2015
NATIONAL STANDARD PLUMBING CODE ILLUSTRATED

PLUMBING-HEATING-COOLING CONTRACTORS ASSOCIATION®
Best People. Best Practices.®

Published By
**PLUMBING-HEATING-COOLING CONTRACTORS—
NATIONAL ASSOCIATION**

All inquiries or questions relating to interpretation should be forwarded to
Code Secretariat
180 S. Washington St., Ste. 100
Falls Church, VA 22046
Email Contact: nspc@naphcc.org

© 2015 Plumbing-Heating-Cooling Contractors–National Association
To order additional books call 1-800-85-IAPMO or
www.iapmostore.org
(if calling from New Jersey call 1-800-652-7422)
(if calling from Maryland call 410-461-5977)

The 2015 National Standard Plumbing Code is a Copyright protected document. This document may not be duplicated, posted, reproduced, or distributed in any form (including, without limitations, electronic, optical, mechanical, or stored in any electronic sharing data base) without advance written permission from PHCC—National Association. Appendix G, excerpts from the Green Plumbing and Mechanical Code Supplement are printed under agreement with the International Association of Plumbing and Mechanical Officials and their Copyright protections also apply.

Illustrations and comments are presented in bold italics or denoted with a border similar to this note and are intended as supplemental information. This information may or may not be adopted by the Authority Having Jurisdiction and as such, acceptance or application should be verified by the user of this document.

Second Printing – March 2015

NATIONAL STANDARD PLUMBING CODE ILLUSTRATED

Title:

National Standard Plumbing Code – Illustrated

Scope:

The development of a recommended code of plumbing practice, design, and installation, including the establishment of performance criteria predicated on the need for protection of health and safety through proper design, installation, and maintenance of plumbing systems. This scope excludes the development of specific standards related to the composition, dimensions, and/or mechanical and physical properties of materials, fixtures, devices, and equipment used or installed in plumbing systems.

Purpose:

To provide practices and performance criteria for the protection of health and safety through proper design and installation of plumbing systems.

Exceptions:

In case of practical difficulty, unnecessary hardship or new developments, exceptions to the literal requirements may be granted by the Authority Having Jurisdiction to permit the use of other devices or methods, but only when it is clearly evident that equivalent protection is thereby secured.

Disclaimer:
PHCC is not responsible or liable for any personal injury, property, or other damages of any nature whatsoever directly or indirectly resulting from the publication, use of, or reliance on the National Standard Plumbing Code or its Supplements. PHCC also makes no guarantee or warranty as to the accuracy or completeness of any information published herein.

FOREWORD

Since its founding as the National Association of Master Plumbers in 1883, the National Association of Plumbing-Heating-Cooling Contractors has maintained a serious interest in plumbing standards, codes and good plumbing design practices.

The Association published the NAMP "Standard Plumbing Code" in 1933 and furnished revised editions until 1942. NAPHCC participated in the development of special standards for war-time plumbing and later was represented on the National Plumbing Code Coordinating Committee, whose work ultimately resulted in the adoption of A40.8 as a standard or model plumbing code in 1955.

NAPHCC served as a sponsor in the early 1960's of the project which attempted to update the 1955 document. This project was operated through the procedures of the American National Standards Institute. However, the A40.8 revision project was not completed because consensus could not be achieved.

In order to provide local and state governments, code administrative bodies and industry with a modern, updated code, NAPHCC published the "National Standard Plumbing Code," in 1971, following the format and sequence of the A40.8 to provide for maximum convenience of users.

With the June 1973 revision, the American Society of Plumbing Engineers joined this effort by co-sponsoring the National Standard Plumbing Code. ASPE maintained its co-sponsorship status until September, 1980. Upon ASPE's withdrawal of co-sponsorship, the Code Committee composition was changed to include not only members of the contracting and engineering communities but also members of the inspection community. Contractors, engineers and inspectors now comprise the National Standard Plumbing Code Committee.

The National Standard Plumbing Code Committee has worked closely with the plumbing industry to maintain a document of minimum requirements for plumbing systems that reflect current practices, materials, and techniques, consistent with public health and safety.

However, the written requirements of a Code can sometimes be interpreted differently by different individuals. For this reason, the NSPC Committee developed the NSPC Illustrated, which includes explanatory comments and illustrations to demonstrate the intent of the various Code Sections.

The Committee also realizes that despite the countless hours of preparation and review, perfection may not have been achieved in this document. For this reason, please send any questions, comments, suggestions, or problems to the Code Secretary, Plumbing-Heating-Cooling Contractors - National Association, 180 S. Washington Street, Ste 100, Falls Church, VA 22046.

Comments on Code Text

The comments following the various Sections of the Code text are intended to explain the intent of that Section of the Code. The comments themselves are not Code requirements but are intended to supplement the Code and provide guidance toward its interpretation.

Figures and Figure Notes

The illustrations (figures) are intended to graphically demonstrate the intent or provide an example of the referenced definition or Code Section. The figures are based only on the referenced definition or Code Section and do not necessarily include all details of the complete installation, such as pipe sizes, specific pipe fittings, required pipe supports, required cleanouts, and other details that are not part of the definition or Code Section being illustrated. The figures must not be used to justify work that does not comply with all requirements of the Code.

The figures are not intended to restrict installations to the arrangement shown. In many cases, the figures show only one example or a typical example of an acceptable arrangement. Any arrangement that meets the intent of the referenced Code Section is acceptable.

This Edition

This edition of the National Standard Plumbing Code – Illustrated includes changes from the 2012 Code, the 2013 Supplement, and the changes that were approved at the July 23, 2014 Public Hearing.

INTRODUCTORY NOTES

The material presented in this Code does not have legal standing unless it is adopted by reference, or by inclusion, in an act of state, county, or municipal government. Therefore, administration of the provisions of this Code must be preceded by suitable legislation at the level of government where it is desired to use this Code.

In some places in this Code, reference is made to "Authority Having Jurisdiction." The identity of an Authority Having Jurisdiction will be established by the act that gives legal standing to the Code provisions.

Proposed changes to the Code can be submitted by any interested party on forms available from PHCC-National Association. Cutoff dates for submitting proposed changes are published by PHCC-National Association, along with the dates for public hearings and final voting by the NSPC Committee. The NSPC is revised and reprinted on a three-year cycle.

Personal appearance before the Committee for a hearing on any Code matter can be had by interested parties after a request in writing.

In the course of revision, certain outdated sections may have been deleted. In order to maintain consistency and perpetuity of the numbering system, those deleted sections and numbers have been removed from this printed text, or placed in reserve.

All changes from the previous edition of this Code are marked by a vertical line in the margin.

2015 National Standard Plumbing Code Committee

J. Richard Wagner, PE
J. Richard Wagner, PE, LLC
Chairman, NSPC Committee

Charles Chalk
Maryland PHCC

James Galvin
Plumbing Manufacturers International

John Heine
New Jersey PHCC

Leon LaFreniere
City of Manchester, NH
Dept. of Buildings

Frank R. Maddalon
F. R. Maddalon Plumbing & Heating

Michael Maloney
Plumbers & Pipefitters Local #9

Thomas C. Pitcherello
NJ Department of Community Affairs
Codes Assistance Unit

Luis A. Rodriguez, CPD, LEED AP
KSi Professional Engineers

Ronald W. Stiegler
Maryland PHCC

Kevin Tindall
PHCC Executive Committee Liaison

Alex Tucciarone
Old Bridge Plumbing Inspector

Jerry Van Pelt, CIPE
GVP Consulting, LLC

Charles R. White
Secretariat & Staff Liaison
Plumbing-Heating-Cooling
Contractors–National Association
800-533-7694
e-mail: white@naphcc.org

Contents

National Standard Plumbing Code - Illustrated		i
Foreword		iii
Introductory Notes		v
National Standard Plumbing Code Committee		vi
Administration		1
Basic Principles		9
Chapter 1	Definitions	13
Chapter 2	General Regulations	67
Chapter 3	Materials	87
Chapter 4	Joints and Connections	115
Chapter 5	Traps, Cleanouts and Backwater Valves	141
Chapter 6	Liquid Waste Treatment Equipment	159
Chapter 7	Plumbing Fixtures, Fixture Fittings and Plumbing Appliances	169
Chapter 8	Hangers and Supports	201
Chapter 9	Indirect Waste Piping and Special Wastes	205
Chapter 10	Water Supply and Distribution	217
Chapter 11	Sanitary Drainage Systems	263
Chapter 12	Vents and Venting	279
Chapter 13	Storm Water Drainage	319
Chapter 14	Special Requirements For Health Care Facilities	333
Chapter 15	Tests and Maintenance	339
Chapter 16	Regulations Governing Individual Sewage Disposal Systems for Homes and Other Establishments Where Public Sewage Systems Are Not Available	341
Chapter 17	Private Potable Water Supply Systems	355
Chapter 18	Mobile Home & Travel Trailer Park Plumbing Requirements	363
Chapter 19	Referenced Standards	371
Appendix A	Sizing Storm Drainage Systems	387
Appendix B	Sizing the Building Water Supply and Distribution Piping Systems	395
Appendix C	Conversions: Metric to U.S. Customary Units	431
Appendix D	Determining the Minimum Number of Required Plumbing Fixtures	433
Appendix E	Special Design Plumbing Systems	439
Appendix F	Requirements of the Adopting Agency	447
Appendix G	Green Plumbing and Mechanical Code Supplement (Excerpts)	449
Appendix H	Combined Building Drains and Building Sewers	511
Appendix I	Fixture Unit Value Curves for Water Closets	513

Appendix K	Flow in Sloping Drains	517
Appendix L	An Acceptable Brazing Procedure for General Plumbing	519
Appendix M	Converting Water Supply Fixture Units (WSFU) to Gallons Per Minute Flow (GPM)	525
Alphabetical Index		531

Administration

ADM 1.1 TITLE

The regulations contained in the following chapters and sections shall be known as the "National Standard Plumbing Code" and may be cited as such, and hereinafter referred to as "this Code".

ADM 1.2 SCOPE

The provisions of this Code shall apply to every installation, including the erection, installation, alteration, relocation, repair, replacement, addition to, use or maintenance of the plumbing system as defined within this Code.

ADM 1.3 PURPOSE

This Code establishes the minimum requirements and standards pertaining to the design, installation, use and maintenance of plumbing systems as defined within this Code.

ADM 1.4 APPLICABILITY

1.4.1 Addition or Repair

Additions, alterations or repairs in compliance to this Code may be made to any existing plumbing system without requiring the existing installation to comply with all the requirements of this Code. Additions, alterations or repairs shall not cause an existing system to become unsafe, insanitary or overloaded.

1.4.2 Existing Plumbing Installation

Plumbing systems that were lawfully installed prior to the adoption of this Code may continue their use, maintenance or repairs, provided the maintenance or repair is in accordance with the original design, location, and no hazard has been created to life, health or property by such plumbing system.

1.4.3 Existing Use

The lawful use of any plumbing installation, appliances, fixtures, fittings and appurtenances may have their use continued, provided no hazards to life, health or property have been created by their continued use.

1.4.4 Maintenance and Repairs

The maintenance of all plumbing systems, materials, appurtenances, devices or safeguards, both existing and new, shall be maintained in a safe and proper condition. The owner, or his designated agent, shall be responsible for the maintenance of the plumbing system. Minor repairs to or replacement of any existing systems are permitted, provided they are made in the same manner and arrangement as the original installation and are approved.

1.4.5 Change of Building Use

The plumbing systems of any building or structure that is proposed for a change in use or occupancy shall comply to all the requirements of this Code for the new use or occupancy.

1.4.6 Moved Buildings or Structures

The plumbing system in any building or structure to be moved into this jurisdiction shall comply with the provisions of this Code for new construction.

1.4.7 Special Historic Buildings

The provisions of this Code relating to the additions, alterations, repair, replacement or restoration of those structures designated as historic buildings shall not be mandatory when such work is deemed to be safe and in the public interest of health, safety and welfare by the Authority Having Jurisdiction.

1.4.8 Appendices

The provisions in the appendices are intended to supplement the requirements of this Code and are considered to be part of this Code when adopted by the Authority Having Jurisdiction.

ADM 1.5 APPROVALS

1.5.1 Alternates

The provisions cited in this Code are not intended to prevent the use of any material or method of installation when it is determined to meet the intent of this Code and approved by the Authority Having Jurisdiction.

1.5.2 Authority Having Jurisdiction

The Authority Having Jurisdiction may approve any such alternate material or method of installation not expressly conforming to the requirements of this Code, provided it finds the proposed material or method of installation is at least the equivalent of that required in this Code. A record of such approval shall be kept and shall be available to the public.

1.5.3 Tests Required

The Authority Having Jurisdiction shall require sufficient evidence to substantiate any claims made regarding the equivalency of any proposed alternate material or method of installation. When the Authority Having Jurisdiction determines that there is insufficient evidence to substantiate the claims, it may require tests to substantiate the claims made by an approved testing agency at the expense of the applicant.

1.5.4 Test Procedure

The Authority Having Jurisdiction shall require all tests be made in accordance with approved standards; but, in the absence of such standards, the Authority Having Jurisdiction shall specify the test procedure.

1.5.5 Retesting

The Authority Having Jurisdiction may require any tests to be repeated if, at any time, there is reason to believe that any material or method of installation no longer conforms to the requirements on which the original approval was based.

ADM 1.6 ORGANIZATION AND ENFORCEMENT

1.6.1 Authority Having Jurisdiction

The Authority Having Jurisdiction shall be the individual official, board, department or agency duly appointed by the jurisdiction as having the authority to administer and enforce the provisions of this Code as adopted or amended.

1.6.2 Deputies

In accordance with the procedures set forth by the jurisdictional authority, the Authority Having Jurisdiction may appoint such assistants, deputies, inspectors or other designated employees to carry out the administration and enforcement of this Code.

1.6.3 Right of Entry

When inspections are required to enforce the provisions of this Code, or there is reasonable cause to believe there exists in any building, structure or premises, any condition or violation of this Code causing the building, structure or premises to be unsafe, insanitary, dangerous or hazardous, the Authority Having Jurisdiction or its designated representative may enter such building, structure or premises at reasonable times to inspect or perform the duties imposed by this Code. When the building, structure or premises are occupied, proper credentials shall be presented to the occupant when entry is required. In the event the building, structure or premises is unoccupied, and entry is required, a reasonable effort shall be made to locate the owner or his agent in charge of such building, structure or premises. In the event the occupant or owner of such building, structure or premises refuses entry, the Authority Having Jurisdiction shall have recourse to the remedies provided by law to gain entry.

1.6.4 Stop Work Order

Upon notice from the Authority Having Jurisdiction, work being done on any building, structure or premises contrary to the provisions of this Code, or in an unsafe and dangerous manner, shall be stopped immediately. The stop work notice shall be in writing, served on the owner of the property, or his agent, or to the person doing such work. It shall state the conditions under which the Authority Having Jurisdiction may grant authorization to proceed with the work.

1.6.5 Authority to Condemn

When the Authority Having Jurisdiction determines that any plumbing system or portion thereof that is regulated by this Code has become insanitary or hazardous to life, health or property, it shall order in writing that such plumbing system or portion thereof be repaired, replaced or removed so as to be in code compliance. The written order shall fix a reasonable time limit for the work to be brought into code compliance, and no person shall use the condemned plumbing system until such work is complete and approved by the Authority Having Jurisdiction.

1.6.6 Authority to Abate

Any plumbing system, or portion thereof that is found to be insanitary or constitute a hazard to life, health or property is hereby declared to be a nuisance. Where a nuisance exists, the Authority Having Jurisdiction shall require the nuisance to be abated and shall seek such abatement in the manner prescribed by law.

1.6.7 Liability

The Authority Having Jurisdiction, or any individual duly appointed or authorized by the Authority Having Jurisdiction to enforce this Code, acting in good faith and without malice, shall not thereby be rendered personally liable for any damage that may occur to persons or property as a result of any act, or by reason of any act or omission in the lawful discharge of his duties. Should a suit be brought against the Authority Having Jurisdiction or duly appointed representative because of such act or omission, it shall be defended by legal counsel provided by this jurisdiction until final termination of the proceedings.

1.6.8 Work Prior to Permit

Where work for which a permit is required by this Code is started prior to obtaining the prescribed permit, the applicant shall pay a double fee. In the event of an emergency where it is absolutely necessary to perform the plumbing work immediately, such as nights, weekends or holidays, said fee shall not be doubled if a permit is secured at the earliest possible time after the emergency plumbing work has been performed.

ADM 1.7 VIOLATIONS AND PENALTIES

1.7.1 Violations

It shall be unlawful for any individual, partnership, firm or corporation to, or cause to, install, construct, erect, alter, repair, improve, convert, move, use or maintain any plumbing system in violation of this Code.

1.7.2 Penalties

Any individual, partnership, firm or corporation who shall violate or fail to comply with any of the requirements of this Code shall be deemed guilty of a _____, and if convicted, shall be punishable by a fine or imprisonment or both as established by this jurisdiction. Each day during which a violation occurs or continues, shall constitute a separate offense.

ADM 1.8 PERMITS

1.8.1 Permits Required

It shall be unlawful for any individual, partnership, firm or corporation to commence, or cause to commence, any installation, alteration, repair, replacement, conversion or addition to any plumbing system, or part thereof, regulated by this Code, except as permitted in Section 1.8.2, without first obtaining a plumbing permit for each separate building or structure, on forms prepared and provided by the Authority Having Jurisdiction.

1.8.2 Permits Not Required for the Following

   a. Permits shall not be required for the following work:

     1. The stoppage of leaks in drains or vent pipes. However, should the defect necessitate removal and replacement with new material, it shall constitute new work and a permit shall be obtained and inspection made as required in this Code.

     2. The clearing of stoppages.

     3. The repairing of leaks in valves or fixtures.

     4. The removing and reinstallation of a water closet for a cleanout opening provided the reinstallation does not require replacement or rearrangement of valves, pipes or new fixtures.

   b. Exemptions from obtaining a permit required by this Code shall not be construed as to authorize any work to be performed in violation of this Code.

ADM 1.9 PROCESS FOR OBTAINING PERMITS

1.9.1 Application

   a. Applications for a permit shall be made in writing by the person, or his agent, proposing to do such work covered by the permit. The applicant shall file such application in writing on a form prepared and provided by the Authority Having Jurisdiction. Every such permit shall:

     1. Describe in detail the work to be done for which the permit was obtained.

     2. Describe in detail the parcel of land on which the proposed work is to be done by legal description, street address or other means to definitely locate the site or building where the work is to be performed.

3. List the type of occupancy or use.
4. Provide plans, drawings, diagrams, calculations or other data as required by Section 1.9.2.
5. Be signed by the person or agent making application.
6. Provide any other information the Authority Having Jurisdiction may require.

1.9.2 Plans

Two or more sets of plans shall be submitted with each permit application. The plans shall contain all the engineering calculations, drawings, diagrams and other data as required for approval. The Authority Having Jurisdiction may also require that the plans, drawings, diagrams and calculations be designed by an engineer and/or architect licensed by the state in which the work is to be performed. Except that the Authority Having Jurisdiction may waive the submission of plans and other data, provided it is determined that the nature of the work covered by the permit does not require plan review to obtain code compliance.

1.9.3 Specifications

All specifications required to be on the plans shall be drawn to scale and sufficiently clear to indicate the nature, location and extent of the proposed work so as to show how it will conform to the requirements of this Code.

1.9.4 Permit Issuance

If, after reviewing the plans and specifications, the Authority Having Jurisdiction finds that they are complete and conform to the requirements of this Code, it shall authorize a permit to be issued upon payment of all the fees specified in Section 1.10.1.

1.9.5 Approved Plans

When the Authority Having Jurisdiction issues a permit and plans were required, it shall endorse, either in writing or stamp the plans "APPROVED", and all work shall be done in accordance with the plans without deviation.

1.9.6 Plans Retention

The Authority Having Jurisdiction shall retain one set of approved plans until final approval of the work contained therein. One set of approved plans shall be returned to the applicant and this set of approved plans shall be kept on the job site at all times until final approval of the work contained therein.

1.9.7 Permit Validity

The issuance of a permit by the Authority Having Jurisdiction is not and shall not be construed to be authorization or approval of any violation of the requirements of this Code. Any presumption of a permit to be authorization to violate or cancel any provisions of this Code shall be invalid. The issuance of a permit based on plans submitted shall not prevent the Authority Having Jurisdiction from requiring the correction of any errors in the plans or preventing the progress of the construction when it is in violation of any provision of this Code.

1.9.8 Permit Expiration

Every permit issued by the Authority Having Jurisdiction, in accordance with the provisions of this Code, shall expire by limitation and become null and void when such work authorized by the permit has not commenced within _____days from the date of issuance or if such work is suspended or abandoned for a period of _____days after commencement of such work. In order for such work to recommence, a new permit shall be obtained and a fee of __ percent of the original permit fee shall be charged, provided no changes have been made or will be made to the original plans as submitted. The Authority Having Jurisdiction may grant an extension to any permit provided the request is in writing by the permittee stating the reason or circumstances that prevented him from completing such work as required by this Code.

1.9.9 Revocation or Suspension

At any time, the Authority Having Jurisdiction may suspend or revoke a permit issued in error or on the basis of incorrect information submitted or in violation of any section of this Code. The suspension or revocation of such permit shall be in written form by the Authority Having Jurisdiction stating the reason or purpose of such suspension or revocation.

ADM 1.10 PERMITS

1.10.1 Fees Schedule

The permit fees for all plumbing work shall be set forth by the Authority Having Jurisdiction of the jurisdiction having authority.

1.10.2 Plan Review Fees

When plans are reviewed as a requirement prior to issuance of a permit, the fee shall be equal to ____ percent of the total permit fee as set forth in Section 1.10.1.

1.10.3 Plan Review Expiration

Permit application and plan review for which no permit is issued shall expire by limitation within ___ days following the date of application. All plan review fees shall be forfeited and the plans may be destroyed by the Authority Having Jurisdiction or returned to the applicant.

l.10.4 Work Without a Permit

When any plumbing work is commenced without first obtaining a permit from the Authority Having Jurisdiction, an investigation of such work shall be made before a permit may be issued. The investigation fee shall be collected whether or not a permit is then or subsequently issued. Any investigation fee shall equal the amount of the permit fee, if a permit were to be issued in accordance with this Code. If the investigation fee is collected, it shall not exempt any person from compliance or penalties set forth in this Code.

1.10.5 Refunding of Fees

Any fee collected by the Authority Having Jurisdiction that was erroneously paid or collected may be refunded, provided not more than____percent of the fee payment shall be refunded when no work has been done. Any request for the refunding of any fee shall be in writing by the applicant no later than____days after the date of fee payment.

ADM 1.11 INSPECTIONS

1.11.1 Required Inspections

All new plumbing systems, and parts of existing systems that require a permit shall be tested and inspected by the Authority Having Jurisdiction prior to being covered or concealed. Where any such work has been covered or concealed, the Authority Having Jurisdiction shall require that such work be exposed for inspection and testing. All equipment, material and labor required for testing the plumbing system shall be furnished by the permittee. The Authority Having Jurisdiction shall not be liable for any expense incurred by the removal or replacement of materials required to permit inspection or testing. Such expense is the responsibility of the permittee. Upon completion of the rough plumbing installation, prior to covering or concealing any such work, the Authority Having Jurisdiction shall inspect the work and any such test, as prescribed hereinafter, to disclose any leaks or defects. After completion of the plumbing system and the plumbing fixtures are set and their traps filled with water, a final inspection shall be conducted as required by this Code. Additional inspections may be required when alternate materials or methods of installation are approved by the Authority Having Jurisdiction.

1.11.2 Exception:

For moved-in or relocated structures, minor installations and repairs, the Authority Having Jurisdiction may make other such inspections or tests as necessary to assure that the work has been performed and is safe for use in accordance with the intent of this Code.

1.11.3 Use of Existing Plumbing

The operation of any plumbing installation to replace existing systems or fixtures serving an occupied portion of any building or structure shall not be considered by the requirements of this Code to prohibit such operation, provided a request for inspection has been made to the Authority Having Jurisdiction within 48 hours of such work and before any such work is covered or concealed.

1.11.4 System Testing

All new plumbing systems and parts of existing systems shall be tested and approved as required elsewhere in this Code.

1.11.5 Requests for Inspection

The Authority Having Jurisdiction shall be notified by the person doing the work, authorized by the permit, that such work has been subjected to the required tests and is ready for inspection. The method of request, whether in writing or by telephone, shall be established by the Authority Having Jurisdiction. It shall be the duty of the permittee doing the work authorized by a permit to provide reasonable access and means for accomplishing proper inspections.

1.11.6 Other Inspections

The Authority Having Jurisdiction may require other inspections, in addition to those required by this Code, of any plumbing work in order to ascertain compliance with the requirements of this Code.

1.11.7 Reinspection Fees

    a. The assessment of a reinspection fee may be required for any of the following:
        1. For any portion of work not completed for which inspection was requested.
        2. For any required corrections that have not been completed and for which reinspection was requested.
        3. For not having the approved plans on site and readily available to the inspector.
        4. Failure to provide access for inspection on the date inspection was requested.
        5. Deviation from the approved plans that would require reapproval of the Authority Having Jurisdiction.
        6. Failure to provide correct address.
    b. This provision is intended to control the practice of calling for inspections prior to having work ready for inspection and not for the first time job rejection for not complying with the installation requirements.
    c. Upon the assessment of a reinspection fee, the applicant shall pay the reinspection fee in accordance with Section 1.10.1 and no additional inspections shall be performed until all fees have been paid.

ADM 1.12 FINAL CONNECTIONS

1.12.1 Energy or Fuel

It shall be unlawful for any person to make, or cause to make, any connection to any source of energy or fuel to any plumbing system or equipment regulated by this Code prior to the approval of the Authority Having Jurisdiction.

1.12.2 Water and Sewer

It shall be unlawful for any person to make, or cause to make, any connection to any water supply or sewer system to any plumbing system or equipment regulated by this Code prior to the approval of the Authority Having Jurisdiction.

1.12.3 Temporary Connection

By authorization of the Authority Having Jurisdiction, a temporary connection may be made to any plumbing equipment to a source of energy or fuel for testing purposes only.

ADM 1.13 UNCONSTITUTIONALITY

Should any chapter, section, subsection, sentence, clause or phrase of this Code be held for any reason as unconstitutional, such decision shall not affect the validity of the remaining chapters, sections, subsections, sentences, clause or phrases of this Code.

Basic Principles

This Code is founded upon certain basic principles of environmental sanitation and safety through properly designed, acceptably installed, and adequately maintained plumbing systems. Some of the details of plumbing construction may vary but the basic sanitary and safety principles desirable and necessary to protect the health of the people are the same everywhere.

The establishment of trade jurisdictional areas is not within the scope of this Code. This inclusion of material, even though indicated as approved for purposes of this Code, does not infer unqualified endorsement as to its selection or serviceability in any or every installation.

As interpretations may be required, and as unforeseen situations arise which are not specifically covered in this Code, the twenty-two principles which follow shall be used to define the intent.

Principle No. 1—ALL OCCUPIED PREMISES SHALL HAVE POTABLE WATER

All premises intended for human habitation, occupancy, or use shall be provided with a supply of potable water. Such a water supply shall not be connected with unsafe water sources, nor shall it be subject to the hazards of backflow.

Principle No. 2—ADEQUATE WATER REQUIRED

Plumbing fixtures, devices, and appurtenances shall be supplied with water in sufficient volume and at pressures adequate to enable them to function properly and without undue noise under normal conditions of use.

Principle No. 3—HOT WATER REQUIRED

Hot water shall be supplied to all plumbing fixtures which normally need or require hot water for their proper use and function.

Principle No. 4—WATER CONSERVATION

Plumbing shall be designed and adjusted to use the minimum quantity of water consistent with proper performance and cleaning.

Principle No. 5—SAFETY DEVICES

Devices for heating and storing water shall be so designed and installed as to guard against dangers from explosion or overheating.

Principle No. 6—USE PUBLIC SEWER WHERE AVAILABLE

Every building with installed plumbing fixtures and intended for human habitation, occupancy, or use, and located on premises where a public sewer is on or passes said premises within a reasonable distance, shall be connected to the public sewer.

Principle No. 7—REQUIRED PLUMBING FIXTURES

Each family dwelling unit shall have at least one water closet, one lavatory, one kitchen-type sink, and one bathtub or shower to meet the basic requirements of sanitation and personal hygiene. All other structures for human habitation shall be equipped with sufficient sanitary facilities. Plumbing fixtures shall be made of durable, smooth, non-absorbent and corrosion resistant material and shall be free from concealed fouling surfaces.

Principle No. 8—DRAINAGE SYSTEM

The drainage system shall be designed, constructed, and maintained to guard against fouling, deposit of solids and clogging, and with adequate cleanouts so arranged that the pipes may be readily cleaned.

Principle No. 9—DURABLE MATERIALS AND GOOD WORKMANSHIP

The piping of the plumbing system shall be of durable material, free from defective workmanship and so designed and constructed as to give satisfactory service for its reasonable expected life.

Principle No. 10—FIXTURE TRAPS

Each fixture directly connected to the drainage system shall be equipped with a liquid seal trap.

Principle No. 11—TRAP SEALS SHALL BE PROTECTED

The drainage system shall be designed to provide an adequate circulation of air in all pipes with no danger of siphonage, aspiration, or forcing of trap seals under conditions of ordinary use.

Principle No. 12—EXHAUST FOUL AIR TO OUTSIDE

Each vent terminal shall extend to the outer air and be so installed as to minimize the possibilities of clogging and the return of foul air to the building.

Principle No. 13—TEST THE PLUMBING SYSTEM

The plumbing system shall be subjected to such tests as will effectively disclose all leaks and defects in the work or the material.

Principle No. 14—EXCLUDE CERTAIN SUBSTANCES FROM THE PLUMBING SYSTEM

No substance that will clog or accentuate clogging of pipes, produce explosive mixtures, destroy the pipes or their joints, or interfere unduly with the sewage-disposal process shall be allowed to enter the building drainage system.

Principle No. 15—PREVENT CONTAMINATION

Proper protection shall be provided to prevent contamination of food, water, sterile goods, and similar materials by backflow of sewage. When necessary, the fixture, device, or appliance shall be connected indirectly with the building drainage system.

Principle No. 16—LIGHT AND VENTILATION

No water closet, urinal, or bidet shall be located in a room or compartment which is not properly lighted and ventilated.

Principle No. 17—INDIVIDUAL SEWAGE DISPOSAL SYSTEMS

If water closets or other plumbing fixtures are installed in buildings where there is no public sewer within a reasonable distance, suitable provision shall be made for disposing of the sewage by some accepted method of sewage treatment and disposal.

Principle No. 18—PREVENT SEWER FLOODING

Where a plumbing drainage system is subject to backflow of sewage from the public sewer or private disposal system, suitable provision shall be made to prevent its overflow in the building.

Principle No. 19—PROPER MAINTENANCE

Plumbing systems shall be maintained in a safe and serviceable condition from the standpoint of both mechanics and health.

Principle No. 20—FIXTURES SHALL BE ACCESSIBLE

All plumbing fixtures shall be so installed with regard to spacing as to be accessible for their intended use and for cleaning.

Principle No. 21—STRUCTURAL SAFETY

Plumbing shall be installed with due regard to preservation of the strength of structural members and prevention of damage to walls and other surfaces through fixture usage.

Principle No. 22—PROTECT GROUND AND SURFACE WATER

Sewage or other waste shall not be discharged into surface or sub-surface water unless it has first been subjected to some acceptable form of treatment.

Blank Page

Chapter 1

Definitions

1.1 GENERAL

For the purpose of this Code, the following terms shall have the meaning indicated in this chapter. No attempt is made to define ordinary words that are used in accordance with their established dictionary meaning, except where it is necessary to define their meaning as used in this Code to avoid misunderstanding.

1.2 DEFINITION OF TERMS

Accessible and Readily Accessible:

Accessible: access thereto without damaging building surfaces, but that first may require the removal of an access panel, door or similar obstructions with the use of tools. ***See Figure 1.2.1***

Readily accessible: access without requiring the use of tools for removing or moving any panel, door or similar obstruction. ***See Figure 1.2.1***

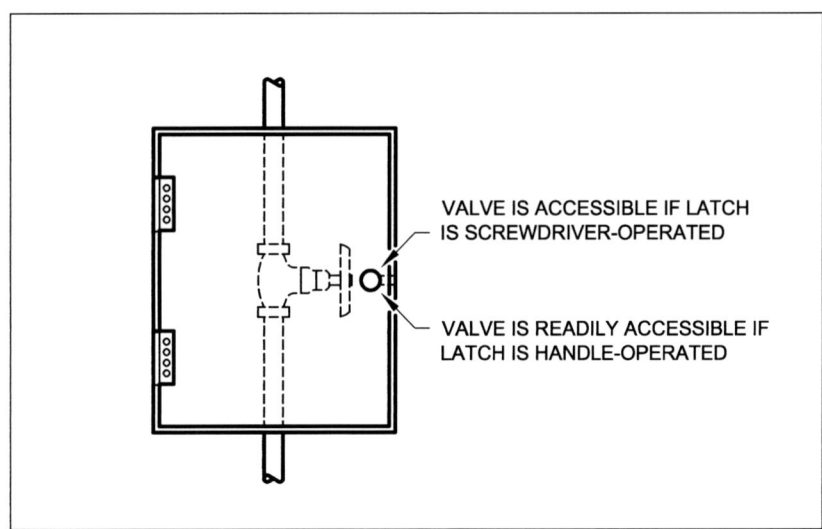

Figure 1.2.1
AN EXAMPLE OF ACCESSIBLE AND READILY ACCESSIBLE

Acid Waste: See "Special Wastes"

Adopting Agency: The agency, board or authority having the duty and power to establish codes and regulations that directly and indirectly affect the plumbing work to be performed in a jurisdiction as administered and enforced by an Authority Having Jurisdiction.

Air Break (drainage system): A piping arrangement in which a drain from a fixture, appliance, or device discharges into a receptor at a point below the flood level rim and above the trap seal of the receptor. ***See Figure and Section 9.1.3***

> *Comment: Air breaks are permitted where backflow cannot occur due to back siphonage.*

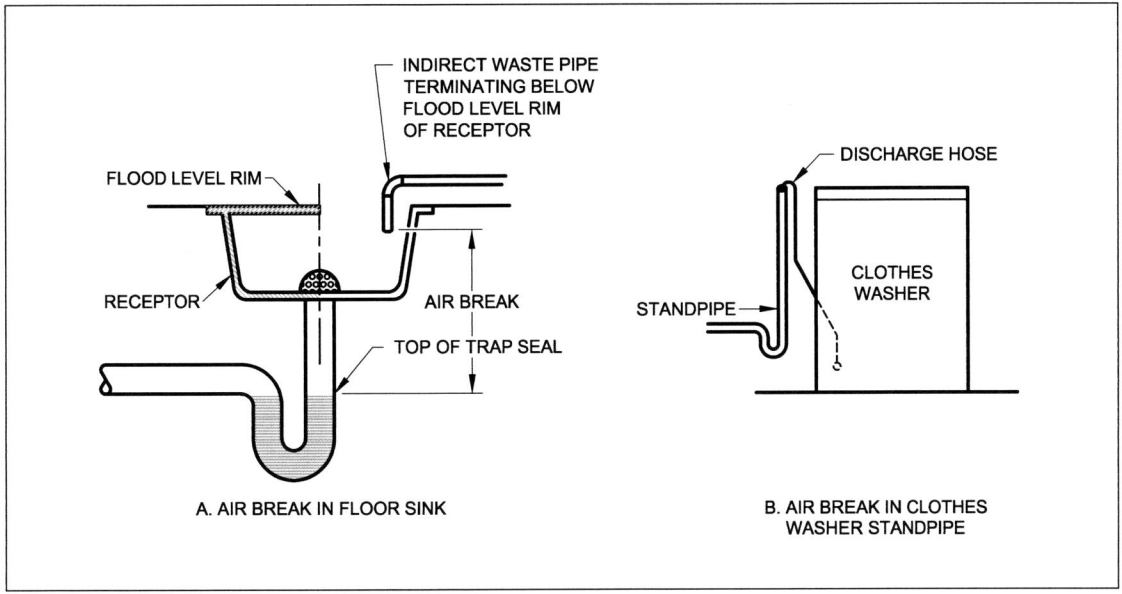

Figure 1.2.2
AIR BREAKS

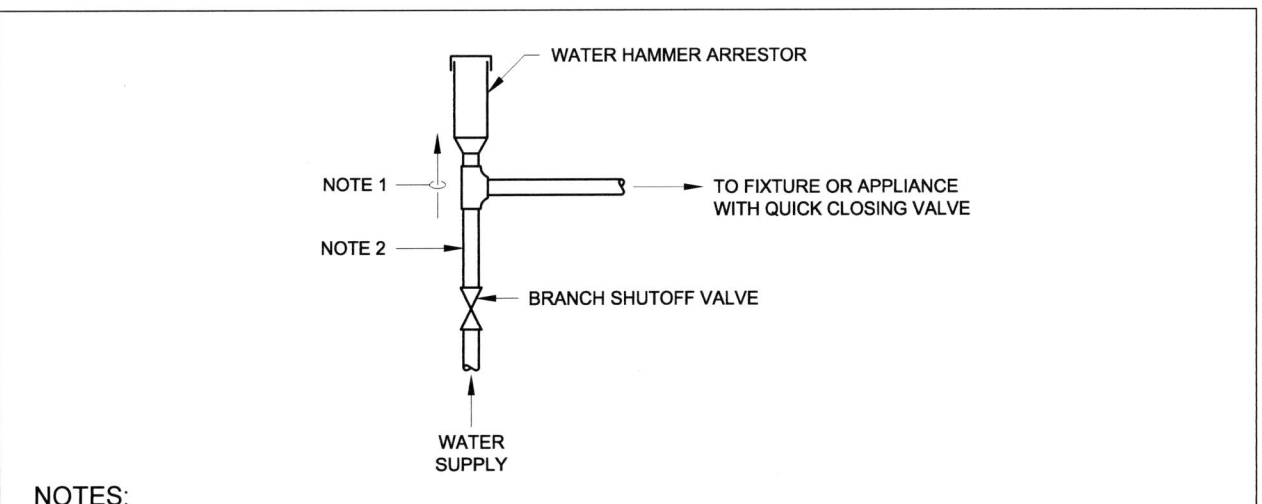

NOTES:
1. For maximum effectiveness, the water hammer arrestor should be oriented so that the direction of the shock wave caused by the quick-closing valve is into the arrestor. Refer to the manufacturer's instructions.
2. The number of elbows upstream from the water hammer arrestor should be minimized. Each elbow represents a point of shock and potential failure.

Figure 1.2.3
A WATER HAMMER ARRESTOR

Air Gap (drainage system): The unobstructed vertical distance through the free atmosphere between the outlet of the waste pipe and the flood level rim of the receptor into which it is discharging. ***See Figure 1.2.4***

Comment: Air gaps are required where backflow can occur due to back siphonage.

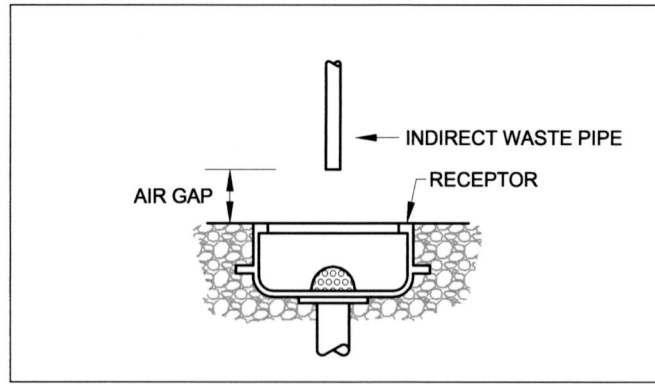

Figure 1.2.4
AN AIR GAP FOR INDIRECT WASTE PIPING

Air Gap (water distribution system): The unobstructed vertical distance through the free atmosphere between the lowest opening from any pipe or faucet supplying water to a tank, plumbing fixture or other device and the flood level rim of the receptor. *See Figure 1.2.5*

Comment #1: The minimum required air gap distance is based on the effective opening of the water supply outlet. The air gap must be increased if the outlet is close to walls or other vertical surfaces. See Section 10.5.2 and Table 10.5.2.

Comment #2: If air is being drawn into the tub spout by a vacuum in the water supply piping, waste water at the flood level rim of the fixture will tend to be lifted upward towards the spout opening by the flow of air. The water will lift higher if the spout opening is close to a wall.

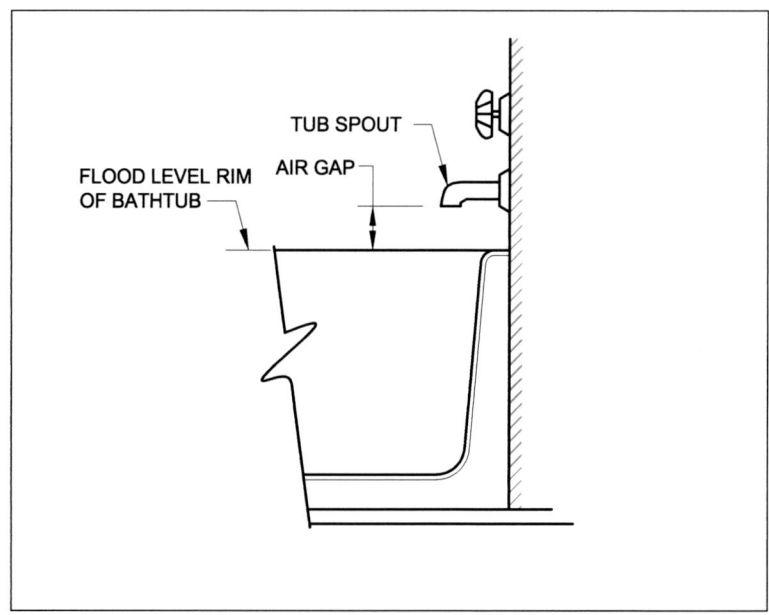

Figure 1.2.5
AN AIR GAP FOR A POTABLE WATER OUTLET

Approved: Accepted or acceptable under an applicable standard stated or cited in this Code, or accepted as suitable for the proposed use under procedures and powers of the Authority Having Jurisdiction as defined in Section 3.12. *See Sections 3.1.1, 3.1.2, 3.1.3, and 3.12*

Area Drain: A receptor designed to collect surface or storm water from an open area. *See Figures 1.2.6 and 13.1.6*

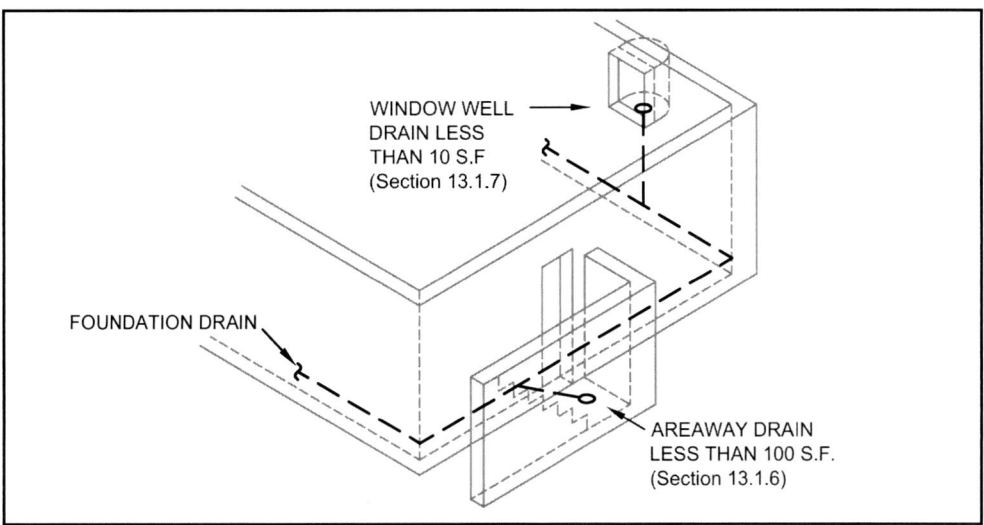

Figure 1.2.6
AREA DRAINS IN WINDOW WELLS AND STAIR WELLS

Aspirator: A fitting or device supplied with water or other fluid under positive pressure that passes through an integral orifice or "constriction" causing a vacuum. *See Figure 1.2.7 and Section 14.13*

Comment: Backflow prevention is required where the fluid supply is potable water.

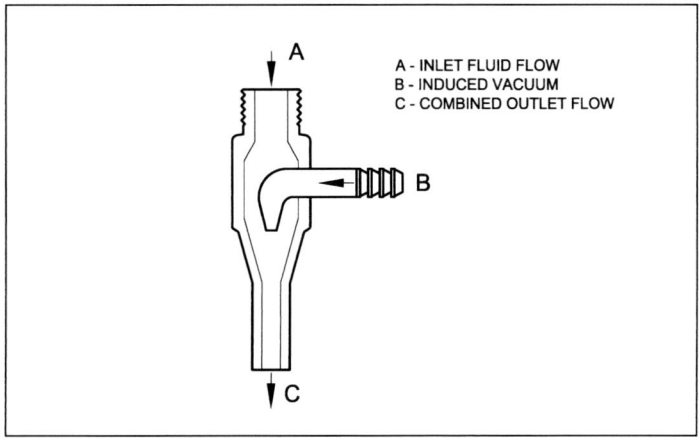

Figure 1.2.7
AN ASPIRATOR FITTING

Authority Having Jurisdiction: An administrative authority. The individual official, board, department or agency established and authorized by a state, county, city or other political subdivision created by law to administer and enforce the provisions of codes and regulations that directly and indirectly affect the plumbing work to be performed in a jurisdiction as adopted and amended by an Adopting Agency.

Automatic Compensating Valve: A temperature control valve for individual shower and tub/shower combinations designed to minimize thermal shock and reduce the risk of scalding. They are installed at the point of use where the user has access to adjust the water flow and discharge temperature. The valves include Type P (pressure balancing), Type T (thermostatic), and Type T/P (combination), and comply with ASME A112.1016/ASSE 1016/CSA B125.16.

Automatic Flushing Device: A device that automatically flushes a fixture after each use without the need for manual activation.

Auxiliary Floor Drain: A floor drain that does not receive the discharge from any indirect waste pipe or other predictable drainage flows. Auxiliary floor drains have no DFU loading.

Back Pressure Backflow: Backflow into potable water piping from a source having a higher pressure than in the potable water piping. *See Figure 1.2.8*

Back Siphonage Backflow: Backflow into potable water piping caused by a vacuum or partial vacuum in the potable water piping. *See Figure 1.2.9*

Backflow Preventer: A device or assembly that prevents backflow into potable water piping caused by back pressure, back siphonage, or both.

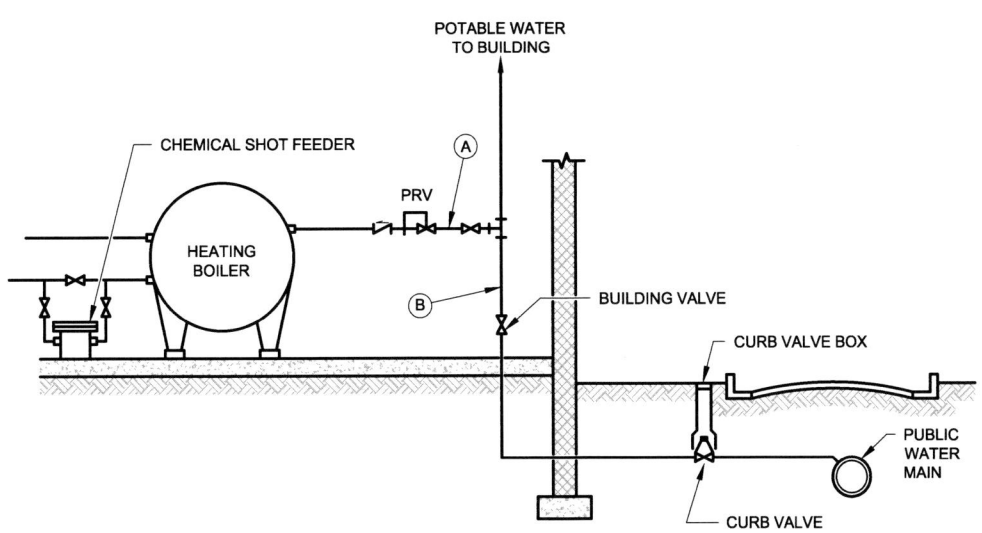

NOTES:
1. Back-pressure backflow is generally caused by water pressure producing equipment within a building.
2. The hot water heating boiler operates at up to 30 psig and has a chemical shot feeder.
3. The pressure in the public water main can drop below 30 psig due to a shutdown for repair or heavy demand for fire fighting operations.
4. The check valve and pressure reducing valve in the water makeup to the heating system are not adequate to prevent back-pressure backflow.
5. The chemical feeder creates a potential "high hazard" that requires a reduced pressure backflow preventer (RP). Otherwise, a double check valve assembly would be adequate.
6. An RP at Point "A" is required to protect the water distribution system within the building from backflow of chemically treated boiler water.
7. An RP at Point "B" would "contain" the building and prevent it from backflowing any potential contamination into the public water system.

Figure 1.2.8
BACKFLOW CAUSED BY BACK-PRESSURE

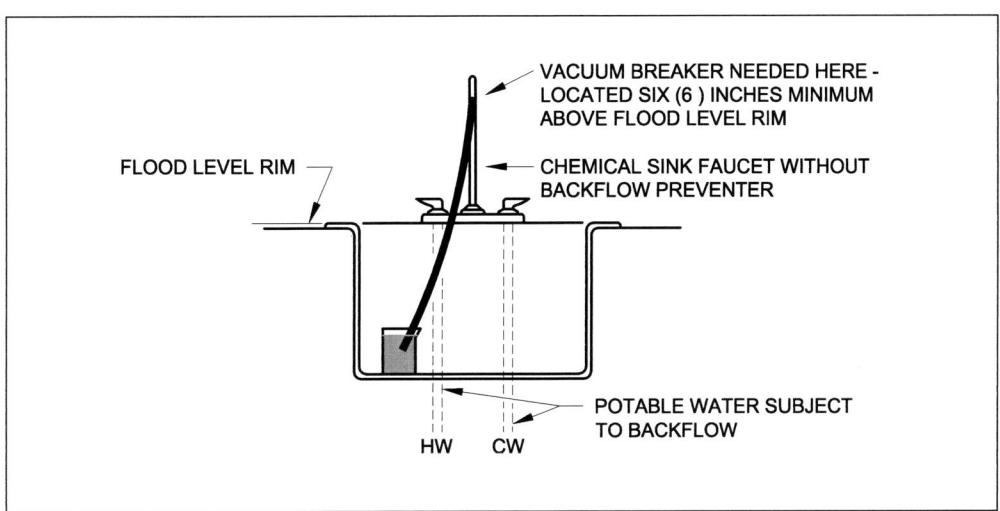

Figure 1.2.9
BACKFLOW CAUSED BY BACK SIPHONAGE

Backwater Valve: A device installed in a drain pipe to prevent reverse flow in the drainage system. *See Figure 1.2.10*

> *Comment: Backwater valves are swing-type check valves that are installed in drainage pipe to prevent the reversal of flow in the piping and overflows due to stoppages, flooding, or other abnormal conditions. Refer to Section 5.5.1 for where backwater valves are required.*

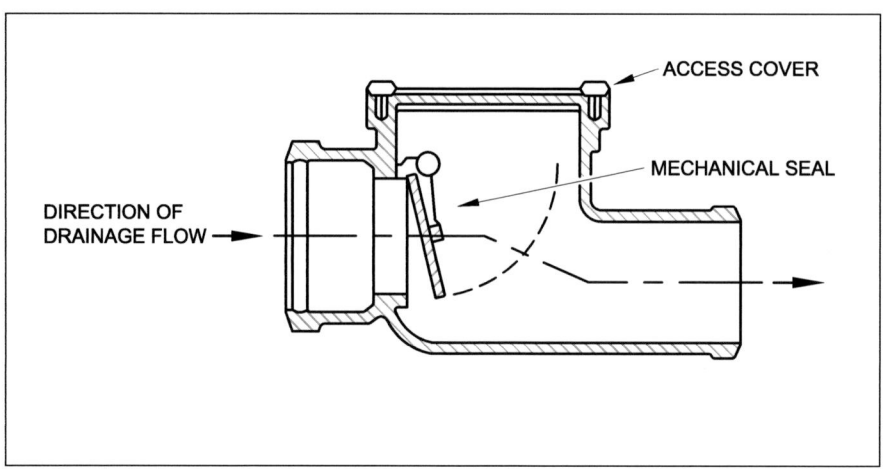

Figure 1.2.10
A BACKWATER VALVE

Baptistery: A tank or pool for baptizing by total immersion.

Bathroom Group: A group of fixtures in a dwelling unit bathroom consisting of one water closet, one or two lavatories, and either one bathtub, one combination bath/shower or one shower stall. Other fixtures within the bathing facility shall be counted separately when determining the total water supply and drainage fixture loads for the bathroom group.

Battery of Fixtures: Any group of two or more similar adjacent fixtures that discharge into a common fixture branch drain. *See Figure 1.2.11*

> *Comment: Batteries of fixtures can be "battery vented" in groups of up to eight fixtures in accordance with Section 12.13. Fixtures that are "battery vented" do not have to be the same type.*

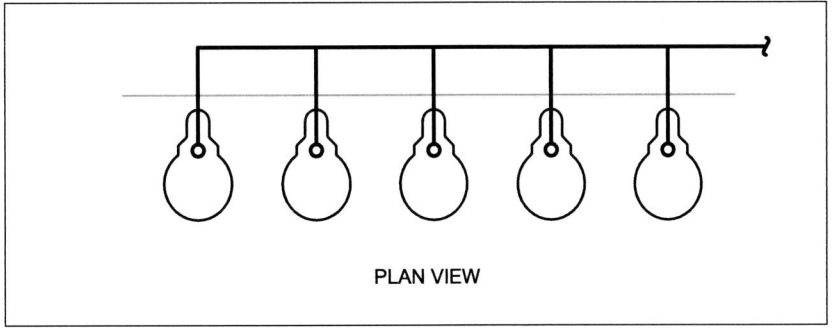

Figure 1.2.11
A BATTERY OF FIXTURES

Bedpan Steamer: A fixture used for scalding bedpans or urinals by direct application of steam. *See Section 14.10*

Boiler Blow-off: An outlet on a boiler to permit emptying or discharge of sediment. *See Figure 1.2.12*

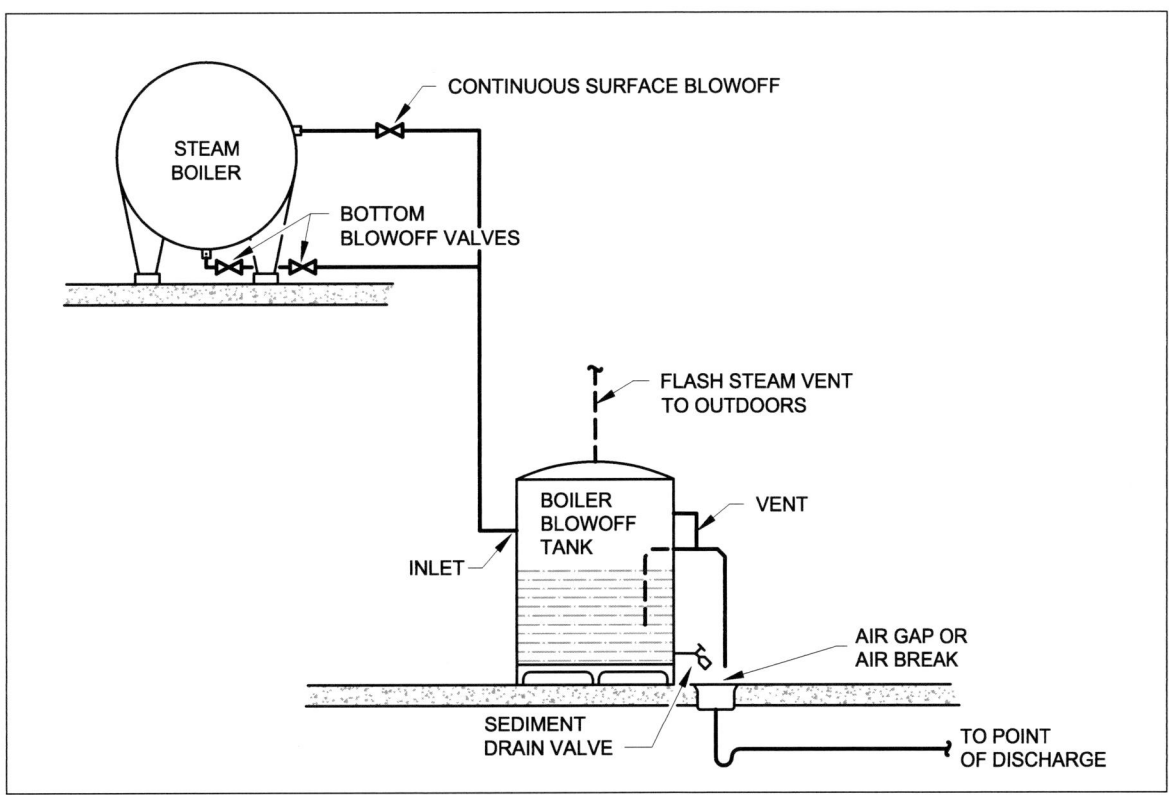

Figure 1.2.12
A BOILER BLOWOFF TANK

Boiler Blow-off Tank: A vessel designed to receive the discharge from a boiler blow-off outlet and to cool the discharge to a temperature that permits its safe discharge to the drainage system. *See Figure 1.2.12*

Comment: Boiler blow-off must be cooled to 140°F or less before being discharged into the drainage system. If potable water is supplied for cooling, the water source must be protected from backflow. An air break can be provided at the discharge from the blow-off tank into the drainage system if the makeup water supply to the boiler is protected against backflow.

Branch Drain: A branch of drain piping, including horizontal fixture branches, horizontal branch drains, horizontal battery-vented drains, and branches of the building drain.

Branch Interval: A distance along a vertical sanitary drain stack corresponding, in general, to a story height, but in no case less than 8 feet within which the horizontal branches from one floor or story of a building are connected to the stack. *See Figure 1.2.14*

Comment: Branch intervals are used to determine the potential drainage load on stacks for the purpose of sizing the stacks and providing pressure relief.

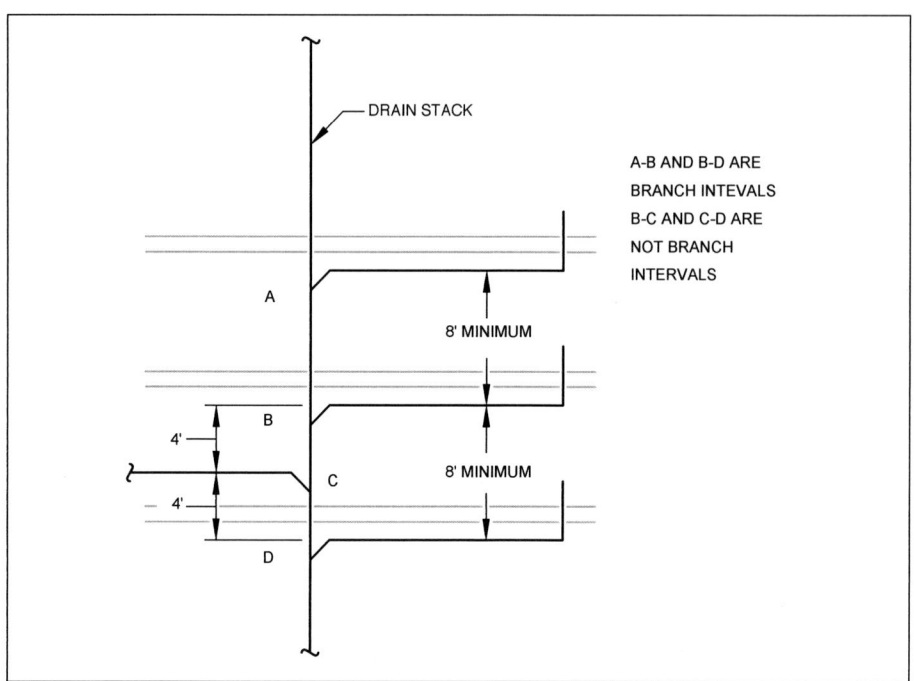

Figure 1.2.14
BRANCH INTERVALS

Branch Vent: See "Vent, Branch"

Building Classification: The arrangement for the designation of buildings according to occupancy based on the applicable building code.

Comment: The building classifications in Table 7.21.1 for the minimum number of required plumbing fixtures include assembly (A), business (B), education (E), factory (F), institutional (I), mercantile (M), residential (R), and storage (S).

Building Drain, Combined: A building drain that conveys both sewage and storm water. *See Figure 1.2.15*

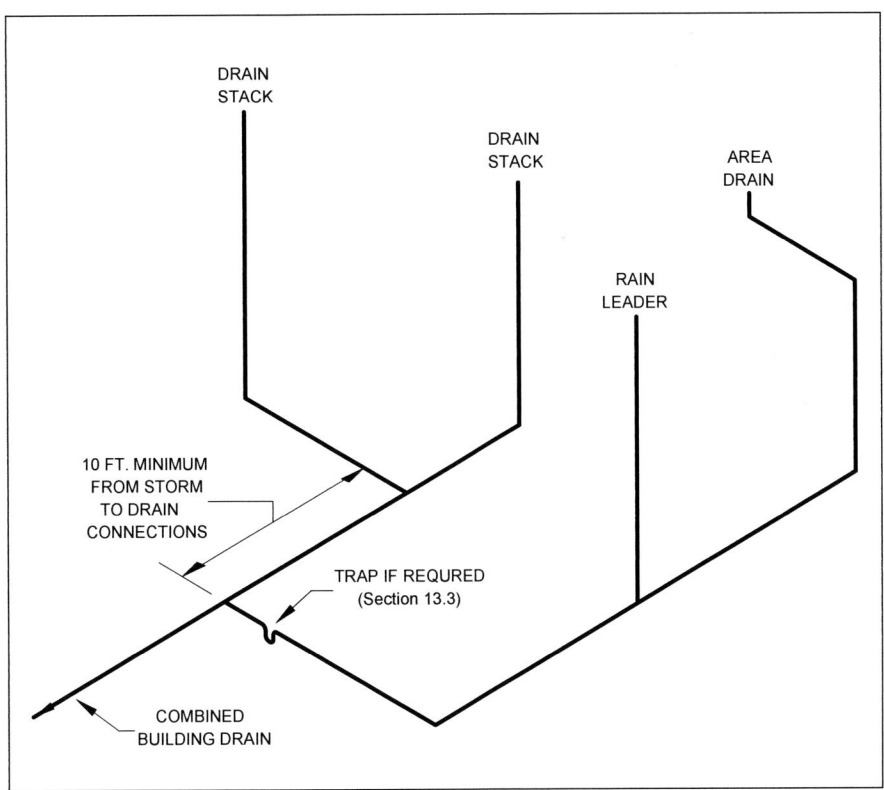

**Figure 1.2.15
COMBINED BUILDING DRAINAGE SYSTEM**

Building Drain, Sanitary: The lowest piping in the building sanitary drainage system that receives the discharge from drain stacks, horizontal branch drains, and fixture drains within the building and conveys the sewage to the building sanitary sewer which begins three (3) feet beyond the outside face of the building. *See Figure 1.2.16*

Building Drain, Storm: A building drain that conveys only storm water from the building to the building storm sewer. *See Figure 1.2.16*

Building Sewer, Combined: A building sewer that conveys both sewage and storm water. *See Figure 1.2.15*

Building Sewer, Sanitary: That part of the building sanitary drainage system that extends from the end of the sanitary building drain and conveys the sewage to a public sewer, an individual sewage disposal system, or other point of disposal. The building sanitary sewer begins at a point three (3) feet beyond the outside face of the building. It does not extend beyond any property lines if discharging to an off-site public sewer or off-site point of disposal. *See Figure 1.2.16*

Building Sewer, Storm: A building sewer that conveys only storm water from the building storm drain to a public storm sewer or other point of disposal. It does not extend beyond any property lines if discharging to an off-site public sewer or off-site point of disposal. *See Figure 1.2.16*

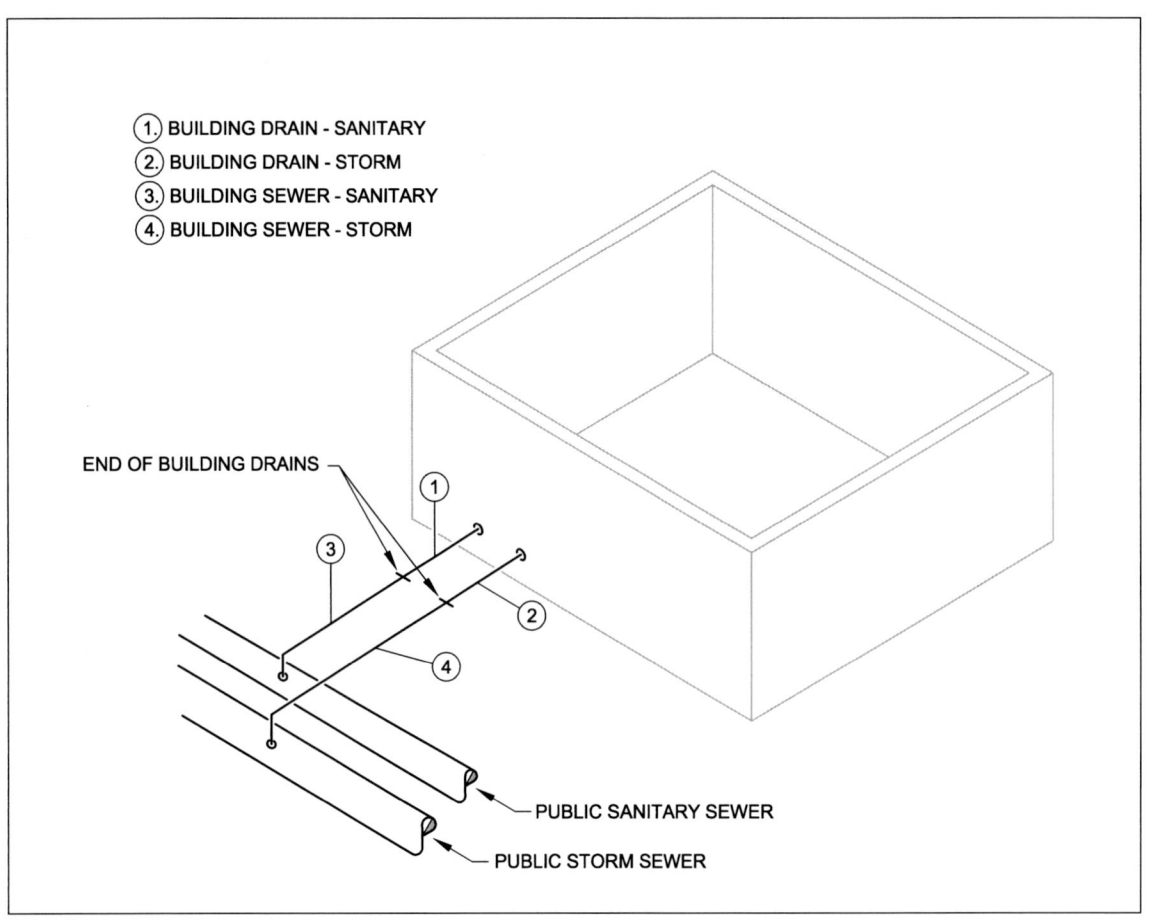

Figure 1.2.16
SEPARATE SANITARY AND STORMWATER BUILDING DRAINS AND SEWERS

Building Subdrain: That portion of a drainage system that does not drain by gravity into the building drain or building sewer. *See Figure 1.2.17*

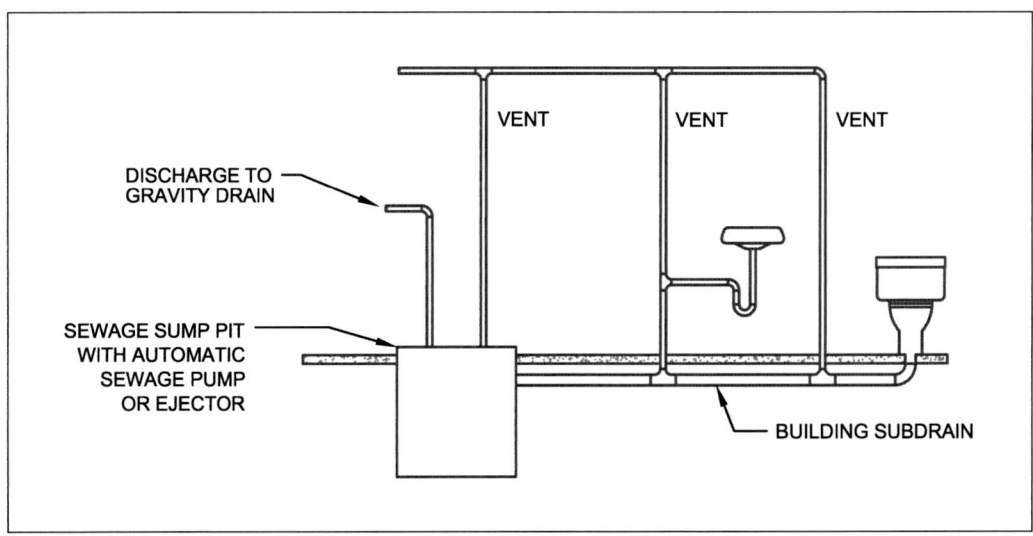

Figure 1.2.17
A BUILDING SUBDRAIN

Building Trap: A device, fitting, or assembly of fittings, installed in the building drain to prevent circulation of sewer gas between the building sewer and the drainage system in the building. *See Figure 1.2.18*

> *Comment: Building traps are currently installed only when required by the Authority Having Jurisdiction.*

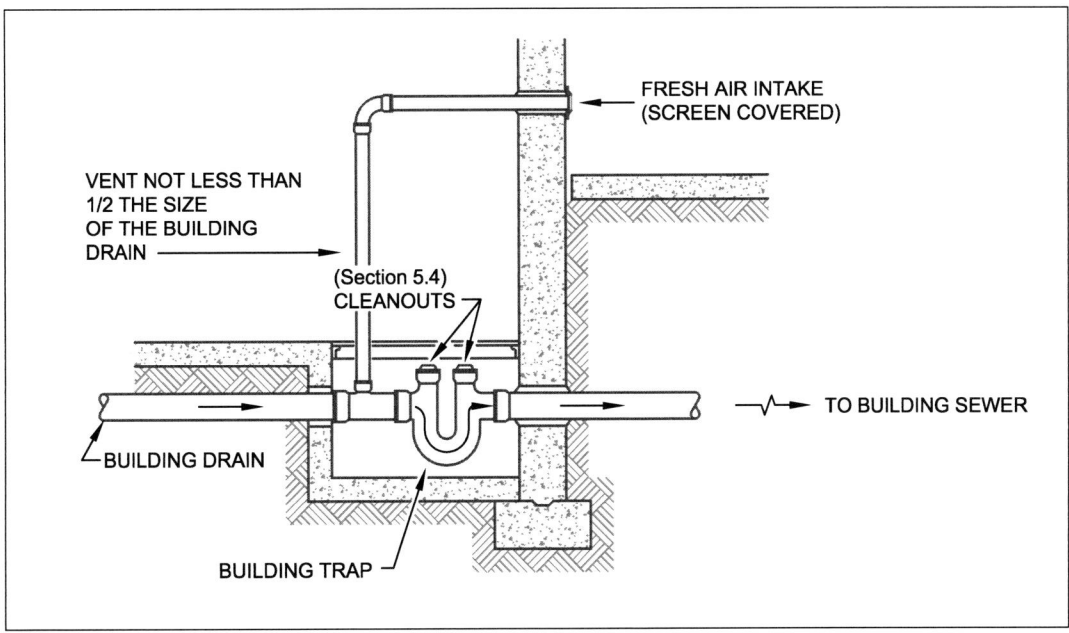

**Figure 1.2.18
A BUILDING TRAP**

Cesspool: A lined and covered excavation in the ground that receives the discharge of domestic sewage or other organic wastes from a drainage system, so designed as to retain the organic matter and solids, but permitting the liquids to seep through the bottom and sides.

> *Comment: Chapter 16 does not permit cesspools or cesspits, into which untreated sewage is discharged and allowed to seep into the ground. Chapter 16 requires septic tanks to retain the sewage until digested and absorption trenches or seepage pits for underground disposal of the effluent.*

Chemical Dispensing System: Equipment that mixes potable water with chemicals to provide the user with a chemical solution that is diluted to a fixed or adjustable amount and is ready for use.

Chemical Waste: Includes industrial liquid waste, process waste, diluted and undiluted acid waste and corrosive and non-corrosive chemical liquid waste. *See Sections 2.10, 3.11, and 9.4*

Circuit Vent: See "Vent, Circuit" *See Figure 12.13.1*

Clear Water Waste: Effluent in which impurity levels are less than concentrations considered harmful by the Authority Having Jurisdiction, such as cooling water and condensate drainage from refrigeration and air conditioning equipment, cooled condensate from steam heating systems, and residual water from ice making processes.

> *Comment: Refer to Section 9.1.8 for whether a clear water waste requires an air gap or an air break at its discharge into the drainage system.*

Clinical Sink: A sink designed primarily to receive wastes from bedpans, having a flushing rim, integral trap with a visible trap seal, and having the same flushing and cleansing characteristics as a water closet. *See Section 14.8*

Combination Fixture: A fixture combining one sink and laundry tray, or a two- or three-compartment sink or laundry tray in one unit. *See Figure 1.2.19*

> *Comment: Combination fixtures with waste outlets not more than 30 inches apart can have one trap.*

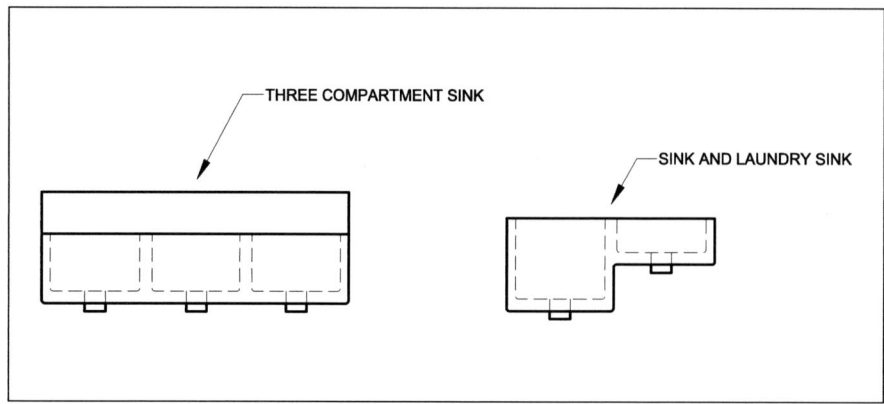

Figure 1.2.19
COMBINATION FIXTURES

Combination Thermostatic/Pressure Balancing Valve: See "Thermostatic/Pressure Balancing Valve, Combined"

Combination Waste and Vent System: A designed system of waste piping embodying the horizontal wet venting of one or more sinks or floor drains by means of a common waste and vent pipe adequately sized to provide free movement of air above the flow line of the drain. *See Figure 1.2.20 and Section 12.17*

> *Comment #1: Combination waste and vent piping systems are permitted where conditions preclude the installation of a conventionally vented system. Such systems are frequently used in exhibition halls and other spaces where long clear spans are required without partitions or pipe chases.*
>
> *Comment #2: Only floor drains, floor receptors, sinks, lavatories, and standpipes can be discharged into a combination waste and vent piping system.*

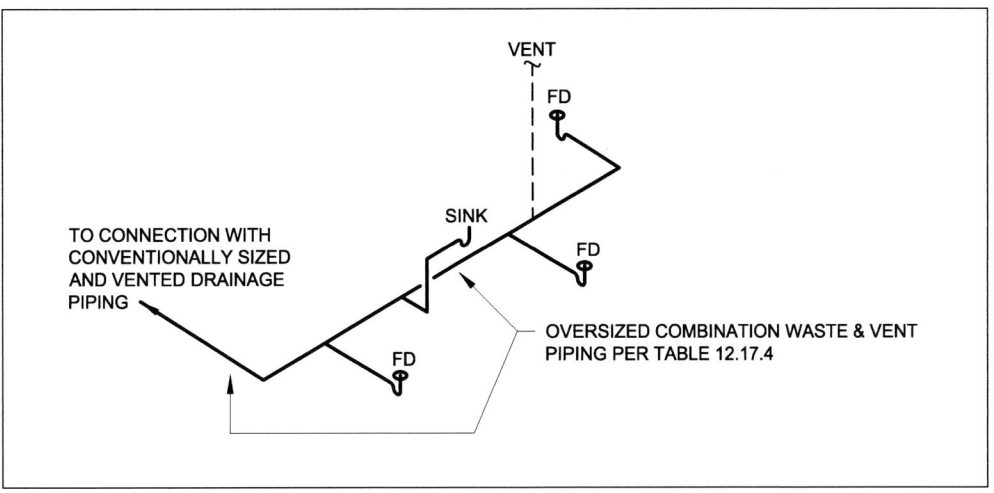

Figure 1.2.20
COMBINATION WASTE AND VENT PIPING

Combined Building Drain: See "Building Drain, Combined"

Combined Building Sewer: See "Building Sewer, Combined"

Commercial Kitchen: One or more rooms in a building that is licensed to prepare food to be served for consumption or process food to be packaged for distribution.

Common Laundry Room: A laundry facility for personal use by multiple individuals in commercial buildings and buildings having more than two dwelling units.

Common Vent: See "Vent, Common"

Conductor: A pipe within a building that conveys stormwater from a roof to its connection to a building storm drain or other point of disposal. *See Figure 1.2.21*

Comment: A vertical stormwater drain pipe on the exterior of a building is a leader.

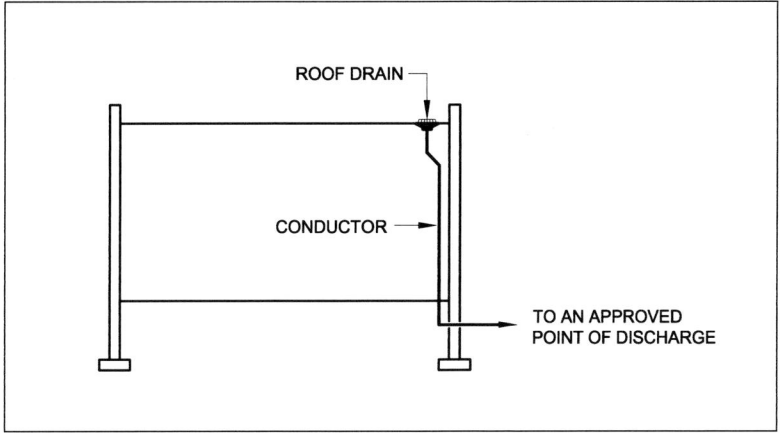

Figure 1.2.21
A STORMWATER CONDUCTOR

26   2015 National Standard Plumbing Code - Illustrated

Contamination: The impairment of the quality of the potable water that creates an actual hazard to the public health through poisoning or through the spread of disease by sewage, industrial fluids or waste. (See the definition of "pollution").

Continuous Vent: See "Vent, Continuous"

Continuous Waste Piping: Drain piping from two or three adjacent lavatories or sinks that connects them to a single trap. *See Figure 1.2.22*

Comment: Continuous waste piping can connect up to three adjacent sinks or lavatories to a single trap if the fixture outlets are no more than 30" apart. See Section 5.1.

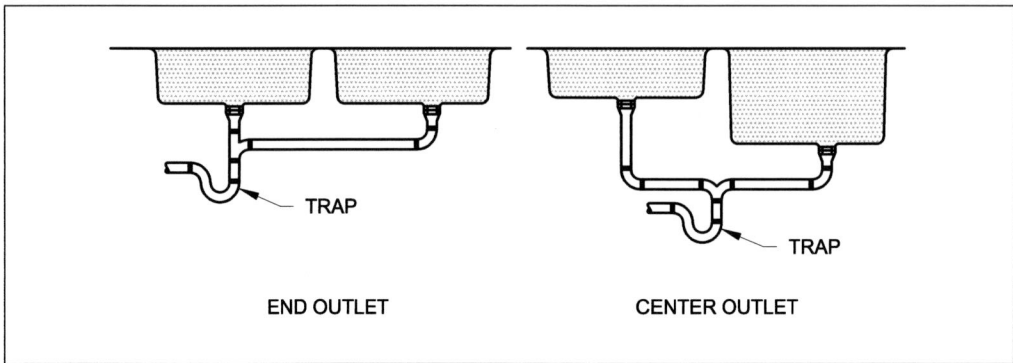

Figure 1.2.22
CONTINUOUS WASTE PIPING

Continuous Water Pressure: With regard to the proper application of backflow preventers, continuous pressure is if they are normally subjected to continuous water supply pressure for periods of more than twelve (12) hours.

Critical Level: The marking on a backflow prevention device or vacuum breaker established by the manufacturer, and usually stamped on the device by the manufacturer, that determines the minimum elevation above the flood level rim of the fixture or receptor served at which the device must be installed. When a backflow prevention device does not bear a critical level marking, the bottom of the vacuum breaker, combination valve, or the bottom of any approved device constitutes the critical level. *See Figure 1.2.23 and Sections 10.5.5.b, c, and e*

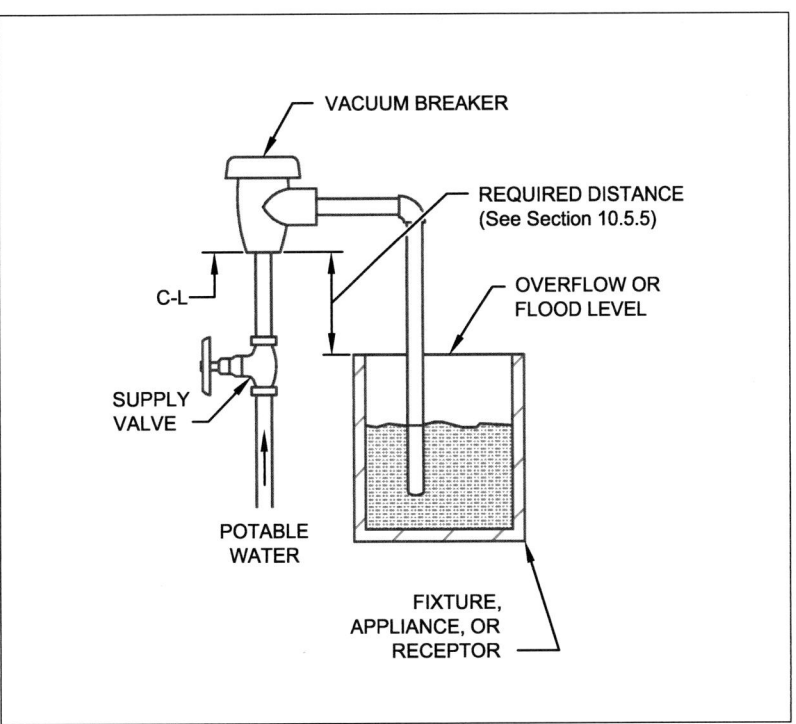

Figure 1.2.23
THE CRITICAL LEVEL (C-L) OF A VACUUM BREAKER

Cross Connection: Any connection or arrangement between two otherwise separate piping systems, one of which contains potable water and the other either water of questionable safety, steam, gas, or chemical, whereby there may be a flow from one system to the other, the direction of flow depending on the pressure differential between the two systems (See "Backflow and Back Siphonage). *See Figures 1.2.8 and 1.2.9*

> *Comment: The backflow of contamination into a potable water system through a cross connection can occur by back siphonage caused by the water system or back pressure from the source of contamination.*

CTS: An abbreviation for "copper tube size".

Day Care Center: A facility for the care and/or education of children ranging from 2-1/2 years of age to 5 years of age.

Day Nursery: A facility for the care of children less than 2-1/2 years of age.

Dead End: Inactive drain or vent piping that does not perform a function in the plumbing system.

Developed Length of Pipe: The length of a pipe line measured along the center line of the pipe and its fittings. Where pipe sizing tables are based on only the developed length of pipe, verify that there has been an allowance for an equivalent length of fittings.'

DFU: See "Fixture Unit (Drainage – DFU)"

DIPS: An abbreviation for "ductile iron pipe size".

Domestic Sewage: Sanitary sewage from assembly, business, educational, institutional, mercantile, and residential facilities. Not industrial or storm water drainage.

Double Check Valve Assembly: A backflow prevention device consisting of two independently acting check valves, internally force loaded to a normally closed position between two tightly closing shut-off valves, and with means of testing for tightness. *See Figure 1.2.25*

Double Offset: See "Offset, Double"

DR: An abbreviation for the "dimension ratio" of OD controlled plastic pipe that is based on the ratio of the outside diameter of the pipe divided by its minimum wall thickness. Lower DR ratings of the same pipe design and material have higher pressure ratings.

> *Comment: DR equals SDR for the same pipe material composition (designation code) with the same outside diameter and minimum wall thickness. The terms are used interchangeably depending on the pipe manufacturer or standards organization. Pipes of the same material composition with the same DR and SDR have the same pressure ratings regardless of pipe size.*

> *Comment: SIDR equals IDR for the same pipe material composition (designation code) with the same inside diameter and minimum wall thickness. The terms are used interchangeably depending on the pipe manufacturer or standards organization. Pipes of the same material composition with the same SIDR and IDR have the same pressure ratings regardless of pipe size.*

Drain Stack: A vertical drain, with or without offsets, that conveys drainage from fixture drains and horizontal fixture branch drains to the building drain or a branch of the building drain.

Drainage Sump: A liquid and air-tight tank that receives sewage and/or liquid waste, located below the elevation of a gravity drainage system, that is emptied by pumping.

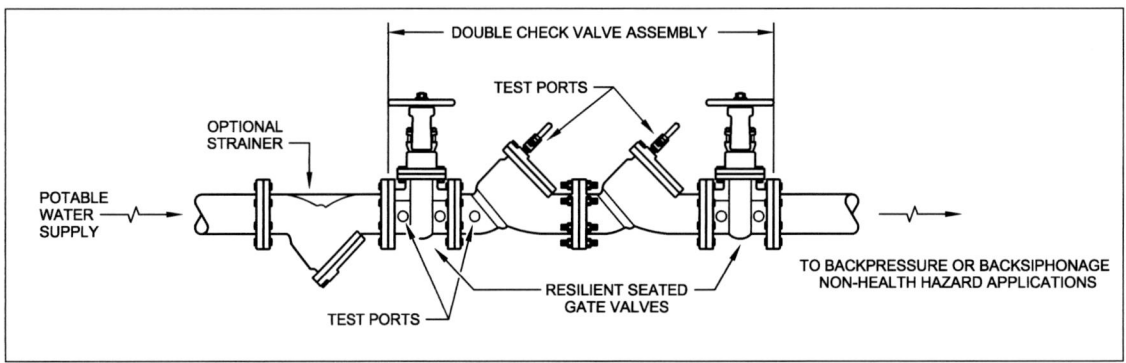

**Figure 1.2.25
A DOUBLE CHECK VALVE ASSEMBLY**

Drainage System, Sub-building: See "Building Subdrain" *See Figure 1.2.17*

Dry Vent: See "Vent, Dry"

Dry Well: See "Leaching Well"

Dual Vent: See "Vent, Common"

Dwelling Unit, Multiple: A room, or group of rooms, forming a single habitable unit with facilities that are used, or intended to be used, for living, sleeping, cooking and eating; and whose sewer connections and water supply, within its own premise, are shared with one or more other dwelling units. Multiple dwelling units include apartments, condominiums, and hotel and motel guest rooms.

Dwelling Unit, Single: A room, or group of rooms, forming a single habitable unit with facilities that are used, or intended to be used, for living, sleeping, cooking and eating; and whose sewer connections and water supply are, within its own premise, separate from and completely independent of any other dwelling.

DWV: An acronym for "drain-waste-vent" referring to sanitary drainage, waste, and venting systems. The term is equivalent to "soil-waste-vent" (SWV).

Effective Opening: The cross-sectional area at the point of water supply discharge, measured or expressed in terms of (1) diameter of a circle, or (2) if the opening is not circular, the diameter of a circle of equivalent cross-sectional area. *See Figures 1.2.26 and 1.2.27*

> *Comment: The required air gap distance for a water supply outlet is based on its effective opening. The air gap must be increased if the outlet is near a wall or other vertical surface. Refer to Table 10.5.2 for minimum air gaps for plumbing fixtures.*

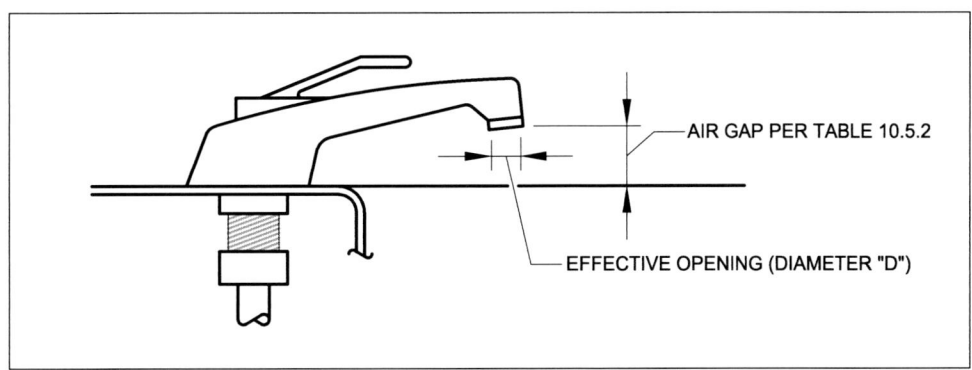

Figure 1.2.26
THE EFFECTIVE OPENING OF A POTABLE WATER OUTLET

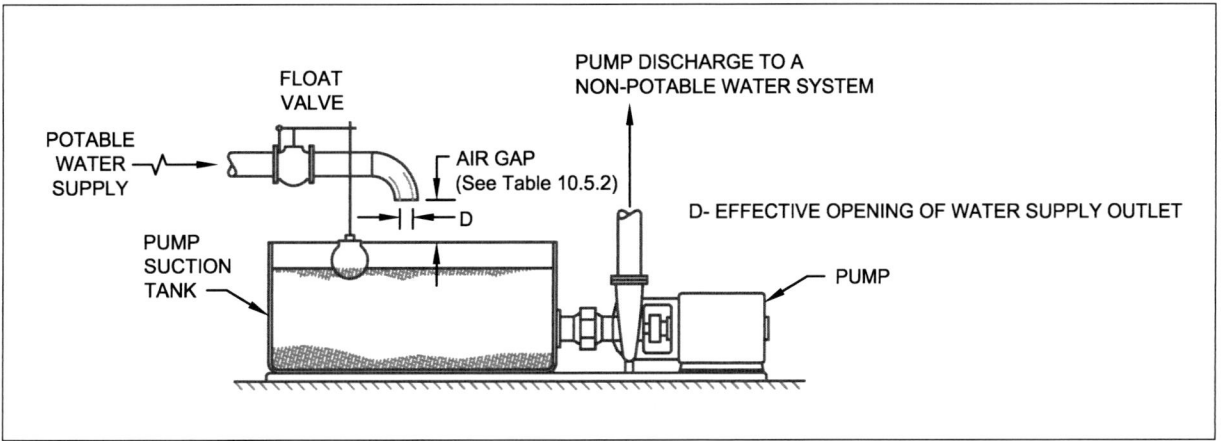

Figure 1.2.27
AN AIR GAP BETWEEN POTABLE AND NON-POTABLE WATER SYSTEMS

Equivalent Length of Fittings and Valves: The equivalent length of straight pipe that would produce the same resistance to flow as a specific fitting or valve of the same pipe size.

Equivalent Length of Pipe, Fittings, and Valves or Total Equivalent Length: The sum of the developed lengths of pipe in a pipe line and the equivalent lengths of the fittings and valves in that line.

Existing Plumbing System: A plumbing system, or any part thereof, that has been installed, was complete, and accepted by the Authority Having Jurisdiction for the adopted edition of the plumbing code under which its permit was issued.

Fixture: See "Plumbing Fixture"

Fixture Branch Drain: A branch drain that receives the discharge from two or more fixture drains.

> *Comment: See Table 11.5.1.B for the maximum number of drainage fixture units (DFU) permitted on each size of horizontal fixture branch.*

Fixture Branch, Supply: A branch of the water distribution system supplying one fixture. *See Figure 1.2.29*

> *Comment: See Table 10.14.2.A for minimum supply fixture branch sizes for various fixtures.*

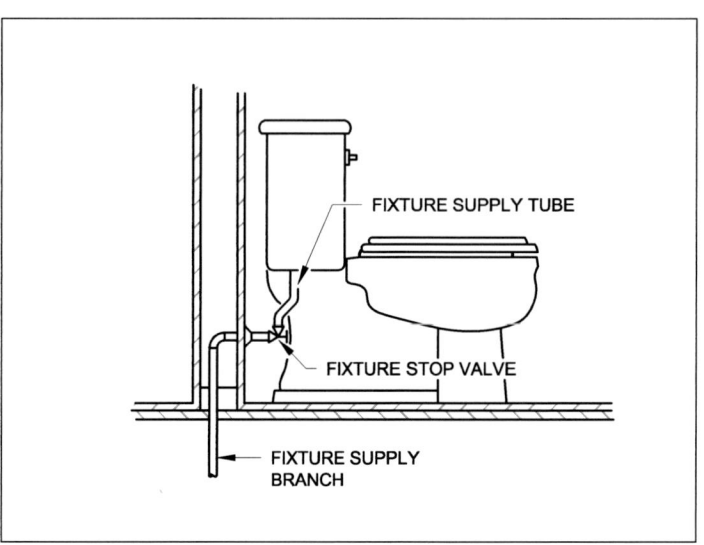

Figure 1.2.29
A FIXTURE SUPPLY BRANCH AND FIXTURE SUPPLY TUBE

Fixture Drain: The drain from the trap of a fixture to the junction of that drain with any other drain pipe.

Fixture Unit (Drainage - DFU): An index number that represents the load of a fixture on the drainage system so that the load of various fixtures in various applications can be combined. The value is based on the volume or volume rate of drainage discharge from the fixture, the time duration of that discharge, and the average time between successive uses of the fixture. One DFU was originally equated to a drainage flow rate of one cubic foot per minute or 7.5 gallons per minute through the fixture outlet. *See Table 11.4.1*

Fixture Unit (Water Supply - WSFU): An index number that represents the load of a fixture on the water supply system so that the load of various fixtures in various applications can be combined. The value is based on the volume rate of supply for the fixture, the time duration of a single supply operation, and the average time between successive uses of the fixture. Water supply fixture units were originally based on a comparison to a flushometer valve water closet, which was arbitrarily assigned a value of 10 WSFU. *See Table 10.14.2A. Also Tables B.5.2 and B.5.3.*

Flexible Water Connector: A connector under continuous pressure in an accessible location that connects a supply fitting, faucet, dishwasher, cloths washer, water heater, water treatment unit, or other fixture or equipment to a stop valve or its water supply branch pipe.

Flood Level Rim: The edge of the receptor or fixture over which water flows if the fixture is flooded. *See Figure 1.2.30*

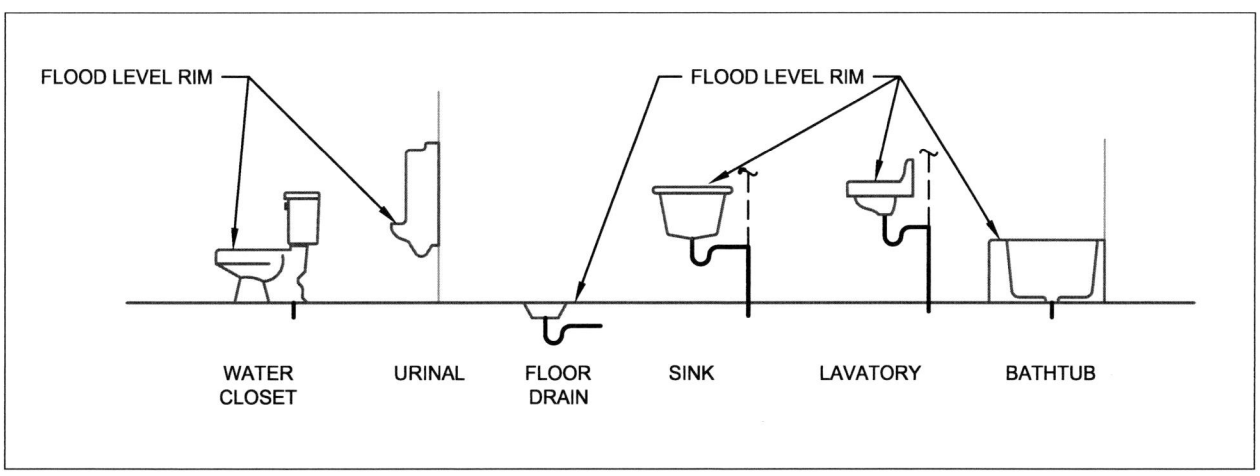

Figure 1.2.30
THE FLOOD LEVEL RIM OF FIXTURES

Flow Pressure: The pressure in the water supply pipe near the faucet or water outlet while the faucet or water outlet is fully open and flowing. *See Figure 1.2.31*

> *Comment: The minimum required flowing water pressure for most fixtures and appliances is 15 psig. Blowout water closets and blowout urinals require 25 psig minimum. Some one-piece water closets require 30 psig minimum and 1/2" supply tubes.*

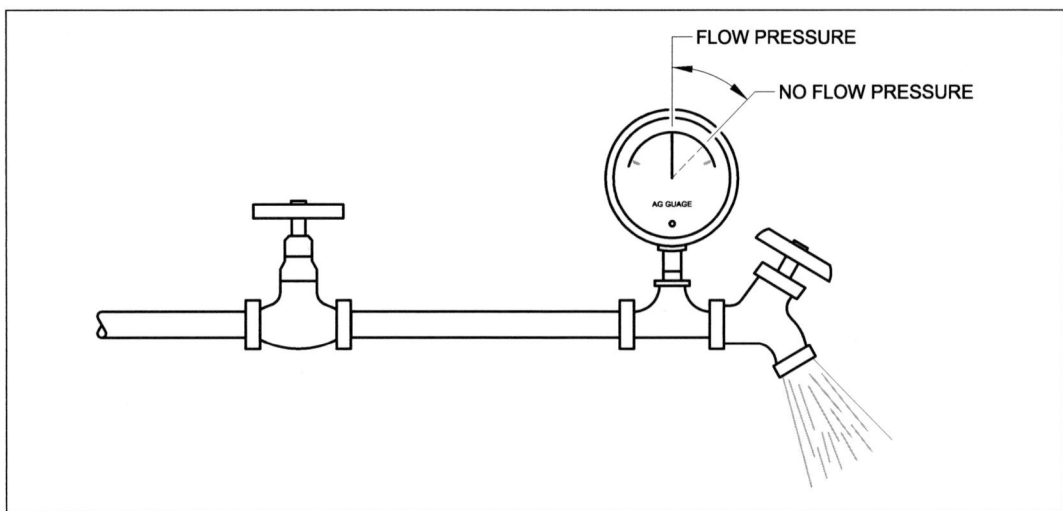

Figure 1.2.31
THE FLOW PRESSURE OF THE WATER SUPPLY TO AN OUTLET

Flush Pipes and Fittings: The pipe and fittings that connect a flushometer valve or elevated flush tank to a water closet, urinal, or bed pan washer.

Flushing Type Floor Drain: A floor drain that is equipped with an integral water supply connection, enabling flushing of the drain receptor and trap. *See Figure 1.2.32*

> *Comment: The water supply to flushing floor drains must be protected from backflow.*

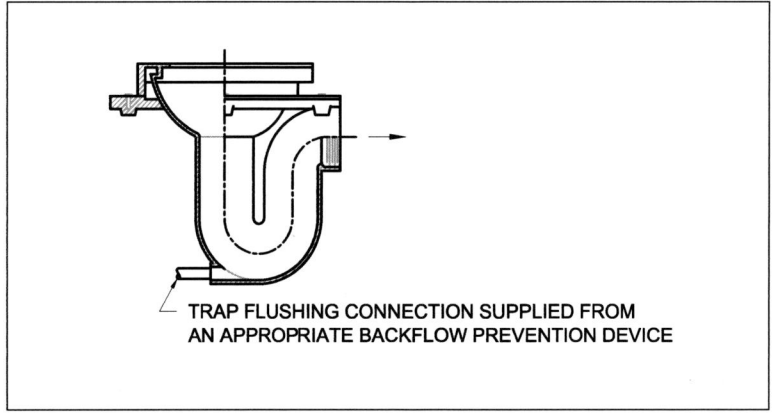

Figure 1.2.32
A FLUSHING TYPE FLOOR DRAIN

Flush Valve: A device located in a water closet flush tank for flushing water closets and similar fixtures. *See Figure 1.2.33*

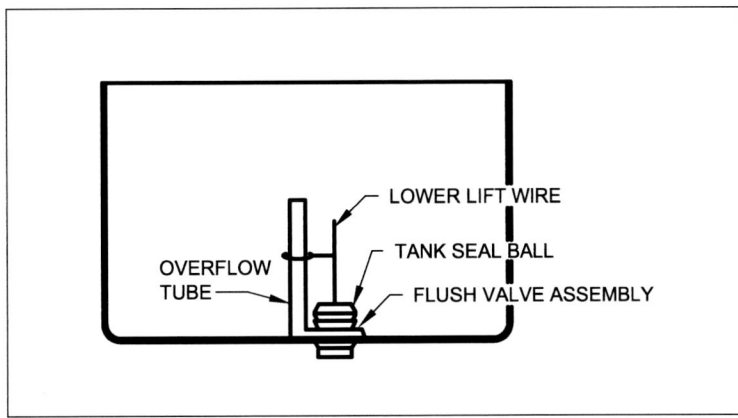

Figure 1.2.33
A FLUSH VALVE IN A WATER CLOSET FLUSH TANK

Flushometer Tank: A water closet flush tank that uses an air accumulator vessel to discharge a predetermined quantity of water into the closet bowl for flushing purposes. *See Figure 1.2.34*

> *Comment: Flushometer tanks are pressure-assisted flush tanks that store water for flushing water closets at the inlet water supply pressure, as opposed to gravity tanks. The discharge rate from flushometer tanks is approximately 35 gallons per minute.*

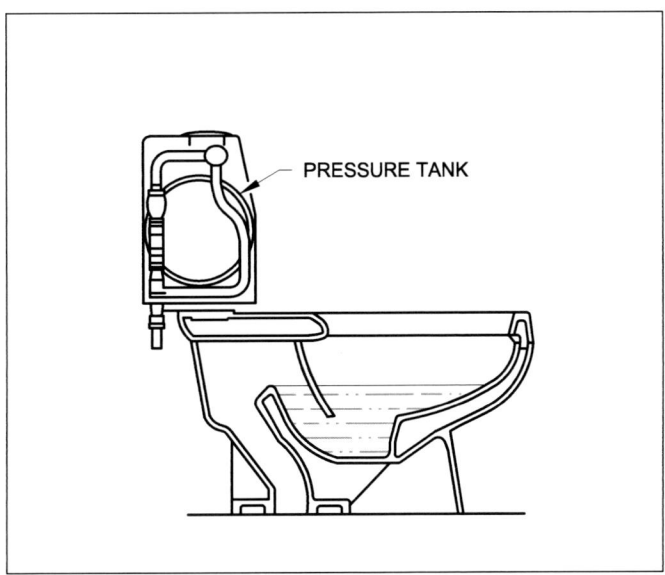

Figure 1.2.34
A PRESSURE-ASSISTED WATER CLOSET WITH A FLUSHOMETER TANK

Flushometer Valve: A device that discharges a predetermined quantity of water to fixtures for flushing and is closed by direct water pressure or other means. *See Figure 1.2.35*

Comment: Flushometer valves are typically used on public water closets and public urinals. They can be manually operated or electronically operated.

Figure 1.2.35
A FLUSHOMETER VALVE

FOG: Fats, oil, and grease in the waste drainage from food handling facilities.

Force Main: A main that delivers waste water under pressure from a sewage ejector or pump to its destination.

Foundation Drain: A drainage system installed near the footings under foundation walls to drain ground water away from the foundation walls to prevent water leakage into building spaces below grade.

Full-way valve: Full-way valves include gate valves, full port ball valves, and other valves that are identified by their manufacturer as full port or full bore.

Grade: The fall (slope) of a line of pipe in reference to a horizontal plane. *See Figure 1.2.36.*

> *Comment: See Table 11.5.1A for sizing sloping building drains, branches of the building drain, and building sewers. See Table 13.6.2 for sizing sloping stormwater drains.*

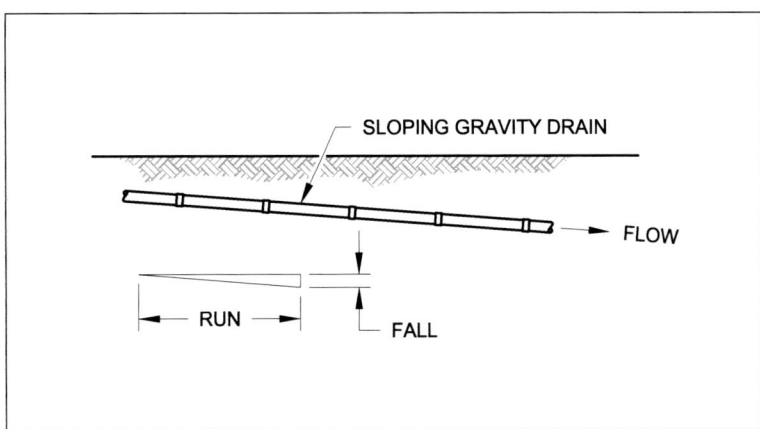

Figure 1.2.36
GRADE ON A SLOPING GRAVITY DRAIN

Graywater: Wastewater from lavatories, bathtubs, showers, and clothes washers that is collected and stored for use as non-potable water in the same facility.

Grease Interceptor: A plumbing appurtenance that is installed in the sanitary drainage system to intercept oily and greasy wastes from wastewater discharges, typically in commercial kitchens and food processing plants. Such equipment has the ability to intercept commonly occurring free-floating fats and oils.

Grease Removal (or Recovery) Device (GRD): A plumbing appurtenance that is installed in the sanitary drainage system to intercept and remove free-floating fats, oils, and grease from wastewater discharges, typically in commercial kitchens and food processing plants. Such equipment operates on a time or event-controlled basis and has the ability to remove the entire range of commonly occurring free-floating fats, oils, and grease automatically without intervention from the user except for maintenance. The removed material is essentially water-free, which allows for recycling of the removed product.

Grease Trap: See "Interceptor"

Grinder Pump: A pump for sewage that shreds or grinds the solids in the sewage that it pumps.

Ground Water: Subsurface water occupying a zone of saturation. (a) Confined ground water - a body of ground water overlaid by material sufficiently impervious to prevent free hydraulic connection with overlying ground water. (b) Free ground water - ground water in the zone of saturation extending down to the first impervious barrier.

Half-Bath: A room that contains one water closet and one lavatory within a dwelling unit.

Harvested Rainwater: Rainwater from the roof of a building that is collected and stored for use as non-potable water in the same facility.

Health Hazard: In backflow prevention, an actual or potential threat of contamination of the potable water supply to the plumbing system of a physical or toxic nature that would be a danger to health. Health hazards include any contamination that could cause death, illness, or spread of disease.

Horizontal Battery-Vented Drain: A horizontal drain that connects battery-vented fixtures to a drain stack, the building drain, or a branch of the building drain.

Horizontal Branch Drain: Horizontal drain piping, with or without offsets, that receives the discharge from two or more fixture drains and conveys the drainage to the building drain or a branch of the building drain. Horizontal branch drains do not connect to drain stacks.

Horizontal Fixture Branch: Horizontal drain piping, with or without offsets, that receives the discharge from two or more fixture drains and conveys the drainage to a vertical drain stack with the reduced allowable DFU loading for the branch as indicated in Table 11.5.1B.

> *Comment: The reduced DFU loading for the branch is to prevent the branch flow from impacting the flow in the stack, causing turbulence in the stack and restricted flow in the branch.*

Horizontal Pipe: Any pipe or fitting that makes an angle of less than 45° with the horizontal. *See Figure 1.2.37*

> *Comment: The sizing of offsets in drain stacks varies depending on whether the offset is horizontal or vertical. See Section 11.6.*

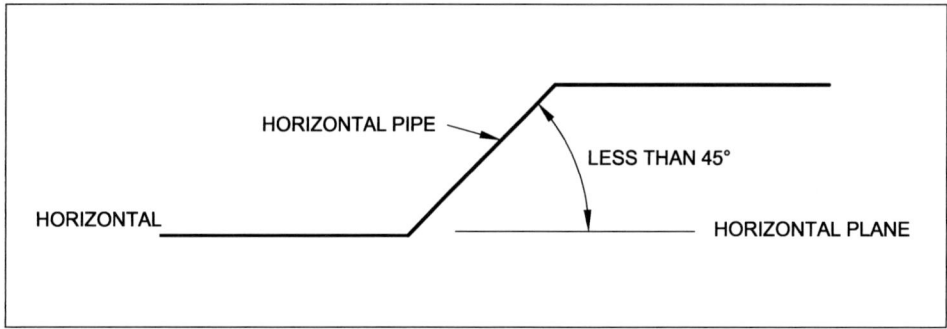

Figure 1.2.37
THE DEFINITION OF "HORIZONTAL PIPE"

Hot Water: Potable water that is heated for domestic use including bathing, washing, dishwashing, clothes washing, cleaning, and maintenance.

IDR: An abbreviation for the "dimension ratio" of ID controlled plastic pipe that is based on the ratio of the inside diameter of the pipe divided by its minimum wall thickness. Lower IDR ratings of the same pipe design and material have higher pressure ratings.

> *Comment: IDR equals SIDR for the same pipe material composition (designation code) with the same inside diameter and minimum wall thickness. The terms are used interchangeably depending on the pipe manufacturer or standards organization. Pipes of the same material composition with the same IDR and SIDR have the same pressure ratings regardless of pipe size.*

Indirect Connection (Waste): The introduction of waste into the drainage system by means of an air gap or air break.

Indirect Waste Pipe: A waste pipe that does not connect directly with the drainage system, but which discharges into the drainage system through an air break or air gap into a trap, fixture, receptor or interceptor. *See Figure 1.2.38*

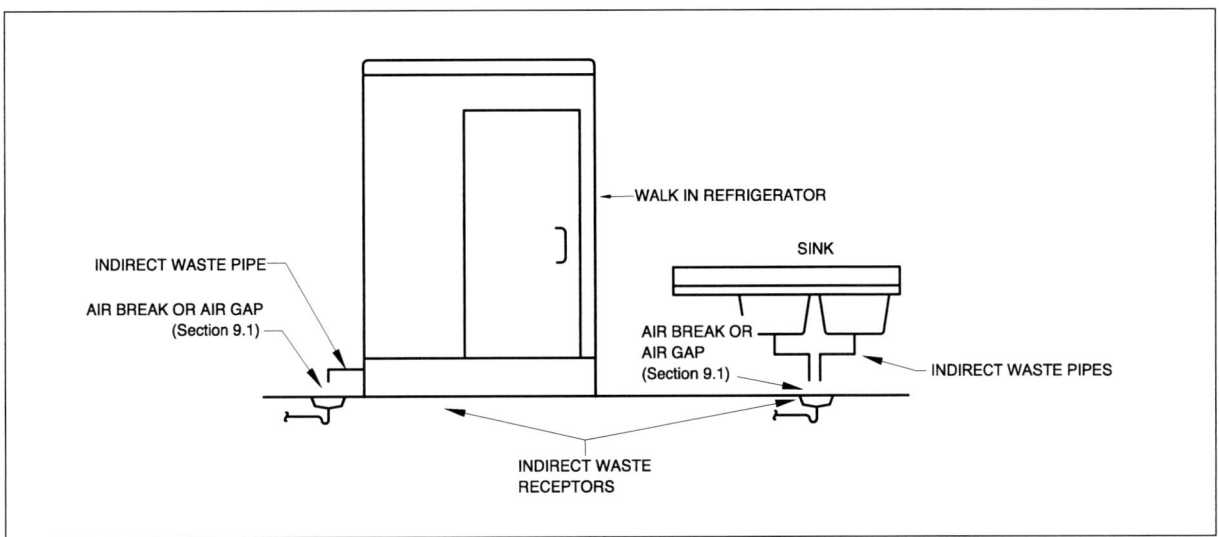

Figure 1.2.38
INDIRECT WASTE PIPES

Individual Vent: See "Vent, Individual"

Industrial Wastes: Liquid or liquid-borne wastes resulting from the processes employed in industrial and commercial establishments.

> *Comment: Industrial wastes must not be discharged into public sewers if they will damage the sewer or interfere with the operation of the sewage treatment plant.*

Interceptor: A device designed and installed so as to separate and retain deleterious, hazardous, or undesirable matter from normal wastes while permitting normal sewage or liquid wastes to discharge into the drainage system by gravity. *See Figure 1.2.39*

> *Comment: Interceptors include grease interceptors, oil/water separators, sand interceptors, solids interceptors, and neutralizing or dilution tanks.*

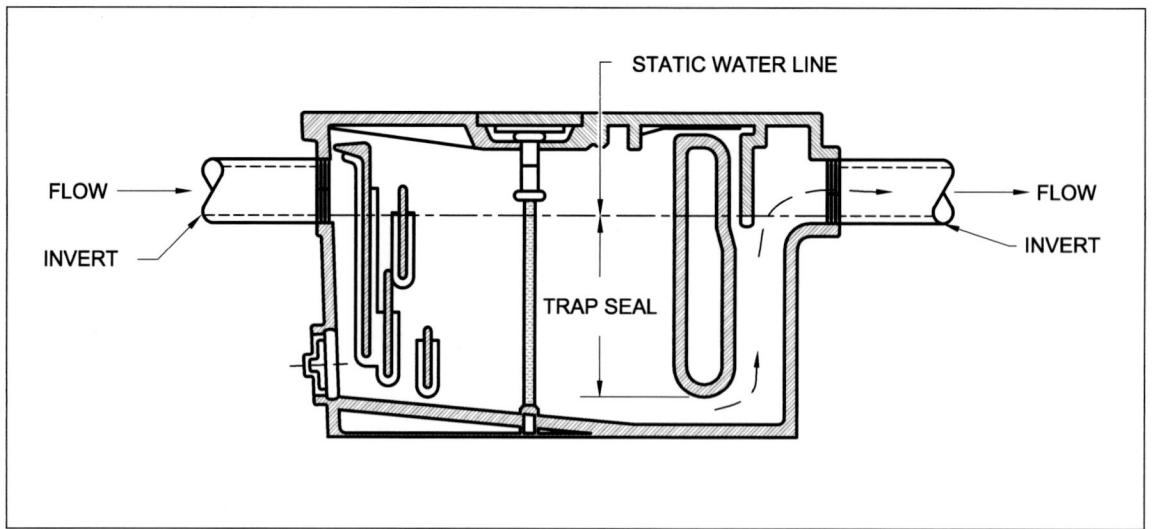

**Figure 1.2.39
AN INTERCEPTOR**

Invert: The lowest portion of the inside of a horizontal pipe. *See Figure 1.2.40*

> *Comment: Invert elevations are used to design and install drainage pipe at the required grade or slope.*

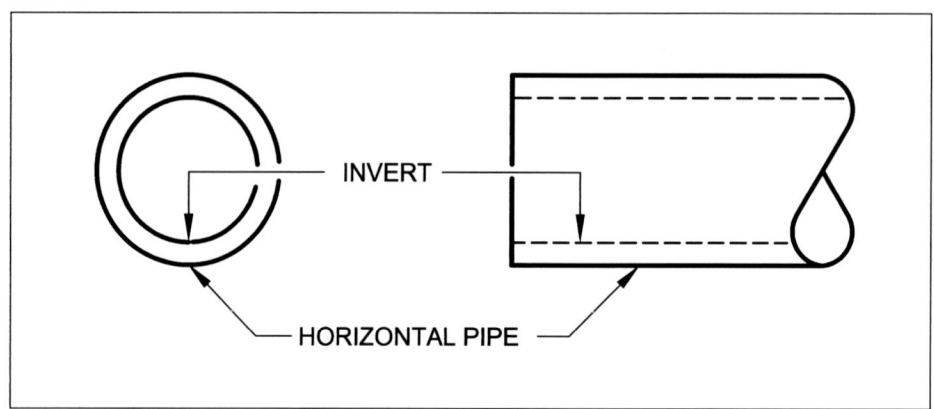

**Figure 1.2.40
THE INVERT OF A DRAIN PIPE**

IPS: An abbreviation for "iron pipe size".

Leaching Well or Pit: A pit or receptor having porous walls that permit the liquid contents to seep into the ground. *See Figure 1.2.41*

> *Comment: Leaching wells or pits are used to disperse the effluent from septic tanks into the ground for secondary treatment. They may supplement or be used in lieu of absorption trenches. Leaching wells or pits can only be used where there is very deep soil of good permeability and considerable depth to groundwater.*

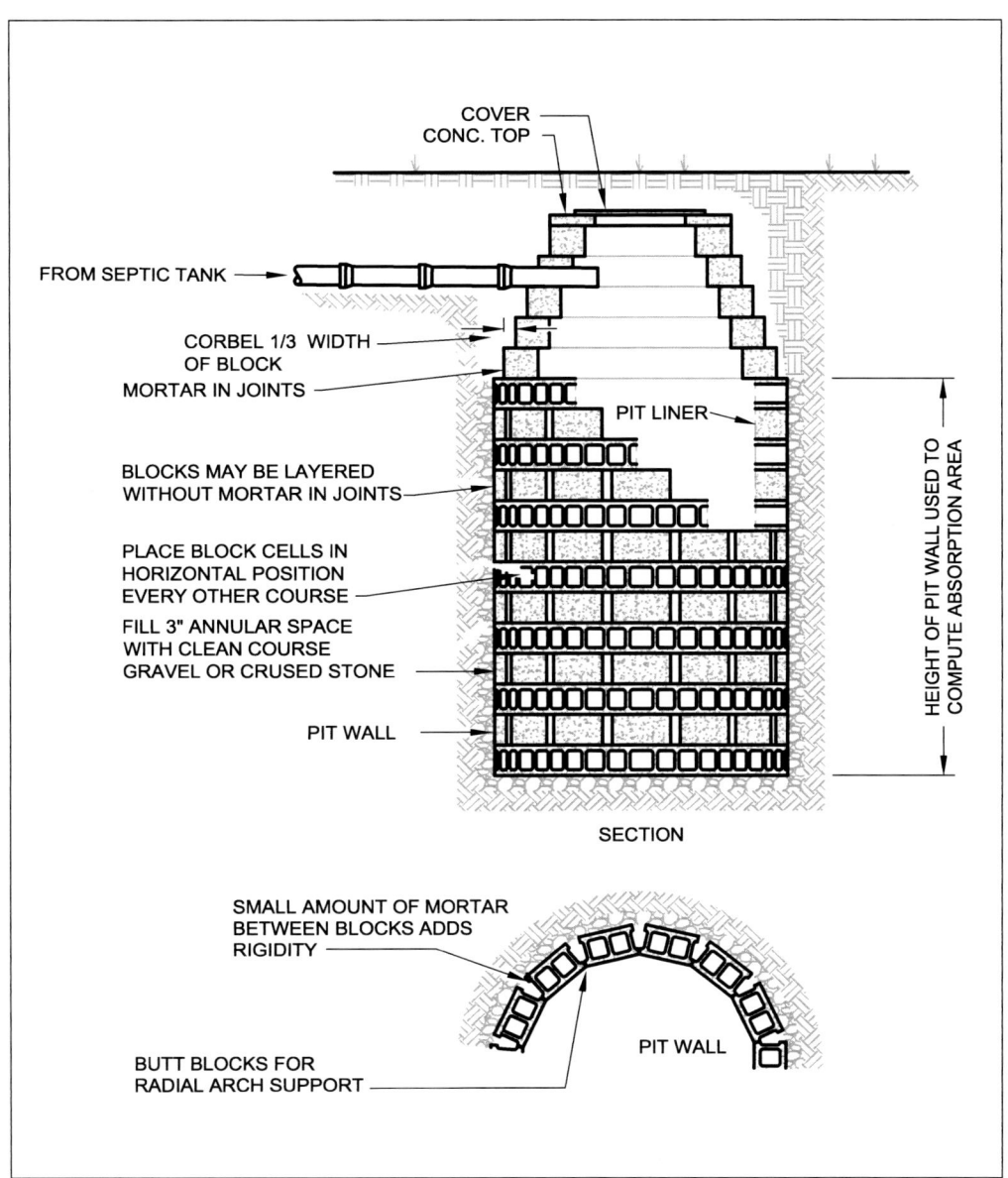

**Figure 1.2.41
A LEACHING OR SEEPAGE WELL OR PIT**

Lead Content:

a. Materials used in potable water supply systems, including piping, faucets, and valves, shall be "lead-free" as defined by current Federal Law.

b. Drinking water system components shall comply with the lead leachate requirements of NSF 61. Refer to Section 3.1.5 for components that are within the scope of NSF 61.

Leader: An exterior vertical drain pipe for conveying storm water from roof or gutter drains.
See Figure 1.2.42

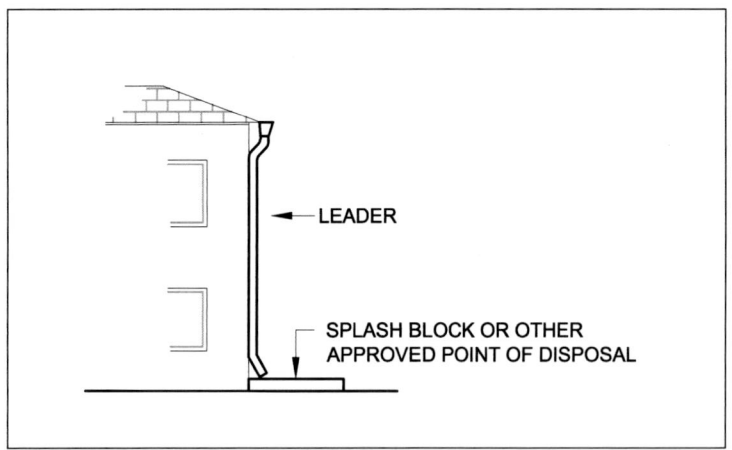

Figure 1.2.42
A STORMWATER LEADER

Loop Vent: See "Vent, Loop"

Macerating Toilet System: A system that collects drainage from a single water closet, lavatory and/or bathtub located in the same room. It consists of a receiving container, a grinder pump, and associated level controls. The system pumps shredded or macerated sewage up to a point of discharge.

May: The word "may" is a permissive term.

Medical Gas System: The complete system used to convey medical gases for direct application from central supply systems (bulk tanks, manifolds and medical air compressors) through piping networks with pressure and operating controls, alarm warning systems, etc., and extending to station outlet valves at use points.

Medical Vacuum Systems: A system consisting of central-vacuum-producing equipment with pressure and operating controls, shut-off valves, alarm warning systems, gauges and a network of piping extending to and terminating with suitable station inlets to locations where suction may be required.

Multiple Dwelling: A building containing two or more dwelling units.

Non-Health Hazard: In backflow prevention, an actual or potential threat to the physical properties or potability of the water supply to the plumbing system, but which would not constitute a health or system hazard.

Non-Potable Water: Water not safe for drinking, personal or culinary use.

Nominal Size: (Pipe or Tube): The industry-recognized size of a plumbing pipe or tube that is not necessarily an actual dimension. It indicates the size of the pipe or tube as indicated in its material standard listed in Table 3.1.3.

Nuisance: Public nuisance at common law or in equity jurisprudence; whatever is dangerous to human life or detrimental to health; whatever building, structure, or premises is not sufficiently ventilated, sewered, drained, cleaned, or lighted in reference to its intended or actual use; and whatever renders the air, human food, drink or water supply unwholesome.

Offset: A combination of elbows or bends that brings one section of the pipe out of line but into a line parallel with the other section. *See Figure 1.2.44 - Parts A and B for single offsets*

Comment: Offsets can occur in vertical and horizontal piping. Offsets can consist of 90-degree, 45-degree, or other angle fittings. Single offsets have one change of alignment.

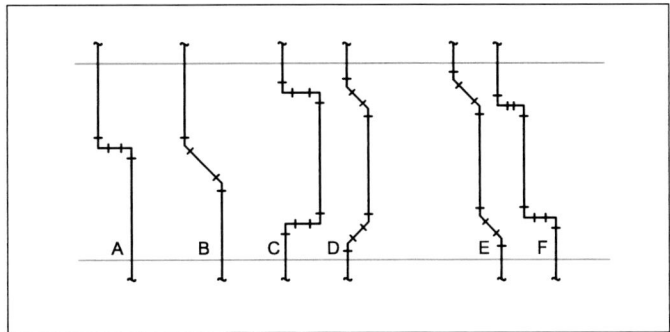

Figure 1.2.44
OFFSETS IN PIPING

Offset, Double: Two offsets installed in succession or series in a continuous pipe. *See Figure 1.2.44 - Parts E and F*

Comment: Double offsets have two offsets, both away from the original alignment of the pipe.

Offset, Return: A double offset installed so as to return the pipe to its original alignment. *See Figure 1.2.44 - Parts C and D*

Comment: Return offsets have two offsets, both in different directions. The second offset does not necessarily bring the pipeline back into exact alignment with the first. The alignment of the piping on both sides of the offset is not necessarily exactly the same laterally.

Oil Interceptor: See "Interceptor"

Pitch: See "Grade"

Plenum: An enclosed portion of the building structure, other than an occupiable space being conditioned, that is designed to allow air movement, and thereby serve as part of an air distribution system.

Plumbing: The practice, materials, and fixtures within or adjacent to any building structure or conveyance, used in the installation, maintenance, extension, alteration and removal of any piping, plumbing fixtures, plumbing appliances, and plumbing appurtenances in connection with any of the following:
    a. Sanitary drainage system and its related vent system
    b. Storm water drainage facilities
    c. Public or private potable water supply systems
    d. The initial connection to a potable water supply upstream of any required backflow prevention devices and the final connection that discharges indirectly into a public or private disposal system
    e. Medical gas and medical vacuum systems
    f. Indirect waste piping including refrigeration and air conditioning drainage
    g. Liquid waste or sewage, and water supply, of any premises to their connection with an approved water supply system or to an acceptable disposal facility
    h. Reclaimed water piping

    i. Graywater systems
    j. Harvested rainwater systems

NOTE: The following are excluded from the definition:

    1. All piping, equipment or material used exclusively for environmental control.
    2. Piping used for the incorporation of liquids or gases into any product or process for use in the manufacturing or storage of any product, including product development.
    3. Piping used for the installation, alteration, repair or removal of automatic sprinkler systems installed for fire protection only.
    4. The related appurtenances or standpipes connected to automatic sprinkler systems or overhead or under-ground fire lines beginning at a point where water is used exclusively for fire protection.
    5. Piping used for lawn sprinkler systems downstream from backflow prevention devices.

Plumbing Appliance: Any one of a special class of plumbing fixture that is intended to perform a special plumbing function. Its operation and/or control may be dependent upon one or more energized components, such as motors, controls, heating elements, or pressure or temperature-sensing elements. Such fixtures may operate automatically through one or more of the following actions: a time cycle, a temperature range, a pressure range, a measured volume or weight; or the fixture may be manually adjusted or controlled by the user or operator.

> *Comment: Plumbing appliances include clothes washers, dishwashers, food-waste-disposal and grinder units, water heaters, water softeners, and similar devices. Refer to Table 3.1.3 - Part VII for listed appliances.*

Plumbing Appurtenance: A manufactured device, a prefabricated assembly, or an on-the-job assembly of component parts, that is an adjunct to the basic piping system and plumbing fixtures. An appurtenance demands no additional water supply, nor does it add any discharge load to a fixture or to the drainage system. It is presumed that an appurtenance performs some useful function in the operation, maintenance, servicing, economy, or safety of the plumbing system.

> *Comment: Some examples of plumbing appurtenances are water filters, backflow prevention devices, backwater valves, interceptors, separators, and neutralizing or dilution tanks. Refer to Table 3.1.3 - Part VIII for listed appurtenances. Plumbing appurtenances do not change the load on the water supply or drainage system.*

Plumbing Code: These regulations and any deletions, modifications, and additions made by the Adopting Agency.

Plumbing Fixture: A receptacle or device connected to the water distribution system of the premises, and demands a supply of water there from; or discharges used water, liquid-borne waste materials, or sewage either directly or indirectly to the drainage system of the premises; or which requires both a water supply connection and a discharge to the drainage system of the premises. Plumbing appliances as a special class of fixture are further defined.

> *Comment: Plumbing fixtures include water closets, urinals, bidets, lavatories, bathtubs, whirlpool baths, showers, sinks, floor drains, and receptors. Refer to Table 3.1.3 - Part V for listed plumbing fixtures.*

Plumbing System: Includes the water supply and distribution piping, plumbing fixtures and traps; drain and vent piping; sanitary and storm drains and building sewers, including their respective connections, devices and appurtenances within the property lines of its site.

Pollution (of Potable Water): An impairment of the quality of the potable water to a degree that does not create a hazard to the public health but that does adversely and unreasonably affect the aesthetic qualities of such potable water for domestic use. (See the definition of "contamination").

Potable Water: Water free from impurities present in amounts sufficient to cause disease or harmful physiological effects and conforming in its bacteriological and chemical quality to the requirements of the Public Health Service Drinking Water Standards or the regulations of the public health authority having jurisdiction.

Powder Room: See "Half-Bath"

Pressure Assisted Water Closet: See "Water Closet, Pressure Assisted"

Pressure Balancing Compensating Valve: An automatic compensating valve that senses its incoming hot and cold water pressures and compensates for fluctuations in order to stabilize its adjusted outlet water temperature. See "Automatic Compensating Valve".

Pressure Relief Connection: For drain stacks that require a vent stack, their pressure relief connection is the lower connection of their vent stack to a point near the base of the drain stack to relieve the increased pressure at that point.

Private Sewage Disposal System: A system for disposal of domestic sewage by means of a septic tank or mechanical treatment, designed for use apart from a public sewer to serve a single establishment or building. *See Chapter 16*

Private Use, Public Use:

Private Use - Plumbing facilities for the private and restricted use of one or more individuals in dwelling units (including hotel and motel guest rooms) and other plumbing facilities that are not intended for public use. Refer to the definition of "Public Use".

Public Use - Plumbing facilities intended for the unrestricted use of more than one individual (including employees) in assembly occupancies, business occupancies, public buildings, transportation facilities, schools and other educational facilities, office buildings, restaurants, bars, other food service facilities, mercantile facilities, manufacturing facilities, military facilities, and other plumbing installations that are not intended for private use. Refer to the definition of "Private Use".

Private Water Supply: A supply, other than an approved public water supply, which serves one or more buildings. *See Figure 1.2.46 and Chapter 17*

Property Line: The recorded boundary of a plot of land.

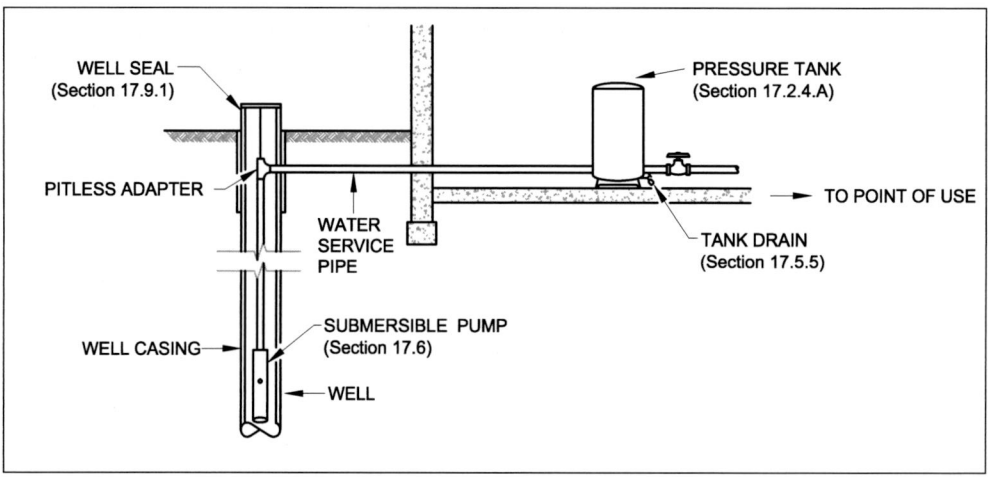

Figure 1.2.46
A TYPICAL PRIVATE WATER SUPPLY SYSTEM

PS: An abbreviation for the relative "pipe stiffness" of plastic pipe that is based on the ratio of wall thickness, mean radius, and flexural modulus of elasticity. Higher PS ratings of the same pipe design and material represent stiffer pipe.

> *Comment: This Code does not cover public sewers. The design and construction of public sewers is regulated by the sewage authority, a public works department, or other Authority Having Jurisdiction.*

Public Toilet Room: Toilet facilities that are intended for public use in accordance with the definition of "Private Use, Public Use").

Public Use: See "Private Use, Public Use").

Pump Assisted Water Closet: See Water Closet, Pump Assisted.

Push-Fit Fittings: A type of mechanical joint used with copper, CPVC and/or PEX that is either permanent or removable and may be used separately or integrated into plumbing fitting devices used in domestic or commercial applications in potable water distribution systems.

Radon Gas: A colorless, odorless, tasteless, radioactive, and chemically inert gas that is naturally produced in the ground through the normal decay of uranium, which decays to radium, which then decays to radon.

Readily Accessible: See "Accessible and Readily Accessible"

Receptor: A fixture or device which receives the discharge from indirect waste pipes. *See Figure 1.2.47*

> *Comment: Receptors include floor drains, standpipes, and certain sinks.*

Reclaimed Non-Potable Water: Effluent from a wastewater treatment facility that complies with the requirements of the Authority Having Jurisdiction for reuse as non-potable water for flushing water closets, urinals, landscape irrigation, and other approved uses.

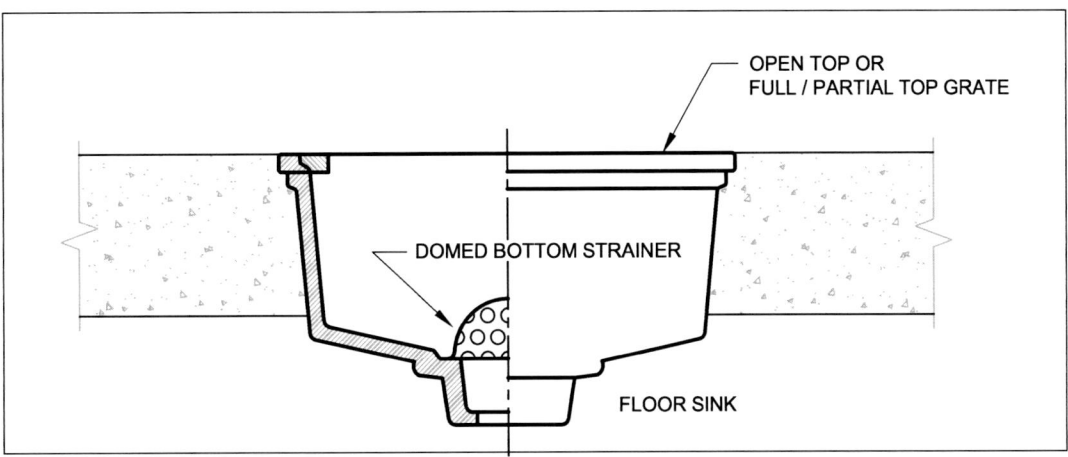

Figure 1.2.47
AN INDIRECT WASTE RECEPTOR

Reduced Pressure Principle Backflow Preventer: A backflow prevention assembly with shutoff valves consisting of two independently acting check valves, internally force loaded to a closed position and separated by an intermediate chamber (or zone), in which there is an automatic relief means of venting to atmosphere, internally loaded to an open position, and with means for testing for tightness of the checks and opening of the relief means. *See Figure 1.2.48*

> *Comment: Reduced pressure principle preventers provide the highest level of protection against back pressure and back siphonage. The shutoff valves on these assemblies are resiliently seated to assure tight close-off for testing and repairing. Provisions must be made to drain any discharge from the relief vent outlet, which may occur due to normal variations in system pressure.*

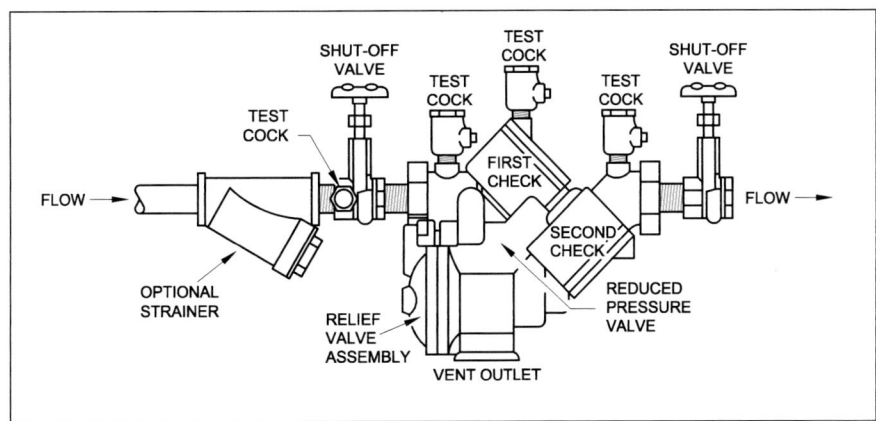

Figure 1.2.48
A REDUCED PRESSURE BACKFLOW PREVENTER ASSEMBLY

Relief Vent: See "Vent, Relief"

Return Offset: See "Offset, Return"

Reuse of Reclaimed Water: Where reclaimed water from a reclaimed water supplier is used for a non-potable water application by an end user.

Revent: An individual vent that is installed for a fixture that is otherwise vented by battery venting or by connection to a drain stack by an oversized drain pipe having limited developed length and slope.

Reventing: The installation of individual vents for fixtures that are also vented by other means.

> *Comment: Reventing is not required in Sections 12.12.2, 12.12.3, and 12.12.4.*

Riser, Water Supply: A water supply pipe that extends vertically one full story or more with branches to convey water to plumbing fixtures and water-supplied equipment. *See Figure 1.2.49*

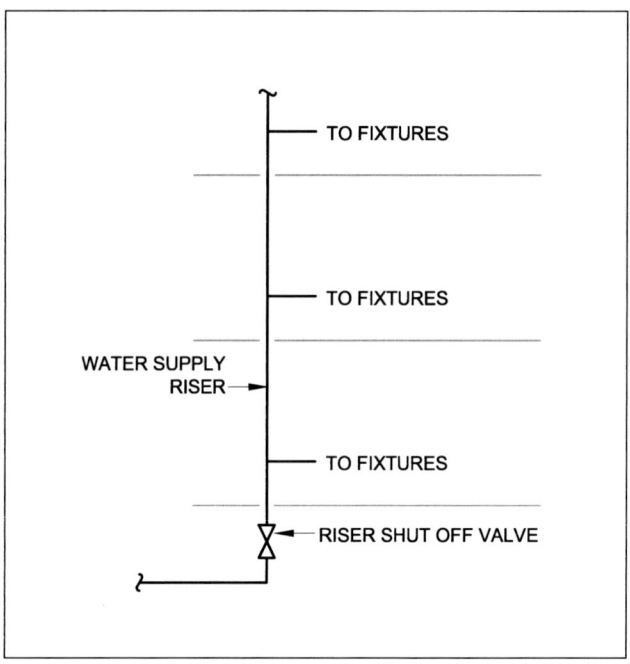

Figure 1.2.49
A WATER SUPPLY RISER

Roof Drain: A drain installed to receive rainwater collecting on the surface of a roof and to discharge it into a conductor or leader. *See Figure 1.2.50*

> *Comment: Roof drains include flat deck drains and scupper drains.*

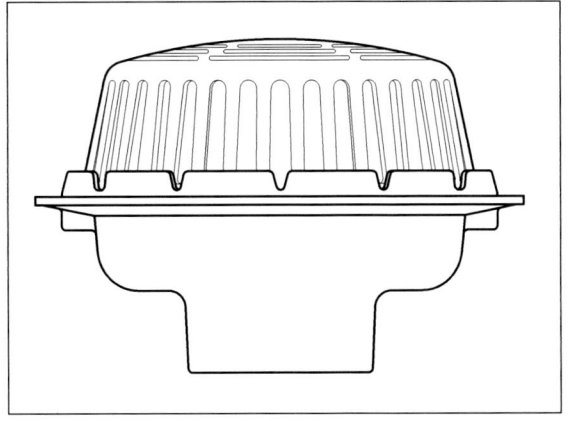

Figure 1.2.50
A GENERAL PURPOSE ROOF DRAIN

Rough Plumbing: The installation of all parts of the plumbing system that can be completed prior to the installation of fixtures. This includes drainage, water supply, and vent piping, and the necessary fixture supports, or any fixtures that are built into the structure.

Sand Filter: A treatment device or structure, constructed above or below the surface of the ground, for removing solid or colloidal material of a type that cannot be removed by sedimentation, from septic tank effluent. *See Figure 1.2.51 and Section 16.12*

> *Comment #1: Sand filters provide additional treatment of septic tank effluent. They are used where the soil depth is shallow and cannot provide sufficient secondary treatment of the effluent.*
>
> *Comment #2: Sand interceptors (or sand traps) are associated with oil/water separators in Chapter 6.*

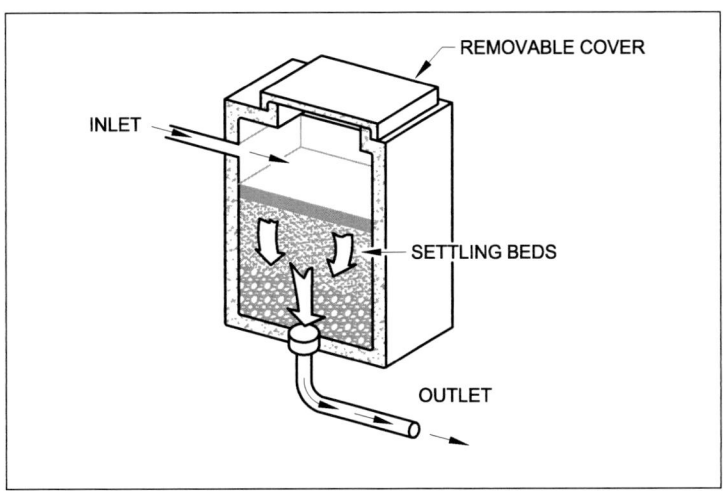

Figure 1.2.51
A SAND FILTER FOR SEPTIC TANK EFFLUENT

Sand Interceptor: See "Interceptor" *See Section 6.4*

> *Comment: Sand interceptors (or sand traps) are associated with oil/water separators in Chapter 6. Sand filters are associated with the effluent from septic tanks in Chapter 16.*

Sanitary Sewer: A system of public or private drain piping or conduits that convey sewage from one or more properties to a point of treatment or disposal.

SDR: An abbreviation for the "standard dimension ratio" of OD controlled plastic pipe that is based on the ratio of the outside diameter of the pipe divided by its minimum wall thickness. Lower SDR ratings of the same pipe design and material have higher pressure ratings.

> *Comment: SDR equals DR for the same pipe material composition (designation code) with the same outside diameter and minimum wall thickness. The terms are used interchangeably depending on the pipe manufacturer or standards organization. Pipes of the same material composition with the same SDR and DR have the same pressure ratings regardless of pipe size.*

Seepage Well or Pit: See "Leaching Well"

Self-draining: With regard to dead ends in vent piping, piping that will drain by gravity and not retain waste or sewage if its associated drainage piping is functioning normally.

Septic Tank: A watertight receptacle that receives the discharge of a building sanitary drainage system or part thereof; and that is designed and constructed so as to separate solids from the liquid, digest organic matter through a period of detention, and allow the liquids to discharge into the soil outside of the tank through a system of open joint or perforated piping, or a seepage pit. *See Figure 1.2.52 and Section 16.6*

> *Comment: Figure 1.2.52 shows baffles at the inlet and outlet connections with two access covers.*

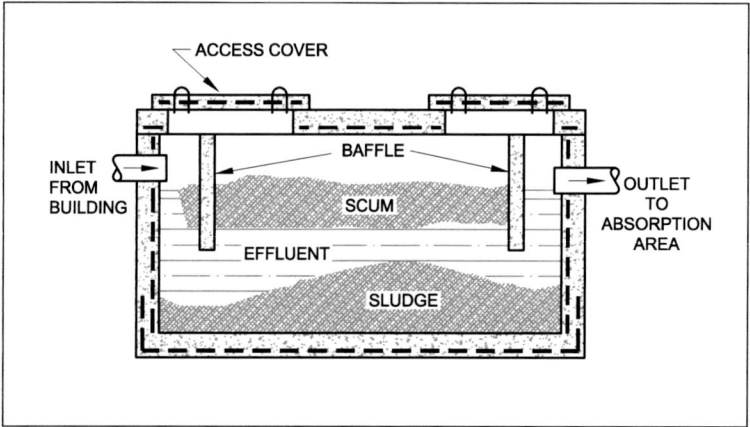

Figure 1.2.52
A TYPICAL SEPTIC TANK

Service Sink: A sink or receptor intended for custodial use that is capable of being used to fill and empty a janitor's bucket. Included are mop basins, laundry sinks, utility sinks, and similar fixtures. *(See Table 7.21.1)*

Sewage: Liquid discharge into the plumbing drainage system from a fixture, appliance, or other connection that contains urine and/or fecal matter.

Sewage Ejector, Pneumatic Type: A unit that uses compressed air to discharge and lift sewage to a gravity sewage system. *See Sections 11.7 and 12.14.3*

Sewage Pump or Pump-Type Ejector: A non-clog or grinder-type sewage pump or ejector. Sewage pumps and pump-type ejectors are either the submersible or vertical type. *See Figure 1.2.54 and Section 11.7*

> *Comment: Figure 1.2.54 shows a pedestal-type sewage pump. Sewage pumps can also be submersible pumps or dry pit centrifugal pumps.*

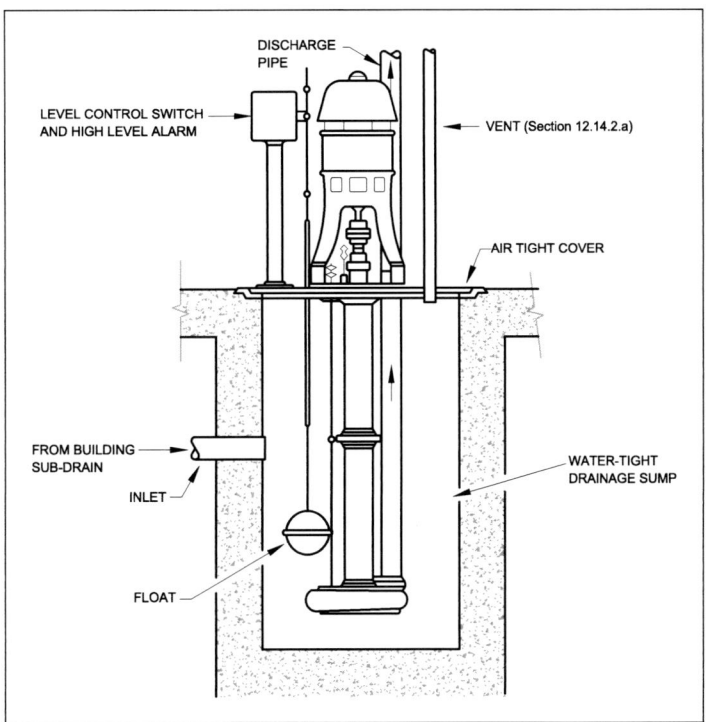

**Figure 1.2.54
A PEDESTAL TYPE SEWAGE PUMP**

Shall: "Shall" is a mandatory term.

Short Term: A period of time not more than 30 minutes.

Shut-off Valve: A ball, gate, or other full way valve located in water supply mains, risers, and branches for the purpose of controlling water supply to a fixture or group of fixtures.

Side Vent: See "Vent, Side"

SIDR: An abbreviation for the "standard inside dimension ratio" of ID controlled plastic pipe that is based on the inside diameter of the pipe divided by its minimum wall thickness. Lower SIDR ratings of the same pipe design and material have higher pressure ratings

> *Comment: SIDR equals IDR for the same pipe material composition (designation code) with the same inside diameter and minimum wall thickness. The terms are used interchangeably depending on the pipe manufacturer or standards organization. Pipes of the same material composition with the same SIDR and IDR have the same pressure ratings regardless of pipe size.*

Sink, Commercial: A sink other than for a domestic application. Commercial sinks include, but are not limited to:
1. pot sinks
2. scullery sinks
3. sinks used in photographic or other processes
4. laboratory sinks

Size of Pipe and Tubing, Incremental: Where relative size requirements are mentioned, the following schedule of sizes is recognized, even if all sizes may not be available commercially: 1/4, 3/8, 1/2, 3/4, 1, 1-1/4, 1-1/2, 2, 2-1/2, 3, 3-1/2, 4, 4-1/2, 5, 6, 7, 8, 10, 12, 15, 18, 21, 24.

Slip Joint: A connection in drainage pipe consisting of a compression nut with a compression washer or ground joint compression ring that permits drainage tubing to be inserted into the joint and secured by tightening the compression nut. Slip joints are typically used in trap connections for lavatories, sinks, and bathtubs. They permit the trap to be removed for cleaning or replacement, and to provide access to the drainage pipe.

Slope: See "Grade"

Soil Pipe: A drain pipe that conveys sewage.

Special Wastes: Wastes that require special treatment before entry into the normal plumbing system. *See Figures 1.2.55 and 9.4.1. Also Section 9.4*

> *Comment: Figure 1.2.55 shows a neutralizing tank for acid waste using limestone or marble chips. Figure 9.4.1 shows an automatic neutralizing tank that could treat either acid or caustic wastes, depending on the neutralizing solution used.*

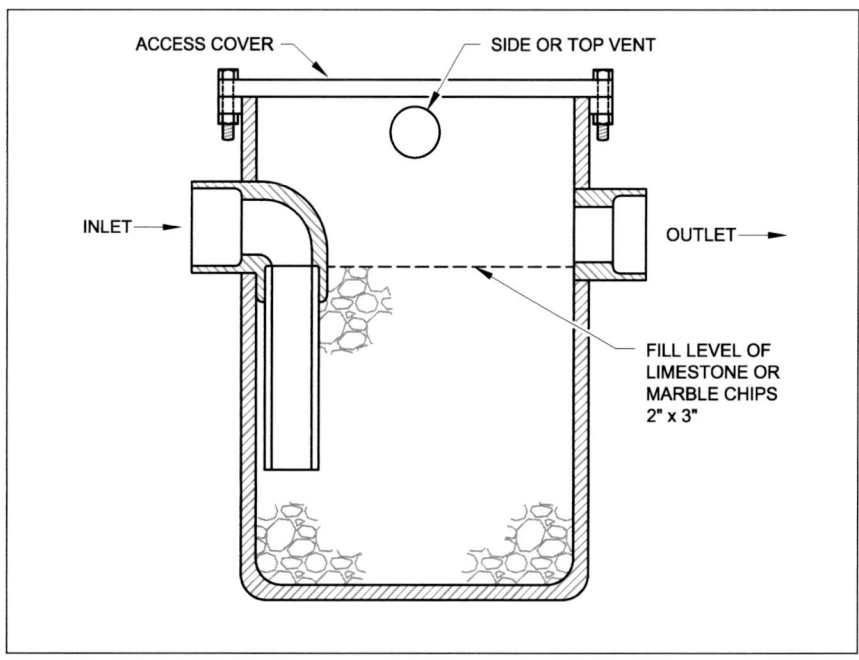

Figure 1.2.55
A NEUTRALIZING OR DILUTION TANK FOR SPECIAL WASTE

Special Waste Pipe: Pipes which convey special wastes. *See Section 3.11 for acceptable piping materials.*

Stack: Vertical drain piping and vent piping, with or without offsets, that serves multiple fixtures on two or more story heights.

Stack Group: A group of fixtures located adjacent to the stack so that by means of proper fittings, vents may be reduced to a minimum. *See Section 12.11 for stack venting groups of fixtures.*

Stack Vent: The extension of the top of a drain stack to a vent header or outdoor vent terminal.

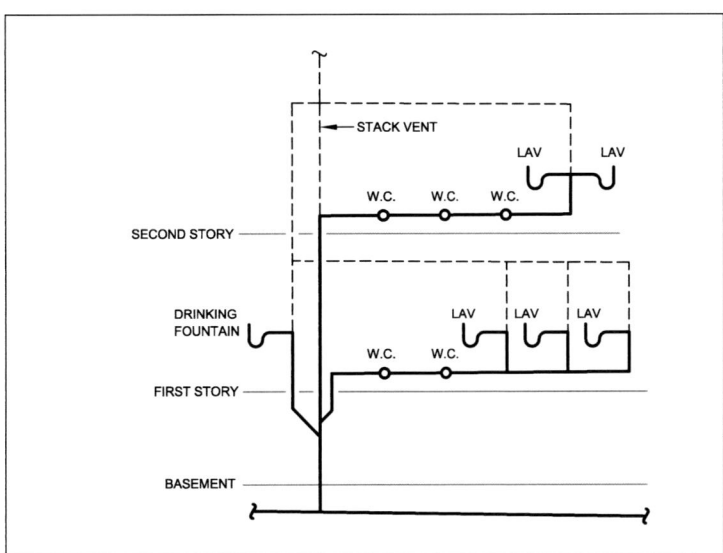

**Figure 1.2.56
A STACK VENT**

Stack Venting: A method of venting a fixture or bathroom group through a stack vent or the vent for a sub-stack. *See Figures 1.2.57, 12.11.1-A&B, 12.11.2-A&B, and Section 12.11*

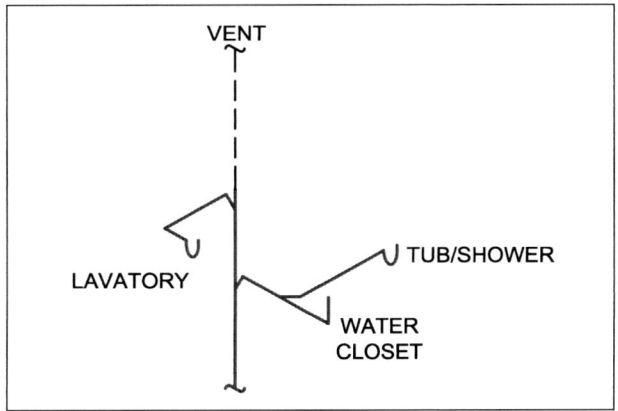

**Figure 1.2.57
STACK VENTING**

Standpipe (indirect waste receptor): A vertical drain pipe that has an open top inlet that provides an air break or air gap for indirect waste discharge.

Storm Drain: A drain that conveys only storm water.

Storm Sewer: An underground drain that conveys only rainwater, surface water, and ground water.

Storm Water: Rainwater, surface water, and ground water.

Sub-Stack: A vertical branch drain from a drain stack that serves a stack group of fixtures on the same floor level. *See Figure 12.11.2-A*

Sump: A tank or pit that receives only liquid wastes, located below the elevation of a gravity discharge, that is emptied by pumping.

Sump, Drainage (sewage): A liquid and air-tight tank that receives sewage and/or liquid waste, located below the elevation of a gravity drainage system, that is emptied by pumping. *See Figure 1.2.54*

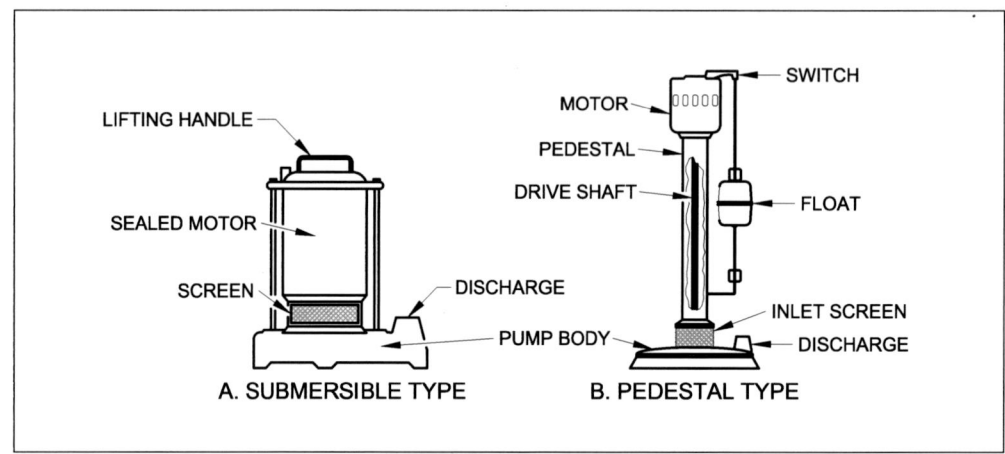

Figure 1.2.60
DIFFERENT TYPES OF SUMP PUMPS

Tempered Water: A mixture of hot and cold water to reach a desired temperature for its intended use, typically 95°F - 105°F.

Thermostatic Compensating Valve: An automatic compensating valve that senses its outlet water temperature and compensates for fluctuations to stabilize its adjusted outlet water temperature. See "Automatic Compensating Valve".

Thermostatic/Pressure Balancing Combination Compensating Valve: An automatic compensating valve that senses its outlet water temperature and its incoming hot and cold water pressures and compensates for fluctuations to stabilize its adjusted outlet water temperature. See "Automatic Compensating Valve".

Toilet Facility: A room or combination of interconnected spaces in other than a dwelling that contains one or more water closets and associated lavatories, with signage to identify its intended use.

Trap: A fitting or device that provides a liquid seal to prevent the emission of sewer gasses without materially affecting the flow of sewage or waste water through it. *See Figure 1.2.62 and Sections 5.1, 5.2, and 5.3*

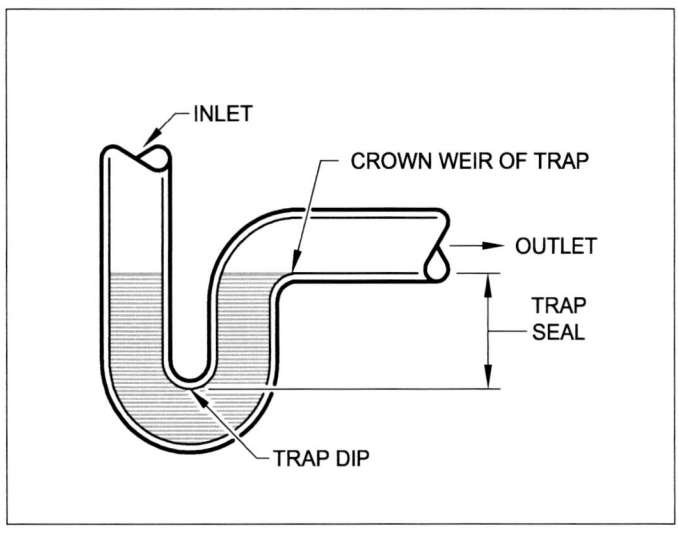

Figure 1.2.62
ELEMENTS OF A FIXTURE TRAP

Trap Arm: That portion of a fixture drain between a trap and its vent. *See Figure 1.2.63 and Section 12.8.1*

Comment: Refer to Table 12.8.1 for the maximum allowable length of trap arms to avoid trap siphonage.

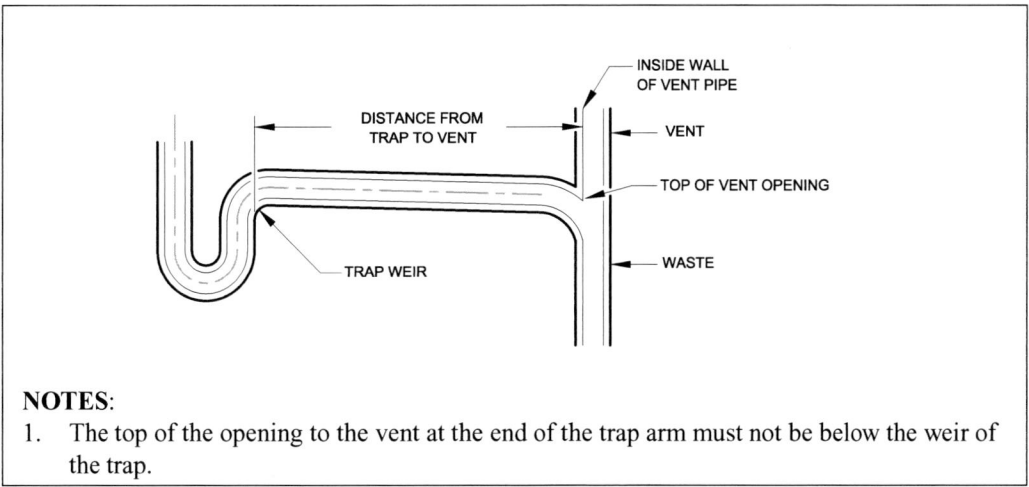

NOTES:
1. The top of the opening to the vent at the end of the trap arm must not be below the weir of the trap.

Figure 1.2.63
THE LENGTH OF A TRAP ARM

Trap Primer: A device or system of piping to maintain a water seal in a trap. *See Figures 1.2.64 and 1.2.65. Also Section 7.16.2*

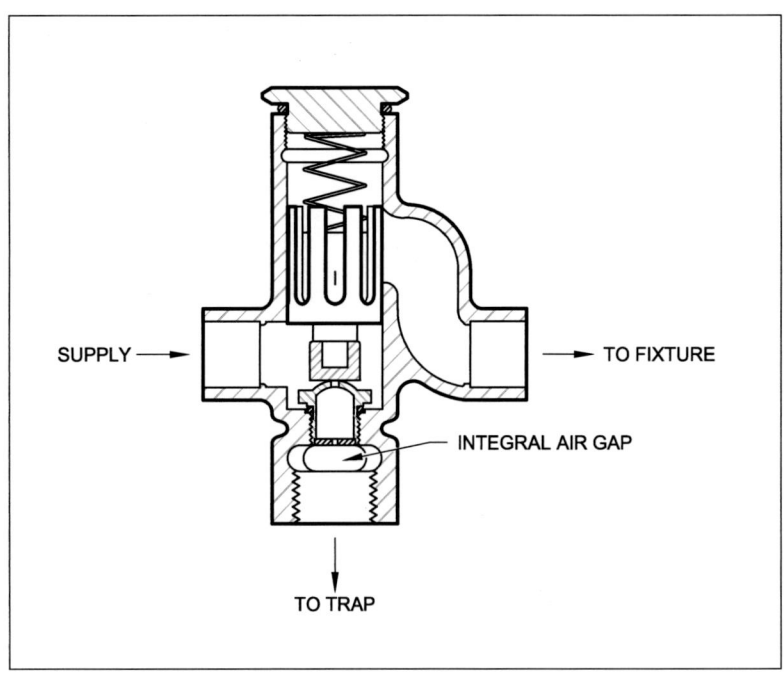

Figure 1.2.64
A FLOW-ACTIVATED TRAP PRIMER

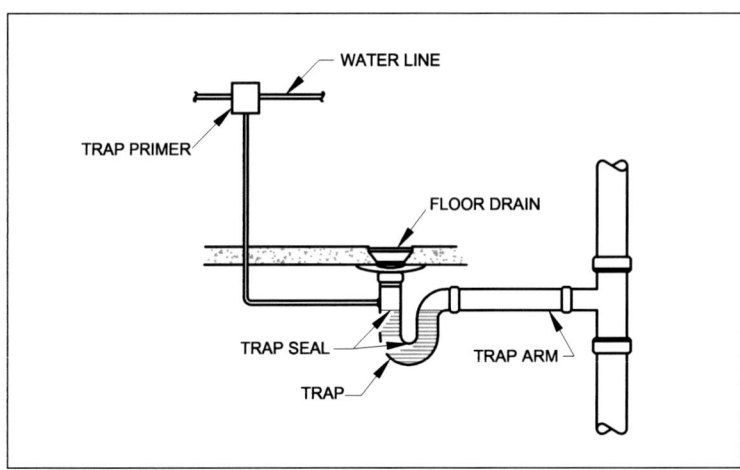

Figure 1.2.65
CONNECTION OF A TRAP PRIMER TO A TRAP

Trap Seal: The maximum vertical depth of liquid that a trap will retain, measured between the crown weir and the top of the dip of the trap. *See Figure 1.2.62 and Section 5.3.2*

Tub/Shower: A bathtub, shower, bath/shower combination, or whirlpool bath.

Vacuum Assisted Water Closet: See Water Closet, Vacuum Assisted.

Vacuum Breaker, Atmospheric Type: A vacuum breaker that is not designed to be subject to continuous line pressure.

Vacuum Breaker, Pressure Type: A vacuum breaker designed to operate under continuous line pressure. *See Figure 1.2.67*

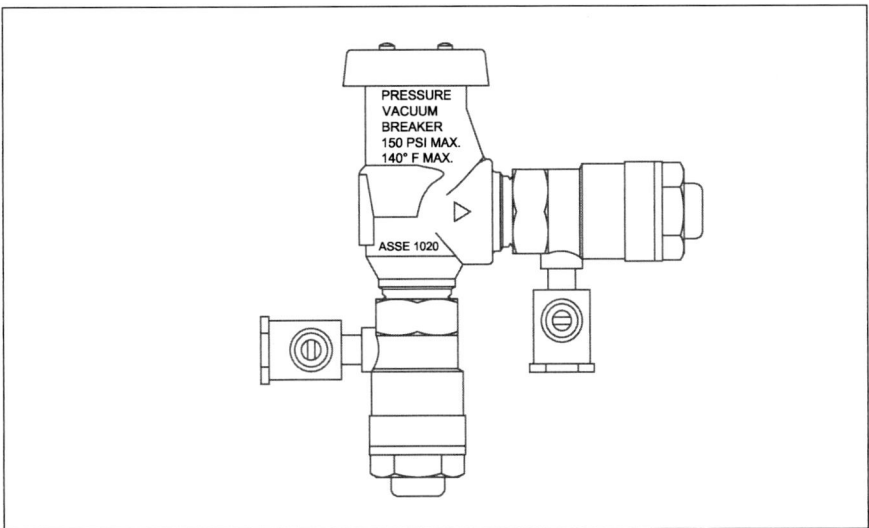

Figure 1.2.67
A PRESSURE TYPE VACUUM BREAKER

Vacuum Breaker, Spill-resistant (SVB): A pressure-type vacuum breaker specifically designed to avoid spillage during operation, consisting of one check valve force-loaded closed and an air inlet vent valve force-loaded open to atmosphere, positioned downstream of the check valve, located between two shutoff valves, and including a means for testing.

> *Comment: SVB vacuum breakers are spill-resistant, not spill-proof.*

Vacuum Relief Valve: A device to prevent vacuum in a pressure vessel. *See Figure 1.2.69 and Section 10.16.7*

> *Comment: Vacuum relief valves are required on storage-type hot water heaters that are located above the fixtures that they serve to prevent the tank from being siphoned dry and damaged by dry-firing.*

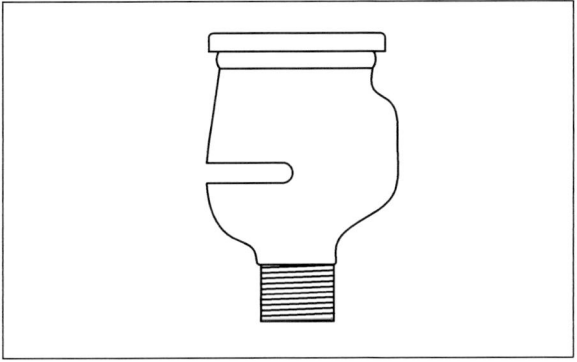

Figure 1.2.69
A VACUUM RELIEF VALVE

Vent, Back: An individual vent or continuous vent.

Vent, Branch: A vent connecting two or more individual fixture vents to a vent riser, vent stack, stack vent, or outdoor vent terminal. *See Figure 1.2.70*

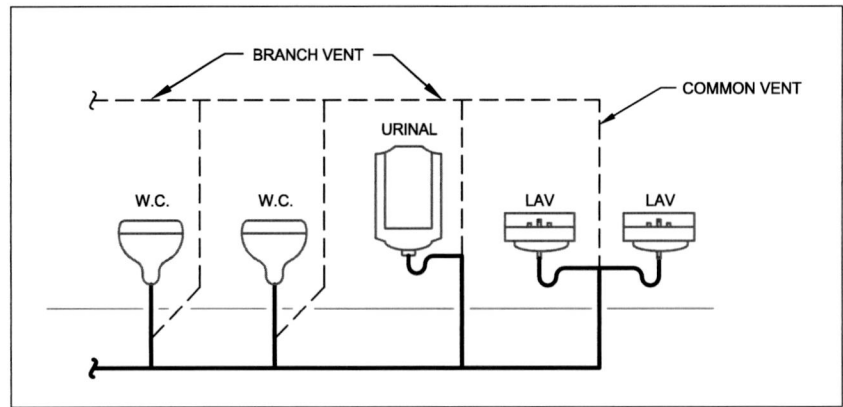

**Figure 1.2.70
A BRANCH VENT**

Vent, Circuit: The vent for a group of battery-vented fixtures that is connected to their horizontal drain branch between the first two upstream fixtures and extends to a vent stack or other vertical vent. *See Figure 1.2.71*

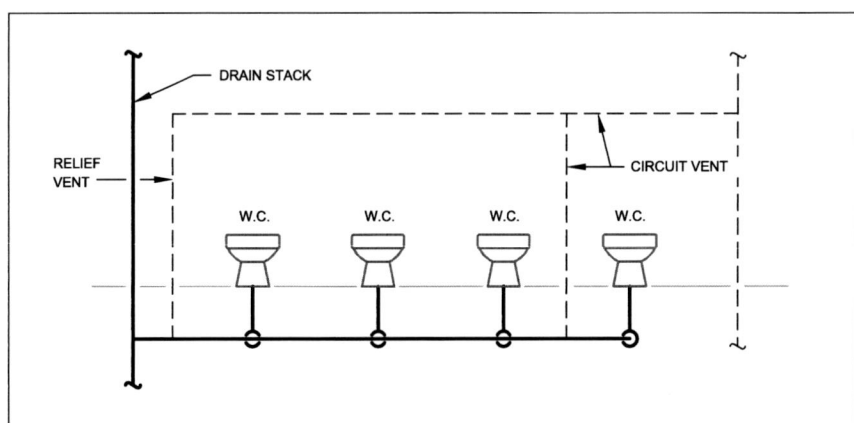

**Figure 1.2.71
A CIRCUIT VENT**

Vent, Common: A vent connected at the common connection of two fixture drains and serving as a vent for both fixtures. *See Figure 1.2.72*

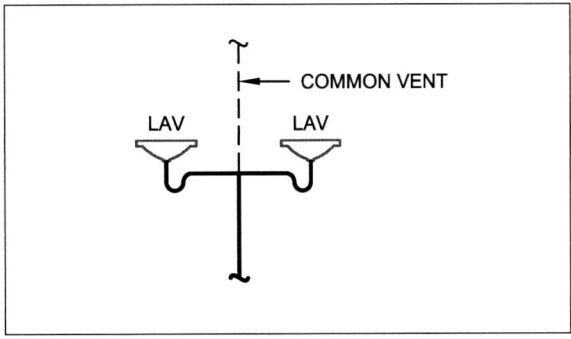

Figure 1.2.72
A COMMON VENT

Vent, Continuous: A vertical vent that is a continuation of the drain to which it connects. *See Figure 1.2.73*

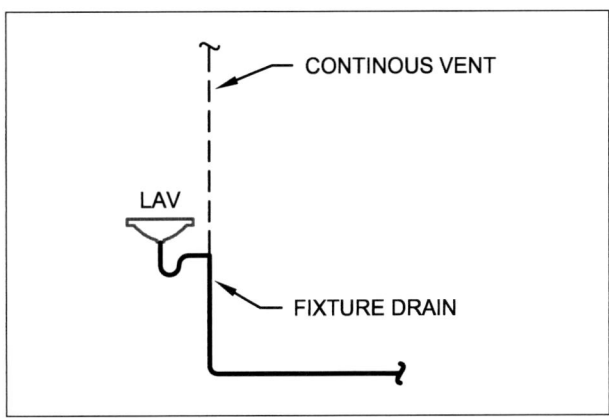

Figure 1.2.73
A CONTINUOUS VENT

Vent, Dry: A vent that does not receive the discharge of any sewage or waste.

Vent Header: Vent piping that connects two or more stack vents or vent stacks to an outdoor vent terminal.

Vent, Individual: A pipe installed to vent a single fixture drain. *See Figure 1.2.74*

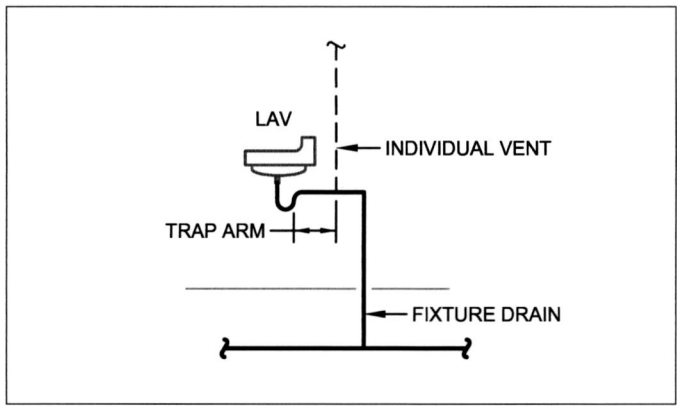

**Figure 1.2.74
AN INDIVIDUAL VENT**

Vent, Loop: A circuit vent that loops back and connects to its stack vent instead of a vent riser or vent stack. *See Figure 12.13.1*

Comment: Loop vents connect to stack vents. Circuit vents connect to vent stacks.

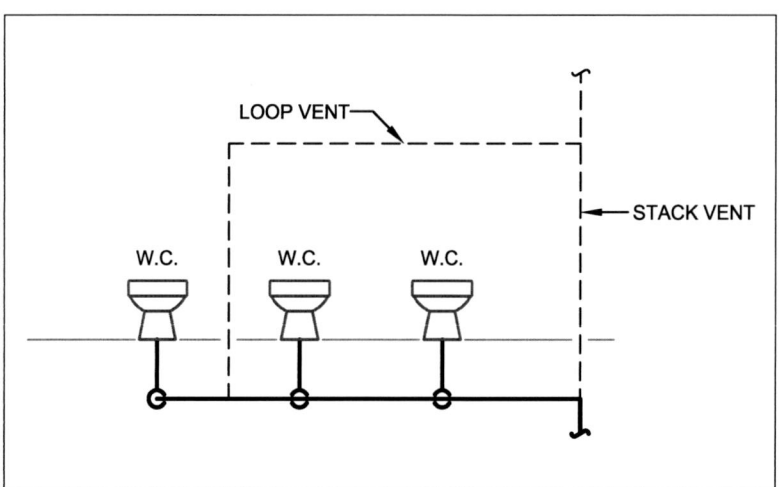

**Figure 1.2.75
A LOOP VENT**

Vent, Relief: An auxiliary vent that permits additional circulation of air in or between a drainage and vent system. *See Figures 12.3.1, 12.3.2, 12.3.3-A, 12.3.3-B, 12.13.1, 12.15.1, and 12.16.2*

Comment: Relief vents are required at points in some vertical drain stacks, at horizontal offsets in some drain stacks, at the base of some stacks, and at some battery vented branch connections to stacks. Refer to Chapter 12.

Vent Riser: A vertical vent pipe with connections from fixture vents, branch vents, and circuit vents on different floor levels that connects to a stack vent, vent stack, or outdoor vent terminal. Not a vent stack.

Vent, Side: A vent connecting to a horizontal or vertical drain pipe at an angle not more than 45° from vertical. *See Figure 1.2.76*

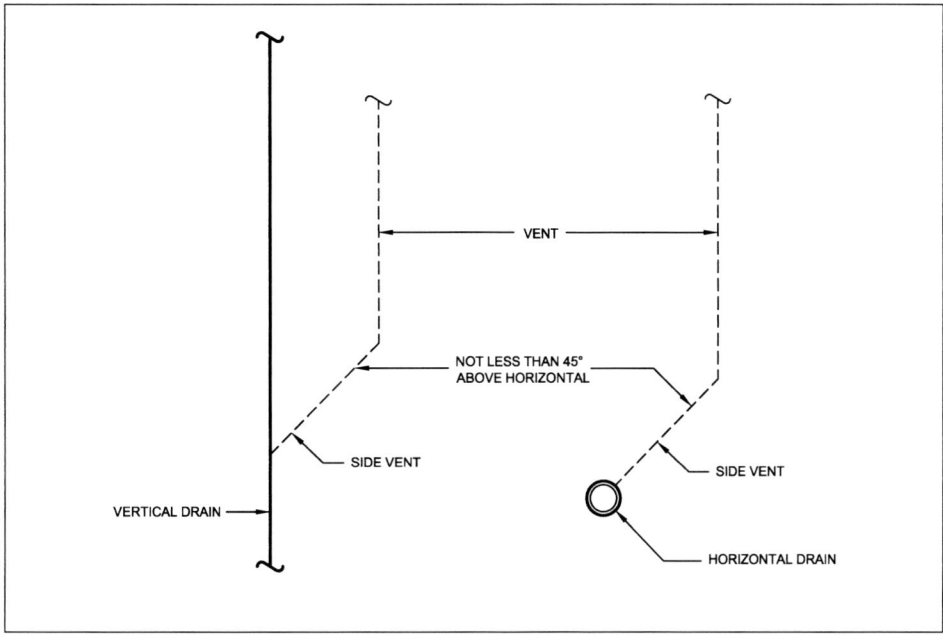

Figure 1.2.76
SIDE VENTS

Vent Stack: A vertical vent pipe that is connected to the base of a drain stack having five or more branch intervals. It extends up and connects to the stack vent or an outdoor vent terminal to relieve the pressure at the base of the drain stack. It is sized according to Section 12.16.4. *See Figures 1.2.78 and 12.3.1*

Vent, Sterilizer: A separate pipe or stack, indirectly drained to the building drainage system at the lower terminal, that receives the vapors from non-pressure sterilizers, or the exhaust vapors from pressure sterilizers, and conducts the vapors directly to the outer air. Sometimes called vapor, steam, atmosphere or exhaust vent. *See Section 14.11*

Vent, Wet: A vent pipe that is sized and arranged in accordance with this Code to receive the discharge of waste from a fixture. *See Figure 1.2.77. Refer to Section 12.10 for other arrangements of wet venting.*

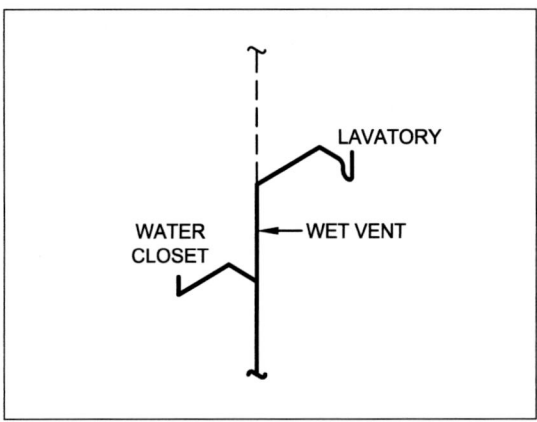

Figure 1.2.77
A WET VENT

Vent, Yoke: A vent pipe connecting up to ten branch intervals of a drain stack to its vent stack to relieve the pressure in the drain stack. *See Figure 1.2.78 and Section 12.3.2*

Comment: A yoke vent is vertical. No portion of a yoke vent can be horizontal.

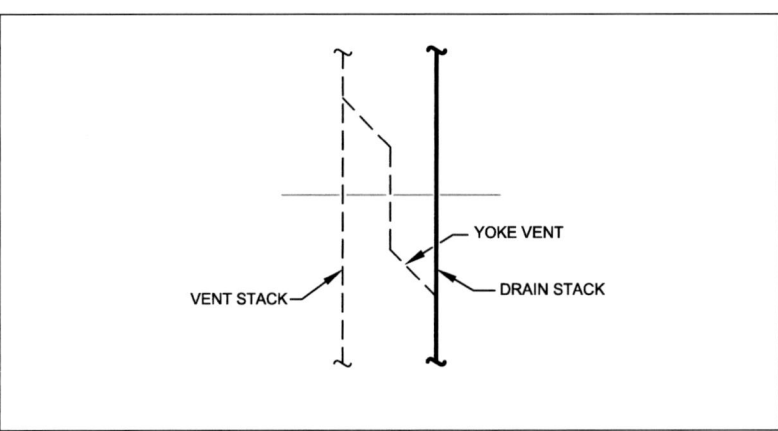

Figure 1.2.78
A YOKE VENT

Vertical Pipe: Any pipe or fitting that makes an angle of 45° or more with the horizontal. *See Figure 1.2.79*

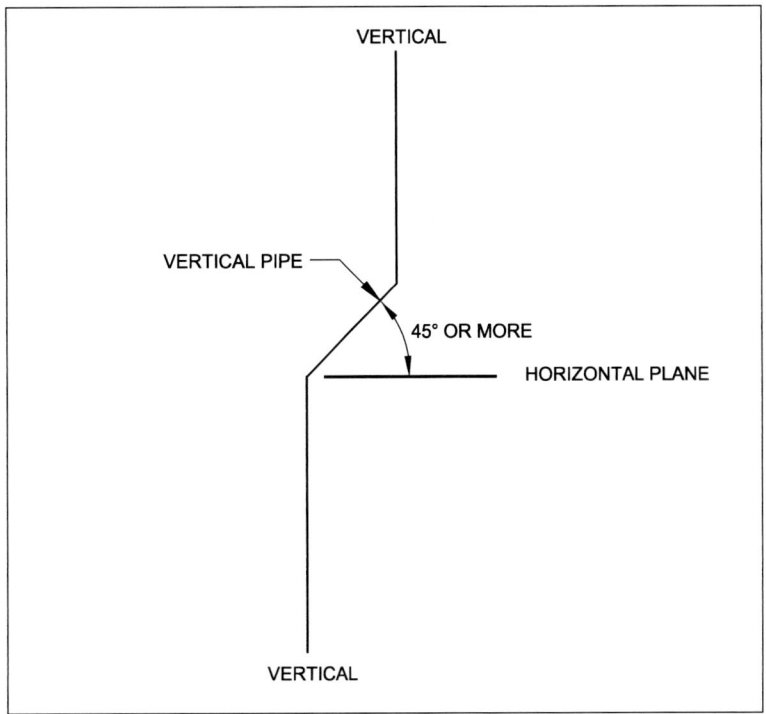

Figure 1.2.79
THE DEFINITION OF "VERTICAL PIPE"

Wall Hung Water Closet: A water closet installed in such a way that no part of it touches the floor. *See Figure 1.2.80*

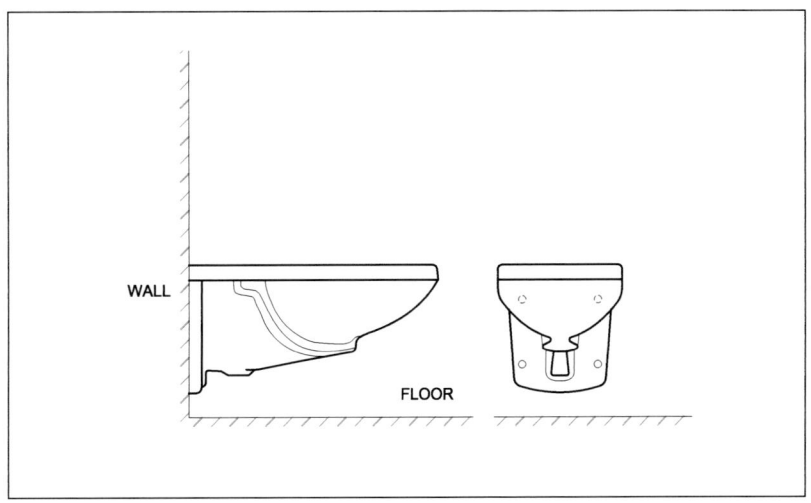

Figure 1.2.80
A WALL HUNG WATER CLOSET

Wash Fountain: A large circular or semi-circular sink that is equivalent to a lavatory except it serves more than one person.

Wash Sink: A long rectangular sink that is equivalent to a lavatory except it serves more than one person.

Waste: Liquid discharge into the plumbing drainage system from a fixture, appliance, or other connection that does not contain urine or fecal matter.

Waste Stack, Pipe or Piping: Pipes that convey the discharge from fixtures (other than water closets), appliances, areas, or appurtenances, that do not contain fecal matter.

Water Bottle Filling Station: A drinking water supply fixture that is connected to the domestic water supply and drain piping and provides for individuals to fill personal-use water bottles. Stations can be recessed or wall-mounted and can include filtration units, self-contained refrigeration units, and drinking fountains.

Water Closet, Pressure Assisted: A low consumption water closet with an air accumulator vessel in the tank that stores water and air under pressure, using the water supply pressure. When flushed, the air produces a high velocity jet of water and air that forces the contents out of the bowl.

Water Closet, Pump Assisted: A low consumption water closet with a fractional horsepower pump in the tank that produces a high velocity jet in the trap way that assists the flushing action.

Water Closet, Vacuum Assisted: A low consumption water closet that uses the falling water level in the tank to induce a vacuum near the outlet of the trap way that assists the flushing action.

Water Distribution Piping: Piping within the building or on the premises that conveys water from the water service pipe to the points of use. *See Figure 1.2.81*

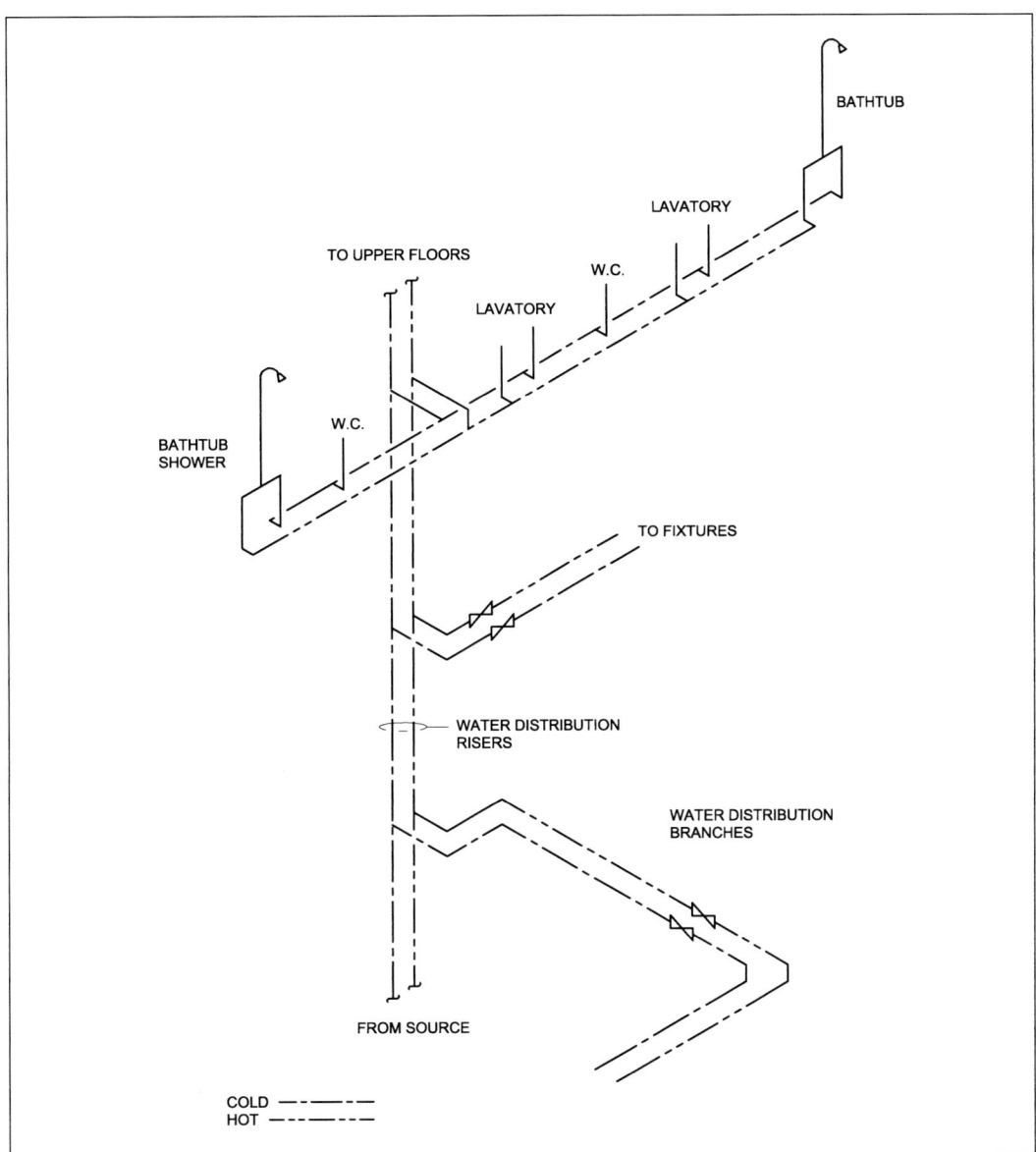

**Figure 1.2.81
WATER DISTRIBUTION PIPING**

Water Hammer Arrestor: A manufactured pre-charged device operating through the compressibility of air.

> *Comment: Water hammer arrestors are either the bellows or piston type and are pre-charged with compressed air or gas. The Code no longer mentions air chambers, which are subject to losing their initial captive air charge through absorption into the system water.*

Water Main: A water supply line that serves the public or a community and is under the jurisdiction of a municipality or water purveyor.

Water Outlet: A discharge opening through which water is supplied to a fixture, into the atmosphere (except into an open tank that is part of the water supply system), to a boiler or heating system, to any devices or equipment requiring water to operate but that are not part of the plumbing system.

Water Service Piping: Piping from a water main or other source of potable water supply to the water distribution system within the building served. *See Figure 1.2.82 and Section 10.6*

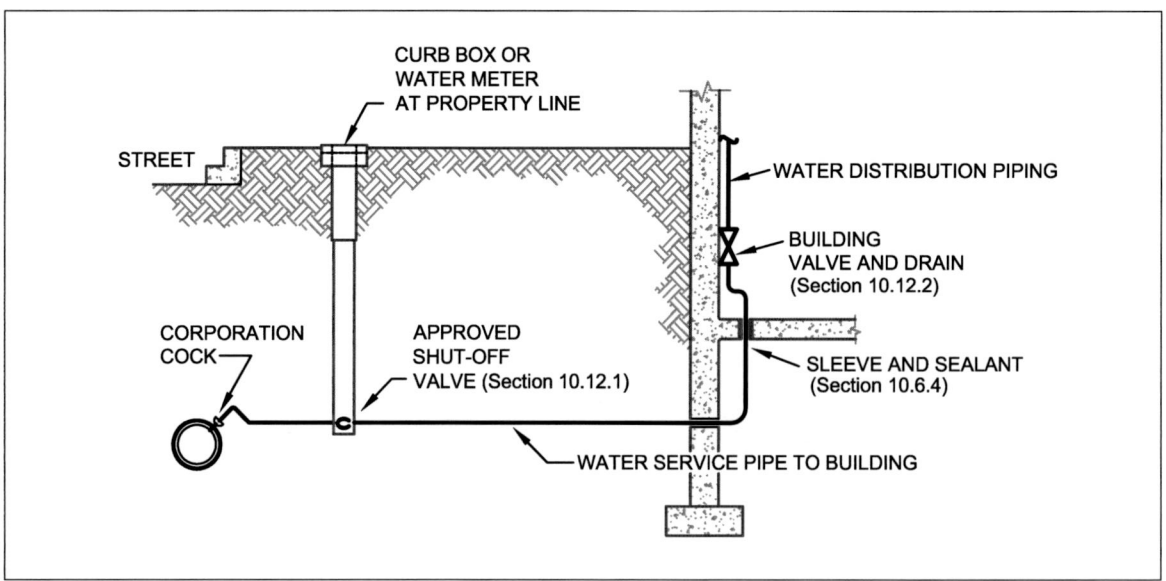

Figure 1.2.82
THE WATER SERVICE PIPE IN A PUBLIC WATER SUPPLY

Wet Vent: See "Vent, Wet"

Whirlpool Bathtub: A plumbing appliance consisting of a bathtub fixture that is equipped and fitted with a circulation piping system, pump, and other appurtenances and is so designed to accept, circulate, and discharge bathtub water upon each use. *See Figure 1.2.83*

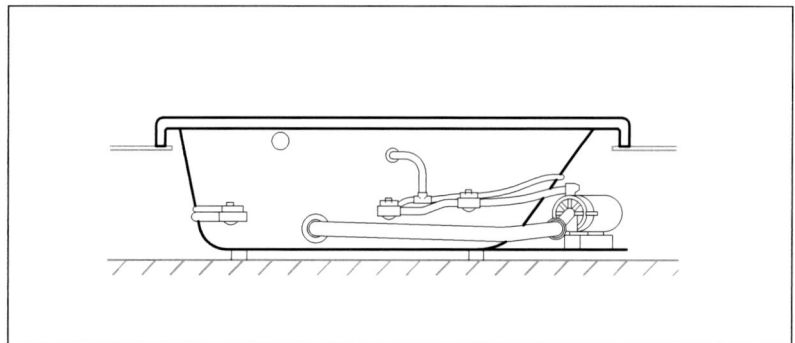

Figure 1.2.83
A WHIRLPOOL BATH

WSFU: See "Fixture Unit (Water Supply – WSFU)"

Blank Page

Chapter 2

General Regulations

2.1 RESERVED

2.2 CAST IRON SOIL PIPE MARKING

All cast iron soil pipe and fittings shall be listed and tested to comply with standards referenced in Table 3.1.3 and marked with country of origin and identification of the original manufacturer in addition to any markings required by those referenced standards.

2.3 CHANGES IN DIRECTION OF DRAINAGE PIPING

2.3.1 Uses for Drainage Fittings

   a. Changes in direction of drainage piping shall be made with long radius drainage fittings except where short radius fittings are required or permitted.
      1. Short radius fittings are required to connect individual fixture trap arms to vertical drain and vent piping.
      2. Short radius fittings may be used in the drain piping for an individual fixture.
      3. Short radius fittings shall not be used in drain piping serving two or more fixtures.
      4. Short radius fittings shall not be used at the base of stacks.
      5. Short radius fittings may be used in vent piping.
   b. Short radius drainage fittings are those having radius or centerline dimensions that are approximately equal to or less than their nominal pipe size. The radius or centerline dimensions of long radius drainage fittings are greater than their nominal pipe size.
   c. Long radius drainage fittings shall not be used to connect fixture trap arms to vertical drain and vent piping. Connections to fixture vents shall be above the top weir of the fixture trap.
   EXCEPTION: Double wyes with 1/8 bends or 45-degree elbows that drain two lavatories into a common drain with a common vent shall be permitted.
See Figures 2.3.1-A, 2.3.1-B, 2.3.1-C, and 2.3.1-D

2.3.2 Double Pattern Fittings

The uses for double pattern drainage fittings shall be the same as for single pattern fittings in Table 2.3.1. EXCEPTION: Double sanitary tees and crosses shall not be used to connect blowout fixtures, back-outlet water closets, back-to-back pressure-assisted water closets, and fixtures or appliances having pumped discharge. *See Figure 2.3.2*

2.3.3 Back-to-Back Back-Outlet Fixtures

Stack fittings, including carriers, for back-outlet fixtures installed back-to-back shall be either the wye pattern, incorporate baffles within the drainage fitting, or otherwise be designed to prevent crossflow or mixing of the discharges from the two fixtures prior to the change in direction. *See Figure 2.3.3*

2.4 FITTINGS AND CONNECTIONS IN DRAINAGE SYSTEMS

2.4.1 Prohibited Fittings

No running threads or saddles shall be used in the drainage or vent system. No drainage or vent piping shall be drilled, tapped, burned or welded.

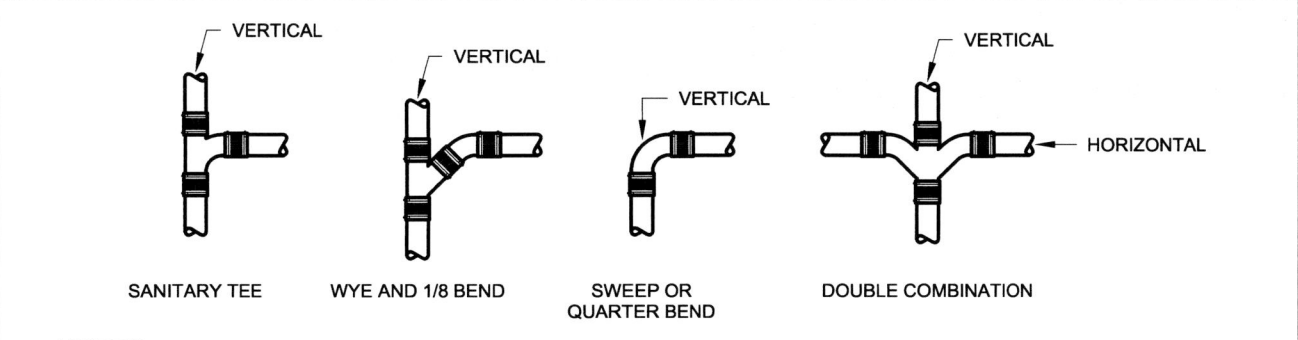

NOTES:
1. Horizontal-to-vertical changes in direction are not as critical as other changes because gravity controls the velocity in vertical drops.
2. No-hub cast iron fittings are shown. See Table 2.3.1 for fitting patterns for other drainage pipe materials.
3. Short radius fittings are permitted in piping for individual fixtures.
4. Long radius fittings are not permitted to connect fixture trap arms to vertical drain vent piping because the vent opening at the end of the trap arm will be below the weir of the trap, causing the trap to self-siphon. Exception: Long radius fittings are permitted to drain two lavatories into a common drain with a common vent per Section 2.3.1 to permit cleaning of the common drain. Self-siphonage is not an issue with the low drain flow rates of lavatories.

Figure 2.3.1 - A

DRAINAGE FITTINGS FOR HORIZONTAL-TO-VERTICAL CHANGES IN DIRECTION

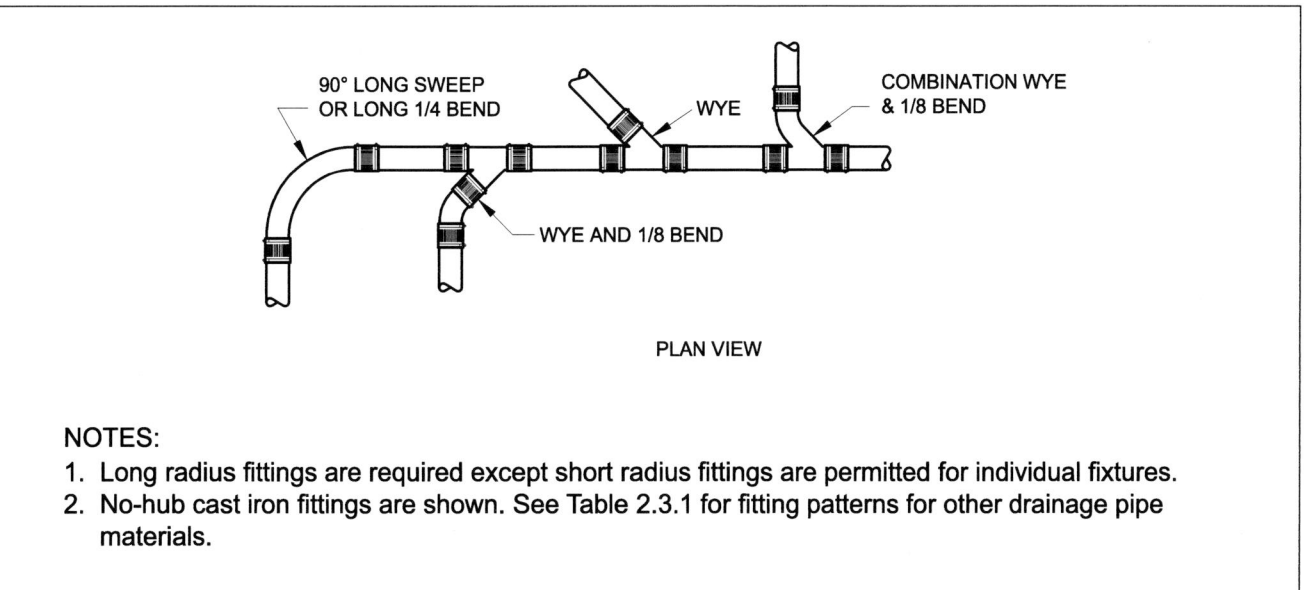

NOTES:
1. Long radius fittings are required except short radius fittings are permitted for individual fixtures.
2. No-hub cast iron fittings are shown. See Table 2.3.1 for fitting patterns for other drainage pipe materials.

Figure 2.3.1 - B

DRAINAGE FITTINGS FOR HORIZONTAL-TO-HORIZONTAL CHANGES IN DIRECTION

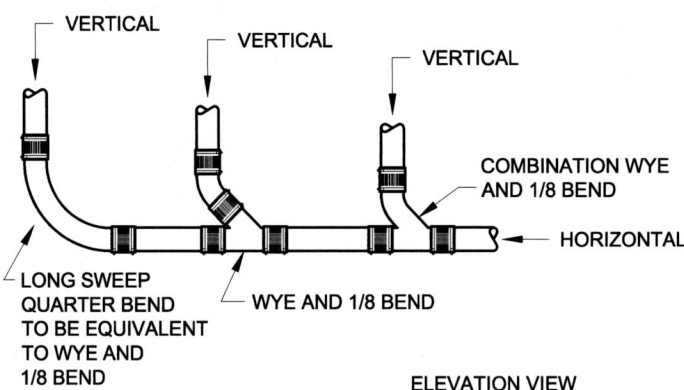

NOTES:
1. Long radius fittings are required at the base of drainage stacks serving two or more fixtures. Short radius fittings are permitted for vertical-to-horizontal drains from individual fixtures.
2. No-hub cast iron fittings are shown. See Table 2.3.1 for fitting patterns for other drainage pipe materials.
3. See Sections 11.11 and 12.15 for restrictions on branch drain connections near the base of stacks that are subject to suds pressure.

Figure 2.3.1 – C
DRAINAGE FITTINGS FOR VERTICAL-TO-HORIZONTAL CHANGES IN DIRECTION

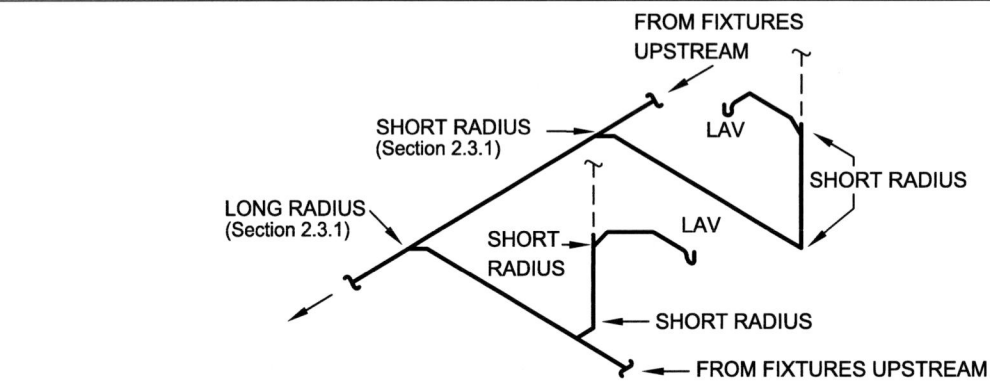

NOTES:
1. Short radius fittings are permitted in drain piping for individual fixtures.
2. Long radius fittings are not permitted to connect fixture trap arms to vertical drain and vent piping because the vent opening at the end of the trap arm will be below the weir of the trap, causing the trap to self-siphon. Exception: Long radius fittings are permitted to drain two lavatories into a common drain with a common vent per Section 2.3.1 to permit cleaning of the common drain. Self-siphonage is not an issue with the low drain flow rates of lavatories.

Figure 2.3.1 – D
USE OF SHORT AND LONG RADIUS DRAINAGE FITTINGS

Table 2.3.1 PERMISSIBLE DRAINAGE FITTINGS FOR CHANGES IN DIRECTION			
PIPE MATERIAL	**CHANGES IN DIRECTION**		
	HORIZONTAL TO HORIZONTAL	**HORIZONTAL TO VERTICAL**	**VERTICAL TO HORIZONTAL**
CAST IRON HUB & SPIGOT **CAST IRON NO-HUB**	long sweep	sanitary tee	eighth bend & wye
	short sweep	eighth bend & wye (2)	combination wye & eighth bend
	wye	combination wye & eighth bend	long sweep
	combination wye & eighth bend	long sweep	short sweep 3" or larger
	fifth bend (72-deg)	short sweep	quarter bend (1)
	sixth bend (60-deg)	quarter bend	short sweep (1)
	eighth bend (45-deg)		
	sixteenth bend (22-1/2 deg)		
	quarter bend		
CAST IRON DRAINAGE (threaded)	extra long turn 90-deg elbow	drainage tee	long turn 90-deg TY
	long turn 90-deg elbow	short turn 90-deg TY	extra long turn 90-deg elbow
	long turn 45-deg elbow	long turn 90-deg TY (2)	long turn 90-deg elbow
	short turn 22-1/2 deg elbow	45-deg elbow or 45-deg Y branch (2)	short turn 90-deg elbow (1)
	short turn 11-1/4 deg elbow	extra long turn 90-deg elbow	45-deg elbow & 45-deg Y branch (1)
	long turn 90-deg TY	long turn 90-deg elbow	
	short turn 45-deg Y branch	short turn 90-deg elbow	
	short turn 90-deg elbow		
	short turn 60-deg elbow		
	short turn 45-deg elbow		
COPPER DWV	DWV 90-deg long radius elbow	DWV tee	DWV long turn T-Y
	90-deg elbow - long radius	DWV 90-deg sanitary tee	45-deg elbow & DWV 45-deg Y
	DWV long turn T-Y	DWV long turn T-Y (2)	DWV 90-deg long radius elbow
	DWV 45-deg Y	45-deg elbow & DWV 45-deg Y (2)	90-deg elbow - long radius
	DWV 90-deg elbow	DWV 90-deg long radius elbow	DWV 90-deg elbow (1)
	DWV 45-deg elbow	90-deg elbow - long radius	
		DWV 90-deg elbow	
PLASTIC DWV	90-deg long turn elbow	sanitary tee	long radius TY
	long sweep 1/4 bend	fixture tee	45-deg elbow and 45-deg wye
	60-deg elbow or 1/6 bend	long radius TY (2)	90-deg long turn elbow
	45-deg elbow or 1/8 bend	45-deg elbow & DWV 45-deg Y (2)	90-deg elbow or 1/4 bend (1)
	22-1/2 deg elbow or 1/16 bend	90-deg long turn elbow	
	long radius TY	90-deg elbow or 1/4 bend	
	45-deg wye		
	90-deg elbow or quarter bend		
STAINLESS STEEL PUSH-FIT DWV	long sweep	tee	eighth bend & wye (2)
	wye	sanitary tee	combination wye & eighth bend
	combination wye & eighth bend	eighth bend & wye (2)	long sweep
	15-degree 1/24 bend	combination wye & eighth bend	90-degree 1/4 bend (1)
	22-1/2 degree 1/16 bend	long sweep	
	30-degree 1/12 bend	90-deg 1/4 bend	
	45-degree 1/8 bend		
	90-deg 1/4 bend		

Footnotes for Table 2.3.1

(1) Short radius fittings shall not be permitted in horizontal offsets in drainage stacks and at the base of stacks.

(2) Long radius fittings shall not be used to connect fixture trap arms to vertical drain and vent piping. Exception: Long radius fittings are permitted to drain two lavatories into a common drain with a common vent per Section 2.3.1.

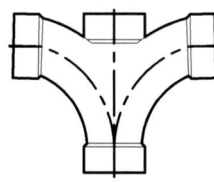
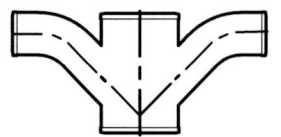

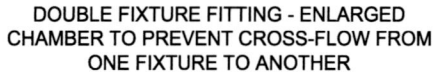

DOUBLE FIXTURE FITTING - ENLARGED CHAMBER TO PREVENT CROSS-FLOW FROM ONE FIXTURE TO ANOTHER

DOUBLE COMBINATION WYE & 1/8 BEND HORIZONTAL DRAINAGE TO VERTICAL DRAINAGE

NOTES:
1. Double fixture fittings are used to connect two horizontal fixture trap arms to a vertical drain and common vent.
2. Long radius fittings are not permitted to connect fixture trap arms to vertical drain and vent piping because the vent opening at the end of the trap arm will be below the weir of the trap, causing the trap to self-siphon.

Figure 2.3.2
DOUBLE PATTERN DRAINAGE FITTINGS

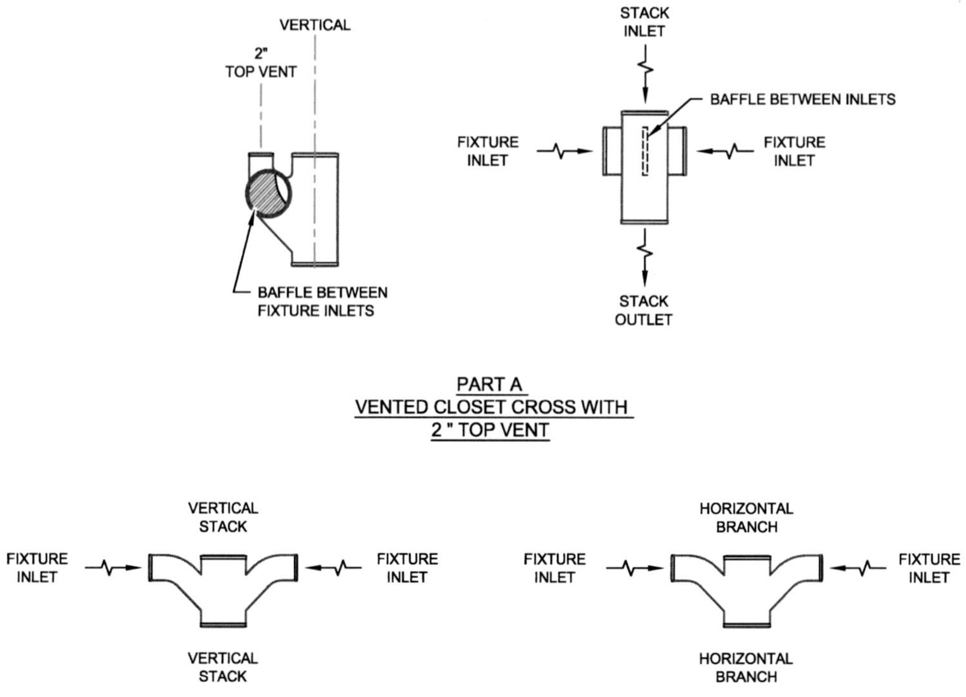

PART A
VENTED CLOSET CROSS WITH
2" TOP VENT

PART B
DOUBLE COMBINATION WYE & 1/8 BEND
HORIZONTAL - TO - VERTICAL

PART C
DOUBLE COMBINATION WYE & 1/8 BEND
HORIZONTAL - TO - HORIZONTAL

NOTES:
1. No-hub cast iron fittings are shown. See Table 2.3.1 for fitting patterns for other drainage pipe materials.
2. In Part B, the fixtures must be vented prior to connection to the stack.
3. In Part C, if not battery vented, the fixtures must be individually vented prior to connection to the branch drain.

Figure 2.3.3
PREVENTING CROSSFLOW BETWEEN BACK-TO-BACK BACK-OUTLET FIXTURES

2.4.2 Heel or Side-Inlet Bends

A heel or side-inlet quarter bend shall not be used as a dry vent when the inlet is placed in a horizontal position or any similar arrangement of pipe or fittings producing a similar effect.

EXCEPTION: When the entire fitting is part of a dry vent arrangement system the heel or side-inlet bend shall be acceptable. *See Figures 2.4.2-A and 2.4.2-B*

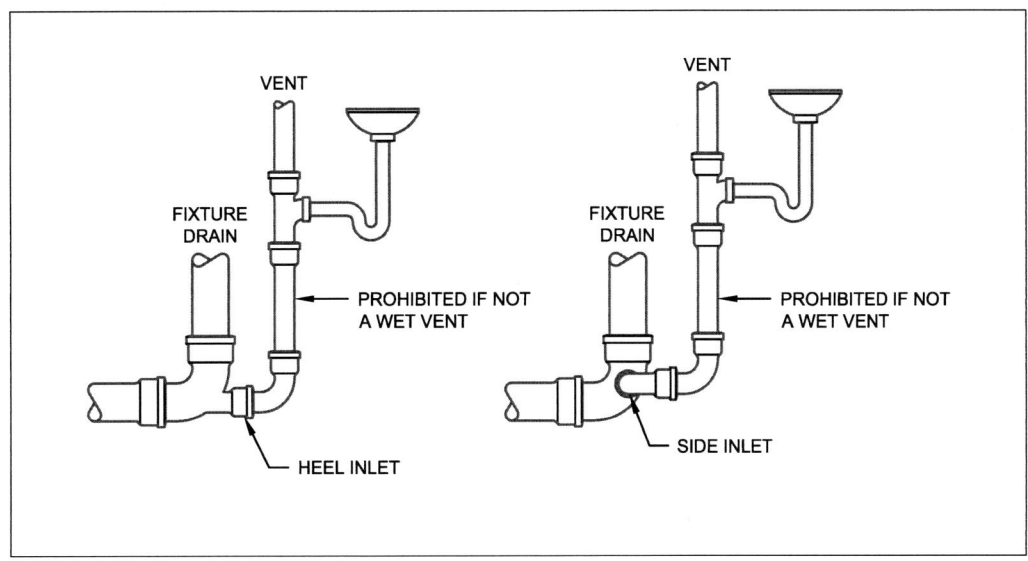

Figure 2.4.2 - A
HORIZONTAL HEEL OR SIDE INLET VENT CONNECTION

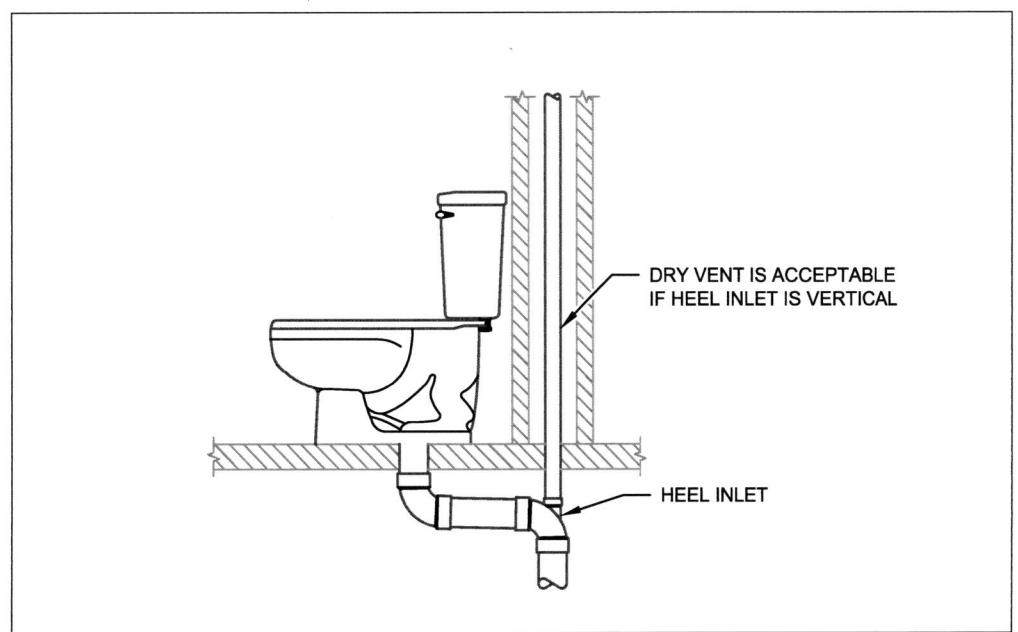

Figure 2.4.2 - B
VERTICAL HEEL INLET VENT CONNECTION

2.4.3 Obstruction to Flow

a. No fitting, connection, device, or method of installation that obstructs or retards the flow of water, wastes, sewage, or air in the drainage or venting systems in an amount greater than the normal frictional resistance to flow, shall be used unless it is indicated as acceptable in this Code.

b. 4x3 closet bends and 4x3 closet flanges shall not be considered as obstructions to flow.

> *Comment #1: This Section does not prohibit double hub fittings for sanitary drain pipe installations since they create no more restriction to flow that would be encountered with a hubless coupling or similar fitting.*
>
> *Comment #2: Failure to ream or deburr drainage piping constitutes an obstruction to flow.*

2.4.4 Prohibited Joints

Cement mortar joints are prohibited.
EXCEPTION: When used for repairs and/or when used for connections to existing lines constructed with such joints.

2.5 HEALTH AND SAFETY

Where a health or safety hazard is found to exist on a premise, the owner or his agent shall be required to make such corrections as may be necessary to abate such nuisance, and bring the plumbing installation within the provisions of this Code.

2.6 TRENCHING, BEDDING, TUNNELING AND BACKFILLING

2.6.1 Trenching and Bedding

a. Trenching and excavation for the installation of underground piping shall be performed in compliance with occupational safety and health requirements. Trenches shall be of sufficient width to permit proper installation of the pipe. Where shoring is required, additional allowance shall be made in the width of the trench to provide adequate clearance.

b. A firm, stable, uniform bedding shall be provided under the pipe for continuous support. Bell holes shall be provided for joints in bell and spigot pipe and for other joints requiring such clearance. Blocking shall not be used to support the pipe.

c. The trench bottom may provide the required bedding when adequate soil conditions exist and when excavated to the proper depth and grade. Where trenches are excavated to depths below the bottom of the pipe, bedding shall be added beneath the pipe as required. Such bedding shall be of clean sand, gravel, or similar select material that is compacted sufficiently to provide the support required under 2.6.1.b.

d. Where rock is encountered in trenching, it shall be removed to a depth of not less than 6 inches below the bottom of the pipe and bedding shall be added as required under 2.6.1.c. The pipe shall not rest on rock at any point, including joints.

Required under Section 2.6.1.c. *See Figures 2.6.1-A and 2.6.1-B. Also Section 2.6.2, 2.6.3, and 2.6.4*

2.6.2 Side-fill

The haunch areas adjacent to the pipe between the bottom of the pipe and its horizontal centerline shall be filled with a clean coarse-grain material such as sand, gravel, or soil. Such side-fill shall be placed by hand, extending to the sides of the trench, and be compacted to provide lateral support for the pipe. *See Figures 2.6.1-A and 2.6.1-B*

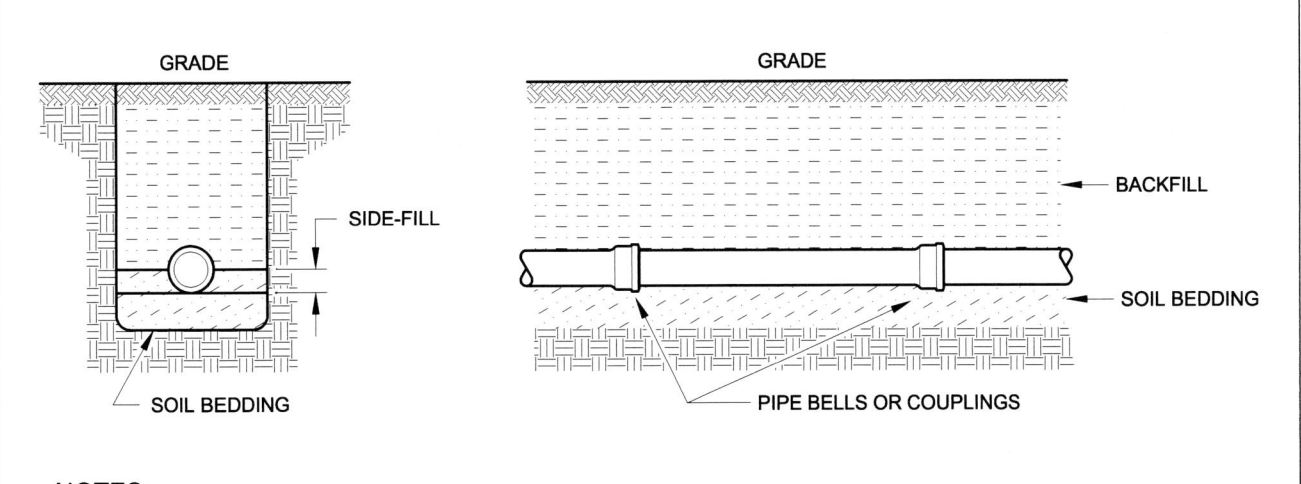

NOTES:
1. The soil under the pipe must be stable. Unstable bedding may damage the pipe, fittings, or joints when settlement occurs.
2. The side-fill should be shoveled under the pipe to make sure that there are no voids in the side-fill.

Figure 2.6.1 - A
UNDERGROUND PIPING BEDDED ON SOIL

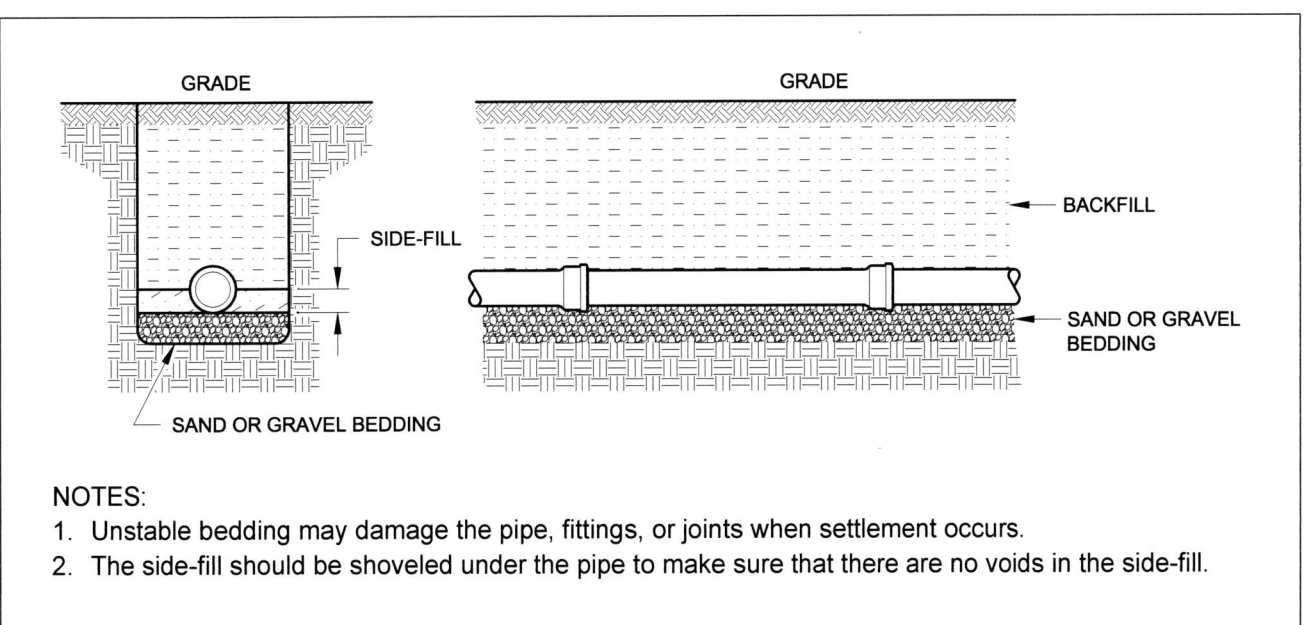

NOTES:
1. Unstable bedding may damage the pipe, fittings, or joints when settlement occurs.
2. The side-fill should be shoveled under the pipe to make sure that there are no voids in the side-fill.

Figure 2.6.1 - B
UNDERGROUND PIPING BEDDED ON SAND OR GRAVEL

2.6.3 Initial Backfill

a. Backfill material shall be sand, gravel, or loose soil that is free of rocks and debris. Maximum particle size shall be 1-1/2 inches.

b. The side-fill adjacent to the pipe shall be backfilled, tamped, and compacted to the top of the pipe.

c. After the side-fill is backfilled, the trench shall be backfilled in not more than 6-inch layers, each tamped and compacted, to a level not less than 2 feet above the top of the pipe.

d. Heavy equipment shall not be used for the initial backfill.

See Figure 2.6.3

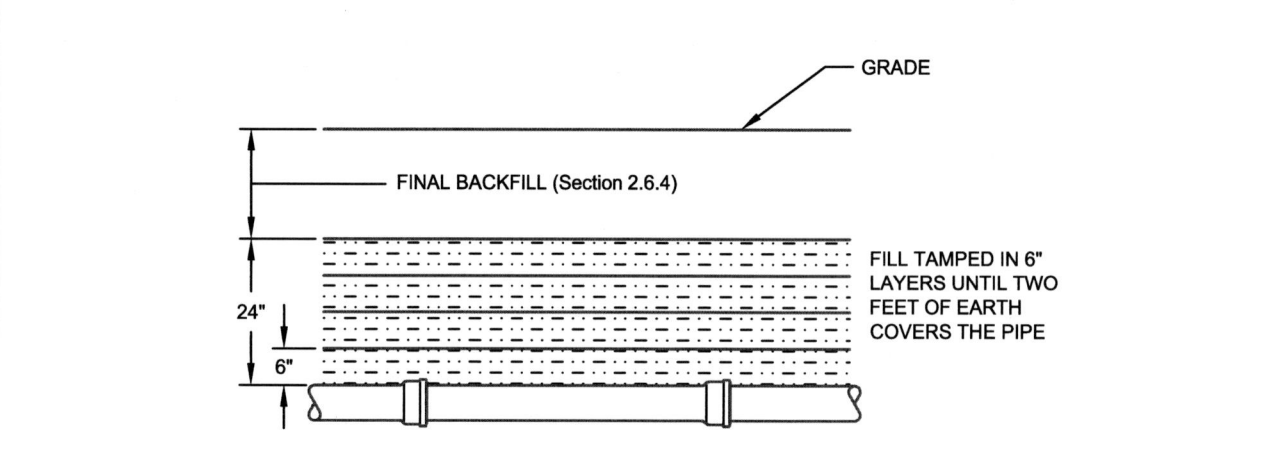

NOTE:
1. The side-fill adjacent to the pipe must be backfilled and tamped to the top of the pipe.
2. The initial backfill must be placed in 6" tamped layers. The use of heavy compacting equipment is prohibited.
3. The initial backfill material must be sand, gravel, or loose soil from the excavation that is free from rocks and debris. Broken concrete, frozen earth, and other solid materials may damage the pipe from point loads.
4. After the pipe is covered with 2 feet of tamped initial backfill, the final backfill to grade can be compacted with heavy equipment.

Figure 2.6.3
TYPICAL BACKFILLING PROCEDURE

2.6.4 Final Backfill

The trench shall be backfilled from the top of the compacted initial backfill to finish grade using suitable material. Heavy compacting equipment may be used for the final backfill. *See Figure 2.6.3*

2.6.5 Tunneling

When pipe is installed in a dug or bored earth tunnel, the space around the pipe between the pipe and the wall of the tunnel shall be completely filled with packed concrete or grout. When pipe is installed in a jacked-in-place conduit or sleeve, the space around the pipe between the pipe and the inside of the conduit or sleeve shall be sealed in an approved manner in accordance with Section 2.12.d.

2.6.6 Underground Plastic Pipe

a. Underground plastic pipe shall be installed in accordance with the requirements of Section 2.6.
EXCEPTIONS:
    (1) The maximum particle size in the side-fill and initial backfill shall be not more than 1/2-inch for pipe 6" size and smaller, and 3/4-inch for pipe 8" and larger.
    (2) For water service piping, refer to ASTM D2774, *Standard Practice for Underground Installation of Thermoplastic Pressure Piping.*
    (3) For gravity-flow drainage pipe, refer to ASTM D2321, *Underground Installation of Thermoplastic Pipe for Sewers and Other Gravity-flow Applications.*

b. An insulated copper tracer wire or other approved conductor shall be installed adjacent to underground non-metallic water service piping and non-metallic force mains, to facilitate locating the piping. One end shall be brought above ground inside or outside the building wall. The tracer wire for the water service shall originate at the curb valve required in Section 10.12.1. The tracer wire for the force main shall originate at the final point of disposal. The tracer wire shall not be less than 18 AWG insulated. The insulation shall not be yellow in color.

2.6.7 Underground Copper Piping

Underground copper piping shall be installed in accordance with the requirements of Section 2.6.
EXCEPTION: The maximum particle size in the side-fill and initial backfill shall be not more than 1/2-inch for pipe 6" size and smaller, and 3/4-inch for pipe 8" and larger.

2.6.8 Safety Precautions

Rules and regulations pertaining to safety and protection of workers, other persons in the vicinity, and neighboring property shall be adhered to where trenching or similar operations are being conducted.

2.6.9 Supervision

Where excavation, bedding or backfilling are performed by persons other than the installer of the underground piping, the pipe installer shall supervise the bedding, side-fill, and initial backfill, and shall be responsible for its conformance to this Code.

2.6.10 Trenchless Pipe Replacement Systems

Trenchless replacement of water and sewer piping shall be performed using equipment and procedures recommended by the equipment manufacturer. Where underground piping beneath paved surfaces or concrete floor slabs is replaced by this method, the manufacturer's recommendations for the specific conditions shall be used. Approved mechanical couplings shall be used to make the connections between new and existing piping.

> *Comment: When piping beneath paving or concrete floors is replaced by the trenchless method, there must be sufficient distance above the pipe to prevent cracking the paving or floor slab when the pipe being replaced bursts.*

2.7 SAFETY

Any part of a building or premise that is changed, altered, or required to be replaced as a result of the installation, alteration, renovation, or replacement of a plumbing system, or any part thereof, shall be left in a safe, non-hazardous condition.

2.8 INSTALLATION PRACTICES

Plumbing systems shall be installed in a manner conforming to this Code and industry installation standards.

2.9 PROTECTION OF PIPES

2.9.1 From Breakage

Pipes passing under or through foundation walls shall be protected from breakage.

2.9.2 From Corrosion

Pipe subject to corrosion by passing through or under corrosive fill, such as, but not limited to, cinders, concrete, or other corrosive material, shall be protected against external corrosion by protective coating, wrapping, or other means that will resist such corrosion.

> *Comment #1: Soil samples should be taken to assure that the soil will not corrode the pipe. Wrappings and coatings reduce contact corrosion, but cathodic protection may be required where stray electric currents exist.*
>
> *Comment #2: Job site debris should not be allowed in the backfill for piping trenches. Material such as metal cans, metal studs, and gypsum board may chemically react with some types of pipe.*

2.9.3 From Weakened Structure

Any structural member weakened or impaired by cutting, notching, or otherwise, shall be reinforced, repaired, or replaced, so as to be left in a safe structural condition in accordance with the requirements of the applicable building code.

2.9.4 From Nails, Screws, and Other Fasteners

    a. Plastic and copper piping run through framing members (wood or metal) to within one inch of the edge of the framing shall be protected by steel nail plates not thinner than 16 gauge. Where such piping penetrates top plates or sole plates of the framing, the nail plate shall extend at least two inches below top plates and two inches above sole plates.

    b. Where plastic and copper piping runs through metal framing members, it shall be protected from abrasion caused by expansion and contraction of the piping or movement of the framing.

2.10 EXCLUSION OF MATERIALS DETRIMENTAL TO THE SEWAGE SYSTEM

2.10.1 General

No material shall be deposited into a building drainage system or sewer that would or could either obstruct, damage, or overload such system; that could interfere with the normal operation of sewage treatment processes; or that could be hazardous to people or property. This provision shall not prohibit the installation of special waste systems when approved by the Authority Having Jurisdiction.

2.10.2 Industrial Wastes

Waste products from manufacturing or industrial operations shall not be introduced into the public sewer system until it has been determined by the Authority Having Jurisdiction that the introduction thereof will not cause damage to the public sewer system or interfere with the functioning of the sewage treatment plant.

> *Comment: Where industrial wastes will be created, the facility should provide the following information to the Authority Having Jurisdiction for the sewerage system: (1) the quantity of water and waste material that will be discharged into the sewer system, (2) the industrial processes that create the waste, (3) the composition and concentration of the chemicals in the waste, (4) the water supply demand for the facility, and (5) the intended design of the pre-treatment or neutralizing system for the wastes prior to its discharge into the sewer system.*

2.11 MATERIALS EXPOSED WITHIN PLENUMS

All piping materials exposed within plenums shall comply with the provisions of other applicable Codes.

2.12 SLEEVES FOR PIPING

a. All piping passing through concrete walls, floors, slabs, and masonry walls shall be provided with sleeves for protection.

EXCEPTION: Sleeves shall not be required for pipes passing through drilled or bored holes. Such holes shall provide 1/2 inch minimum clearance around the pipe and any thermal insulation.

b. Sleeves shall be sized so there is a minimum of 1/2-inch clearance around the pipe and/or insulation.

c. Piping through concrete or masonry walls shall not be subject to any load from building construction.

d. The annular space between sleeves and pipes shall be filled or tightly caulked with coal tar, asphaltum compound, lead, or other material found equally effective and approved as such by the Authority Having Jurisdiction.

e. All penetrations of construction required to have a fire resistance rating shall be protected in accordance with the applicable building regulations.

See Figure 2.12

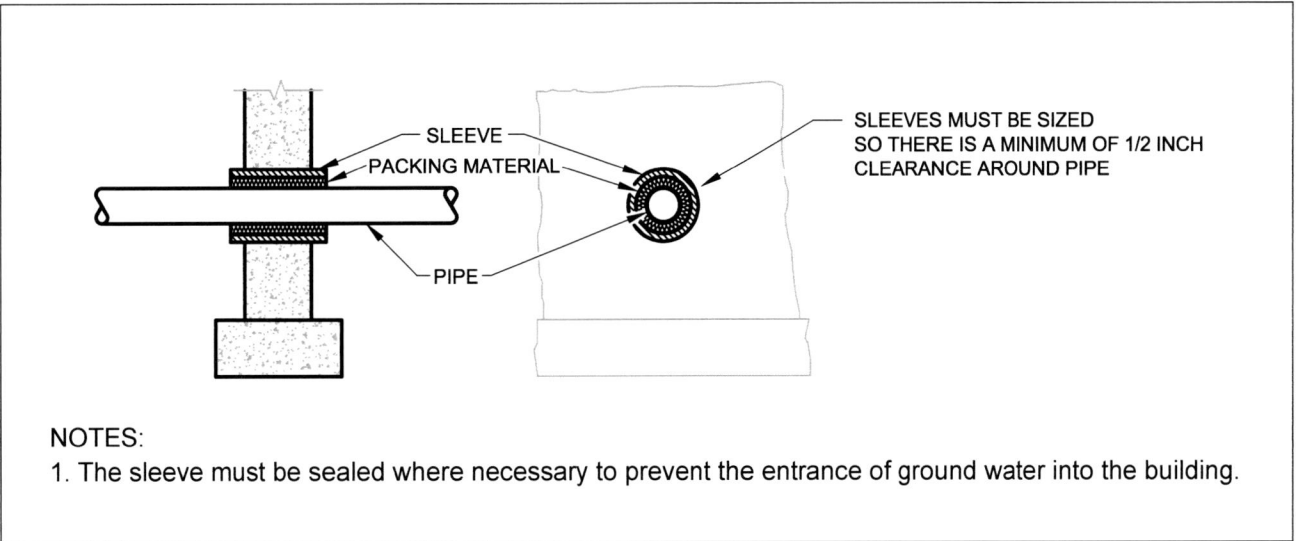

Figure 2.12
A PIPE SLEEVE THROUGH A FOUNDATION WALL

2.13 OPENINGS FOR PIPING

a. Openings for plumbing piping shall be sealed as required to maintain the integrity of the wall, floor, ceiling, or roof that has been penetrated.

b. Collars or escutcheon plates shall be provided to cover the openings around pipes where the piping penetrates walls, floors, or ceilings in finished areas that are exposed to view.

2.14 USED MATERIAL OR EQUIPMENT

Used plumbing material or equipment that does not conform to the standards and regulations set forth in this Code shall not be installed in any plumbing system.

2.15 CONDEMNED EQUIPMENT

Any plumbing equipment condemned by the Authority Having Jurisdiction because of wear, damage, defects or sanitary hazards, shall not be used for plumbing purposes.

2.16 FREEZING OR OVERHEATING

a. The plumbing system shall be protected from freezing or overheating. The following conditions shall be met:

1. Exterior water piping shall be installed below recorded frost lines. Minimum earth cover above the top of the pipe shall be _____ inches.

2. Minimum earth cover above the top of exterior building drains and building sewers that connect to public sewage systems shall be _____ inches. Minimum earth cover above the top of exterior building drains and building sewers that connect to individual sewage disposal systems shall be _____ inches.

3. In systems that are used seasonally, water piping shall have provisions to be drained.

4. Piping shall be installed so that the contents will not be heated due to close proximity to any heat source or from direct solar radiation.

5. In areas with seasonal freezing outdoor temperatures, all drain piping and water piping installed in exterior walls, attics, and other areas exposed to outdoor temperatures shall be protected from freezing. In heated spaces, the piping shall be installed on the heated side of the building insulation.

See Figure 2.16

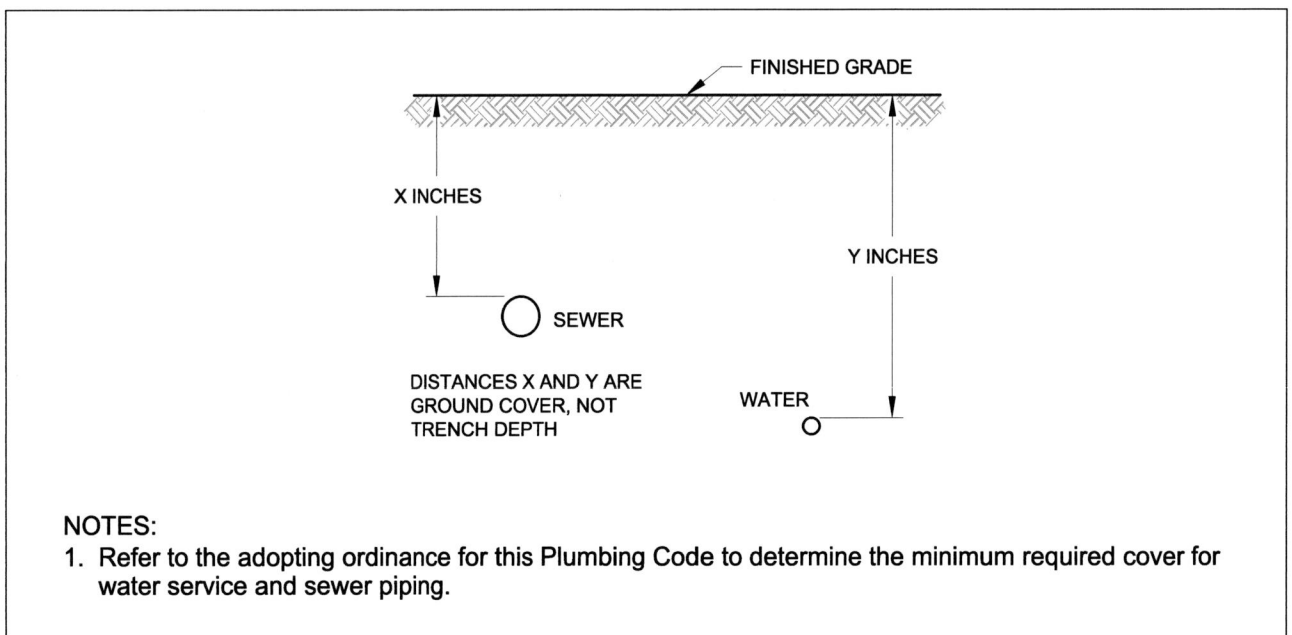

Figure 2.16
DEPTH OF COVER FOR WATER SERVICE AND SEWER PIPING

2.17 PROTECTING FOOTINGS

Trenching parallel to and below the bottom of footings or walls shall not penetrate a 45° plane extending outward from the bottom corner of the footing or wall, unless the soil type is approved by the Authority Having Jurisdiction for a different angle of repose. *See Figure 2.17*

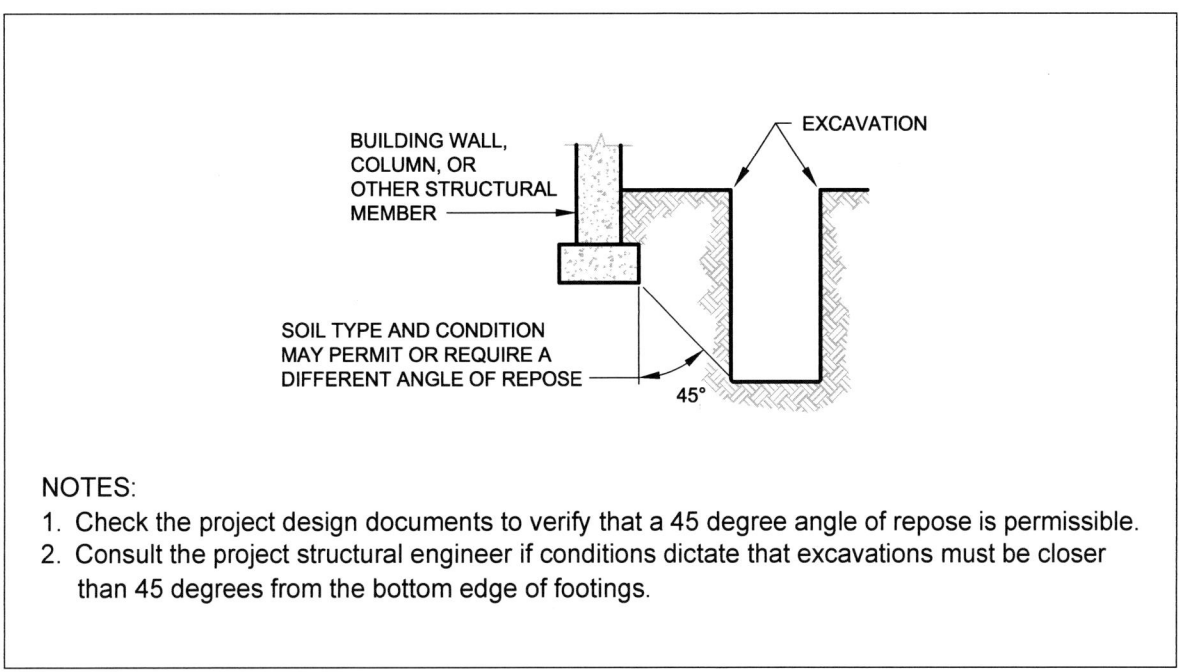

Figure 2.17
PROTECTING STRUCTURAL FOOTINGS FROM BEING UNDERMINED

2.18 CONNECTIONS TO PLUMBING SYSTEMS REQUIRED

Every plumbing fixture, drain, appliance, or appurtenance thereto that is to receive or discharge any liquid waste or sewage shall discharge to the sanitary drainage system of the building in accordance with the requirements of this Code.

2.19 CONNECTION TO WATER AND SEWER SYSTEMS

2.19.1 Availability of Public Water and Sewer

The water distribution and drainage systems of any building in which plumbing fixtures are installed shall be connected to a public water supply and sewer system respectively if the public water supply and/or public sewer is within _____ feet of any property line on the premises, or other reasonable distance as determined by the Authority Having Jurisdiction. *See Figure 2.19.1*

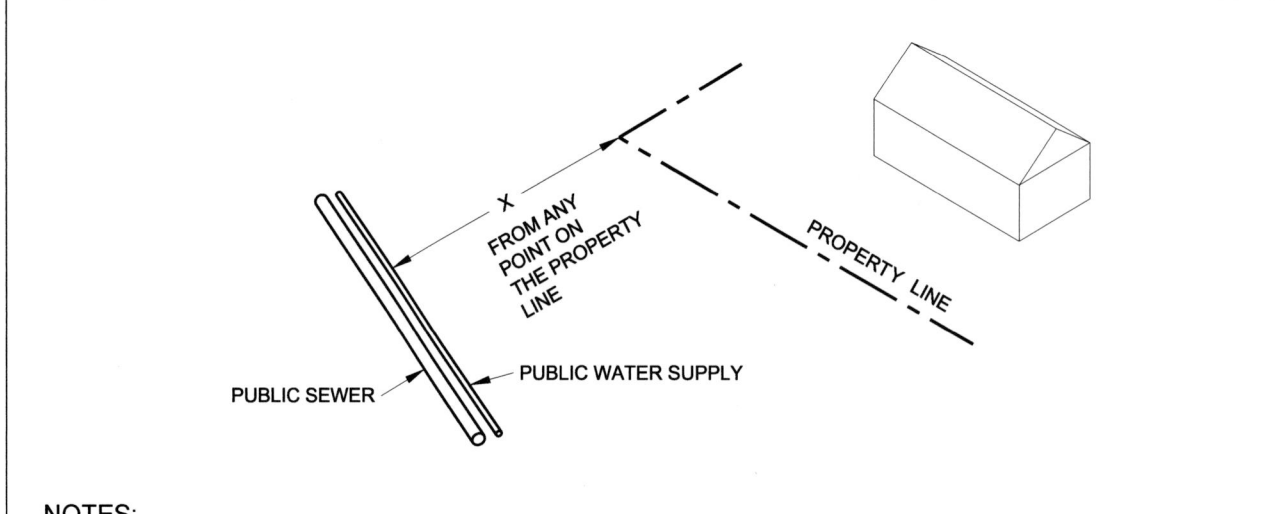

NOTES:
1. Refer to the adopting ordinance for this Plumbing Code to determine the distance from which properties must be connected to public water and sewer systems.
2. Connections to public water and sewer systems are more desirable than private systems, from the standpoint of public health and convenience.

Figure 2.19.1
REQUIRED CONNECTION OF PROPERTIES TO PUBLIC WATER AND SEWER SYSTEMS

2.19.2 Private Systems

Where either a public water supply or sewer system, or both, are not available, a private individual water supply or individual sewage disposal system, or both, shall be provided, and the water distribution system and drainage system shall be connected thereto. Such private systems shall meet the standards for installation and use established by the Health Department or other agency having jurisdiction. (See Chapters 16 and 17.) *See Figure 2.19.1. Also the definition of "Private Sewage Disposal System" and "Private Water Supply".*

Comment: Plumbing in buildings connected to private water or sewage systems must comply with all applicable requirements of this Code or the Authority Having Jurisdiction.

2.20 REQUIREMENTS FOR WASHROOMS, TOILET ROOMS & BATHROOMS

2.20.1 Light and Ventilation

Light and ventilation shall be provided as required by other applicable codes.

2.20.2 Location of Piping and Fixtures

Piping, fixtures, or equipment shall not be located in such a manner as to interfere with the normal operation of windows, doors, or other exit openings.

2.21 PIPING MEASUREMENTS

Except where otherwise specified in this Code, all measurements shall be made to the center lines of the pipes.

2.22 WATER CLOSET CONNECTIONS

a. Three-inch bends may be used on water closets or similar connections provided a 4-inch by 3-inch flange is installed to receive the closet fixture horn.

b. Four-inch by three-inch closet bends shall be permitted.

See Figure 2.22

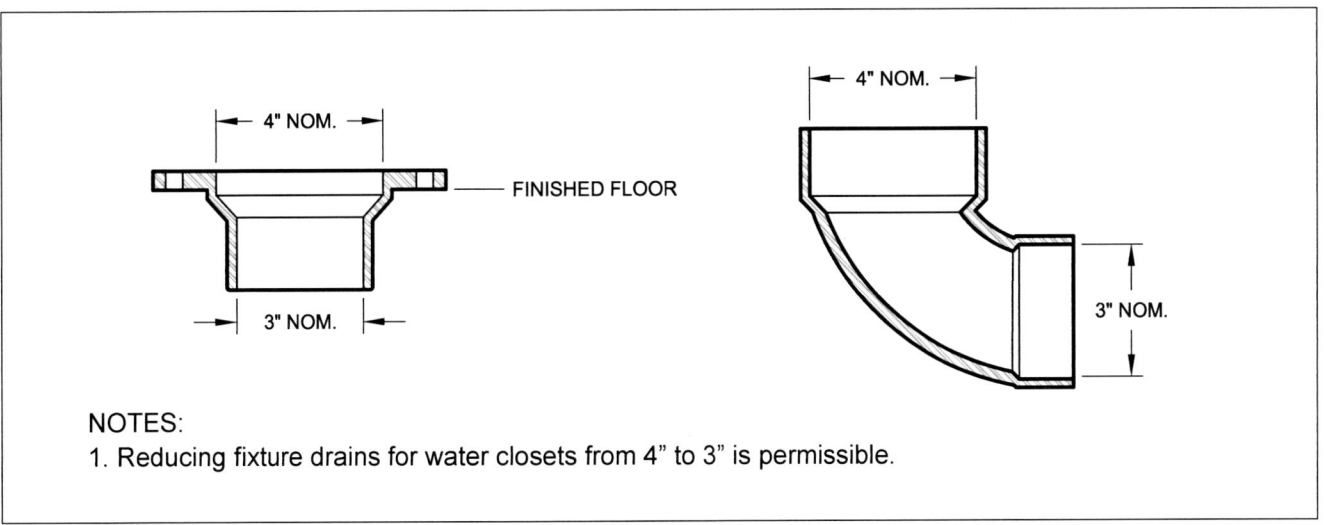

Figure 2.22
WATER CLOSET DRAIN CONNECTIONS

2.23 DEAD ENDS

2.23.1 Drain Piping

a. In the installation of drain piping, dead ends shall not be permitted.
EXCEPTION: Piping to make cleanouts accessible.
See Figure 5.4.5 - B
b. Rough-ins for future fixtures shall not be considered dead ends.
c. If an existing fixture, appliance, or equipment is removed, any open ends in the drain piping shall be capped or plugged watertight.

2.23.2 Vent Piping

In the installation, removal, or disconnecting of any vent piping, any open ends shall be capped or plugged watertight. Dead ends that are not self-draining shall not be permitted.

2.24 TOILET FACILITIES FOR CONSTRUCTION WORKERS

Suitable toilet facilities shall be provided and maintained in a sanitary condition for the use by workers during construction. Non-sewer type toilet facilities for construction workers shall conform to ANSI Z4.3.

2.25 FOOD HANDLING ESTABLISHMENTS AND FOOD HANDLING AREAS WITHIN BUILDINGS

2.25.1 General Area Protection

a. All food and drink, while being stored, prepared, displayed, served, or sold in food handling establishments and food handling areas in buildings shall be protected against contamination from drainage overflow, flooding, backflow, or leakage.
EXCEPTION: Seating areas.

b. Food or drink shall not be stored, prepared or displayed beneath overhead drain or vent piping unless such pipes are protected against leakage or condensation reaching the food or drink.

c. In new or remodeled construction, drain and vent piping shall not be located above food preparation, storage, display, or serving areas where possible.
EXCEPTION: Seating areas.

d. Where drain and vent piping must be installed above such areas, the amount of piping and the number of pipe joints shall be minimized.

e. Where plumbing fixtures are located above such areas, all openings through floors, including those for piping, cleanouts, and other plumbing work shall be provided with sleeves securely bonded to the floor construction and projecting not less than 3/4 inches above the top of the finished floor with the space between the penetration and the sleeve sealed waterproof. *See Figure 2.25 - A*

f. Except for bathtubs or whirlpool baths, plumbing fixtures installed on the floor above such areas shall be the wall-mounted or back-outlet type.

g. Floor drains and shower drains installed on the floor above such areas shall be flashed and equipped with integral seepage pans.

h. The waste and overflow connections for bathtubs and whirlpool baths installed on the floor above such areas shall be made above the floor and piped through a single sleeved floor opening to the fixture trap below the floor. No floor openings, other than the sleeve for the waste pipe to the fixture trap, and a sleeved opening for a vent pipe, if required, shall be permitted above such areas. Pipes shall be sealed watertight to their sleeves.

i. Drain and vent piping above such areas shall be subjected to a standing water test of not less than 25 feet or 10 psig.

j. Piping subject to operation at temperatures that may cause condensation on the external surfaces of the pipe shall be thermally insulated to prevent condensation.

k. Where drain and vent piping is installed above finished ceilings in such areas, the ceiling panels shall be the removable type or sufficient access panels shall be provided to permit complete inspection of the piping.

> *Comment: The seating areas in restaurants, cafeterias, and other dining areas are not subject to the requirements of Section 2.25.1.*

2.25.2 Food Service Equipment and Fixtures

a. All food and/or drink service equipment, including sinks, dishwashers, ice machines, brewers, and dispensers, shall be indirectly discharge to the drainage system through an air gap or air break in accordance with Chapter 9.
EXCEPTION: If a properly vented floor drain is installed adjacent to a food service fixture that is properly trapped and vented, the food service fixture shall be permitted to connect directly to the drainage system on the sewer side of the floor drain.

b. Where multi-compartment sinks are drained indirectly, each compartment shall discharge separately into a sanitary floor sink that is capable of draining all of the compartments simultaneously.

2.25.3 Exposed Food Products

Wherever unwrapped or unpacked food products are prepared, displayed, or sold for human consumption, all fixtures, equipment, devices, utensils, tableware, and apparatus involved in the food service process shall be protected against backflow, cross-connection, and flooding from the drainage system by indirect connections to the drainage system through air gaps in accordance with Chapter 9.

2.25.4 Food Display Equipment

Display cases for refrigerated and frozen food and drink products and other equipment and appliances that produce clear water waste shall be indirectly connected to the drainage system through an air gap or air break in accordance with Chapter 9.

2.25.5 Sanitary Floor Sinks

Floor sinks in food handling areas shall be of the sanitary design with smooth, corrosion-resistant surfaces that can be readily cleaned.

    a. Where sanitary floor sinks serve as waste receptors in food handing areas, they shall not be located where subject to foot traffic or other loads on their grates, if provided.

    b. Sanitary floor sinks shall comply with ASME A112.6.7 and shall be of the sanitary design with smooth, corrosion-resistant surfaces that can be readily cleaned. They shall include a dome strainer over the outlet opening to prevent clogging the drain or a sediment bucket to collect solids.

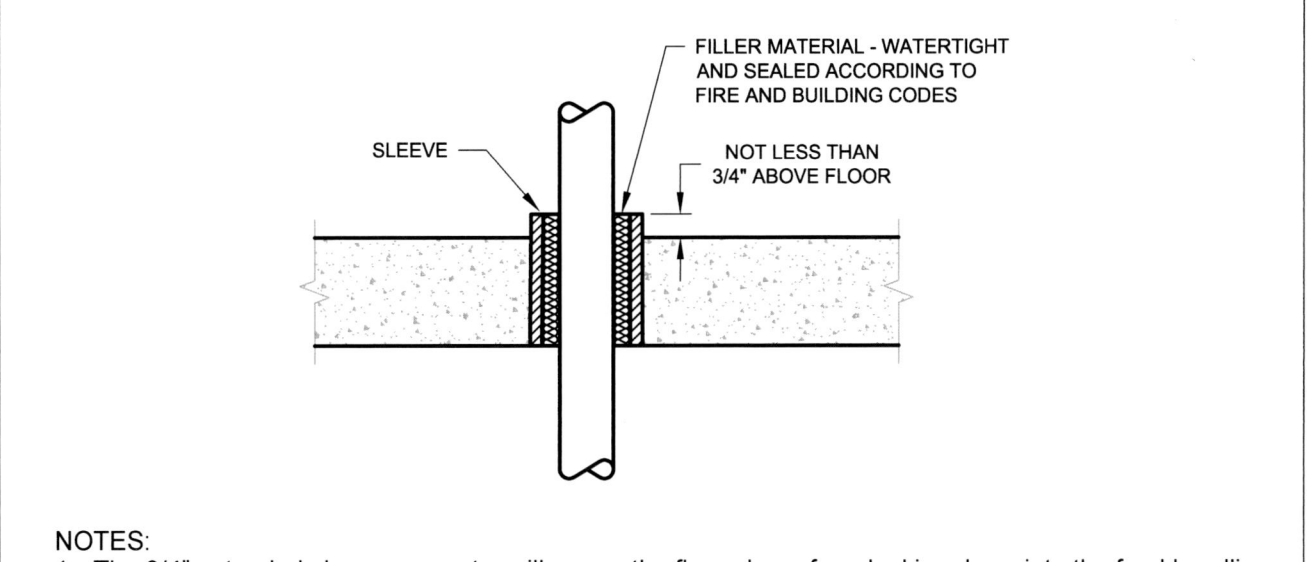

NOTES:
1. The 3/4" extended sleeve prevents spillage on the floor above from leaking down into the food handling area and contaminating the food products.

Figure 2.25 - A

A PIPE PENETRATION OF A FLOOR ABOVE A FOOD HANDLING AREA

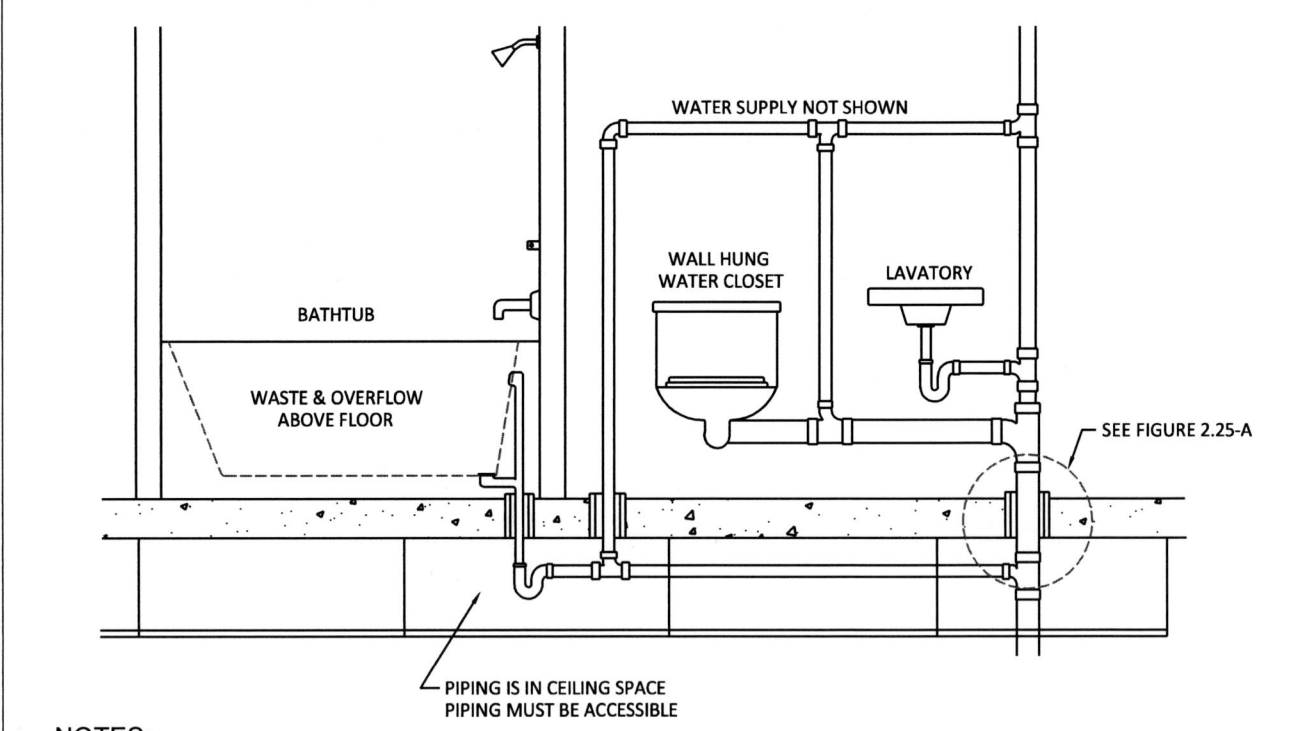

NOTES:
1. All pipe penetrations through the floor above must have extended pipe sleeves. See Figure 2.25 - A. The number of floor penetrations should be minimized.
2. Floor-outlet water closets are not permitted above a food handling area.
3. Bathtubs and showers must have above-the-floor waste outlets with piping extended to a point where an extended pipe sleeve can be installed.
4. Piping run in a ceiling space above a food handling area must be fully accessible.

Figure 2.25 - B
PLUMBING FIXTURES ABOVE A FOOD HANDLING AREA

2.25.6 Vacuum Condensate Drainage Systems

a. Vacuum condensate drainage systems to lift condensate from refrigerated food and drink display cases shall be designed by the manufacturer specifically for supermarkets and similar applications. Systems shall comply with the requirements of this Code for protection against backflow, cross connections, and contamination of food and drink products.

b. Systems shall be installed in accordance with the manufacturer's requirements for piping material, arrangement of the piping and equipment, pipe sizing, and pipe support. Drain pipe connections to display cases shall include swing check valves and isolation valves when recommended by the system manufacturer.

c. The waste discharge from vacuum condensate drainage systems shall be connected to the building gravity drainage system through an air gap or air break. The gravity drainage system shall be sized for the maximum waste discharge rate from the vacuum system.

2.26 REQUIREMENTS FOR ELEVATORS

2.26.1 General

 a. The plumbing requirements for elevator shafts, pits, and equipment rooms shall be in accordance with the applicable elevator code.

 b. Piping shall not be installed in elevator machine rooms that is prohibited by the applicable elevator code.

 c. Foundation drains and other sources of ground water shall not be connected to elevator pits.

 d. Where elevators have Phase II firefighter override, drainage shall be provided for their elevator pit.

2.26.2 Where Elevator Pit Drainage is Provided

 a. Drainage from elevator pit drains or pumps shall discharge to an approved location. Such drainage shall not be connected to storm drain piping.

 b. Where the drainage discharge is outdoors, the location shall be approved by the Authority Having Jurisdiction and be marked "ELEVATOR PIT DISCHARGE".

 c. An indoor discharge from elevator pit drains or pumps shall extend from the pit to an indirect waste receptor or be connected to gravity drain piping indirectly through an air gap or air break. The point of discharge shall be marked "ELEVATOR PIT DISCHARGE".

 d. The controls for sump pumps serving hydraulic elevators shall include automatic oil sensing with pump cutoff or there shall be oil separation for pump operation.

 e. Sump pits for drains or pumps shall have a removable or operable grate-type cover that is secured and level with the elevator pit floor.

 f. The discharge piping from elevator sump pumps shall include a check valve and a manual quarter-turn shutoff valve with visual indication of its full open and full closed positions from within the elevator pit.

 g. Where elevators have Phase II firefighter override, the design flow capacity of the required drains or pumps shall be not less than 3000 gph (50 gpm) per elevator. Pumps for hydraulic elevators shall have oil separation for pump operation.

2.27 IDENTIFICATION OF SYSTEM PIPING

 a. Where plumbing system piping is identified, the labeling shall comply with the applicable requirements of ASME A13.1 or its equivalent.

 b. Piping for non-potable water used for typical potable water applications shall be identified in accordance with Section 10.21.

2.28 NFPA 13D MULTIPURPOSE RESIDENTIAL FIRE SPRINKLER SYSTEMS

Where approved by the Authority Having Jurisdiction, NFPA 13D multipurpose residential fire sprinkler systems that provide both domestic cold water distribution and fire sprinkler protection for one- and two-family dwellings with a combination piping system shall comply with Section 10.20.

2.29 RADON GAS SYSTEMS

Radon systems and their components shall be designed to comply with the laws, ordinances, codes, and regulations of relevant jurisdictional authorities, including applicable mechanical, electrical, building, plumbing, energy, and fire prevention codes.

2.30 SWIMMING POOLS, WADING POOLS, SPAS, AND HOT TUBS

 a. The water supply shall be protected from backflow by a fixed air gap above the overflow level of the pool, spa, or hot tub, or by an approved backflow preventer. Where the water supply connection is below the overflow level, an ASSE 1013 reduced pressure principle backflow preventer shall be provided.

 b. The drainage from swimming pools and wading pools shall be indirect through an air gap in accordance with Section 9.1.11.

Chapter 3

Materials

3.1 MATERIALS

3.1.1 Minimum Standards

The standards cited in this chapter shall control all materials, systems, and equipment used in the construction, installation, alteration, repair, or replacement of plumbing or drainage systems or parts thereof.
EXCEPTIONS:

(1) The Authority Having Jurisdiction shall allow the extension, addition, or relocation of existing water, drain, and vent piping with materials of like grade or quality as permitted in Section 3.12.2.

(2) Materials not covered by the standards cited in this chapter may be used with the approval of the Authority Having Jurisdiction as permitted in Section 3.12.2.

3.1.2 General Requirements

a. Materials, fixtures, or equipment used in the installation, repair or alteration of any plumbing system shall conform at least to the standards listed in this chapter, except as otherwise approved by the Authority Having Jurisdiction under the authority contained in Section 3.12.

b. Materials installed in plumbing systems shall be handled and installed as to avoid damage so that the quality of the material will not be impaired.

c. No defective or damaged materials, equipment or apparatus shall be installed or maintained. (See Sections 2.14 and 2.15)

d. All materials used shall be installed in strict accordance with the standards under which the materials are accepted and approved, including the appendices of the standards, and in strict accordance with the manufacturer's instructions. Where the provisions of material standards or manufacturer's instructions conflict with the requirements of this Code, this Code shall prevail.

e. Marking of cast iron soil pipe and fittings: Each length of cast iron soil pipe and fittings used in the plumbing system shall be marked by the manufacturer's name or registered trademark to enable the end user to readily identify the manufacturer. The marking shall be done during the time of manufacture. Field marking shall not be permitted.

f. Certification of cast iron soil pipe and fittings: Where cast iron soil pipe and fittings are being installed, the Authority Having Jurisdiction shall be furnished, when requested, certification by the manufacturer of compliance to the product standards. Resellers of cast iron soil pipe and fittings manufactured by others and using third party certifications or inspections to support proof of compliance to the product standard shall, in addition to the manufacturer's certification, provide, when requested, copies of the third party reports to the Authority Having Jurisdiction.

3.1.3 Standards Applicable to Plumbing Materials

A material shall be considered approved if it is listed or certified by a recognized certification body as complying with one or more of the standards cited in Table 3.1.3, and in the case of plastic pipe, fittings and solvent cement also NSF 14. Materials not listed in Table 3.1.3 shall be used only as provided for in Section 3.12.2 or as permitted elsewhere in this Code.

3.1.4 Identification of Materials

Materials shall be identified as provided in the standard to which they conform.

3.1.5 Health Effects of Drinking Water Components

The health effects of drinking water components shall comply with NSF 61. Components within the scope of NSF 61 have surfaces in contact with potable water. Components include, but are not limited to, plastic materials, plastic and metallic pipe, pipe-related products, protective materials, joining and sealing materials, process media, treatment devices, transmission devices, distribution devices, and end-point devices.

3.2 SPECIAL MATERIALS

3.2.1 Miscellaneous Materials

Sheet and tubular copper and brass for the following uses shall be not less than:
- a. General use - 12 oz. per square foot
- b. Flashing for vent pipes - 8 oz. per square foot
- c. Fixture traps and trap arms - 17 gauge or ASME A112.18.2/CSA B125.2
- d. Fixture tailpieces - 20 gauge or ASME A112.18.2/CSA B125.2
- e. Tailpieces with dishwasher connections - ASME A112.18.2/CSA B125.2
- f. Continuous wastes - 20 gauge or ASME A112.18.2/CSA B125.2

3.2.2 Lead

See Table 3.1.3. Sheet lead shall be not less than the following:

a. Shower pans—not less than 4 pounds per square foot (psf) and be coated with an asphalt paint or equivalent.

b. Flashings of vent terminals—not less than 3 pounds per square foot (psf).

c. Lead bends and lead traps shall not be less than 1/8" wall thickness.

3.2.3 Plastic

- a. Trap and tailpiece fittings— minimum 0.062" wall thickness.
- b. Piping— see specific application — Sections 3.4 to 3.11.
- c. Shower pans—approved plastic sheeting material.

> *Comment: Plastic shower pans should not be covered with asphalt unless recommended by the manufacturer.*

3.3 FITTINGS, FIXTURES, APPLIANCES & APPURTENANCES

3.3.1 Drainage Fittings

See Sections 2.3 and 2.4 for fittings and connections in drainage systems.

3.3.2 Cleanout Plugs and Caps

a. Cleanout plugs shall be of brass, plastic, stainless steel, or other approved materials and shall have raised or countersunk square heads, except that where raised heads will cause a tripping hazard, countersunk heads shall be used. *See Figure 3.3.2*

b. Cleanout caps shall be of brass, plastic, reinforced neoprene, cast-iron, or other approved material and shall be readily removable.

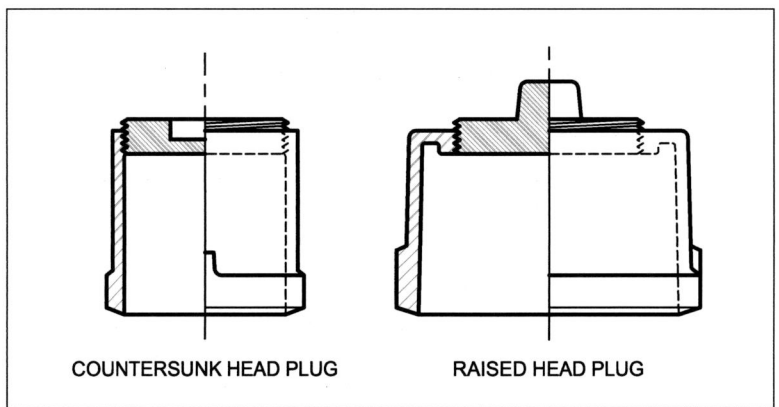

Figure 3.3.2
CLEANOUT PLUGS

3.3.3 Fixtures

a. Plumbing fixtures shall be constructed from approved materials having smooth, non-absorbent surfaces and be free from defects, and except as permitted elsewhere in this Code, shall conform to the standards listed in Table 3.1.3.

b. Refer to Chapter 7 for the requirements for specific plumbing fixtures.

c. Materials for special use fixtures not otherwise covered in this Code shall be constructed of materials especially suited to the use for which the fixture is intended.

> *Comment #1: The standards for plumbing fixtures that are listed in Table 3.1.3 reflect consensus agreement between manufacturers, industry representatives, and consumer groups.*
>
> *Comment #2: Slip resisting surfaces in bathtubs and showers do not constitute a violation of the requirement for smooth surfaces.*

3.3.4 Floor Flanges and Mounting Bolts

a. Floor flanges for water closets or similar fixtures shall be not less than 1/8" thick for brass, 1/4" thick and not less than 1-1/2" caulking depth for cast-iron or galvanized malleable iron. Approved copper and plastic flanges may be used.

b. If of hard lead, they shall weigh not less than 1 lb. 9 oz. and be composed of lead alloy with not less than 7.75 percent antimony by weight. Flanges shall be soldered to lead bends or shall be caulked, soldered, or threaded into other metal.

c. Closet screws and bolts shall be corrosion-resisting.

d. Connections between drain piping and floor outlet water closets shall be made by means of an approved flange that is attached to the drain piping in accordance with the provisions of this chapter. The floor flange shall be set on and securely anchored to the finished floor.

3.3.5 Flush Pipes and Fittings

Flush pipes and fittings shall be of nonferrous material. When of brass or copper tube, the material shall be at least 0.0375" in thickness (No. 20 U.S. gauge).

3.3.6 Plumbing Supply Fittings

Plumbing supply fittings covered under the scope of NSF 61 shall comply with its requirements. Refer to Section 3.1.5 for components that are within the scope of NSF 61.

3.3.7 Interceptors, Separators, Grease Recovery Devices

Interceptors, separators, and grease recovery devices shall meet the requirements of Chapter 6.

3.3.8 Pressure Tanks and Vessels

a. Hot water storage tanks shall meet construction requirements of ASME, CSA, or UL as appropriate. (See Table 3.1.3)

b. ASME pressure vessels shall be designed and constructed in accordance with the requirements of the American Society of Mechanical Engineers (ASME), Rules for Construction of Heating Boilers, Section IV Part HLW and/or Rules for Construction of Pressure Vessels, Section VIII. Any pressure vessel that exceeds any of the following shall meet the requirements of ASME and shall be stamped ASME: A heat input rating of 200,000 BTU per hour; or a water temperature of 210 degrees Fahrenheit; or a nominal water capacity of 120 gallons; or any other thresholds of ASME that apply.

c. Storage tanks less in volume than those requirements specified by ASME shall be of durable materials and constructed to withstand 125 p.s.i. with a safety factor of 2.

3.3.9 Roof Drains

Roof drains shall be of cast iron, copper, lead, plastic, or other approved corrosion-resisting materials. *See Section 13.5.1*

3.3.10 Safety Devices for Pressure Tanks

Safety devices shall meet the requirements of the American National Standards Institute, American Society of Mechanical Engineers, or the Underwriters Laboratories. Listing by Underwriters Laboratories, Canadian Standards Association, or National Board of Boiler and Pressure Vessel Inspectors shall constitute evidence of conformance with these standards. Where a device is not listed by any of these organizations, it shall have certification by an approved laboratory as having met these requirements. *See Section 10.16*

3.3.11 Tanks

a. Plans for all septic tanks shall be submitted to the Authority Having Jurisdiction for approval. Such plans shall show all dimensions, reinforcing, structural calculations, and such other pertinent data as may be required.

b. Septic tanks shall be constructed of sound durable materials, not subject to excessive corrosion or decay and shall be watertight. *See Sections 16.6.5 and 16.6.6*

3.3.12 Carriers and Supports

Carriers and supports for plumbing fixtures shall comply with ASME A112.6.1M, ASME A112.6.2, or ASME A112.19.12.

3.4 POTABLE WATER PIPING

3.4.1 Materials

3.4.1.1 Plastic Piping

Plastic piping materials used for the conveyance of potable water shall comply with NSF 14 and be marked accordingly.

3.4.1.2 Stainless Steel Piping

a. Stainless steel pipe shall comply with ASTM A312, TP316, or TP316L.

b. Stainless steel fittings shall be the press connect or approved joint type.

3.4.2 Water Service Piping

Water service pipe and pipe fittings to the point of entrance into the building through a foundation wall or floor shall be of materials listed in Table 3.4 and shall be water pressure rated not less than 160 psi at 73 deg F. See Table 3.4.2. Water service pipe and pipe fittings shall comply with NSF 61.

3.4.3 Water Distribution Piping

Water piping for the distribution of hot or cold water within buildings shall be of material listed in Table 3.4, and shall be water pressure rated for not less than 100 psi at 180 deg F and 160 psi at 73 deg F. Plastic piping used for hot water distribution shall be installed in accordance with the requirements of Section 10.15.8. Water distribution pipe and pipe fittings shall comply with NSF 61.

NOTE: The working pressure rating of certain approved plastic piping materials varies depending on the pipe size, material composition, wall thickness, and methods of joining. See Table 3.4.3.

3.4.4 Fittings

a. Fittings for water supply piping shall be compatible with the pipe material used.

b. Insert fittings for plastic tubing shall be the metallic or plastic type that comply with standards listed in Table 3.1.3.

3.4.5 Material Ratings and Installation

a. Piping used for domestic water shall be suitable for the maximum temperature, pressure, and velocity that may be encountered, including temporary increases and surges.

b. Relief valve temperature and pressure relief settings shall not exceed the pipe, tubing, or fitting manufacturer's recommendations.

c. Pipe and fittings shall be installed in accordance with the manufacturer's installation instructions and the applicable material standards, recognizing any limitations in use.

> *Comment: Refer to the standards listed in Table 3.1.3 to determine the pressure and temperature ratings of the various piping materials.*

3.4.6 Limit on Lead Content

a. Materials used in potable water supply systems, including piping, faucets, and valves, shall be "lead-free" as defined by current Federal Law.

b. Drinking water system components shall comply with the lead leachate requirements of NSF 61 and NSF 372. Refer to Section 3.1.5 for components that are within the scope of NSF 61 and NSF 372.

3.4.7 Shutoff Valves

All gate valves, ball valves, butterfly valves, globe valves, and other shutoff valves in water service piping and water distribution piping shall comply with the requirements of NSF 61.

3.5 SANITARY DRAIN PIPING

3.5.1 Aboveground Piping - Soil, Waste and Indirect Waste

Aboveground sanitary drain piping within buildings shall be of materials listed in Table 3.5.

3.5.2 Underground Building Sanitary Drains

Underground building drains and other underground sanitary drain piping within buildings shall be of materials listed in Table 3.5.

3.5.3 Building Sanitary Sewer

Sanitary sewer piping outside of buildings shall be of materials listed in Table 3.5. Joints shall be watertight and root proof.

3.5.4 Plastic Piping

a. Pipe and fittings classified by standard dimension ratio that are underground outside of buildings shall be SDR 35 or heavier (lower SDR number). Pipe and fittings within buildings shall be SDR 26 or heavier (lower SDR number).

b. Pipe and fittings classified by pipe stiffness that are underground outside of the buildings shall be PS-46 or heavier (higher PS number). Pipe and fittings within buildings shall be PS-100 minimum.

3.5.5 Vitrified Clay Pipe

Vitrified clay pipe shall be joined using compression joints or couplings. Vitrified clay pipe installed underground within buildings shall be extra strength and shall have 12" minimum earth cover.

3.5.6 Fittings

Fittings in drainage systems shall be compatible with the pipe used and shall have no ledges, shoulders, or reductions that can retard or obstruct flow. Threaded fittings shall be the recessed drainage type.

3.6 VENT PIPING

3.6.1 Aboveground Piping

Aboveground vent piping in buildings serving sanitary drainage or storm drainage systems shall be of materials listed in Table 3.6.

3.6.2 Underground Piping

Vent piping installed underground shall be of materials listed in Table 3.6.

3.6.3 Plastic Piping

a. Pipe and fittings classified by standard dimension ratio that are underground outside of buildings shall be SDR 35 or heavier (lower SDR number). Pipe and fittings within buildings shall be SDR 26 or heavier (lower SDR number).

b. Pipe and fittings classified by pipe stiffness that are underground outside of buildings shall be PS-46 or heavier (higher PS number). Pipe and fittings within buildings shall be PS-100 minimum.

3.6.4 Fittings

Fittings in vent piping shall be compatible with the pipe material used. Where threaded pipe is used, fittings shall be either the drainage or pressure type, galvanized or black.

3.7 STORM DRAIN PIPING

3.7.1 Reserved

3.7.2 Interior Conductors

Stormwater drain piping installed aboveground in buildings shall be of materials listed in Table 3.7.

3.7.3 Underground Building Storm Drains

Underground building storm drains and other underground stormwater piping within buildings shall be of materials listed in Table 3.7.

3.7.4 Building Storm Sewer

Building storm sewer piping outside of buildings shall be of materials listed in Table 3.7.

3.7.5 Plastic Piping

a. Pipe and fittings classified by standard dimension ratio that are underground outside of buildings shall be SDR 35 or heavier (lower SDR number). Pipe and fittings within buildings shall be SDR 26 or heavier (lower SDR number).

b. Pipe and fittings classified by pipe stiffness that are underground outside of buildings shall be PS-46 or heavier (higher PS number). Pipe and fittings within buildings shall be PS-100 minimum.

3.7.6 Vitrified Clay Pipe

Vitrified clay pipe shall be joined using compression joints or couplings. Vitrified clay pipe installed underground within buildings shall be extra strength and shall have 12" minimum earth cover.

3.7.7 Fittings

Fittings in drainage systems shall be compatible with the pipe used and shall have no ledges, shoulders, or reductions that can retard or obstruct flow. Threaded fittings shall be the recessed drainage type.

3.8 FOUNDATION DRAINS AND SUBSOIL DRAINAGE

Piping for foundation drains and other subsoil drainage shall be of materials listed in Table 3.8.

3.9 AIR CONDITIONING CONDENSATE DRAIN PIPING

Indirect waste piping from air conditioning unit drains to the point of disposal shall be of a material approved for potable water, sanitary drainage or storm drainage.
EXCEPTION: Flexible plastic tubing shall be permitted to be used for the discharge from small, fractional horsepower condensate pumping units.

3.10 CONDENSATE DRAINS FROM COMBUSTION PROCESSES

Piping used to convey condensate from combustion processes (such as from flues and chimneys) shall conform to the equipment manufacturer's instructions.

3.11 CHEMICAL WASTE PIPING SYSTEMS

3.11.1 Where Required

Drainage and vent piping systems for other than sanitary drainage shall be separate from building sanitary drainage and vent piping and shall be compatible with the wastes being handled.

3.11.2 Drain piping

Drain piping materials, including the type of joints and methods of support, shall be listed by the manufacturer as suitable for the wastes being handled.

3.11.3 Vent Piping

Vent piping for chemical wastes shall be the same as the drain piping and shall be run separately to the outdoor air, independent from any vent piping for sanitary drainage.

3.11.4 Separate Systems

Where separate chemical waste drainage systems are required to handle different chemical wastes, the drainage and vent piping for each system shall also be separate.

3.12 ALTERNATE MATERIALS AND METHODS

3.12.1 Existing Buildings

a. Plumbing work performed in existing buildings shall conform to the requirements of this Code, unless the Authority Having Jurisdiction finds that such conformance would result in an undue hardship.

b. The Authority Having Jurisdiction may grant a variation to the extent necessary to relieve the undue nature of the hardship.

c. A record, open to the public, shall be kept of each variation granted under this section.

3.12.2 Approval

a. The Authority Having Jurisdiction may approve the use of any material or method not expressly conforming to the requirements of this Code provided all of the following conditions are met:

1. The material or method is not expressly prohibited by this Code.

2. The material or method is determined to be of such design or quality as to appear suitable for the proposed use.

3. A record of such approval is kept and shall be available to the public.

3.12.3 Tests

When there is insufficient evidence to verify claims for alternate materials, the Authority Having Jurisdiction may require tests of compliance as proof of suitability. Such tests shall be made by an approved testing agency at the expense of the applicant.

3.12.4 Test Procedure

Tests shall be made in accordance with applicable standards; but in the absence of such standards, the Authority Having Jurisdiction shall specify the test procedure.

3.12.5 Repeated Tests

The Authority Having Jurisdiction may require tests to be repeated if, at any time, there is reason to believe that an alternate material no longer conforms to the requirements on which its approval was based.

Table 3.1.3
APPROVED MATERIALS

Table 3.1.3 - Part I FERROUS PIPE AND FITTINGS

#	Description	Standard
1	Cast iron threaded drainage fittings	ASME B16.12
2	Cement mortar lining for ductile iron pipe and fittings	AWWA C104/A21.4
3	Ductile iron compact water service fittings	AWWA C153/A21.53
4	Ductile iron culvert pipe	ASTM A716
5	Ductile iron gravity sewer pipe	ASTM A746
6	Ductile iron pressure pipe	ASTM A377 AWWA C151/A21.51
7	Ductile iron and gray iron fittings	AWWA C110/A21.10
8	Ferrous pipe flanges and flanged fittings	ASME B16.5
9	Gray iron threaded fittings, class 125 & 250	ASME B16.4
10	Hub & spigot cast iron soil pipe and fittings	ASTM A74
11	Hubless cast iron soil pipe and fittings	CISPI 301 ASTM A888
12	Malleable iron threaded fittings, class 150 & 300	ASME B16.3
13	Stainless steel DWV pipe and fittings, 304 & 316L	ASME A112.3.1
14	Stainless steel pipe, seamless, welded, austenitic	ASTM A312
15	Steel pipe, galvanized, welded, seamless	ASTM A53
16	Threaded ferrous pipe plugs, bushings, and locknuts	ASME B16.14

Table 3.1.3 - Part II NON-FERROUS PIPE AND FITTINGS

#	Description	Standard
1	Cast bronze threaded fittings, class 125 & 250	ASME B16.15
2	Cast copper fittings for flared copper tube	ASME B16.26
3	Cast copper pipe flanges and flanged fittings	ASME B16.24
4	Cast copper solder joint drainage fittings	ASME B16.23
5	Cast copper solder joint pressure fittings	ASME B16.18
6	Copper drainage tube (DWV)	ASTM B306
7	Copper pipe, threadless	ASTM B302
8	Copper water tube (K, L, M)	ASTM B88
9	Press-connect pressure fittings, copper and copper alloy	ASME B16.51
10	Push-fit fittings for copper water tube	ASSE 1061
11	Red brass pipe, seamless	ASTM B43
12	Wrought copper braze joint pressure fittings	ASME B16.50
13	Wrought copper solder joint drainage fittings	ASME B16.29
14	Wrought copper solder joint pressure fittings	ASME B16.22

Table 3.1.3 - Part III NON-METALLIC PIPE AND FITTINGS

#	Description	Standard
1	ABS plastic DWV pipe with cellular core, schedule 40	ASTM F628
2	ABS plastic DWV pipe and fittings	CSA B181.1
3	ABS plastic DWV pipe and fittings, schedule 40	ASTM D2661

4	ABS/PVC plastic composite sewer pipe	ASTM D2680
5	Clay drain tile	ASTM C4
6	CPVC plastic DWV pipe and fittings	CSA B181.2
7	CPVC plastic pipe, schedule 40 & 80	ASTM F441
8	CPVC plastic pipe, SDR-PR	ASTM F442
9	CPVC plastic pipe and fittings, chemical waste drainage	ASTM F2618
10	CPVC plastic pipe fittings, socket-type, schedule 40	ASTM F438
11	CPVC plastic pipe fittings, socket-type, schedule 80	ASTM F439
12	CPVC plastic pipe fittings, threaded, schedule 80	ASTM F437
13	CPVC plastic pipe, tubing, and fittings for HW/CW distribution (copper tube size)	ASTM D2846
14	CPVC plastic pipe, tubing, and fittings for HW/CW distribution (schedule 40 & 80)	CSA B137.6
15	CPVC-AL-CPVC plastic tubing, (copper tube size)	ASTM F2855
16	Cold expansion fittings with metal compression sleeves for PEX pipe	ASTM F2080
17	Cold expansion fittings with PEX reinforcing rings for PEX tubing	ASTM F1960
18	Concrete culvert, storm drain, and sewer pipe, non-reinforced	ASTM C14
19	Concrete culvert, storm drain, and sewer pipe, reinforced	ASTM C76
20	Concrete drain tile	ASTM C412
21	Concrete perforated pipe	ASTM C444
22	Corrugated PE plastic tubing and fittings	ASTM F405
23	Fiberglass (GFR) pipe fittings, non-pressure	ASTM D3840
24	Fiberglass (GFR) pressure pipe	ASTM D3517
25	Fiberglass (GFR) sewer pipe	ASTM D3262
26	Fiberglass pressure pipe	AWWA C950
27	Flexible pre-insulated piping	ASTM F2165
28	Insert fittings for PEX tubing, mechanical cold expansion, with compression sleeve	ASTM F1865
29	Mechanical cold flare compression fittings with disc spring for PEX tubing	ASTM F1961
30	Metal insert fittings for PE-AL-PE and PEX-AL-PEX plastic pressure pipe	ASTM F1974
31	Metal insert fittings with copper crimp ring for SDR-9 PEX and PE-RT tubing	ASTM F1807
32	Metal insert fittings with copper crimp ring for SDR-9 PEX and PEX-AL-PEX tubing	ASTM F2434
33	PE plastic electro-fusion fittings for OD controlled PE and PEX pipe and tubing	ASTM F1055
34	PE plastic fittings for OD controlled PE pipe and tubing, socket type	ASTM D2683
35	PE plastic fittings for PE pipe and fittings, butt heat fusion	ASTM D3261
36	PE plastic pipe, DR-PR, OD controlled	ASTM F714
37	PE plastic pipe, SDR-PR, OD controlled	ASTM D3035
38	PE plastic pipe, SIDR-PR, ID controlled	ASTM D2239
39	PE plastic pipe for drainage and waste disposal absorption fields, smooth wall	ASTM F810
40	PE plastic tubing, CTS	ASTM D2737
41	PE plastic water service pipe and tubing	AWWA C901

	Table 3.1.3 - PART III NON-METALLIC PIPE AND FITTINGS (continued)	
42	PE-RT plastic HW/CW distribution tubing	ASTM F2769
43	PE-AL-PE composite plastic pressure pipe	ASTM F1282 CSA B137.9
44	PE-AL-PE composite plastic water service pipe	AWWA C903
45	PEX plastic pipe, OD controlled, 3"–54"	ASTM F2788
46	PEX plastic HW/CW distribution systems	ASTM F877
47	PEX plastic tubing	ASTM F876
48	PEX plastic water service pressure pipe	AWWA C904
49	PEX-AL-PEX composite plastic pressure pipe	ASTM F1281 CSA B137.10
50	PEX-AL-PEX composite plastic tubing, OD controlled, SDR-9	ASTM F2262
51	PEX-AL-PEX composite plastic water service pressure pipe	AWWA C903
52	Plastic DWV fitting patterns	ASTM D3311
53	Plastic insert fittings for PE plastic pipe	ASTM D2609
54	Plastic insert fittings for PEX and PE-RT SDR-9 plastic tubing	ASTM F2735
55	Plastic insert fittings for PEX and PE-RT SDR-9 plastic tubing, with copper crimp ring	ASTM F2159
56	Plastic piping components, physical, performance, and health effect requirements	NSF 14
57	PP plastic piping systems, pressure-rated	ASTM F2389
58	Push-fit fittings for copper, CPVC, PEX, and PE-RT tubing	ASSE 1061
59	PVC plastic gasketed sewer pipe fittings	ASTM F1336
60	PVC plastic DWV pipe and fittings	CSA B181.2
61	PVC plastic DWV pipe and fittings, schedule 40	ASTM D2665
62	PVC plastic DWV pipe and fittings, 3.25" OD	ASTM D2949
63	PVC plastic pipe, coextruded, with cellular core, non-pressure, IPS, schedule 40	ASTM F891
64	PVC plastic pipe, coextruded, with cellular core, non-pressure, PS & sewer-drain	ASTM F891
65	PVC plastic pipe, coextruded, reprocessed-recycled, non-pressure, IPS, schedule 40	ASTM F1760
66	PVC plastic pipe, coextruded, reprocessed-recycled, non-pressure, IPS, PS	ASTM F1760
67	PVC plastic pipe, coextruded, reprocessed-recycled, non-pressure, sewer-drain	ASTM F1760
68	PVC plastic pipe, schedules 40, 80, 120	ASTM D1785
69	PVC plastic pipe, SDR, pressure-rated	ASTM D2241
70	PVC plastic pipe fittings, socket-type, schedule 40	ASTM D2466
71	PVC plastic pipe fittings, socket-type, schedule 80	ASTM D2467
72	PVC plastic pipe fittings, threaded, schedule 80	ASTM D2464
73	PVC plastic pressure pipe and fabricated fittings	AWWA C900
74	PVC plastic sewer pipe and fittings	ASTM D2729
75	PVC plastic sewer pipe and fittings, type PSM	ASTM D3034
76	SR plastic drain pipe and fittings	ASTM D2852
77	Stainless steel clamps for insert fittings for SDR-9 PEX tubing	ASTM F2098
78	Vitrified clay pipe, standard, extra strength, and perforated	ASTM C700

Table 3.1.3 - Part IV PIPE JOINTS, JOINING MATERIALS, COUPLINGS, GASKETS

#	Description	Standard
1	Compression joints for vitrified clay pipe and fittings	ASTM C425
2	Couplings for dissimilar DWV pipe and fittings, aboveground, shielded transition	ASTM C1460
3	Couplings for hubless cast iron soil pipe and fittings	CISPI 310 ASTM C1277 FM 1680
4	Couplings for hubless cast iron soil pipe and fittings, shielded, heavy duty	ASTM C1540
5	Couplings for joining hubless cast iron soil pipe and fittings	ASTM A1056
6	Couplings, mechanical, gasketed, for grooved and plain-end pipe	ASTM F1476
7	Dielectric pipe unions	ASSE 1079
8	Elastomeric seals (gaskets) for joining plastic pipe	ASTM F477
9	Electrofusion PE fittings for PE and PEX OD-controlled pipe and tubing	ASTM F1055
10	Extruded tee connections for piping, non-reinforced	ASTM F2014
11	Filler metals for brazing and braze welding	AWS A5.8
12	Flexible transition couplings for underground drain and sewer piping	ASTM C1173
13	Fluxes for soldering copper and copper alloy tube, liquid and paste	ASTM B813
14	Grooved and shouldered joints	AWWA C606
15	Joints for concrete pipe and manholes using rubber gaskets	ASTM C443
16	Joints for drain and sewer plastic pipes using elastomeric seals	ASTM D3212
17	Joints for IPS PVC plastic pipe using solvent cements	ASTM D2672
18	Joints for plastic pressure pipe using flexible elastomeric seals	ASTM D3139
19	Lead, refined	ASTM B29
20	Mechanical couplings for DWV and sewer pipe	CSA B602
21	Mechanical couplings using TPE gaskets for DWV piping above and below ground	ASTM C1461
22	Pipe threads, general purpose, inch	ASME B1.20.1
23	Plastic fittings for connecting water closets to drain piping	ASME A112.4.3
24	Primers for solvent cement joints in PVC plastic pipe and fittings	ASTM F656
25	Push-fit joints for PEX, copper, CPVC, and PE-RT tubing	ASSE 1061
26	Rubber gaskets for cast iron soil pipe and fittings	ASTM C564
27	Rubber-gasket joints for ductile iron pressure pipe and fittings	AWWA C111/A21.11
28	Rubber sheet gaskets	ASTM D1330
29	Solder metal	ASTM B32
30	Solvent cement for ABS plastic pipe and fittings	ASTM D2235
31	Solvent cements for CPVC plastic pipe and fittings	ASTM F493
32	Solvent cements for PVC plastic pipe and fittings	ASTM D2564
33	Solvent cements for SR plastic pipe and fittings	ASTM D3122
34	Solvent cements for transition joints between ABS and PVC non-pressure piping	ASTM D3138

Table 3.1.3 - Part V   PLUMBING FIXTURES

#	Item	Standard
1	Area drains	ASME A112.6.3
2	Bathtubs and whirlpool baths with pressure sealed doors	ASME A112.19.15
3	Ceramic plumbing fixtures	ASME A112.19.2 CSA B45.1
4	Deck and balcony drains	ASME A112.6.4
5	Drinking water coolers	UL 399
6	Electro-hydraulic, tank-type water closets	ASME A112.19.2 CSA B45.1 ASME A112.19.14
7	Emergency eyewash and shower equipment	ANSI/ISEA Z358.1
8	Enameled cast iron plumbing fixtures	ASME A112.19.1 CSA B45.2
9	Enameled steel plumbing fixtures	ASME A112.19.1 CSA B45.2
10	Fabricated stainless steel security water closets	ASME A112.19.3 CSA B45.4
11	Floor drains	ASME A112.6.3
12	Hydraulic requirements for water closets and urinals	ASME A112.19.2 CSA B45.1
13	Hydromassage bathtubs	ASME A112.19.7 CSA B45.10
14	Non-water urinals, vitreous china	ASME A112.19.19
15	Non-water urinals, plastic	IAPMO Z124 CSA B45.5
16	Plastic plumbing fixtures	IAPMO Z124 CSA B45.5
17	Plumbing fixtures with pumped waste	ASME A112.3.4 CSA B45.9
18	Prefabricated plastic spa shells	IAPMO/ANSI Z124.7
19	Roof drains	ASME A112.6.4
20	Sanitary floor sinks	ASME A112.6.7
21	Stainless steel plumbing fixtures	ASME A112.19.3 CSA B45.4
22	Trench drains	ASME A112.6.3
23	Vitreous china plumbing fixtures	ASME A112.19.2 CSA B45.1
24	Water closets, dual flushing, six liter	ASME A112.19.14
25	Whirlpool bathtubs	ASME A112.19.7

Table 3.1.3 - Part VI   PLUMBING FIXTURE TRIM AND ACCESSORIES

#	Item	Standard
1	Anti-siphon fill valves for water closet tanks	ASSE 1002 CSA B125.3
2	Automatic compensating valves for showers and tub/shower combinations	ASSE 1016 ASME A112.1016 CSA B125.16

	Table 3.1.3 - Part VI PLUMBING FIXTURE TRIM AND ACCESSORIES (continued)	
3	Automatic compensating valves for wall-mounted shower systems	ASME A112.18.1 CSA B125.1
4	Bidet sprays	ASME A112.4.2
5	Carriers and waste drains for adjustable lavatories, sinks, and shampoo bowls	ASME A112.19.12
6	Dual flush devices for water closets	ASME A112.19.10
7	Faucets for sinks and lavatories	ASME A112.18.1 CSA B125.1
8	Flexible water connectors	ASME A112.18.6 CSA B125.6
9	Flush valves and spuds for water closet bowls, tanks, and urinals	ASME A112.19.5 CSA B45.15 ASSE 1037
10	Plastic liners for bathtubs and showers	IAPMO/ANSI Z124.8
11	Plumbing fittings	CSA B125.3
12	Plumbing supply fittings	ASME A112.18.1 CSA B125.1
13	Plumbing waste fittings (2" and smaller)	ASME A112.18.2 CSA B125.2
14	Protectors/insulators for exposed piping on accessible fixtures	ASME A112.18.9
15	Safety vacuum release systems (SVRS) for pools, spas, and hot tubs	ASME A112.19.17
16	Shower heads, hand-held showers, and body sprays	ASME A112.18.1 CSA B125.1
17	Supports for off-the-floor public plumbing fixtures	ASME A112.6.1M
18	Supports for off-the-floor water closets with concealed tanks	ASME A112.6.2
19	TAFR valves for individual fixture fittings	ASSE 1062
20	Tubular plastic traps and waste fittings	ASTM F409
21	Water closet seats	ASME A112.4.2
22	Water closet seats, plastic	IAPMO/ANSI Z124.5
	Table 3.1.3 - Part VII PLUMBING APPLIANCES	
1	Clothes washers, commercial, electric	UL 1206
2	Clothes washers, household	AHAM HLW-1
3	Clothes washers, general public, electric	UL 2157
4	Dishwashers, commercial, electric	UL 921
5	Dishwashers, household	AHAM DW-1 UL 749
6	Drinking water treatment system	CSA B483.1
7	Drinking water treatment systems, reverse osmosis	NSF 58
8	Drinking water treatment systems, UV microbiological	NSF 55
9	Drinking water treatment units - aesthetic effects	NSF 42
10	Drinking water treatment units - health effects	NSF 53
11	Food waste disposers, domestic	AHAM FWD-1 ASSE 1008 UL 430

12	Hot water dispensers, household, electric, storage type	ASSE 1023
13	Warewashing equipment, commercial	NSF 3
14	Water heaters, storage type, electric, commercial	UL 1453
15	Water heaters, storage type, electric, household	UL 174
16	Water heaters, storage type, gas-fired, 75,000 BTUH or less	ANSI Z21.10.1 CSA 4.1
17	Water heaters, storage type, gas-fired, above 75,000 BTUH	ANSI Z21.10.3 CSA 4.3
18	Water heaters, storage type, oil-fired	UL 732
19	Water heaters, tankless, electric	UL 499
20	Water heaters, tankless, gas-fired	ANSI Z21.10.3 CSA 4.3
21	Water softeners, cation exchange, residential	NSF 44
Table 3.1.3 - Part VIII VALVES AND APPURTENANCES		
1	Automatic compensating valves for showers and bath/showers	ASME A112.1016 ASSE 1016 CSA B125.16
2	Automatic temperature control mixing valves for discharge water	ASSE 1069
3	Backwater valves	ASME A112.14.1 CSA B181.1 (ABS) CSA B181.2 (PVC) CSA B181.2 (CPVC)
4	Ball valves	ASME A112.4.14
5	Ball valves, threaded, solder joint, grooved, flared, socket-welding	MSS SP-110
6	Bronze gate, globe, angle, and check valves	MSS SP-80
7	Cleanouts	ASME A112.36.2
8	Drain tubes for water heater relief valves	ASME A112.4.1
9	Expansion tanks for water, pre-pressurized	IAPMO/ANSI Z1088
10	Floor drain trap seal protection devices, barrier type	ASSE 1072
11	Gate valves, metal-seated, for water service	AWWA C500
12	Gray iron gate valves, threaded, flanged	MSS SP-70
13	Gray iron swing check valves, threaded, flanged	MSS SP-71
14	Pressure balancing valves for individual fixtures, in-line	ASSE 1066
15	Pressure reducing valves for water distribution	ASSE 1003
16	Relief valves for hot water supply, temperature or temperature-pressure actuated	ANSI Z21.22 CSA 4.4
17	Temperature actuated mixing valves for emergency equipment	ASSE 1071
18	Temperature actuated mixing valves for HW distribution	ASSE 1017 CSA B125.3
19	Trap seal priming valves, potable water supplied	ASSE 1018
20	Trap seal priming devices, drainage and electronic types	ASSE 1044
21	Water hammer arrestors	ASSE 1010 PDI WH 201
22	Water shutoff valves, manual, quarter-turn	ASME A112.4.14
23	Water temperature limiting devices for hot water supply to fixtures	ASSE 1070 CSA B125.3

Table 3.1.3 - Part IX BACKFLOW PREVENTION

1	Air gaps	ASME A112.1.2
2	Air gap fittings	ASME A112.1.3
3	Backflow preventers for beverage dispensing equipment	ASSE 1022
4	Backflow preventers with intermediate atmospheric vent	ASSE 1012
5	Backflow prevention in plumbing fixture fittings	ASME A112.18.3
6	Backflow prevention devices for hand-held showers	ASSE 1014
7	Backflow prevention for commercial dishwashing machines	ASSE 1004
8	Backflow prevention for chemical dispensers	ASSE 1055
9	Domestic dishwasher discharge air gap fittings	ASSE 1021
10	Double check backflow preventers	ASSE 1015
11	Double check detector fire protection backflow preventers	ASSE 1048
12	Double check fire protection backflow preventers	ASSE 1015
13	Dual check backflow preventers	ASSE 1024
14	Dual check backflow preventers for post-mix carbonated beverage dispensers	ASSE 1032
15	Hose connection backflow preventers	ASSE 1052
16	Hose connection vacuum breakers	ASSE 1011
17	Laboratory faucet backflow preventers	ASSE 1035
18	Outdoor enclosures for backflow preventers	ASSE 1060
19	Reduced pressure backflow preventers	ASSE 1013
20	Reduced pressure detector fire protection backflow preventers	ASSE 1047
21	Reduced pressure fire protection backflow preventers	ASSE 1013
22	Vacuum breakers, atmospheric type	ASSE 1001
23	Vacuum breakers, pressure type	ASSE 1020
24	Vacuum breakers, spill resistant	ASSE 1056
25	Wall hydrants with backflow prevention	ASSE 1019
26	Wall hydrants with dual check backflow preventer	ASSE 1053
27	Yard hydrants with backflow prevention	ASSE 1057

	Table 3.1.3 - Part X MISCELLANEOUS MATERIALS	
1	FOG disposal systems for fats, oils, greases	ASME A112.14.6
2	Grease interceptors	ASME A112.14.3 PDI G 101
3	Grease interceptors, gravity, pre-fabricated	IAPMO/ANSI Z1001
4	Grease removal devices (GRD)	ASME A112.14.4
5	Macerating toilet systems	ASME A112.3.4 CSA B45.9
6	Pipe hangers and supports	MSS SP-58
7	Waste disposal systems, non-sewered	ANSI Z4.3
8	Water containment membrane, concealed, CPE plastic	ASTM D4068
9	Water containment membrane, concealed, PVC plastic	ASTM D4551
	Table 3.1.3 - Part XI INSTALLATION PROCEDURES AND PRACTICES	
1	Ductile iron water mains	AWWA C600
2	Electro fusion joining PE and PEX piping	ASTM F1290
3	Heat fusion joining PE piping	ASTM F2620
4	Safe handling of solvent cements, primers, and cleaners	ASTM F402
5	Soldered joints in copper tubing	ASTM B828
6	Solvent cemented joints in PVC piping	ASTM D2855
7	Thermoplastic pipe and corrugated tubing for septic tank leach fields	ASTM F481
8	Underground thermoplastic pressure piping	ASTM D2774
9	Underground thermoplastic drain and sewer pipng	ASTM D2321
10	Vitrified clay piping	ASTM C12

Table 3.4 MATERIALS FOR POTABLE WATER PIPING

	HOT WATER DISTRIBUTION (2) (5)	COLD WATER DISTRIBUTION (2) (5)	WATER SERVICE (1) (4)		PIPE	FITTINGS
1	A	A	A	Brass Pipe, seamless, red brass	ASTM B43	ASME B16.15 (threaded), ASME B16.24 (flanged) brazed, compression, flare, transition
2	A	A	A	Copper Tube, seamless, Type K, L, or M	ASTM B88	ASME B16.18 (cast, solder joint) ASME B16.22 (wrought, solder joint) ASME B16.24 (cast, flanged) ASME B16.26 (cast, flared) ASME B16.50 (wrought, braze joint) ASME B16.51 (cast or wrought press-connect for hard drawn tube) ASSE 1061 (push-fit) ASME Mechanically crimped (pressed) per Section 4.2.6 Split couplings per Section 4.2.17
3	A	A	A	CPVC Plastic Pipe, schedule 40	ASTM F441	ASTM F438 (socket), ASTM F1970 (special engineered)
4	A	A	A	CPVC Plastic Pipe, schedule 80	ASTM F441	ASTM F437 (threaded), ASTM F439 (threaded, socket), ASTM F1970 (special engineered)
5	A	A	A	CPVC Plastic Pipe, SDR	ASTM F442	ASTM F438 (sched 40 socket), ASTM F439 (sched 80 socket) ASTM F1970 (special engineered)
6	A	A	A	CPVC Plastic HW/CW Tubing	ASTM D2846	ASTM F1960, ASTM F1961, ASTM F1807, ASTM F2098, ASTM F2159, ASTM F2434, ASTM F2735, ASTM F1970 (special engineered)
7	A	A	A	CPVC-AL-CPVC Plastic Composite Tubing	ASTM F2855	AWWA C800 (couplings, adapters)
8	A	A	X	Ductile Iron Pipe, cement-mortar lined	AWWA C151/A21.51	AWWA C110/A21.10, AWWA C153/A21.53 (compact)
9	A	X	X	Fiberglass (GFR) Pressure Pipe	ASTM D3517	ASTM D3517
10	A	X	X	Fiberglass Pressure Pipe	AWWA C950	AWWA C950
11	(4)	(5)	(5)	Flexible Pre-Insulated Piping	ASTM F2165	ASTM D2239, ASTM D3035, ASTM F714, ASTM F876, ASTM F877, ASTM F1281, ASTM F1282
12	A	A	A	Galvanized Steel Pipe and Fittings	ASTM A53	ASME B16.3 (malleable, threaded), ASME B16.4 (gray iron, threaded), ASME B16.5 (cast, forged, flanged), Split couplings per Section 4.2.17
13	X	A	X	PE Plastic Pipe, DR	ASTM D3035	ASTM D2683 (socket fusion), ASTM D3261 (butt heat fusion), ASTM F1055 (electro-fusion), IPS OD fittings and adapters for ASTM D3035

Table 3.4 MATERIALS FOR POTABLE WATER PIPING (continued)

#	Material				Standards
14	PE Plastic Pipe, DR	A	X		ASTM F714	ASTM D2683 (socket fusion), ASTM D3261 (butt heat fusion), ASTM F1055 (electro-fusion)
15	PE Plastic Pipe, SIDR	A	X		ASTM D2239	ASTM D2609 (plastic insert) IPS ID fittings and adapters for ASTM D2239
16	PE Plastic Water Service Pipe & Tubing	A	X		AWWA C901	ASTM D3261 (butt heat fusion), ASTM F1055 (electro-fusion)
17	PE Plastic Tubing, SDR (CTS)	A	X		ASTM D2737	ASTM D3261 (butt heat fusion), ASTM F1055 (electro-fusion) CTS OD fittings and adapters for ASTM D2737
18	PE-RT Plastic HW/CW Tubing	A		A	ASTM F2769	ASTM F1807 (crimped metal insert) ASTM F2159 (crimped plastic insert) ASTM F2098 (SS clamps for ASTM F1807 & F2159 fittings) ASTM F2735 (crimped plastic insert)
19	PE-AL-PE Composite Pipe	A		A	ASTM F1282	ASTM F1974 (metal insert), connectors marked F1282 or F1281/2
20	PE-AL-PE Composite Water Service Pipe	A	X		AWWA C903	AWWA C800 (metal insert)
21	PEX Crosslinked Water Service Pipe	A	X		AWWA C904	ASTM F1807 (crimped metal insert) ASTM F1960 (metal cold expansion insert) ASTM F2080 (metal cold expansion compression) ASTM F2098 (SS clamps for ASTM F1807 fittings)
22	PEX Plastic Tubing	A		A	ASTM F876	ASTM F1807 (crimped metal insert) ASTM F1865 (mechanical cold expansion metal insert) ASTM F1960 (metal cold expansion insert) ASTM F1961 (metal mechanical cold flare compression) ASTM F2080 (metal cold expansion compression) ASTM F2098 (SS clamps for ASTM F1807 and ASTM F2159 fittings) ASTM F2159 (crimped plastic insert) ASTM F2434 (crimped metal insert) ASTM F 2735 (crimped plastic) ASSE 1061 (push-fit)
23	PEX Plastic HW/CW Tubing	A		A	ASTM F877	ASTM F1807 (crimped metal insert) ASTM F1865 (mechanical cold expansion metal insert) ASTM F1960 (metal cold expansion insert) ASTM F1961 (metal mechanical cold flare compression) ASTM F2080 (metal cold expansion compression) ASTM F2098 (SS clamps for ASTM F1807 and ASTM F2159 fittings) ASTM F2159 (crimped plastic insert) ASTM F2434 (crimped metal insert) ASTM F 2735 (crimped plastic) ASSE 1061 (push-fit)
24	PEX-AL-PEX Crosslinked Pipe	A	A		ASTM F1281	ASTM F1974 (metal insert), connectors marked F1281 or F1281/2
25	PEX-AL-PEX Crosslinked Pipe	A	X		AWWA C903	AWWA C903 (metal insert)

Table 3.4 MATERIALS FOR POTABLE WATER PIPING (continued)

26	PEX-AL-PEX Plastic Tubing	A	A	A	ASTM F2262	ASTM F2434 (crimped metal insert)
27	PP Plastic Piping, IPS schedule 80	A	X	X	ASTM F2389	ASTM F2389 (socket weld), electro-fusion
28	PVC Plastic Pipe, schedule 40	A	X	X	ASTM D1785	ASTM D2466 (socket, threaded) ASTM F1970 (special engineered)
29	PVC Plastic Pipe, schedule 80	A	X	X	ASTM D1785	ASTM D2464 (threaded), ASTM D2467 (socket, threaded) ASTM F1970 (special engineered)
30	PVC Plastic Pipe, schedule 120	A	X	X	ASTM D1785	
31	PVC Plastic Pipe, SDR	A	X	X	ASTM D2241	ASTM D2466 (socket), ASTM F1970 (special engineered)
32	PVC Plastic Water Pipe & Fittings	A	X	X	AWWA C900	AWWA C900
	Approved	A	A	A		
	Disapproved	X	X	X		

NOTES FOR TABLE 3.4
(1) Piping for water service shall be water pressure rated for not less than 160 psi at 73°F.
(2) Piping for hot and cold water distribution shall be water pressure rated for not less than 160 psi at 73°F and 100 psi at 180°F.
(3) Plastic piping materials shall comply with NSF 14 and NSF 61.
(4) See Table 3.4.2 for plastic water service piping.
(5) See Table 3.4.3 for plastic hot and cold water distribution piping.

Table 3.4.2
PLASTIC WATER SERVICE PIPING (1) (2) (3) (4)
(water pressure rated for not less than 160 psi at 73 deg F)

MATERIAL	COMPOSITION	DIMENSIONS	JOINTS	PIPE SIZES
CPVC (ASTM D2846)	CPVC 4120	SDR 11 (CTS)	not threaded	all sizes
CPVC (ASTM F441)	CPVC 4120	Schedule 40	not threaded	up through 8"
		Schedule 80	threaded	up through 4"
		Schedule 80	not threaded	up through 16"
CPVC (ASTM F442)	CPVC 4120	SDR 26 or lower	not threaded	up through 12"
CPVC-AL-CPVC (ASTM F2855)	CPVC-AL-CPVC	ASTM F2855	not threaded	all sizes
PE (ASTM D2239)	PE 1404	none	none	none
	PE 2708 PE 3608 PE 4608	SIDR 9 or lower	not threaded	all sizes
	PE 4710	SIDR 11.5 or lower	not threaded	all sizes
PE (ASTM D2737)	PE 2708 PE 3608 PE 4608 PE 4710	SDR 11 or lower	not threaded	all sizes
PE (ASTM D3035)	PE 1404	none	none	none
	PE 2708 PE 3608 PE 4608	DR 11 or lower	not threaded	all sizes
	PE 4710	DR 13.5 or lower	not threaded	all sizes
PE (ASTM F714) IPS/DIPS	PE 2708 PE 3608 PE 4608	DR 11 or lower	not threaded	all sizes
	PE 4710	DR 13.5 or lower	not threaded	all sizes
PE (AWWA C901) SIDR ID-Controlled IPS Pipe	PE 2708 PE 3608	SIDR 9 or lower	not threaded	1/2" - 3"
	PE 4710	SIDR 11.5 or lower	not threaded	3/4" - 3"
PE (AWWA C901) SDR OD-Controlled IPS Pipe	PE 2708 PE 3608	SDR 11 or lower	not threaded	1/2" - 3"
	PE 4710	SDR 13.5 or lower	not threaded	3/4" - 3"
PE (AWWA C901) SDR OD-Controlled CTS Tubing	PE 2708 PE 3608 PE 4710	SDR 11 or lower	not threaded	3/4" - 2"
PE-RT (ASTM F2769)	PE-RT	SDR 9	not threaded	all sizes
PE-AL-PE (ASTM F1282)	PE-AL-PE	ASTM F1282	not threaded	all sizes
PE-AL-PE (AWWA C903)	PE-AL-PE	AWWA C903	not threaded	all sizes
PEX (ASTM F876)	PEX	SDR 9	not threaded	all sizes
PEX (ASTM F877)	PEX	SDR 9	not threaded	all sizes
PEX (AWWA C904)	PEX 1006	SDR 9	not threaded	all sizes
PEX-AL-PEX (ASTM F1281)	PEX-AL-PEX	ASTM F1281	not threaded	all sizes
PEX-AL-PEX (ASTM F2262)	PEX-AL-PEX	SDR 9	not threaded	all sizes
PEX-AL-PEX (AWWA C903)	PEX-AL-PEX	AWWA C903	not threaded	all sizes
PP (ASTM F2389) IPS	PP-R	Schedule 80	not threaded	up through 3"

Table 3.4.2 (continued)
PLASTIC WATER SERVICE PIPING (1) (2) (3) (4)
(water pressure rated for not less than 160 psi at 73 deg F)

MATERIAL	COMPOSITION	DIMENSIONS	JOINTS	PIPE SIZES
PVC (ASTM D1785)	PVC 1120	Schedule 40	not threaded	up through 8"
	PVC 1220	Schedule 80	threaded	up through 4"
	PVC 2120	Schedule 80	not threaded	up through 24"
		Schedule 120	threaded	up through 12"
		Schedule 120	not threaded	up through 12"
	PVC 2110	Schedule 40	not threaded	up through 1-1/2"
		Schedule 80	threaded	up through 1"
		Schedule 80	not threaded	up through 4"
		Schedule 120	threaded	up through 1"
		Schedule 120	not threaded	up through 12"
	PVC 2112	Schedule 40	not threaded	up through 3"
		Schedule 80	threaded	up through 1-1/4"
		Schedule 80	not threaded	up through 6"
		Schedule 120	threaded	up through 1-1/2"
		Schedule 120	not threaded	up through 12"
	PVC 2116	Schedule 40	not threaded	up through 5"
		Schedule 80	threaded	up through 2"
		Schedule 80	not threaded	up through 24"
		Schedule 120	threaded	up through 5"
		Schedule 120	not threaded	up through 12"
PVC (ASTM D2241)	PVC 1120 PVC 1220 PVC 2120	SDR 26 or lower	not threaded	all sizes
	PVC 2110	SDR 13.5	not threaded	all sizes
	PVC 2112	SDR 17 or lower	not threaded	all sizes
	PVC 2116	SDR 21 or lower	not threaded	all sizes
PVC (AWWA C900)	PVC 1120	DR 14 (4)	not threaded	all sizes

NOTES FOR TABLE 3.4.2

(1) The application of a pipe material for water service piping and its required water pressure rating of not less than 160 psi at 73 deg F shall be indicated in the manufacturer's data.
(2) Refer also to the manufacturer's recommendations, instructions, and limitations.
(3) Lower SDR, SIDR, IDR, and DR numbers for the same material composition have heavier wall thicknesses and higher pressure ratings.
(4) AWWA C900 pipe shall be rated by FM pressure class.

Table 3.4.3
PLASTIC HOT AND COLD WATER DISTRIBUTION PIPING (1)(2)
(water pressure rated for not less than 100 psi at 180 deg F and 160 psi at 73 deg F)

MATERIAL	COMPOSITION	DIMENSIONS	JOINTS	PIPE SIZES
CPVC (ASTM D2846)	CPVC 4120 (23447)	SDR 11	not threaded	all sizes
CPVC (ASTM F441)	CPVC 4120 (23447)	Schedule 40	not threaded	up through 1"
		Schedule 80	threaded	up through 1/2"
		Schedule 80	not threaded	up through 2-1/2"
CPVC (ASTM F442)	CPVC 4120 (23447)	SDR 11	not threaded	all sizes
CPVC-AL-CPVC (ASTM F2855)	CPVC-AL-CPVC	ASTM F2855	not threaded	all sizes
PE-AL-PE (ASTM F1282)	PE-AL-PE	ASTM F1282	not threaded	all sizes
PE-RT (ASTM F2769)	PE-RT	SDR 9	not threaded	all sizes
PEX (ASTM F876)	PEX	SDR 9	not threaded	all sizes
PEX (ASTM F877)	PEX	SDR 9	not threaded	all sizes
PEX-AL-PEX (ASTM F1281)	PEX-AL-PEX	ASTM F1281	not threaded	all sizes
PEX-AL-PEX (ASTM F2262)	PEX-AL-PEX	SDR 9	not threaded	all sizes

NOTES FOR TABLE 3.4.3

(1) The application of a pipe material for hot and cold water distribution and its required pressure ratings of not less than 100 psi at 180 deg F and 160 psi at 73 deg F shall be indicated in the manufacturer's data.

(2) Refer also to the manufacturer's recommendations, instructions, and limitations.

Table 3.5 MATERIALS FOR SANITARY WASTE AND DRAIN (3)

	ABOVE GROUND WITHIN BUILDINGS	UNDERGROUND WITHIN BUILDINGS	SEWERS OUTSIDE OF BUILDINGS		PIPE	FITTINGS
1	A	A	A	ABS Plastic Pipe and Fittings	ASTM D2661 (DWV, schedule 40 IPS)	ASTM D2661 (socket, threaded)
2	A	A	A	ABS Cellular Core Plastic Pipe	ASTM F628 (DWV, schedule 40 IPS)	ASTM F628, ASTM D2661 (socket, threaded)
3	A	X	A	ABS and PVC Composite Sewer Pipe	ASTM D2680	ASTM D2680 (socket), PVC (gasket)
4	A	A	A	Cast Iron Soil Pipe and Fittings (bell & spigot)	ASTM 74	ASTM A74
5	A	A	A	Cast Iron Soil Pipe and Fittings (no-hub)	CISPI 301, ASTM A888	CISPI 301, ASTM A888
6	A	X	X	Concrete Sewer Pipe, Non-reinforced	ASTM C14	ASTM C14
7	A	X	X	Concrete Sewer Pipe, Reinforced	ASTM C76	ASTM C76
8	A	A	A	Copper Drainage Tube, DWV (7)	ASTM B306	ASME B16.23 (cast copper solder joint) ASME B16.29 (wrought copper solder joint)
9	A	A	A	Copper Water Tube, K, L, M (7)	ASTM B88	ASME B16.23 (cast copper solder joint) ASME B16.29 (wrought copper solder joint)
10	A	X	X	Fiberglass Sewer Pipe	ASTM D3262	ASTM D3840
11	X	X	A	Galvanized Steel Pipe	ASTM A53	ASME B16.12 (cast iron drainage)
12	A	X	X	PE Plastic Pipe, SDR-PR	ASTM F714 (6)	
13	A	A	A	PVC Plastic Pipe and Fittings, DWV	ASTM D2665	ASTM D2665 (socket), ASTM F1866 (socket)
14	X	A	A	PVC Plastic Pipe and Fittings, 3.25" OD	ASTM D2949	ASTM D2949
15	A	X	X	PVC Sewer and Drain Pipe (cellular core)	ASTM F891	ASTM D2665 (socket), ASTM F1866 (socket)
16	A	A	X	PVC Sewer and Drain Pipe (cellular core)	ASTM F891 (4)	ASTM D2665 (socket), ASTM F1866 (socket)
17	A	A	A	PVC IPS Schedule 40 (cellular core)	ASTM F891	ASTM D2665 (socket), ASTM F1866 (socket)
18	A	X	X	PVC Plastic Sewer Pipe (PSM) and Fittings	ASTM D3034 (1)	ASTM D3034 (1)
19	A	A	A	Stainless Steel DWV Systems - Type 316L (5)	ASME A112.3.1	ASME A112.3.1
20	A	X	X	Stainless Steel DWV Systems - Type 304 (5)	ASME A112.3.1	ASME A112.3.1
21	A	X	X	Vitrified Clay Pipe, Standard Strength	ASTM C700	ASTM C700
22	A	X	X	Vitrified Clay Pipe, Extra Strength	ASTM C700	ASTM C700
23	A	X	X	Ductile Iron Gravity Sewer Pipe	ASTM A746	AWWA C110, AWWA 153 (push-on, mechanical)
24	A	X	X	Ductile Iron Culvert Pipe	ASTM A716	AWWA C110, AWWA 153 (push-on, mechanical)
25	A	A	X	CPVC Chemical Waste Systems	ASTM F2618 (schedule 40 IPS)	ASTM F2618 socket
26	A	A	X	PVC Non-Pressure, Reprocessed-Recycled Content	ASTM F1760, Sewer-Drain	ASTM D3034, ASTM F1336
27	A	A	A	PVC Non-Pressure, Reprocessed-Recycled Content	ASTM F1760, IPS, Schedule 40	ASTM D2466, ASTM D2665
	A	A	A	Approved		
	X	X	X	Not Approved		

NOTES FOR TABLE 3.5
(1) Plastic waste and drain piping classified by standard dimension ratio shall be SDR 26 or heavier (lower SDR number).
(2) Plastic sewer pipe classified by pipe stiffness shall be PS-46 or stiffer (higher PS number).
(3) Piping shall be applied within the limits of its listed standard and the manufacturer's recommendations.
(4) PS-100 pipe or stiffer (higher PS number) is required for underground cellular core PVC sewer and drain pipe underground within buildings.
(5) The alloy shall be marked on stainless steel DWV piping systems.
(6) Minimum DR-17 is required for trenchless sewer replacement systems.
(7) Copper tube shall not be used for drainage from non-water urinals to the point where it connects to piping containing drainage from one or more water closets.

Table 3.6 MATERIALS FOR VENT PIPING (1)

		ABOVEGROUND	UNDERGROUND	PIPE	FITTINGS
1	ABS Plastic Pipe and Fittings	A	A	ASTM D2661 (DWV, schedule 40 IPS)	ASTM D2661 (socket, threaded)
2	ABS Cellular Core Plastic Pipe	A	A	ASTM F628 (DWV, schedule 40 IPS)	ASTM F628, ASTM D2661 (socket, threaded)
3	Cast Iron Soil Pipe and Fittings (bell & spigot)	A	A	ASTM A74	ASTM A74
4	Cast Iron Soil Pipe and Fittings (no-hub)	A	A	CISPI 301, ASTM A888	CISPI 301, ASTM A888
5	Copper Drainage Tube, DWV	A	A	ASTM B306	ASME B16.23 (cast copper solder joint) ASME B16.29 (wrought copper solder joint)
6	Copper Water Tube, K, L, M	A	A	ASTM B88	ASME B16.23 (cast copper solder joint) ASME B16.29 (wrought copper solder joint)
7	CPVC Chemical Waste Systems	A	A	ASTM F2618 (sched 40 pipe)	ASTM F2618 socket
8	Galvanized Steel Pipe	A	X	ASTM A53	ASME B16.12 (cast iron drainage)
9	PVC Non-Pressure, Reprocessed-Recycled Content	A	A	ASTM F1760, IPS, Schedule 40	ASTM D2466, ASTM D2665
10	PVC Plastic Pipe and Fittings, DWV	A	A	ASTM D2665	ASTM D2665 (socket), ASTM F1866 (socket)
11	PVC Plastic Pipe and Fittings, DWV, 3.25" OD	A	A	ASTM D2949	ASTM D2949
12	PVC Sewer and Drain Pipe (cellular core)	A	X	ASTM F891 (3), (4)	ASTM D2665 (socket), ASTM F1866 (socket)
13	PVC IPS Schedule 40 (cellular core)	A	A	ASTM F891	ASTM D2665 (socket), ASTM F1866 (socket)
14	Stainless Steel DWV Systems - Type 316L (2)	A	A	ASME A112.3.1	ASME A112.3.1
15	Stainless Steel DWV Systems - Type 304 (2)	A	X	ASME A112.3.1	ASME A112.3.1
16	Vitrified Clay Pipe, Extra Strength	A	X	ASTM C700	ASTM C700
	Approved	A	A		
	Not Approved	X	X		

NOTES FOR TABLE 3.6

(1) Piping shall be applied within the limits of its listed standard and the manufacturer's recommendations.
(2) The alloy shall be marked on stainless steel DWV piping systems.
(3) ASTM F891 PVC sewer and drain piping shall be PS 100 or PS 50.
(4) ASTM F891 PVC sewer and drain piping underground within buildings shall be PS-100.

Table 3.7 MATERIALS FOR STORM DRAIN PIPING (1)

		ABOVE GROUND WITHIN BUILDINGS	UNDERGROUND WITHIN BUILDINGS	SEWERS OUTSIDE OF BUILDINGS	PIPE	FITTINGS
1	ABS Plastic Pipe and Fittings	A	A	A	ASTM D2661 (DWV, schedule 40 IPS)	ASTM D2661 (socket, threaded)
2	ABS Cellular Core Plastic Pipe	A	A	A	ASTM F628 (DWV, schedule 40 IPS)	ASTM F628, ASTM D2661 (socket, threaded)
3	ABS and PVC Composite Sewer Pipe	A	X	X	ASTM D2680	ASTM D2680 (socket), PVC (gasket)
4	Cast Iron Soil Pipe and Fittings (bell & spigot)	A	A	A	ASTM 74	ASTM A74
5	Cast Iron Soil Pipe and Fittings (no-hub)	A	A	A	CISPI 301, ASTM A888	CISPI 301, ASTM A888
6	Concrete Sewer Pipe, Non-reinforced	A	X	X	ASTM C14	ASTM C14
7	Concrete Sewer Pipe, Reinforced	A	X	X	ASTM C76	ASTM C76
8	Copper Drainage Tube, DWV	A	A	A	ASTM B306	ASME B16.23 (cast copper solder joint) ASME B16.29 (wrought copper solder joint)
9	Copper Water Tube, K, L, M	A	A	A	ASTM B88	ASME B16.23 (cast copper solder joint) ASME B16.29 (wrought copper solder joint)
10	Ductile Iron Culvert Pipe	A	X	X	ASTM A716	AWWA C110, AWWA 153 (push-on, mechanical)
11	Ductile Iron Gravity Sewer Pipe	A	X	X	ASTM A746	AWWA C110, AWWA 153 (push-on, mechanical)
12	Fiberglass Sewer Pipe	A	X	X	ASTM D3262	ASME D3840
13	Galvanized Steel Pipe	X	X	X	ASTM A53	ASME B16.12 (cast iron drainage)
14	PE Plastic Pipe, SDR-PR	A	X	X	ASTM F714 (4)	ASTM D2949
15	PVC IPS Schedule 40 (cellular core)	A	A	A	ASTM F891	ASTM D2665 (socket), ASTM F1866 (socket)
16	PVC Non-Pressure, Reprocessed-Recycled Content	A	X	X	ASTM F1760, Sewer-Drain, PS 46	ASTM D3034, ASTM F1336
17	PVC Non-Pressure, Reprocessed-Recycled Content	A	X	X	ASTM F1760, IPS, Schedule 40	ASTM D2466, ASTM D2665
18	PVC Plastic Pipe and Fittings, DWV	X	X	X	ASTM D2665	ASTM D2665 (socket), ASTM F1866 (socket)
19	PVC Plastic Pipe and Fittings, DWV, 3.25" OD	X	A	A	ASTM D2949	ASTM D2949
20	PVC Sewer and Drain Pipe (cellular core)	A	X	X	ASTM F891(5)	ASTM D2665 (socket), ASTM F1866 (socket)
21	PVC Sewer and Drain Pipe (cellular core) (2)	X	A	A	ASTM F891 (2)	ASTM D2665 (socket), ASTM F1866 (socket)
22	Stainless Steel DWV Systems - Type 316L (3)	A	X	A	ASME A112.3.1	ASME A112.3.1
23	Stainless Steel DWV Systems - Type 304 (3)	X	X	A	ASME A112.3.1	ASME A112.3.1
24	Type PSM PVC Sewer Pipe and Fittings	A	X	X	ASTM D3034 (6)	ASTM D3034
25	Vitrified Clay Pipe, Standard Strength	A	A	A	ASTM C700	ASTM C700
26	Vitrified Clay Pipe, Extra Strength	A	A	A	ASTM C700	ASTM C700
	Approved	A	A	A		
	Not Approved	X	X	X		

NOTES FOR TABLE 3.7

(1) Piping shall be applied within the limits of its listed standard and the manufacturer's recommendations.
(2) PS-100 pipe or stiffer (higher PS number).
(3) The alloy shall be marked on stainless steel DWV piping systems.
(4) Minimum SDR-17 is required for trenchless sewer replacement systems.
(5) ASTM F891 PVC sewer piping shall be PS 100 or PS 50.
(6) ASTM D3034 PVC sewer drain piping shall be SDR 26, SDR 35, or SDR 41.

	Table 3.8 MATERIALS FOR FOUNDATION DRAINS AND SUB-SOIL DRAINAGE	
1	Clay Drain Tile (ASTM C4)	A
2	Concrete Drain Tile (ASTM C412)	A
3	Perforated Concrete Pipe (ASTM C444)	A
4	Corrugated Polyethylene Tube (ASTM F405)	A
5	SR Plastic Drain Pipe and Fittings, Perforated (ASTM D2852)	A
6	PVC Sewer Pipe and Fittings, Perforated (ASTM D2729)	A
7	Stainless Steel DWV Systems – Type 316L (ASME A112.3.1)	A
8	Vitrified Clay Pipe, Perforated, Standard and Extra Strength (ASTM C700)	A
	Approved	A

Chapter 4

Joints and Connections

4.1 GENERAL REQUIREMENTS

4.1.1 Tightness

Joints and connections in the plumbing system shall be gas tight and watertight for the pressure required by test, with the exceptions of those portions of perforated or open joint piping that are installed for the purpose of collecting and conveying ground or seepage water to the underground storm drains.

> *Comment: Perforated and open-joint piping is also used for the piping in absorption trenches for private sewage disposal systems. See Section 16.9.5.*

4.1.2 Joint Standards

    a. Pipe and tube shall be cut 90° or perpendicular to the pipe center lines.

    b. The inside diameter of pipe and tube ends shall be reamed, filed, or smoothed to size of bore and all chips removed. All burrs on the outside of the pipe and butt ends shall be removed before the installation.

    c. Pipe and tube shall engage into fittings the full manufacturer's design depth of the fitting socket.

    d. Male pipe threads shall be made of sufficient length to ensure the proper engagement.

    e. Pipe shall not extend into a fitting or other pipe to such a depth that it will impede or restrict the design flow.

    f. Joints made by bonding, welding, brazing, solvent cementing, soldering, burning, fusion or mechanical means shall be free from grease or other substances not specifically required to achieve a satisfactory joint.

    g. Pipe sealing or lubricating compound required for threaded pipe joints shall be applied to the male pipe end only and shall be chemically compatible with the pipe and fitting, insoluble, and nontoxic.

> *Comment: For industry standards for the various joining methods, refer to Table 3.1.3 - Part IV Pipe Joints, Joining Materials, Couplings, Gaskets.*

4.1.3 Expansion Joints

Mechanical type expansion joints requiring or permitting adjustment shall be accessible for adjustment and/or replacement.

4.1.4 Increasers and Reducers

Where different sizes of pipes or pipes and fittings are to be connected, increaser and reducer fittings or bushings shall be used. (See Section 2.4.3)

4.2 TYPES OF JOINTS FOR PIPING MATERIALS

4.2.1 Caulked

4.2.1.1 Cast-Iron Soil Pipe

Lead caulked joints for cast-iron hub and spigot soil pipe shall be firmly packed with oakum or hemp and filled with molten lead not less than 1 inch deep and not to extend more than 1/8 inch below the rim of the hub. No paint, varnish, or other coatings shall be permitted on the jointing material until after the joint has been tested and approved. Lead shall be run in one pouring and shall be caulked tight. *See Figure 4.2.1.1*

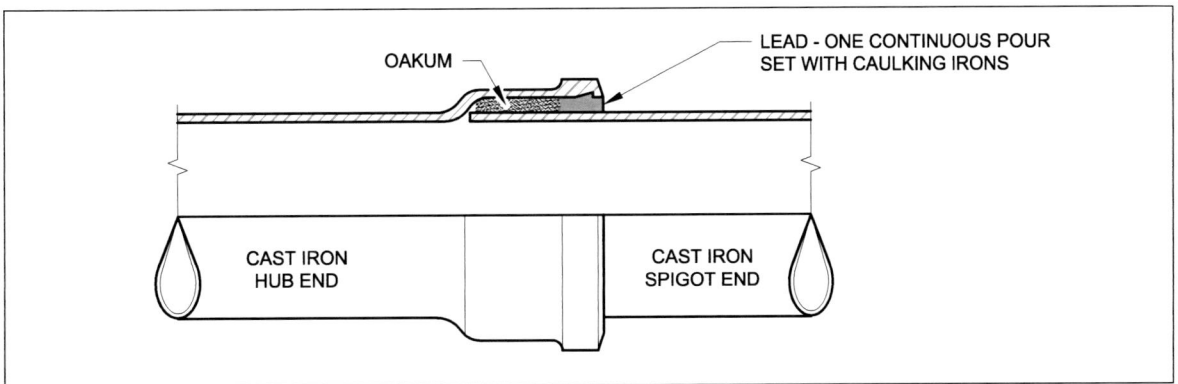

Figure 4.2.1.1
A LEAD CAULKED JOINT IN CAST IRON SOIL PIPE

4.2.1.2 Cast-Iron Water Pipe

Lead caulked joints for cast-iron bell and spigot water pipe shall be firmly packed with clean dry jute, or treated paper rope packing. The remaining space in the hub shall be filled with molten lead according to the following schedule:

Pipe Size	Depth of Lead
Up to 20 inches	2-1/4 inches
24, 30, 36 inches	2-1/2 inches
Larger than 36 inches	3 inches

Lead shall be run in one pouring and shall be caulked tight. *See Figure 4.2.1.2*

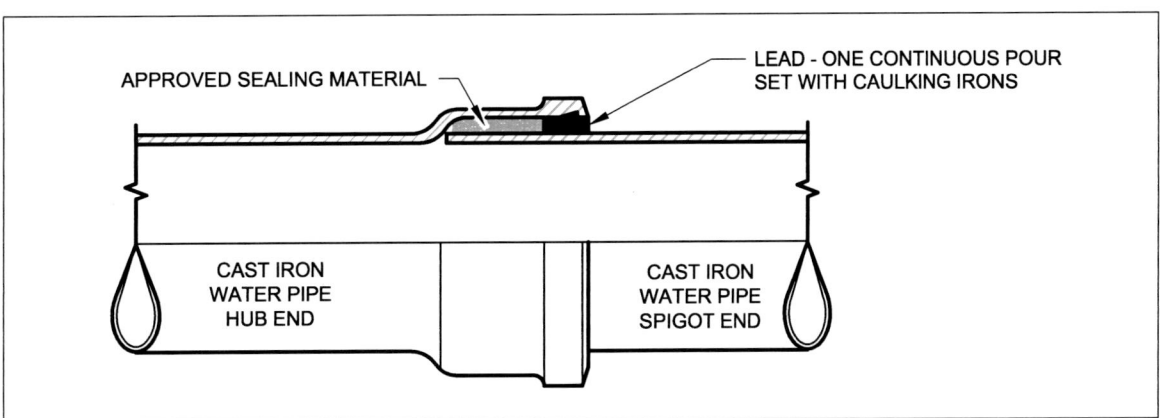

Figure 4.2.1.2
A LEAD CAULKED JOINT IN IRON WATER PIPE

4.2.2 Threaded

The threads in tapered general purpose pipe joints shall conform to ASME B1.20.1. Pipe ends shall be reamed or filed out to the full size of the bore and all chips removed. Thread seal tape, pipe joint compound, or other thread lubricant shall be applied only to the male threads. *See Figure 4.2.2*

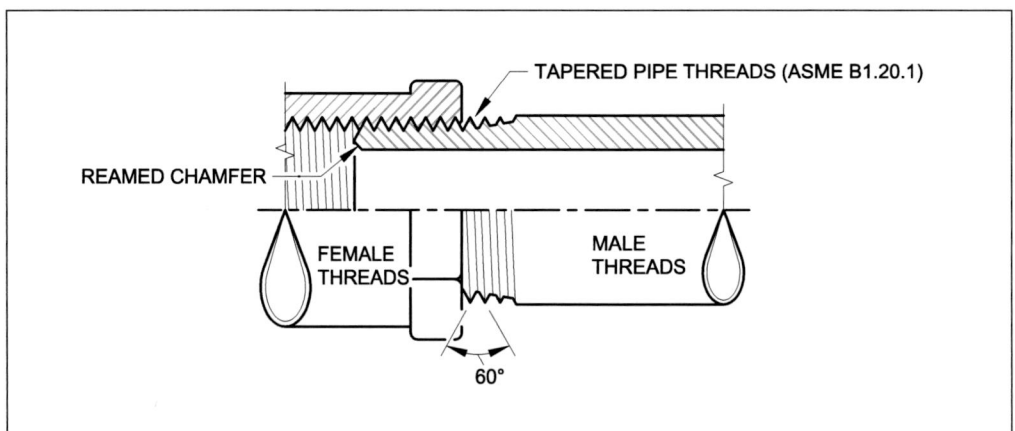

**Figure 4.2.2
A THREADED PIPE JOINT**

4.2.3 Wiped

Joints in lead pipe or fittings, or between lead pipe or fittings and brass or copper pipe, ferrules, solder nipples, or traps, shall be full wiped joints. Wiped joints shall have an exposed surface on each side of a joint not less than 3/4 inch and at least as thick as the material being jointed. Wall or floor flange lead-wiped joints shall be made by using a lead ring or flange placed behind the joints at wall or floor. Joints between lead pipe and cast-iron, steel, or wrought iron shall be made by means of a caulking ferrule, soldering nipple, or bushing. *See Figure 4.2.3*

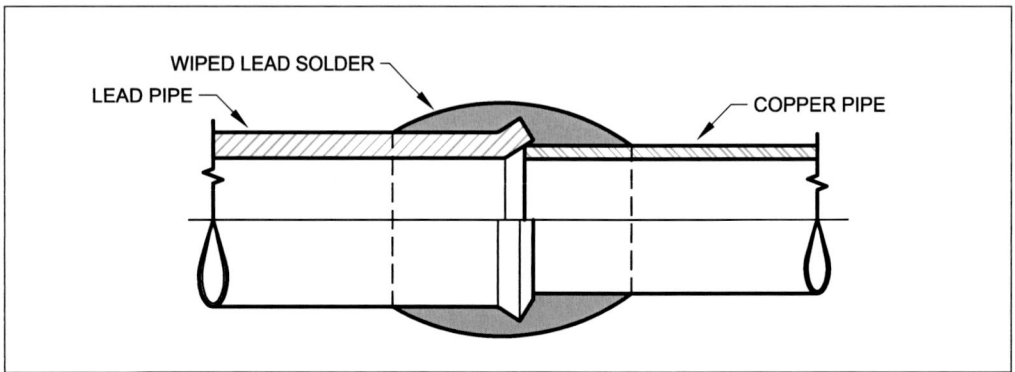

**Figure 4.2.3
A WIPED JOINT ON LEAD PIPE**

4.2.4 Soldered

   a. Soldered joints in copper water piping shall be made using wrought copper pressure fittings complying with ASME B16.22, cast copper alloy pressure fittings complying with ASME B16.18, or cast copper alloy flanges complying with ASME B16.24.

NOTE: Short-cup brazing fittings complying with ASME B16.50 and bearing the mark "BZ" and solder joint fittings with reduced insertion depths for brazing shall not be used where joints are soldered.

   b. Soldered joints in copper drain and vent piping shall be made with wrought copper drainage fittings complying with ASME B16.29 or cast copper alloy drainage fittings complying with ASME B16.23.

   c. Soldered joints shall be made in accordance with ASTM B828.

d. Solder shall comply with ASTM B32. Flux shall comply with ASTM B813. Flux residue shall be noncorrosive and non-toxic, inside and outside the completed system, after soldering potable water systems, in accordance with ASTM B813.

e. Solder and flux for joints in potable water piping shall contain not more than 0.2% lead.

See Figure 4.2.4

Comment: The limit on lead in solder filler metal is to reduce the level of lead in drinking water in accordance with the EPA Safe Drinking Water Act.

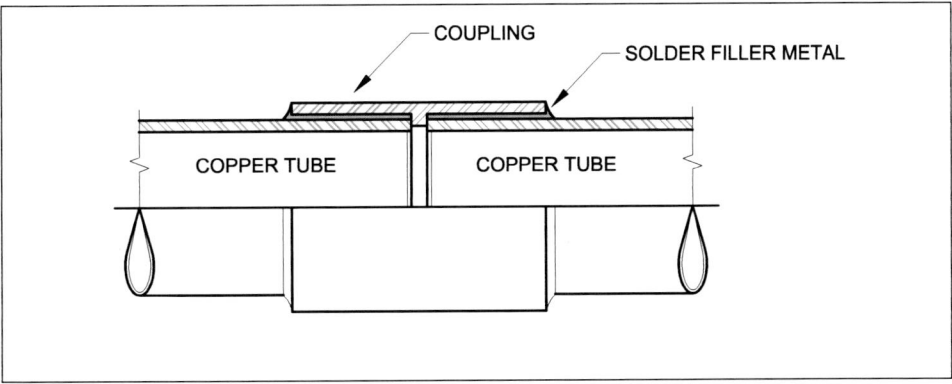

**Figure 4.2.4
A SOLDERED JOINT**

4.2.5 Flared

Flared joints for copper water tube shall be made with fittings complying with ASME B16.26. The tube shall be reamed and then expanded with an approved flaring tool. *See Figure 4.2.5*

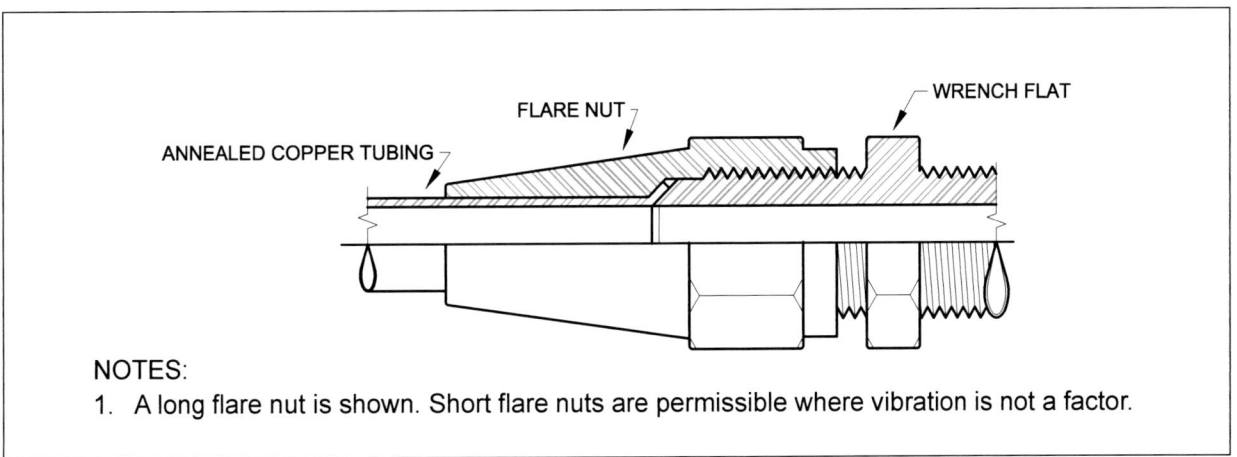

**Figure 4.2.5
A FLARED COPPER JOINT**

4.2.6 Mechanically Crimped (Pressed) Joints

a. Copper fittings for water supply and distribution, designed for mechanically crimped (pressed) connections to ASTM B88 hard drawn copper water tube, shall include an O-ring gasket complying with NSF 61 for potable water.

EXCEPTION: Mechanically crimped (pressed) joints shall be permitted with annealed copper water tube when such use is included in the fitting manufacturer's technical data and installation instructions.

b. The fittings shall comply with the material and sizing requirements of ASME B16.22 (wrought copper or copper alloy fittings) or ASME B16.18 (cast copper alloy fittings).

c. During installation, the tube end shall be deburred and depth-marked to permit visual verification of full insertion of the tube into the fitting socket.

d. The joint shall be crimped (pressed) using a tool approved by the manufacturer of the fitting.

e. The joints shall be rated by the manufacturer for not less than 200 psig at 180 deg F.

f. The fittings shall be permitted to be installed in concealed locations.

4.2.7 Push-Fit Joints

a. Fittings for water supply and distribution, designed for manual push-fit connections to copper, CPVC, and/or PEX tubing, shall comply with ASSE 1061.

b. Fittings shall be marked with those tubing materials that they are intended to be used with unless they are suitable for use with all three tubing materials (copper, CPVC, and PEX).

c. Fittings will be installed in accordance with their manufacturer's instructions

d. During installation, the tube end shall be de-burred and depth-marked to permit visual verification of full insertion of the tube into the fitting socket.

e. The fittings shall be rated by their manufacturer for not less than 125 psig at 180 deg F as listed by ASSE 1061.

f. The fittings shall be permitted to be installed in concealed locations.

4.2.8 Brazed

4.2.8.1 General

a. Brazed joints in copper tubing shall be made in accordance with accepted industry practice. See Appendix L for an accepted practice for general plumbing.

b. Brazed joints in medical gas and vacuum piping shall be made in accordance with NFPA 99.

4.2.8.2 Fittings

a. Fittings in copper tubing with brazed joints shall be wrought solder joint fittings complying with ASME B16.22 or short-cup brazing fittings complying with ASME B16.50. Short-cup brazed joint fittings shall be clearly marked by the manufacturer to differentiate them from solder-joint fittings and avoid their being used in piping with soldered joints.

b. Fittings for medical gas and vacuum piping shall be as required by NFPA 99.

4.2.8.3 Mechanically Formed Tee Branches

a. Mechanically formed tee branches shall be permitted in copper tubing in water distribution systems. The branch connections shall be formed with appropriate tools and joined by brazing. The branch tube end shall be notched and dimpled with two sets of double dimples. The first dimples shall act as depth stops to prevent the branch tube from being inserted beyond the depth of the branch collar. The second dimples shall be 1/4" above the first dimples and provide a visual means of verifying that the branch connection has been properly fitted. The dimples in the branch tube shall be in line with the run of the main. The joints shall be brazed in accordance with Section 4.2.8.1 and ASTM F2014.

b. Mechanically former tee branches shall not be permitted in drain piping.

4.2.9 Cement Mortar

Where permitted as outlined in Section 2.4.4, cement mortar joints shall be made in the following manner:

a. A layer of jute or hemp shall be inserted into the annular joint space and packed tightly to prevent mortar from entering the interior of the pipe or fitting.

b. Not more than 25 percent of the depth of the annular space shall be used for jute or hemp.

c. The remaining depth shall be filled in one continuous operation with a thoroughly mixed mortar composed of one part cement and two parts sand, with only sufficient water to make the mixture workable by hand.

d. Additional mortar of the same composition shall then be applied to form a one-to-one slope with the barrel of the pipe.

e. The bell or hub of the pipe shall be left exposed for inspection.

f. When necessary, the interior of the pipe shall be swabbed to remove any mortar or other material that may have found its way into the pipe.

> *Comment: Cement mortar joints have been virtually eliminated by the use of alternate joining methods and materials.*

4.2.10 Burned Lead (Welded)

Burned (welded) joints shall be made in such a manner that the two or more sections to be joined shall be uniformly fused together into one continuous piece. The thickness of the weld shall be at least as thick as the lead being joined.

> *Comment: The added welding material must be of the same composition as the lead pipe being joined.*

4.2.11 Mechanical (Flexible or Slip Joint)

4.2.11.1 Stainless Steel DWV Systems

a. Joints in stainless steel DWV systems shall be made with an elastomeric O-ring of a material that is suitable for the intended service.

b. Joints between stainless steel drainage systems and other piping materials shall be made with an approved adapter coupling.

4.2.11.2 Cast-Iron Soil Pipe

a. Hubless pipe: Joints for hubless cast-iron soil pipe and fittings shall be made with an approved elastomeric sealing sleeve and corrosion resistant metallic shielded coupling.

b. Hub and Spigot: Joints for hub and spigot cast-iron soil pipe and fittings, designed for use with a compressed gasket, may be made using a compatible compression gasket that is compressed when the spigot is inserted into the hub of the pipe.

See Figures 4.2.11.2-A and 4.2.11.2-B

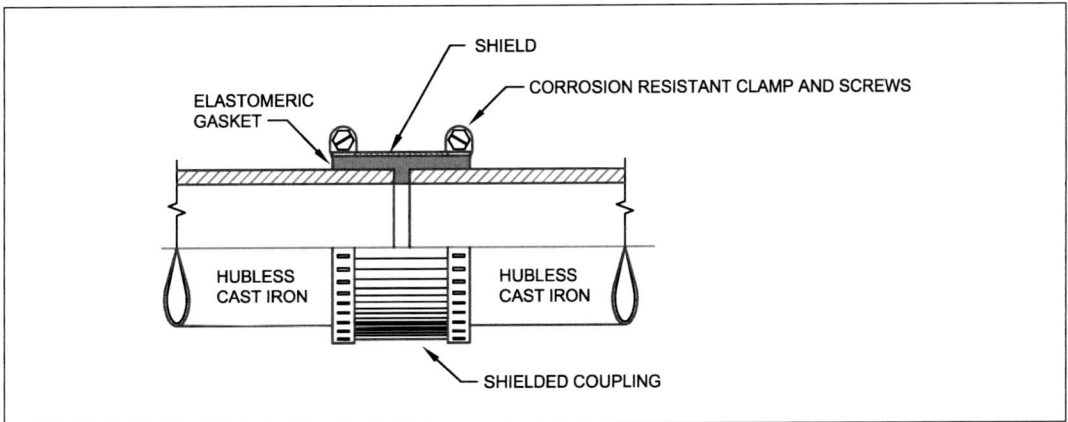

Figure 4.2.11.2-A
A SHIELDED COUPLING ON HUBLESS CAST IRON SOIL PIPE

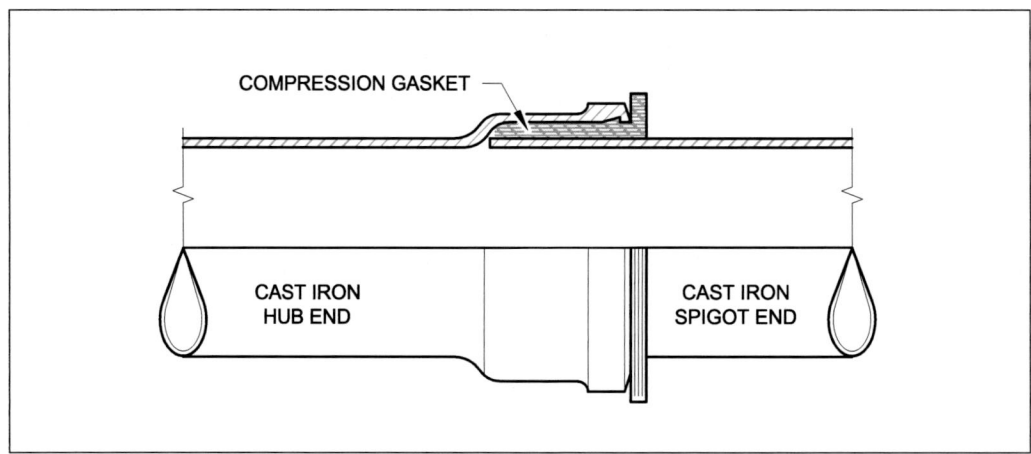

**Figure 4.2.11.2-B
A CAST IRON HUBBED JOINT WITH A COMPRESSION GASKET**

4.2.11.3 Cast Iron Water Pipe

Mechanical joints in cast-iron water pipe shall be made with a flanged collar, a rubber ring gasket, and the approved number of securing bolts. *See Figure 4.2.11.3*

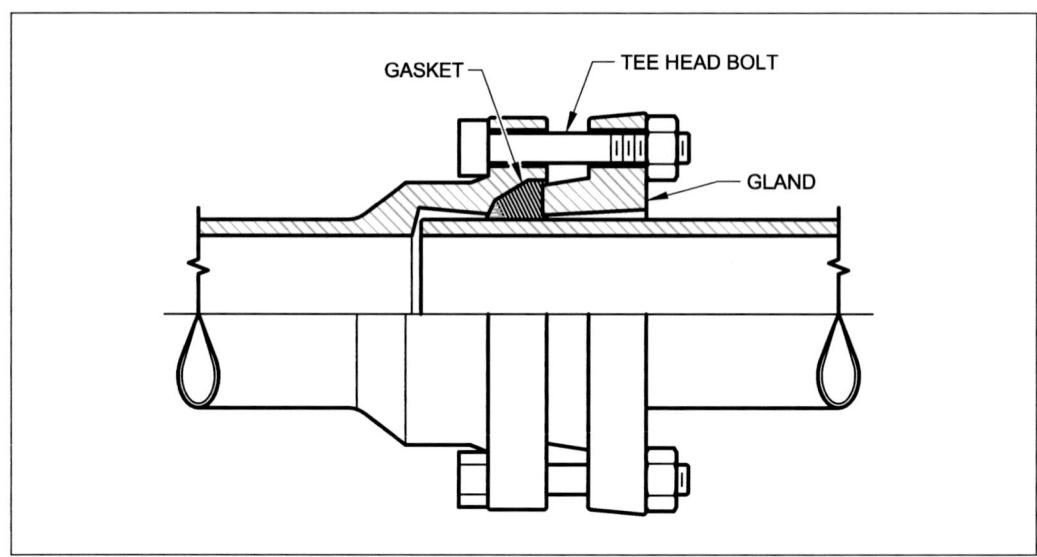

**Figure 4.2.11.3
A MECHANICAL JOINT ON IRON WATER PIPE**

4.2.11.4 Clay Pipe

Joints in piping and/or fittings shall be made using flexible compression joints. *See Figure 4.2.11.4*

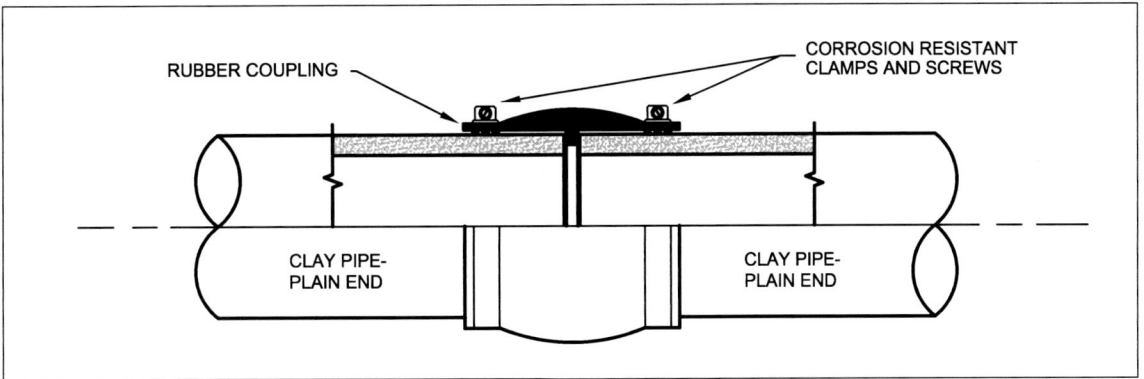

Figure 4.2.11.4
AN UNSHIELDED COUPLING ON PLAIN END CLAY PIPE

4.2.11.5 Concrete Pipe

Flexible joints between lengths of concrete pipe may be made using approved compression type joints or elastomeric materials on the spigot end and in the bell (or hub) end of the pipe. ***See Figure 4.2.11.5***

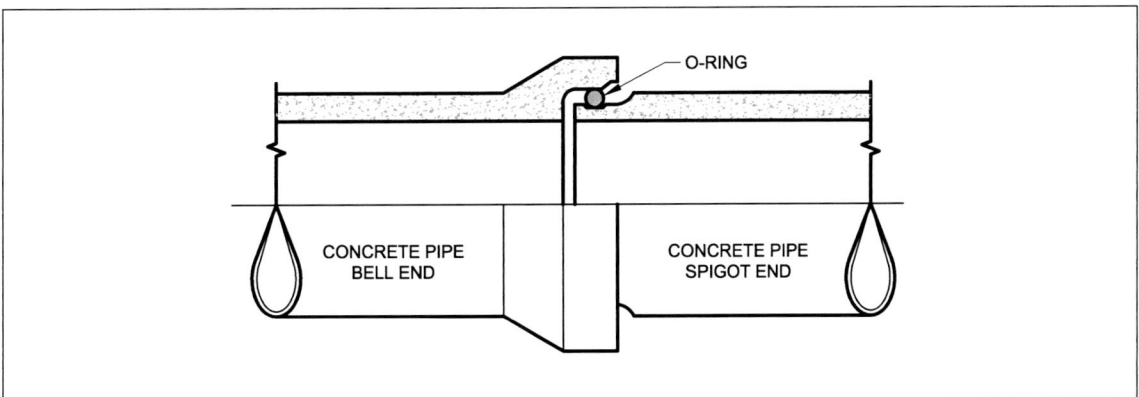

Figure 4.2.11.5
A JOINT IN BELL AND SPIGOT CONCRETE PIPE

4.2.11.6 Elastomeric Sleeves

Mechanical joints on drain pipes below ground shall be made with an elastomeric seal conforming to ASTM D3212, CSA B602 or ASTM C1173. Joints shall be installed in accordance with the manufacturer's instructions.

4.2.12 Ductile Iron Pressure Piping

a. Ductile iron pressure pipe and fittings shall comply with standards listed in Table 3.1.3 and shall be applied and installed in accordance with the manufacturer's instructions. Joints shall be the push-on type, mechanical type, grooved and shouldered, or flanged. Rubber gaskets shall be suitable for the maximum service temperature.

b. Flexible, unshielded couplings shall not be used for joints in ductile iron pressure piping.

c. With push-on and mechanical joints, spigot ends shall be inserted into bells, as recommended by the manufacturer, rather than bells on to spigot ends.

d. Transitions with plastic or other pressure piping shall be made with transition fittings using push-on or mechanical joints sized for the piping being joined.

4.2.13 Reserved

4.2.14 Plastic

4.2.14.1 General

a. Joints in plastic piping shall be made by one of the following methods where appropriate:
  1. solvent cement
  2. heat fusion
  3. couplings with elastomeric sleeves and corrosion resisting metal screw clamps
  4. approved insert fittings
  5. approved mechanical fittings, or
  6. threads according to approved standards

b. Joints shall be made in accordance with the manufacturer's instructions for the method used.

See Figures 4.2.14.1-A through-F

4.2.14.2 Solvent Cement Joints in PVC Piping

Primers and solvent cements shall be suitable for joints in PVC piping. Primers shall be purple in color and solvent cements shall not be purple in color.

4.2.14.3 Solvent Cement Joints in CPVC Piping

Primers (where used) and solvent cements shall be suitable for joints in CPVC piping. Primer shall be purple in color and solvent cements shall be orange in color. Primers (where used) shall not be orange in color. Single-step solvent cement used without a primer on piping ½ inch through 2 inch in diameter shall be yellow in color. One step solvent cement used for joining CPVC fire sprinkler pipe shall be listed for fire protection use.

4.2.14.4 UV Detectable Clear Primers

Where a clear primer that is detectable by ultra-violet light is used as a substitute for a colored primer, the installer shall make a UV light detection device available for inspection of the joints for final acceptance.

4.2.14.5 Heat Fusion Joints in PE Water Piping

Heat fusion joints in PE water piping complying with ASTM F714, ASTM D3035, AWWA C901, or AWWA C906 shall be made in accordance with the applicable requirements of ASTM F2620.

4.2.14.6 Threaded Joints in Plastic Piping

Tapered pipe threads for leak-tight joints in thermoplastic pipe and fittings shall comply with ASTM F1498.

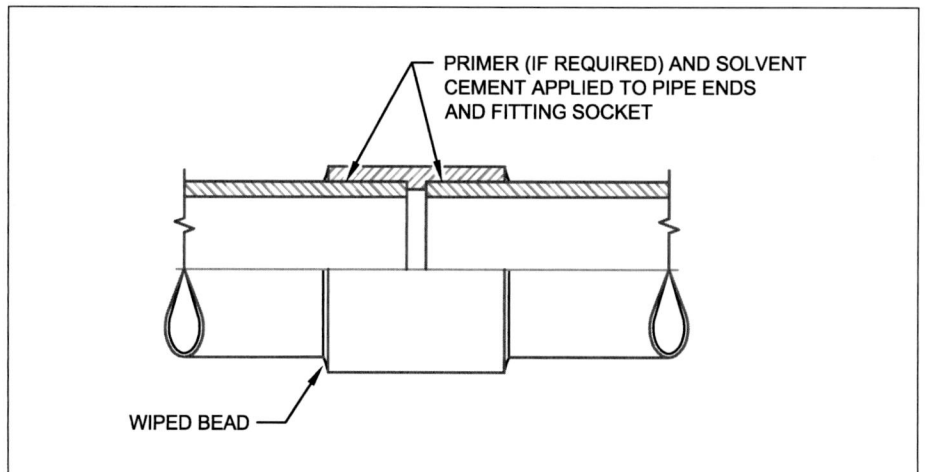

Figure 4.2.14-A
A SOLVENT CEMENT JOINT IN PLASTIC DWV OR WATER PIPING

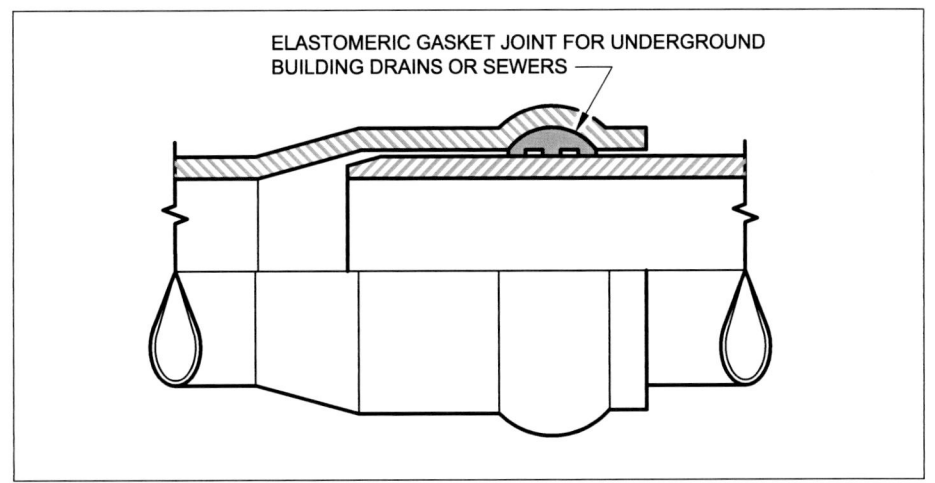

Figure 4.2.14-B
AN ELASTOMERIC GASKET JOINT FOR UNDERGROUND PLASTIC DWV PIPING

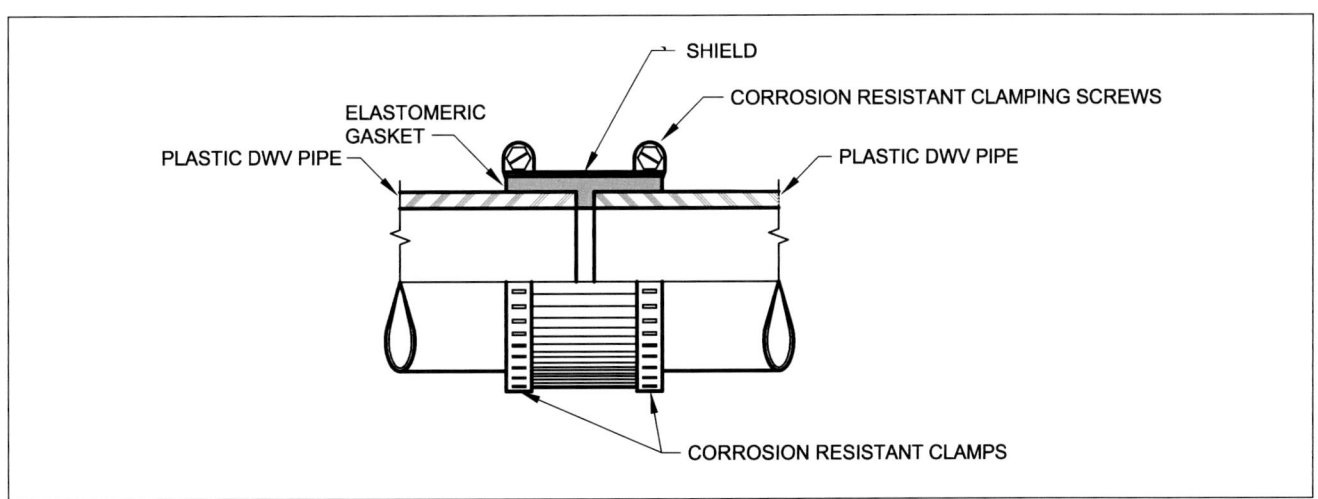

Figure 4.2.14-C
A SHIELDED COUPLING ON PLASTIC DWV PIPING

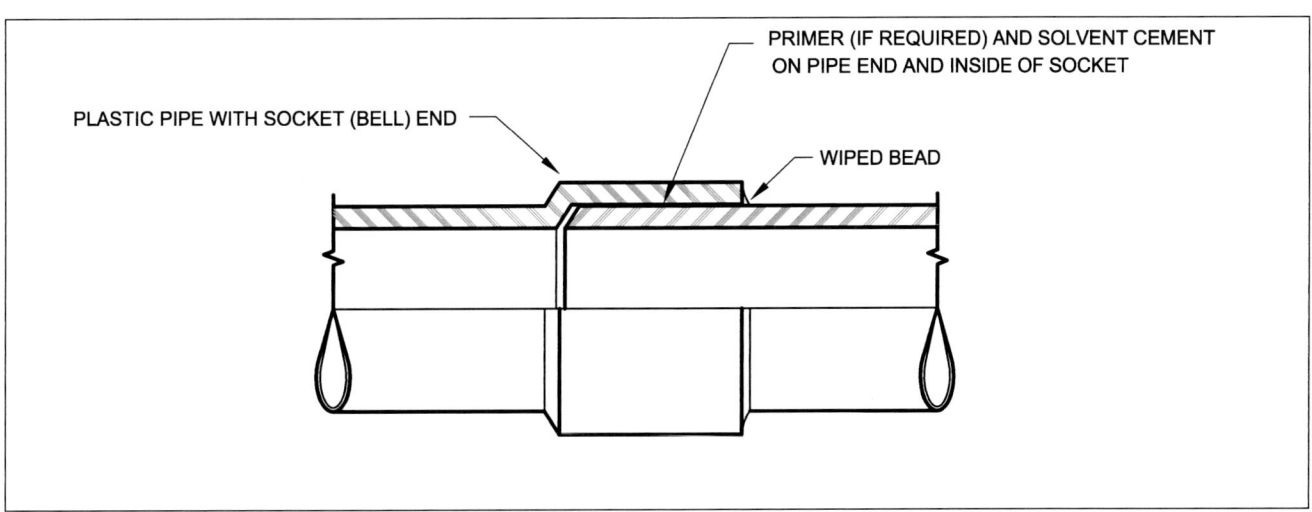

Figure 4.2.14-D
A SOLVENT CEMENT JOINT IN SOCKET (BELL) END PLASTIC PRESSURE PIPE

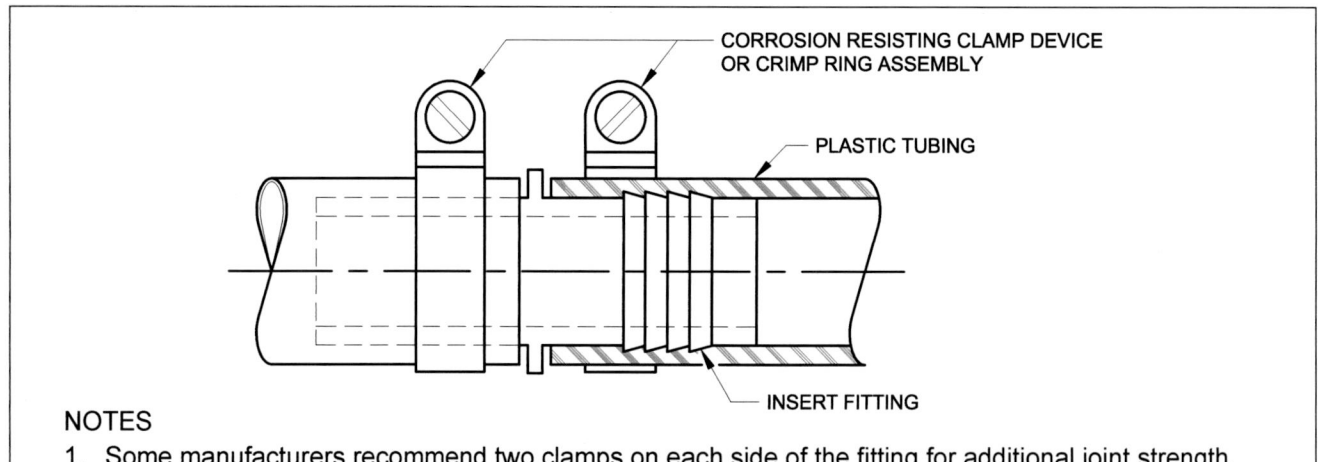

Figure 4.2.14-E
AN INSERT FITTING JOINT IN PLASTIC TUBING

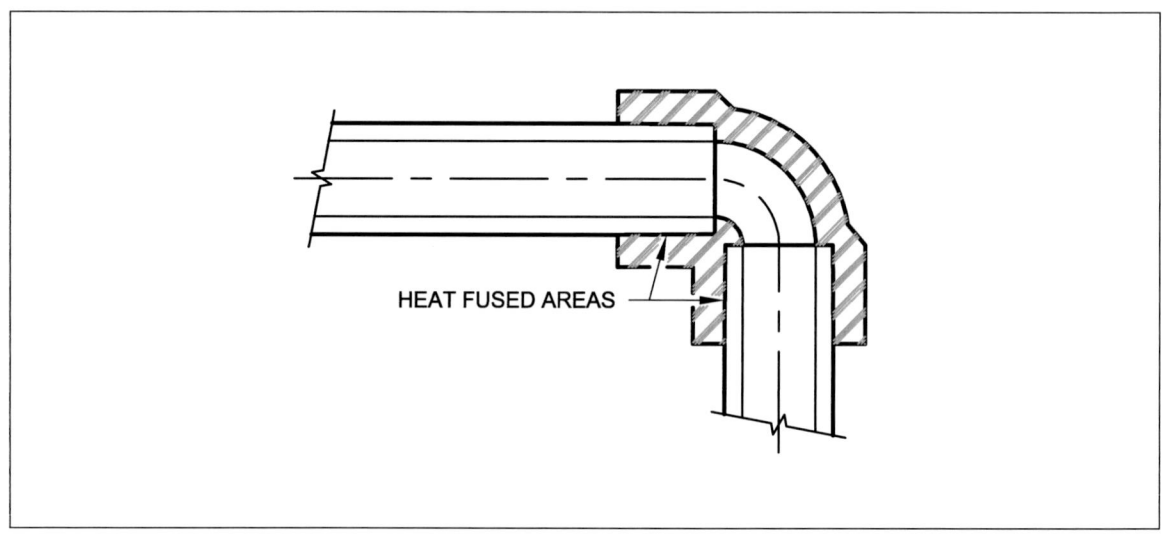

Figure 4.2.14-F
A HEAT FUSED JOINT IN PLASTIC WATER PIPING

4.2.15 Slip

Slip joints using washers or approved packing or gasket material, when installed in concealed locations, shall be provided with an access panel. Slip joints using approved ground joint brass compression rings that allow adjustment of tubing but provide a rigid joint when made up, shall not be considered as slip joints that require access. EXCEPTION: Slip joints in tub waste and overflows and their related traps shall not require an access panel. *See Figures 4.2.15-A and-B*

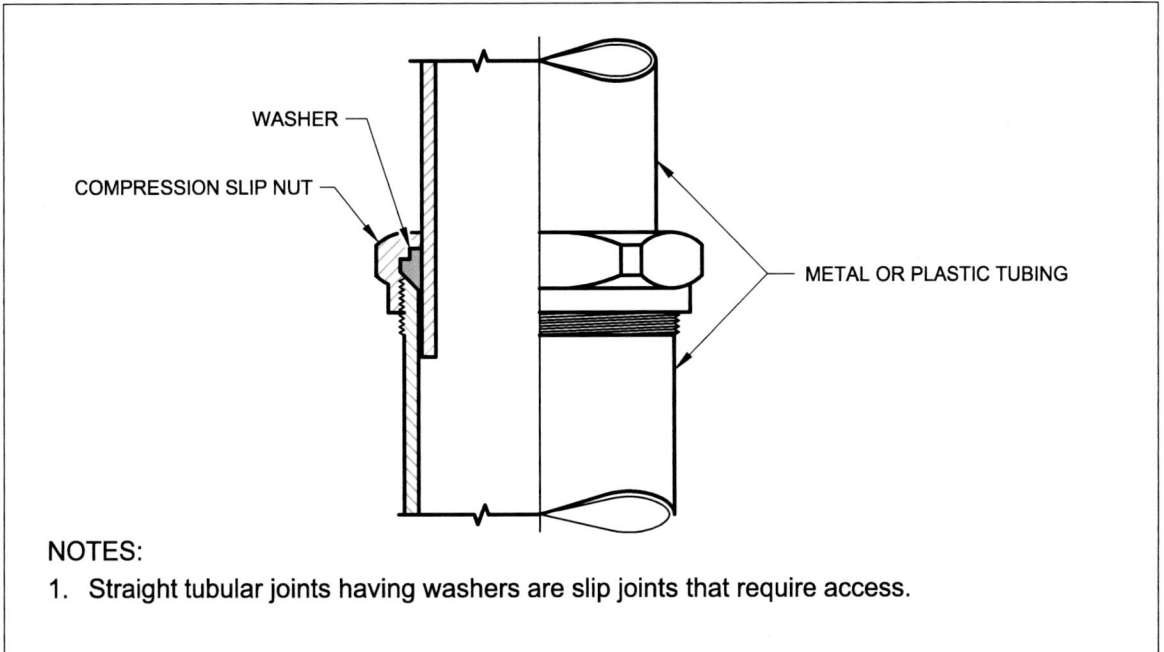

Figure 4.2.15-A
SLIP JOINT IN DRAIN TUBING

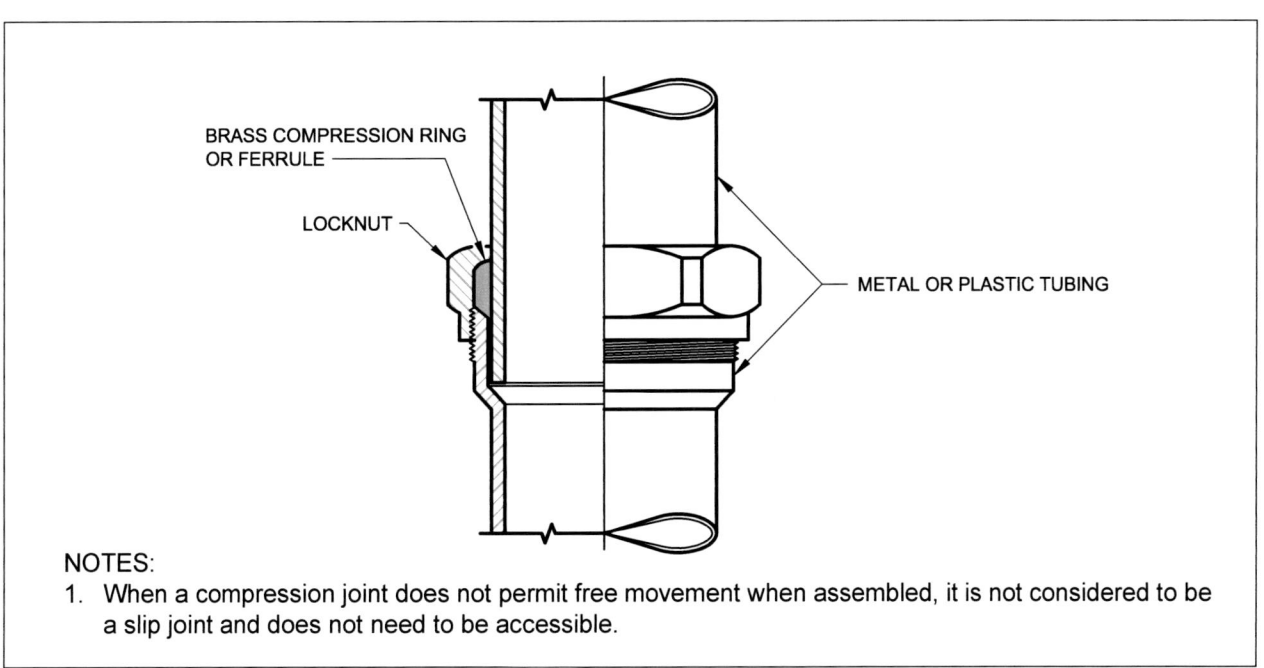

Figure 4.2.15-B
A COMPRESSION JOINT IN DRAIN TUBING

4.2.16 Expansion

Expansion joints shall be of approved type and its material shall conform with the type of piping in which it is installed. *See Figures 4.2.16-A, 4.2.16-B, and 4.2.16-C*

Comment: Expansion joints must to be accessible for adjustment or replacement. Refer to the definition of "accessible".

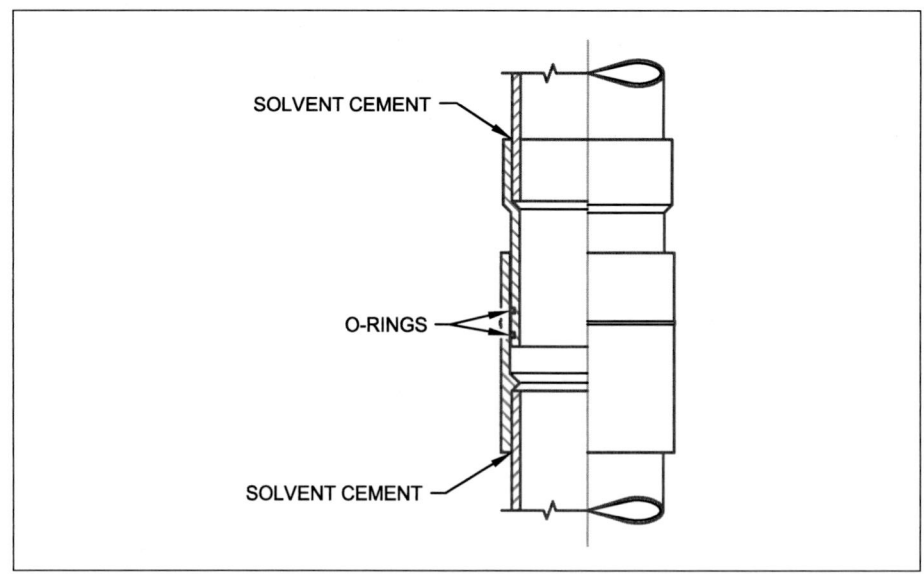

Figure 4.2.16-A
AN EXPANSION JOINT FOR PLASTIC DWV PIPING

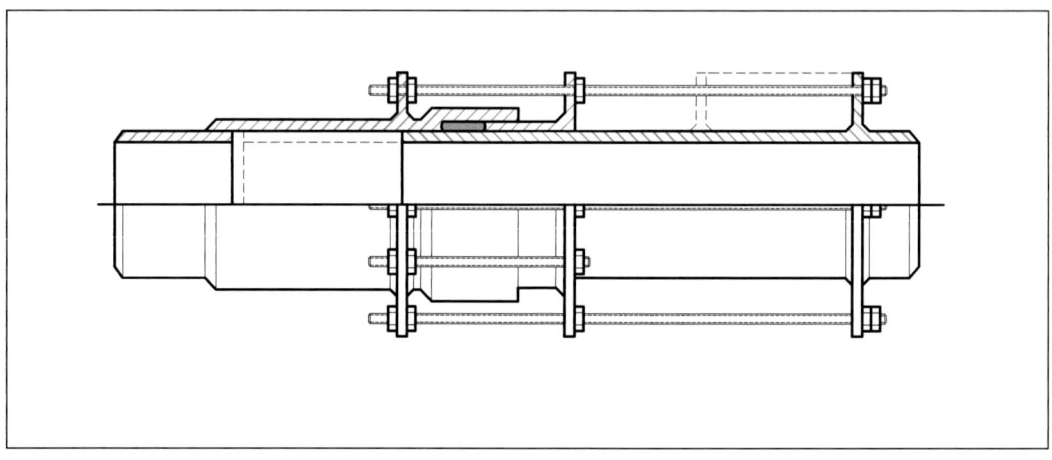

Figure 4.2.16-B
A MECHANICAL EXPANSION JOINT IN PRESSURE PIPING

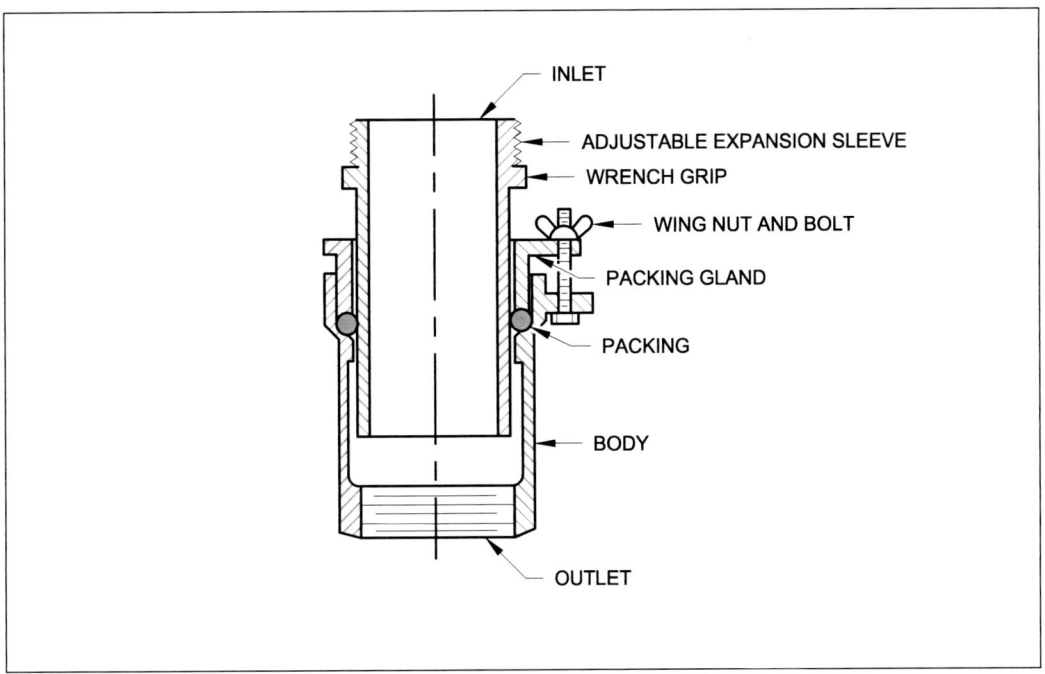

Figure 4.2.16-C
AN EXPANSION JOINT FOR ROOF DRAINS

4.2.17 Split Couplings

a. Split couplings consisting of two or more parts and a compression gasket, designed for use with grooved or plain end pipe and fittings, shall be permitted to be used for water service piping, hot and cold water distribution piping, storm water piping, and sump pump discharge piping. The complete joint assembly shall be suitable for the intended use and comply with a standard listed in Table 3.1.3.

b. Galvanized steel pipe may be jointed using rolled or cut grooves. Other interior coated pipe shall not be joined using rolled grooves.

See Figure 4.2.17

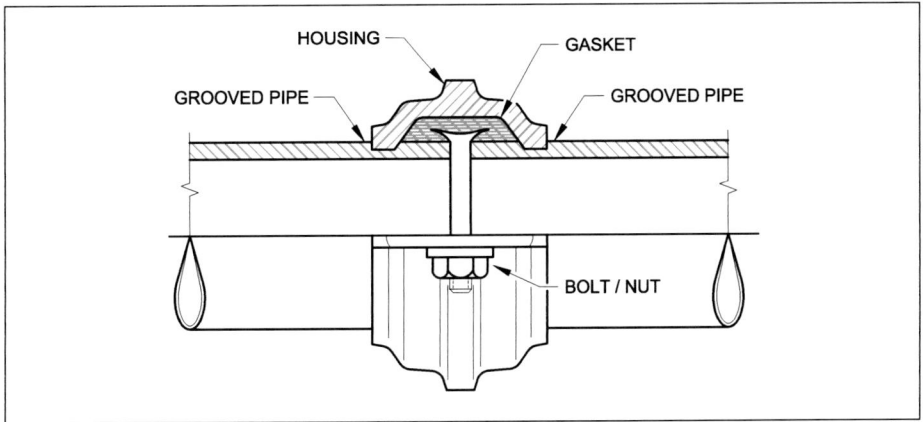

Figure 4.2.17
A GROOVED PIPE JOINT

4.2.18 Heat Fusion

a. ASTM F714 high-density polyethylene (HDPE) pipe and ASTM D3261 butt fittings shall be joined by butt heat fusion in accordance with ASTM D2657.

b. Heat fusion joints in polyethylene (PE) piping, including socket fusion, butt fusion, and saddle fusion, shall be made according to ASTM F2620.

4.2.19 Bending

Changes in direction in copper water tube shall be permitted to be made by the use of factory or field bends. Field bends shall be made in accordance with Table 4.2.19. Bends shall be made only with bending equipment and procedures intended for that purpose. Hard drawn tubing shall not be bent with tubing benders intended for only annealed (soft) tube. All bends shall be smooth and free from buckling, cracks, and other evidence of mechanical damage.

Table 4.2.19: BENDING COPPER TUBE			
Nominal Tube Size -in.	**Tube Type**	**Temper**	**Min. Bend Radius, in.**
1/4 inch	K,L	Annealed (soft)	3/4"
3/8 inch	K,L	Annealed (soft)	1-1/2"
3/8 inch	K,L,M	Drawn (hard)	1-3/4"
1/2 inch	K,L	Annealed (soft)	2-1/4"
1/2 inch	K,L,M	Drawn (hard)	2-1/2"
3/4 inch	K,L	Annealed (soft)	3"
3/4 inch	K,L	Drawn (hard)	3"
1 inch	K,L	Annealed (soft)	4"
1-1/4 inches	K,L	Annealed (soft)	9"

4.3 TYPES OF JOINTS BETWEEN DIFFERENT PIPING MATERIALS

4.3.1 Vitrified Clay to Other Material

Joints between vitrified clay and other piping materials shall be made with an approved joint.
See Figures 4.3.1-A and 4.3.1-B

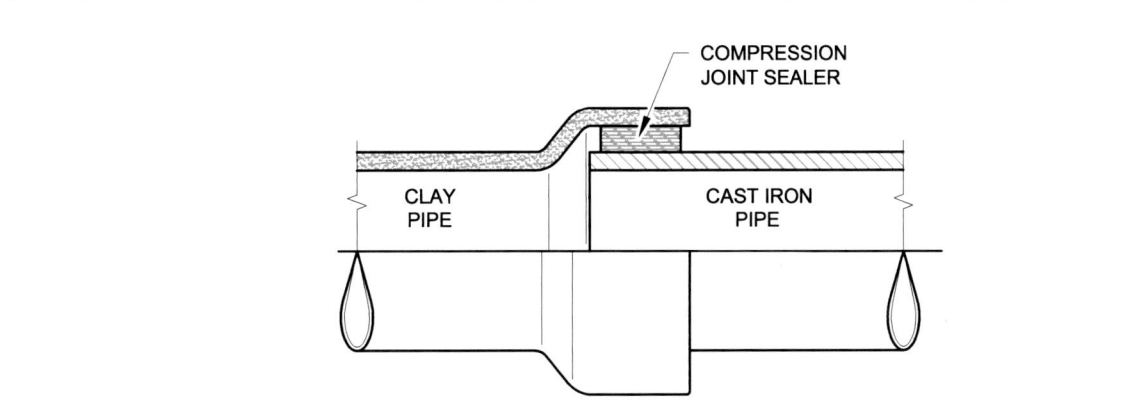

NOTES:
1. The rubber rings must be sized to adapt to the different outside diameters (O.D.) of the different piping materials being joined.
2. There are hundreds of different adapters available for joining different materials, different pipe sizes, and different wall thickness, as well as clay pipe from different manufacturers.

Figure 4.3.1-A
A RUBBER RING TRANSITION JOINT TO VITRIFIED CLAY PIPE

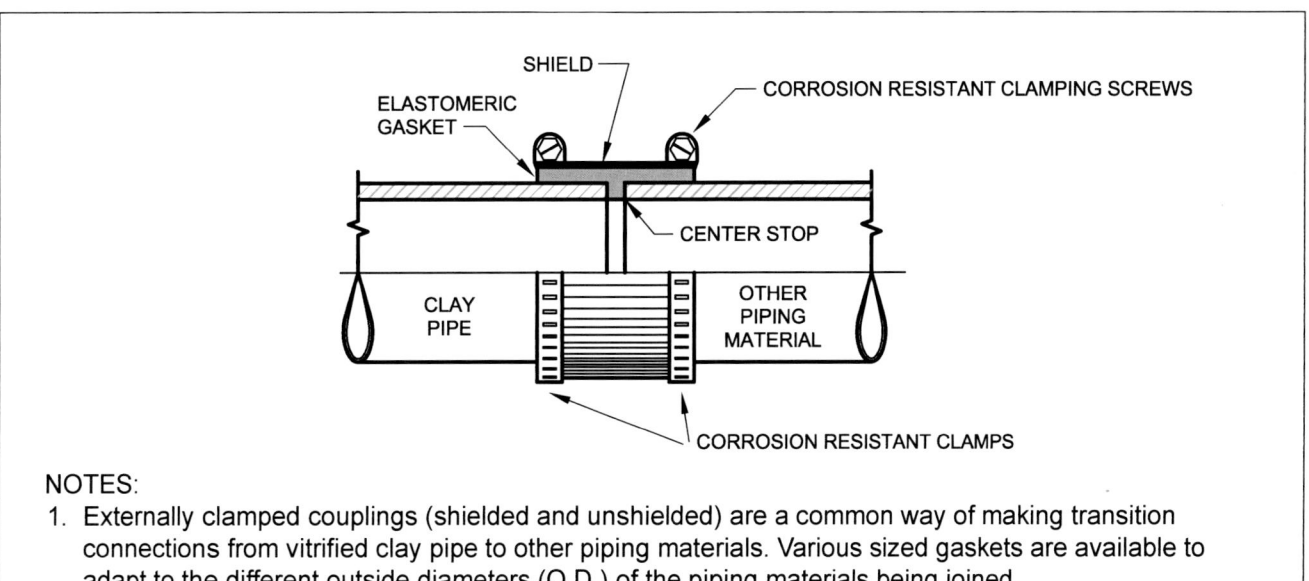

NOTES:
1. Externally clamped couplings (shielded and unshielded) are a common way of making transition connections from vitrified clay pipe to other piping materials. Various sized gaskets are available to adapt to the different outside diameters (O.D.) of the piping materials being joined.

Figure 4.3.1-B
AN EXTERNALLY CLAMPED TRANSITION COUPLING JOINT TO VITRIFIED CLAY PIPE

4.3.2 Reserved

4.3.3 Reserved

4.3.4 Threaded Pipe to Cast-Iron

Joints between steel or brass and cast-iron pipe shall be either caulked or threaded or shall be made with approved adapter fittings. *See Figure 4.3.4*

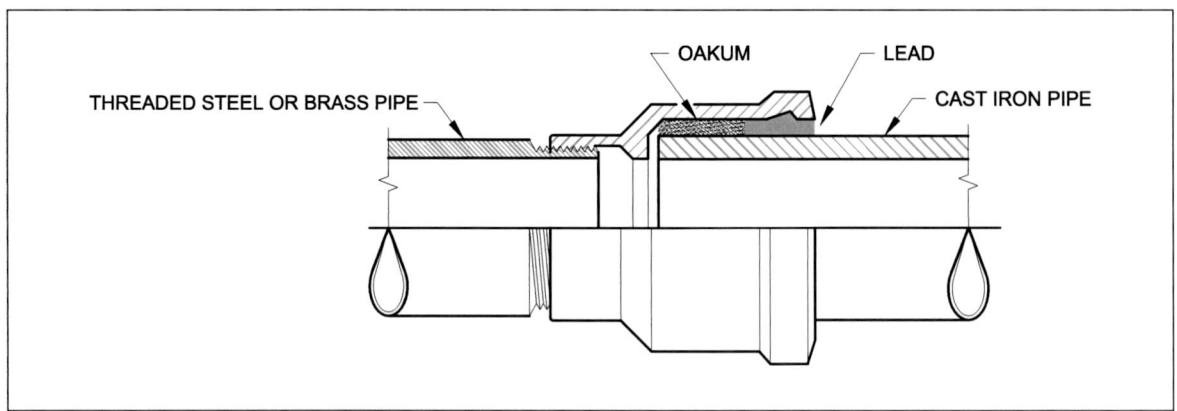

Figure 4.3.4
A THREADED PIPE TO CAST IRON ADAPTER FITTING

4.3.5 Lead to Cast-Iron or Steel

Joints between lead and cast-iron or steel pipe shall be made by means of wiped joints to a caulking ferrule, soldering nipple, bushing, or by means of a mechanical adapter. *See Figure 4.3.5*

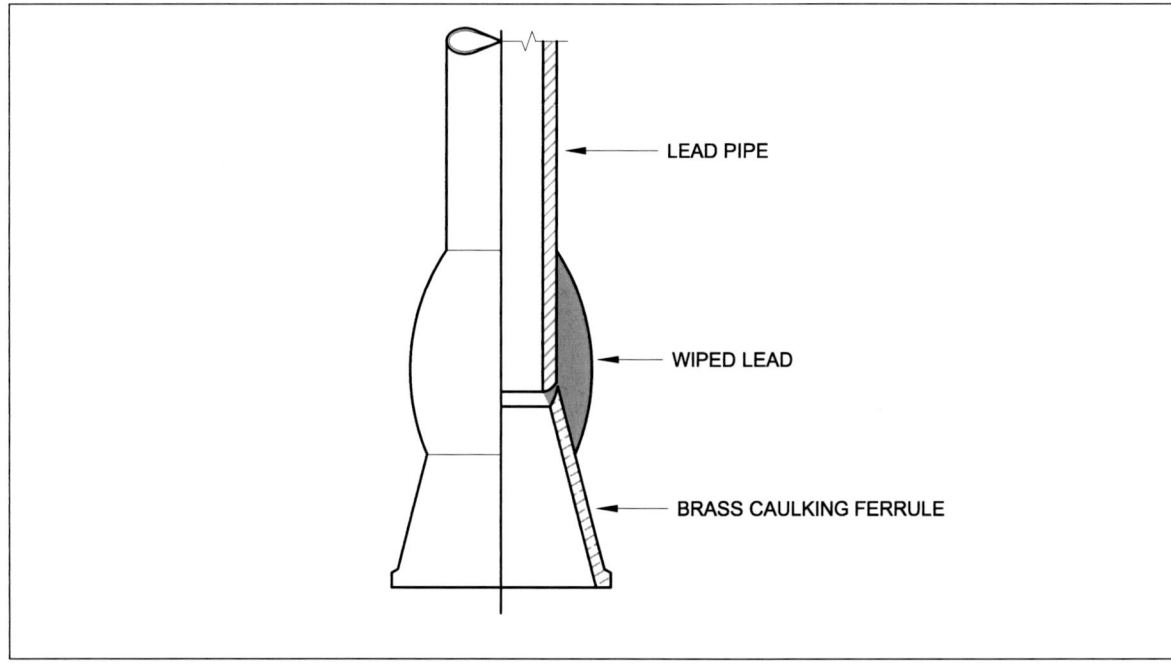

Figure 4.3.5
A WIPED LEAD TRANSITION JOINT

4.3.6 Cast-Iron to Copper Tube

Joints between cast-iron and copper tube shall be made by using an approved brass or copper caulking ferrule and by properly soldering the copper tube to the ferrule. *See Figure 4.3.6*

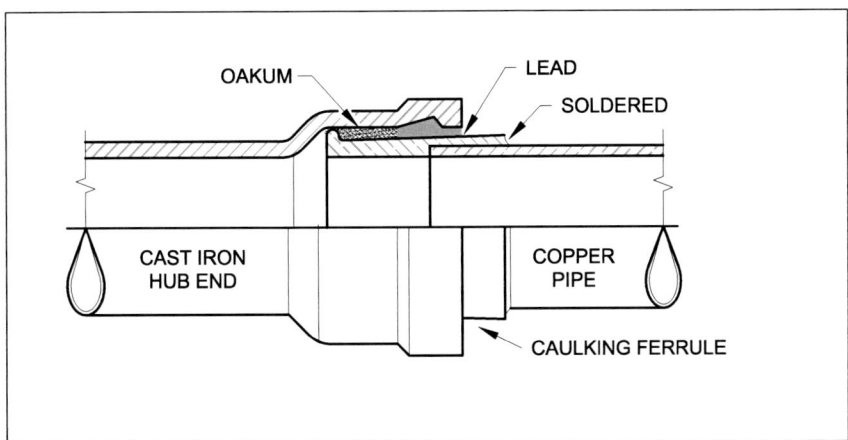

Figure 4.3.6
A CAULKING FERRULE FOR CAST IRON TO SWEAT COPPER TRANSITION

4.3.7 Copper Tube to Threaded Pipe Joints

a. Joints from copper tube to threaded pipe shall be made as follows:

   1. DWV Systems: with copper or brass threaded adapters.

   2. Water Systems and Galvanized Steel Pipe: cast brass threaded adapters, dielectric pipe unions conforming to ASSE 1079, dielectric flanges or dielectric waterway fittings that comply with IAPMO PS 66. EXCEPTION: Dielectric pipe unions shall not be installed on connections to water heaters when not recommended by the water heater manufacturer.

   3. To any Non-Ferrous Piping: copper or brass threaded adapter.

b. The adapter fitting shall be connected to the tubing by approved methods, and the threaded section assembled with tapered national pipe threads (NPT).

See Figures 4.3.7-A and 4.3.7-B

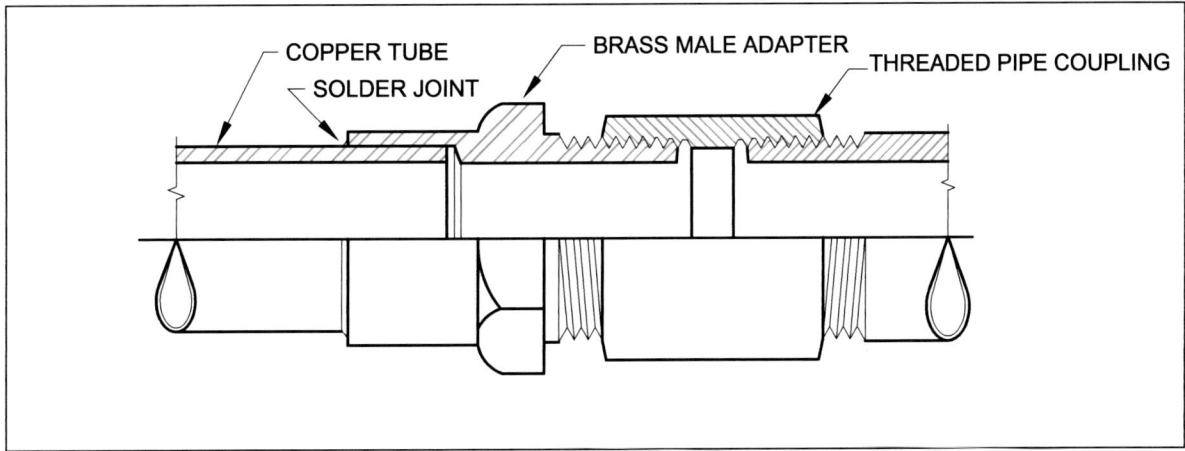

Figure 4.3.7-A
A SWEAT COPPER TO THREADED PIPE TRANSITION

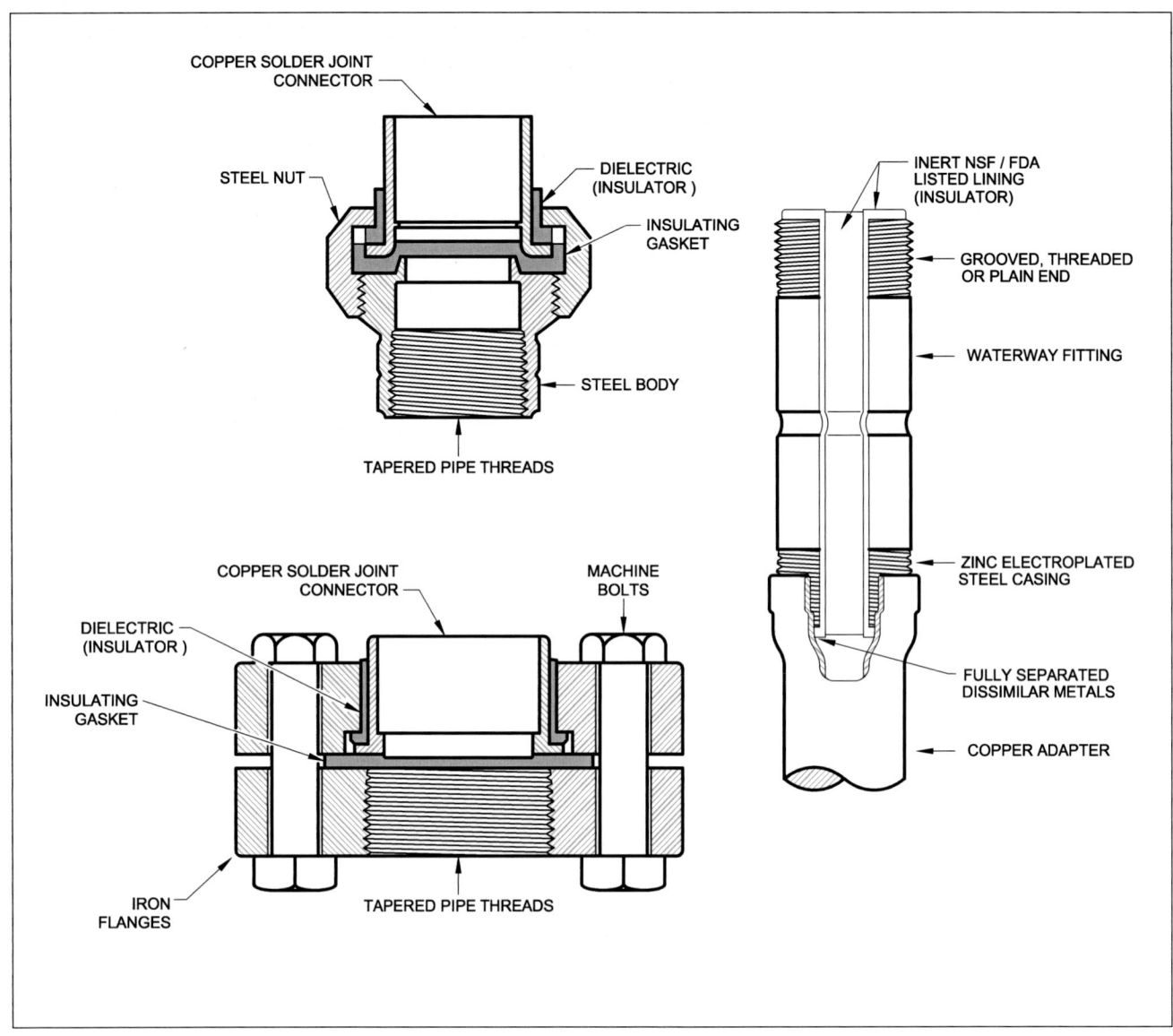

Figure 4.3.7-B
DIELECTRIC UNION, FLANGE, AND WATERWAY FITTING

4.3.8 Special Joints and Couplings for Drain Piping

a. Joints between two different drain piping materials or between different size piping, of the same or different material, shall be made using fittings or mechanical couplings that are designed for the specific application, including adapter fittings, hubless pipe couplings, slip-on couplings, transition couplings, and repair couplings. Installation shall comply with the coupling manufacturer's instructions and intended use.

b. Shielded couplings shall consist of a flexible elastomeric sealing sleeve, a protecting and supporting continuous metal shield or shear ring, and metal screw clamping bands. All metal parts shall be corrosion resisting. Shielded couplings shall be capable of withstanding a shear test and shall be permitted to partially support the pipe being joined when such installation is recommended by the manufacturer's instructions.

c. Mechanical unshielded couplings using thermoplastic elastomer gaskets shall consist of a rigid or semi-rigid sealing sleeve and corrosion-resisting metal screw clamping bands. Mechanical unshielded couplings using thermoplastic elastomer gaskets shall not be installed where the operating internal or external temperatures exceed 130°F (54°C) or are below 0°F (-18°C). The pipe shall be supported on both sides of the coupling within 18 inches of the centerline of the coupling. Mechanical unshielded couplings using thermoplastic elastomer gaskets shall not be installed in construction that has a fire rating that restricts the use of flammable materials or be installed in through-penetrations or plenums without additional fire resistance protection.

EXCEPTION: Mechanical unshielded couplings using thermoplastic elastomer gaskets shall not be used as a joining method for hubless cast-iron soil pipe and fittings.

d. Flexible unshielded couplings shall consist of an elastomeric sealing sleeve and corrosive-resisting metal screw clamping bands. The use of flexible unshielded couplings shall be limited to joints in underground sewer, drain, and vent piping. Where eccentric flexible unshielded couplings are used to maintain the invert of a drain line, the couplings shall align the inverts of the inside walls of the pipes being joined to prevent an obstruction to free flow.

e. The installation of couplings shall comply with the following.

1. Couplings that comply with the following standards may be installed aboveground or underground:
   ASTM A1056: cast iron couplings for hubless cast iron soil pipe and fittings
   ASTM C1277: shielded couplings for hubless cast iron soil pipe and fittings
   ASTM C1461: mechanical couplings with thermoplastic gaskets for DWV piping
   ASTM C1540: heavy duty shielded couplings for hubless CISP and fittings
   CISPI 310: cast iron couplings for hubless cast iron soil pipe and fittings
   FM 1680-Class 1: couplings for hubless cast iron soil pipe and fittings
2. Couplings that comply with the following standard shall only be installed aboveground.
   ASTM C1460: shielded transition couplings for dissimilar DWV piping
3. Couplings that comply with the following standard shall only be installed underground.
   ASTM C1173: flexible transition couplings for underground piping

f. Where eccentric flexible unshielded couplings are used to maintain the invert of a drain line, the couplings shall align the inverts of the inside walls of the pipes being joined to prevent an obstruction to free flow.

g. Center stops shall not be required for couplings that join pipes with different inside diameters and for slip-on repair couplings used for repair or rework.

See Figures 4.3.8-A through 4.3.8-D

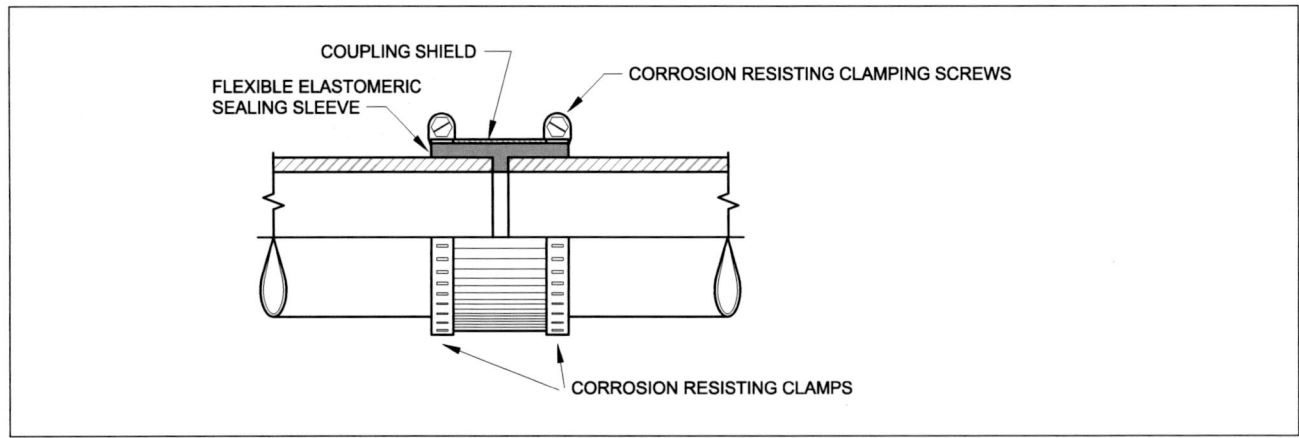

Figure 4.3.8-A
A SHIELDED COUPLING ON DRAINAGE PIPING

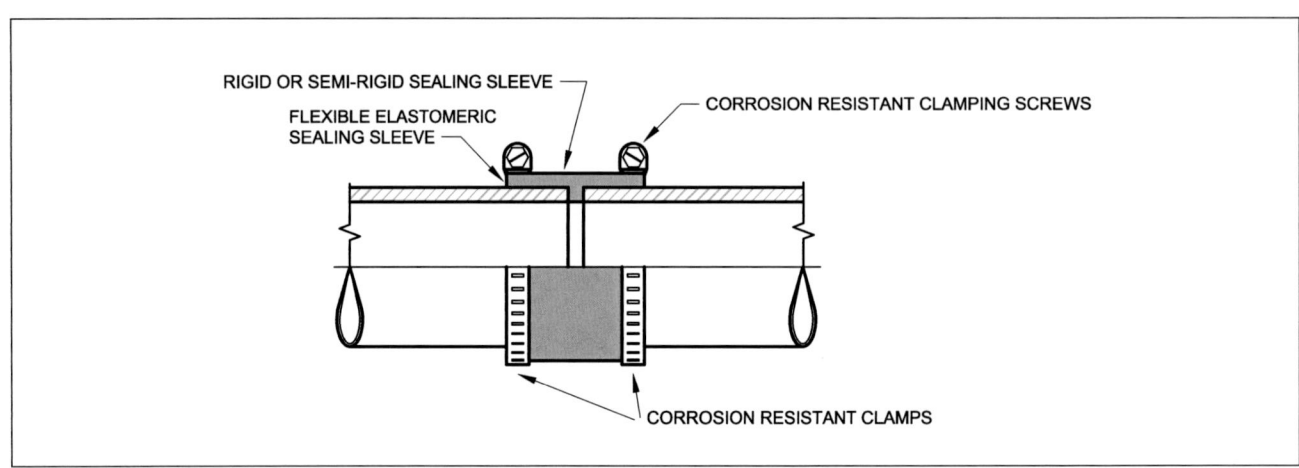

Figure 4.3.8-B
A RIGID UNSHIELDED COUPLING ON DRAINAGE PIPING

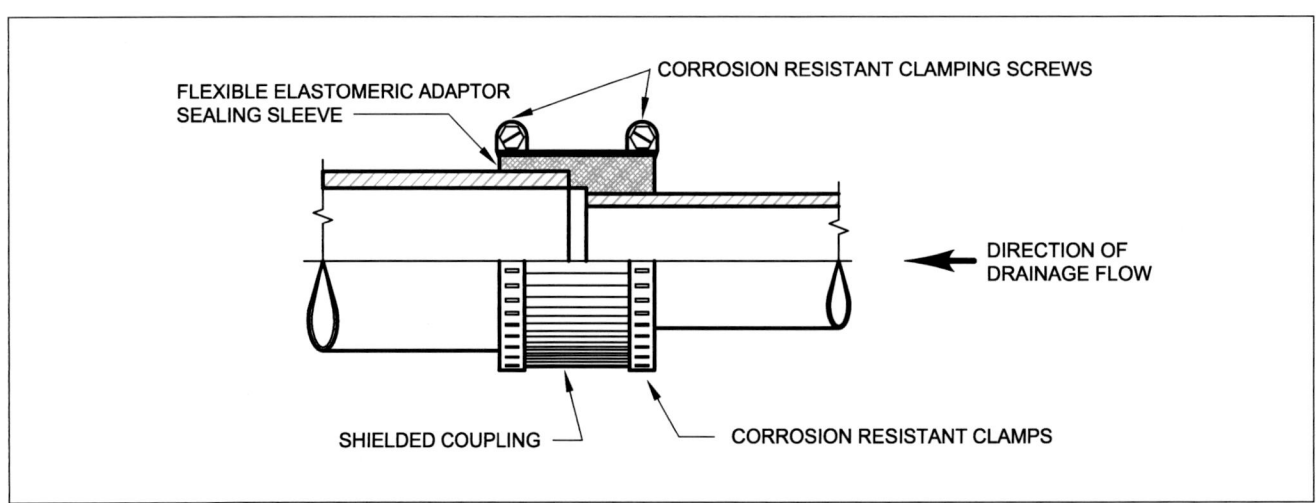

Figure 4.3.8-C
A TRANSITION CONNECTION USING A SHIELDED COUPLING

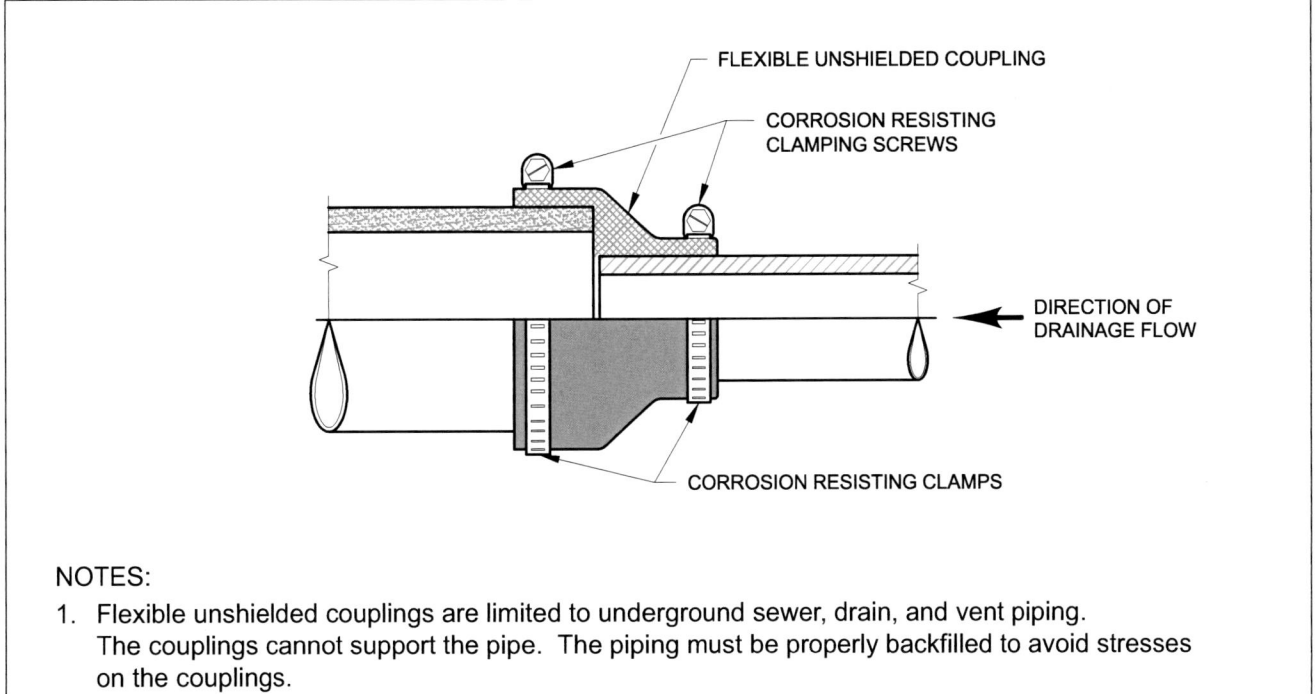

NOTES:
1. Flexible unshielded couplings are limited to underground sewer, drain, and vent piping. The couplings cannot support the pipe. The piping must be properly backfilled to avoid stresses on the couplings.

Figure 4.3.8-D
A TRANSITION CONNECTION USING A FLEXIBLE UNSHIELDED COUPLING

4.3.9 Plastic DWV Pipe to Other Materials

a. Joints between plastic DWV pipe and piping of other materials shall be made in accordance with the manufacturer's instructions for the pipe and fittings unless prohibited by this Code.

b. Threaded Joints: Threaded joints for connecting plastic drain piping to other materials shall be made with proper male or female threaded plastic adapters. Joints shall not be over-tightened. After hand tightening, one-half to one full turn shall be made with a strap wrench. *See Figure 4.3.9-A*

c. Solid Wall PVC Schedule 40 DWV Plastic Pipe to Cast-Iron Hub Ends: Joints shall be made by caulking the plastic pipe into the hub end with molten lead and oakum or by use of a compression gasket that is compressed when the plastic pipe is inserted into the hub end. Joints shall be permitted to be made with or without a hub end plastic adapter. Adapters without a caulking bead shall be permitted. *See Figures 4.3.9-B through 4.3.9-D*

d. Cellular Core PVC Schedule 40 DWV Plastic Pipe to Cast-Iron Hub Ends: Joints shall be made by caulking a solid plastic adapter into the cast-iron hub end with molten lead and oakum or by use of a compression gasket that is compressed when the plastic pipe is inserted into the hub end. Adapters without a caulking bead shall be permitted. Cellular core plastic pipe shall not be lead caulked.

e. Plastic Pipe to Galvanized Steel, Copper or Stainless Steel DWV Tube, or Cast-iron Spigot Ends: Joints between plastic pipe and the listed materials shall be made with proper transition fittings.

f. ABS and PVC Plastic Pipe: Solvent cemented non-pressure joints between ABS and PVC DWV piping systems shall be made with an ASTM D3138 solvent cement intended for such transition joints. Transition cement shall not be permitted to be used within buildings. Transition joints shall be a minimum of 3 feet outside of buildings.

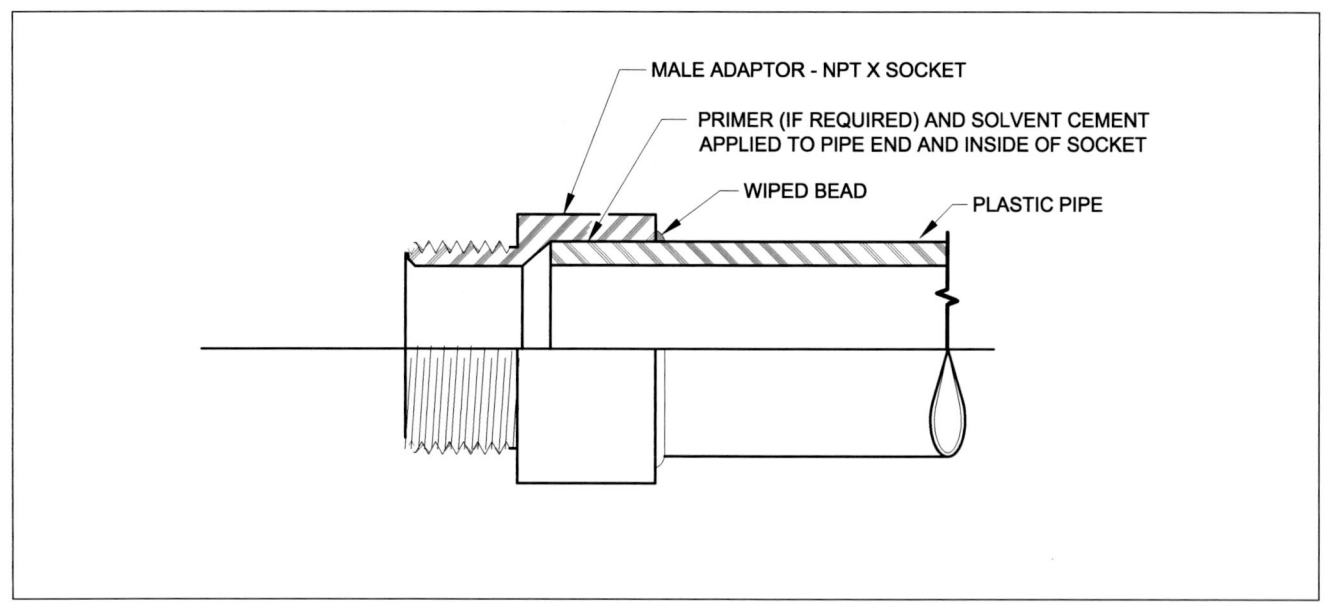

Figure 4.3.9-A
A PLASTIC DWV THREADED MALE ADAPTER

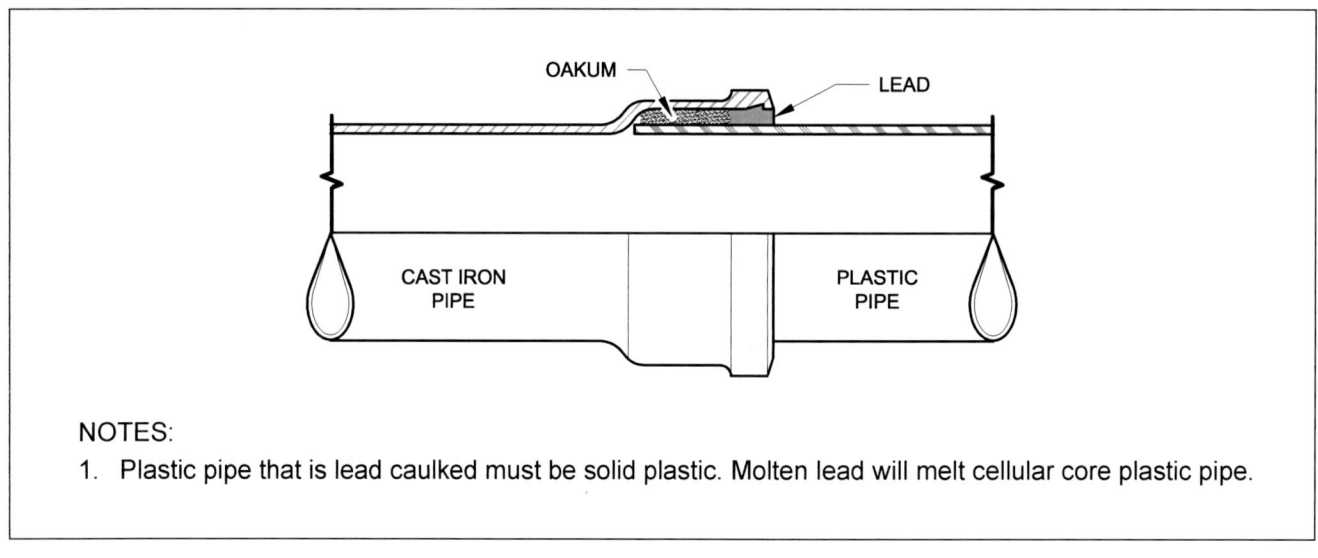

NOTES:
1. Plastic pipe that is lead caulked must be solid plastic. Molten lead will melt cellular core plastic pipe.

Figure 4.3.9-B
A LEAD CAULKED JOINT FOR PLASTIC DWV TO CAST IRON HUB

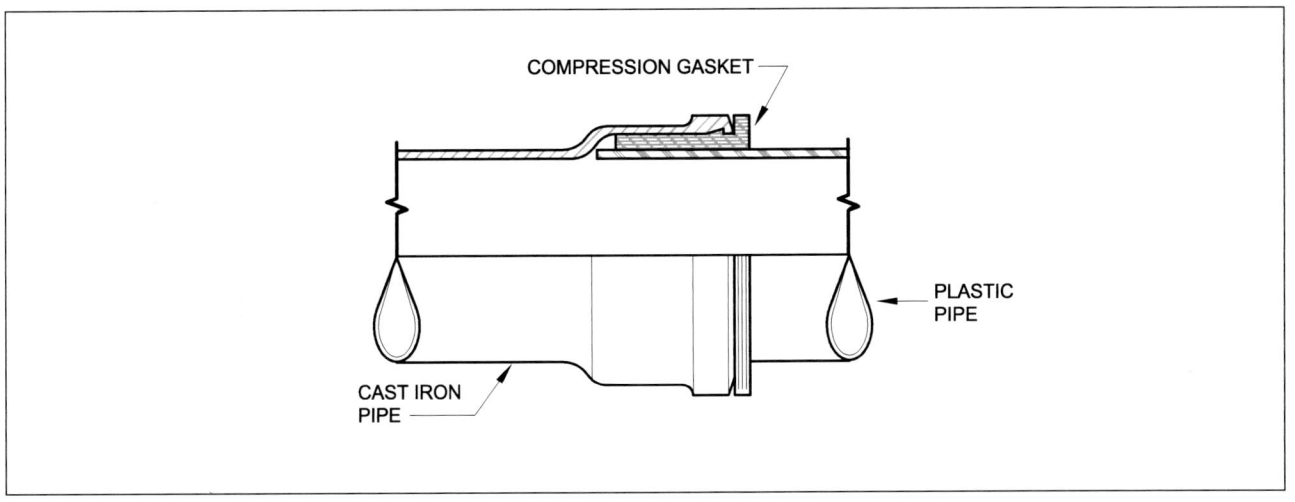

Figure 4.3.9-C
A COMPRESSION GASKETED JOINT FOR PLASTIC DWV TO CAST IRON HUB

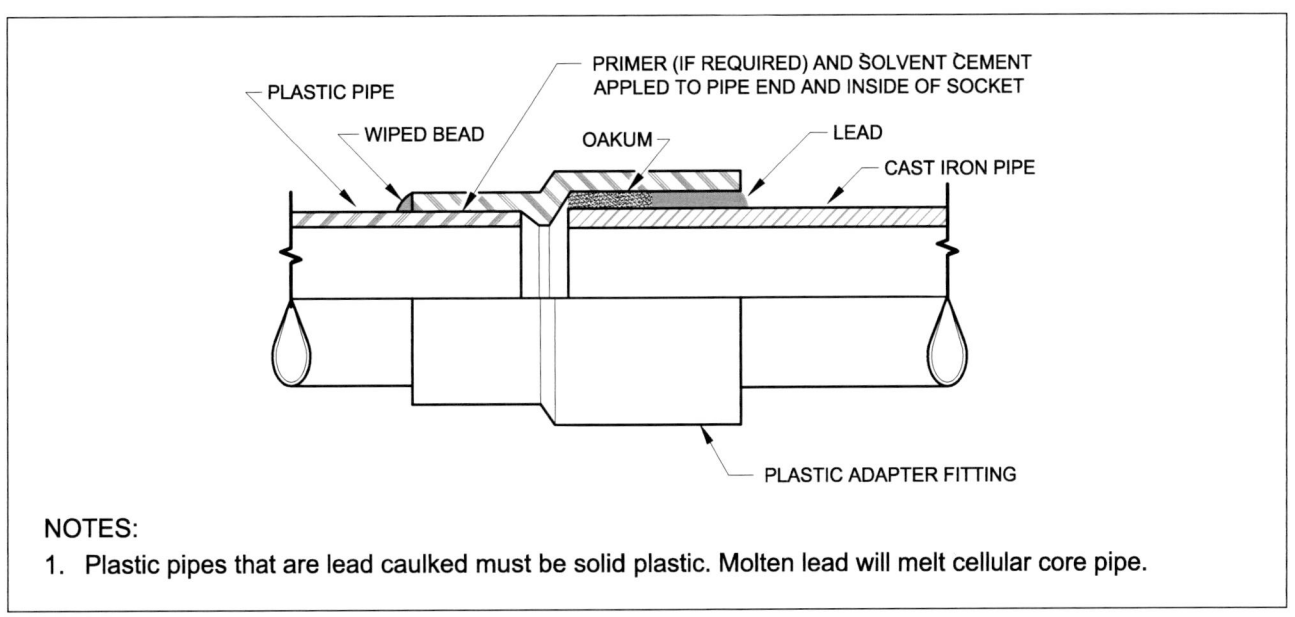

NOTES:
1. Plastic pipes that are lead caulked must be solid plastic. Molten lead will melt cellular core pipe.

Figure 4.3.9-D
A LEAD CAULKED PLASTIC SOCKET JOINT TO CAST IRON PIPE

4.4 CONNECTIONS BETWEEN DRAIN PIPING AND CERTAIN FIXTURES

a. Connections between drain piping and floor outlet plumbing fixtures shall be made by means of an approved flange that is attached to the drain piping in accordance with the provisions of this chapter. The floor flange shall be set on and securely anchored to the building structure.

b. Connections between drain piping and wall hung water closets shall be made by means of an approved extension nipple or horn adapter.

c. Connections shall be bolted to the flange or carrier using corrosion resisting bolts or screws, or assemblies recommended by the manufacturer.

4.5 WATERPROOFING OF OPENINGS

a. Joints around vent pipes at the roof shall be made watertight by the use of lead, copper, aluminum, plastic, or other approved flashing or flashing materials. See Section 12.4.7.

b. Exterior wall openings shall be made watertight.

4.6 JOINTS IN LARGE SIZE CAST IRON SOIL PIPE

Cast-iron soil pipe and fittings (hub & spigot and hubless) five inches and larger shall be braced to prevent horizontal and vertical movement. Support shall be provided at every branch connection and change of direction by the use of braces, blocks, rodding, or other effective method to prevent movement and joint separation.

Blank Page

Chapter 5

Traps, Cleanouts and Backwater Valves

5.1 SEPARATE TRAPS FOR EACH FIXTURE

a. Plumbing fixtures shall be separately trapped by a liquid seal trap placed as close as possible to the fixture outlet.

b. The vertical distance from the fixture outlet to the trap weir shall not exceed 24 inches.

c. Fixtures shall not be double-trapped unless a relief vent is provided between the two traps.

EXCEPTIONS:

(1) Fixtures that have integral traps.

(2) Interceptors in Chapter 6 that provide the required trap seal. Interceptors in Chapter 6 that do not provide the required trap seal shall be provided with a separate trap.

(3) A combination plumbing fixture may be installed on one trap provided the waste outlets are not more than 30 inches from the trap inlet.

(4) One trap may be installed for up to three (3) sinks or lavatories in the same room with their waste outlets no more than 30 inches from the trap, and where the trap is centrally located when three such fixtures are installed.

(5) No clothes washer or laundry sink shall be discharged to a trap serving a kitchen sink.

(6) As otherwise permitted by this Code.

See Figures 5.1-A through 5.1-D

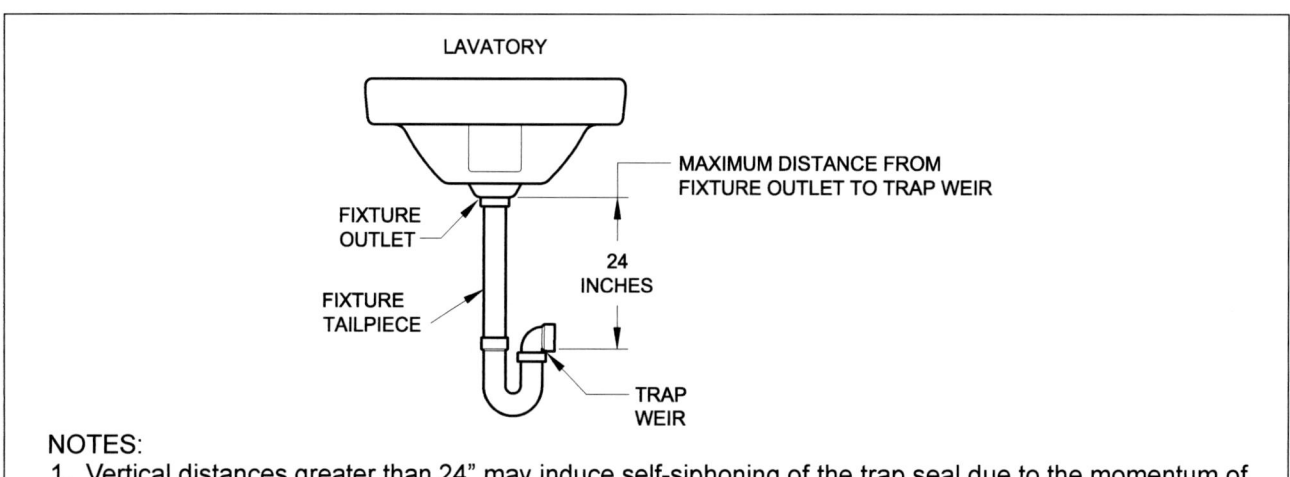

Figure 5.1 - A
THE VERTICAL DISTANCE FROM A FIXTURE OUTLET TO ITS TRAP WEIR

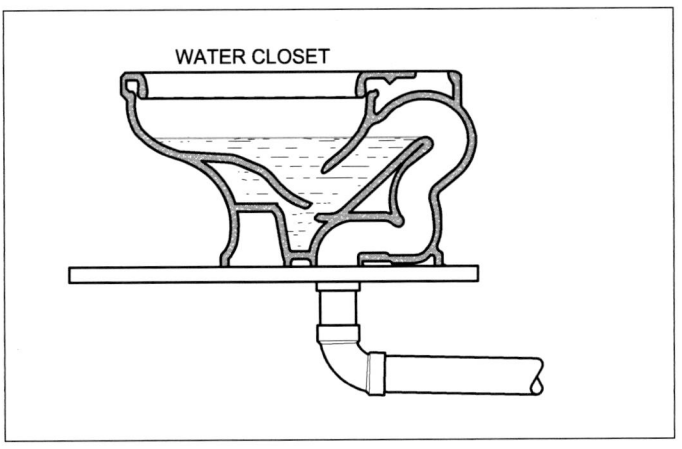

Figure 5.1 - B
A FIXTURE WITH AN INTEGRAL TRAP

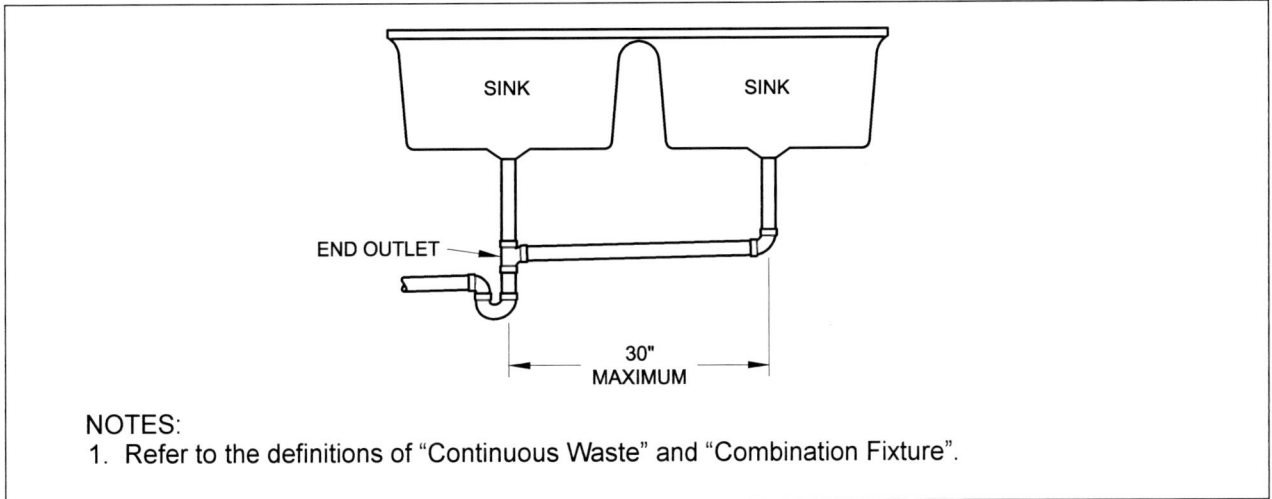

NOTES:
1. Refer to the definitions of "Continuous Waste" and "Combination Fixture".

Figure 5.1 - C
A CONTINUOUS WASTE WITH END OUTLET

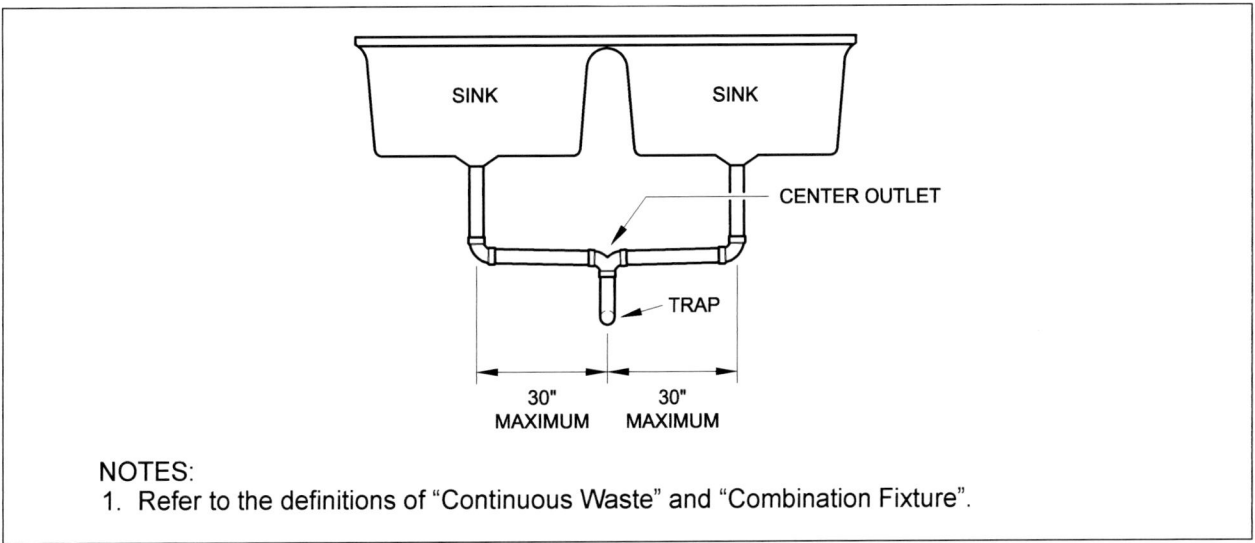

NOTES:
1. Refer to the definitions of "Continuous Waste" and "Combination Fixture".

Figure 5.1 - D
A CONTINUOUS WASTE WITH CENTER OUTLET

5.2 SIZE OF FIXTURE TRAPS

Fixture trap size (nominal diameter) shall be sufficient to drain the fixture rapidly and in no case less than given in Table 5.2. No trap shall be larger than the drain pipe into which it discharges. Integral traps shall conform to appropriate standards.

> *Comment: A drain pipe that is smaller than the trap to which it is connected could create an obstruction to flow through the trap.*

Table 5.2
MINIMUM SIZE OF NON-INTEGRAL TRAPS

Plumbing Fixture	Trap Size in inches
Bathtub (with or without overhead shower)	1-1/2
Bidet	1-1/4
Clothes washing machine standpipe	2
Combination kitchen sink, domestic dishwasher, and food waste disposer	1-1/2
Dental unit or cuspidor	1-1/4
Dental lavatory	1-1/4
Drinking fountain	1-1/4
Dishwasher, commercial	2
Dishwasher, domestic (non-integral trap)	1-1/2
Floor drain	2
Food waste disposer, commercial use	2
Food waste disposer, domestic use	1-1/2
Kitchen sink, domestic, with food waste disposer unit	1-1/2
Kitchen sink, domestic	1-1/2
Lavatory, common (private and public)	1-1/4
Lavatory (barber shop, beauty parlor or surgeon's)	1-1/2
Laundry tray (1 or 2 compartments)	1-1/2
Shower stall or shower drain (single shower head)	1-1/2
Shower stall or shower drain (multiple shower heads)	2
Sink (surgeon's)	1-1/2
Sink (flushing rim type, flush valve supplied)	3
Sink (service type with floor outlet trap standard)	3
Sink (service type with P trap)	2
Sink, commercial (pot, scullery, or similar type)	2
Sink, commercial (with food disposer unit)	2
Wash Fountain (4 rated users or less)	1-1/2
Wash Fountain (8 rated users or less)	2
Wash Sink	1-1/2

5.3 GENERAL REQUIREMENTS FOR TRAPS

5.3.1 Design of Traps

Fixture traps shall be self-scouring and shall have no interior partitions except where such traps are integral with the fixture or where corrosion resistant materials of plastic or glass are used. Solid connections, slip joints, or couplings may be used on the trap inlet, trap outlet, or within the trap seal.
See Figures 5.3.1-A through 5.3.1-C. Refer to Sections 4.2.15 and 7.3.7 for slip joints and their required access.

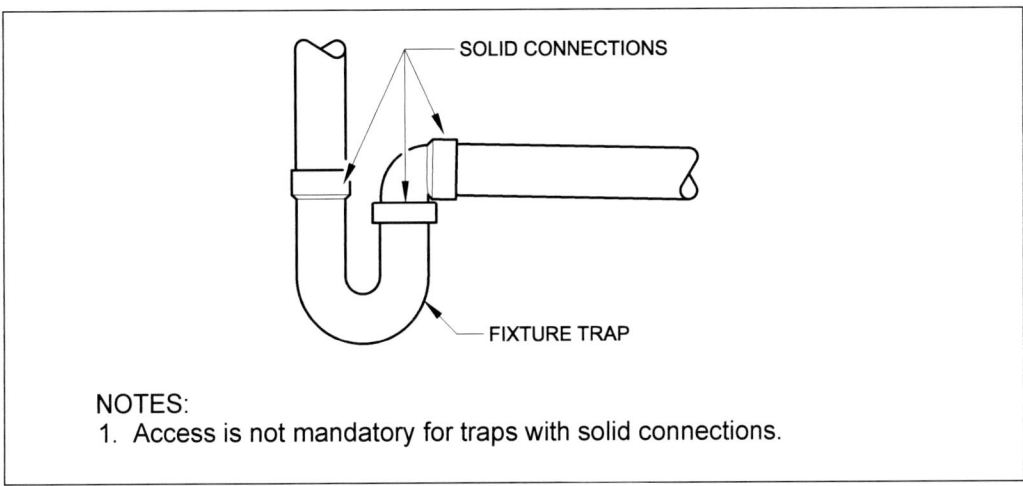

Figure 5.3.1 - A
A FIXTURE TRAP WITH SOLID CONNECTIONS

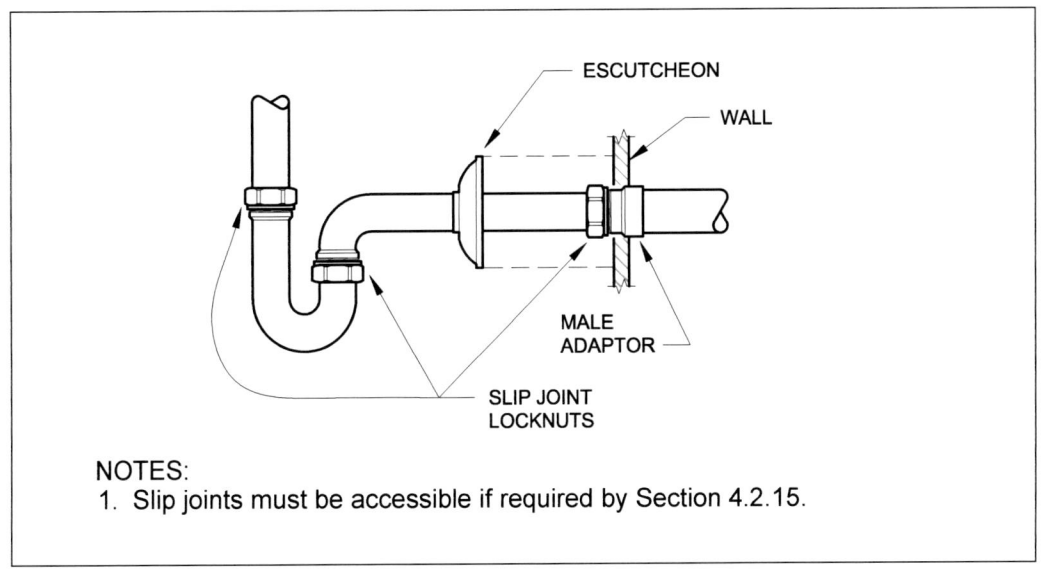

Figure 5.3.1 - B
A FIXTURE TRAP WITH SLIP JOINTS

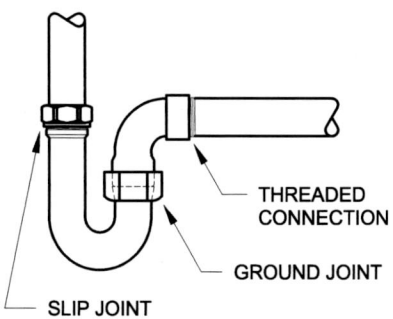

NOTES:
1. Only the slip joint must be accessible if required by Section 4.2.15.

Figure 5.3.1 - C
A FIXTURE TRAP WITH THREE DIFFERENT TYPES OF JOINTS

5.3.2 Trap Seals

Each fixture trap shall have a liquid seal of not less than two inches and not more than four inches.
EXCEPTIONS:
 (1) Interceptors in Chapter 6 that provide the required trap seal.
 Note: Interceptors in Chapter 6 that do not provide the required trap seal shall be provided with a separate trap.
 (2) Special conditions such as accessible fixtures, a deeper seal may be required by the Authority Having Jurisdiction.
See Figure 5.3.2

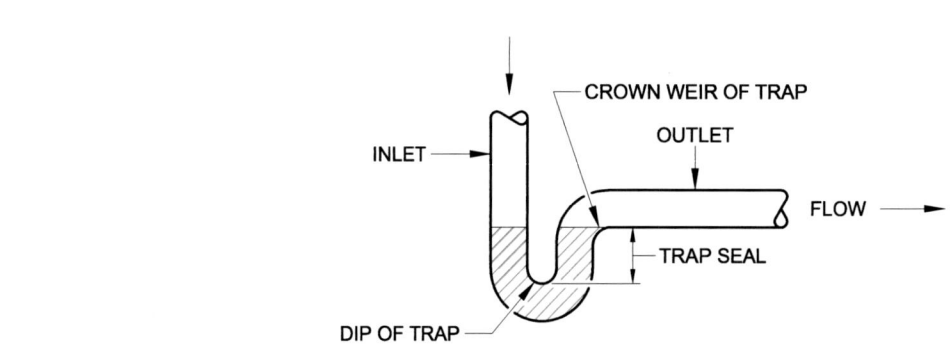

NOTES:
1. The minimum trap seal depth of 2" is based on the design criteria for the vent piping. Trap seals deeper than 4" will tend to trap solids and create a breeding ground for bacteria.

Figure 5.3.2
TRAP SEAL DEPTH

5.3.3 Trap Setting and Protection

Traps shall be set level with respect to their water seals and, where necessary, shall be protected from freezing. *See Figure 5.3.3*

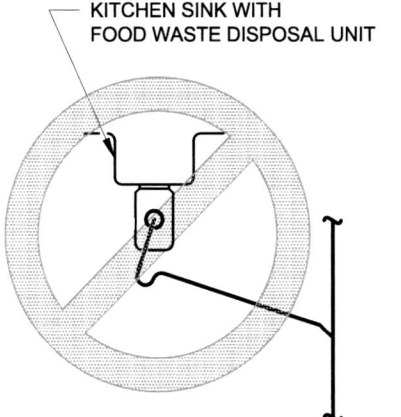
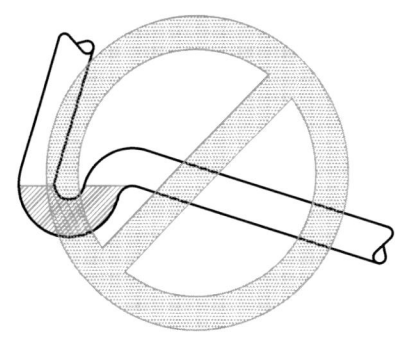

NOTES:
1. Tilted traps may self-siphon because of the reduction in the effective depth of the trap seal.
2. Where freeze protection is necessary, thermal insulation alone will not prevent freezing, only delay it. Traps should be located in heated spaces.
3. Where seasonal facilities are shutdown during the winter, fixture traps can be filled with an anti-freeze solution that will not damage the traps or drain piping.

Figure 5.3.3
TRAPS MUST BE LEVEL AND PROTECTED FROM FREEZING

5.3.4 Building Traps

Building traps shall not be installed except where required by the Authority Having Jurisdiction. Each building trap when installed shall be provided with a cleanout and with a relieving vent or fresh air intake on the inlet side of the trap that shall be at least one-half the diameter of the drain to which it connects. Such relieving vent or fresh air intake shall be carried above grade and terminate in a screened outlet located outside the building.

Comment: Building traps are no longer installed in modern plumbing systems. They provide a collection point for solids and other waste that may cause line stoppages. Building traps also prevent the building vent system from venting the building sewer. The Authority Having Jurisdiction may require building traps if the sewer gas in the municipal sewage system is particularly corrosive or aggressive and could damage the drain, waste, and vent piping in the building.

5.3.5 Prohibited Traps

a. The following types of traps shall be prohibited:
   1. Traps that depend upon moving parts to maintain their seal.
   2. Bell traps.
   3. Crown vented traps.
   4. Separate fixture traps that depend on interior partitions for their seal, except if made from plastic, glass or other corrosion resistant materials.
   5. "S" traps, of uniform internal dimension.
   6. Drum traps, except drum trap solids separators that are permitted in Section 5.3.7.

See Figure 5.3.5

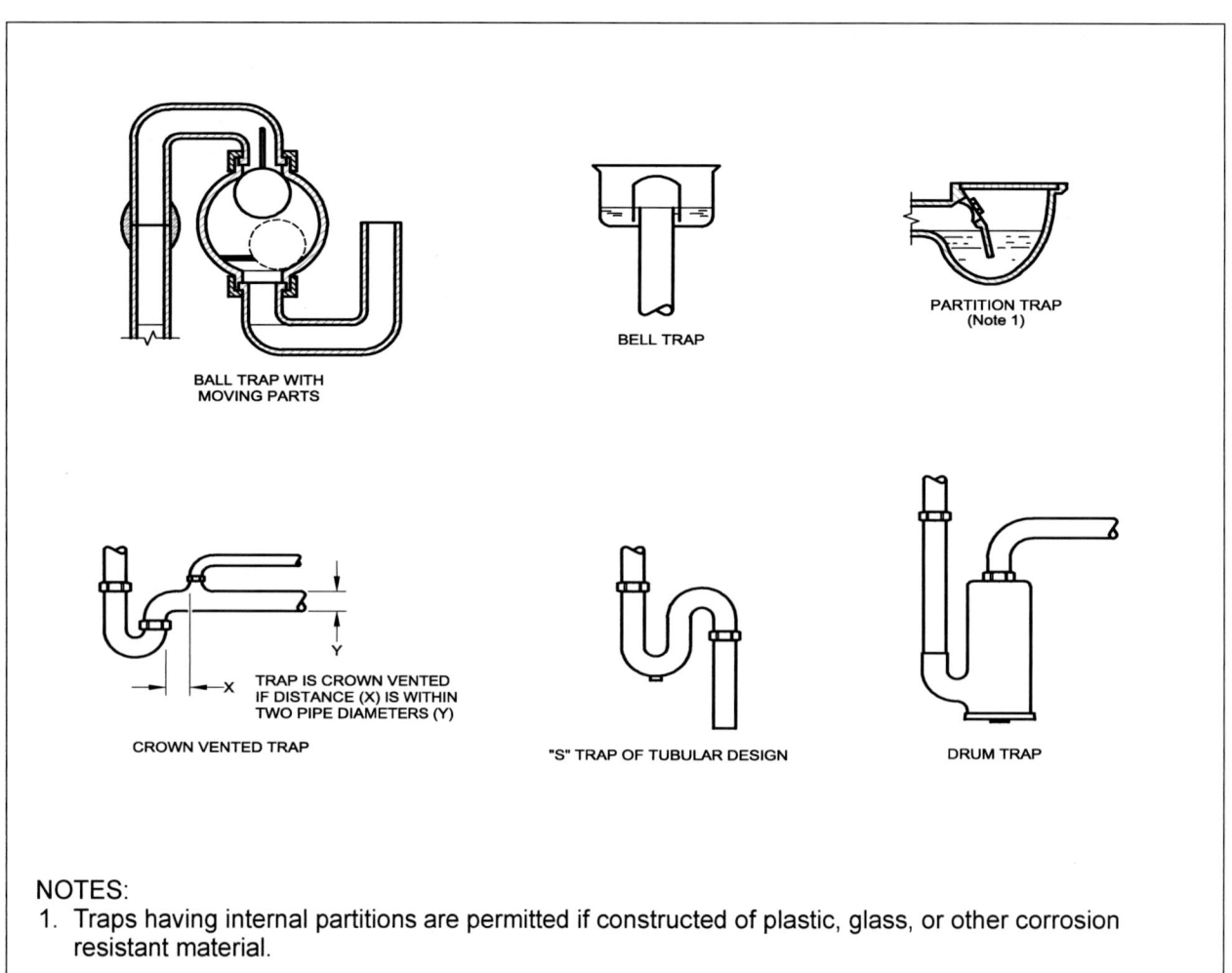

**Figure 5.3.5
PROHIBITED TRAPS**

5.3.6 Trap Seal Maintenance

a. Traps that could lose their seal due to evaporation because of infrequent use shall have accessible means to replenish the trap seal or be connected to a trap primer conforming to ASSE 1018 or ASSE 1044.

b. Floor drains that comply with ASME A112.6.3 shall be permitted to be fitted with a barrier type floor drain trap seal protection device complying with ASSE 1072. The device shall have an application designation marking for the type of floor and environment it has been designed and tested for (AF, AF-GW, SF, CF, CT, WF).

5.3.7 Drum Trap Solids Interceptors

a. Drum trap solids interceptors shall be permitted for lavatories and sinks in beauty salons, dental offices, jewelry stores, hotel and motel guest rooms, dormitories and other areas that require interception of hair, solids, and valuables.

b. Solids interceptors of the drum trap type shall comply with Section 6.5.

c. The depth of trap seals shall comply with Section 5.3.2.

5.4 DRAIN PIPE CLEANOUTS

5.4.1 Cleanout Spacing

a. Cleanouts in horizontal drainage lines shall be spaced at intervals not exceeding the following values:
   4" pipe size or less: 75 feet
   5" size and larger: 100 feet

b. The distances referred to in Section 5.4.1.a shall include the developed length from a cleanout opening to the connection of the next downstream cleanout or cleanout pipe. *See Figure 5.4.1*

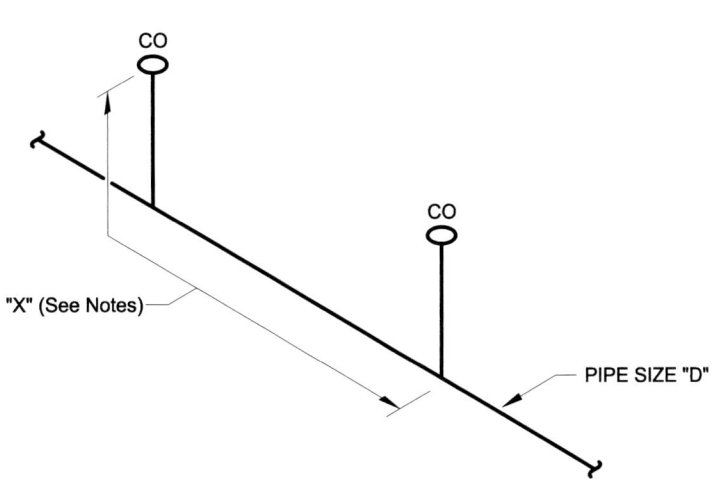

NOTES:
1. The maximum spacing "X" for pipes 4" and smaller is 75 feet, including the developed length from the cleanout opening.
2. The maximum spacing "X" for pipes 5" and larger is 100 feet, including the developed length from the cleanout opening.
3. Manholes may be used for cleanouts in building drains or building sewers (or branches thereof) that are 8" size or larger. The maximum spacing is 300 feet. Refer to Section 5.4.10.

Figure 5.4.1
CLEANOUT SPACING ON HORIZONTAL DRAIN LINES

5.4.2 Building Sewer

Cleanouts, when installed on an underground building sewer, shall be extended vertically to or above the finished grade level. *See Figure 5.4.2*

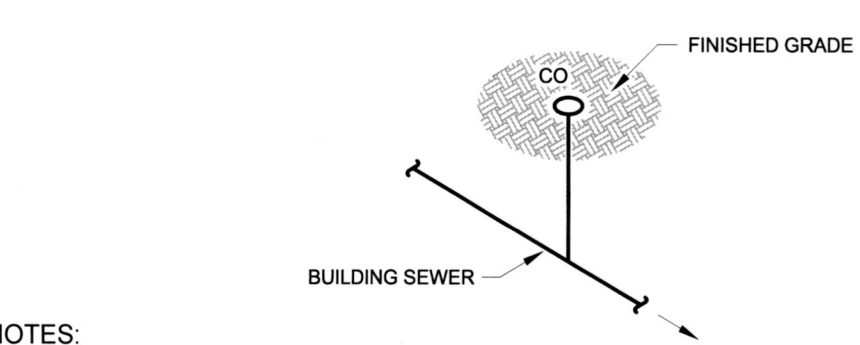

NOTES:
1. Cleanouts must have flush top covers if located in a walkway or other paved area.
2. The length of the vertical extension to grade must be included in the developed length between cleanouts.

Figure 5.4.2
A SEWER CLEANOUT EXTENDED TO FINISHED GRADE

5.4.3 Change of Direction

a. Cleanouts shall be installed at changes of direction in drain piping made with piping made with 60°, 72°, and 90° fittings.
EXCEPTION: Where there are multiple changes of direction in a run of piping, one cleanout shall be permitted for up to forty total feet of offset piping. *See Figure 5.4.3*

b. Cleanouts shall not be required where changes of direction are made with one or more 22-1/2° or 45° fittings.

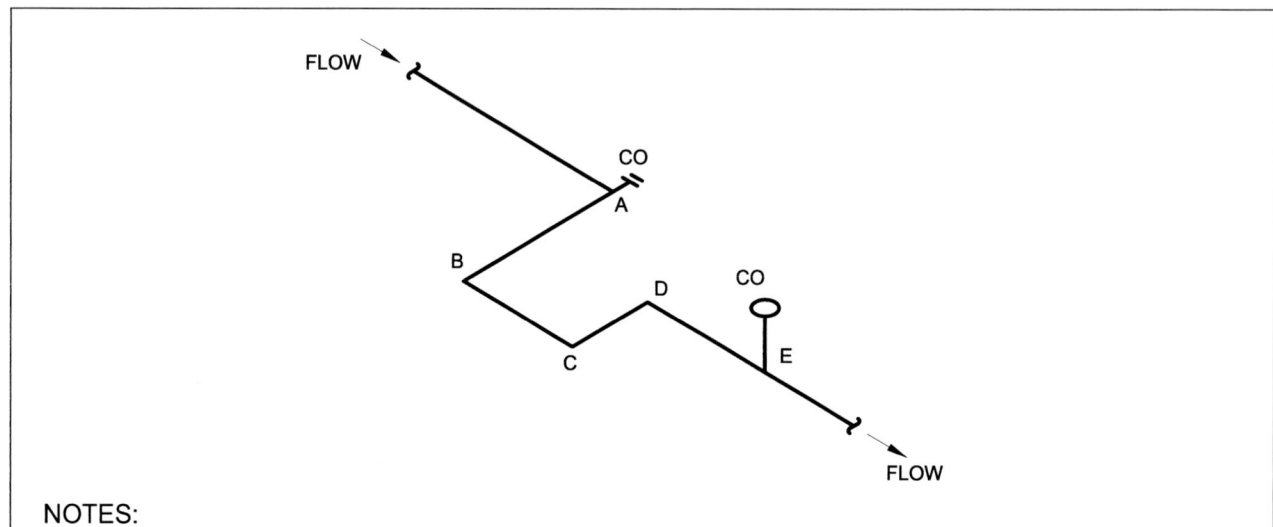

NOTES:
1. A cleanout is not required at Points "B", "C", & "D" if the distance from Point "A" to Point "D" is not more than 40 feet.

Figure 5.4.3
WHERE CLEANOUTS ARE REQUIRED AT CHANGES IN DIRECTION

5.4.4 Cleanouts for Concealed Piping

Cleanouts for concealed piping in finished walls or ceilings shall be accessible through access doors or cover plates which provide clearance for removal of the cleanout plug and insertion of the drain cleaning equipment. *See Figure 5.4.4*

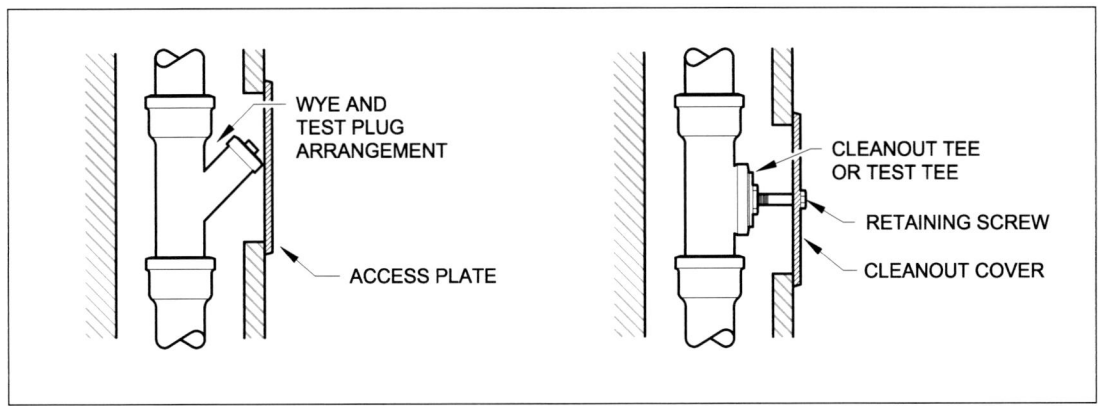

Figure 5.4.4
ACCESS TO CONCEALED CLEANOUTS

5.4.5 Base of Stacks

a. A cleanout shall be provided near the base of each vertical sanitary drain stack and located 6 inches above the flood level rim of the lowest fixture on the lowest floor. If there are no fixtures installed on the lowest floor, the cleanout shall be installed at the base of the stack.

b. For buildings with a floor slab, a crawl space of less than 18 inches clear height, or where a stack cleanout is not accessible from within the building, the cleanout shall be installed not more than five feet outside the building wall.

c. Rain leaders and conductors connected to a building storm drain or storm sewer shall have a cleanout installed at or above the base of the outside leader or inside conductor before it connects to the building drain or sewer.

See Figures 5.4.5-A through 5.4.5-C

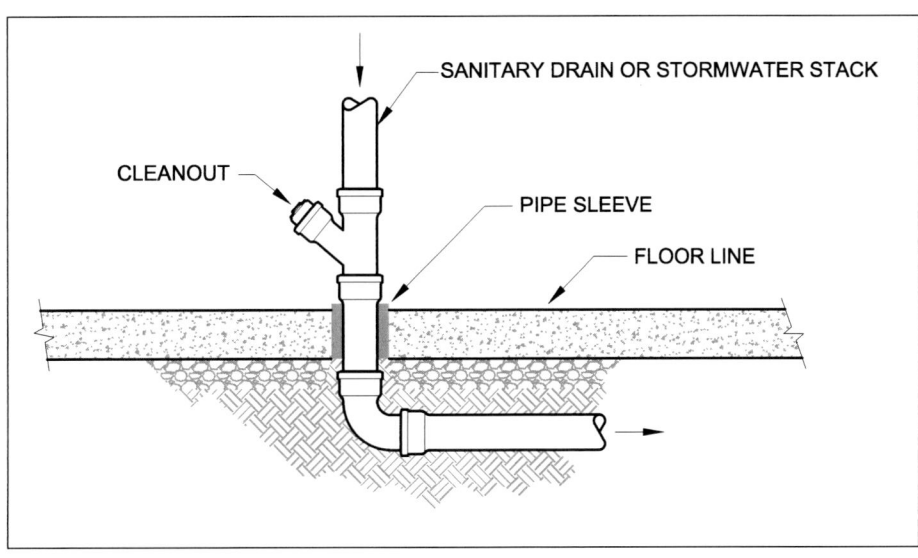

Figure 5.4.5 - A
A CLEANOUT NEAR THE BASE OF A STACK

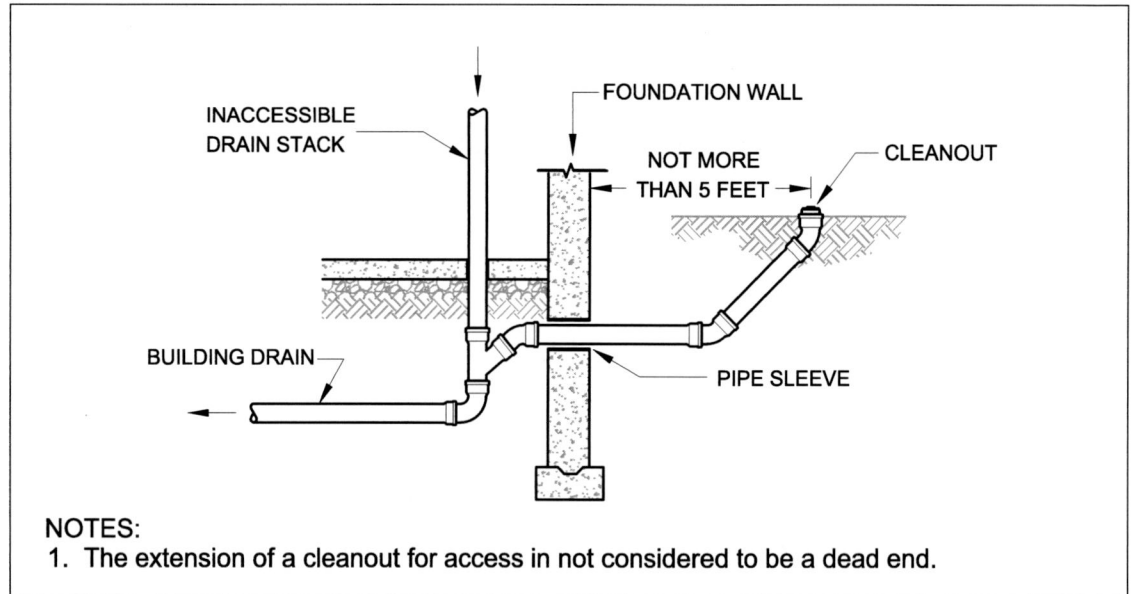

Figure 5.4.5 - B
A CLEANOUT EXTENDED TO OUTSIDE OF THE BUILDING FOR ACCESS

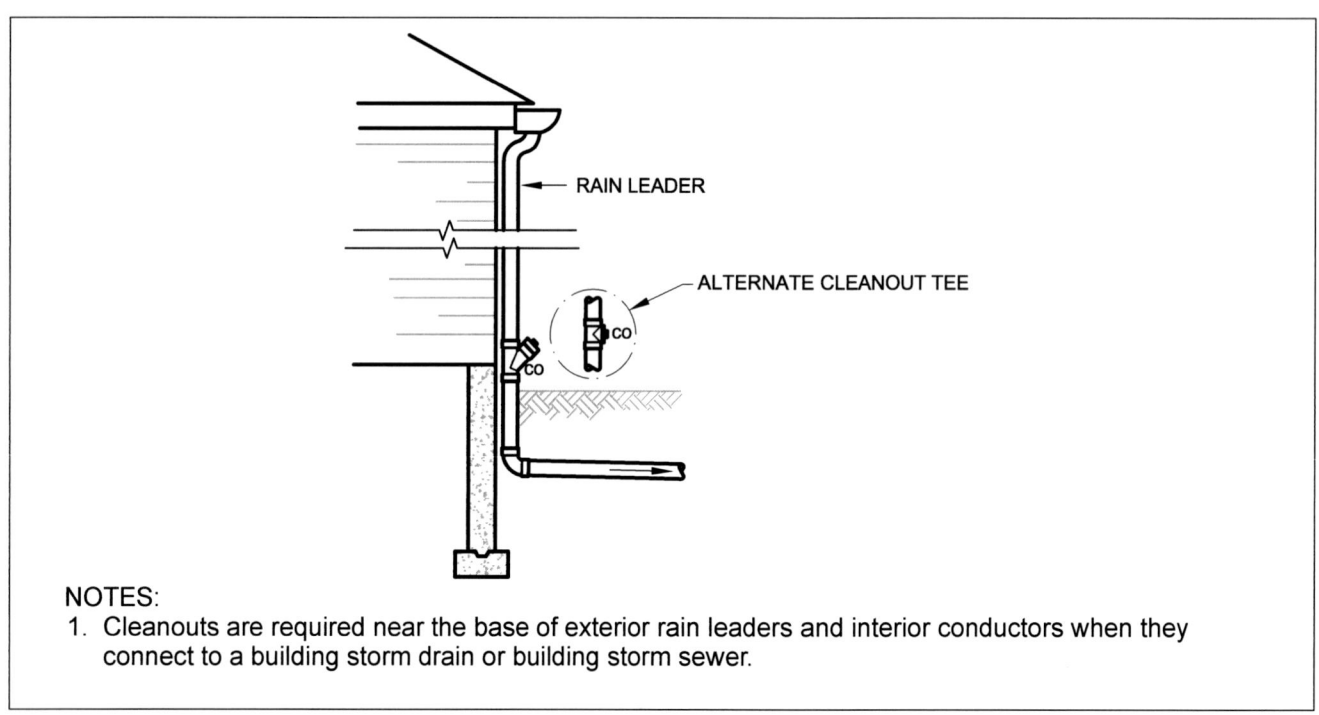

Figure 5.4.5 - C
A CLEANOUT NEAR THE BASE OF A RAIN LEADER

5.4.6 Building Drain and Building Sewer Junctions and the Property Line

a. There shall be a cleanout near the junction of a building drain and building sewer either inside or outside the building wall.

b. Cleanouts shall be placed in the building sanitary sewer and the building storm sewer at the property line and brought to the surface in accordance with the requirements of the Authority Having Jurisdiction. *See Figures 5.4.6-A and 5.4.6-B*

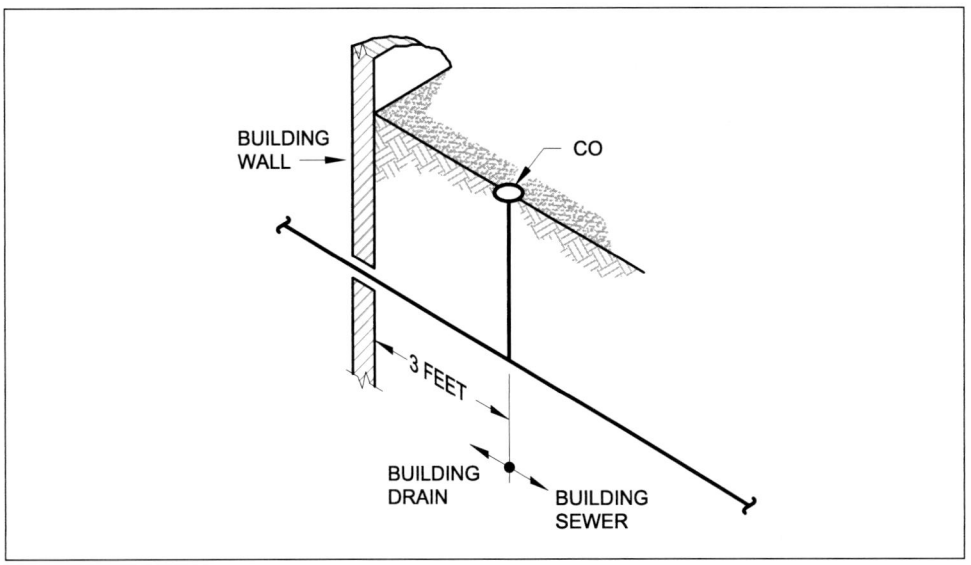

Figure 5.4.6 - A
A CLEANOUT AT THE JUNCTION OF THE BUILDING DRAIN AND BUILDING SEWER

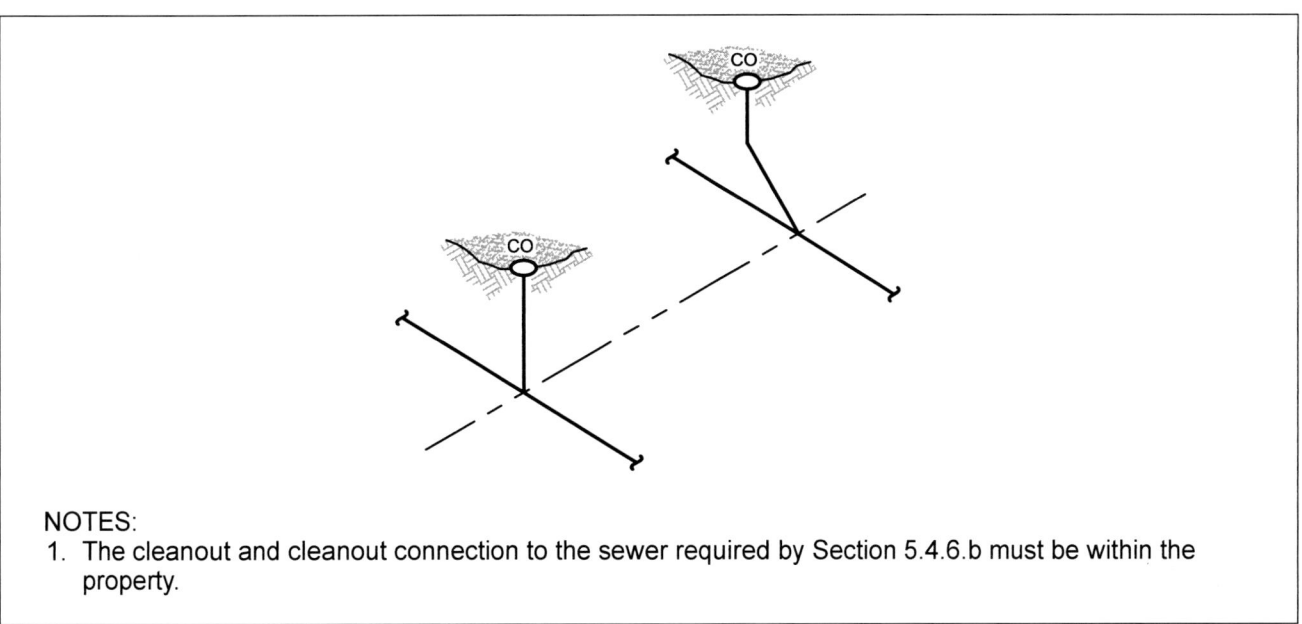

NOTES:
1. The cleanout and cleanout connection to the sewer required by Section 5.4.6.b must be within the property.

Figure 5.4.6 - B
A CLEANOUT AT THE PROPERTY LINE

5.4.7 Direction of Flow

Cleanouts shall be installed so that the cleanout opens in the direction of the flow of the drainage line or at right angles thereto. *See Figure 5.4.7*

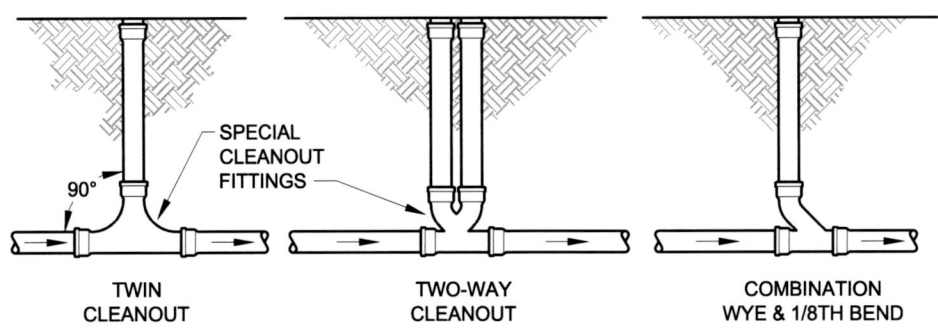

Figure 5.4.7
THE FLOW DIRECTION OF CLEANOUTS

5.4.8 Connections to Cleanouts Prohibited

a. Cleanout plug openings in other than drainage pattern fittings shall not be used for the installation of new fixtures or floor drains.

b. If a cleanout fitting or cleanout plug opening is removed from a drainage pattern fitting in order to extend the drain, another cleanout of equal access and capacity shall be provided in the same location.

5.4.9 Cleanout Size

Cleanout size shall conform with Table 5.4.9.

Table 5.4.9 SIZE OF CLEANOUTS	
Nominal Drain Pipe Size (inches)	**Nominal Size of Cleanout (inches)**
1-1/4	1-1/4
1-1/2	1-1/2
2	2
3	3
4, 5, 6	4
8 & 10	6
12 & 15	8

NOTES FOR TABLE 5.4.9
(1) See Section 5.4.10 for sizes 12" or larger for building sewers.
(2) See Section 5.4.13 for cleanout equivalents.

5.4.10 Manholes for Large Pipes

a. Manholes shall be provided as cleanouts for building sewers 12" size and larger. Manholes shall be provided at every change of size, alignment, direction, grade, or elevation. The distance between manholes shall not exceed 300 feet.

b. Manholes may be provided in lieu of cleanouts in underground building sewers, building drains, and branches thereof, 8" size and larger. Such manholes shall comply with the requirements of Section 5.4.10.a.

c. If manholes are installed indoors, they shall have a bolted, gas-tight cover.

d. Manhole construction shall comply with the standards of the Authority Having Jurisdiction.

See Figure 5.4.10

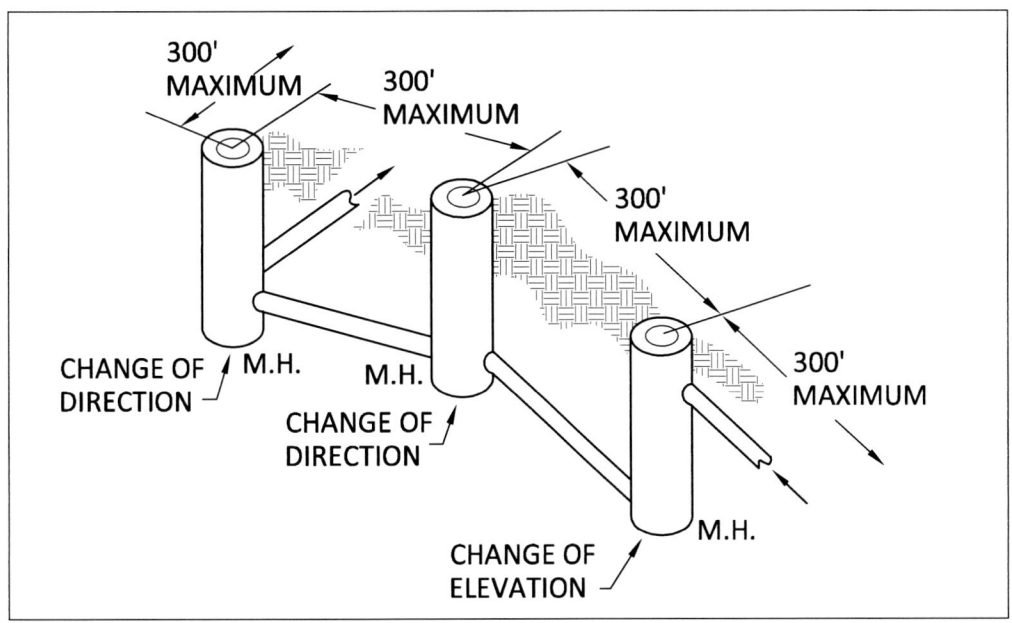

Figure 5.4.10
LOCATION AND SPACING OF MANHOLES

5.4.11 Cleanout Clearances

Cleanouts on 3" or larger pipes shall be so installed that there is a clearance of not less than 18" for the purpose of rodding. Cleanouts smaller than 3 inches shall be so installed that there is a 12" clearance for rodding. *See Figures 5.4.11-A and 5.4.11-B*

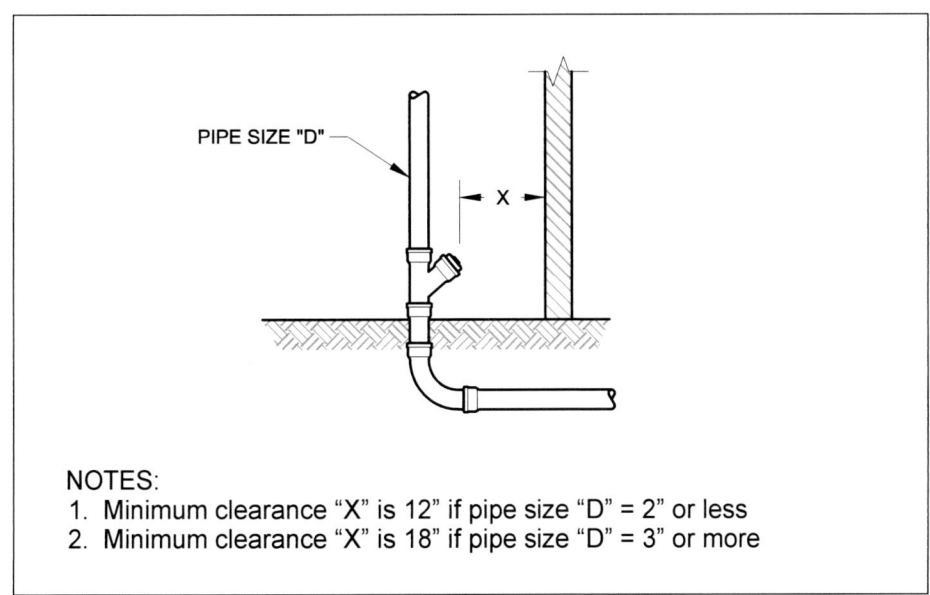

Figure 5.4.11 - A
CLEANOUT CLEARANCE ABOVE THE BASE OF A STACK

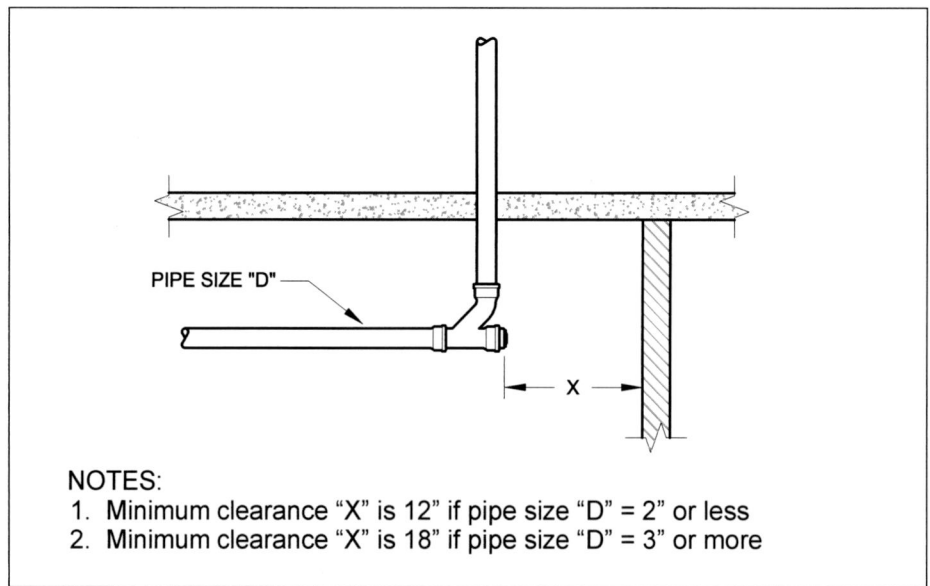

Figure 5.4.11 - B
CLEANOUT CLEARANCE AT THE BASE OF A STACK

5.4.12 Cleanouts to be Kept Uncovered

Cleanout plugs shall not be covered with cement, plaster, or any other permanent finishing material. Where it is necessary to conceal a cleanout plug, a covering plate or access door shall be provided that will permit access to the plug.

5.4.13 Cleanout Equivalent

Where the piping is concealed, a fixture trap or a fixture with integral trap, removable without disturbing concealed roughing work, shall be accepted as a cleanout equivalent, provided the opening to be used as a cleanout opening is the size required by Table 5.4.9.
EXCEPTIONS:
(1) The trap arm of a floor drain with a removable strainer.
(2) Fixtures with removable traps not more than one pipe size smaller than the drain served shall be permitted.
See Figure 5.4.13

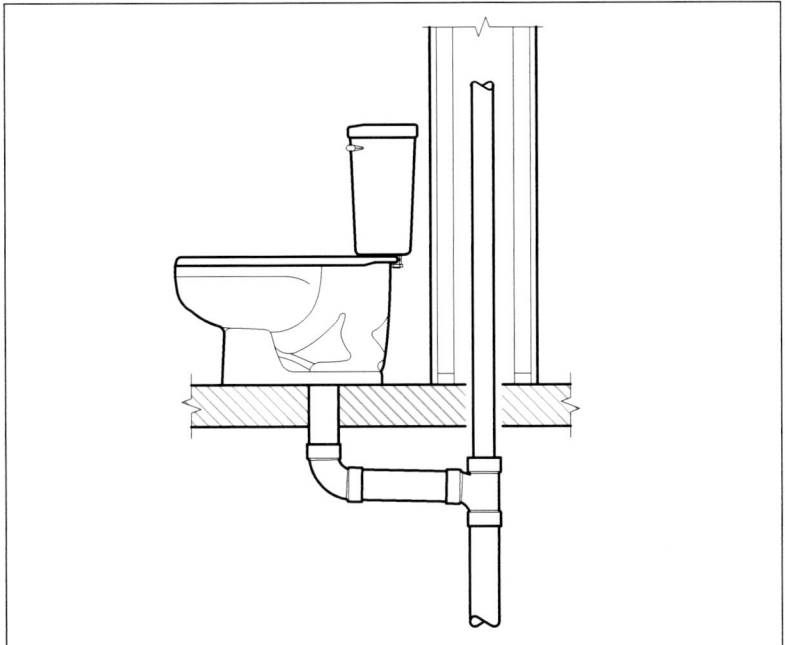

Figure 5.4.13
A WATER CLOSET AS A CLEANOUT EQUIVALENT

5.4.14 Cleanouts for Floor Drains

A cleanout shall be provided immediately downstream from a floor drain whose strainer is not removable.

5.5 BACKWATER VALVES

5.5.1 Where Required

a. Fixtures and/or drain inlets subject to backflow and overflow from blocked or restricted public sewers shall be protected by a backwater valve.

b. Such situations include those where the flood level rim of fixtures and/or drain inlets are below the overflow level of the first upstream manhole in the public sewer that will overflow due to a blockage or flow restriction in the public sewer.

c. Backwater valves shall be installed in drain piping that receives flow only from fixtures and/or drains that are subject to backflow from public sewers.

d. Other portions of the drainage system not subject to such backflow shall not drain through a backwater valve. EXCEPTION: Existing buildings without backwater valves that have fixtures that are below the overflow level of the public sewer shall be permitted to install backwater valves in piping that also serves fixtures that are above the overflow level of the sewer if necessary to protect the fixtures that are below the overflow level.

See Figure 5.5.1.

5.5.2 Material Standard and Accessibility

Backwater valves shall conform to ASME A112.14.1 and be installed so that their internal working parts are accessible for periodic cleaning, repair or replacement.

5.5.3 Notice of the Installation of Backwater Valves

When backwater valves are installed in building sanitary drainage systems, a notice shall be posted at the building water service shutoff valve(s) describing where backwater valves are located.

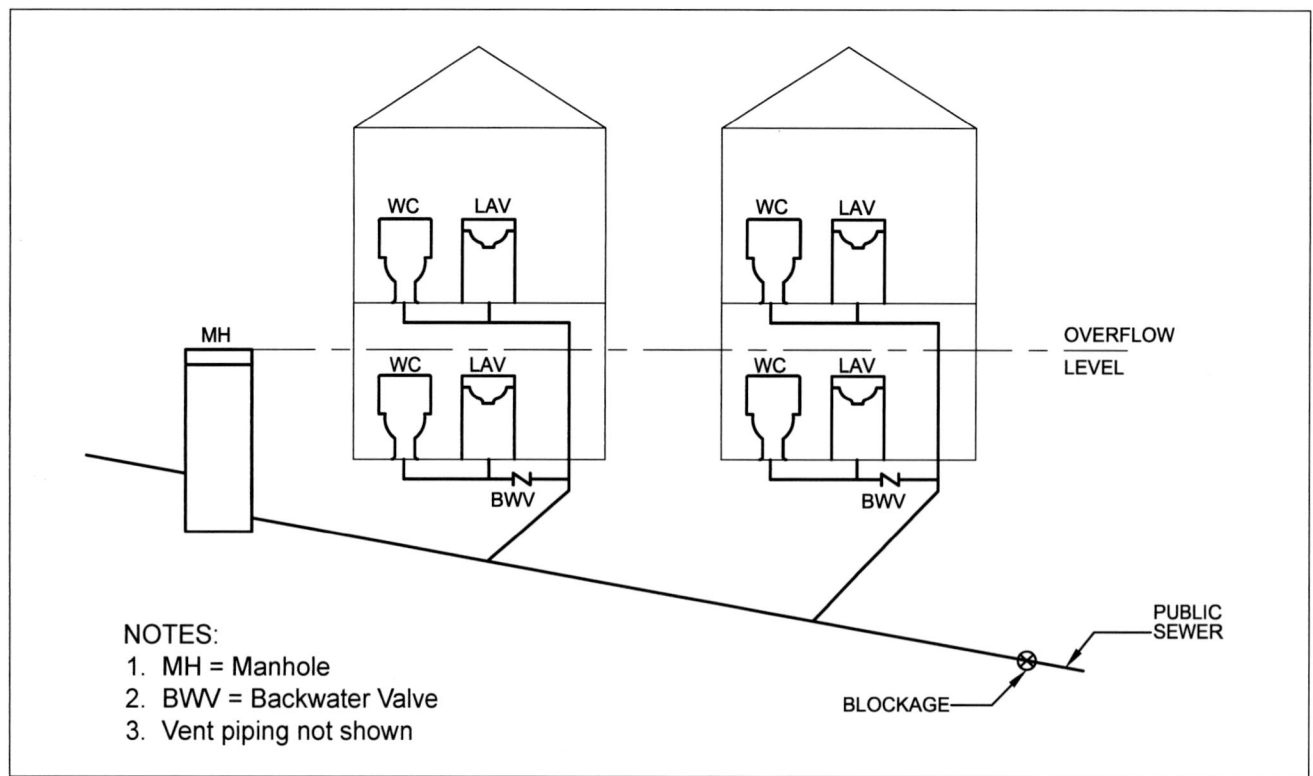

**Figure 5.5.1
BACKWATER VALVES**

Blank Page

Chapter 6

Liquid Waste Treatment Equipment

6.1 GENERAL

6.1.1 Where Required

Interceptors, separators, neutralizers, dilution tanks, or other means shall be provided where required to prevent liquid wastes containing fats, oils, greases, flammable liquids, sand, solids, acid or alkaline waste, chemicals, or other harmful substances from entering a building drainage system, a public or private sewer, or sewage treatment plant or process. ***See Figure 6.1.1***

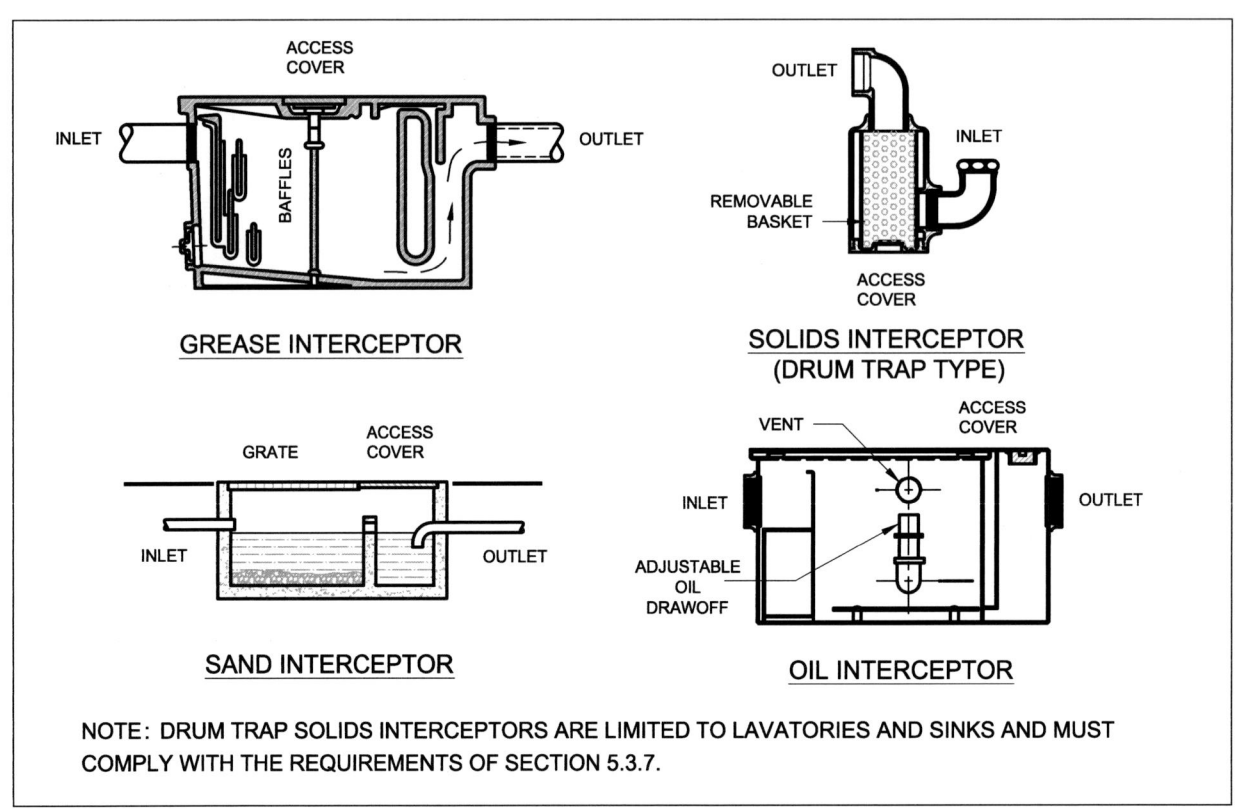

Figure 6.1.1
DIFFERENT TYPES OF INTERCEPTORS

2015 National Standard Plumbing Code - Illustrated

6.1.2 Design

The size and type of liquid waste treatment equipment shall be based on the maximum volume and rate of discharge of the plumbing fixtures and equipment being drained.

6.1.3 Exclusion of Other Liquid Wastes

Only wastes from fixtures and equipment requiring treatment or separation shall be discharged into treatment equipment.
EXCEPTION: Non-grease discharges into grease interceptors that are permitted in Section 6.2.1.e.

6.1.4 Approval

6.1.4.1 General

The type, size, capacity, design, arrangement, construction, and installation of liquid waste treatment devices shall be as required by the Authority Having Jurisdiction.

6.1.4.2 Grease Interceptors and Grease Removal (or Recovery) Devices (GRD)

Grease interceptors rated for up to 100 gallons per minute shall be certified according to PDI Standard G101 or ASME A112.14.3. Grease removal (or recovery) devices rated for up to 100 gallons per minute shall be certified according to ASME A112.14.4.

6.1.4.3 Mechanical Equipment

Each installation of a manufactured liquid waste treatment device employing pumps, filters, drums, collection plates, or other mechanical means of operation shall be certified by the manufacturer to provide effluent meeting the environmental requirements of the sewer or other approved point to which it discharges.

6.1.5 Venting

Liquid waste treatment equipment shall be so designed that they will not become air-bound if tight covers are used. Equipment shall be properly vented if loss of its trap seal is possible.

6.1.6 Accessibility

    a. Liquid waste treatment equipment shall be so installed that it is accessible for the removal of covers and the performance of necessary cleaning, servicing and maintenance.

    b. The need to use ladders or move bulky objects in order to service interceptors and other liquid waste treatment equipment shall constitute a violation of accessibility.

6.1.7 Point of Discharge

Connections to sewers or other points of discharge for the effluent from liquid waste treatment equipment shall be as permitted by the Authority Having Jurisdiction.

6.2 GREASE INTERCEPTORS

6.2.1 General

    a. Grease interceptors shall comply with the requirements of the Adopting Agency.
    b. Grease interceptors include the following types:
        1. Hydro-mechanical interceptors
        2. Grease removal (or recovery) devices (GRD)
        3. Gravity interceptors
        4. FOG (fats, oils, and greases) disposal systems

c. Grease interceptors shall be provided to receive the waste discharges from fixtures in food handling areas that introduce grease into the sanitary drainage system. Fixtures include, but are not limited to pot washing sinks, utensil soak sinks, pre-rinse sinks at dishwashers, wok range stations, drains from washdown ventilation hoods, can washing drains, mop sinks, floor drains and floor sinks in areas around grease producing fixtures, and similar fixtures.

d. Water closets, urinals, and other fixtures that discharge human waste shall not discharge through a grease interceptor.

e. Non-grease drainage from handwashing sinks, lavatories, food preparation sinks, ice machines, ice bins, and similar fixtures shall be permitted to discharge through a grease interceptor along with grease producing fixtures.

f. Ice machines, ice bins, food preparation sinks, and other food containing equipment shall drain indirectly through an air gap.

6.2.1.1 Hydro-Mechanical Grease Interceptors

a. Hydro-mechanical interceptors shall comply with the performance, testing, and installation requirements of ASME A112.14.3 or PDI Standard G101 and Section 6.2.

b. These interceptors shall be sized according to Section 6.2.10.

c. A calibrated, non-adjustable flow control device shall be provided on the inlet side of each interceptor to prevent the waste flow (gpm) from exceeding the rated flow capacity of the interceptor. The flow control device shall be vented in accordance with Section 6.2.4.

6.2.1.2 Grease Removal (or Recovery) Devices (GRD)

a. Hydro-mechanical interceptors that are capable of automatically removing free-floating grease, fats, and oils from their waste discharge without intervention of the user except for maintenance shall comply with ASME A112.14.4 and Section 6.2.

b. These interceptors shall be sized according to Section 6.2.10.

c. A calibrated, non-adjustable flow control device shall be provided on the inlet side of each interceptor to prevent the waste flow (gpm) from exceeding the rated flow capacity of the interceptor. The flow control device shall be vented in accordance with Section 6.2.4.

6.2.1.3 Gravity Grease Interceptors

a. Outdoor underground gravity grease interceptors serving commercial kitchens shall be sized and designed by a registered design professional who is licensed to practice in the particular jurisdiction.

b. Gravity grease interceptors shall comply with the requirements of Section 6.2.10.b or the Adopting Agency, including materials of construction, arrangement, retention time, storage factor for floatable FOG and settled solids, and minimum size.

c. Where drain piping and a gravity grease interceptor are provided for the future installation of a commercial kitchen, the design plans shall indicate the maximum permitted future drainage load in either gallons per minute (excluding diversity) or drainage fixture units (DFU).

d. Prefabricated gravity grease interceptors shall comply with IAPMO/ANSI Z1001.

6.2.1.4 FOG (Fats, Oils, and Greases) Disposal Systems

a. FOG disposal systems shall be designed to:
   1. remove FOG from the effluent,
   2. retain the separated FOG,
   3. internally dispose of the retained FOG by means of mass and volume reduction by thermal, chemical, electrical, biological or other approved processes.

b. FOG disposal systems shall comply with ASME A112.14.6 and be installed according to the manufacturer's instructions.

6.2.2 Compliance for Hydro-Mechanical Grease Interceptors

a. Hydro-mechanical grease interceptors shall comply with ASME A112.14.3 and Section 6.2.1.1, and be installed in accordance with the recommendations of PDI Standard G101 and the manufacturer's instructions. They shall have a grease retention capacity not less than two pounds for each gpm of rated flow.

b. Grease interceptors that include automatic grease removal or recovery (GRD) shall comply with ASME A112.14.4 and Section 6.2.1.2.

6.2.3 Fixture Traps

a. Fixtures that discharge into a hydro-mechanical or GRD grease interceptor shall be trapped and vented between the fixture and the interceptor.
EXCEPTION: A hydro-mechanical or GRD grease interceptor with the required flow control device shall be permitted to serve as a trap for an individual fixture if the developed length of the drain between the fixture and the interceptor does not exceed four feet horizontally and 30 inches vertically.

b. Where one or more fixtures discharge into a hydro-mechanical or GRD grease interceptor, the required vented flow control device shall be installed in the drain line between the fixture(s) and the interceptor.
See Figures 6.2-A and 6.2-B

6.2.4 Fixture Venting

a. Trapped fixtures draining to grease interceptors shall be vented in accordance with the manufacturer's instructions and the applicable provisions of Chapter 12, including combination waste and vent venting.

b. Where a grease interceptor is permitted to serve as the trap for a single fixture that does not have a trap, the air intake for the flow control device between the fixture and the grease interceptor shall be vented by a return bend that is open to the space at least 6 inches above the flood level rim of the fixture being served.

c. When the fixture connected to the grease interceptor has a separate vented trap, the air intake on the flow control device shall be connected to the vent piping system.
See Figures 6.2-A and 6.2-B

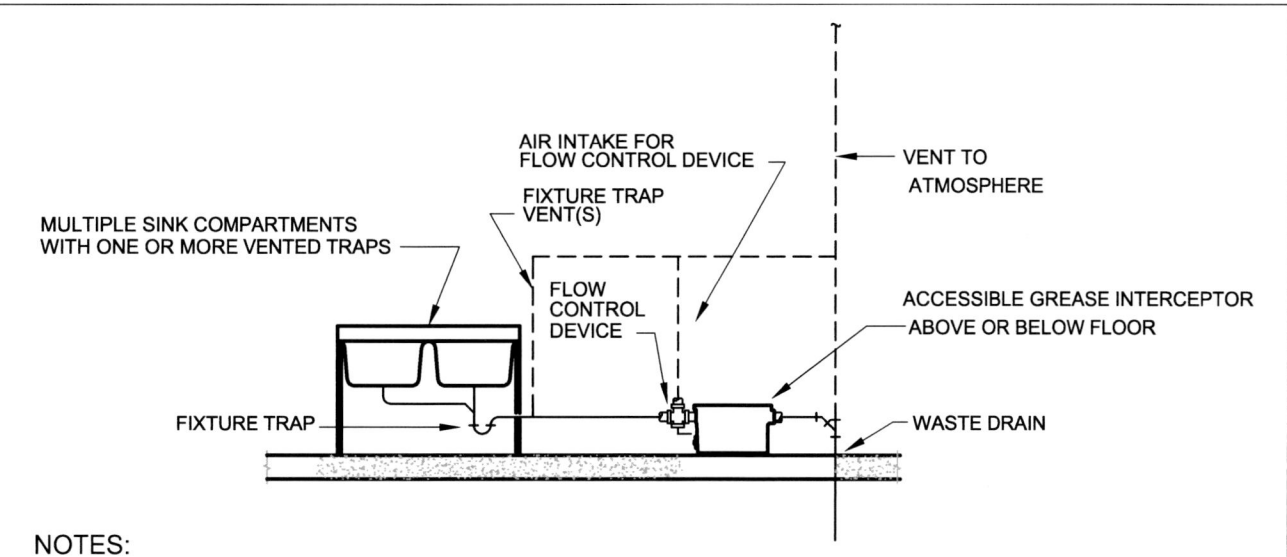

NOTES:
1. Flow control devices are necessary to prevent the waste flow from exceeding the design flow rating of the grease interceptor.
2. When the fixtures connected to the grease interceptor have traps, the air intake on the flow control device must be connected to the fixture vent piping system.
3. Air intake through the flow control device aerates the grease-laden waste, which is essential to the separation process.

**Figure 6.2-A
A GREASE INTERCEPTOR SERVING A TRAPPED AND VENTED FIXTURE**

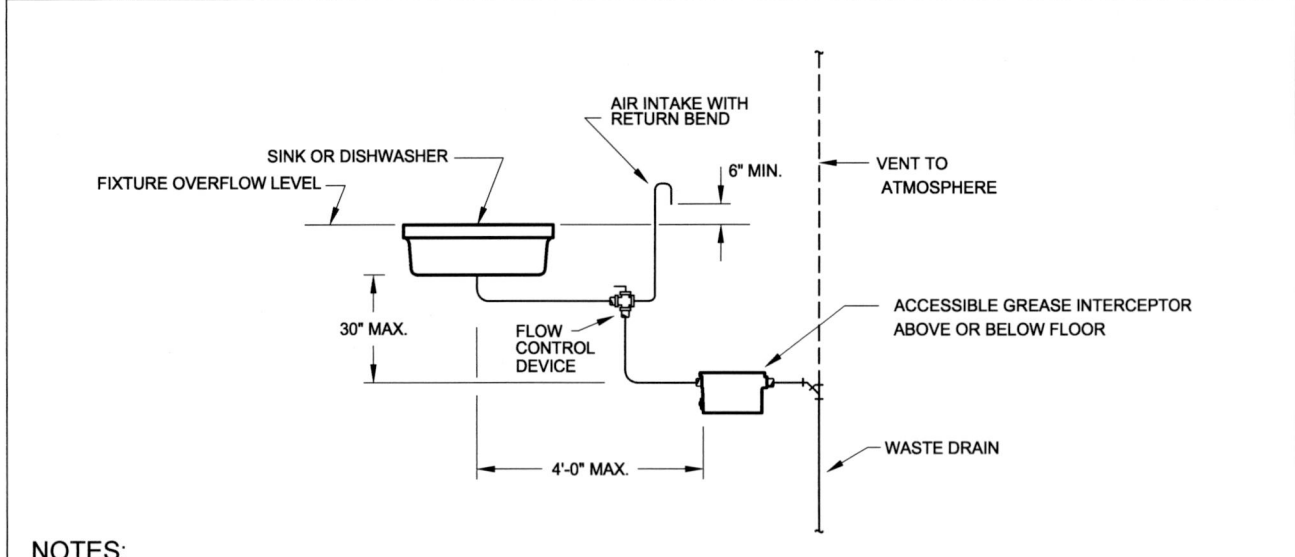

NOTES:
1. When a fixture that is connected to a grease interceptor is within 4 feet horizontally and 30 inches vertically from the inlet to the interceptor, a fixture trap is not required. If the fixture is not trapped, the air intake for the flow control device must not be connected to the vent piping system. It must be terminated at a 180 degree return bend with its inlet at least 6 inches above the flood level of the fixture.

Figure 6.2-B
A GREASE INTERCEPTOR SERVING AS A FIXTURE TRAP

6.2.5 Food Waste Disposers

a. Where food waste disposers discharge through a hydro-mechanical or GRD grease interceptor, a solids separator shall be installed either in the drain line from the food waste disposer or upstream of the grease interceptor to prevent food waste particles from entering the grease interceptor.

b. Solids separators shall not be required where food waste disposers discharge to a gravity grease interceptor.
See Figure 6.2.5

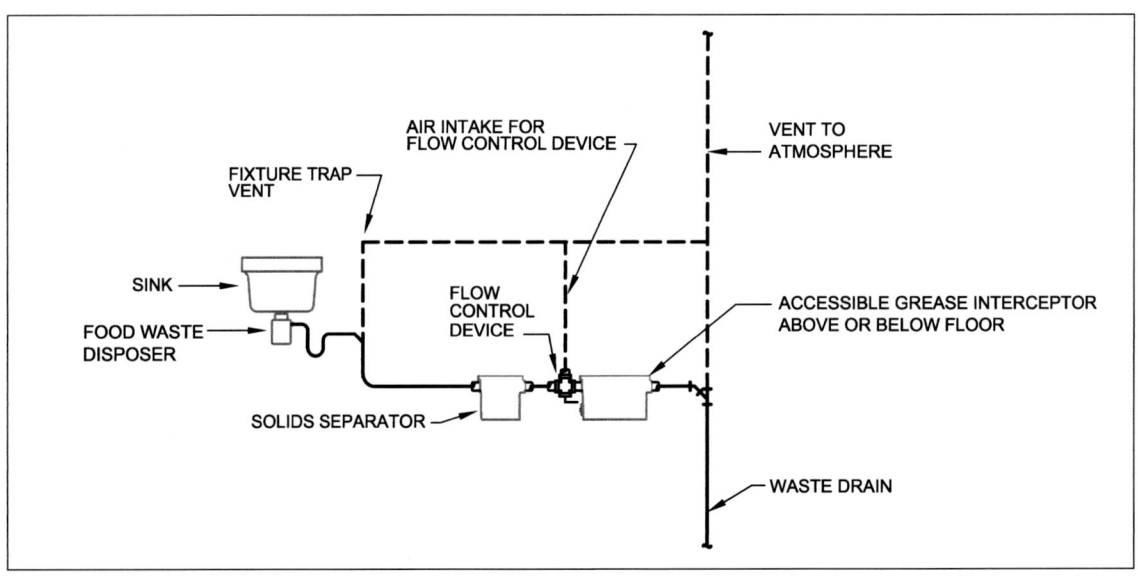

Figure 6.2.5
A FOOD WASTE DISPOSER WITH SOLIDS SEPARATOR AND GREASE INTERCEPTOR

6.2.6 Commercial Dishwashers

a. Commercial dishwashers shall be permitted to discharge through a grease interceptor.

b. Where the discharge rate of a commercial dishwasher in gallons per minute is converted to drainage fixture units (DFU), each 7.5 GPM of discharge shall be equated to one (1) DFU, with the total rounded up to the next whole DFU.

6.2.7 Location

Hydro-mechanical and GRD interceptors may be installed within buildings if permitted by the Authority Having Jurisdiction. Where gravity grease interceptors or holding tanks are remote from the fixtures served, the drain piping between the fixtures and the interceptor or holding tank shall be as direct as possible and shall include provisions for periodic cleaning.

6.2.8 Prohibited Interceptors

The installation of water-cooled grease interceptors is prohibited.

6.2.9 Chemicals - Where Prohibited

a. The use of enzymes, emulsifiers, or similar chemicals in hydro-mechanical grease interceptors, GRD grease interceptors, and gravity interceptors is prohibited.

b. Sinks or sink compartments used for sanitizing pots or other ware shall not be drained through a grease interceptor.

6.2.10 Interceptor Sizing

a. Where hydro-mechanical interceptors and grease removal devices (GRD) serve one or more individual fixtures, they shall be sized for the total drainage flow rate from the fixtures served in accordance with PDI G101, Table 6.2.10, or the manufacturer's instructions. All compartments of multi-compartment sinks shall be considered to drain simultaneously, except that sanitizing compartments shall not be drained through a grease interceptor.

b. Gravity interceptors for commercial kitchens shall be sized based on the inlet pipe flowing half-full according to Appendix K, a 30-minute retention time, and an additional 25% storage factor for floatable FOG and settled solids, or as required by the Adopting Agency.

Table 6.2.10 FIXTURE DRAINAGE FLOW RATES FOR SIZING HYDRO-MECHANICAL AND GRD GREASE INTERCEPTORS	
FIXTURE	**FLOW**
1-1/4" Sink Drain Outlet (each)	7.5 GPM
1-1/2" Sink Drain Outlet (each)	15 GPM
2" Sink Drain Outlet (each)	22.5 GPM
Floor Drain without Indirect Waste	0 GPM
Floor Drain or Floor Sink with Indirect Waste	(1)
Commercial Dishwasher	(2)

NOTES FOR TABLE 6.2.10

(1) The GPM drain load shall be the total indirect waste flow in GPM.

(2) The GPM drain load for a commercial dishwasher shall be not less than the manufacturer's peak rate of drain flow with a full tank.

6.2.11 Individual Dwelling Units

Grease interceptors shall not be required in individual dwelling units or any private living quarters.

6.2.12 Combination Systems

A combination of hydro-mechanical and exterior gravity grease interceptors shall be allowed in order to meet separation needs of the Adopting Agency when space or existing physical constraints of existing buildings necessitates such installations.

6.3 OIL/WATER SEPARATORS

6.3.1 Where Required and Approved Point of Discharge

    a. Liquid waste containing grease, oil, solvents, or flammable liquids shall not be directly discharged into any sanitary sewer, storm sewer, or other point of disposal. Such contaminants shall be removed by an appropriate separator.

    b. Sand interceptors and oil separators shall be provided wherever floors, pits or surface areas subject to accumulation of grease or oil from service or repair operations are drained or washed into a drainage system. Such locations include, but are not limited to, car or truck washing facilities, engine cleaning facilities, and similar operations. The drainage or effluent from the interceptors and separators in such locations shall be connected to the sanitary sewer.

    c. Drains shall not be required in service or repair garages that employ dry absorbent cleaning methods; however, if any drains are located in such areas, they shall discharge to the sanitary sewer through a sand interceptor and oil interceptor.

    d. Drains shall not be required in parking garages unless the garage, or portions thereof, has provisions for either washing vehicles or rinsing the floor. Where such cleaning facilities are provided, the area subject to waste drainage shall be provided with one or more floor drains, complete with sand interceptor and oil interceptor, and the effluent from the oil separator shall be connected to the sanitary sewer. Any storm water shall be drained separately and directly to the storm sewer.

    e. Where parking garages without provisions for vehicle washing or floor rinsing require storm water drainage, drains shall be permitted to connect to the storm sewer without a sand interceptor and oil separator. Such drainage, including rainwater, melting snow and ice, or rainwater runoff from vehicles, shall not be connected to the sanitary sewer.

    f. Where oil separators include a waste holding tank, it shall not be used to store or contain any other waste oil (e.g., motor oil) or hazardous fluid.

6.3.2 Design of Oil Separators

    a. Where oil separators are required in garages and service stations, they shall have a minimum volume of six cubic feet for the first 100 square feet of area drained, plus one cubic foot for each additional 100 square feet of area drained. Oil separators in other applications shall be sized according to the manufacturers rated flow.

    b. Field-fabricated oil separators shall have a depth of not less than two feet below the invert of the discharge outlet. The outlet opening shall have a water seal depth of not less than 18 inches.

    c. Manufactured oil separators shall be sized according to gallons per minute of rated flow. They shall include a flow control device and adjustable oil draw-off.

    d. Oil separators shall have a 3-inch minimum discharge line and a 2-inch minimum vent to atmosphere. The discharge line shall have a full-size cleanout extended to grade or otherwise be accessible.

    e. The oil draw-off or overflow piping from oil separators shall be connected to an approved waste oil tank that is installed and permitted according to the environmental requirements of the Authority Having Jurisdiction. The waste oil from the separator shall flow by gravity or may be pumped to a higher elevation by an automatic pump. Pumps shall be adequately sized, explosion-proof, and accessible. Waste oil tanks shall have a 2-inch minimum pump-out connection and a 1-1/2 inch minimum vent to atmosphere.

f. Where oil separators are subject to backflow from a sewer or other point of disposal, their discharge line shall include a backwater valve, installed in accordance with Section 5.5.

g. Where oil separators are installed in parking garages and other areas where the waste flow will include sand, dirt or similar solids, a sand interceptor shall be provided upstream from the oil separator. Sand interceptors shall comply with Section 6.4.

h. Oil interceptors, waste oil tanks, oil pump-out connections, backwater valves and atmospheric vent piping shall be permanently identified by suitable labels or markings.

6.3.3 Vapor Venting

The atmospheric vents from oil separators and their waste holding tanks shall be separate from other plumbing system vents and shall be extended to an approved location at least 12 feet above grade or the surrounding area.

6.3.4 Combination Oil Separator and Sand Separator

A combination oil separator and sand separator meeting the functional requirements of Sections 6.3 and 6.4 shall be permitted to be installed.

6.4 SAND INTERCEPTORS

6.4.1 Where Required

a. A sand interceptor shall be installed upstream from an oil separator if required in Section 6.3.2.g.

b. A sand interceptor shall be provided downstream from any drain whose discharge may contain sand, sediment, or similar matter on a continuing basis that would tend to settle and obstruct the piping in the drainage system. Multiple floor drains shall be permitted to discharge through one sand interceptor.

6.4.2 Construction and Size

a. Sand interceptors shall be constructed of concrete, brick, fabricated coated steel, or other watertight material, and shall be internally baffled to provide an inlet section for the accumulation of sediment and a separate outlet section.

b. The outlet pipe of a sand interceptor shall be the same size as the drain served.
EXCEPTION: If serving an oil separator, the outlet from the sand interceptor shall be the same size as the inlet to the oil separator.

c. The inlet baffle shall have two top skimming openings, each the same size as the outlet pipe and at the same invert as the outlet opening. The openings in the baffle shall be offset to prevent straight-line flow through the interceptor from any of its inlets to its outlet.

d. The inlet to the interceptor shall be at the same elevation as, or higher than, the outlet. The bottom of the inlet section shall be at least 24 inches below the invert of the outlet pipe.

e. The bottom of the inlet section shall be at least two feet wide and two feet long for flow rates up to 20 gallons per minute. The bottom of the inlet section shall be increased by one square foot for each 5 gpm of flow or fraction thereof over 20 gpm. The bottom of the outlet section shall be not less than 50% of the area of the bottom of the inlet section.

f. A solid removable cover shall cover the outlet section. An open grating suitable for the traffic in the area in which it is located shall cover the inlet section. Covers shall be set flush with the finished floor.

6.4.3 Water Seal

When a sand interceptor is used separately without also discharging through an oil separator, its outlet pipe shall be turned down inside the separator below the water level to provide a six-inch minimum water seal. A cleanout shall be provided for the outlet line.

6.4.4 Alternate Design

Alternate designs for construction of, or baffling in, sand interceptors shall comply with the intent of this Code and be submitted to the Authority Having Jurisdiction for approval.

6.5 SOLIDS INTERCEPTORS

a. Solids interceptors shall be provided where necessary to prevent harmful solid materials from entering the drainage system on a continuing basis. Such harmful materials include, but are not limited to, aquarium gravel, barium, ceramic chips, clay, cotton, denture grindings, dental silver, fish scales, gauze, glass particles, hair, jewels, lint, metal grindings, plaster, plastic grindings, precious metal chips, sediment, small stones, and solid food particles.

b. Solids interceptors shall separate solids by gravity, trapping them in a removable bucket or strainer.

c. Solids interceptors shall be sized according to their drain pipe size or by the required flow rate.

d. Drum trap solids separators shall comply with Section 5.3.7.

6.6 NEUTRALIZING AND DILUTION TANKS

a. Neutralizing or dilution tanks shall be provided where necessary to prevent acidic or alkaline waste from entering the building drainage system. Such waste shall be neutralized or diluted to levels that are safe for the piping in the drainage and sewer systems.

b. Vents for neutralizing or dilution tanks shall be constructed of acid-resistant piping and shall be independent from sanitary system vents.

6.7 SPECIAL APPLICATIONS

6.7.1 Laundries

Commercial laundries shall be equipped with one or more lint interceptors having wire baskets or similar devices, removable for cleaning, that will prevent passage into the drainage system of solids 1/2 inch or larger in size, strings, rags, buttons, lint, and other materials that would be detrimental to the drainage system.

6.7.2 Bottling Establishments

Bottling plants shall discharge their process wastes into a solids separator that will retain broken glass and other solids, before discharging liquid wastes into the drainage system.

6.7.3 Slaughter Houses

Drains in slaughtering rooms and dressing rooms shall be equipped with separators or interceptors, approved by the Authority Having Jurisdiction, that will prevent the discharge into the drainage system of feathers, entrails, and other waste materials that are likely to clog the drainage system.

6.7.4 Barber Shops and Beauty Parlors

Shampoo sinks in barbershops, beauty parlors, and other grooming facilities shall have hair interceptors installed in lieu of regular traps.

Chapter 7

Plumbing Fixtures, Fixture Fittings and Plumbing Appliances

7.1 FIXTURE STANDARDS

Plumbing fixtures, plumbing fixture trim, and plumbing appliances shall comply with the standards listed in Table 3.1.3. Plumbing supply fittings covered under the scope of NSF 61 shall comply with the requirements of NSF 61.

7.2 FIXTURES FOR ACCESSIBLE USE

a. Plumbing fixtures for accessible use and their installation shall conform to the local accessibility code or regulations.

> *Comment: Requirements for accessible plumbing fixtures are in ADA Standards for Accessible Design and ICC ANSI A117.1. Refer to local codes and regulations.*

b. Exposed waste and water supply piping for accessible sinks and lavatories shall be covered with protectors or insulators that comply with ASME A112.18.9.

c. The support and drain systems for adjustable sinks that facilitate accessible use shall comply with ASME A112.19.12.

7.3 INSTALLATION

7.3.1 General

Plumbing fixtures, fixture trim, and plumbing appliances shall be installed in accordance with the requirements of this Code and the manufacturer's instructions and recommendations.

7.3.2 Minimum Clearances

For other than accessible applications, minimum clearances between plumbing fixtures and from fixtures to adjacent walls shall be in accordance with Figure 7.3.2.

7.3.3 Access for Cleaning

Plumbing fixtures shall be so installed as to provide access for cleaning the fixture and the surrounding area.

7.3.4 Securing Floor-Mounted Fixtures

Floor-mounted fixtures shall be securely supported by the floor or floor/wall structure. No strain shall be transmitted to the connecting piping. Fastening screws or bolts shall be corrosion-resisting.

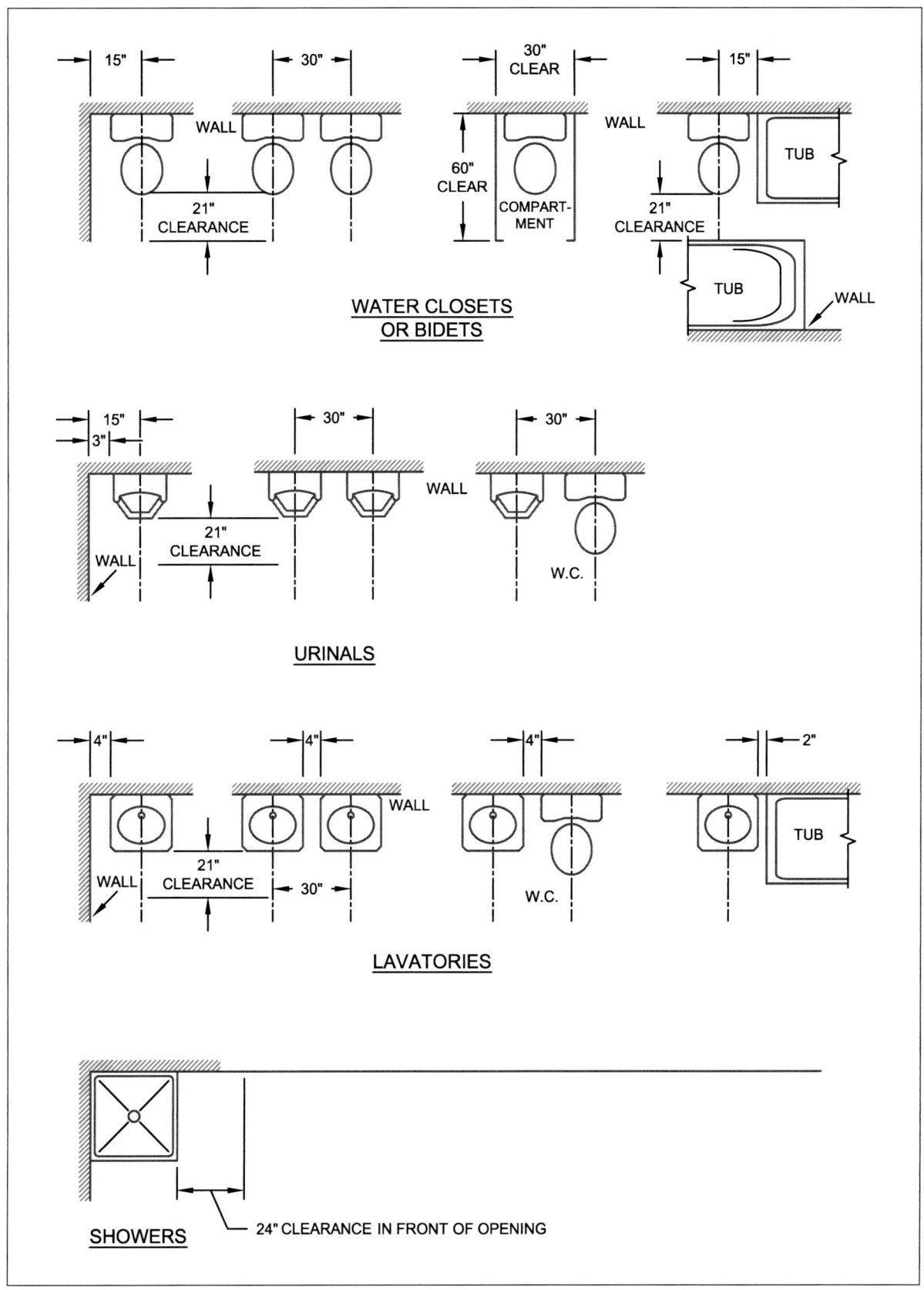

Figure 7.3.2
MINIMUM FIXTURE CLEARANCES
(for other than accessible applications)

7.3.5 Supporting Wall-Hung Fixtures

a. Wall-hung water closets shall be supported by concealed metal carriers that transmit the entire weight of the fixture to the floor and place no strain on the wall or connecting piping. Supports of this design shall comply with ASME A112.6.1M.

b. Supports for off-the-floor water closets with concealed tanks shall comply with ASME A112.6.2.

c. Lavatories, urinals, and other wall-hung fixtures shall be supported by a concealed or exposed finished wall hanger plate or equivalent that transmits the weight of the fixture to the wall structure, if adequate, or to the floor without placing strain on the piping. Such supports shall comply with ASME A112.6.1M. In addition to the wall support brackets, pedestals or legs may provide additional support for pedestal lavatories. *See Figure 7.3.5*

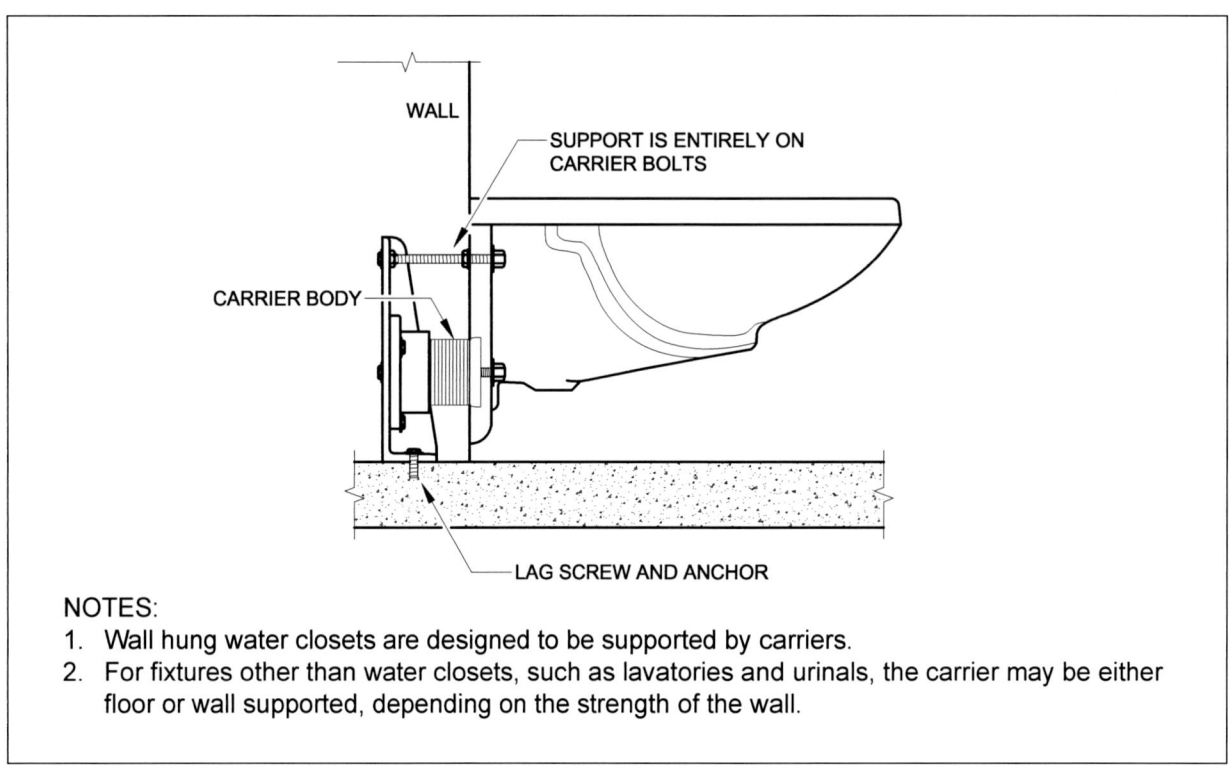

Figure 7.3.5
A CARRIER FOR A WALL HUNG WATER CLOSET

7.3.6 Orientation and Operation of Faucets

Where fixtures are supplied with both hot and cold water, the faucet(s) and supply piping shall be installed so that the hot water is controlled from the left side of the fixture or faucet when facing the controls during fixture use.

EXCEPTION: Single handle and single control valves for showers and tub/shower combinations where the hot and cold temperature orientation is marked on the fitting surface.

7.3.7 Access to Concealed Connections

Where fixtures have drains with concealed slip joint connections or incorporate a cleanout plug, a means of access shall be provided for inspection and repair. Such access is not required for connections that are soldered, threaded, solvent cemented, or equivalently secured.

Comment: Access to concealed slip joints and cleanouts can require the use of tools.

7.3.8 Joints with Walls and Floors

Joints where fixtures contact walls and floors shall be caulked or otherwise made water-tight.
See Figure 7.3.8

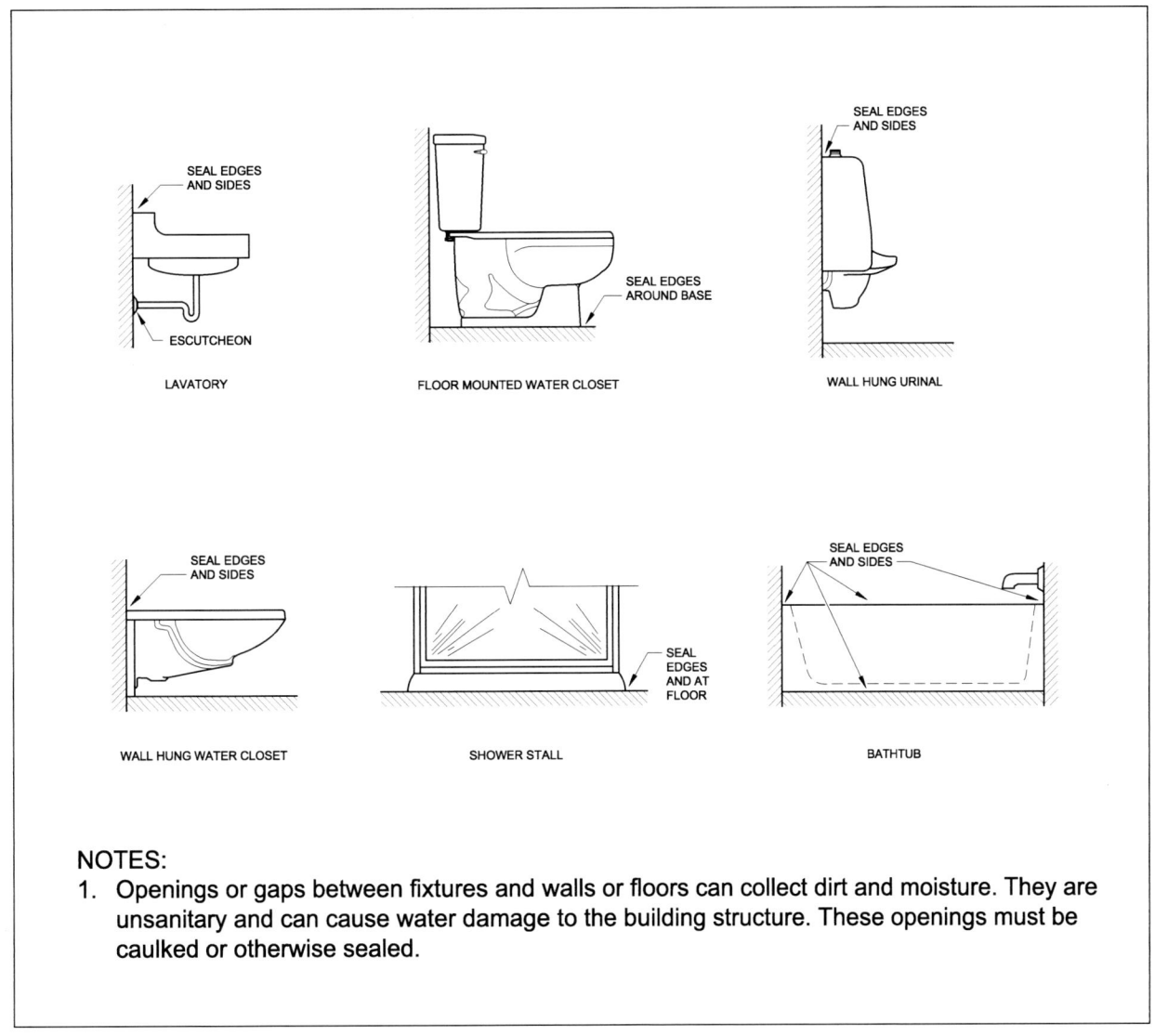

Figure 7.3.8
SEALING FIXTURES TO WALLS AND FLOORS

7.4 WATER CLOSETS

7.4.1 Compliance

Vitreous china water closets shall comply with ASME A112.19.2/CSA B45.1. Plastic water closets shall comply with CSA B45.5/IAPMO Z124.

7.4.2 Water Conservation

Water closets, whether operated by flush tank, flushometer tank, or flushometer valve, shall be the low-consumption type having an average consumption of not more than 1.6 gallons per flush when tested in accordance with ASME A112.19.2/CSA B45.1 or CSA B45.5/IAPMO Z124.
EXCEPTION: Blow-out water closets and clinical sinks.

7.4.3 Type of Bowls and Seats

Water closets shall have elongated bowls with open-front seats.
EXCEPTIONS:
   (1) Water closets having closed-front seats and either round or elongated bowls are permitted in single dwelling units, apartments, condominiums, private office toilet rooms intended for the exclusive use of one individual, and in pre-school and kindergarten facilities.
   (2) Closed-front seats are permitted in hotel and motel guest rooms.
   (3) Closed-front seats that have automatic seat cover protection are permitted in lieu of open-front seats.

See Figure 7.4.3

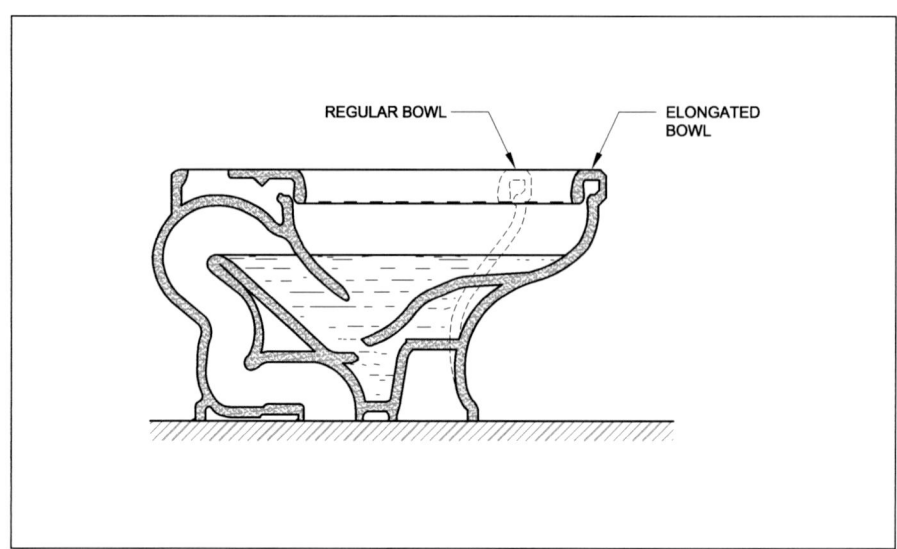

Figure 7.4.3
AN ADULT-HEIGHT ELONGATED WATER CLOSET BOWL
COMPARED TO A ROUND BOWL

7.4.4 Bowl Height

The height of water closet bowls shall be a minimum of 13-1/2" from the finished floor to the top of the rim.
EXCEPTION:
   Bowls intended for children's use (5 years and younger) are permitted to be 9-1/2" to 10-1/2" high to the rim and juvenile use (6-12 years) are permitted to be 10-1/2" to 13-1/2" high to the rim.

7.4.5 Water Closet Seats

Seats for water closets shall be of smooth, non-absorbent materials, be properly sized to fit the water closet bowl. Plastic toilet seats shall comply with IAPMO/ANSI Z124.5.
EXCEPTION: Water closets in single dwelling units are not required to comply with IAPMO/ANSI Z124.5.

7.4.6 Hotels, Motels, Dormitories, and Boarding Houses

Water closets in hotels, motels, dormitories, boarding houses and similar occupancies shall be the elongated type with open-front seats.
EXCEPTIONS:
   (1) Closed-front seats shall be permitted in hotel and motel guest rooms.
   (2) Closed-front seats that are provided with automatic seat cover protection.

7.4.7 Prohibited Water Closets

Water closets not having a visible trap seal or having either unventilated spaces or walls that are not washed at each discharge shall be prohibited.

7.4.8 Water Closets with Pumps

Water closets with pumps shall comply with ASME A112.3.4/CSA B45.9.

7.5 URINALS

7.5.1 Compliance

a. Water-fed vitreous china urinals shall comply with ASME A112.19.2/CSA B45.1.

b. Non-water vitreous china urinals shall comply with ASME A112.19.19.

c. Plastic urinals shall comply with CSA B45.5/IAPMO Z124.

d. Non-water urinals shall have a liquid trap seal as required by Section 5.3.2.

e. Water-fed urinals with liquid trap seals that are not visible in the bowl shall be supplied by an automatic flushing device.

7.5.2 Water Conservation

Water-fed urinals shall be the low-consumption type having an average water consumption of not more than 1.0 gallon per flush when tested in accordance with ASME A112.19.2/CSA B45.1 or CSA B45.5/IAPMO Z124.

7.5.3 Surrounding Surfaces

Urinals shall not be installed where wall and floor surfaces are not waterproof and do not have a smooth, readily cleanable, non-absorbent surface extending not less than four feet above the floor, one foot to each side of the urinal, and one foot in front of the lip of the urinal. *See Figure 7.5.3*

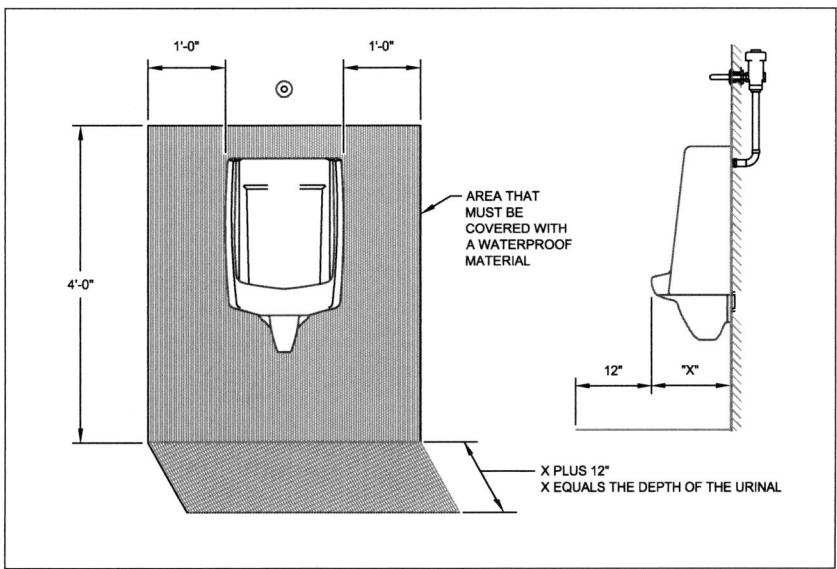

Figure 7.5.3
PROTECTING SURROUNDING SURFACES AT URINALS

7.5.4 Prohibited Urinals

Trough urinals and urinals having walls that are not washed at each discharge shall be prohibited.
EXCEPTION: Non-water urinals.

7.5.5 Reserved

7.5.6 Non-water Urinals

a. Non-water urinals shall not be installed in facilities that do not have provisions for necessary maintenance.

b. Drain piping shall be sloped not less than 1/4 inch per foot to the point where it connects to piping with drainage from one or more water closets.

c. Drain piping shall not be copper to the point where it connects to piping with drainage from one or more water closets.

7.6 LAVATORIES

7.6.1 Compliance

a. Lavatories shall comply with the following standards:
 1. Ceramic, non-vitreous; ASME A112.19.2/CSA B45.1.
 2. Enameled cast-iron; ASME A112.19.1/CSA B45.2.
 3. Enameled steel; ASME A112.19.1/CSA B45.2.
 4. Plastic; CSA B45.5/IAPMO Z124.
 5. Stainless steel; ASME A112.19.3/CSA B45.4.
 6. Vitreous china; ASME A112.19.2/CSA B45.1

7.6.2 Water Conservation

a. Except as required under Section 7.6.2.c, lavatory faucets shall be designed and manufactured so that they will not exceed a water flow rate of 2.2 gallons per minute when tested in accordance with ASME A112.18.1/ CSA B125.1.

b. Public lavatory faucets, other than the metering type, shall be designed and manufactured according to ASME A112.18.1/CSA B125.1.

c. Self-closing or self-closing/metering faucets shall be installed on lavatories intended to serve the transient public, such as those in, but not limited to, service stations, train stations, airport terminals, restaurants, and convention halls. Metering faucets shall deliver not more than 0.25 gallon of water per use when tested in accordance with ASME A112.18.1/CSA B125.1. Self-closing faucets shall be designed and manufactured so that they will not exceed a water flow rate of 0.5 gallon per minute when tested in accordance with ASME A112.18.1/CSA B125.1.

7.6.3 Waste Outlet

The waste outlet pipe on individual lavatories shall be not less than 1-1/4" nominal size. A strainer, pop-up stopper, crossbar grid, or other device shall be provided to protect the waste outlet.

7.6.4 Integral Overflow

Where lavatories include an integral overflow drain, the waste fitting shall be designed and installed so that standing water in the bowl of the fixture cannot rise in the overflow channel when the drain is closed, nor shall any water remain in the overflow channel when the bowl is empty. The overflow shall drain to the inlet side of the fixture trap. *See Figure 7.6.4*

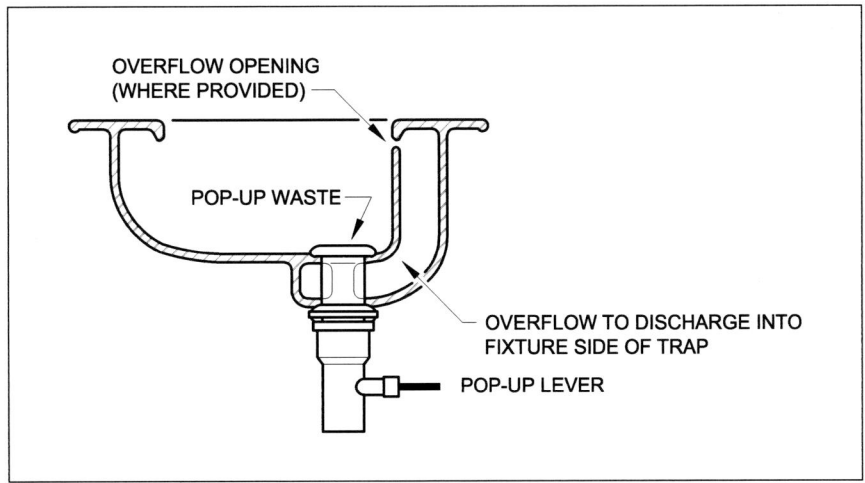

**Figure 7.6.4
A LAVATORY OVERFLOW**

Comment: Overflows are not provided in all lavatories. Lavatories with overflows are not used in hospitals because of the potentially unsanitary condition caused by inaccessible surfaces exposed to waste.

7.6.5 Wash Fountains and Wash Sinks

a. Circular and semi-circular wash fountains and wall-mounted and free-standing wash sinks shall comply with the requirement of Section 7.6. Mixed water temperature control shall comply with Section 10.15.6.

b. The water supply to wash fountains shall not exceed 0.75 gpm for each of the rated number of users. The rated number of users for wash fountains shall be as specified by the manufacturer but shall not be more than one person per 18-inches of usable length of the rim having an available water spray.

c. The faucets for wash sinks shall comply with Section 7.6.2 for lavatories. The rated number of users for wash sinks shall equal the number of faucets.

7.6.6 Control of Mixed Water Temperature

The control of mixed water temperature for public-use hand washing facilities shall comply with Section 10.15.6.

7.6.7 Lavatory Equivalents

Where group-type wash fountains or wash sinks are used to satisfy the number of lavatories required by Section 7.21.1, each rated user shall be considered as the equivalent of one lavatory. ***See Figure 7.6.7***

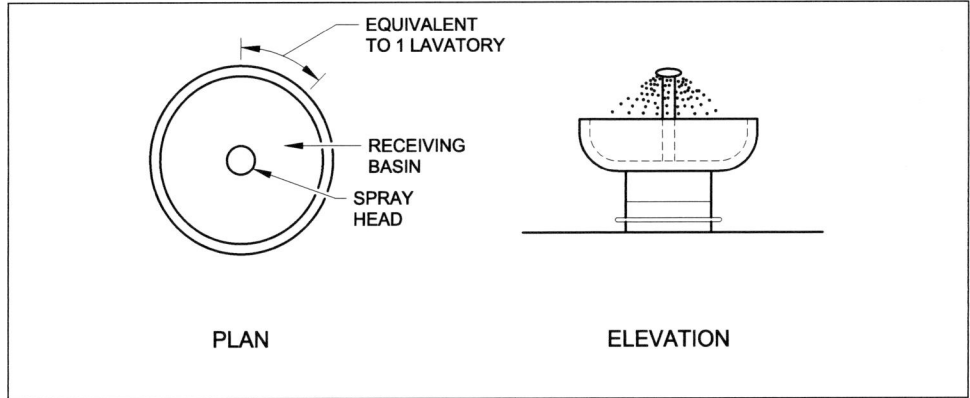

**Figure 7.6.7
A CIRCULAR WASH FOUNTAIN**

7.7 BIDETS

7.7.1 Compliance

Bidets shall comply with ASME A112.19.2/CSA B45.1. Bidet faucets shall comply to ASME A112.18.1/CSA B125.1. Bidet sprays shall comply with ASME A112.4.2.

7.7.2 Backflow Prevention

Bidets having integral flushing rims shall have a vacuum breaker assembly on the mixed water supply to the fixture. Bidets without flushing rims shall have an over-the-rim supply fitting providing the air gap required by Chapter 10.

7.7.3 Integral Overflow

Where bidets include an integral overflow drain, the waste fitting shall be designed and installed so that standing water in the bowl of the fixture cannot rise in the overflow channel when the drain is closed, nor shall any water remain in the overflow channel when the bowl is empty. The overflow shall drain to the inlet side of the fixture trap.

7.7.4 Control of Mixed Water Temperature

The control of mixed water temperature for bidets shall comply with Section 10.15.6.

7.8 BATHTUBS

7.8.1 Compliance

   a. Bathtubs shall comply with the following standards:
     1. Plastic, cultured marble and other synthetic products or finishes; CSA B45.5/IAPMO Z124
     2. Enameled cast-iron; ASME A112.19.1/CSA B45.2
     3. Enameled steel; ASME A112.19.1/CSA B45.2
     4. Bathtubs with pressure sealed doors; ASME A112.19.15

7.8.2 Waste and Overflow

Bathtubs shall have waste outlet and overflow pipes not less than 1-1/2" nominal size. Waste outlets shall be equipped with a pop-up waste, chain and stopper, or other type of drain plug. *See Figure 7.8.2*

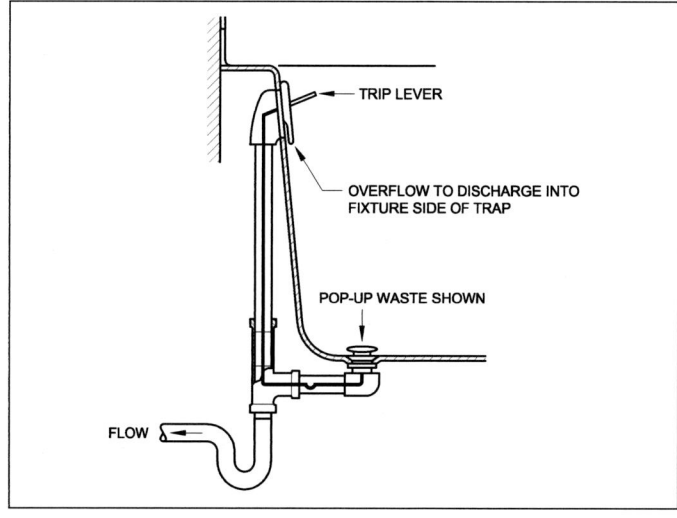

Figure 7.8.2
A BATHTUB WASTE AND OVERFLOW

7.8.3 Combination Bath/Showers

Shower heads, including the hand-held type, shall be designed and manufactured so that they will not exceed a water supply flow rate of 2.5 gallons per minute when tested in accordance with ASME A112.18.1/ CSA B125.1. The control of mixed water temperatures to bath/shower combinations shall comply with Section 10.15.6. Surrounding wall construction shall be in accordance with Section 7.10.5.e. Riser pipes to shower heads shall be secured in accordance with Section 7.10.7.

7.8.4 Backflow Prevention

Unless equipped with an ASSE 1001 atmospheric vacuum breaker, there shall be an air gap between the outlet of the bath spout and the flood level rim of the tub that complies with Section 10.5.2 and Table 10.5.2.

7.8.5 Control of Mixed Water Temperature

The control of mixed water temperature for bathtubs shall comply with Section 10.15.6.

7.9 WHIRLPOOL BATHS

7.9.1 General

The requirements of Section 7.8 for bathtubs shall also apply to whirlpool baths. The provisions for wet venting in Section 12.10 shall also apply to whirlpool baths.

7.9.2 Compliance

Whirlpool bathtubs and their suction fittings shall comply with ASME A112.19.7/CSA B45.10

7.9.3 Drainage

The arrangements of circulating piping and pumps shall not be altered in any way that would prevent the pump and associated piping from draining after each use of the fixture.

7.9.4 Access

One or more removable panels shall be provided where required for access to pumps, heaters, and controls, as recommended by the fixture manufacturer. ***See Figure 7.9.4***

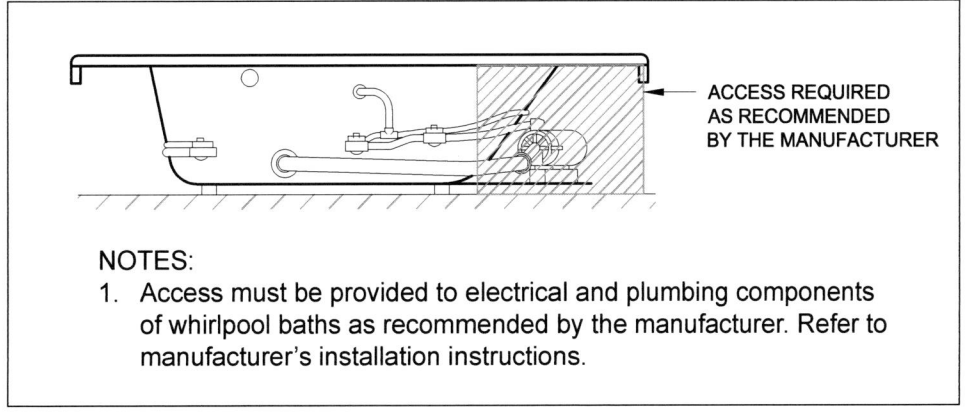

Figure 7.9.4
ACCESS REQUIREMENTS FOR A WHIRLPOOL BATH

7.9.5 Control of Mixed Water Temperature

The control of mixed water temperature for whirlpool baths shall comply with Section 10.15.6.

7.10 SHOWERS

7.10.1 Compliance

Plastic shower receptors and stalls shall comply with CSA B45.5/IAPMO Z124.

7.10.2 Water Conservation

Shower heads shall be designed and manufactured so that they will not exceed a water supply rate of 2.5 gallons per minute when tested in accordance with ASME A112.18.1/CSA B125.1.
EXCEPTION: Emergency safety showers.

7.10.3 Control of Mixed Water Temperature

a. The control of mixed water temperatures shall comply with Section 10.15.6.

b. Where shower heads are rated for maximum water flow rates of less than 2.5 gallons per minute, the mixed water temperature control valve that supplies the shower head shall be rated for the maximum flow rate of the shower head or less.

7.10.4 Shower Waste Outlets

a. For a shower with a single 2.5 gpm shower head, the waste outlet connection shall be sized for a drain not less than 1-1/2 inches nominal pipe size. Combination bath/showers with 1-1/2" drains can have two 2.5 gpm shower heads.

b. Showers with multiple shower heads shall have a 2" outlet pipe size for up to 10 gpm and a 3" outlet pipe size for up to 20 gpm.

c. In group showers where each shower is not provided with an individual waste outlet, the waste outlet(s) shall be located and the floor so pitched that waste from one shower does not flow over the area serving another shower.

d. Waste outlets shall be securely connected to the drainage piping system.

e. Outlet covers/strainers shall be removable and not less than 4 inches in diameter with 1/4" minimum openings.

7.10.5 Shower Compartments

a. The minimum outside rough-in dimension for shower bases and prefabricated shower compartments shall be 32 inches, except where a shower receptor has a minimum overall dimension of 30 inches (750 mm) in width and 60 inches (1500 mm) in length.

b. The minimum rough-in depth for prefabricated tub/shower combinations shall be 30 inches.

c. Where shower compartments have glass enclosures or field-constructed tile walls, the compartment shall provide clearance for a 30 inch diameter circle with the door closed.

d. The walls in shower compartments and above built-in bathtubs having installed shower heads shall be constructed of smooth non-corrosive, non-absorbent waterproof materials that extend to a height of not less than 68 inches above the fixture drain.

e. The joints between walls and with bathtubs and shower compartment floors shall be water-tight.

7.10.6 Shower Floors and Shower Pan Liners

a. Adequate structural support shall be provided under shower floors.

b. Finished shower floor surfaces shall be smooth and water-proof.
EXCEPTION: Grouted shower floor surfaces shall be smooth and water-resistant.

c. Manufactured shower pans, shower bases, and shower receptors shall be installed in accordance with this Code and the manufacturer's instructions.

d. The edges of shower pans, bases, and receptors shall include flanges or other means of making a waterproof joint with the walls of the shower enclosure.

e. Shower pan liners shall be provided beneath shower floors that are water-resistant (not water-proof) and also if required by the manufacturer's installation instructions for pre-fabricated or pre-finished shower pans, bases, or receptors.

EXCEPTION: Shower pan liners shall not be required under water-resistant shower floors if they are installed on shower pans, bases, and receptors that provide for the drainage of water seepage through the floor finish.

f. Shower pan liners shall slope to the shower drain outlet and be sealed to the weep holes in the drain fitting. The liner shall provide a water-tight basin up to the overflow elevation of the shower floor.

g. Shower pan liners shall be tested for leakage. There shall be no leakage indicated on the floor surrounding the shower pan for a period of not less than 15 minutes.
See Figure 7.10.6

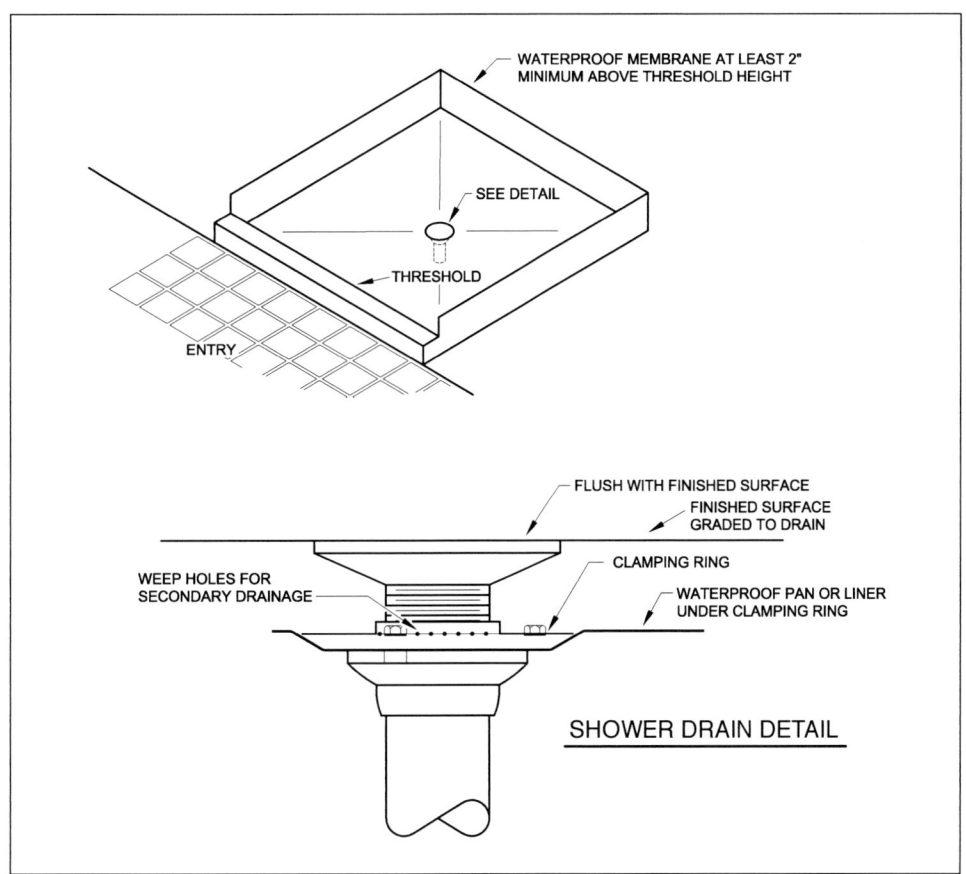

**Figure 7.10.6
A SHOWER PAN AND DRAIN**

7.10.7 Water Supply Riser

Whether exposed or concealed, the water supply riser pipe from the shower control valve(s) to the shower head outlet shall be secured to the wall structure. *See Figure 7.10.7*

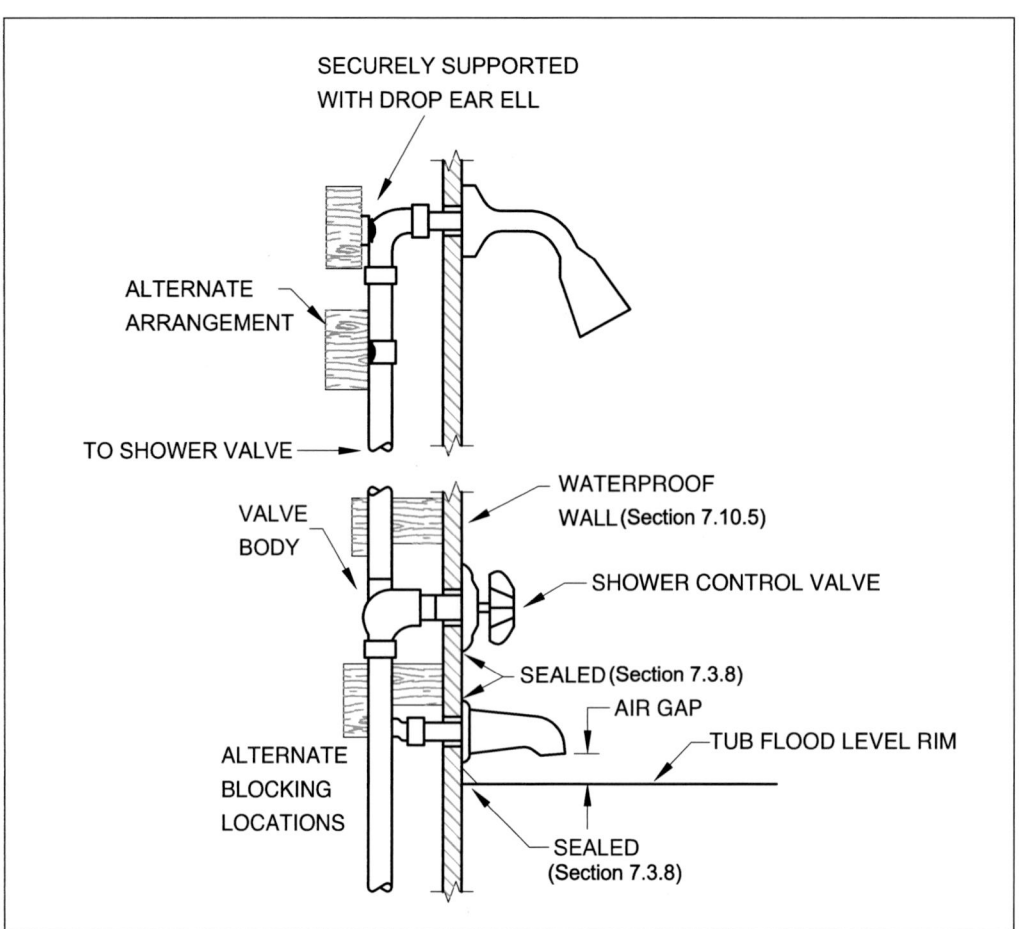

Figure 7.10.7
A SHOWER HEAD SUPPLY RISER PIPE

7.11 SINKS

7.11.1 Compliance

   a. Sinks shall comply with the following standards:
      1. Enameled cast-iron; ASME A112.19.1/CSA B45.2
      2. Enameled steel; ASME A112.19.1/CSA B45.2
      3. Stainless steel; ASME A112.19.3/CSAB45.4
      4. Plastic; CSA B45.5/IAPMO Z124
      5. Ceramic; ASME A112.19.2/CSA B45.1
   b. Fixture supply fittings, including faucets and supply stops, shall comply with ASME A112.18.1/CSA B125.1.
   c. Fixture waste outlet fittings shall comply with ASME A112.18.2/CSA B125.2.

7.11.2 Domestic Kitchen Sinks and Bar Sinks

   a. Each compartment in a kitchen sink or bar sink shall have an outlet suitable for either a domestic food waste disposer or a basket strainer. The waste outlet pipe for each compartment shall be 1-1/2" nominal size. Outlet fittings shall have crossbars or other provisions for protecting the drain outlet and shall include a means of closing the drain outlet.
   b. Faucets for kitchen sinks and bar sinks shall be designed and manufactured so that they will not exceed a water flow rate of 2.2 gallons per minute when tested in accordance with ASME A112.18.1/CSA B125.1.
See Figure 7.11.2

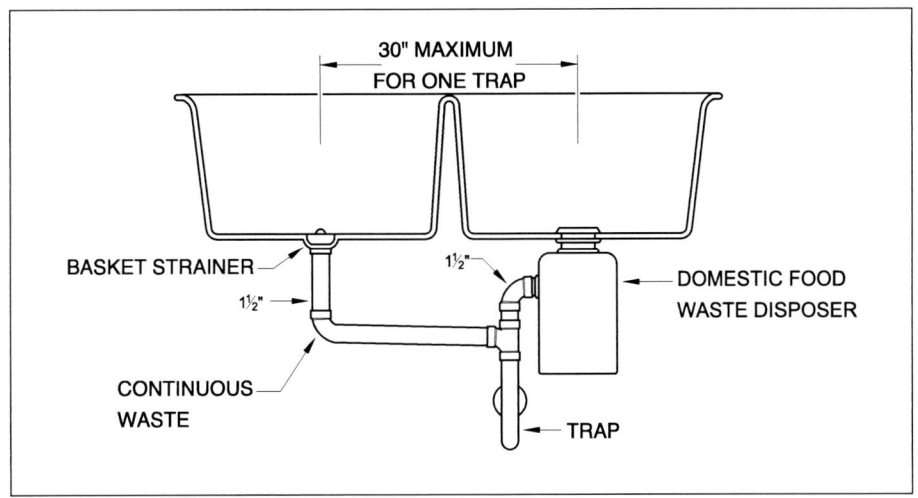

Figure 7.11.2
KITCHEN SINK DRAIN OUTLETS

7.11.3 Laundry Sinks

a. Sinks for laundry use shall be not less than 12 inches deep with a strainer and waste outlet connection not less than 1-1/2" nominal size.

b. Utility faucets for laundry sinks shall comply with ASME A112.18.1/CSA B125.1.

7.11.4 Service Sinks and Mop Receptors

a. Service sinks and mop receptors shall have removable strainers and waste outlet connections not less than 2" nominal size.

b. Service sinks and mop receptors shall not be installed where walls and floors are not waterproof and do not have a smooth, readily cleanable surface at least one foot in front of the sink or receptor, at least one foot on each side, and up to a point one foot above the faucet height.

7.11.5 Sanitary Floor Sinks

Sanitary floor sinks shall comply with ASME A112.6.7.

7.11.6 Backflow Prevention

a. Sink faucets having a hose thread or other means of attaching a hose to the outlet shall be protected from back siphonage by either an integral vacuum breaker, an atmospheric vacuum breaker attached to the outlet, or pressure type vacuum breakers on the fixture supply lines.

b. Sink faucets without provisions for connecting hoses shall be protected from back siphonage by a fixed air gap complying with ASME A112.1.2 or Table 10.5.2.

7.12 DRINKING WATER FACILITIES

7.12.1 Compliance

a. Drinking water facilities shall comply with the "lead-free" requirements of Section 3.4.6.

b. Drinking water facilities for accessible use shall comply with Section 7.2.

c. Drinking fountains shall comply with the following plumbing fixture standards:

    1) Enameled cast iron or steel: ASME A112.19.1/CSA B45.2
    2) Ceramic: ASME A112.19.2/CSA B45.1
    3) Stainless steel: ASME A112.19.3/CSA B45.4

d. Drinking water coolers with self-contained refrigeration units shall comply with UL 399 and ASHRAE Standard 18.

e. Water bottle filling stations shall comply with applicable standards. Stations with self- contained refrigeration units for chilled water shall comply with UL 399 and ASHRAE Standard 18.

7.12.2 Required Drinking Water Facilities

The required drinking water facilities shall be in accordance with Section 7.21.

7.12.3 Prohibited Locations

Drinking water facilities shall not be located in public toilet rooms. Convertible lavatory faucets that provide a discharge stream similar to a drinking fountain shall only be permitted in bathrooms in dwelling units.

7.12.4 Outdoor Drinking Fountain

Freeze-resistant drinking fountains shall be the sanitary type. Weep hole drains that form a cross connection between ground water and the potable water supply shall not be permitted.

7.13 AUTOMATIC CLOTHES WASHERS

7.13.1 Compliance

a. Automatic clothes washers for domestic use or use by the general public shall comply with AHAM HLW-1 or UL 2157 and shall have an air gap incorporated in the internal tub fill line.

b. Commercial electric clothes washers shall comply with UL 1206 and include backflow prevention.

> *Comment: Automatic clothes washers that comply with the referenced appliance standards include an internal air gap in the water fill line and do not require the installation of an external backflow prevention device.*

7.14 FOOD WASTE DISPOSERS

7.14.1 Compliance

Domestic food waste disposer units shall comply with UL 430 and either AHAM FWD-1 or ASSE 1008.

7.14.2 Domestic Units

Domestic food waste disposer units shall have not less than a 1-1/2" nominal waste connection to the drainage system. Such units may connect to a kitchen sink drain outlet, as permitted under Section 7.11.2.a.

7.14.3 Commercial Units

Commercial food waste disposer units shall be connected to the drainage system and be separately trapped from any sink compartment or other fixture. The waste pipe size for such fixtures shall be of sufficient size to serve the fixture but shall be not less than 2-inch nominal size. ***See Figure 7.14.3***

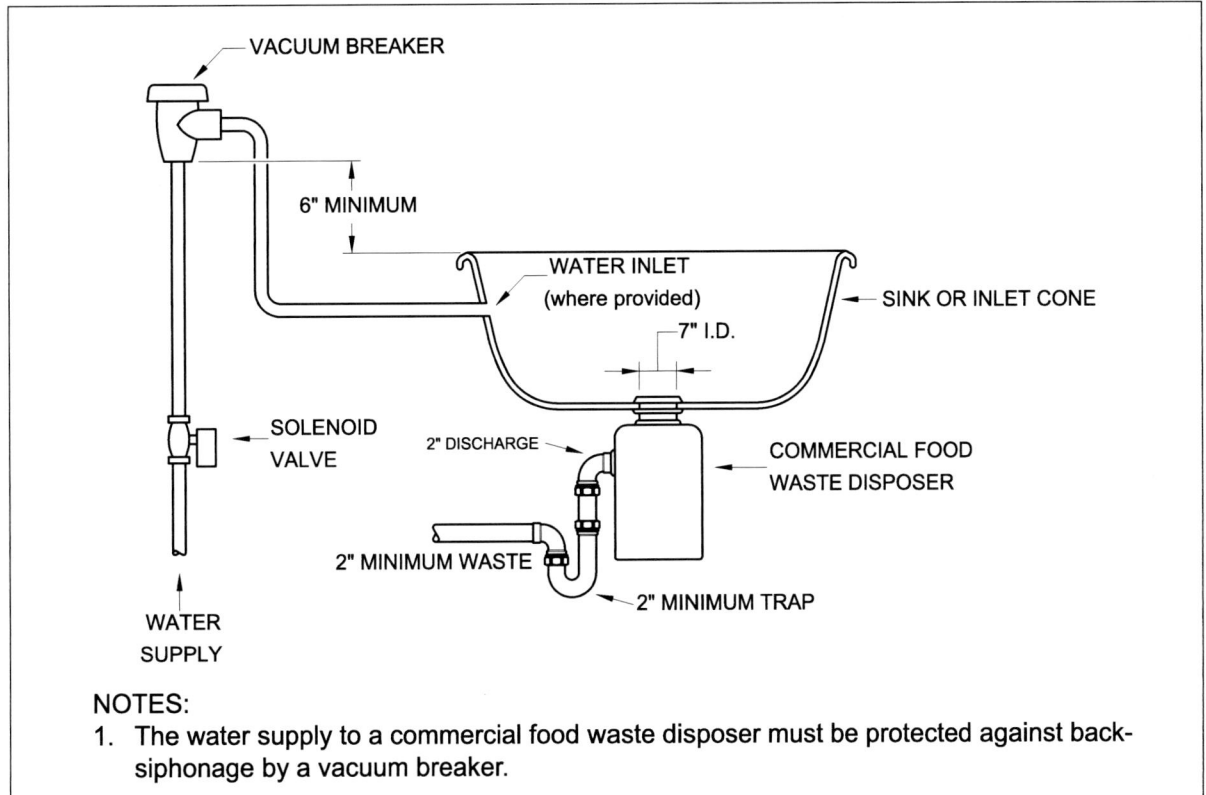

Figure 7.14.3
A COMMERCIAL FOOD WASTE DISPOSER INSTALLATION

7.14.4 Water Supply

An adequate supply of water shall be provided for proper operation of food waste disposers.

7.15 DISHWASHING MACHINES

7.15.1 Compliance

a. Domestic dishwashing machines shall comply with UL 749 and AHAM DW-1. The water supply to domestic dishwashing machines shall be protected from back siphonage by an integral air gap, an other internal means, or a vacuum breaker. Air gap fittings for drain discharge shall comply with ASSE 1021.

b. Commercial dishwashing machines shall comply with UL 921 and NSF 3. Backflow prevention shall comply with ASSE 1004.

7.15.2 Residential Sink and Dishwasher

The discharge from a residential kitchen sink and dishwasher may discharge through a single 1-1/2" trap. The discharge line from the dishwasher shall be not less than the size recommended by the dishwasher manufacturer. It shall either be looped up and securely fastened to the underside of the counter or be connected to a deck-mounted dishwasher air gap fitting. The discharge shall then be connected to a branch inlet wye fitting between the sink waste outlet and the trap inlet. The discharge may also drain indirectly into a trapped standpipe, or receptor. ***See Figure 7.15.2***

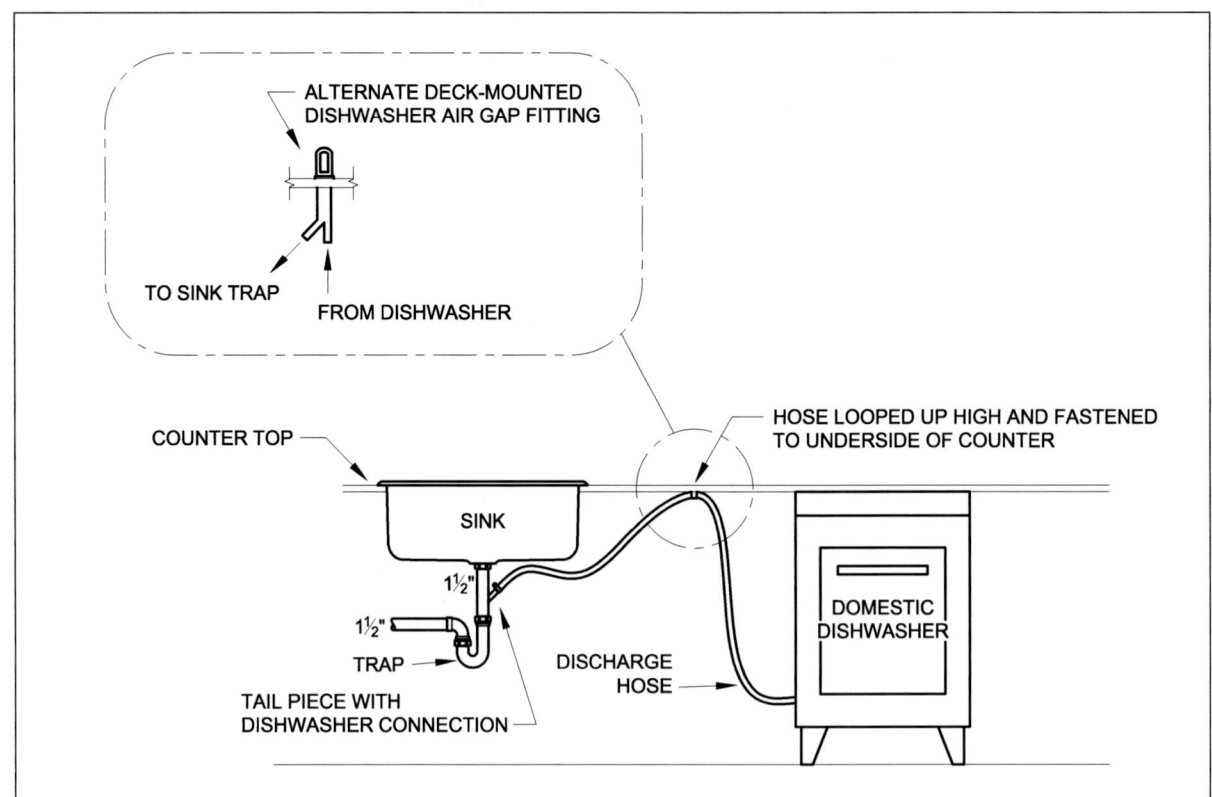

Figure 7.15.2
A RESIDENTIAL KITCHEN SINK AND DISHWASHER

7.15.3 Residential Sink, Dishwasher, and Food Waste Disposer

The discharge from a residential kitchen sink, dishwasher, and food waste disposer may discharge through a single 1-1/2" trap. The discharge line from the dishwasher shall be not less than the size recommended by the dishwasher manufacturer. It shall either be looped up and securely fastened to the underside of the counter or be connected to a deck-mounted dishwasher air gap fitting. The discharge shall then be connected either to the chamber of the food waste disposer or to a branch inlet wye fitting between the food waste disposer outlet and the trap inlet. The discharge may also drain indirectly into a trapped standpipe or receptor.
See Figure 7.15.3

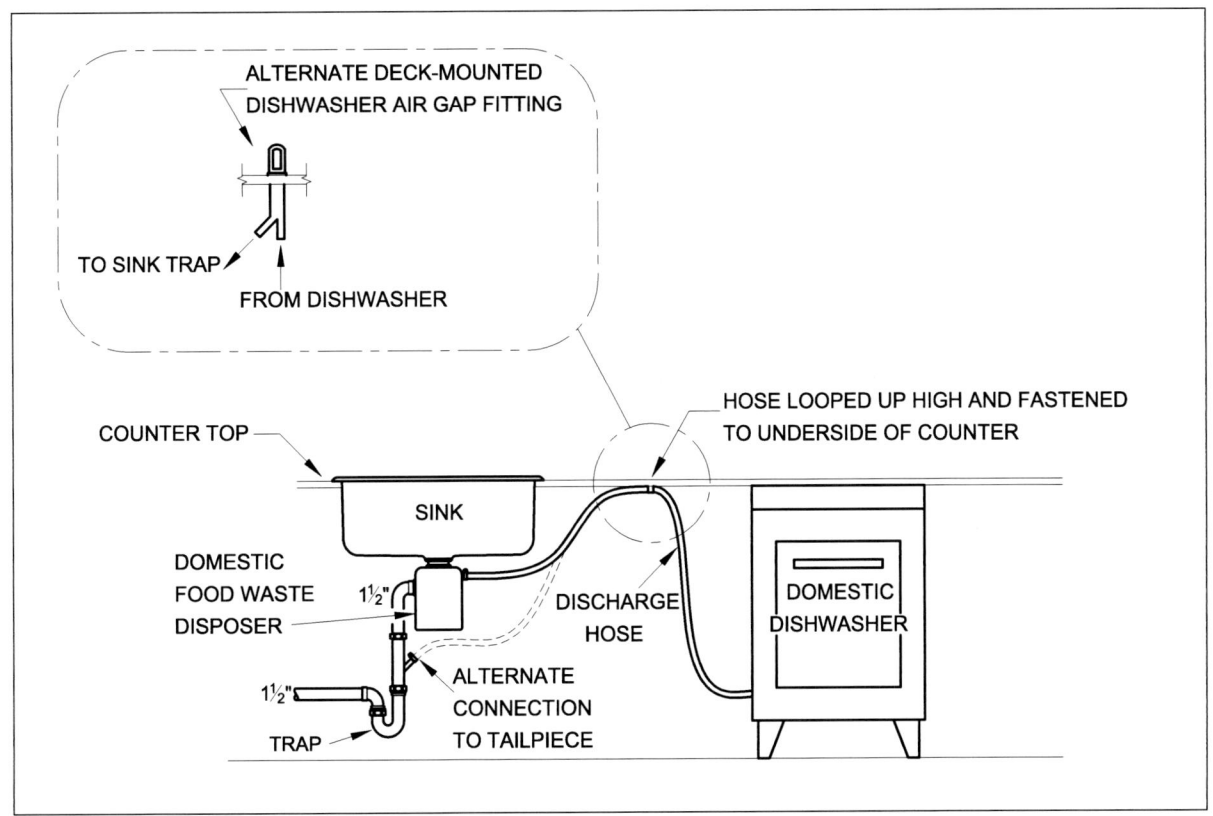

Figure 7.15.3
A RESIDENTIAL KITCHEN SINK, DISHWASHER, AND FOOD WASTE DISPOSER

7.15.4 Commercial Dishwashing Machine

a. Commercial dishwashing machines shall be indirectly connected to the drainage system through either an air gap or an air break. When the machine is within 5 feet developed length of a trapped and vented floor drain, an indirect waste pipe from the dishwasher may be connected to the inlet side of the floor drain trap. *See Figure 7.15.4*

b. Commercial dishwashers shall be permitted to discharge through a grease interceptor in accordance with Section 6.2.6.a.

7.16 FLOOR DRAINS, AREA DRAINS, AND TRENCH DRAINS

7.16.1 Compliance

Floor drains, area drains, and trench drains shall comply with ASME A112.6.3 or, if stainless steel, ASME A112.3.1.

7.16.2 Grate Free Area and Strainer or Sediment Basket

a. The free area of the top grate for floor drains, area drains, and trench drains shall comply with ASME A112.6.3 based on their outlet connection size. Where floor drains and area drains receive other than clear-water waste, they shall include an internal outlet strainer or removable metal sediment basket to retain solids.

b. The trap seal for floor drains shall be not less than 2 inches deep.

c. Where infrequently used floor drains are subject to evaporation of their trap seals, they shall either 1) be provided with a 4-inch deep trap seal and have an accessible means to replenish, or 2) be fed from an automatic trap seal priming device complying with ASSE 1018 or ASSE 1044, or 3) be fitted with a barrier type floor drain trap seal protection device complying with ASSE 1072.

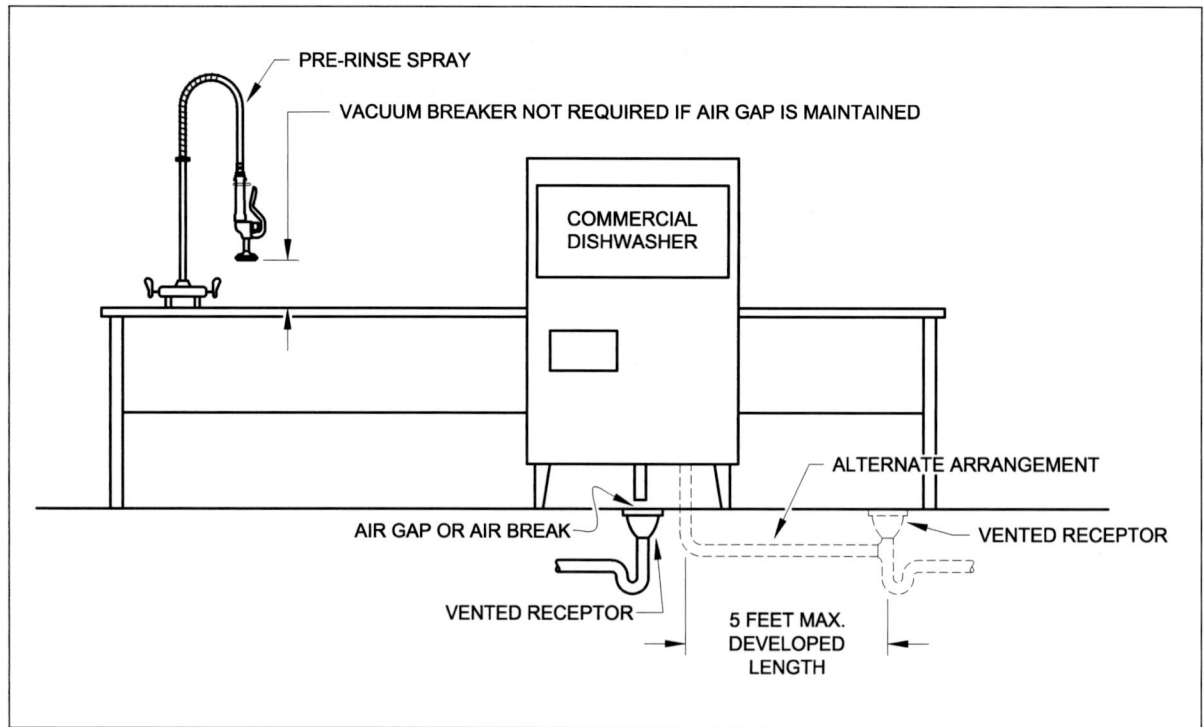

**Figure 7.15.4
A COMMERCIAL DISHWASHING MACHINE**

d. Where traps are automatically primed, the priming connection shall be above the weir of the trap.
See Figures 7.16.2-A through -C

7.16.3 Size of Floor Drains

a. Floor drains and their discharge piping shall be sized in accordance with Section 11.4.4. Floor drains shall be not less than 2" pipe size.

7.16.4 Required Locations for Floor Drains

a. Floor drains shall be provided in the following areas:
   1. Toilet rooms containing either one water closet and one wall hung urinal, two or more water closets, or two or more wall hung urinals, except in a dwelling unit.
   2. Commercial kitchens.
   3. Common laundry rooms.

7.16.5 Walk-in Coolers and Freezers

Floor drains located in walk-in coolers and walk-in freezers where food or other products for human consumption are stored shall be indirectly connected to the drainage system in accordance with Section 9.1.6.

7.16.6 Floor Slope

Where floor drains receive indirect waste or other drainage on a regular or frequent basis, the elevation of the floor drain shall be set so that the floor within a 2 foot radius can be sloped to the drain.

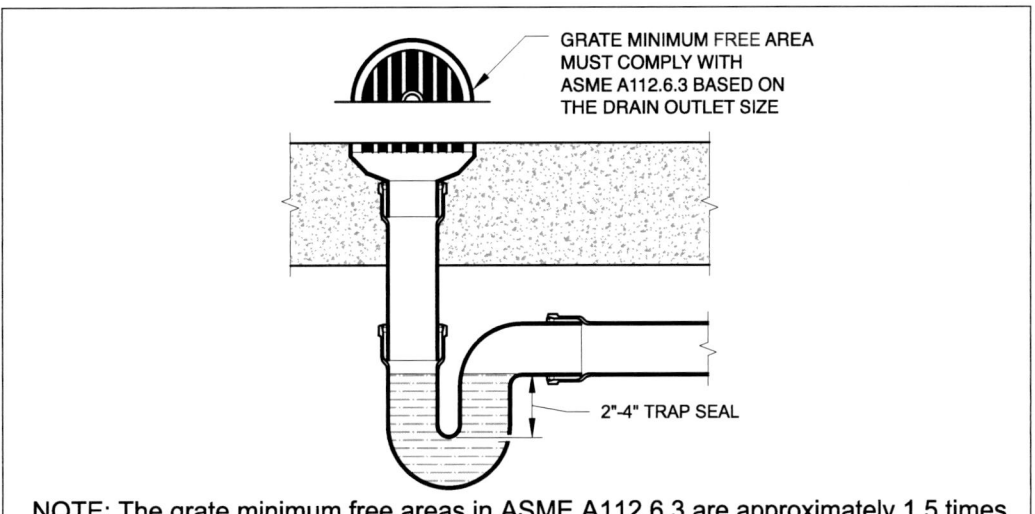

Figure 7.16.2 - A
A FLOOR DRAIN TRAP AND STRAINER

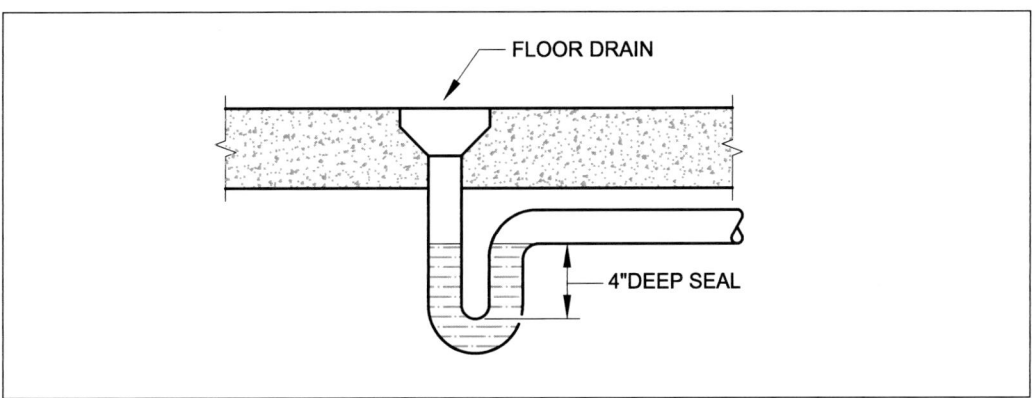

Figure 7.16.2 - B
A DEEP SEAL TRAP

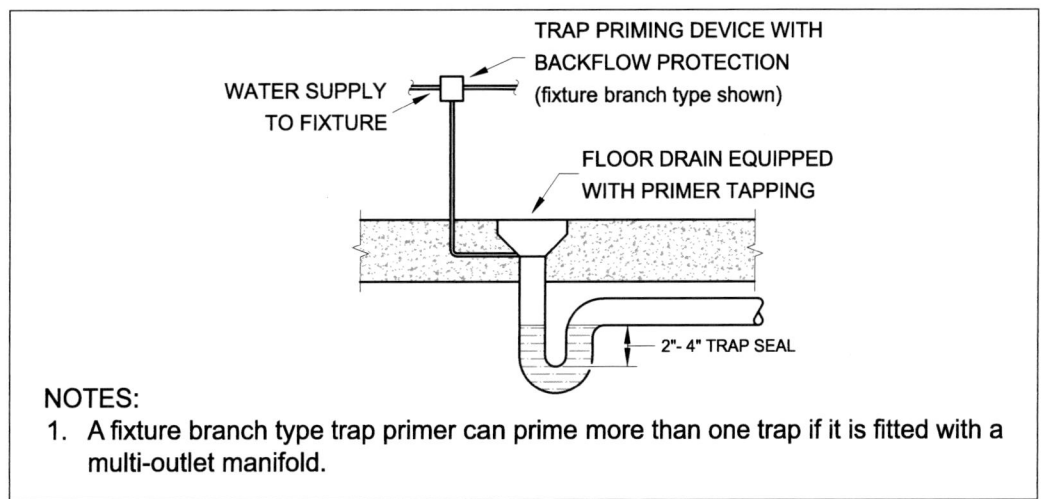

NOTES:
1. A fixture branch type trap primer can prime more than one trap if it is fitted with a multi-outlet manifold.

Figure 7.16.2 - C
A FLOOR DRAIN WITH TRAP PRIMER

7.17 GARBAGE CAN WASHERS

Garbage can washers shall include a removable basket or strainer to prevent large particles of garbage from entering the drainage system. The water supply connection shall be protected from back siphonage in accordance with Chapter 10. Garbage can washers shall be trapped and vented as required for floor drains.
See Figure 7.17

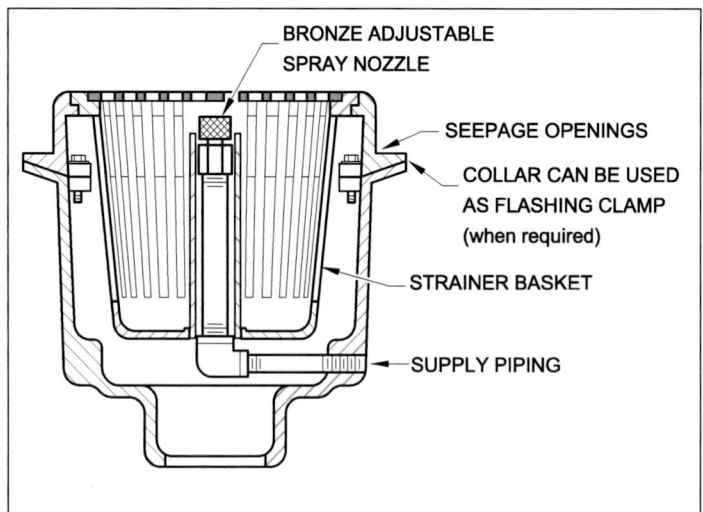

**Figure 7.17
A GARBAGE CAN WASHER**

7.18 SPECIAL INSTALLATIONS

7.18.1 Protection of Water Supply

The water supply to special installations shall be protected from backflow in accordance with Sections 10.4.3 and 10.5. Examples of such special installations include decorative fountains, ornamental pools and waterfalls, swimming and wading pools, baptisteries, and similar custom-built equipment.

7.18.2 Approval

Special installations requiring water supply and/or drainage shall be submitted to the Authority Having Jurisdiction for approval.

7.19 FLUSHING DEVICES FOR WATER CLOSETS AND URINALS

7.19.1 General

Appropriate flushing devices shall be provided for water closets, urinals, clinical sinks, and other fixtures that depend on trap siphonage to discharge the contents of the fixture.

7.19.2 Separate Devices

A separate flushing device shall be provided for each fixture.
EXCEPTION: A single device may be used to automatically flush two or more urinals.

7.19.3 Flush Tanks: Gravity, Pump Assisted, Vacuum Assisted.

a. Flush tanks shall have ballcocks or other means to refill the tank after each discharge and to shutoff the water supply when the tank reaches the proper operating level. Ballcocks shall be the anti-siphon type and comply with ASSE 1002.

b. Except in approved water closet and flush tank designs, the seat of the tank flush valve shall be at least 1 inch above the flood level rim of the fixture bowl.

c. The flush valve shall be designed so that it will close tightly if the tank is flushed when the fixture drain is clogged or partly restricted, so that water will not spill continuously over the rim of the bowl or backflow from the bowl to the flush tank.

d. Flush tanks shall include a means of overflow into the fixture served having sufficient capacity to prevent the tanks from overflowing with normal flow through the fill valve.
See Figure 7.19.3

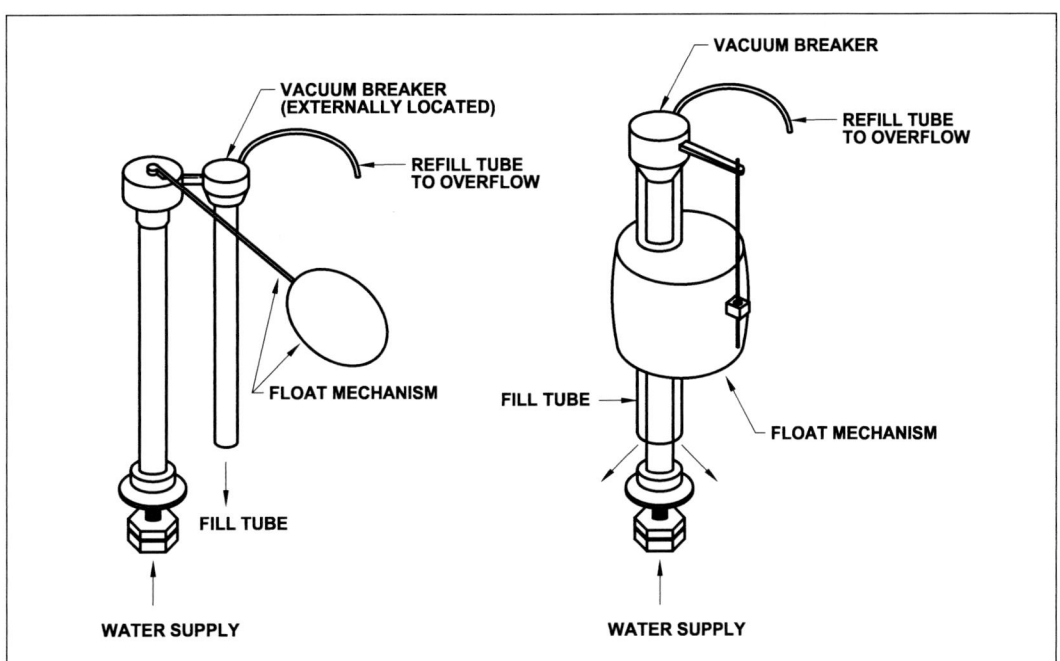

Figure 7.19.3
BALLCOCKS FOR WATER CLOSET FLUSH TANKS

7.19.4 Flushometer Tanks (Pressure Assisted)

Flushometer tanks (pressure assisted) shall comply with ASSE 1037 or CSA B125.3 and shall include built-in pressure regulation and backflow prevention devices.

7.19.5 Flushometer Valves

Flushometer valves shall comply with ASSE 1037 or CSA B125.3 and include a vacuum breaker assembly and means of flow adjustment. Flushometer valves shall be accessible for maintenance and repair.
See Figure 7.19.5

7.19.6 Required Water Pressure

The available water supply pressure shall be adequate for proper operation of the particular flushing devices used, as recommended by the manufacturer.

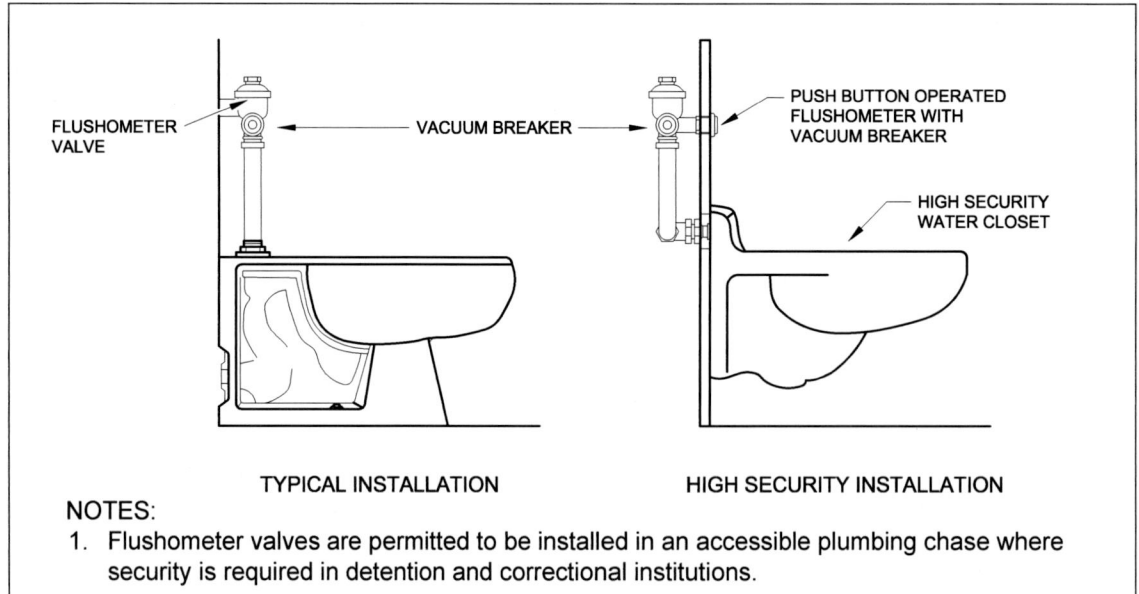

Figure 7.19.5
FLUSHOMETER VALVE INSTALLATIONS

NOTE: Some one-piece tank-type water closets require 30 psig flowing pressure and 1/2" supplies for proper operation.

7.20 FIXTURES FOR DETENTION AND CORRECTIONAL INSTITUTIONS

Special design fixtures for use in detention and correctional institutions shall comply with the requirements of this Code except that fixtures may be fabricated from welded seamless stainless steel and be equipped with necessary security devices. Water closets shall be the elongated type with integral or separate seats. Urinals shall have a continuous flushing rim that washes all four walls of the fixture.

7.21 MINIMUM NUMBER OF REQUIRED FIXTURES

7.21.1 Number of Fixtures

Plumbing fixtures shall be provided for the type of building occupancy and in the numbers not less than those shown in Table 7.21.1.

7.21.2 Occupant Load

    a. The minimum number of plumbing fixtures shall be based on the number of persons to be served by the fixtures, as determined by the person responsible for the design of the plumbing system.

    b. Where the occupant load is not established and is based on the egress requirements of a building code, the number of occupants for plumbing purposes shall be permitted to be reduced to two-thirds of that for fire or life safety purposes.

    c. Wherever both sexes are present in approximately equal numbers, the total occupant load shall be multiplied by 50 percent to determine the number of persons of each sex to be provided for, unless specific information concerning the percentage of male and female occupants is available.

    d. Plans for plumbing systems, where required, shall indicate the maximum number of persons to be served by the facilities.

    e. In occupancies having established seating, such as auditoriums and restaurants, the number of occupants for plumbing purposes shall not be less than the number of seats.

7.21.3 Access to Fixtures

a. In multi-story buildings, accessibility to the required fixtures shall not exceed one story.

b. Fixtures accessible only to private offices shall not be counted to determine compliance with this section.

c. The lavatories required by Tables 7.21.1 for employee and public toilet facilities shall be located within the same toilet facility as their associated water closets and urinals.

7.21.4 Separate Facilities

a. Separate toilet facilities shall be provided for each sex.
EXCEPTIONS:

(1). Residential installations.

(2). In occupancies serving 15 or fewer people, one toilet facility, designed for use by no more than one person at a time, shall be permitted for use by both sexes.

(3). In business occupancies with a total floor area of 1500 square feet or less, one toilet facility, designed for use by no more than one person at a time, shall satisfy the requirements for serving customers and employees of both sexes.

(4). In mercantile occupancies with a net occupiable floor area of 1500 square feet or less that is accessible to customers, one toilet facility designed for use by no more than one person at a time shall satisfy the requirements for serving customers and employees of both sexes.

7.21.5 Substitution and Omission of Fixtures

a. Urinals: In male toilet rooms, urinals may be substituted for up to 50% of the required number of water closets.

b. Drinking Water Facilities: A kitchen or bar sink shall be considered as meeting the requirements for drinking water facilities for employees who have access to the facilities.

c. Provisions for Clothes Washers: Apartment buildings and condominiums without laundry rooms for their occupants shall not require provisions for centralized clothes washers for use by multiple occupants.

d. Service Sinks: Service sinks may be omitted when the Authority Having Jurisdiction determines that they are not necessary for proper cleaning of the facility.

7.21.6 Fixture Requirements for Special Occupancies

a. Additional fixtures may be required when unusual environmental conditions or special activities are encountered.

b. In food preparation areas of commercial food establishments, fixture requirements may be dictated by the health and/or sanitary codes. Fixtures, fixture compartments and appliances used for rinsing or sanitizing equipment or utensils, processing or preparing food for sale or serving, shall be installed in accordance with Section 9.1.1 to ensure the required protection from backflow and flooding.

c. Types of occupancies not shown in Table 7.21.1 shall be considered individually by the Authority Having Jurisdiction.

d. Where swimming pools operated by an apartment building, condominium, or similar multi-family dwelling unit are restricted to the use of residents and guests of residents of dwelling units in the immediate vicinity of the pool, the minimum required toilet facilities for bathers within the pool compound shall be one (1) male toilet room and one (1) female toilet room, each consisting of a water closet and lavatory as a minimum.

e. Hand washing facilities shall be provided in each examination room in a doctor's office or medical office, dental treatment room, massage treatment room, tattoo parlor or any other facility where skin to skin contact is necessary for performing a service and if required by the use and occupancy regulations for a facility.

7.21.7 Facilities in Mercantile and Business Occupancies Serving Customers

a. Within a single establishment, facilities for customers and employees shall be permitted to be met with a single set of restrooms accessible to both groups. The required number of fixtures shall be based on the total number of customers and employees.

b. Fixtures for customer and employee use shall be permitted to be met by providing centrally located facilities accessible to multiple establishments. The maximum distance of entry to these facilities shall not exceed 500 feet from the entrance of any establishment served.

c. Drinking water facilities shall not be required for customers where normal occupancy is short term as defined in Section 1.2.

d. For establishments less than 1500 square feet in total floor area, one water closet and one lavatory in a restroom shall be permitted to provide the requirements for serving the both customers and employees.

7.21.8 Food Service Establishments

a. Restaurants and other food service establishments with a seated occupant load exceeding 100 customers shall be provided with separate toilet facilities for employees and customers. Customer and employee toilet facilities may be combined for seated customer loads of 100 or less and fixtures shall be provided for the total number of customers and employees. For employees of 15 or less, one employee toilet facility, designed for use by no more than one person at a time, shall be permitted for use by both sexes.

b. Where food service establishments and food courts are open to public areas in malls and transportation terminals, toilet facilities that serve both customers and employees may be centrally located and serve multiple establishments. The maximum distance of entry to central toilet facilities shall be 500 feet from any establishment or food court served.

c. Drinking water facilities shall not be required in restaurants or other food service establishments if drinking water service is provided or available on request.

7.21.9 Family and Assisted-Use Toilet Rooms

a. Accessible family or assisted-use toilet rooms containing one water closet and one lavatory shall be provided as required by the applicable building code.

b. In assembly and mercantile occupancies, the plumbing fixtures within family and assisted-use toilet rooms may be counted as part of the minimum required fixtures for males or for females.

7.21.10 Fractions of the Minimum Number of Required Plumbing Fixtures

a. In Table 7.21.1, where the number of persons of each sex is less than 100% of a particular group, the number of fixtures required for that group may be reduced by the fraction of the persons in that group. The fraction of the required fixtures shall be rounded up to the next whole number.

b. Where a facility includes two or more different occupancy classifications with access to the same installed plumbing fixtures, the fraction of the required fixtures for each classification may be added and the total combined fraction of the required fixtures shall be rounded up to the next whole number

7.22 WATER TREATMENT SYSTEMS

Water softeners, reverse osmosis water treatment units, and other drinking water treatment systems shall meet the requirements of the appropriate standards referenced in Section 10.18.1. Waste discharge from such equipment shall enter the drainage system through an air gap. Discharge piping shall be of a material approved for potable water, sanitary drainage, or storm drainage.

7.23 SAFETY FEATURES FOR SPAS AND HOT TUBS

7.23.1 Spas and Hot Tubs

Spas and hot tubs shall comply with the requirements of subsections 7.23.2, 7.23.3, and 7.23.4.

7.23.2 Entrapment Avoidance

There shall be nothing in the spa or hot tub that can cause the user to become entrapped underwater. Types of entrapment can include, but not be limited to, rigid, non-giving protrusions, wedge-shaped openings, and any arrangement of components that could pinch and entrap the user.

7.23.3 Outlets Per Pump

There shall be a minimum of two (2) suction inlets for each pump in the suction inlet system, separated by at least 3 feet or located on two (2) different planes, such as one on the bottom and one on a vertical wall, or one on each of two vertical walls. The suction inlets shall be piped so that water is drawn through the inlets simultaneously by a common suction line to the pump. Blocking one suction inlet shall not create excessive suction at other suction inlets.

7.23.4 Obstructions and Entrapment Avoidance

Where vacuum cleaning fittings are provided, they shall be located outside of the spa or hot tub and shall not be accessible to the spa or hot tub user.

7.24 PLUMBED EMERGENCY EYEWASH AND SHOWER EQUIPMENT

a. Emergency eyewash, eye/face wash, and shower equipment shall comply with the requirements of ANSI/ISEA Z358.1

b. The installer shall assemble and install the equipment in accordance with the manufacturer's instructions.

c. The location of the emergency equipment shall be determined by the facility designer to provide the required access for protection from the specific hazards within the facility.

d. The emergency equipment shall be designed and manufactured to provide the discharge patterns and flow rates required by ANSI/ISEA Z358.1 for the hazard(s) for which protection is being provided.

e. The temperature of the flushing fluid shall be from 60°F to 100°F unless the facility designer has indicated that other specific temperatures are required to provide protection from a particular hazard.

f. Temperature actuated mixing valves for plumbed emergency equipment shall comply with ASSE 1071.

g. The minimum required water supply shall be 0.4 gpm for eyewashes, 3.0 gpm for eye/face washes, and 20 gpm for showers in accordance with ANSI/ISEA Z358.1.

h. Water shutoff valves shall have provisions to prevent unauthorized closure.

i. Equipment with waste pipe connections shall be connected to the building drain piping. Floor drains shall be provided for showers and equipment without waste connections.

j. Where necessary for the hazards that the emergency equipment is being provided for, the drainage shall be neutralized or otherwise treated before discharge into the building drainage system. Acid waste shall be connected to an acid waste disposal system if available.

7.25 COMMERCIAL DISHWASHING PRE-RINSE SPRAY VALVES

Commercial dishwashing pre-rinse spray valves shall have a maximum flow rate of 1.6 gallons per minute at 60 psi.

Table 7.21.1 see (1)
MINIMUM NUMBER OF REQUIRED PLUMBING FIXTURES - Page 1

Classification	Occupancy Group see (14)	No. of Persons of Each Sex see (4), (16)	Water Closets see (5) for urinals		Lavatories		Bath or Shower	Drinking Water Facilities	Other
			Male	Female	Male	Female			
Assembly (A)	A-1: Theaters and other facilities for the viewing of the performing arts or motion pictures, typically with fixed seats. See Notes: (6), (9)	100 or less	2	2	1	1		1 per 500 people	1 service sink per floor (21)
		101 - 200	add 1	add 1	add 1	add 1			
		201-300	add 1	add 2	add 0	add 1			
		ea. add'l. 300 over 300	add 1	add 2	add 1	add 1			
	A-2: Night clubs, bars, taverns, and similar facilities where drink and/or food is consumed. See Notes: (3), (6), (8), (9), (13)	25 or less	1	1	1	1		1 per 250 people (2)	1 service sink per floor (21)
		26 - 50	add 1	add 1	add 1	add 1			
		51 - 100	add 1	add 1	add 1	add 1			
		ea. add'l. 200 over 100	add 1	add 1	add 1	add 1			
	A-2: Restaurants, banquet halls, food courts, carry-outs with seating. See Notes: (3), (6), (8), (9), (13)	50 or less	1	1	1	1		1 per 250 people (2)	1 service sink per floor (21)
		51 - 100	add 1	add 1	add 1	add 1			
		101 – 200	add 1	add 1	add 0	add 0			
		201 - 300	add 1	add 1	add 1	add 1			
		ea. add'l. 200 over 300	add 1	add 1	add 1	add 1			
	A-3: Museums, art galleries, exhibition halls, lecture halls, libraries. See Notes: (6), (9)	50 or less	1	1	1	1		1 per 500 people	1 service sink per floor (21)
		51 - 150	add 1	add 1	add 1	add 1			
		151 - 300	add 1	add 1	add 1	add 1			
		ea. add'l. 300 over 300	add 1	add 1	add 1	add 1			
	A-3: Passenger terminals, transportation facilities. See Notes: (6), (9)	100 or less	1	1	1	1		1 per 500 people	1 service sink per floor (21)
		101 - 250	add 2	add 2	add 1	add 1			
		ea. add'l. 250 over 250	add 2	add 2	add 1	add 1			
	A-3: Places of worship and other religious services. See Notes: (6), (9)	25 or less	1	1	1	1		1 per 500 people	1 service sink per floor (21)
		26 - 150	add 1	add 1	add 1	add 1			
		ea. add'l. 150 over 150	add 1	add 1	add 1	add 1			
	A-3: Gymnasiums (without spectator seating), health spas. See Notes: (6), (9), (20)	50 or less	1	1	1	1		1 per 200 people	1 service sink per floor (21)
		51 - 100	add 1	add 1	add 1	add 1			
		101 - 200	add 1	add 1	add 1	add 1			
		ea. add'l. 200 over 200	add 2	add 2	add 2	add 2			
	A-4: Coliseums, arenas, skating rinks, pools, and tennis courts for indoor sporting events and activities. See Notes: (6), (9), (20)	1 - 100	2	2	2	2		1 per 500 people	1 service sink per floor (21)
		101 - 200	add 2	add 2	add 2	add 2			
		201 - 400	add 2	add 2	add 2	add 2			
		ea. add'l. 400 up to 2400	add 2	add 4	add 1	add 2			
		ea. add'l. 600 over 2400	add 2	add 4	add 1	add 2			

Table 7.21.1 see (1)
MINIMUM NUMBER OF REQUIRED PLUMBING FIXTURES - Page 2

Classification	Occupancy Group see (14)	No. of Persons of Each Sex see (4), (16)	Water Closets see (5) for urinals		Lavatories		Bath or Shower	Drinking Water Facilities	Other
			Male	Female	Male	Female			
Assembly (A) (continued)	A-5: Stadiums, amusement parks, bleachers and grandstands for outdoor sporting events and activities. See Notes: (6), (9), (20)	1 - 100	2	2	2	2		1 per 1000 people	1 service sink per floor (21)
		101 - 200	add 2	add 2	add 2	add 2			
		201 - 400	add 2	add 2	add 2	add 2			
		ea. add'l. 400 up to 2400	add 2	add 4	add 1	add 2			
		ea. add'l. 600 over 2400	add 2	add 4	add 1	add 2			
Business (B)	B: Buildings for the transaction of business, professional services, office buildings, banks, and similar uses. See Notes: (6), (9), (12), (15)	1 - 15	1	1	1	1		1 per 100 people (2)	1 service sink per floor (21)
		16 - 50	add 1	add 1	add 0	add 0			
		ea. add'l. 50 over 50	add 1	add 1	add 1	add 1			
Education (E)	E-1: Day Care, Preschool, Kindergarten. See Notes: (6), (9)	15 or less	1	1	1	1		1 per 100 people	1 service sink per floor (21)
		16 - 30	add 1	add 1	add 1	add 1			
		31 - 50	add 1	add 1	add 1	add 1			
		51 - 75	add 1	add 1	add 1	add 1			
		76 - 100	add 1	add 1	add 1	add 1			
		ea. add'l. 50 over 100	add 1	add 1	add 1	add 1			
	E-2: Elementary Grades 1 – 6 See Notes: (6), (9)	25 or less	1	1	1	1		1 per 100 people	1 service sink per floor (21)
		26 - 50	add 1	add 1	add 1	add 1			
		51 - 75	add 1	add 1	add 1	add 1			
		76 - 100	add 1	add 1	add 1	add 1			
		ea. add'l. 50 over 100	add 1	add 1	add 1	add 1			
	E-3: Secondary Grades 7 – 12 and higher education facilities. See Notes: (6), (9), (17)	30 or less	1	1	1	1		1 per 100 people	1 service sink per floor (21)
		31 - 60	add 1	add 1	add 1	add 1			
		61 - 100	add 1	add 1	add 1	add 1			
		ea. add'l. 50 over 100	add 1	add 1	add 1	add 1			
Factory (F)	F-1 and F-2: Structures in which occupants are engaged in work fabricating, assembling, or processing products, materials, food, etc. See Notes: (6), (9)	25 or less	1	1	1	1	1 emergency shower per 10 people exposed to skin contamination	1 per 250 people	1 service sink per floor (21)
		26 - 50	add 1	add 1	add 1	add 1			
		51 - 100	add 1	add 1	add 1	add 1			
		ea. add'l. 100 over 100	add 1	add 1	add 1	add 1			
Institutional (I)	I-1: Residential care, assisted living facilities (24-hour). See Notes: (6), (9), (11)	1 - 10	1	1	1	1	1 per 8 occupants	1 per 100 occupants	1 service sink per floor (21)
		ea. add'l. 10 over 10	add 1	add 1	add 1	add 1			

Table 7.21.1 see (1)
MINIMUM NUMBER OF REQUIRED PLUMBING FIXTURES - Page 3

Classification	Occupancy Group see (14)	No. of Persons of Each Sex see (4), (16)	Water Closets see (5) for urinals		Lavatories		Bath or Shower	Drinking Water Facilities	Other
			Male	Female	Male	Female			
Institutional (I) (continued)	I-2: Patients in hospitals, nursing homes, etc.		1 water closet per patient room		1 lavatory per patient room		1 per patient room		
	I-2: Employees in hospitals, etc. See Notes: (6), (9)	25 or less	1	1	1	1		1 per 100 people	1 service sink per floor per wing (21)
		26 - 50	add 1	add 1	add 0	add 0			
		51 - 100	add 1	add 1	add 1	add 1			
		ea. add'l. 100 over 100	add 1	add 1	add 1	add 1			
	I-2: Visitors in hospitals, etc. See Notes: (6), (9)	50 or less	1	1	1	1		1 per 500 people	
		51 - 100	add 1	add 1	add 1	add 1			
		ea. add'l. 100 over 100	add 1	add 1	add 1	add 1			
	I-3: Prisoners in prisons.		1 per cell		1 per cell		1 per 15 prisoners	1 per 100 prisoners	
	I-3: Detainees in reformatories, detention centers, and correctional centers.	15 or less	1	1	1	1	1 per 15 detainees	1 per 100 detainees	
		16 - 30	add 1	add 1	add 1	add 1			
		31 - 50	add 1	add 1	add 1	add 1			
		51 - 75	add 1	add 1	add 1	add 1			
		76 - 100	add 1	add 1	add 1	add 1			
		ea. add'l. 50 over 100	add 1	add 1	add 1	add 1			
	I-3: Employees in prisons, reformatories, etc. See Note: (9)	25 or less	1	1	1	1		1 per 100 people	1 service sink per floor per wing (21)
		26 - 50	add 1	add 1	add 0	add 0			
		51 - 100	add 1	add 1	add 1	add 1			
		ea. add'l. 50 over 100	add 1	add 1	add 1	add 1			
	I-3: Visitors in prisons, reformatories, etc. See Note: (9)	50 or less	1	1	1	1		1 per 500 people	
		51 - 100	add 1	add 1	add 1	add 1			
		ea. add'l. 100 over 100	add 1	add 1	add 1	add 1			
	I-4: Adult day care, child day care. See Note: (9)	15 or less	1	1	1	1		1 per 100 people	1 service sink per floor (21)
		16 - 30	add 1	add 1	add 1	add 1			
		31 - 50	add 1	add 1	add 1	add 1			
		51 - 75	add 1	add 1	add 1	add 1			
		76 - 100	add 1	add 1	add 1	add 1			
		ea. add'l. 50 over 100	add 1	add 1	add 1	add 1			
Mercantile (M)	M: Stores, shops, markets, shopping centers, sales rooms, service stations, etc. See Notes: (6), (7), (9), (12), (15)	50 or less	1	1	1	1		1 per 500 people (2)	1 service sink per floor (21)
		51 - 150	add 1	add 1	add 0	add 0			
		151 - 400	add 1	add 1	add 1	add 1			
		ea. add'l. 400 over 400	add 1	add 1	add 1	add 1			

Table 7.21.1 see (1)
MINIMUM NUMBER OF REQUIRED PLUMBING FIXTURES - Page 4

Classification	Occupancy Group see (14)	No. of Persons of Each Sex see (4), (16)	Water Closets see (5) for urinals		Lavatories		Bath or Shower	Drinking Water Facilities	Other
			Male	Female	Male	Female			
Residential (R)	R-1: Hotels, motels, boarding houses (transient)		1 water closet per guest room		1 lavatory per guest room		1 per guest room		1 service sink per floor (21)
	R-2: Dormitories, fraternities, sororities, boarding houses (not transient) See Note: (9)	10 or less	2	2	2	2	1 per 8 people	1 per 100 people	1 service sink per floor (21)
		11 - 20	add 1	add 1	add 1	add 1			
		ea. add'l. 20 over 20	add 2	add 2	add 2	add 2			
	R-2: Apartment houses, condominiums.		1 water closet per unit		1 lavatory per unit		1 per unit		1 kitchen sink per unit provisions for 1 clothes washer per 20 units (10)
	R-2: One- and two-family dwellings.		1 water closet per unit		1 lavatory per unit		1 per unit		1 kitchen sink per unit provisions for 1 clothes washer per unit
	R-3: Congregate living facilities with 16 or fewer persons. See Note: (9)	10 or less	1	1	1	1	1 per 8 people	1 per 100 people	1 service sink per floor (21)
		ea. add'l. 10 over 10	add 1	add 1	add 1	add 1			
	R-4: Residential care, assisted living facilities. See Note: (9)	10 or less	1	1	1	1	1 per 8 residents	1 per 100 people	1 service sink per floor (21)
		ea. add'l. 10 over 10	add 1	add 1	add 1	add 1			
Storage (S)	S-1 & S-2: Structures for the storage of goods, warehouses, storehouses, and freight depots. See Notes: (6), (9), (18), (19)	50 or less	1	1	1	1		1 per 500 people	1 service sink per floor (21)
		51 - 100	add 1	add 1	add 0	add 0			
		ea. add'l. 100 over 100	add 1	add 1	add 1	add 1			

Notes for Table 7.21.1 (where indicated in the Table):

(1) Plumbing fixtures shall be provided in numbers not less than those shown in this Table for the type of building occupancy (7.21.1). For accessible requirements, see local, state, and national codes. Additional fixtures may be required where environmental conditions or special activities may be encountered.

(2) Drinking water facilities for customers are not required in restaurants or other food service establishments if drinking water service is available (7.21.8.c). Drinking water facilities are not required for customers in mercantile or business establishments where normal occupancy is short term (7.21.7.c). Kitchen sinks and bar sinks may be used for employee drinking water facilities (7.21.5.b).

(3) In food preparation areas, fixture requirements may be dictated by local Health Codes (7.21.6.b).

(4) Wherever both sexes are present in approximately equal numbers, multiply the total census by 50% to determine the number of persons of each sex to be provided for (7.21.2.c). This regulation applies only where specific information that would otherwise affect the fixture count is not provided.

(5) Not more than 50% of the required number of water closets for males may be urinals (7.21.5.a).

(6) In buildings with multiple floors, access to fixtures shall not exceed one vertical story (7.21.3.a).

(7) Fixtures for customers and employees as required by this Table may be met by providing centrally located facilities accessible to several stores (7.21.7.b). The maximum distance from the entry for any store to these facilities shall not exceed 500 feet (7.21.7.b).

(8) Where food service establishments and food courts are open to public areas in malls and transportation terminals, toilet facilities for both customers and employees may be centrally located and serve multiple establishments (7.21.8.b).

(9) Fixtures accessible only to private offices shall not be counted to determine compliance with this Table (7.21.3.b).

(10) Facilities without laundry rooms for their occupants shall not require provisions for clothes washers.

(11) In residential care facilities where water closets and lavatories are provided in individual patient rooms, the minimum number of fixtures for employees and visitors shall be as required for hospitals.

(12) Requirements for employees and customers may be met with a single set of restrooms (7.21.7.a).

(13) If the design number of customers in food handling establishments exceeds 100, separate facilities for employees and customers are required (7.21.8.a).

(14) Occupancy groups shall be as described in the local building code.

(15) See 7.21.4 and 7.21.7 for toilet facilities for occupancies with a total floor area of 1500 square feet or less.

(16) In determining the number of required fixtures for numbers of persons that fall in the "each additional (xx) over (xx)" listings, the requirement applies to fractions of the listed group (7.21.10).

(17) Laboratories in higher education facilities shall have safety showers if required by ANSI/ISEA Z358.1 and the facility design.

(18) Warehouse storage area requirements shall be permitted to be met by providing a facility centrally located within the storage area. The maximum travel distance to the facility shall not exceed 500 feet.

(19) The requirements for multiple individual self-storage areas shall be permitted to be met by fixtures located in the facility's administration building. The administration office must be accessible during normal business hours.

(20) Showers may be omitted in recreational facilities without locker rooms when approved by the Authority Having Jurisdiction.

(21) Service sinks may not be required on floor levels if the AHJ determines that housekeeping is not required (7.21.5.d). Service sinks shall be permitted to serve two adjacent floors (one above and one below) where there is service elevator access.

Chapter 8

Hangers and Supports

8.1 GENERAL

a. Hangers and anchors shall be securely attached to the building construction at sufficiently close intervals to support the piping and its contents.

b. Fixtures, appliances and equipment shall be connected to support the weight of the device and any additional probable loads that may impact on the device.

c. Fixtures shall be rigidly supported so that no strain is transmitted in the piping connections.

8.2 VERTICAL PIPING

Vertical piping of the following materials shall be supported according to manufacturer's recommendations, but at no greater than the distances listed below:

1. Cast-iron soil pipe—at base and at each story height not to exceed 15 feet.
2. Steel threaded pipe—at every other story height.
3. Copper tube—at each story height but not more than 10-foot intervals.
4. Lead pipe—four-foot intervals.
5. Plastic pipe—see Section 8.7.
6. Flexible plastic tubing—each story height and at mid-story.
7. Stainless steel drainage pipe—at each story height.

8.3 HORIZONTAL PIPING

Horizontal piping of the following materials shall be supported according to manufacturer's recommendations, but at no greater than the distances listed below:

1. Cast-iron soil pipe—minimum of one hanger per pipe length located within 18 inches of each joint (up to 10-foot maximum pipe length), at changes in direction, and at branch connections. Where pipe is suspended by non-rigid hangers more than 18 inches long, provide lateral support at 40-foot maximum spacing for pipe sizes 5 inches and larger. Lateral support shall consist of either 1) a sway brace or 2) either a change in direction or a branch connection that provides the required lateral support.
2. Steel threaded pipe—3/4 inch size and smaller—10-foot intervals. One-inch size and larger—12-foot intervals.
3. Copper tube (1-1/4 inch size and smaller)—6-foot intervals.
4. Copper tube (1-1/2 inch size and larger)—10-foot intervals.
5. Lead pipe—on continuous metal or wood strips for its entire length.
6. Plastic pipe—see Section 8.7.
7. Flexible plastic tubing—32 inches.
8. Stainless steel drainage pipe—10-foot intervals, changes of direction and branch connections.

8.4 MATERIAL

a. Hangers, anchors, and supports shall be of metal or other material of sufficient strength to support the piping and its contents.

b. Piers shall be of concrete, brick, or other masonry construction.

8.5 STRAIN AND STRESS IN PIPE

a. Piping in the plumbing system shall be installed so as to prevent strains and stresses that will exceed the allowable stress for the pipe. Provision shall be made for expansion and contraction of the piping material installed. (See Sections 4.1.3 and 4.2.16).

b. Pipe guides shall be provided on each side of expansion loops and expansion compensators.
See Figure 8.5-A, 8.5-B, and 8.5-C

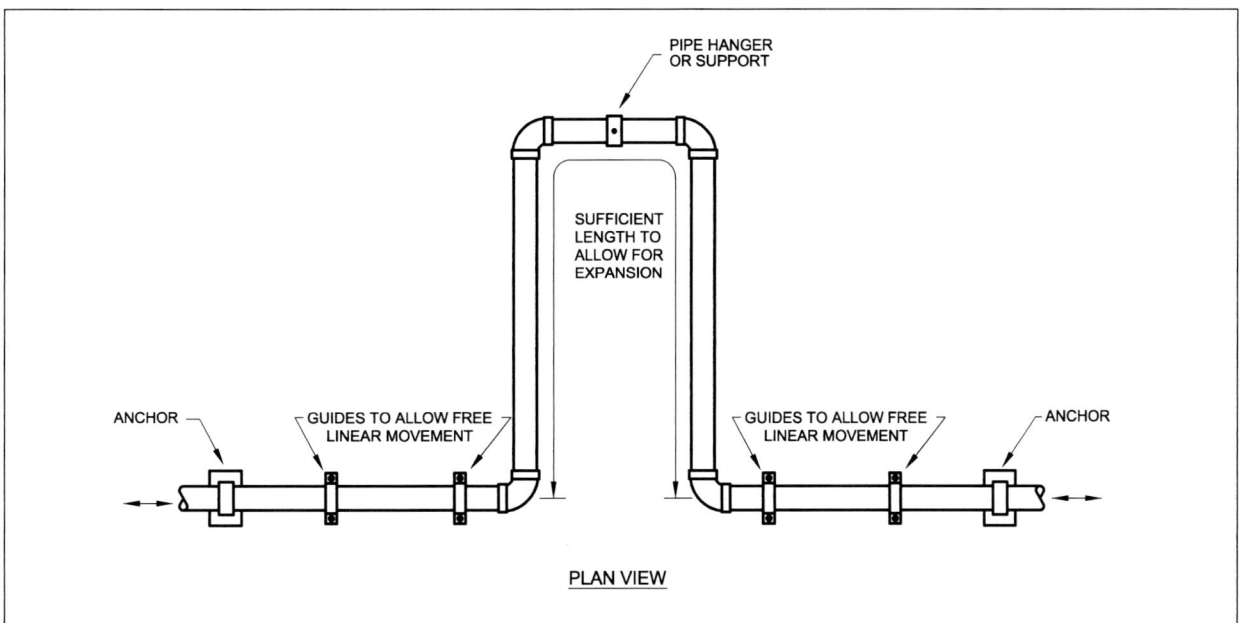

Figure 8.5 - A
AN EXPANSION LOOP IN A HOT WATER PIPE LINE

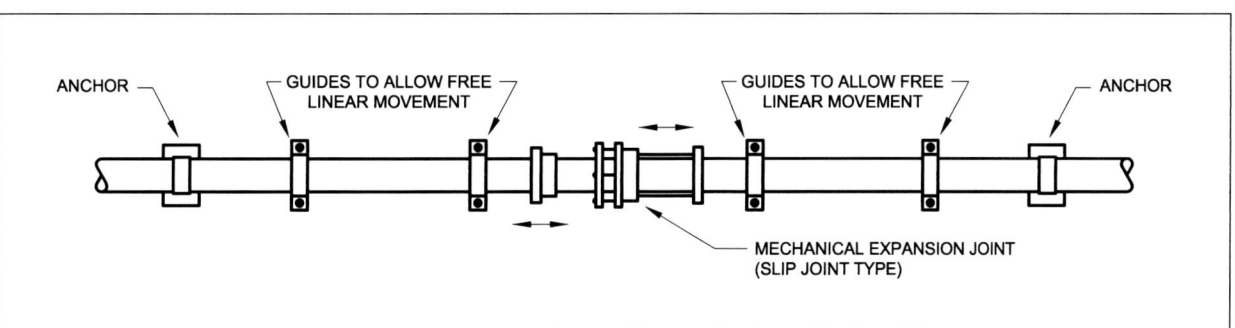

Figure 8.5 - B
AN EXPANSION COMPENSATOR IN A HOT WATER PIPE LINE

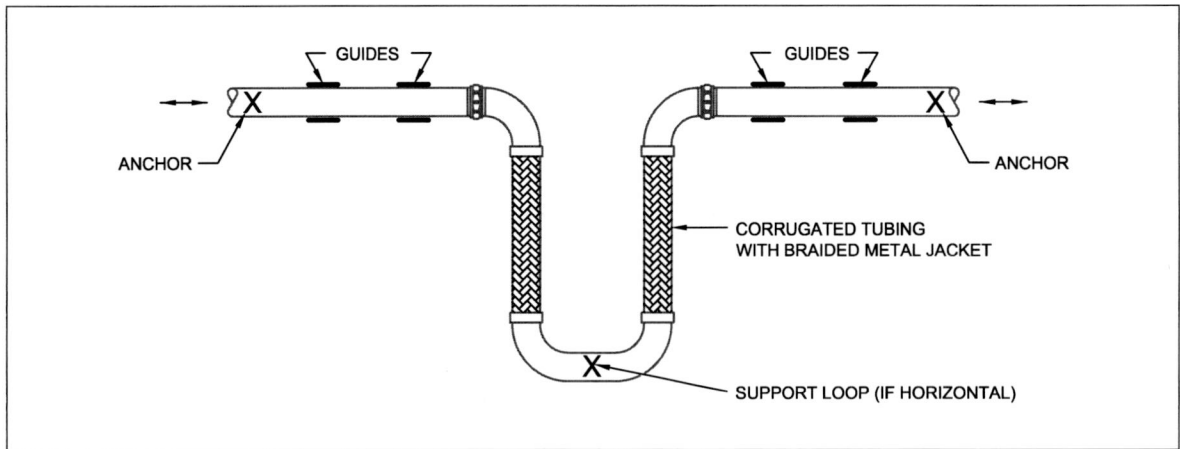

**Figure 8.5 - C
A FLEXIBLE EXPANSION LOOP TO ABSORB THERMAL EXPANSION**

8.6 BASE OF STACKS

Bases of cast-iron stacks shall be supported on concrete, brick laid in cement mortar, metal brackets attached to the building construction, or by other methods approved by the Authority Having Jurisdiction. Other piping material shall be so anchored as to support the stack at the base.

8.7 SUPPORT OF PLASTIC PIPE

a. Plastic drain, waste, vent, and pressure pipe shall be installed and supported as recommended by the manufacturer's instructions.

b. Maximum horizontal support spacing shall be based on the pipe schedule or wall thickness, the pipe size, the system operating temperature, the ambient temperature, and any concentrated loads.

c. Vertical pipe shall be maintained in straight alignment with supports at each story height. Intermediate supports shall be provided where required for stability.

d. Pipe shall also be supported at changes of direction or elevation.

e. Supports shall not compress, distort, cut, or abrade the piping and shall allow free movement.

f. Provisions shall be made for expansion and contraction of the piping.

g. Fixture trap arms longer than three feet shall be supported as close as possible to the trap.

8.8 UNDERGROUND INSTALLATION

See Section 2.6.

8.9 SEISMIC SUPPORTS FOR PIPING

Where earthquake loads are applicable in accordance with the adopted building code, plumbing piping supports shall be designed and installed for the seismic forces in accordance with the adopted building code.

8.10 ALTERNATE PIPE HANGER AND SUPPORT SPACING

In lieu of the pipe hanger and support spacing required by Section 8.2 for vertical piping and Section 8.3 for horizontal piping, the pipe support spacing shall be permitted to comply with MSS SP-58 and the pipe manufacturer's recommendations. The spacing for water piping, waste and sanitary drain piping, and storm water piping shall be based on full water. The spacing for vent piping shall be permitted to be based on full vapor.

Blank Page

Chapter 9

Indirect Waste Piping and Special Wastes

9.1 INDIRECT WASTES

9.1.1 General

Drains from fixtures, fixture compartments, equipment, appliances, appurtenances, and other devices requiring protection against contamination from backflow or flooding from the drainage system or other source shall not be directly connected to the drainage system. Such drains shall discharge separately and indirectly to the drainage system through an air gap or, where permitted, an air break. Indirect wastes shall be trapped if required by Section 9.2.3.

9.1.2 Air Gaps

The clear air gap between a drain outlet or indirect waste pipe and the flood level rim of an indirect waste receptor or other point of disposal shall be not less than twice the diameter of the effective opening of the drain served, but not less than one inch.

9.1.3 Air Breaks

Air breaks shall be permitted where the waste pipe being indirectly drained is not subject to back siphonage. The waste pipe shall be permitted to terminate below the flood level rim of its receptor but shall maintain an air space above the receptor's drain outlet or above the top of its trap seal. Such indirect waste pipes shall not block or restrict the drain outlet of the receptor. They shall be permitted to connect to the inlet side of the receptor's trap above the trap seal. ***See Figure 1.2.2***

9.1.4 Where Indirect Wastes Are Required

Indirect wastes shall be provided for food-handling or food-storage equipment, medical or other sterile equipment, clear-water wastes or discharges, and other drains as required herein.

9.1.5 Food Handling Areas

   a. Fixtures and appliances used for the storage, processing, preparation, serving, dispensing, or sale of food shall be drained indirectly. Examples of such fixtures include refrigerated cases, steam kettles, dishwashing machines, culinary sinks and/or sink compartments used for rinsing, sanitizing, soaking or washing food, ice machines, ice storage bins, drink dispensers, and similar equipment or appliances. A separate indirect waste pipe shall be provided for each fixture and/or compartment drain and each shall discharge separately through an air gap or air break into a trapped and vented receptor.

   b. Where bar sinks, glass-washing sinks, or other counter sinks cannot be vented according to the requirements of Chapter 12, they shall be permitted to each discharge separately to a trapped and vented receptor through indirect waste pipes providing either an air break or an air gap.
See Figure 9.1.5-A through -C

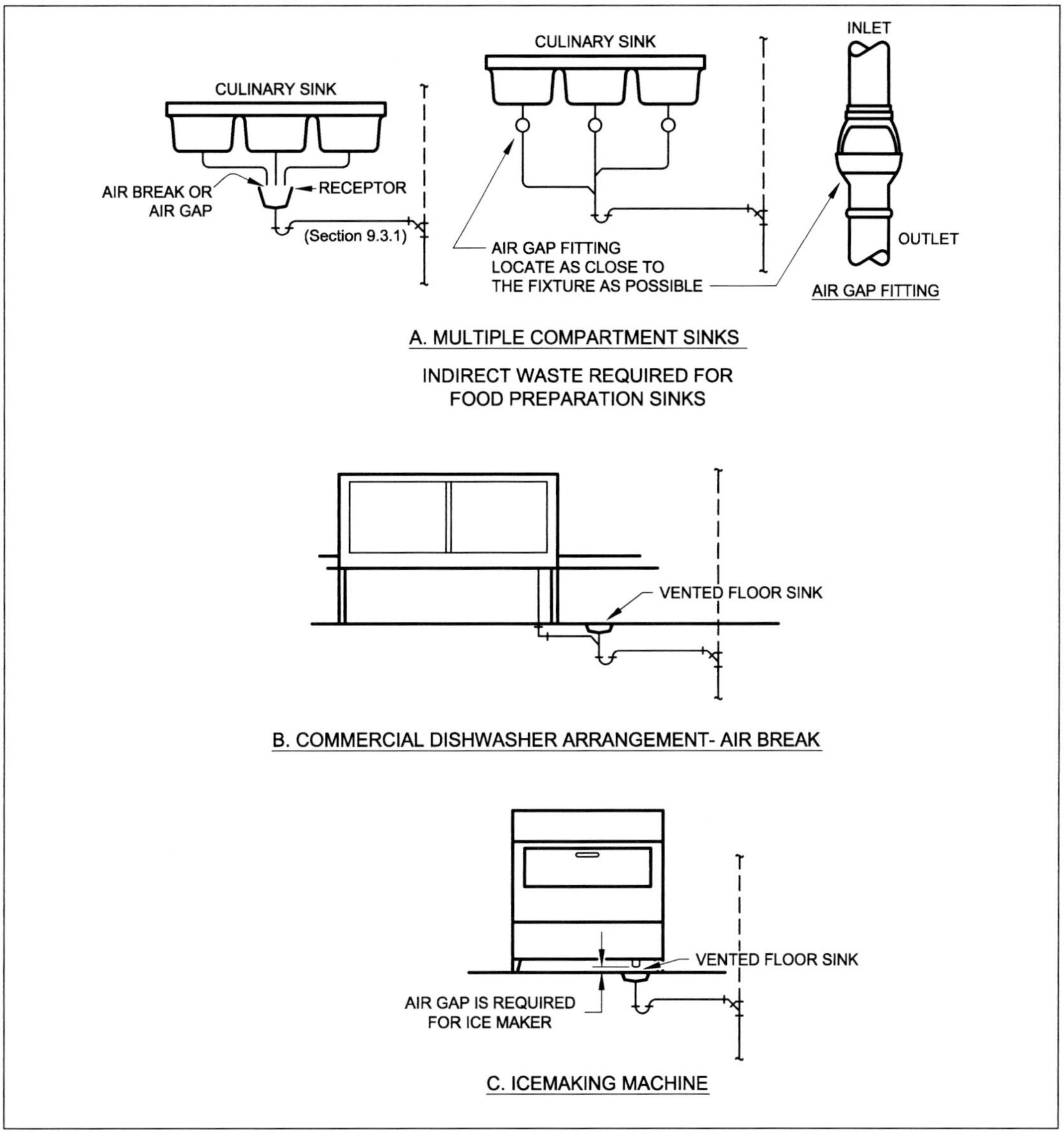

Figure 9.1.5 - A
INDIRECT WASTE PIPING IN A FOOD HANDLING AREA

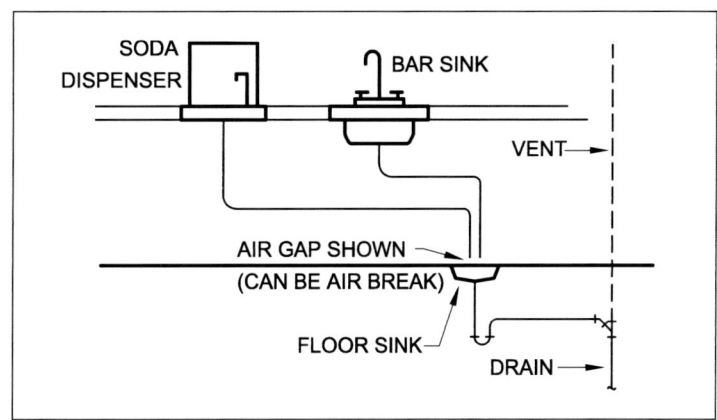

Figure 9.1.5 - B
INDIRECT WASTES FOR COUNTER SINKS AND COUNTER-MOUNTED EQUIPMENT

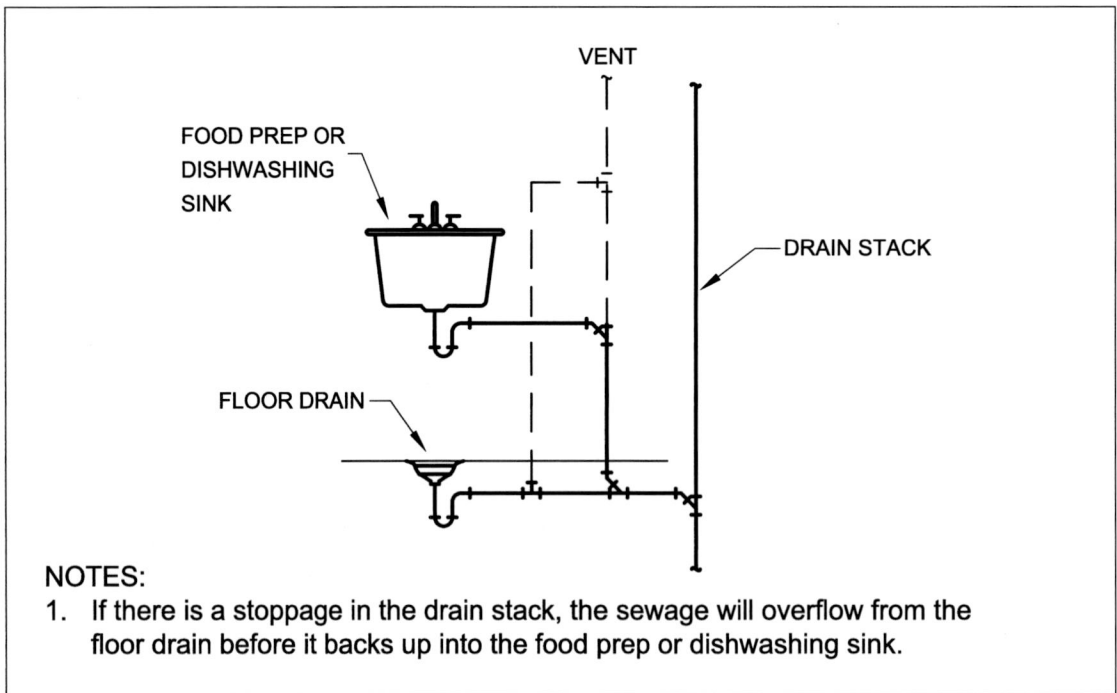

**Figure 9.1.5 - C
PROTECTING A FOOD PREP OR
DISHWASHING SINK FROM SEWAGE BACKUP**

EXCEPTIONS:

(1) For multi-compartment commercial sinks, only the compartment used for washing pots, tableware, kitchenware, and utensils shall discharge to the drainage system through a grease interceptor in accordance with Sections 6.1.1 and 6.2.

(2) The rinsing and sanitizing compartments of multi-compartment commercial sinks shall be drained indirectly, in accordance with Section 9.1.1.

(3) If a properly vented floor drain is installed immediately adjacent to a sink used for dishwashing or food prep, a properly trapped and vented sink or sink compartment shall be permitted to connect directly to the drainage system, on the sewer side of the floor drain trap.

(4) Indirect drains shall not be required for domestic kitchen sinks or domestic dishwashers.

9.1.6 Walk-in Coolers and Freezers

a. If floor drains are located in walk-in coolers or walk-in freezers used for the storage of food or other products for human consumption, they shall be indirectly connected to the sanitary drainage system.

b. Separate indirect waste pipes shall be provided for the floor drains from each cooler or freezer, and each shall discharge separately through an air gap or air break into a trapped and vented receptor.

c. Traps shall be provided in the indirect waste pipe when required under Section 9.2.3.

d. Indirectly connected floor drains may be located in freezers or other spaces where freezing temperatures are maintained, provided that traps are not required under Section 9.2.3. Otherwise, the floor of the freezer shall be sloped to a floor drain located outside the storage compartment.

e. The above requirements do not apply to refrigerated food preparation areas or work rooms.

9.1.7 Medical and Other Sterile Equipment

Stills, sterilizers, and other sterile equipment requiring drainage shall each discharge separately through an air gap into a trapped and vented receptor.

9.1.8 Potable Clear-Water Wastes

Discharges of potable water from the water distribution system, water storage or pressure tanks, water heaters, water pumps, water treatment equipment, boilers, relief valves, backflow preventer, and other potable water sources shall be indirect through an air gap.

EXCEPTION: An air break shall be permitted where the potable water supply to boilers, water-cooled equipment, heating and air-conditioning systems, and similar cross-connections is protected from backflow in accordance with Section 10.5.

9.1.9 Drinking Fountains and Water Coolers

Drinking fountains and water coolers shall be permitted to discharge indirectly through an air break or air gap. Where such fixtures are connected to a dedicated drain stack, the fixtures may connect directly to the stack and the stack shall terminate with an air break or air gap.

9.1.10 Air Conditioning Equipment

Where condensate or other drainage from air conditioning or cooling equipment discharges to a drainage system, it shall discharge indirectly to a trapped and vented receptor through an air break or air gap.
EXCEPTION: An air break shall not be permitted where the drain connects to a point in the air conditioning equipment that operates at a pressure below atmospheric.
See Figure 9.1.10

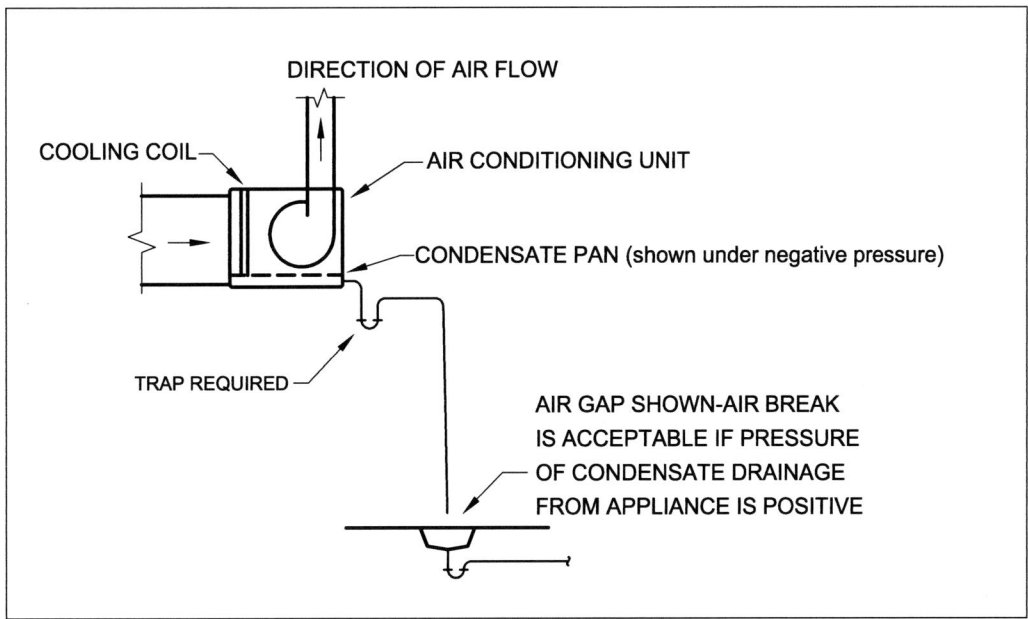

Figure 9.1.10
DRAINING AIR CONDITIONING CONDENSATE

9.1.11 Swimming Pools

Drainage from swimming pools or wading pools, including pool drains, filter backwash, overflows, and pool deck drains, shall discharge indirectly through an air gap to a trapped and vented receptor.

9.1.12 Relief Valve Discharge Piping

Discharge piping from relief valves and any associated indirect waste piping shall be in accordance with Section 10.16.6.

9.2 INDIRECT WASTE PIPING

9.2.1 Materials and Installation

Indirect waste piping shall be of materials approved for sanitary drainage under Section 3.5.

9.2.2 Pipe Size

Indirect waste piping shall be not less than the nominal size of the drain outlet on the fixture or equipment served.

9.2.3 Fixture Traps

Traps shall be provided at fixtures and equipment connections where the developed length of indirect waste piping exceeds ten feet.
EXCEPTION: Drain lines used for clear-water wastes.

9.2.4 Provisions for Cleaning

Indirect waste piping shall be installed in a manner to permit access for flushing and cleaning. Where necessary, cleanouts shall be provided in accordance with Section 5.4.

9.3 INDIRECT WASTE RECEPTORS

9.3.1 General

a. Receptors for indirect wastes shall be properly trapped and vented floor drains, floor sinks, standpipes, open-hub drains, air gap fittings, or other approved fixtures.

b. Receptors shall be of such size, shape, and capacity as required to prevent splashing or flooding by the discharge from any and all indirect waste pipes served by the receptor.

c. Plumbing fixtures that are used for domestic or culinary purposes shall not be used as receptors for indirect wastes, except as follows:

    1. In a dwelling unit, a kitchen sink trap or food waste disposer shall be permitted to receive the discharge from a dishwasher.

    2. In a dwelling unit, a laundry sink, provided that an air gap is maintained for any potable clear-water waste, shall be an acceptable receptor for:

        a) Air conditioning condensate.
        b) Automatic clothes washer.
        c) Water treatment unit.
        d) Water heater relief valve discharge
        e) Condensing furnace condensate.

    3. A service sink or mop basin shall be an acceptable receptor for air conditioning and condensing furnace condensate and any infrequent potable clear-water waste if the required air gap is provided for potable clear-water wastes.

See Figures 7.15.2, 7.15.3, 7.15.4, 9.1.5-A, 9.1.5-B, and 9.1.10

9.3.2 Strainers or Baskets

Floor sinks and floor drains that handle other than clear-water wastes shall include an internal strainer or a removable metal basket to retain solids. ***See Figure 9.3.2***

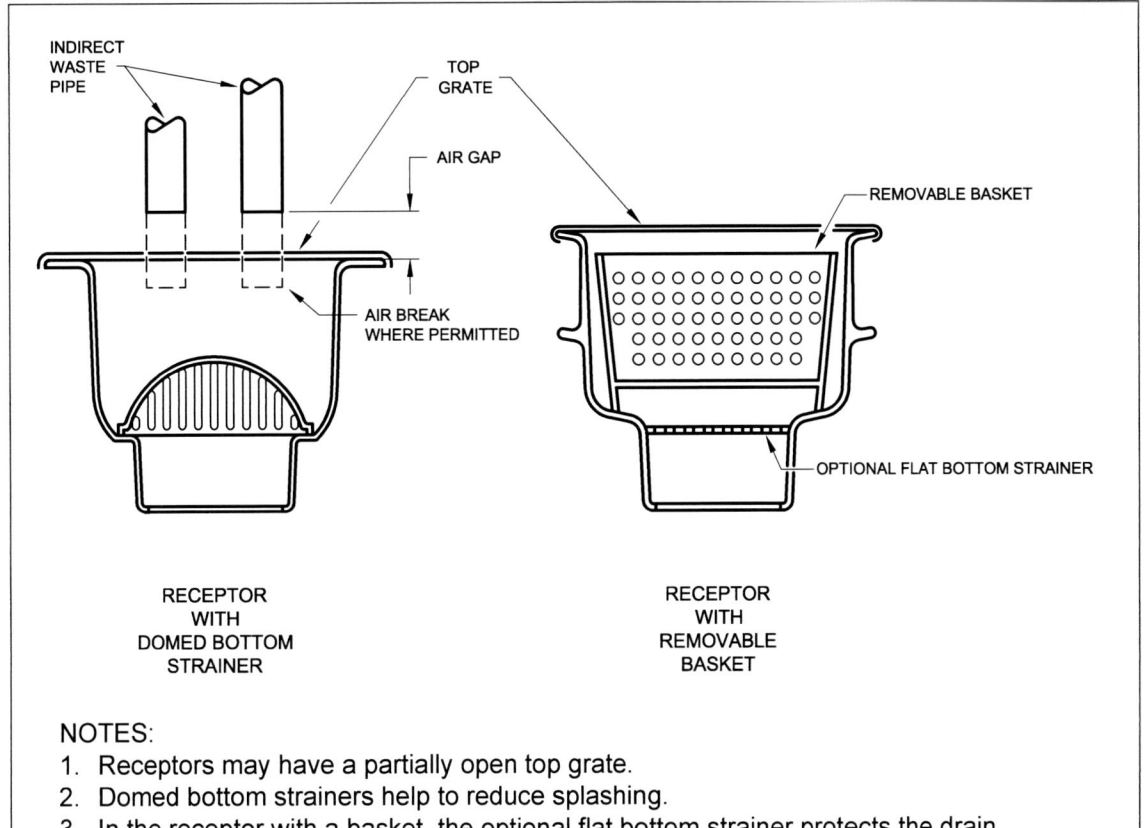

Figure 9.3.2
FLOOR SINKS WITH STRAINER OR BASKET

9.3.3 Prohibited Locations

Receptors for indirect wastes shall not be located in a toilet room or in any confined, concealed, inaccessible, or unventilated space.

EXCEPTION: Air conditioning condensate in dwellings shall be permitted to drain to a tub waste and overflow or lavatory tailpiece in accordance with Section 9.4.3.c.5.

9.3.4 Standpipes

A standpipe, 2-inch minimum pipe size and extending not more than 48 inches nor less than 18 inches above its trap, shall be permitted to serve as a receptor for a domestic clothes washer. In a dwelling, a laundry sink shall be permitted to drain into the standpipe.

See Figure 9.3.4

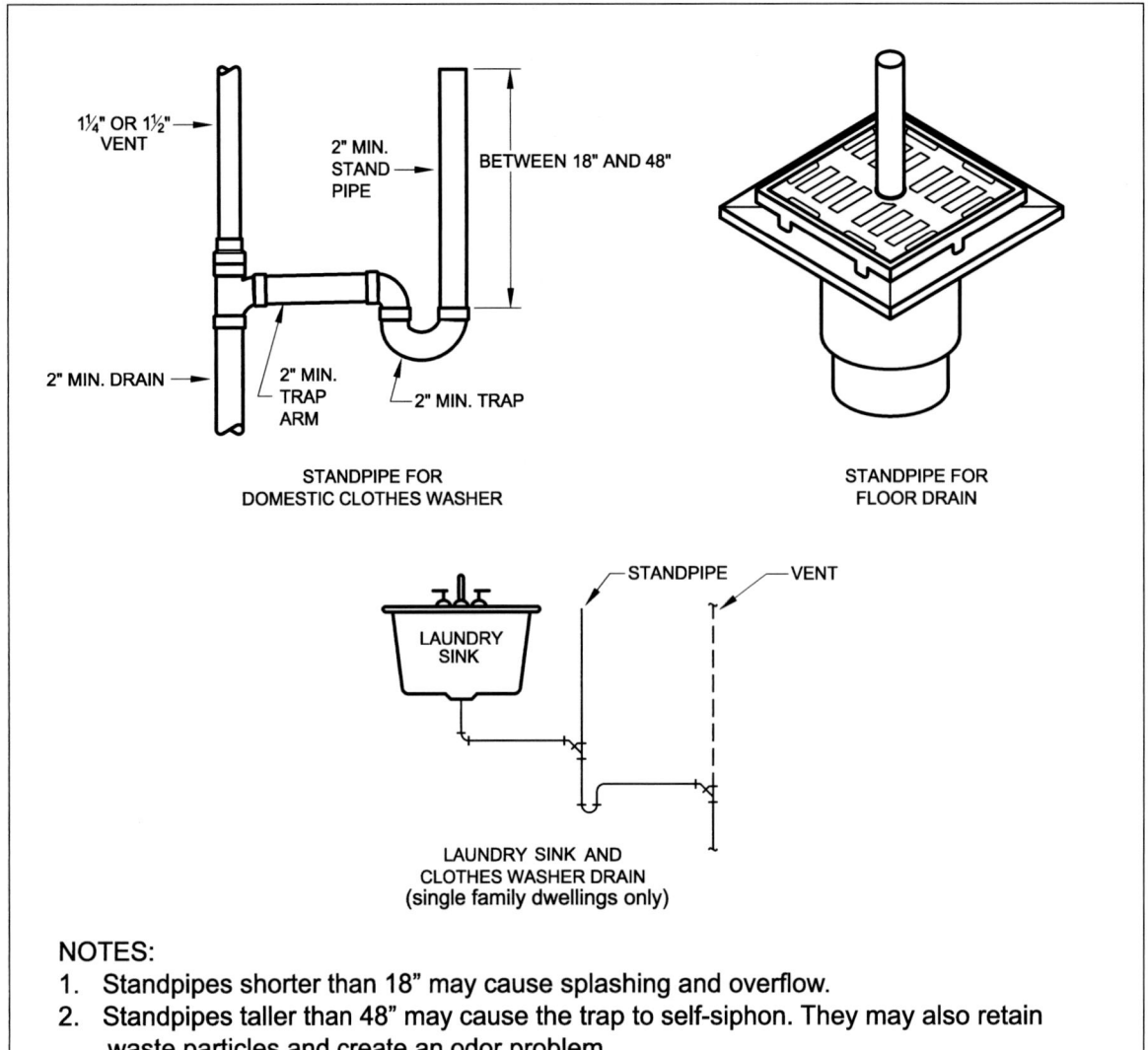

**Figure 9.3.4
STANDPIPE RECEPTORS**

9.3.5 Open-hub Drains

A trapped and vented open-end drain pipe extending not less than 2 inches above the surrounding floor shall be permitted to serve as a receptor for clear-water wastes.

9.3.6 Minimum Receptor Pipe Size

a. The minimum drain pipe size for an indirect waste receptor shall be at least one pipe size larger than the indirect waste pipe that it serves.

EXCEPTION: A laundry sink receiving the discharge from an automatic clothes washer under Section 9.3.1.c(2).b.

b. Where a receptor receives indirect drainage from two or more fixtures, the cross-sectional area of the receptor drain shall be not less than the aggregate cross-sectional area of all indirect waste pipes served by the receptor. For the purposes of this requirement, 1-1/4" pipe = 1.2 in^2, 1-1/2" pipe = 1.8 in^2, 2" pipe = 3.1 in^2, 2-1/2" pipe = 4.9 in^2, 3" pipe = 7.1 in^2, 4" pipe = 12.6 in^2, 5" pipe = 19.6 in^2, and 6" pipe = 28.3 in^2.
See Figure 9.3.6

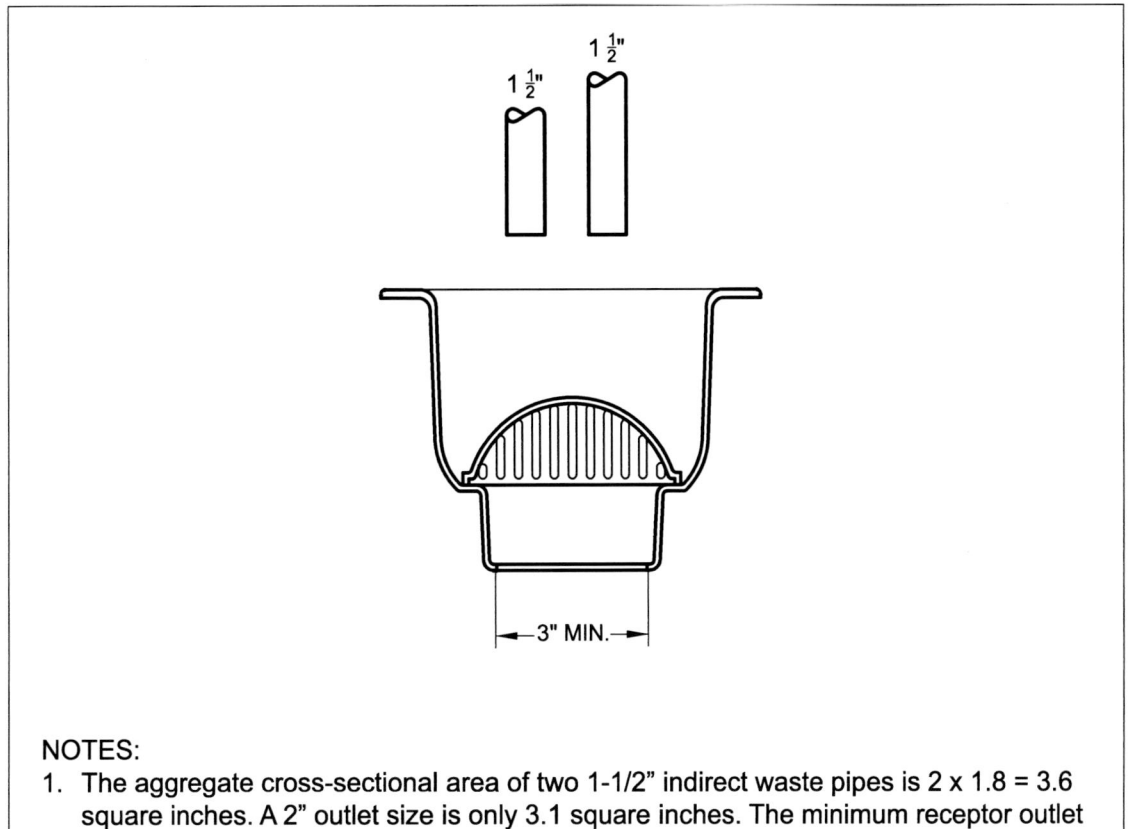

NOTES:
1. The aggregate cross-sectional area of two 1-1/2" indirect waste pipes is 2 x 1.8 = 3.6 square inches. A 2" outlet size is only 3.1 square inches. The minimum receptor outlet pipe size is 3" with 7.1 square inches.

Figure 9.3.6
MINIMUM RECEPTOR OUTLET PIPE SIZE

9.3.7 Drainage Fixture Unit (DFU) Values

The drainage fixture unit values used to combine the loading of indirect waste receptors with other fixtures shall be the sum of the DFU values for all fixtures that are indirectly drained into the receptor.

9.4 SPECIAL WASTES

9.4.1 Treatment of Corrosive Wastes

a. Corrosive liquids, spent acids, or other harmful chemicals that may damage a drain, sewer, or sanitary drain pipe; create noxious or toxic fumes; or interfere with sewage treatment processes shall not be discharged into the plumbing system without being thoroughly neutralized or treated by passing through a properly constructed and approved neutralizing device. Such devices shall be provided automatically with a sufficient supply of neutralizing medium, so as to make its contents non-injurious before discharge into the drainage system. The nature of the corrosive or harmful waste and proposed method of its treatment shall be submitted to and approved by the Authority Having Jurisdiction prior to installation. ***See Figures 9.4.1 and 1.2.55***

b. Vent piping for corrosive waste drain piping shall be separate from vent piping for sanitary waste and drain piping and shall terminate outdoors in accordance with Section 12.4.

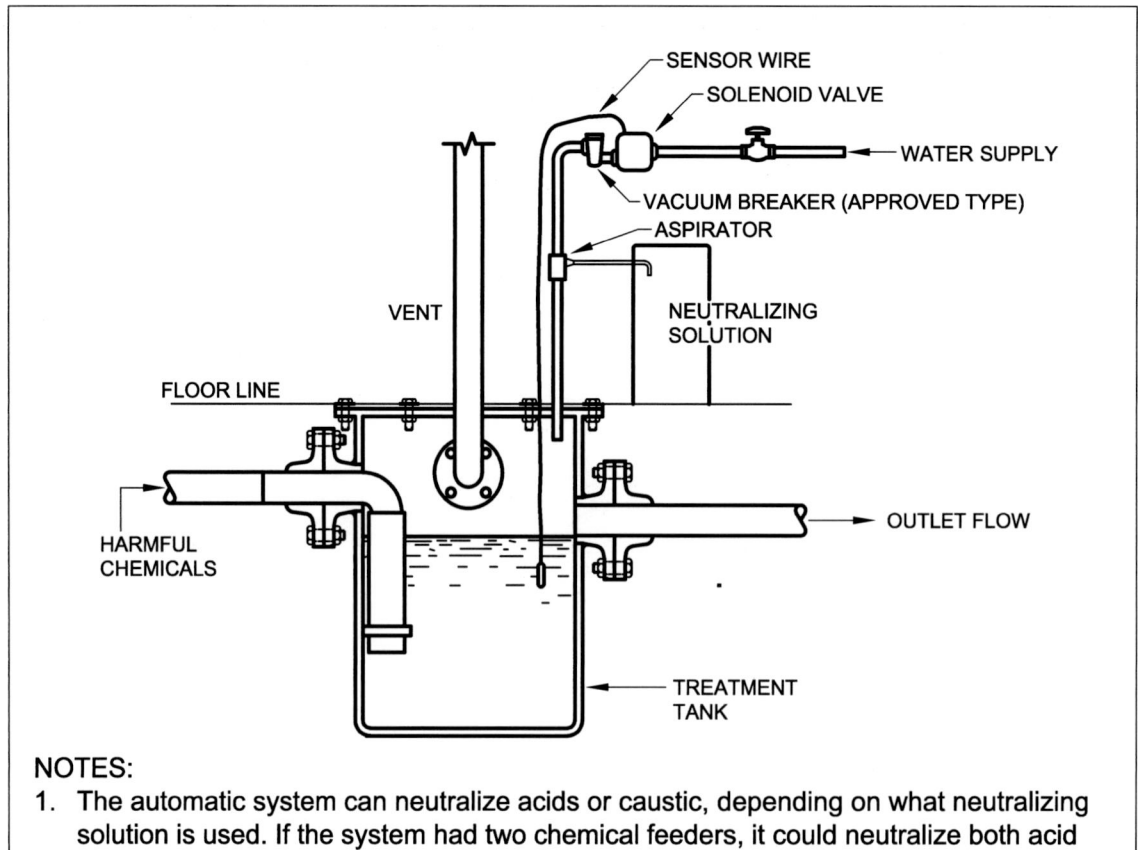

Figure 9.4.1
AN AUTOMATIC WASTE TREATMENT NEUTRALIZING TANK

9.4.2 High Temperature Wastes

No waste at temperatures above 140°F shall be discharged directly into any part of a drainage system. Such wastes shall be discharged to an indirect waste receptor and a means of cooling shall be provided where necessary.

9.4.3 Air Conditioning Condensate

a. Indirect waste piping from air conditioning units shall be sized according to the condensate-generating capacity of the units served. Branches from individual units shall be no smaller than the drain opening or drain connection on the unit. Traps shall be provided at each air conditioning unit or cooling coil to maintain atmospheric pressure in the waste piping.

b. Gravity condensate waste piping shall be sloped not less than 1/8" per foot. Drainage fittings shall be used in sizes 1-1/4" and larger. Minimum pipe sizes for gravity flow shall be as follows:

3/4" pipe size through 3-ton cooling capacity
1" pipe size through 20-ton cooling capacity
1-1/4" pipe size through 100-ton cooling capacity
1-1/2" pipe size through 300-ton cooling capacity
2" size pipe through 600-ton cooling capacity

c. Discharge of air conditioning condensate shall not be permitted to create a nuisance such as by flowing across the ground or paved surfaces. Unless expressly prohibited by the Authority Having Jurisdiction, the point of indirect discharge for air conditioning condensate shall be one of the following:

1. The building sanitary drainage system.
2. The building storm drainage system.
3. A sump pump pit.
4. A subsurface absorption pit or trench.
5. Within dwellings, a tub waste and overflow, lavatory tailpiece, or laundry sink within the same dwelling.

See Figures 9.4.3-A and 9.4.3-B

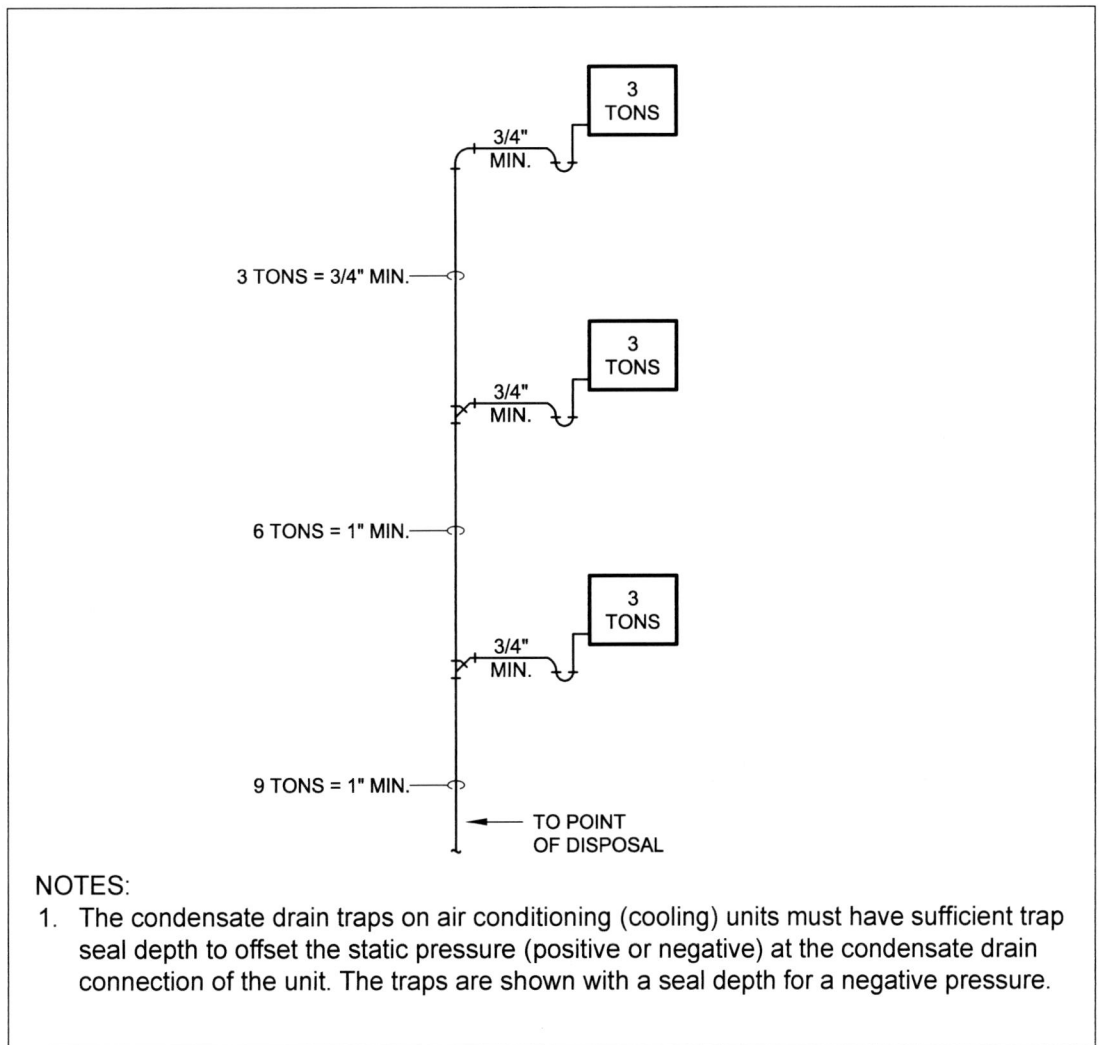

Figure 9.4.3 - A
SIZING INDIRECT WASTE PIPING FOR AIR CONDITIONING CONDENSATE

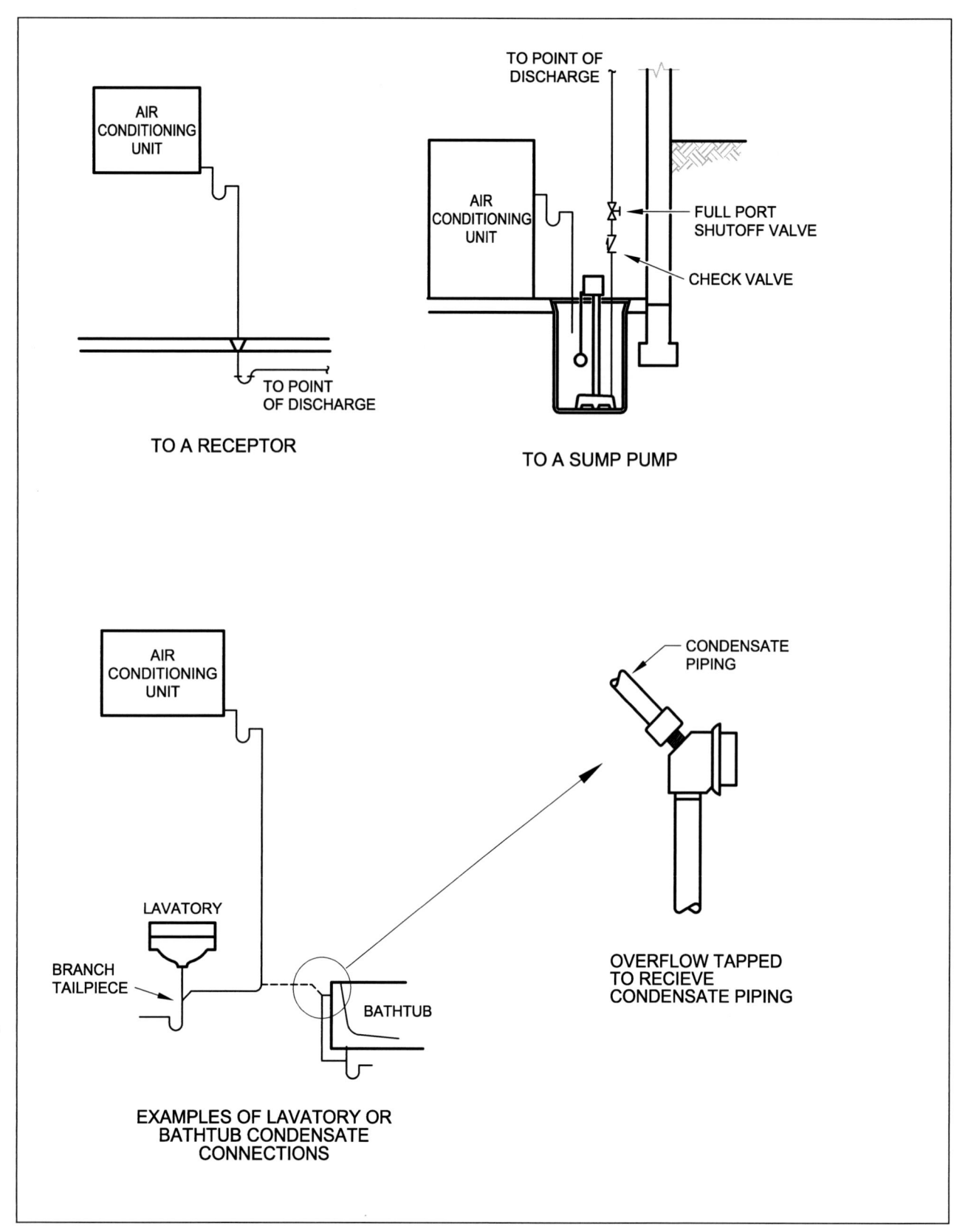

Figure 9.4.3 - B
DISCHARGING AIR CONDITIONING CONDENSATE

Chapter 10

Water Supply and Distribution

10.1 QUALITY OF WATER SUPPLY

a. Only potable water shall be supplied to plumbing fixtures used for drinking, bathing, culinary use or the processing of food, medical or pharmaceutical products.

b. Reclaimed non-potable water shall be permitted in other than single dwelling units for flushing water closets and urinals, landscape irrigation, and other reuse applications in accordance with the requirements of the Authority Having Jurisdiction.

c. Graywater in single dwelling units shall be limited to subsurface landscape irrigation in accordance with the requirements of the Authority Having Jurisdiction.

d. Graywater in other than single dwelling units that is filtered and disinfected shall be permitted to be used for flushing water closets and urinals in accordance with the requirements of the Authority Having Jurisdiction.

e. Harvested rainwater in other than single dwelling units that is filtered shall be permitted to be used for flushing water closets and urinals and subsurface landscape irrigation in accordance with the requirements of the Authority Having Jurisdiction.

f. Reclaimed non-potable water, gray water, and harvested rain water, including the collection piping, distribution piping, storage and pumping equipment, shall be part of the plumbing system as covered by this Code regardless of their location on the property.

10.2 RESERVED

10.3 WATER REQUIRED

Plumbing fixtures shall be provided with a supply of water in the amounts and at the pressures specified in this Chapter.

10.4 PROTECTION OF POTABLE WATER SUPPLY

10.4.1 General

A potable water supply shall be designed, installed and maintained to prevent contamination from non-potable liquids, solids or gases by cross connections.

10.4.2 Interconnections

Interconnections between two or more public water supplies shall be permitted only with the approval of the Authority Having Jurisdiction.

10.4.3 Cross Connection Control

Potable water supplies shall be protected in accordance with the cross connection control program of the Authority Having Jurisdiction and the provisions of this Code. Cross connection control shall be provided at individual outlets, and where required, by containment of the premises. Each potential cross connection within the premises shall be protected. Where containment is required, the potable water supply shall be protected by a backflow prevention device installed immediately downstream of the meter or between the service shutoff valve and the first outlet or branch connection.

See Figures 10.4.3-A and 10.4.3-B

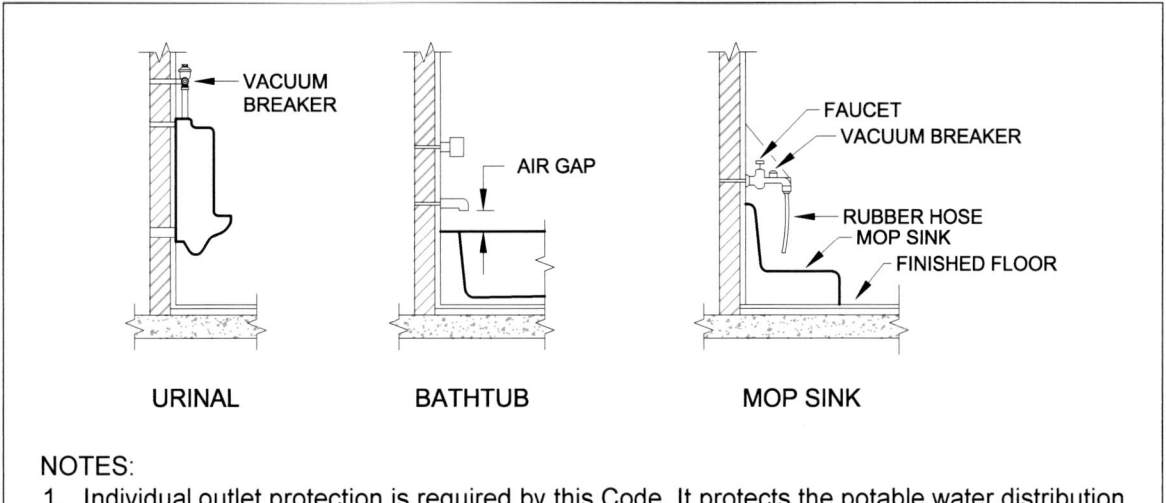

Figure 10.4.3 - A
CROSS CONNECTION CONTROL BY INDIVIDUAL OUTLET PROTECTION

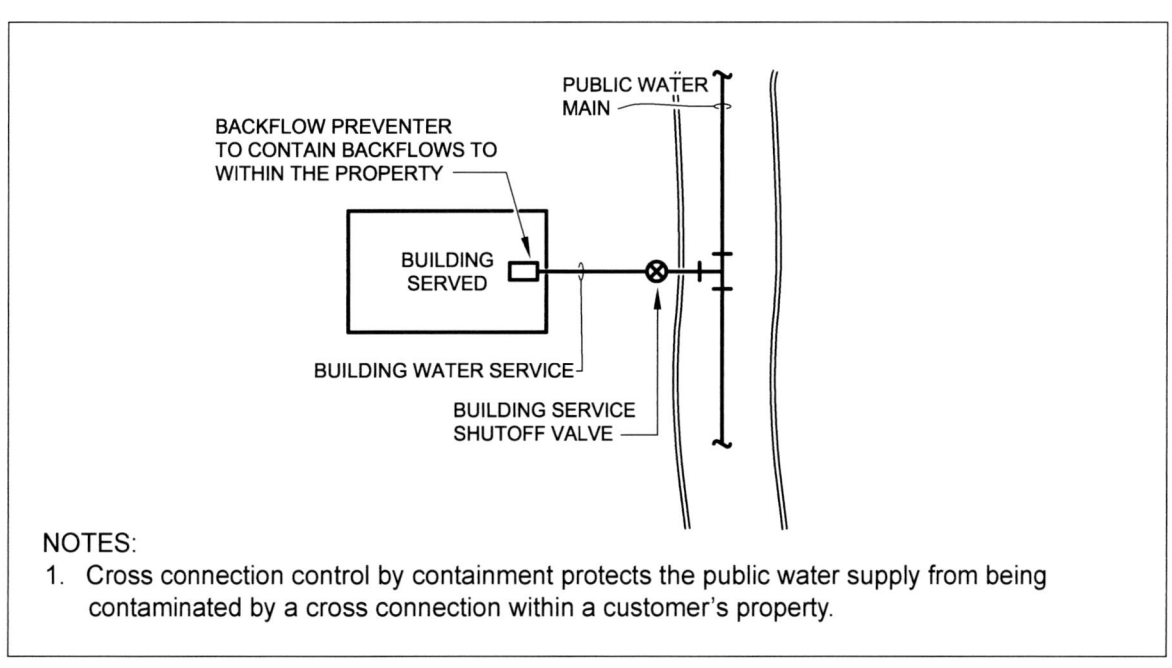

Figure 10.4.3 - B
CROSS CONNECTION CONTROL BY CONTAINMENT OF THE BUILDING

10.4.4 Private Supplies

a. Private potable water supplies (i.e., wells, cisterns, lakes, streams) shall require the same backflow prevention that is required for a public potable water supply.

b. Cross connection between a private potable water supply and a public potable water supply shall not be made unless specifically approved by the Authority Having Jurisdiction.

> *Comment: Interconnections between private water supplies and public water supplies are generally prohibited because private supplies are usually not monitored continuously for water quality.*

10.4.5 Toxic Materials

a. Piping conveying potable water shall be constructed of non-toxic material.

b. The interior surface of a potable water tank shall not be lined, painted, or repaired with any material that will affect the taste, odor, color or potability of the water supply when the tank is placed in or returned to service.

> *Comment: The toxicity rating of a piping material can be found in the material standard listed in Table 3.1.3. The piping materials listed in Table 3.4 are non-toxic and are suitable for conveying potable water.*

10.4.6 Reserved

10.4.7 Reserved

10.4.8 Used Materials

Materials that have been used for any purpose other than conveying potable water shall not be used for conveying potable water.

10.4.9 Water As a Heat-Transfer Fluid

Potable water may be used as a heat-transfer fluid provided that the potable water system is protected against cross connection.

10.5 BACKFLOW PREVENTION

10.5.1 Air Gaps

a. The water supply outlets for plumbing fixtures and other discharges shall be protected from back siphonage by a fixed air gap or a required backflow preventer.

b. Air gaps shall comply with ASME A112.1.2 or Table 10.5.2. Air gap fittings shall comply with ASME A112.1.3.

10.5.2 Requirements for Air Gaps

a. How Measured: Air gaps shall be measured vertically from the lowest opening of the water supply outlet to either (1) the flood level rim of the fixture or receptor served, or (2) the maximum elevation of the source of contamination.

b. Minimum size (distance): The minimum required air gap shall be in accordance with Table 10.5.2 based on the opening of the water supply outlet and the affect of any nearby vertical surfaces (walls).

Table 10.5.2
MINIMUM AIR GAPS FOR FIXTURES, APPLIANCES, AND WATER SUPPLY OUTLETS

Water Supply Outlet Size	Minimum Air Gap		
	Not Affected by Near Walls	Affected by One Near Wall	Affected by Two Near Walls
Outlets not greater than 1/2-inch diameter, including lavatory faucets	1 inch	1-1/2 inches	2 inches
Outlets not greater than 3/4-inch diameter, including sink faucets	1-1/2 inches	2-1/4 inches	3 inches
Outlets not greater than 1-inch diameter, including bathtub spouts	2 inches	3 inches	4 inches
Outlets greater than 1-inch diameter	2 times diameter	3 times diameter	4 times diameter

NOTES FOR TABLE 10.5.2:
1. The listed outlet diameters in Table 10.5.2 are 1/2-inch, 3/4-inch, 1-inch, and greater than 1-inch. They should be used to determine the affect of near walls (vertical surfaces) in Notes 2, 3, and 4 and the minimum required air gap.
2. The minimum air gaps for water supply outlets are not affected by near walls if a single wall is more than 3 times the listed outlet diameter clearance from an edge of the outlet. If there are two walls, the clearances from both must be more than 4 times the listed outlet diameter for the air gap to be not affected.
3. If a water supply outlet has one near wall and the clearance is less than 3 times the listed outlet diameter from an edge of the outlet, the air gap must be sized as affected by one near wall.
4. If a water supply outlet has two near walls that both have clearances that are less than 4 times the listed outlet diameter from an edge of the outlet, the air gap must be sized as affected by two near walls.

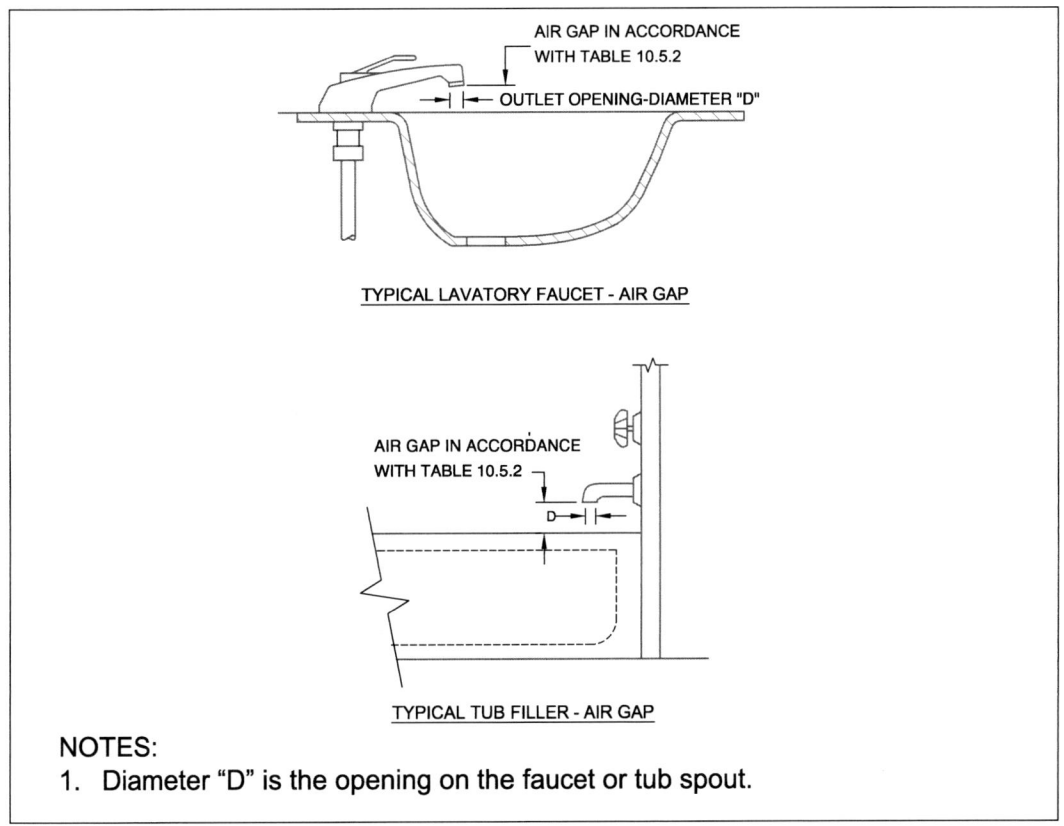

Figure 10.5.2
MINIMUM REQUIRED AIR GAPS

10.5.3 Backflow Prevention Assemblies and Devices

The following requirements shall apply:

A. Back Siphonage Only, Non-Continuous Water Pressure, Non-Health Hazards
 1. Hose connection vacuum breaker - ASSE 1011
 2. Backflow prevention devices for hand-held showers - ASSE 1014
 3. Wall hydrant with backflow prevention and freeze resistance - ASSE 1019
 4. Backflow preventer for beverage dispensing equipment - ASSE 1022
 5. Laboratory faucet backflow preventer - ASSE 1035
 6. Hose connection backflow preventer - ASSE 1052
 7. Dual check backflow preventer wall hydrant - freeze resistant - ASSE 1053
 8. Freeze resistant sanitary yard hydrants with backflow protection - ASSE 1057 (Types I – V)

B. Back Siphonage Only, Non-Continuous Water Pressure, Health and Non-Health Hazards
 1. Air gaps in plumbing systems - ASME A112.1.2
 2. Air gap fittings for use in plumbing fixtures, appliances, and appurtenances - ASME A112.1.3
 3. Atmospheric type vacuum breaker - ASSE 1001

C. Back Siphonage Only, Continuous Water Pressure, Health or Non-Health Hazards
 1. Pressure vacuum breaker assembly - ASSE 1020
 2. Spill resistant vacuum breaker - ASSE 1056 (SVB)

D. Back Siphonage and Back Pressure, Continuous Water Pressure, Non-Health Hazards
 1. Backflow preventer with intermediate atmospheric vent - ASSE 1012
 2. Double check backflow prevention assembly - ASSE 1015 (DC)
 3. Double check fire protection backflow prevention assembly - ASSE 1015 (DCF)
 4. Dual check backflow preventer - ASSE 1024
 5. Double check detector fire protection backflow prevention assembly - ASSE 1048 (DCDA)
 6. Double check detector fire protection backflow prevention assembly - ASSE 1048 Type II (DCDA-II)

E. Back Siphonage and Back Pressure, Continuous Water Pressure, Health and Non-Health Hazards
 1. Reduced pressure principle backflow preventer - ASSE 1013 (RP)
 2. Reduced pressure principle fire protection backflow preventer - ASSE 1013 (RPF)
 3. Reduced pressure detector fire protection backflow prevention assembly - ASSE 1047 (RPDA)
 4. Reduced pressure detector fire protection backflow prevention assembly - ASSE 1047 Type II (RPDA-II)

F. Restrictions
 1. Backflow preventers shall be applied within the ratings of their standard and their manufacturer's instructions.
 2. Backflow preventers with atmospheric vents shall not be located where discharge from their vent will cause water damage.
 3. The following devices must not be subjected to more than 10 feet of water back pressure from elevated hoses:
  a. ASSE 1011 hose connection vacuum breakers
  b. ASSE 1019 wall hydrants with backflow prevention and freeze resistance
  c. ASSE 1052 hose connection backflow preventers
  d. ASSE 1053 dual check backflow preventer wall hydrant - freeze resistant
  e. ASSE 1057 freeze resistant sanitary yard hydrants with backflow prevention
 4. ASSE 1035 laboratory faucet backflow preventers must not be subjected to more than 6 inches of water back pressure from elevated discharge hoses.

10.5.4 Approval of Devices

Backflow prevention devices shall be listed or certified by a recognized certification body as complying with the appropriate standards in Table 3.1.3 - Part IX.

10.5.5 Installation of Backflow Preventers

a. All Types: All backflow preventers shall be accessible for testing (if testable), maintenance, repair, and replacement. Clearances shall be as recommended by the manufacturer. Backflow preventers having atmospheric vents shall not be installed in pits, vaults, or similar potentially submerged locations. Vacuum breakers and other devices with vents shall not be located within fume hoods. Where outdoor enclosures are provided for backflow prevention assemblies, they shall comply with ASSE 1060.

b. Atmospheric Vacuum Breakers: Pipe-applied atmospheric vacuum breakers shall be installed with the critical level not less than six inches above the flood level rim or highest point of discharge of the fixture being served. Approved deck-mounted and pipe-applied vacuum breakers and vacuum breakers within equipment, machinery and fixtures where the critical level is a specified distance above the source of contamination shall be installed in accordance with manufacturer's instructions with the critical level not less than one inch above the flood level rim. Such devices shall be installed on the discharge side of the last control valve to the fixture and no shut-off valve or faucet shall be installed downstream of the vacuum breaker. Vacuum breakers on urinals shall be installed with the critical level not less than six inches above the highest part of the urinal. *See Figures 10.5.5-A through 10.5.5-F.*

c. Pressure Type Vacuum Breakers: Pressure type vacuum breakers shall be installed with the critical level at a height of at least 12 inches above the flood level rim for ASSE 1020 devices and with the critical level at least six inches above the flood level rim or highest point of discharge of the fixture being served for ASSE 1056 devices. Deck-mounted and pipe-applied pressure type (ASSE 1056) vacuum breakers within equipment, machinery and fixtures where the critical level is a specified distance above the source of contamination shall be installed in accordance with manufacturer's instructions with the critical level not less than one inch above the flood level rim. *See Figure 10.5.5-G.*

d. Double Check Valves and Reduced Pressure Principle Valves: Such devices shall be installed at not less than 12 inches above the floor or permanent platform with the maximum of 60 inches above floor or permanent platform. *See Figures 10.5.5-H and -I.*

e. Spill-resistant Vacuum Breakers: Approved deck-mounted and pipe-applied spill-resistant vacuum breakers within equipment, machinery and fixtures where the critical level is a specified distance above the source of contamination shall be installed in accordance with manufacturer's instructions with the critical level not less than one inch above the flood level rim.

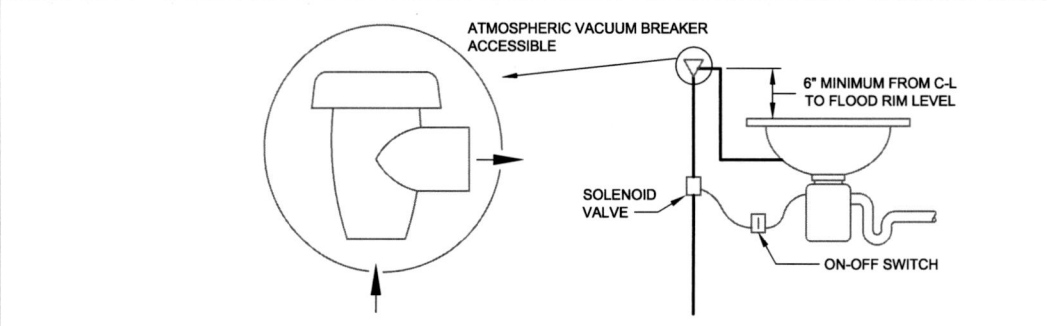

NOTES:
1. If the critical level (C-L) is not marked on the vacuum breaker body, the bottom of the valve is considered to be the C-L reference.
2. Atmospheric vacuum breakers are not rated for periods of more than 12 hours of continuous water pressure.

Figure 10.5.5 - A
THE POTABLE WATER SUPPLY TO A COMMERCIAL FOOD WASTE DISPOSER PROTECTED BY AN ATMOSPHERIC VACUUM BREAKER

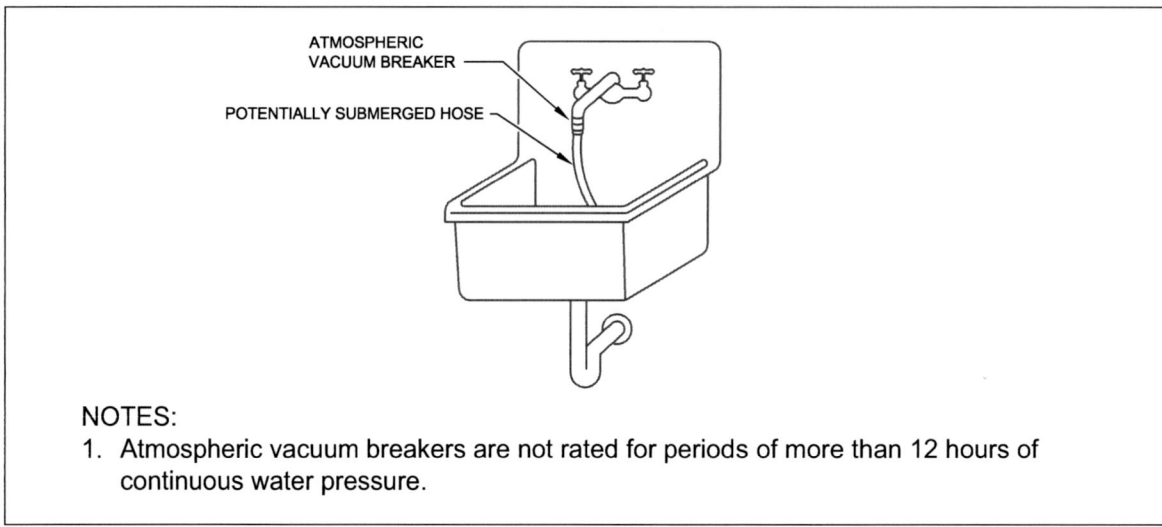

NOTES:
1. Atmospheric vacuum breakers are not rated for periods of more than 12 hours of continuous water pressure.

Figure 10.5.5 - B
THE POTABLE WATER SUPPLY TO A SERVICE SINK

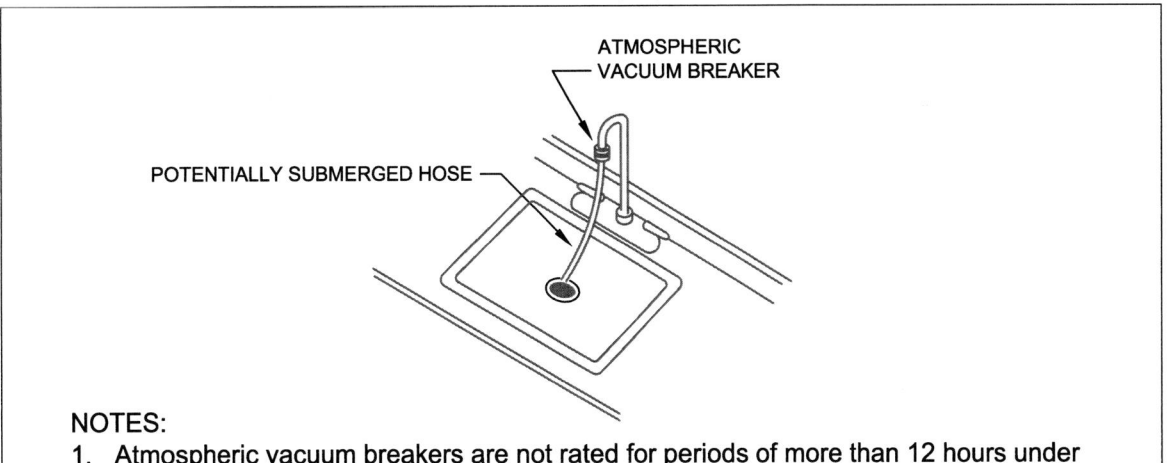

NOTES:
1. Atmospheric vacuum breakers are not rated for periods of more than 12 hours under continuous water pressure.

Figure 10.5.5 - C
THE POTABLE WATER SUPPLY TO A LAB SINK
PROTECTED BY AN ATMOSPHERIC VACUUM BREAKER

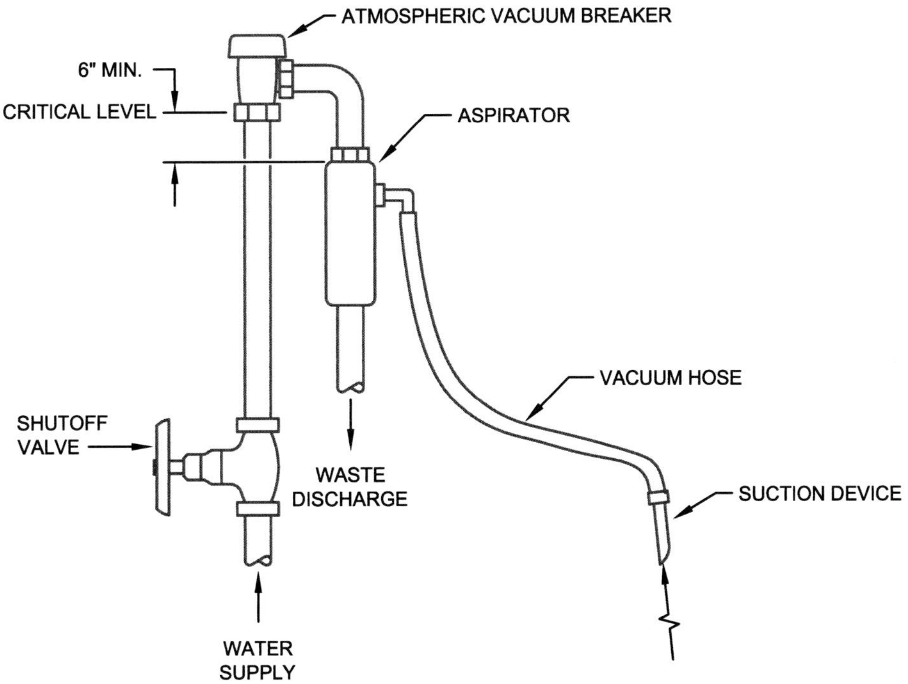

NOTES:
1. Atmospheric vacuum breakers are not rated for periods of more than 12 hours of continuous water pressure.

Figure 10.5.5 - D
THE POTABLE WATER SUPPLY TO AN ASPIRATING DEVICE
PROTECTED BY AN ATMOSPHERIC VACUUM BREAKER

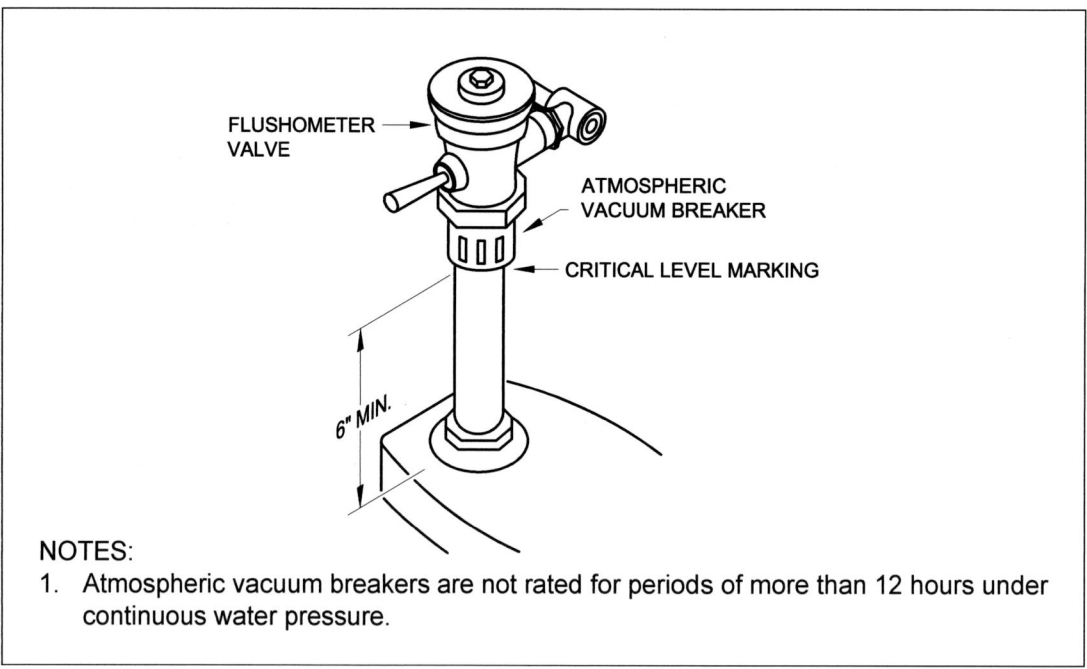

Figure 10.5.5 - E
THE POTABLE WATER SUPPLY TO A WATER CLOSET OR URINAL PROTECTED BY AN ATMOSPHERIC VACUUM BREAKER

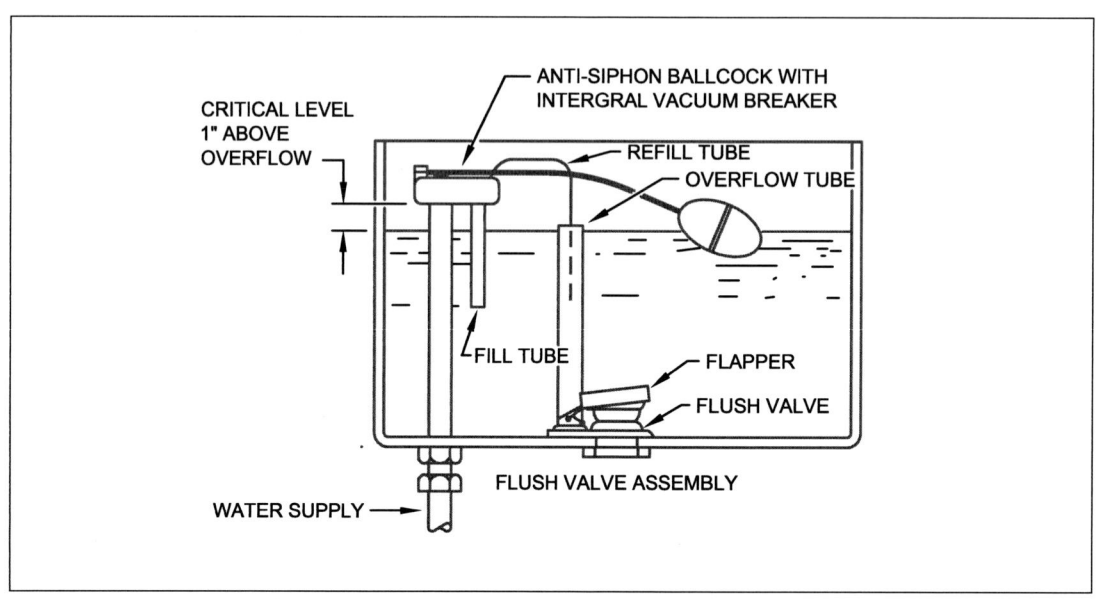

Figure 10.5.5 - F
THE POTABLE WATER SUPPLY TO A WATER CLOSET GRAVITY FLUSH TANK PROTECTED BY AN ANTI-SIPHON BALL COCK WITH INTEGRAL VACUUM BREAKER

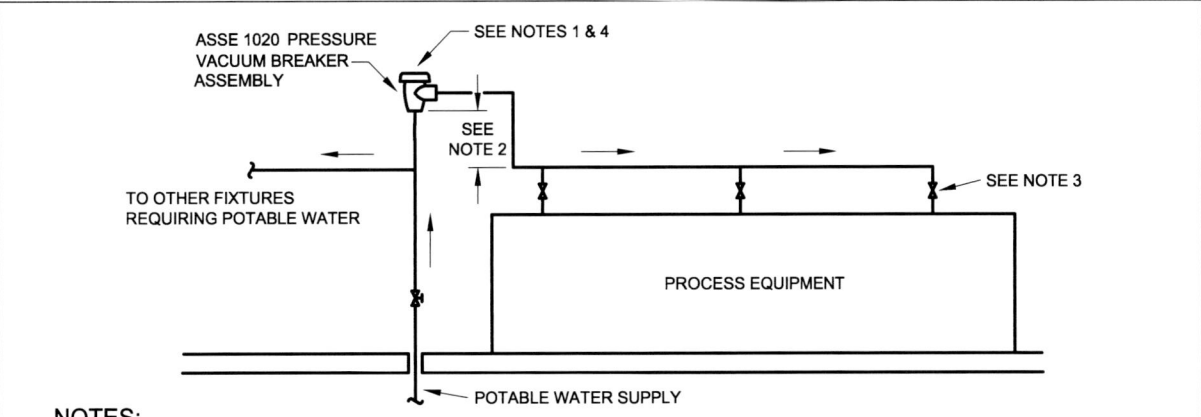

NOTES:
1. ASSE 1020 anti-siphon pressure vacuum breaker assemblies can be located where subjected to continuous pressure.
2. The critical level (C-L) of the vacuum breaker must be at least 12" higher than the highest downstream piping.
3. Shutoff valves are permitted downstream from pressure vacuum breakers.
4 ASSE 1020 pressure vacuum breaker assemblies should not be located where water spillage from the atmospheric vent will cause damage or create a nuisance. If necessary, use ASSE 1056 spill-resistant vacuum breaker assemblies.

Figure 10.5.5 - G
THE POTABLE WATER SUPPLY TO INDUSTRIAL PROCESS EQUIPMENT PROTECTED BY A PRESSURE VACUUM BREAKER

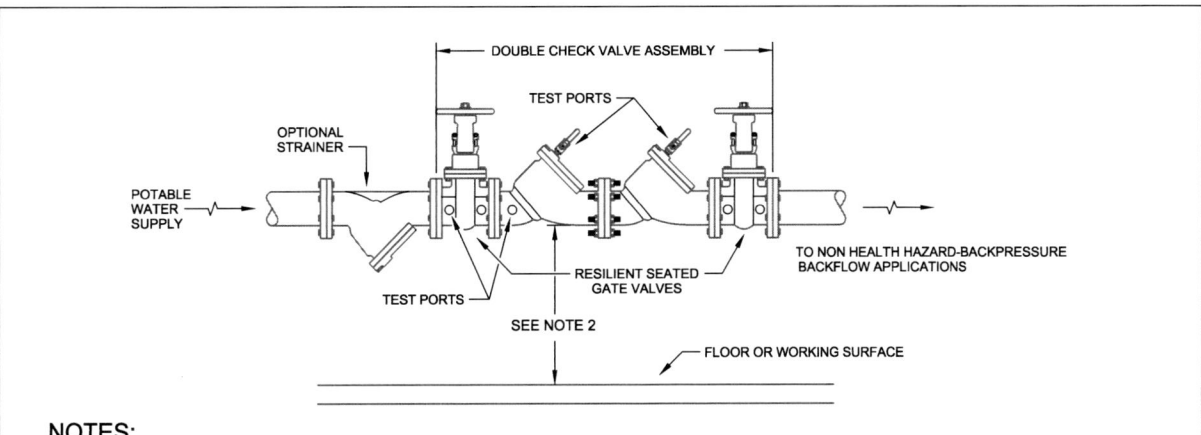

NOTES:
1. Double check valve assemblies or reduced pressure backflow preventers are required for back pressure applications.
2. Double check valve assemblies must be installed between 12" and 60" above the floor or other working surface to provide sufficient access for periodic testing and maintenance.
3. Double check valve assemblies subject to back-siphonage require an intermediate vacuum breaker and a relief vent.
4. Double check valve assemblies with an intermediate vacuum breaker and relief vent must not be located in pit or other area subject to flooding. The relief vent will create a cross-connection if submerged.

Figure 10.5.5 - H
THE POTABLE WATER SUPPLY TO A NON-HEALTH HAZARD BACKFLOW APPLICATION PROTECTED BY A DOUBLE CHECK VALVE ASSEMBLY

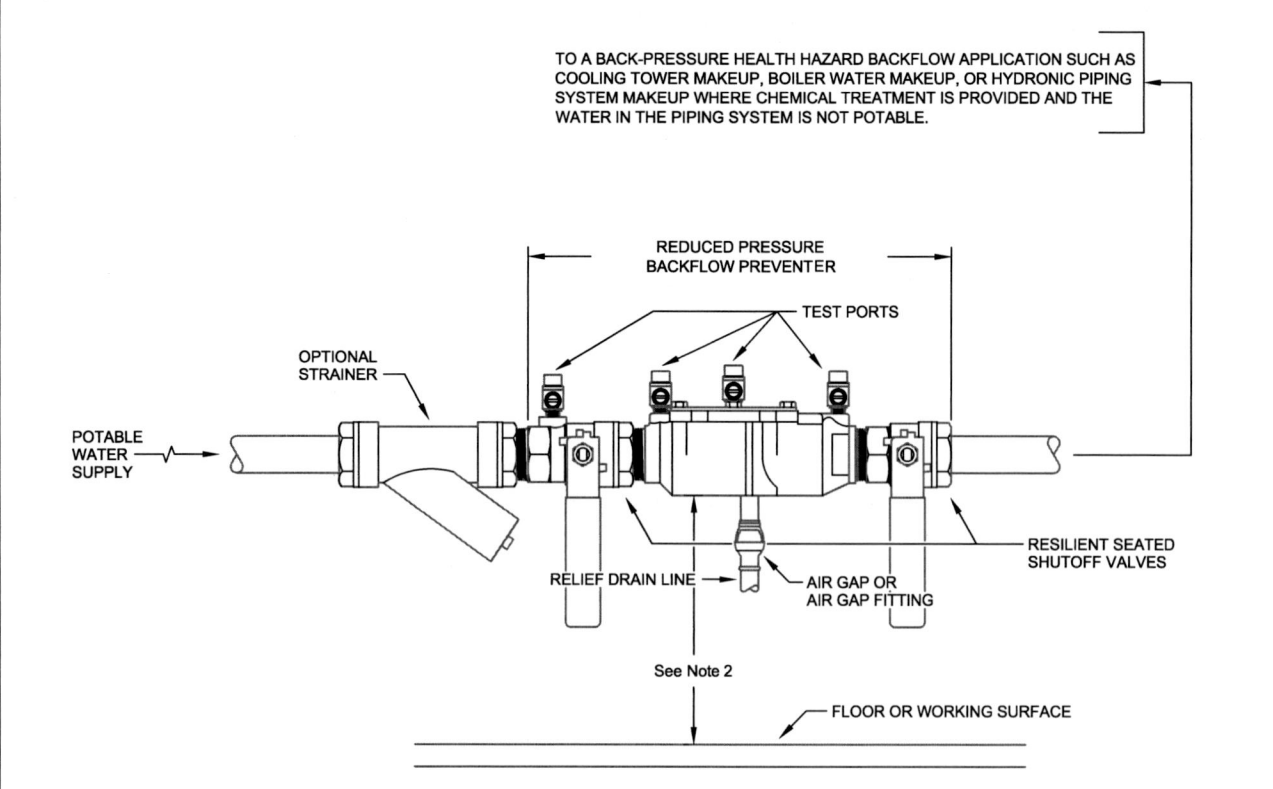

NOTES:
1. Double check valve assemblies are not permitted for health hazard applications.
2. Reduced pressure backflow preventers must be installed between 12" and 60" above the floor or other working surface to provide sufficient access for periodic testing and maintenance.
3. RP devices must not be installed in pits or other area subject to flooding. The relief vent will create a cross-connection if submerged.
4. If occasional spillage from the relief vent will cause damage or be a nuisance, the vent must be equipped with an air gap fitting and indirectly drained to an acceptable point of disposal.

Figure 10.5.5 - I
THE POTABLE WATER SUPPLY TO A HEALTH HAZARD BACKFLOW APPLICATION PROTECTED BY A REDUCED

10.5.6 Testing and Maintenance of Backflow Prevention Assemblies

a. Assemblies that are designed to be field tested shall be tested prior to final inspection of the initial installation and once each year thereafter.

b. Assemblies installed in a building potable water supply distribution system for protection against backflow shall be maintained in good working condition and be repaired when necessary.

c. Testable assemblies are those backflow prevention assemblies having test cocks or test procedures and include, but are not limited, to the following:

1. Reduced pressure principle backflow preventers (ASSE 1013)
2. Reduced pressure fire protection principle backflow preventers (ASSE 1013)
3. Double check backflow prevention assemblies (ASSE 1015)
4. Double check fire protection backflow prevention assemblies (ASSE 1015)
5. Pressure vacuum breaker assemblies (ASSE 1020)
6. Reduced pressure detector fire protection backflow prevention assemblies (ASSE 1047)
7. Double check detector fire protection backflow prevention assemblies (ASSE 1048)
8. Spill resistant vacuum breakers (ASSE 1056)

d. Where tests indicate that the assembly is not functioning properly, it shall be serviced or repaired in accordance with the manufacturer's instructions and be retested.

e. Testing and repair of assemblies shall be performed by certified individuals approved by an agency acceptable to the Authority Having Jurisdiction.

f. Certification for testing and repair shall be in accordance with the appropriate sections of ASSE 5000 Cross-Connection Control Professional Qualification Standard. Testing shall be in accordance with ASSE 5110 Backflow Prevention Assembly Testers. Repair shall be in accordance with ASSE 5130 Backflow Prevention Assembly Repairers.

g. Copies of test reports for the initial installation shall be sent to the Authority Having Jurisdiction and the water supplier. Copies of annual test reports shall be sent to the water supplier.

h. Where a continuous water supply is critical and cannot be interrupted for the periodic testing of a backflow prevention assembly, multiple backflow prevention assemblies or other means of maintaining a continuous supply shall be provided that does not create a potential cross connection.

10.5.7 Tanks and Vats – Below Rim Supply

a. Where a potable water inlet terminates below the rim of the tank or vat and the tank or vat has an overflow of a diameter not less than given in Table 10.8.3, the overflow pipe shall be provided with an air gap as close to the tank as possible.

b. The potable water supply inlet to the tank or vat shall terminate a distance not less than 1-1/2 times the height to which water can rise in the tank above the top of the overflow. This level shall be established at the maximum flow rate of the supply to the tank or vat and with all outlets closed except the air-gapped overflow outlet.

c. An alternate to 10.5.7.b is a vacuum breaker on the water supply inlet pipe above the rim of the tank or vat.

See Figure 10.5.7. Also Sections 10.8.3, 10.8.4, 10.8.5, and 10.8.6

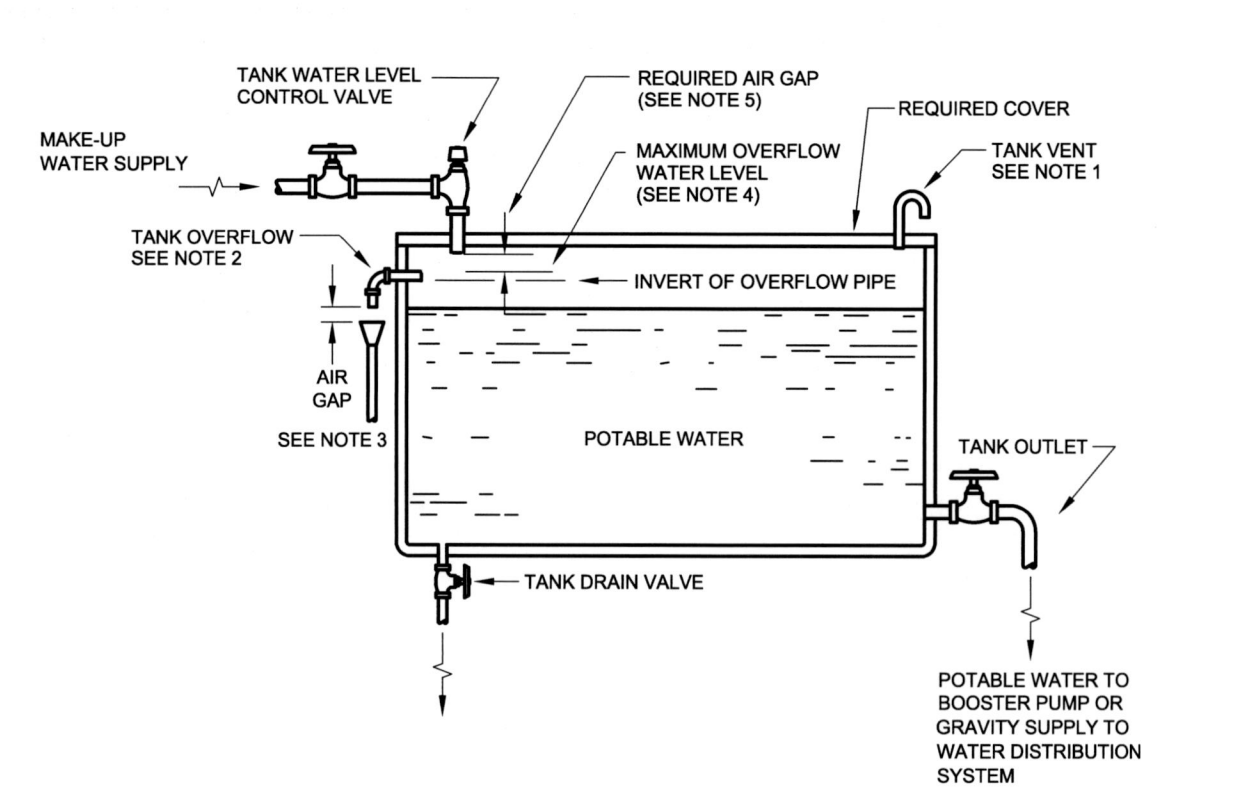

NOTES:
1. The covers of potable water tanks or vats must have screened vent pipes with internal areas not less than that of the outlet pipe.
2. The size of the overflow pipe must comply with Table 10.8.3, based on the maximum makeup water flow rate.
3. The air gap at the tank overflow discharge must comply with Table 10.5.2 based on the inside diameter of the overflow pipe.
4. The maximum overflow water level must be established based on the size of the overflow pipe and the maximum available makeup water flow. Calculated required static head or rise above the invert of the overflow should be confirmed by operational tests under actual full flow conditions with the tank outlet(s) closed.
5. The required air gap at the water inlet pipe to the tank or vat must comply with Table 10.5.2 based on the inside diameter of the water inlet pipe and the maximum water level in the tank under maximum overflow conditions as determined in Note 4.

Figure 10.5.7
A POTABLE WATER TANK OR VAT WITH ITS INLET BELOW THE OVERFLOW RIM

10.5.8 Protection from Equipment in Food Establishments

a. Food preparation and service equipment in food establishments that is connected to potable water piping or drained to waste piping shall be protected from backflow to the potable water system and from contamination of equipment, food, and utensils.

b. Fixtures that require an air gap in their water supply above the flood level rim of the fixture include, but are not limited to, sink faucets, dipper wells, pot fillers, and spray arms in dish areas.

c. Equipment that requires an air gapped waste line includes, but is not limited to, walk-in coolers, food prep sinks, dipper wells, ice machines, ice bins, steam tables, bains-marie, Chinese ranges (woks), combination steamer/ovens, rethermalizers, steamers, kettles, non-evaporating reach-in coolers, soda gun holsters, and water treatment systems.

d. Equipment that requires a pressure vacuum breaker or spill resistant vacuum breaker in its water supply includes, but is not limited to, Chinese ranges (woks), rethermalizers, hoses with a pressure nozzle, and pasta cookers.

e. Equipment that requires an atmospheric vacuum breaker in its water supply includes, but is not limited to, garbage disposals, automatic dishwashers, dish machine automatic detergent feeders, detergent feeders for 3-compartment sinks, wall-mounted chemical feeders, garbage can washers, mop sinks, steam tables with submerged inlets, and dish troughs with submerged inlets.

f. Equipment that requires a vented double check valve or equivalent in its water supply includes, but is not limited to, untreated boilers such as expresso machines, the oven in combination steamer/ovens, carbonated beverage machines, beverage carbonators, non-carbonated beverage machines, coffee machines, iced tea machines, hot chocolate machines, cappuccino machines, instant hot water dispensers, slushy machines, and any other beverage equipment.

g. The water supply to a post-mix carbonated beverage dispenser shall be protected against backflow with an integral backflow preventer conforming to ASSE 1022 or an air gap.

1. Post-mix carbonated beverage dispensers and carbonated beverage systems with an integral dual check backflow preventer without an atmospheric vent (ASSE 1032) or without an integral backflow preventer conforming to ASSE 1022 or without an integral air gap shall have the water supply connection to the dispenser protected by a double check valve with atmospheric vent port conforming to ASSE 1022.

2. When an ASSE 1022 device must be installed in the water supply piping external to the carbonated beverage dispenser, the piping from the device to the beverage dispenser shall be acid-resistant and not copper.

3. ASSE 1012 backflow preventers with intermediate atmospheric vents are not constructed for use with carbonated beverages and shall not be used for backflow prevention from post-mix carbonated beverage dispensers.

h. All backflow prevention assemblies and devices shall be installed in accordance with their manufacturer's instructions for the particular application.

10.5.9 Protection from Fire Systems

a. Potable water supplies to water-based fire protection systems, including but not limited to standpipes and automatic sprinkler systems, shall be protected from back pressure and back siphonage by one of the following testable assemblies:

1. double check fire protection backflow prevention assembly – ASSE 1015 (DCF)

2. double check detector fire protection backflow prevention assembly – ASSE 1048 (DCDA or DCDA-II)

3. reduced pressure fire protection principle backflow prevention assembly – ASSE 1013 (RPF)

4. reduced pressure detector fire protection backflow prevention assembly – ASSE 1047 (RPDA or RPDA-II).

EXCEPTIONS:

(1) ASSE 1024 dual check backflow preventers shall be permitted in stand-alone residential fire sprinkler systems that comply with NFPA 13D or NFPA 13R, do not supply plumbing fixtures, and do not include a fire department connection.

(2) Backflow preventers shall not be required in NFPA 13D multipurpose or network residential fire sprinkler systems that supply both plumbing fixtures and residential fire sprinklers. The piping in such systems shall be approved for potable water. Such systems shall not have a fire department connection.

(3) ASSE 1024 dual check backflow preventers shall be permitted in limited area fire sprinkler systems that comply with NFPA 13 and do not have a fire department connection.

(4) Where fire protection systems include a fire department connection, ASSE 1013 reduced pressure fire protection principle backflow preventers (RPF) or ASSE 1047 reduced pressure detector fire protection backflow prevention assemblies (RPDA or RPDA-II) shall be required.

(5) Where fire protection systems are filled with solutions that are considered to be health hazards as defined in Section 1.2, ASSE 1013 reduced pressure fire protection principle backflow preventers (RPF) or ASSE 1047 reduced pressure detector fire protection backflow prevention assemblies (RPDA or RPDA-II) shall be required.

b. Whenever a backflow prevention assembly is installed in a potable water supply to a fire protection system, the hydraulic design of the fire protection system shall account for the pressure drop through the backflow prevention assembly.

c. If a backflow prevention assembly is retrofitted for an existing fire protection system, the hydraulics of the fire protection system shall be checked to verify that there will be sufficient water pressure available for satisfactory operation of the fire protection system with the backflow prevention assembly in place.

10.5.10 Protection from Lawn Sprinklers and Irrigation Systems

a. Potable water supplies to systems having no pumps or connections for pumping equipment, and no chemical injection or provisions for chemical injection, shall be protected from backflow by one of the following:

1. Atmospheric type vacuum breaker – ASSE 1001 (for non-continuous pressure)
2. Pressure vacuum breaker assembly – ASSE 1020
3. Spill resistant vacuum breaker (SVB) – ASSE 1056
4. Reduced pressure principle backflow preventer – ASSE 1013 (RP)

b. Where lawn sprinkler and irrigation systems have pumps, connections for pumping equipment, auxiliary air tanks or are otherwise capable of creating back pressure, the potable water supply shall be protected by the following type of device if the backflow prevention device is located upstream from the source of back pressure.

1. Reduced pressure principle backflow preventer – ASSE 1013 (RP)

c. Where systems have a backflow preventer installed downstream from a potable water supply pump or a potable water supply pump connection, the preventer shall be one of the following:

1. Atmospheric type vacuum breaker – ASSE 1001 (for non-continuous pressure)
2. Pressure vacuum breaker assembly (PVB) – ASSE 1020
3. Spill resistant vacuum breaker (SVB) – ASSE 1056
4. Reduced pressure principle backflow preventer – ASSE 1013 (RP)

d. Where systems include a chemical injector or any provisions for chemical injection, the potable water supply shall be protected by the following:

1. Reduced pressure principle backflow preventer – ASSE 1013 (RP)

10.5.11 Water Heat Exchangers

a. Heat exchangers used for heat transfer, heat recovery, or solar heating shall protect the potable water system from being contaminated by the heat transfer medium, in accordance with either subparagraph b or c below.

b. Single-wall heat exchangers shall be permitted if they satisfy all of the following requirements:

1. The heat transfer medium is either potable water or contains only substances that are recognized as safe by the U.S. Food and Drug Administration.

2. The pressure of the heat transfer medium is maintained less than the normal minimum operating pressure of the potable water system.

EXCEPTION: Steam complying with subparagraph b.1.

3. The equipment is permanently labeled to indicate that only additives recognized as safe by the FDA shall be used in the heat transfer medium.

c. Double-wall heat exchangers shall separate the potable water from the heat transfer medium by providing a space between the two walls that is vented to the atmosphere.

10.5.12 Protection from Hose Connections

A pressure-type or atmospheric-type vacuum breaker or permanently attached hose connection vacuum breaker shall protect hose bibbs, sill-cocks, wall hydrants and other outlets having a hose connection. The backflow preventer shall not be subjected to continuous pressure and shall be limited to low-head back pressure of not more than 10 feet of water from an elevated hose.

EXCEPTIONS:

(1) Water heater and boiler drain valves that are provided with hose connection threads and that are intended only for tank or vessel draining shall not be required to be equipped with a backflow preventer.

(2) This section shall not apply to water supply valves intended for connection to clothes washing machines where backflow prevention is otherwise provided or is integral with the machine.

10.5.13 Protection from Special Equipment

The water supply for any equipment or device that creates a cross-connection with the potable water supply shall be protected against backflow as required in Section 10.5. Such equipment and devices includes, but is not limited to, chemical dispensers, portable cleaning equipment, sewer and drain cleaning equipment, and dental pump equipment.

a. Chemical Dispensing Systems

Chemical dispensing systems with connections to the potable water distribution system shall protect the water distribution systems from backflow in accordance with ASSE 1055, which requires an air gap, atmospheric vacuum breaker, pressure vacuum breaker, spill resistant vacuum breaker, or reduced pressure principle assembly.

EXCEPTION: Atmospheric vacuum breakers shall not be used where there are shutoff valves or other shutoff devices downstream or where they are subject to continuous pressure for more than 12 hours in a 24 hour period.

b. Portable Cleaning Equipment

Where the water distribution system connects to portable cleaning equipment, the water supply system shall be protected against backflow in accordance with Section 10.5, which allows for an atmospheric vacuum breaker, pressure vacuum breaker, double check valve, or a reduced pressure principle assembly.

EXCEPTION: Atmospheric vacuum breakers shall not be used where there are shutoff valves or other shutoff devices downstream or where they are subject to continuous pressure for more than 12 hours at a time.

c. Dental Pump Equipment

Where the water distribution system connects to dental pumping equipment, the water supply system shall be protected against backflow in accordance with Section 10.5, which allows for an atmospheric vacuum breaker, pressure vacuum breaker, double check valve, or a reduced pressure principle assembly.

EXCEPTION: Atmospheric vacuum breakers shall not be used where there are shutoff valves or other shutoff devices downstream or where they are subject to continuous pressure for more than 12 hours at a time.

d. Water Powered Back-up Sump Pumps

1. The potable water supply to water powered back-up sump pumps shall be protected against backflow by an ASSE 1013 (RP) reduced pressure principle backflow preventer. Vacuum breakers shall not be permitted.

2. Pump control valves, ejectors, and discharge piping shall be installed in accordance with the manufacturer's instructions.

e. Laboratory Sink Faucets

Faucets on laboratory sinks shall be protected against backflow from back siphonage and back pressure by the attachment of an ASSE 1035 laboratory faucet backflow preventer. The backflow preventer shall not be subjected to constant supply pressure and there shall be no shutoff devices downstream. The backflow preventer shall be limited to back pressure of not more than 6 inches of water elevation beyond its outlet.

10.6 WATER SERVICE

10.6.1 Separation of Water Service Piping from Drain Piping and Non-Potable Water Piping.
Underground water service piping shall be separated by not less than one foot horizontally from building drain piping, building sewer piping, and non-potable water piping, including reclaimed water, gray water, and harvested rainwater. *See Figure 10.6.1-A and 10.6.1-B*

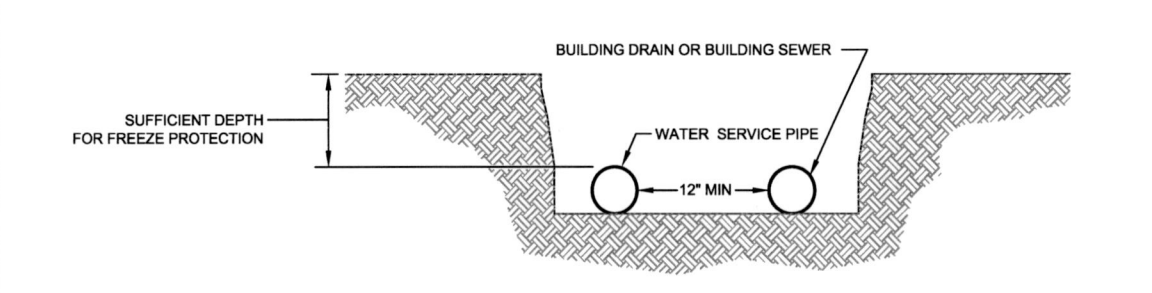

NOTES:
1. Improvements in the integrity of underground drainage system pipe joints has permitted the separation from potable water piping to be reduced to 12 inches.

Figure 10.6.1 - A
THE MINIMUM DISTANCE BETWEEN AN UNDERGROUND WATER SERVICE AND A BUILDING DRAIN OR BUILDING SEWER

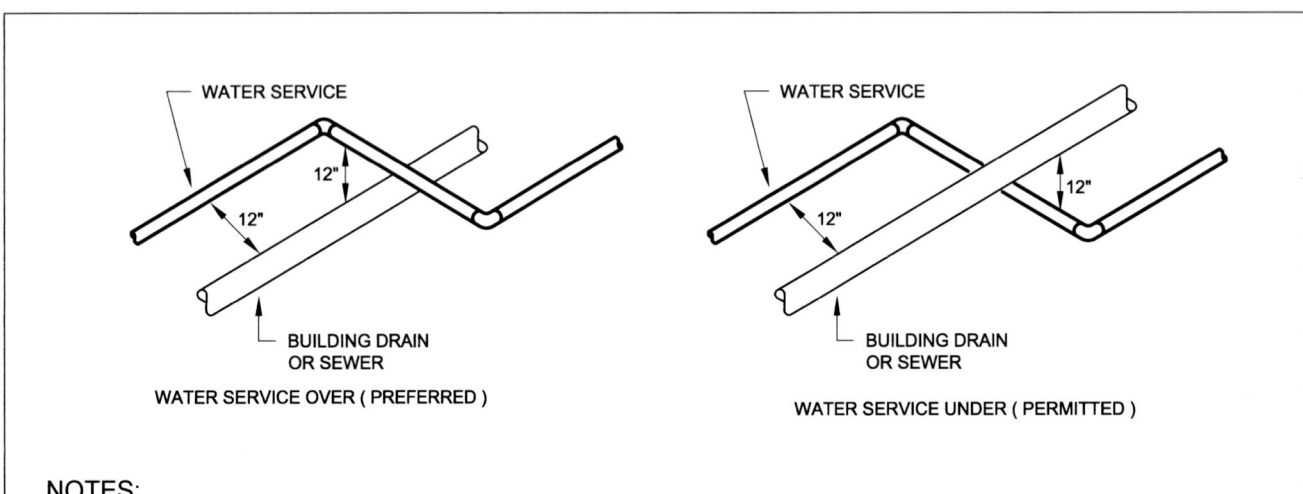

NOTES:
1. The minimum clearance is 12 inches.
2. Where possible, the water service should cross over the building drain or building sewer.
3. If necessary, the water service can cross under the building drain or building sewer.
4. The 12" minimum separation prevents direct contact between the pipe lines, even with considerable settlement.

Figure 10.6.1 - B
CROSSING A WATER SERVICE OVER (OR UNDER) A BUILDING DRAIN OR BUILDING SEWER

10.6.2 Water Service Near Sources of Pollution

Potable water service piping shall not be located in, under, or above cesspools, septic tanks, septic tank drainage fields, or drainage pits. A separation of ten feet shall be maintained from such systems. When a water line parallels or crosses over or under a sewer, a minimum clearance of 12 inches in all directions shall be maintained. *See Figures 10.6.2, 10.6.1-A, and 10.6.1-B*

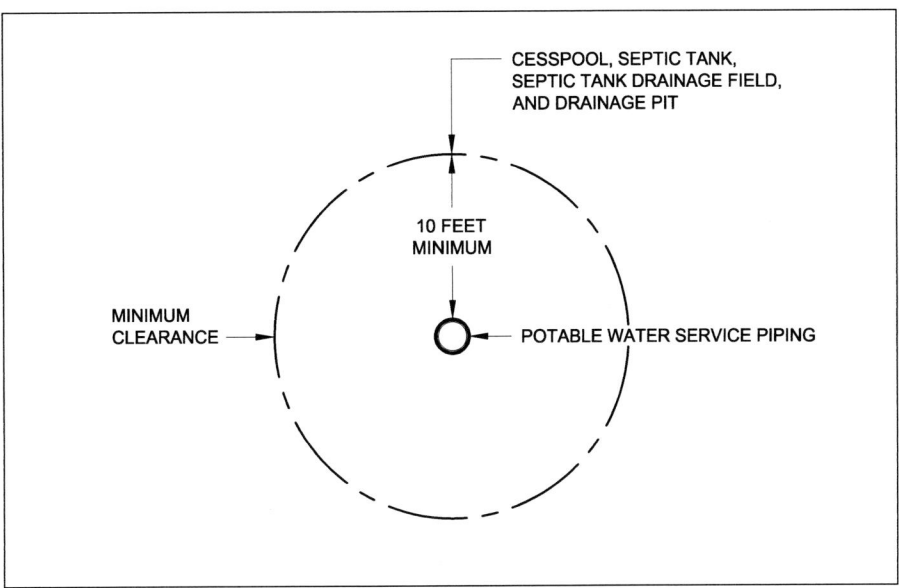

Figure 10.6.2
MINIMUM CLEARANCE BETWEEN A WATER SERVICE PIPE AND SOURCES OF CONTAMINATION

10.6.3 Stop-and-Waste Valves Prohibited

Combination stop-and-waste valves or cocks shall not be installed underground in water service piping.

> *Comment: Open waste outlets on underground water service shutoff valves would permit ground water to contaminate the potable water supply.*

10.6.4 Water Service Pipe Sleeves

a. Pipe sleeves shall be provided where water service pipes penetrate foundation walls or floor slabs to protect against corrosion of the pipe and allow clearance for expansion, contraction and settlement. The sleeve shall form a watertight bond with the wall or floor slab. The annular space between the pipe and the sleeve shall be resiliently sealed watertight.

b. Where water service piping is plastic, a wall sleeve shall be not less than five feet long extended outside beyond the wall to undisturbed earth or other equivalent support.

See Figure 10.6.4

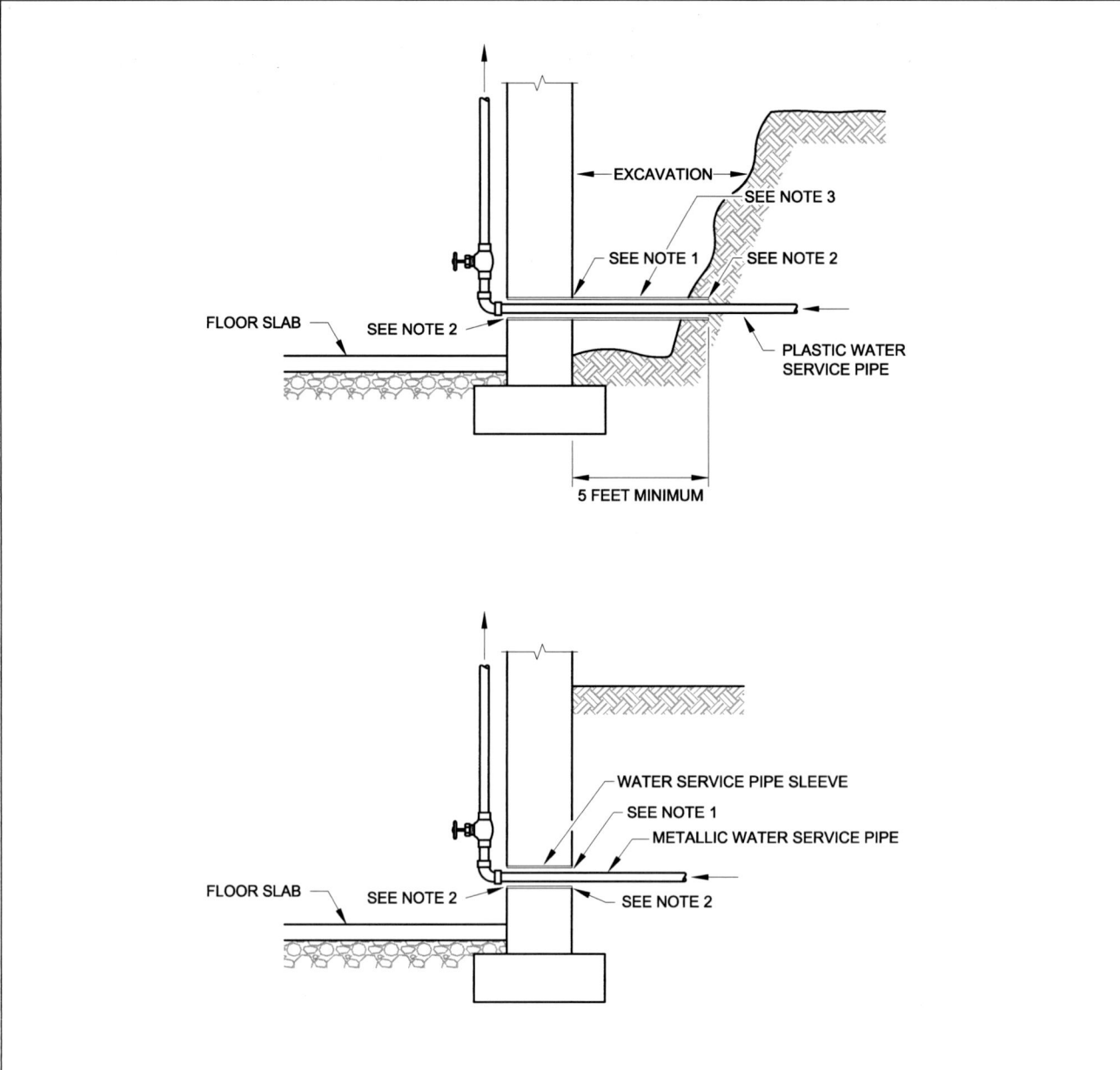

NOTES:
1. The pipe sleeve must be sealed watertight to the wall or floor slab.
2. A 1/2" minimum annular space around the water pipe, between the pipe and the sleeve, is required and must be sealed watertight.
3. Sleeves for plastic water service pipe must extend outside beyond foundation walls to undisturbed earth and be at least 5 feet long.

Figure 10.6.4
WATER SERVICE PIPE SLEEVES FOR METALLIC AND PLASTIC PIPING

10.6.5 Water Service Sizing

The water service pipe shall be of sufficient size to furnish water to the building in the quantities and at the pressures required elsewhere in this Code. The pipe size shall not be less than 3/4 inch nominal.

10.7 WATER PUMPING AND STORAGE EQUIPMENT

10.7.1 Pumps and Other Appliances

Water pumps, filters, softeners, tanks and other appliances and devices used to handle or treat potable water shall be protected against contamination as per Section 10.5.

10.7.2 Prohibited Location of Potable Supply Tanks

Potable water gravity tanks or manholes of potable water pressure tanks shall not be located directly under any sanitary drain piping.

10.8 WATER PRESSURE BOOSTER SYSTEMS

10.8.1 Water Pressure Booster Systems Required

a. When the water pressure in the public water main or individual water supply system is insufficient to supply the potable peak demand flow to plumbing fixtures and other water needs freely and continuously with the minimum pressure and quantities specified in Section 10.14.3, or elsewhere in this Code, and in accordance with good practice, the rate of supply shall be supplemented by one of the following methods:

    1. An elevated water tank.

    2. A hydro-pneumatic pressure booster system.

    3. A water pressure booster pump.

10.8.2 Reserved

10.8.3 Overflows for Water Supply Tanks

Gravity or suction water supply tanks shall be provided with an overflow having a diameter not less than that shown in Table 10.8.3. The overflow outlet shall discharge above and within not less than 6 inches of a roof or roof drain, floor or floor drain, or over an open water-supplied fixture. The overflow outlet shall be covered by a corrosion-resistant screen of not less than 16 x 20 mesh to the inch and by 1/4 inch hardware cloth, or it shall terminate in a horizontally installed 45° angle-seat check valve. Drainage from overflow pipes shall be directed so as not to freeze on roof walkways. *See Figure 10.5.7*

Table 10.8.3 Size of Overflow Pipes for Water Supply Tanks								
Maximum Makeup Water Flow – gpm	13	55	100	165	355	640	1040	>1040
Overflow Pipe Size – inches	1-1/2	2	2-1/2	3	4	5	6	8

10.8.4 Covers

All water supply tanks shall be covered to keep out unauthorized persons, dirt, and vermin. The covers of gravity tanks shall be vented with a return bend vent pipe having an area not less than the area of the down feed riser pipe and the vent shall be screened with corrosion resistant screen having not less than 14 and not more than 20 openings per linear inch. *See Figure 10.5.7*

10.8.5 Potable Water Inlet Control and Location

Potable water inlets to gravity water tanks shall be controlled by a level control valve or other automatic supply valve so installed as to control the water level in the tank and prevent the tank from overflowing. The inlet shall be terminated so as to provide an air gap above the overflow level of the tank in accordance with Table 10.5.2. *See Figure 10.5.7*

10.8.6 Tank Drain Pipes

Each tank shall be provided at its lowest point with a valved pipe to permit emptying the tank.
See Figure 10.5.7

10.8.7 Low Pressure Cut-Off Required on Booster Pumps

Booster pumps shall be protected by a low pressure cut-off switch to shut-off the pump(s) if the suction pressure drops to an unsafe value.

10.8.8 Pressure Tanks—Vacuum Relief

Domestic water pressure tanks shall be provided with vacuum relief if required by Section 10.16 .7 or the tank manufacturer.

10.8.9 Pressure Tanks—Pressure Relief

All water pressure tanks shall be provided with approved pressure relief valves set at a pressure not in excess of the tank working pressure.

10.9 FLUSHING AND DISINFECTING POTABLE WATER SYSTEMS

10.9.1 Flushing

The water service piping and distribution piping to all fixtures and outlets shall be flushed until the water runs clear and free of debris or particles. Faucet aerators or screens shall be removed during flushing operations.

10.9.2 Disinfecting

   a. Water service piping and hot and cold water distribution piping in new or renovated potable water systems shall be disinfected after flushing and prior to use.
EXCEPTION: Single dwelling units with public water supply.
   b. The procedure used shall be as follows or an approved equivalent:
      1. All water outlets shall be posted to warn against use during disinfecting operations.
      2. Disinfecting shall be performed by persons experienced in such work.
      3. The water supply to the piping system or parts thereof being disinfected shall be valved-off from the normal water source to prevent the introduction of disinfecting agents into a public water supply or portions of a system that are not being disinfected.
      4. The piping shall be disinfected with a water-chlorine solution. During the injection of the disinfecting agent into the piping, each outlet shall be fully opened several times until a concentration of not less than 50 parts per million chlorine is present at every outlet. The solution shall be allowed to stand in the piping for at least 24 hours.
      5. An acceptable alternate to the 50 ppm/24-hour procedure described in Section 10.9.2.b.4 shall be to maintain a level of not less than 200 parts per million chlorine for not less than three hours. If this alternate procedure is used, the heavily concentrated chlorine shall not be allowed to stand in the piping system for more than 6 hours. Also, special procedures shall be used to dispose of the heavily concentrated chlorine in an environmentally acceptable and approved manner.
      6. At the end of the required retention time, the residual level of chlorine at every outlet shall be not less than five parts per million. If the residual is less than five parts per million, the disinfecting procedure shall be repeated until the required minimum chlorine residual is obtained at every outlet.

7. After the required residual chlorine level is obtained at every outlet, the system shall be flushed to remove the disinfecting agent. Flushing shall continue until the chlorine level at every outlet is reduced to that of the incoming water supply.

8. Any faucet aerators or screens that were removed under Section 10.9.1 shall be replaced.

9. A certification of performance and laboratory test report showing the absence of coliform organisms shall be submitted to the Authority Having Jurisdiction upon satisfactory completion of the disinfecting operations.

10.9.3 Additions to Existing Systems

Where new piping is required to be flushed and disinfected, if a new addition or additions to an existing domestic water piping system can be completely isolated from the existing system for the purpose of flushing and disinfecting, only the new piping shall be required to be flushed and disinfected.

10.10 WATER SUPPLY SYSTEM MATERIALS

See Sections 3.4.2 and 3.4.3.

10.11 ALLOWANCE FOR CHARACTER OF SOIL AND WATER

When selecting the material and size for water service supply pipe, tube, or fittings, due consideration shall be given to the action of the water on the interior of the pipe and of the soil, fill or other material on the exterior of the piping.

> *Comment: The chemical composition of the service water should be considered when selecting the water service pipe material. Refer to Appendix Section B.2.3. The aggressive nature of the soil or fill material around the water service pipe should be evaluated when considering if additional corrosion protection is need for the exterior surfaces of the pipe.*

10.12 WATER SUPPLY CONTROL VALVES

10.12.1 Curb Valve

On the water service from the street main to the building, an approved gate valve or ground key stopcock or ball valve shall be installed near the curb line between the property line and the curb. This valve or stopcock shall be provided with an approved curb valve box. *See Figure 1.2.82*

10.12.2 Building Valve

a. The building water service shall be provided with a readily accessible gate valve or other full-way shutoff valve located inside the building near the point where the water service enters. *See Figure 1.2.82*

b. Provisions shall be made to drain the piping immediately downstream from the building valve for service or maintenance. Drains shall be capped or plugged to prevent them from being inadvertently discharged. Bleed valves shall be permitted to satisfy this requirement.

c. If a building has two or more water services that are interconnected within the building, full-size check valves shall be installed near the outlets of the building valves to prevent backflow from one water service to another. Pressure test connections shall be provided on both sides of the check valves.

EXCEPTION: Check valves shall not be required if the water services include backflow prevention for containment of the premises.

10.12.3 Water Supply Tank Valve

A shutoff valve shall be provided at the outlet of any tank serving as a water supply source, either by gravity or pressure. *See Figure 10.5.7*

10.12.4 Valves in Dwelling Units

a. In single dwelling units, the building valve required by Section 10.12.2 shall shutoff the water supply to all fixtures and outlets.

b. In multiple dwelling units, one or more shutoff valves shall be provided in the main supply or main branches in each dwelling unit so that the water supply to any plumbing fixture or group of fixtures in that dwelling unit can be shut off without stopping the water supply to fixtures in other dwelling units. These valves shall be accessible in the dwelling unit that they control.

c. Except as permitted in Section 10.12.4.d, individual fixture shutoff or stop valves shall be provided for water closets, lavatories, kitchen sinks, laundry trays, bar sinks, bidets, clothes washing machines, sill cocks, wall hydrants, appliances, and equipment connected to the water supply system. Valves for fixtures, appliances, and equipment shall be accessible without having to move the appliance or equipment.

EXCEPTION: Appliances shall be permitted to be moved for access to shutoff valves for ice makers.

d. Unless individual fixture shutoff valves are provided for fixtures in powder rooms or bathroom groups in accordance with Section l0.12.4.c, shutoff valves may be provided for the each powder room or bathroom group in lieu of individual fixture shutoff valves. In individual dwelling units where powder rooms or bathroom groups are located adjacent to each other or one directly above the other, they may be considered as a single group and shall be permitted to have a single set of shutoff valves.

e. Self-piercing and needle-type saddle valves shall be prohibited.

See Figure 10.12.4-A and 10.12.4-B

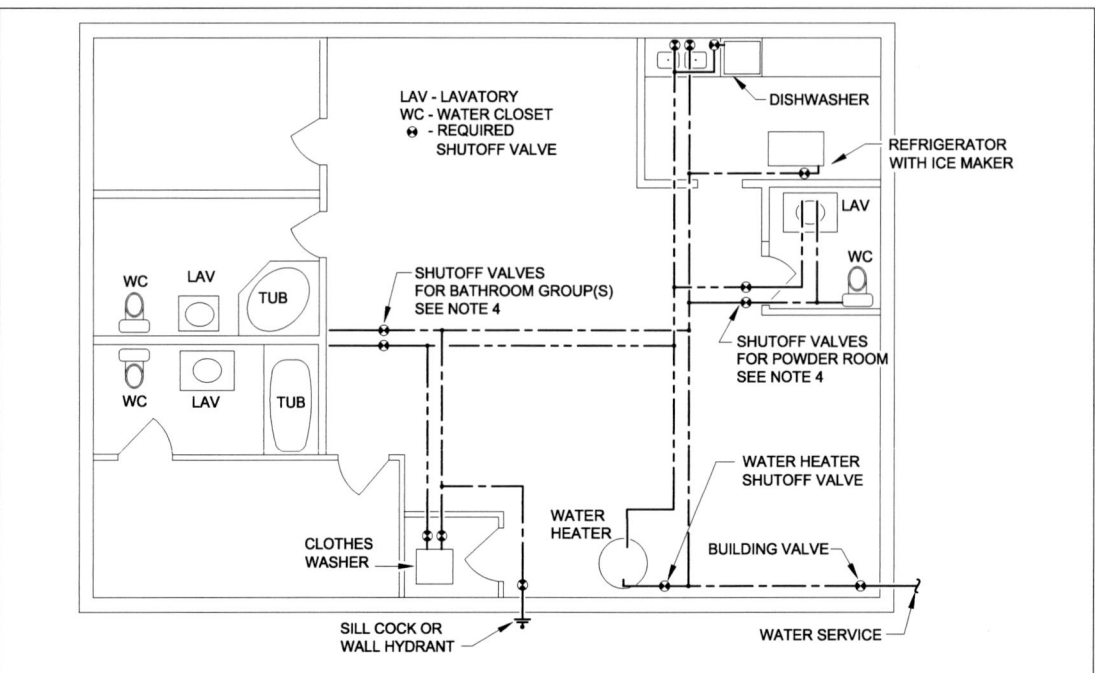

Figure 10.12.4 - A
REQUIRED WATER SHUTOFF VALVES IN SINGLE DWELLING UNITS

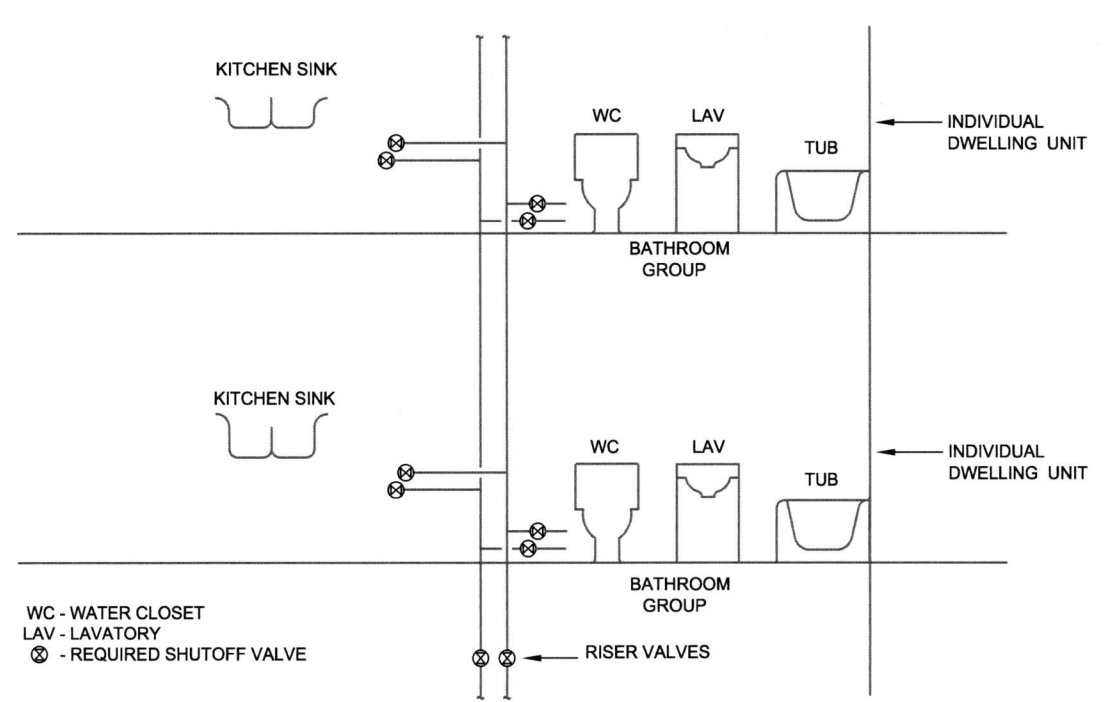

NOTES:
1. Riser shutoff valves are required where risers serve dwelling units on two or more floors.
2. Main shutoff valves are mandatory for each dwelling unit.
3. Shutting off the water supply to one dwelling unit must not stop the water supply to another dwelling unit.
4. Shutoff valves for powder rooms and bathroom groups are not mandatory if all fixtures within the group have individual shutoff valves. One set of shutoff valves can serve two bathroom groups within the same dwelling unit if they are piped as one group of fixtures without separate branch pipes for each group.
5. Water closets and lavatories usually have individual supply stop valves where the fixture supply tube connects to the fixture branch supply pipe.
6. The main shutoff valves for each dwelling unit must be accessible within the dwelling unit that they serve.

Figure 10.12.4 - B
REQUIRED WATER SHUTOFF VALVES IN MULTI-DWELLING UNITS

10.12.5 Riser Valves

Shutoff valves shall be provided for isolating each water supply riser serving fixtures on two or more floors. EXCEPTION: Risers within individual dwelling units that serve fixtures only in that unit. *See Figure 1.2.49*

10.12.6 Individual Fixture Valves

a. In a building used or intended to be used for other than dwelling purposes, the water distribution pipe to each fixture or other piece of equipment shall be provided with a valve or fixture stop to shut off the water to the fixture or to the room in which it is located. These valves shall be accessible. Each sill cock and wall hydrant shall be separately controlled by a valve inside the building.

b. Self-piercing and needle-type saddle valves shall be prohibited.

> *Comment: In buildings or portions thereof that are used for other than dwelling purposes, each fixture or equipment connection must have an individual shutoff means. Shutoff means in such systems cannot shutoff a group of fixtures that do not have individual shutoff means. The guest rooms in hotels and motels are considered to be dwelling units.*

10.12.7 Water Heating Equipment Valve

A shutoff valve shall be provided in the cold water supply to each water heater. If a shutoff valve is also provided in the hot water supply from the heater, it shall not isolate any safety devices from the heater or storage tank. Shutoff valves for water heaters shall be the gate, ball, plug, or butterfly type. ***See Figure 10.12.7***

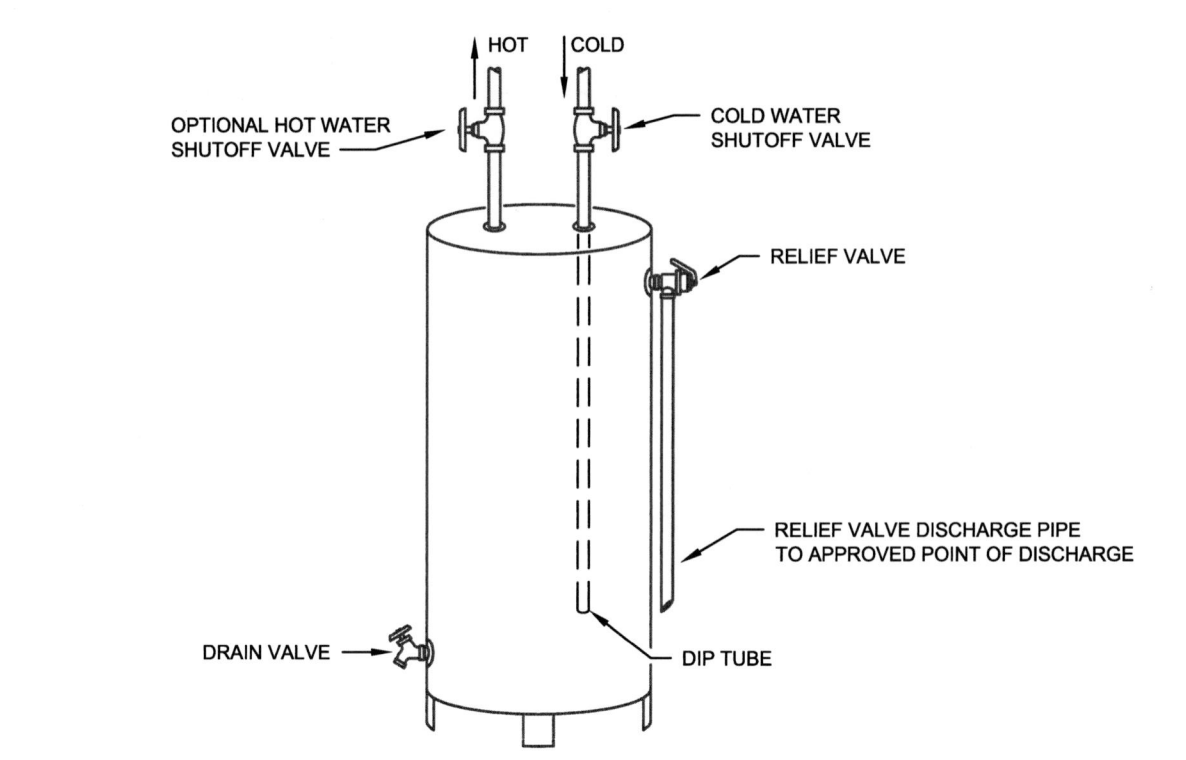

NOTES:
1. The cold water shutoff valve permits the water heater to be serviced or replaced, but requires that the hot water distribution piping be drained.
2. The optional hot water shutoff valve permits the water heater to be serviced or replaced without draining the hot water distribution piping.
3. The water heater relief valve must not be isolated from the water heater tank by any shutoff valves.
4. Water heater shutoff valves must be the full-flow type. Globe valves are not permitted.

**Figure 10.12.7
WATER HEATER SHUTOFF VALVE(S)**

10.12.8 Meter Valve

A gate valve or other full-way valve shall be installed in the line on the discharge side of each water meter. The valve shall not be less in size than the building water service pipe. ***See Figures 10.12.8-A and 10.12.8-B***

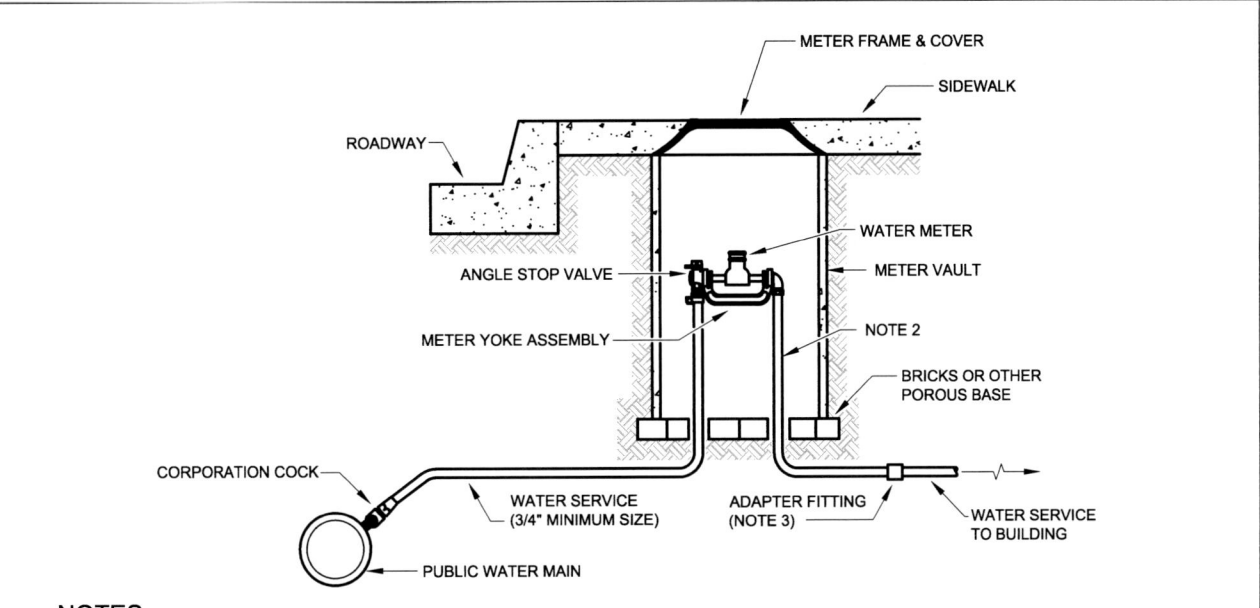

NOTES:
1. The stop valve on the inlet side of the water meter is normally provided by the water service utility. It can be the curb valve required by Section 10.12.1.
2. The distinction between small and large water services is dictated by the Authority Having Jurisdiction. For small water services, the shutoff valve on the outlet side of the water meter can be the building valve required by Section 10.12.2. See Figure 10.12.8 - B for larger water services.
3. Adapter fittings are required where the pipe material for the water service to the building is different from the pipe material provided by the water utility.

Figure 10.12.8 - A
METER SHUTOFF VALVES IN A SMALL WATER SERVICE

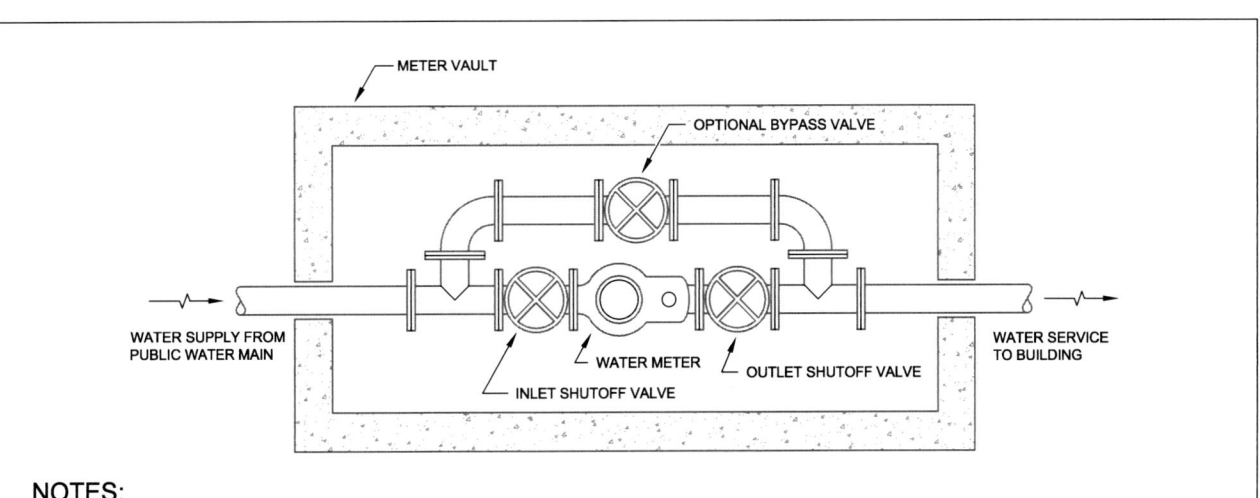

NOTES:
1. The shutoff valve on the inlet side of the water meter is normally provided by the water utility.
2. The bypass valve is optional. It can be used to provide continuous water service while the meter is being repaired or replaced.
3. The bypass valve and shutoff valve on the meter outlet permit the meter to be repaired or replaced without interrupting the water service to the facility.

Figure 10.12.8 - B
METER SHUTOFF VALVES IN A LARGE WATER SERVICE

10.12.9 Valve Accessibility

Water supply shutoff valves shall be placed so as to be accessible for use, service and maintenance.

> *Comment: Access to water supply shutoff valves can require the use of tools to remove access panels or doors.*

10.13 FLEXIBLE WATER CONNECTORS

Flexible water connectors exposed to continuous pressure shall conform to ASME A112.18.6/CSA B125.6. Access shall be provided to all flexible water connectors.

10.14 MINIMUM REQUIREMENTS FOR WATER DISTRIBUTION SYSTEMS

10.14.1 Maximum Velocity (See Appendix B.6)

Water distribution piping within buildings shall be sized for a maximum velocity of 8 feet per second at the design flow rate unless the pipe manufacturer's sizing recommendations call for the maximum velocity to be less than 8 feet per second. The maximum velocity of hot water in copper tubing shall be 5 feet per second.

10.14.2 Size of Individual Fixture Supply Branches

a. Individual fixture supply branch pipe sizes shall be based on the minimum available flowing water pressure at the point of connection to the water distribution system, any elevation difference between that connection and the fixture, and the allowable pressure loss in the fixture supply branch. The minimum fixture supply branch pipe sizes shall be as indicated in Table 10.14.2A. For design purposes, the required pressure at each fixture inlet shall be 15 psig minimum with flow for all fixtures, except 20 psig flowing for flushometer valves on siphon jet water closets and 25 psig flowing for flushometer valves on blowout water closets and blowout urinals. Flushometer tank (pressure assisted) water closets require a minimum of 25 psig static pressure. The following water flow rates shall be used for the purpose of sizing individual fixture supply branch pipes:

    5.0 gpm for hose bibbs and wall hydrants;

    4.0 gpm for bath faucets and clothes washers;

    0.75 gpm for drinking fountains and water coolers;

    2.2 gpm for sink faucets;

    2.5 gpm for showers;

    2.2 gpm for lavatory faucets;

    3.0 gpm for water closets other than the flushometer valve type;

    12.0 gpm for flushometer valve urinals;

    30.0 gpm for flushometer valve water closets

b. Fixture supply branches shall extend from the distribution system to within 30 inches of the point of connection to the fixture or device served and be within the same area and physical space as the point of connection to the fixture or device. Fixture supply tubes and flexible water connectors shall be not less than the size recommended by the manufacturer of the fixture, faucet, appliance or device served.

Table 10.14.2A
WATER SUPPLY FIXTURE UNITS (WSFU) AND
MINIMUM FIXTURE BRANCH PIPE SIZES

	MINIMUM BRANCH PIPE SIZE	INDIVIDUAL DWELLING UNITS	SERVING 3 OR MORE DWELLING UNITS	OTHER THAN DWELLING UNITS	HEAVY-USE ASSEMBLY
BATHROOM GROUPS HAVING 1.6 GPF WATER CLOSETS OTHER THAN THE FLUSHOMETER VALVE TYPE					
Half-Bath or Powder Room		3.5	2.5		
1 Bathroom Group		5.0	3.5		
1-1/2 Bathroom Groups		6.0	4.0		
2 Bathroom Groups		7.0	4.5		
2-1/2 Bathroom Groups		8.0	5.0		
3 Bathroom Groups		9.0	5.5		
Each Additional Half-Bath		0.5	0.5		
Each Additional Bathroom Group		1.0	1.0		
BATHROOM GROUPS HAVING 3.5 GPF (or higher) GRAVITY TANK WATER CLOSETS					
Half-Bath or Powder Room		4.0	3.0		
1 Bathroom Group		6.0	5.0		
1-1/2 Bathroom Groups		8.0	5.5		
2 Bathroom Groups		10.0	6.0		
2-1/2 Bathroom Groups		11.0	6.5		
3 Bathroom Groups		12.0	7.0		
Each Additional Half-Bath		0.5	0.5		
Each Additional Bathroom Group		1.0	1.0		
OTHER GROUPS OF FIXTURES					
Bathroom Group with 1.6 GPF Flushometer Valve		6.0	4.0		
Bathroom Group with 3.5 GPF (or higher) Flushometer Valve		8.0	6.0		
Kitchen Group with Sink and Dishwasher		2.0	1.5		
Laundry Group with Sink and Clothes Washer		5.0	3.0		

Table 10.14.2A (continued)
WATER SUPPLY FIXTURE UNITS (WSFU) AND MINIMUM FIXTURE BRANCH PIPE SIZES

INDIVIDUAL FIXTURES	MINIMUM BRANCH PIPE SIZE	INDIVIDUAL DWELLING UNITS	SERVING 3 OR MORE DWELLING UNITS	OTHER THAN DWELLING UNITS	HEAVY-USE ASSEMBLY
Bar Sink	3/8"	1.0	0.5		
Bathtub or Combination Bath/Shower	1/2"	4.0	3.5		
Bidet	1/2"	1.0	0.5		
Clothes Washer, Domestic	1/2"	4.0	2.5	4.0	
Dishwasher, Domestic	1/2"	1.5	1.0	1.5	
Drinking Fountain or Water Cooler	3/8"			0.5	0.75
Hose Bibb (first)	1/2"	2.5	2.5	2.5	
Hose Bibb (each additional)	1/2"	1.0	1.0	1.0	
Kitchen Sink, Domestic	1/2"	1.5	1.0	1.5	
Laundry Sink	1/2"	2.0	1.0	2.0	
Lavatory	3/8"	1.0	0.5	1.0	1.0
Service Sink or Mop Basin	1/2"			3.0	
Shower	1/2"	2.0	2.0	2.0	
Shower, continuous use	1/2"			5.0	
Urinal, 1.0 GPF	3/4"			4.0	5.0
Urinal, greater than 1.0 GPF	3/4"			5.0	6.0
Wash Fountain (per rated user)				0.5	
Wash Sink (1 or 2 faucets, per faucet)				1.0	
Wash Sink (additional faucets, per faucet)				0.5	
Water Closet, 1.6 GPF Gravity Tank	1/2"	2.5	2.5	2.5	4.0
Water Closet, 1.6 GPF Flushometer Tank	1/2"	2.5	2.5	2.5	3.5
Water Closet, 1.6 GPF Flushometer Valve	1"	5.0	5.0	5.0	8.0
Water Closet, 3.5 GPF Gravity Tank	1/2"	3.0	3.0	5.5	7.0
Water Closet, 3.5 GPF Flushometer Valve	1"	7.0	7.0	8.0	10.0
Whirlpool Bath or Combination Bath/Shower	1/2"	4.0	4.0		

NOTES FOR TABLE 10.14.2A:
1. A Bathroom Group, for the purposes of this Table, consists of not more than one water closet, up to two lavatories, and either one bathtub, one bath/shower combination, or one shower stall. Other fixtures within the bathing facility shall be counted separately to determine the total water supply fixture unit load.
2. A Half-Bath or Powder Room, for the purposes of this Table, consists of one water closet and one lavatory.
3. For unlisted fixtures, refer to a listed fixture having a similar flow and frequency of use.
4. The listed fixture unit values for Bathroom Groups and Individual Fixtures represent their load on the cold water service. The separate cold water and hot water fixture unit values for fixtures having both cold and hot water connections shall each be taken as 3/4 of the listed total value for the individual fixture.
5. When WSFU values are added to determine the demand on the water distribution system or portions thereof, round the sum to the nearest whole number before referring to Table 10.14.2B for the corresponding gallons per minute (gpm) flow. WSFU values of 0.5 or more should be rounded up to the next higher whole number (9.5 = 10 WSFU). Values of 0.4 or less should be rounded down to the next lower whole number (9.4 = 9 WSFU).
6. The listed minimum supply branch pipe sizes for individual fixtures are the nominal (I.D.) pipe size in inches.
7. "Other Than Dwelling Units" applies to business, commercial, industrial, and assembly occupancies other than those defined under "Heavy-Use Assembly." Included are the public and common areas in hotels, motels, and multi-dwelling buildings.
8. "Heavy-Use Assembly" applies to toilet facilities in occupancies that place heavy, but intermittent, time-based demands on the water supply system, such as schools, auditoriums, stadiums, race courses, transportation terminals, theaters, and similar occupancies where queuing is likely to occur during periods of peak use.
9. For fixtures or supply connections likely to impose continuous flow demands, determine their required flow in gallons per minute (gpm) and add it separately to the demand (in gpm) for the distribution system or portion thereof.

Table 10.14.2B
TABLE FOR CONVERTING DEMAND IN WSFU TO GPM[1]

WSFU	GPM Flush Tanks[2]	GPM Flush Valves[3]	WSFU	GPM Flush Tanks[2]	GPM Flush Valves[3]
3	3		120	49	74
4	4		140	53	78
5	4.5	22	160	57	83
6	5	23	180	61	87
7	6	24	200	65	91
8	7	25	225	70	95
9	7.5	26	250	75	100
10	8	27	300	85	110
11	8.5	28	400	105	125
12	9	29	500	125	140
13	10	29.5	750	170	175
14	10.5	30	1000	210	210
15	11	31	1250	240	240
16	12	32	1500	270	270
17	12.5	33	1750	300	300
18	13	33.5	2000	325	325
19	13.5	34	2500	380	380
20	14	35	3000	435	435
25	17	38	4000	525	525
30	20	41	5000	600	600
40	25	47	6000	650	650
50	29	51	7000	700	700
60	33	55	8000	730	730
80	39	62	9000	760	760
100	44	68	10,000	790	790

NOTES FOR TABLE 10.14.2B:
1. This table converts water supply demands in water supply fixture units (WSFU) to required water flow in gallons per minute (GPM) for the purpose of pipe sizing.
2. This column applies to portions of piping systems where the water closets are the flush tank type (gravity or pressure) or there are no water closets, and to hot water piping.
3. This column applies to portions of piping systems where the water closets are the flush valve type.
4. Refer to Appendix M for WSFU values that are interpolated between the values listed in Table 10.14.2B to the degree that the WSFU increments are close enough that they do not produce GPM differences that would be significant in sizing water supply distribution piping.

10.14.3 Sizing Water Distribution Piping

a. The supply demand in gallons per minute in the building hot and cold water distribution system shall be determined on the basis of the load in terms of water supply fixture units (WSFU) as shown in Table 10.14.2A and the relationship between the load in WSFU and the supply demand in gallons per minute (GPM) as shown in Table 10.14.2B. Refer to Appendix M for a more detailed table of WSFU and equivalent GPM. For fixtures having both hot water and cold water connections, the separate hot and cold water loads shall be taken as 75% of the listed WSFU value.

b. Main risers and branches of the water distribution system shall be sized based on the minimum available water pressure at the source, any elevation differences between the source and the fixtures, pressure losses in the distribution system, and the pressure (with flow) required at each connection to the fixture supply branches. See Section 10.14.2.

10.14.4 Inadequate Water Pressure

Whenever water pressure from the street main or other sources of supply is insufficient to provide flow pressures at fixture outlets as required under Section 10.14.2, a booster pump and pressure tank or other approved means shall be installed on the building water supply system.

10.14.5 Variable Street Pressures

Where street water main pressures fluctuate, the building water distribution system shall be designed for the minimum pressure available.

10.14.6 Excessive Pressures

a. Pressure reducing valves complying with ASSE 1003 shall be provided if required to limit the water supply pressure at any fixture appliance, appurtenance, or outlet to not more than 80 psi under no-flow conditions.

b. The requirement of Section 10.14.6.a above shall not prohibit supply pressures higher than 80 psi from water pressure booster systems under Section 10.14.4 or in high pressure distribution systems, provided that the pressure at the fixtures served is subsequently reduced to 80 psi maximum. Where operating water pressures exceed 80 psi, the working pressure rating of materials and equipment shall be suitable for the maximum pressure that may be encountered, including temporary increases or surges.

c. Where pressure reducing valves are installed and the downstream piping is not rated for the maximum upstream pressure, a pressure relief valve shall be installed downstream from the pressure reducing valve. The relief valve shall be set not higher than the working pressure rating of the downstream piping and sized for not less than the flow capacity of the pressure reducing valve. Relief valves shall discharge in accordance with Section 10.16.6.

d. When a pressure reducing valve is installed, a gauge port or pressure gauge with pressure range of 0-150 psi shall be installed within 24 inches downstream from the reducing valve.

EXCEPTION: In dwelling units, gauge ports or pressure gauges shall not be required if there is a hose bibb or hose-end drain valve to which a pressure test gauge can be connected.

See Figures 10.14.6-A and 10.14.6-B

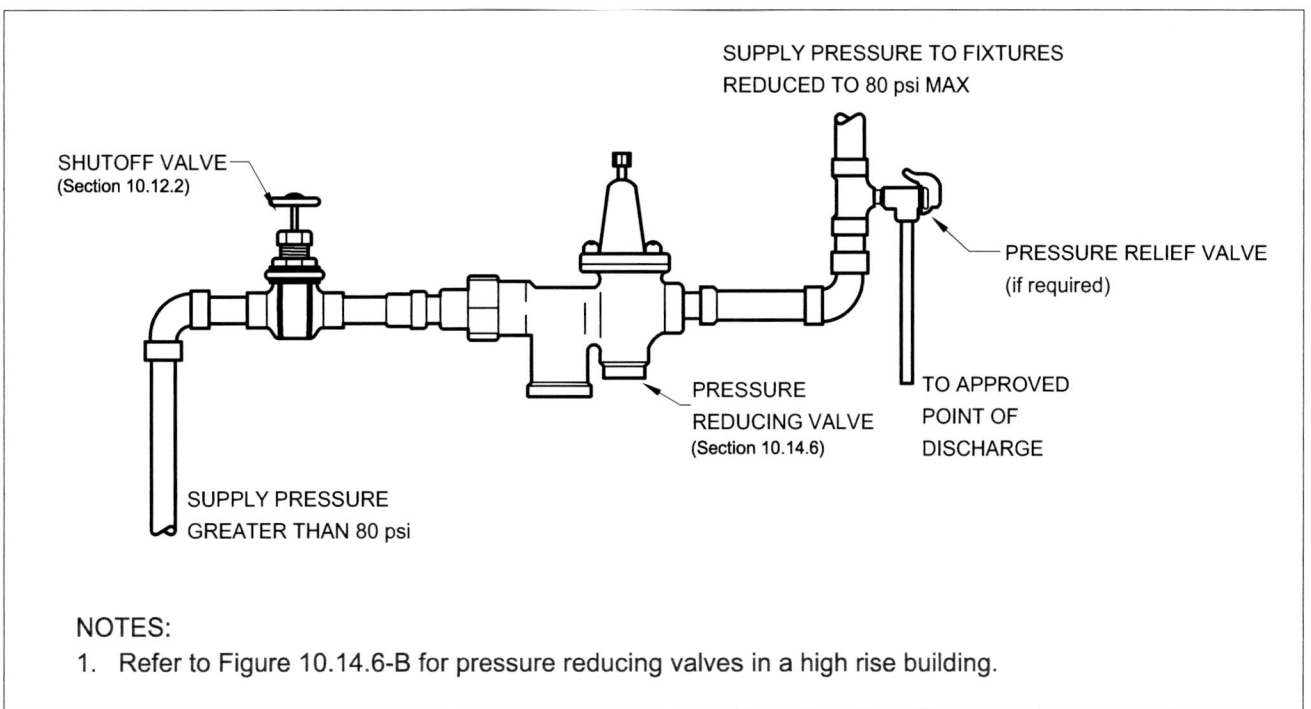

Figure 10.14.6 - A
INSTALLATION OF A WATER PRESSURE REDUCING VALVE WHERE THE WATER SUPPLY PRESSURE EXCEEDS 80 PSIG

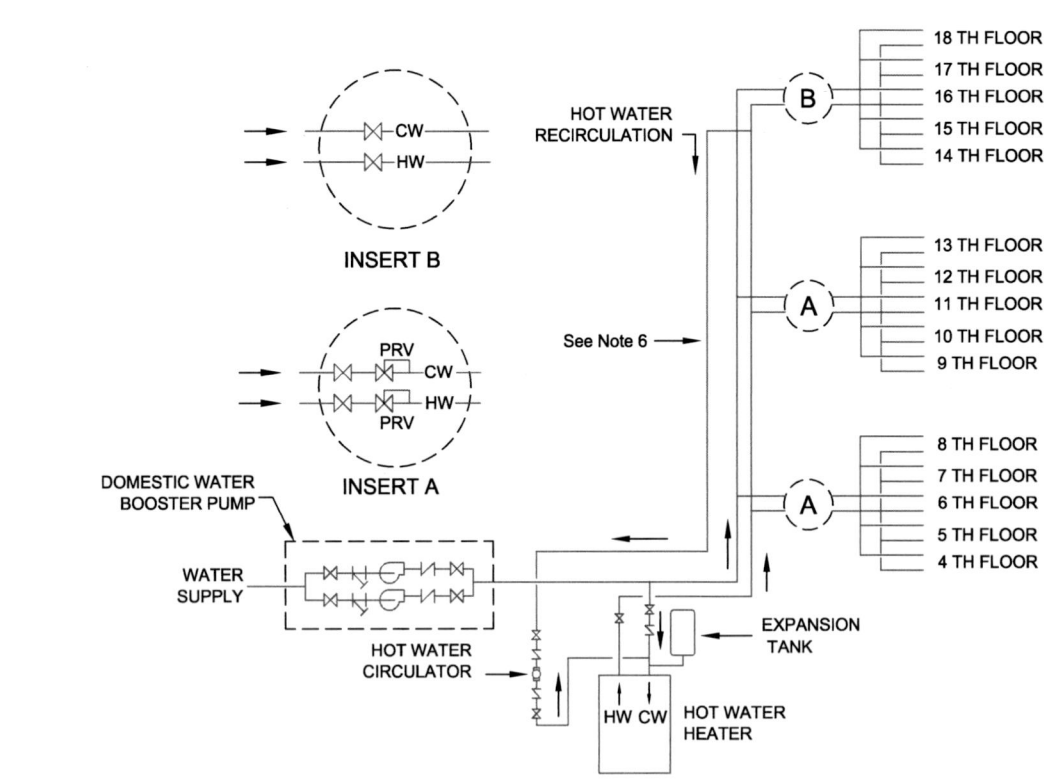

NOTES:
1. Example: A pressure booster pump is required to deliver domestic water to the 18th floor.
2. The 1st, 2nd, and 3rd Floors are supplied with street pressure and a separate hot water heater.
3. Without pressure reducing valves, the water supply pressure on the 4th through 13th Floors would exceed 80 psig.
4. Generally, up to five floors can be zoned with one set of pressure reducing valves without having excessive pressure differences between the top and bottom floors of the zone.
5. The pressure in the hot water recirculation line is essentially the same as in the adjacent hot water supply riser.
6. This figure shows only the hot water riser recirculated. If hot water must be circulated from the individual floors, pressure booster pumps would be required to pump the recirculated water from each reduced pressure zone back into the higher-pressure return riser (the downcomer).

Figure 10.14.6 - B
ONE POSSIBLE ARRANGEMENT OF PRESSURE REDUCING VALVES IN A HIGH-RISE BUILDING

10.14.7 Water Hammer

a. Approved water hammer arresters, complying with ASSE 1010 or PDI WH 201 shall be installed on water distribution piping in which quick closing valves are installed.

EXCEPTION: Single lever faucets, domestic clothes washers, and domestic dishwashers.

b. Water hammer arresters shall be placed as close as possible to the quick acting valve, at the end of long piping runs, or near batteries of fixtures.

c. Arresters shall be accessible for replacement.

See Figure 10.14.7

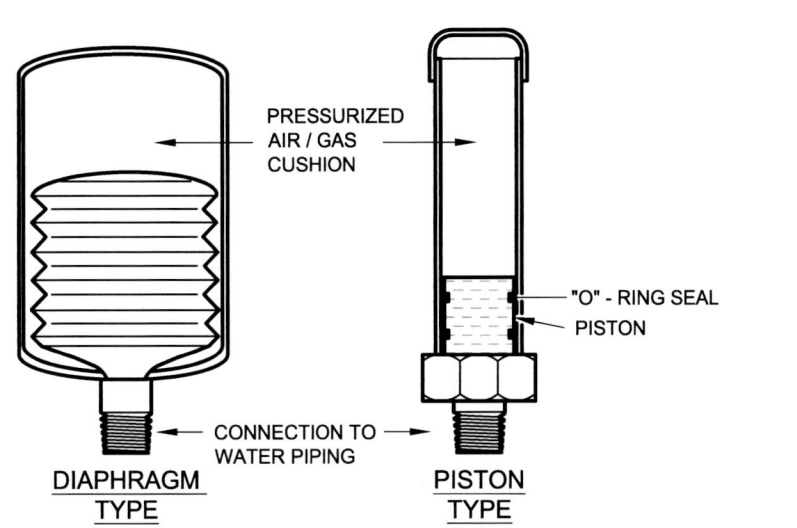

NOTES:
1. Water hammer arrestors should be sized and installed according to the manufacturer's instructions.
2. PDI Standard WH201 establishes PDI Sizes "A" through "F" with recommendations on the number of fixture units served by each size. Most manufacturers rank their products to the PDI Sizes.
3. Water hammer arrestors can be installed vertically or horizontally. They should be installed on the run of a tee fitting so that the unobstructed shock path is directly into the water hammer arrestor. See Figure 1.2.3.
4. The number of elbows upstream from a water hammer arrestor should be minimized because they create points of shock before the arrestor.
5. Water hammer arrestors must be accessible and have a means of shutoff to permit replacement if necessary. The means of shutoff can be the shutoff valve for a fixture or group of fixtures.

Figure 10.14.7
WATER HAMMER ARRESTORS

10.15 HOT WATER

10.15.1 Hot Water Supply System

In residences and buildings intended for human occupancy, hot water shall be supplied to all plumbing fixtures, appliances, and equipment that require hot water for their use.
EXCEPTION: In buildings other than dwelling units, tempered water supply systems shall be permitted to supply fixtures that deliver only tempered water.

10.15.2 Temperature Maintenance Where Required

a. Where the developed length of the hot water supply piping exceeds 100 feet from the hot water source to the farthest hot water outlet, the system shall maintain the temperature of the hot water to within 25 feet developed length from every hot water outlet served.

b. Where required by Section 10.15.2.a, the hot water temperature within the piping shall be maintained by heat tracing or recirculation of the hot water. The temperature of the hot water in the piping shall be maintained by automatic controls with manual auto-off.

c. The requirements of this section for temperature maintenance also apply to tempered water supply piping.

See Figure 10.15.2

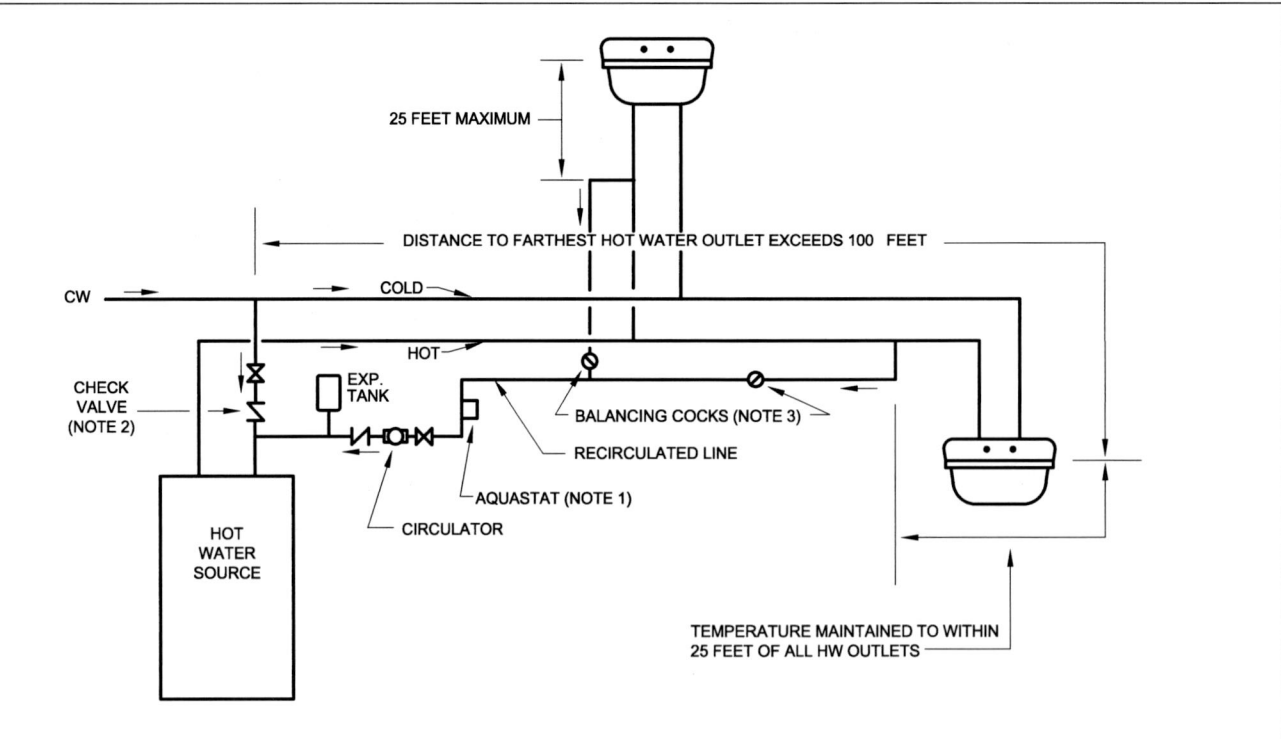

Figure 10.15.2
HOT WATER TEMPERATURE MAINTAINENCE BY RECIRCULATION

10.15.3 Minimum Requirements for Hot Water Storage Tanks

a. Hot water storage tanks shall be adequate in size, when combined with the BTUH input of the water heating equipment, to provide the rise in temperature necessary.

b. Water heaters and storage tanks shall be sized to provide sufficient hot water to provide both daily requirements and hourly peak loads of the occupants of the building.

c. Storage tanks shall be protected against excessive temperatures and pressure conditions as specified in this Code. (See Sections 3.3.8 and 3.3.10)

10.15.4 Drainage of Hot Water Storage Tanks

Hot water storage tanks shall be equipped with a valve capable of draining the tank completely.
See Figure 10.12.7

10.15.5 Pressure Marking of Hot Water Storage Tanks

Hot water storage tanks shall be permanently marked in an accessible place with the maximum allowable working pressure, in accordance with the applicable standard as listed in Table 3.1.3.

10.15.6 Mixed Water Temperature Control

a. Hot Water Supply Sources: The temperature control devices for water heaters and other hot water supply sources shall not be permitted to be used to meet this Section's requirements for mixed water temperature control.

b. Hot Water Distribution Temperature Control: Where temperature-actuated mixing valves are installed to control the hot water supply temperature in the water distribution system, they shall comply with ASSE 1017. Such devices shall be installed at the hot water source and alone shall not supersede the other requirements of Section 10.15.6 for mixed water temperature control.

c. Application of Water Temperature Control and Limiting Devices: The inlet hot and cold water temperatures for temperature control and limiting devices shall be within their operating ranges and have sufficient differential above and below their discharge set point.

d. Where Check Valves Are Required: Where a water temperature control control or temperature limiting device supplies one or more outlets that can be shutoff downstream from the device, the device shall include integral check valves or check valves shall be provided in the hot and cold water supplies to the device or at its inlets to prevent cross flow through the device when there is no flow through the outlet or outlets that it supplies.

e. Showers and Bath/Shower Combinations: The water discharged from shower heads, wall or ceiling mounted hand held showers, body sprays, and tub spouts shall be controlled to a temperature no higher than 120°F by a Type P, Type T, or Type P/T automatic compensating valve complying with ASME A112.1016/ASSE 1016/CSA B125.16.

f. Multiple Showers: Where multiple (gang) showers are supplied by a one pipe tempered water supply system, the water temperature shall be controlled to a temperature no higher than 105°F by an automatic temperature control mixing valve complying with ASSE 1069.

g. Multiple Lavatories: Where multiple lavatories are supplied by a one-pipe tempered water supply system, the water temperature shall be controlled to a temperature no higher than 110°F by a water temperature limiting device complying with ASSE 1070 or CSA B125.3.

h. Bathtubs and Whirlpool Baths: The hot water supply to the faucets for bathtubs and whirlpool baths without showers and with or without deck-mounted hand sprays, shall be controlled to a temperature no higher than 120°F by a water temperature limiting device complying with ASSE 1070 or CSA B125.3. EXCEPTION: A water temperature limiting device shall not be required if the fixture is supplied by an ASSE 1016 automatic compensating valve.

i. Bidets: The hot water supply to the faucet on bidet plumbing fixtures shall be controlled to a temperature no higher than 110°F by a water temperature limiting device complying with ASSE 1070 or CSA B125.3. Where bidets are incorporated into toilet seats or consist of a heated water tank and nozzle, their controls shall limit the discharge temperature to no more than 110°F.

j. Hand Washing Facilities: The hot water supply to the following hand washing fixtures shall be controlled to a temperature no higher than 110°F by a water temperature limiting device complying with ASSE 1070 or CSA B125.3:

    1. in public toilet rooms
    2. in hotel and motel guest rooms
    3. in hospital patient rooms
    4. in medical and clinical treatment rooms
    5. wash fountains
    6. wash sinks

k. Commercial Hair/Shampoo Sink Sprays: The hot water supply to the faucets and controls for commercial hair/shampoo sink sprays and pedicure basins shall be limited to a temperature no higher than 110°F by a temperature limiting device complying with ASSE 1070 or CSA B125.3.

l. Animal Washing Fixtures: The hot water supply to the faucet or control device for an animal washing fixture shall be controlled to a temperature no higher than 110°F by a water temperature limiting device complying with ASSE 1070 or CSA B125.3.

m. Temperature Actuated Flow Reduction (TAFR) Devices: Where temperature actuated flow reduction (TAFR) devices are installed to limit the maximum mixed water temperature to 120°F for individual fixture fittings, such devices shall comply with ASSE 1062. These devices shall not supersede the other requirements of Section 10.15.6.

n. In-Line Pressure Balancing Valves: Where in-line pressure balancing valves are installed to compensate for water pressure fluctuations to stabilize the temperature discharges from their individual faucet or fixture fitting, such devices shall comply with ASSE 1066. These devices shall be installed in an accessible location and alone shall not supersede the other requirements of Section 10.15.6.

o. Devices installed for mixed water temperature control, temperature limiting, flow reduction, or pressure balance shall be field-adjusted in accordance with the manufacturer's instructions.

10.15.7 Thermal Expansion Control

a. Where a water pressure regulator (with or without an internal thermal expansion bypass), a backflow preventer, or a check valve is installed such that a closed system is created to the water supply source for the hot water heating equipment, a device for controlling thermal expansion shall be provided.
EXCEPTIONS: (1) Instantaneous water heaters. (2) Well systems with water pressure tanks.

b. Thermal expansion tanks shall be the adjustable pre-charged type for potable water, steel construction with a flexible bladder or bellows, rated for not less than 125 psig and 200°F, and sized to limit the increased water system pressure to no higher than 100 psig. Tanks shall be sized and installed in accordance with the manufacturer's instructions.

c. Thermal expansion control devices shall be connected to the water supply piping between the hot water heating equipment and its shutoff valve.

10.15.8 Plastic Piping

a. Plastic piping used for hot water distribution shall conform to the requirements of Section 3.4.3 and Table 3.4.3. Piping shall be water pressure rated for not less than 100 psi at 180°F and 160 psi at 73°.

NOTE: The working pressure rating for certain approved plastic piping materials varies depending on material composition, pipe size, wall thickness and method of joining. See Table 3.4.3.

b. Plastic pipe or tube shall not be used downstream from instantaneous water heaters, immersion water heaters or other heaters not having approved temperature safety devices.

c. Piping within six inches of flue or vent connectors shall be approved metallic pipe or tube.

d. The normal operating pressure in water distribution piping systems utilizing approved plastic pipe or tube for hot water distribution shall be not more than 80 psi. Where necessary, one or more pressure reducing valves shall be provided to regulate the hot and cold water supply pressure to not more than 80 psi.

e. The pressure in the hot water distribution piping shall be limited by a pressure relief valve set no higher than 150 psi. When the water heater is protected by a pressure relief valve or combination pressure temperature relief valve having a pressure setting higher than 150 psi, a separate pressure relief valve shall be provided to protect the piping. The relief valve for the piping shall comply with Section 10.16.2 except that it shall be set no higher than 150 psi. Thermal expansion shall be controlled as required under Section 10.15.7.

10.15.9 Drip Pans

10.15.9.1 Where Required

Where tank-type water heaters or hot water storage tanks are installed in locations where leakage will cause structural damage to the building, the tank or water heater shall be installed in a drip pan in accordance with Section 10.15.9.2 and 10.15.9.3.

10.15.9.2 Construction

a. Drip pans shall be watertight and constructed of corrosion-resistant materials. Galvanized steel pans shall be 24 gauge (0.0276-inch) minimum thickness. Aluminum pans shall be 20 gauge (0.0320-inch) minimum thickness. Non-metallic pans shall be 0.0625-inch minimum thickness. Pans shall be not less than 1-1/2" deep but shall not be deeper than the bottom of the water heater tank or hot water storage tank. They shall be of sufficient size to hold the heater or tank without interfering with drain valves, burners, controls, and any required access.

b. High impact plastic pans shall be permitted under gas-fired water heaters where the heater is listed for zero clearance for combustible floors and the application is recommended by the pan manufacturer.

10.15.9.3 Drainage

a. Drip pans shall have drain outlets not less than 3/4" size, with drain pipes extending to an approved point of discharge, a suitably located indirect waste receptor, or to within 2 to 6 inches above the adjacent floor.

b. Discharge from a relief valve into a water heater drip pan shall be permitted if the drain size for the drip pan is not less than the relief valve discharge pipe size and the discharge pipe extends to within 2 to 6 inches above the bottom of the pan.

c. For drip pans installed under water heaters that are located above ceilings, the drain pipe from the drip pan shall extend to a point of disposal or indirect waste that is readily observable in an area below the heater.

10.15.10 Water Heaters Used for Space Heating

a. Water heaters used for space heating shall be listed for such use.

b. Piping and components connected to a water heater for space heating application shall be suitable for use with potable water.

c. Where required, a water temperature control valve shall be installed in every combination water/space heating system application to limit domestic hot water temperature to 140°F. The temperature control device shall be an ASSE 1017 listed device.

10.15.11 Water Heaters

a. Water heaters shall be applied, sized, and installed in accordance with the manufacturer's recommendations and instructions.

b. Gas-fired storage tank water heaters with input ratings of 75,000 Btuh or less shall comply with ANSI Z21.10.1/CSA 4.1.

c. Gas-fired storage tank water heaters with input ratings above 75,000 Btuh shall comply with ANSI Z21.10.3/CSA 4.3.

d. Gas-fired tankless water heaters shall comply with ANSI Z21.10.3/CSA 4.3.

e. Oil-fired storage tank water heaters shall comply with UL 732.

f. Household electric storage tank water heaters up to 120 gallons and 12 KW capacity shall comply with UL 174.

g. Commercial electric storage tank water heaters over 120 gallons and 12 KW capacity shall comply with UL 1453.

h. Tankless electric water heaters shall comply with UL 499.

> *Comment: Tankless water heaters include whole-home, multi-fixture, and point-of-use heaters.*

10.16 SAFETY DEVICES FOR PRESSURE VESSELS

10.16.1 Tank Protection

a. Pressure vessels used for heating water or storing water at pressures above atmospheric shall be protected by approved safety devices in accordance with one of the following methods:

    1. A separate pressure relief valve and a separate temperature relief valve; or

    2. A combination pressure and temperature relief valve; or

    3. Either "1" or "2" above and an energy cut-off device.

4. Tank construction conforming to a standard that does not require a temperature or pressure safety or relief valve.

> *Comment: If a pressure relief valve, temperature relief valve, or combination pressure/ temperature relief valve is installed for a water heater storage tank, there must not be any shutoff valves between the relief valve(s) and the tank that they protect. Also, there must not be any shutoff valves in any relief valve discharge piping.*

10.16.2 Pressure Relief Valves

a. Pressure relief valves shall comply with the applicable standards listed in Table 3.1.3.

b. The valves shall have a relief setting of not more than the pressure rating of the tank, or 150 psig maximum, and shall be installed either directly in a tank tapping or in the hot or cold water piping close to the tank. Pressure relief valves installed in hot water piping shall be rated not less than 180 degrees F.

c. There shall be no shutoff valve between the pressure relief valve and the tank.

d. The pressure relief valve shall be set to open at not less than 25 psig above the street main pressure or not less than 25 psig above the setting of any building water pressure regulating valve.
See Figure 10.16.6

10.16.3 Temperature Relief Valves

a. Temperature relief valves shall have an adequate relief rating, expressed in BTU/HR, for the equipment served.

b. The CSA temperature steam rating shall be used when a water heater has a heat input rating of 200,000 BTU/HR or less, a water temperature of 210°F or less, a nominal water-containing capacity of 120 gallons or less, is not ASME rated, and does not exhibit an ASME "HLW" symbol or a National Board "NB" number.

c. The ASME pressure steam rating shall be used if a water heater is ASME rated and exhibits an ASME "HLW" symbol or a National Board "NB" number.

d. The valves shall be installed so that the temperature sensing element is in the hottest water within the top 6 inches of the tank.

e. The valves shall be set to open when the stored water temperature reaches a maximum of 210°F. (See Section 3.3.10.)

f. The valves shall conform to an approved standard and shall be sized so that when the valve opens, the water temperature cannot exceed 210°F with the water heating equipment operating at maximum input.
See Figure 10.16.6

10.16.4 Combination Pressure-Temperature Relief Valves

Combination pressure-temperature relief valves shall comply with all the requirements of the separate pressure and temperature relief valves. *See Sections 10.16.2 and 10.16.3*

10.16.5 Reserved

10.16.6 Relief Valve Piping

a. There shall be no shut-off valve, check valve, or other restricting device between a relief valve and the pressure vessel or piping system being protected.

b. Piping from the outlet of a relief valve to the point of disposal shall be of a material suitable for potable water (see Section 3.4). Discharge pipes from temperature relief valves and combination pressure-temperature relief valves shall be listed in Table 3.4 for hot water distribution, and shall be suitable for conveying water at 210°F to an open discharge. The pressure rating of the pipe at 210 deg F is not required to equal or exceed the pressure setting of the relief valve.

c. The discharge pipe shall be no smaller than the outlet size of its relief valve and shall extend to a point of disposal without valves, traps, or rises that would prevent the discharge piping from draining by gravity. The discharge end of the pipe shall not be threaded.

d. A visible air gap shall be provided where relief valves discharge into an indirect waste pipe, floor drain, trench drain, service sink, mop basin, laundry sink, standpipe, or other approved receptor. The minimum size of fixture drains or waste pipes that receive the discharge from relief valves shall be as indicated in Table 10.16.6.

e. Where relief valves discharge to the floor, the discharge pipe shall terminate not more than 6 inches nor less than 2 inches above the floor.

f. If the point of discharge is not within the space or room in which the relief valve is located, an indirect gravity drain shall be provided from the room or space to the point of disposal. Indirect waste pipes shall be sized according to Table 10.16.6 and shall be of a material approved for potable water, sanitary drainage, or storm drainage (see Tables 3.4, 3.5, and 3.7). A visible air gap shall be provided in the room or space in which the relief valve is located.

g. Where water heaters are located above ceilings, the relief valve discharge pipe shall extend to a point of disposal or indirect waste that is readily observable in an area below the heater.

h. Where two or more relief valves serving independent systems are located in the same area, each shall be discharged separately.

See Figure 10.16.6

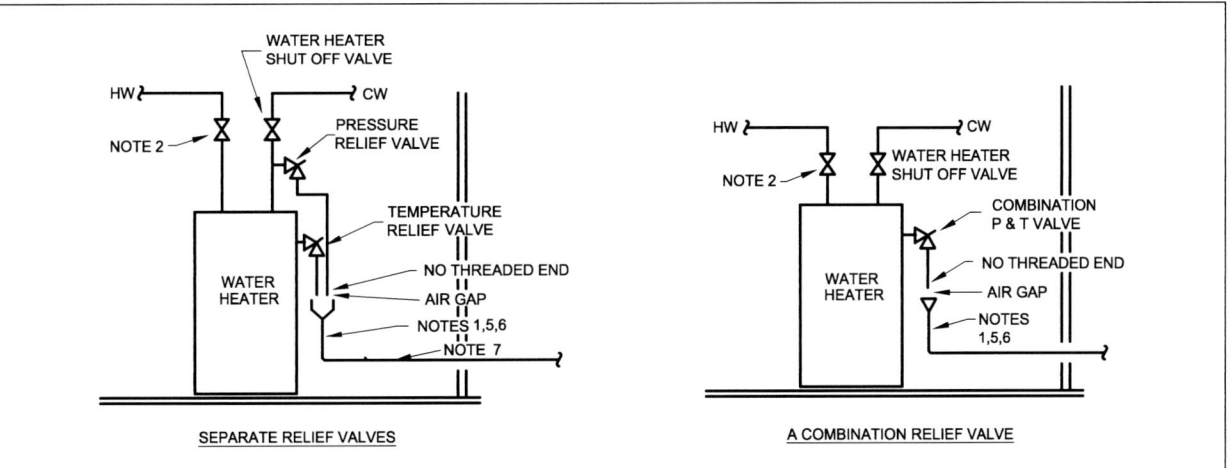

Figure 10.16.6
WATER HEATER RELIEF VALVES AND DISCHARGE PIPING

10.16.7 Vacuum Relief Valves for Water Heaters

a. Where water distribution piping can siphon water from a water heater and cause dry-firing, a vacuum relief valve shall be installed on the cold water inlet piping to the water heater.

b. Vacuum relief valves shall comply with ANSI Z21.22/CSA 4.4 and be rated for not less than 210°F.

c. Valves shall be the full size of the water heater inlet pipe and be installed at an elevation above the top of the heater tank.

See Figure 10.16.7

Table 10.16.6
SIZE OF DRAINS OR WASTE PIPES RECEIVING RELIEF VALVE DISCHARGE

Discharge Pipe Size	Minimum Drain or Indirect Waste Size
3/4"	2"*
1"	3"
1-1/2"	4"
2"	4"
2-1/2"	6"

*EXCEPTION: A laundry sink with 1-1/2" waste pipe.

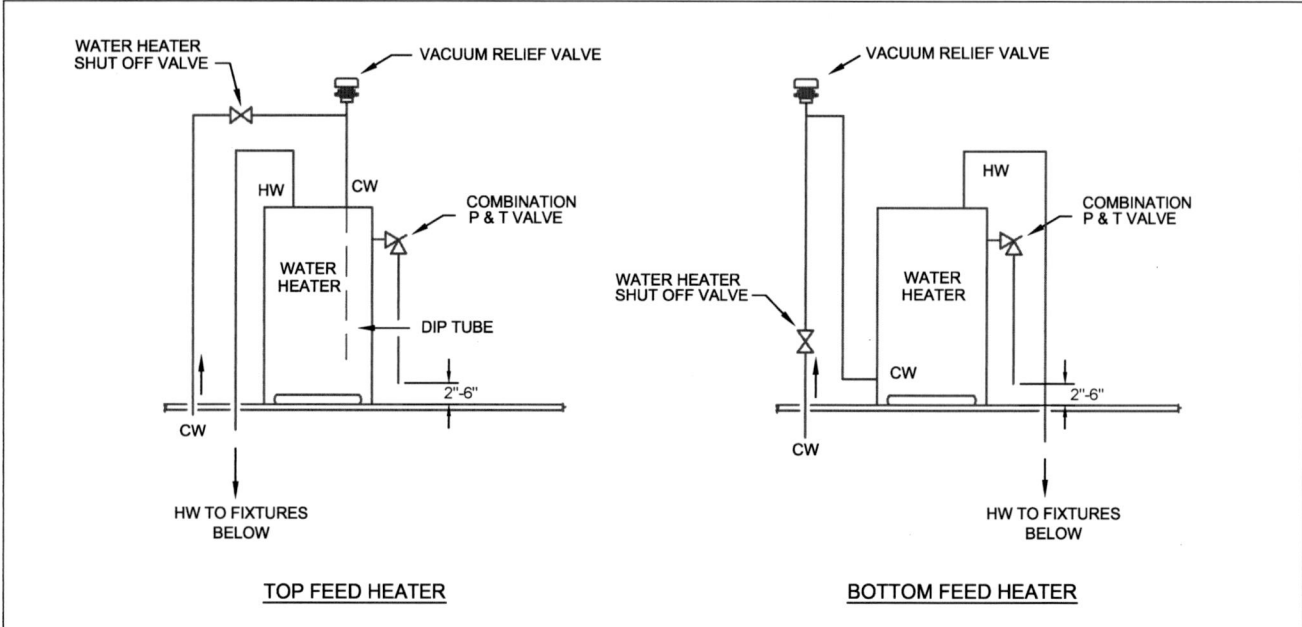

NOTES:
1. Vacuum relief valves at water heaters prevent vacuum conditions that could siphon water from a tank, causing it to be damaged from dry firing or collapse.
2. Vacuum relief valves are tested and rated under ANSI Z21.22 - Relief Valves for Hot Water Supply Systems.
3. Backflow prevention vacuum breakers are not intended for use as vacuum relief valves.
4. Vacuum relief valves must be mounted higher than the tank being protected.

Figure 10.16.7
VACUUM RELIEF VALVES ON OVERHEAD WATER HEATER TANKS

10.16.8 Replacement of Relief Valves

a. Relief valves shall be maintained in proper working order and shall be replaced when necessary.

b. Whenever a water heater is replaced, its temperature relief valve and pressure relief valve, or combination temperature-pressure relief valve shall also be replaced and shall not be reused.

10.17 MANIFOLD-TYPE PARALLEL WATER DISTRIBUTION SYSTEMS

10.17.1 General

a. Parallel water distribution systems shall provide individual hot and cold water supply lines from a manifold to each fixture served.

b. Manifolds shall be specifically designed and manufactured for parallel water distribution.

c. Manufacturer's of such systems shall provide complete sizing and installations instructions, including any limitations or restrictions on use.

d. Piping materials shall be as recommended by the system manufacturer and be listed in Table 3.4 for hot and cold water distribution.

10.17.2 Sizing

See Appendix B.12 for sizing manifolds and distribution lines. Distribution line sizes shall be as recommended by the system manufacturer to provide the fixture flow rates listed in Section 10.14.2.a. The minimum line size shall be 3/8" nominal.

10.17.3 Valving

a. Each manifold outlet that is equipped with a shutoff valve shall identify the fixture being supplied. Additional shut-off or stop valves at the fixtures shall be provided as required by Section 10.12.4.2.a.

EXCEPTION: Additional shut-off or stop valves at the fixtures shall not be required if a manifold with shutoff valves is located within the same room as the fixtures, or in an adjacent closet.

b. Manifolds having shutoff valves shall be readily accessible.

10.17.4 Support

a. Tube bundles for manifold systems shall be supported in accordance with Chapter 8 of this Code.

b. Supports at changes in direction shall be in accordance with the system manufacturer's recommendations.

10.17.5 Combined Distribution Systems

Manifold-type parallel water distribution systems shall be permitted to be combined with conventional main/branch piping systems that serve one or more fixture(s) through common main and branch piping.

10.18 DRINKING WATER TREATMENT UNITS

10.18.1 Compliance with Standards

Drinking water treatment units shall comply with the appropriate standards listed below.

a. NSF 42 Drinking Water Treatment Units - Aesthetic Effects
b. NSF 44 Water Softeners (cation exchange)
c. NSF 53 Drinking Water Treatment Units - Health Effects
d. NSF 55 Ultraviolet Microbiological Water Treatment Systems
e. NSF 58 Reverse Osmosis Drinking Water Systems
f. CSA B483.1 Drinking Water Treatment Systems

10.19 SIZING OF RESIDENTIAL WATER SOFTENERS

Residential-use water softeners shall be sized per Table 10.19.

Table 10.19 SIZING OF RESIDENTIAL WATER SOFTENERS	
Required Size of Softener Connection (in.)	Number of Bathroom Groups Served[1]
3/4	Up to 2[2]
1	Up to 4[3]

NOTES FOR TABLE 10.19:
1. The number of bathroom groups served may include a kitchen sink, dishwasher, laundry sink, and automatic clothes washer.
2. An additional water closet and lavatory shall be permitted without an increase in sizing.
3. Over four Bathroom Groups, the softener shall be engineered for the specific installation.

10.20 NFPA 13D MULTIPURPOSE RESIDENTIAL FIRE SPRINKLER SYSTEMS

10.20.1 Where Permitted

NFPA 13D multipurpose residential fire sprinkler systems shall be permitted where approved by the Authority Having Jurisdiction for plumbing under this Code and by the Authority Having Jurisdiction for fire protection systems.

10.20.2 General

a. The plumbing requirements for NFPA 13D multipurpose residential fire sprinkler piping systems, which provide both domestic cold water distribution and fire sprinkler protection for one- and two-family dwellings from a combination piping system, shall comply with the applicable requirements of this Code.

b. The fire protection requirements for NFPA 13D multipurpose residential fire sprinkler systems shall comply with the requirements of NFPA 13D and the Authority Having Jurisdiction for fire protection systems.

c. NFPA 13D multipurpose piping systems include network systems where each fire sprinkler is supplied from at least three separate paths.

d. NFPA 13D piping systems shall not be multipurpose if they require the use of antifreeze.

e. NFPA 13D multipurpose piping systems shall not include domestic hot water distribution to plumbing fixtures. Piping for domestic hot water distribution shall comply with Section 3.4.3.

f. The design of the plumbing portions of NFPA 13D multipurpose piping systems shall comply with the requirements of this Code.

g. The design of the fire sprinkler portions of NFPA 13D multipurpose piping systems shall comply with the requirements of NFPA 13D.

h. The installation of the plumbing portions of NFPA 13D multipurpose piping systems, which consist of the water supply to the multipurpose system and the branch piping that supplies cold water to plumbing fixtures and end-use devices, shall be performed under the plumbing permit issued by the Authority Having Jurisdiction for plumbing.

i. The installation of the fire sprinkler portions of NFPA 13D multipurpose piping systems, which consist of the fire sprinklers and their supply piping, shall be performed under the fire sprinkler permit issued by the Authority Having Jurisdiction for fire sprinkler systems.

10.20.3 Water Supply

a. The water supply to an NFPA 13D multipurpose piping system shall:

1. not include a fire department connection.
2. not include an alternate supply of non-potable water.
3. not require a backflow prevention device.

b. A single shutoff valve shall be provided to shutoff the water supply to the multipurpose piping system and the cold water supply to the domestic hot water source for the plumbing fixtures served with cold water by the multipurpose piping system. Separate shutoff valves shall be provided for plumbing fixtures beyond the multipurpose piping as required by Section 10.12.4.

10.20.4 Materials for Combined System Piping

a. Piping materials shall be in accordance with this Section and be listed, as defined in NFPA 13D, for both residential fire sprinkler service and potable water distribution.

b. Copper piping shall be ASTM B88 copper water tube, Type L or K. Fittings shall be the solder joint type, ASME B16.22 wrought or ASME B16.18 cast. Solder shall comply with ASTM B32. Flux shall comply with ASTM B813. Solder and flux shall contain no more than 0.2% lead. Soldered joints shall comply with ASTM B828.

c. Plastic piping shall be CPVC or PEX in accordance with Table 3.4.3. Plastic piping shall be rated for 160 psig at 73 deg F, 130 psig at 120 deg F, and 100 psig at 180 deg F.

10.20.5 Pipe Sizing

a. Piping shall be sized to satisfy the fire sprinkler requirements of NFPA 13D and the domestic water distribution requirements of this Code.

b. The pipe sizing for fire sprinkler performance shall include 4 GPM of domestic water flow. The 4 GPM of domestic water flow shall be to any single clothes washer connection or exterior hose bibb that is in the path of the design sprinkler water flow.

c. After the pipe sizing is established for fire sprinkler performance, the sizing of the piping shall be increased if necessary for domestic water distribution with no demand for fire sprinkler flow.

d. The minimum pipe size for combination piping shall be 3/4" except where NFPA 13D permits 1/2" pipe size in network systems. The minimum pipe size for individual plumbing outlets shall be in accordance with Section 10.14.2.

10.20.6 System Flushing

a. Multipurpose piping shall be flushed in accordance with Section 10.9.1 through each plumbing outlet that it serves.

b. Multipurpose piping shall not be disinfected.

10.20.7 Warning: Signs

In accordance with NFPA 13D, a warning sign for the multipurpose piping system shall be installed at the main building water shutoff valve and at any other valves that will shutoff the water supply to the fire sprinkler system.

10.21 NON-POTABLE WATER PIPING

a. Where non-potable water distribution systems are installed for irrigation, flushing water closets or urinals, or other non-potable water use in typical potable water applications, the following shall apply:

1. The use of non-potable water shall be approved by the Authority Having Jurisdiction.
2. The piping shall comply with Section 3.4 for potable water.
3. Non-potable water piping shall be visibly identified to differentiate it from potable water piping.
4. Non-potable water pipe may be purple colored.
5. Potable water pipe shall not be purple colored.

b. Non-potable water piping shall be identified by distinctive labels marked NON-POTABLE WATER that comply with ASME A13.1 and have a safety purple background color and white letter color. One or more labels shall be visible from all points of observation. Piping that passes through walls shall be labeled on both sides of the wall. Where piping is concealed within ceilings. walls, or other construction, at least one label shall be visible from any possible point of access, including penetration of the concealment. The labeling on piping that will be concealed shall be inspected prior to being concealed.

c. Each outlet from a non-potable water piping system that could be mistakenly used for drinking, culinary uses, or domestic purposes shall be posted: DANGER (red background) - UNSAFE WATER - DO NOT DRINK.

Blank Page

Chapter 11

Sanitary Drainage Systems

11.1 MATERIALS

See Section 3.1.

11.2 BUILDING SEWERS AND BUILDING DRAINS

11.2.1 Sewer or Drain in Filled Ground

Building sewers or building drains that are installed in filled or unstable ground shall be installed in accordance with Section 2.6.

11.2.2 Existing Building Sewers and Building Drains

Existing building sewers and building drains may be used in connection with new building sewer and drainage systems only when found by examination to conform to the new system in quality of material prescribed by this Code.

11.2.3 Building Sewer and Building Drain Size

The size of the building sewer and the size of the building drain shall be determined by fixture unit loads connected in accordance with Table 11.5.1A.

11.3 DRAIN PIPING INSTALLATION

11.3.1 Slope of Horizontal Drain Piping

a. Horizontal drain piping shall be installed in uniform alignment at uniform slopes not less than 1/4 inch per foot for 2-inch size and smaller, and not less than 1/8 inch per foot for 3-inch size and larger.
EXCEPTION: Horizontal drain piping for non-water urinals shall be sloped not less than 1/4 inch per foot to the point where it connects to piping with drainage for one or more water closets.
b. Where conditions do not permit building drains and sewers to be laid with slope as great as that specified, a lesser slope may be permitted by the Authority Having Jurisdiction. *See Appendix K for approximate discharge rates and flow velocities in drains at various slopes.*

11.4 FIXTURE UNITS

11.4.1 Load on Drain Piping

The load on drainage system piping shall be computed in terms of drainage fixture unit values in accordance with Table 11.4.1 and Section 11.4.2.

11.4.2 Conversion of Flow in GPM to DFU

Where the discharge rate of fixtures or equipment is expressed in gallons per minute (GPM), two (2) drainage fixture units (DFU) shall be allowed for each gallon per minute (GPM) of flow.

Table 11.4.1
DRAINAGE FIXTURE UNIT (DFU) VALUES

	INDIVIDUAL DWELLING UNITS	SERVING 3 OR MORE DWELLING UNITS	OTHER THAN DWELLING UNITS	HEAVY-USE ASSEMBLY
BATHROOM GROUPS HAVING 1.6 GPF GRAVITY-TANK WATER CLOSETS				
Half-Bath or Powder Room	3	2		
1 Bathroom Group	5	3		
1-1/2 Bathrooms	6	3.5		
2 Bathrooms	7	4.5		
2-1/2 Bathrooms	8	5		
3 Bathrooms	9	5.5		
Each Additional Half-Bath	0.5	0.5		
Each Additional Bathroom Group	1	1		
BATHROOM GROUPS HAVING 1.6 GPF PRESSURE-TANK WATER CLOSETS				
Half-Bath or Powder Room	3.5	2.5		
1 Bathroom Group	5.5	3.5		
1-1/2 Bathrooms	6.5	4		
2 Bathrooms	7.5	5		
2-1/2 Bathrooms	8.5	5.5		
3 Bathrooms	9.5	6		
Each Additional Half-Bath	0.5	0.5		
Each Additional Bathroom Group	1	1		
BATHROOM GROUPS HAVING 3.5 GPF (or higher) GRAVITY TANK WATER CLOSETS				
Half-Bath or Powder Room	3	2		
1 Bathroom Group	6	4		
1-1/2 Bathrooms	8	5.5		
2 Bathrooms	10	6.5		
2-1/2 Bathrooms	11	7.5		
3 Bathrooms	12	8		
Each Additional Half-Bath	0.5	0.5		
Each Additional Bathroom Group	1	1		
BATHROOM GROUP WITH 1.6 GPF FLUSHOMETER VALVE	5	3		
BATHROOM GROUP WITH 3.5 GPF (or higher) FLUSHOMETER VALVE	6	4		

Table 11.4.1 (Continued)
DRAINAGE FIXTURE UNIT (DFU) VALUES

INDIVIDUAL FIXTURES	INDIVIDUAL DWELLING UNITS	SERVING 3 OR MORE DWELLING UNITS	OTHER THAN DWELLING UNITS	HEAVY-USE ASSEMBLY
Bathtub or Combination Bath/Shower, 1-1/2" Trap	2	2		
Bidet, 1-1/4" Trap	1	1		
Clothes Washer, Domestic, 2" Standpipe	3	3	3	
Dishwasher, Domestic, with Independent Drain, 1-1/2" minimum trap	2	2	2	
Drinking Fountain or Watercooler			0.5	
Food Waste Disposer, Commercial, 2" Min Trap			3	
Floor Drain, Auxiliary			0	
Kitchen Sink, Domestic, with One 1-1/2" Trap	2	2	2	
Kitchen Sink, Domestic, with Food Waste Disposer	2	2	2	
Kitchen Sink, Domestic, with Dishwasher	3	3	3	
Kitchen Sink, Domestic, with Disposer and Dishwasher	3	3	3	
Laundry Sink, One or Two Compartments, 1-1/2" Waste	2	2	2	
Laundry Sink, with Discharge from Clothes Washer	2	2	2	
Lavatory, 1-1/4" Waste	1	1	1	1
Mop Basin, 3" Trap			3	
Service Sink, 3" Trap			3	
Shower Stall, 1-1/2" Trap	2	2	2	
Shower Stall, 2" Trap	2	2	2	
Showers, Group, per Head (Continuous Use)			5	
Sink, 1-1/2" Trap	2	2	2	
Sink, 2" Trap	3	3	3	
Sink, 3" Trap			5	
Trap Size, 1-1/4" (Other)	1	1	1	
Trap Size, 1-1/2" (Other)	2	2	2	
Trap Size, 2" (Other)	3	3	3	
Trap Size, 3" (Other)			5	
Trap Size, 4" (Other)			6	
Urinal, 1.0 GPF			4	5
Urinal, Greater Than 1.0 GPF			5	6
Wash Fountain, 1-1/2" Trap			2	
Wash Fountain, 2" Trap			3	
Wash Sink, 1-1/2" Trap			2	

Table 11.4.1 (Continued)
DRAINAGE FIXTURE UNIT (DFU) VALUES

	INDIVIDUAL DWELLING UNITS	SERVING 3 OR MORE DWELLING UNITS	OTHER THAN DWELLING UNITS	HEAVY-USE ASSEMBLY
Water Closet, 1.6 GPF Gravity or Pressure Tank	3	3	4	6
Water Closet, 1.6 GPF Flushometer Valve	3	3	4	6
Water Closet, 3.5 GPF Gravity Tank	4	4	6	8
Water Closet, 3.5 GPF Flushometer Valve	4	4	6	8
Whirlpool Bath or Combination Bath/Shower, 1-1/2" Trap	2	2		

NOTES FOR TABLE 11.4.1:
1. A Bathroom Group, for the purposes of this Table, consists of not more than one water closet, up to two lavatories, and either one bathtub, one bath/shower combination, or one shower stall. Other fixtures within the bathing facility shall be counted separately to determine the total drainage fixture unit load.
2. A Half-Bath or Powder Room, for the purposes of this Table, consists of one water closet and one lavatory.
3. For unlisted fixtures, refer to a listed fixture having a similar flow and frequency of use.
4. When drainage fixture unit (DFU) values are added to determine the load on the drainage system or portions thereof, round the sum to the nearest whole number before referring to Tables 11.5.1A, 11.5.1B, or 12.16.6A for sizing the drainage and vent piping. Values of 0.5 or more should be rounded up to the next higher whole number (9.5 = 10 DFU). Values of 0.4 or less should be rounded down to the next lower whole number (9.4 = 9 DFU).
5. "Other Than Dwelling Units" applies to business, commercial, industrial, and assembly occupancies other than those defined under "Heavy-Use Assembly." Included are the public and common areas in hotels, motels, and multi-dwelling buildings.
6. "Heavy-Use Assembly" applies to toilet facilities in occupancies that place heavy, but intermittent, time-based loads on the drainage system, such as; schools, auditoriums, stadiums, race courses, transportation terminals, theaters, and similar occupancies where queuing is likely to occur during periods of peak use.
7. Where other than water-supplied fixtures discharge into the drainage system, allow 2 DFU for each gallon per minute (gpm) of flow. (See Section 11.4.2.)

11.4.3 Diversity Factors

In certain structures such as hospitals, laboratory buildings, and other special use or special occupancy buildings where the ratio of the number of plumbing fixtures to the number of occupants is proportionally more than required by Table 7.21.1 for Business Occupancy and in excess of 1,000 drainage fixture units, the Authority Having Jurisdiction may permit the use of a diversity factor for sizing drain branches, drain stacks, building drains, and building sewers.

> *Comment: The Authority Having Jurisdiction may permit the use of a diversity factor in systems where the number of fixtures per person is higher than normal. A hospital is such an example where toilet facilities are provided in each patient room for the convenience of the patients. The load on the drainage system is created by the number of persons served, not by the number of plumbing fixtures that are installed.*

11.5 DETERMINING DRAIN PIPE SIZES

11.5.1 Selecting the Size of Drain Piping

Pipe sizes shall be determined from Table 11.5.1A and 11.5.1B on the basis of the drainage fixture unit load (DFU) computed from Table 11.4.1 and Section 11.4.1. Sanitary drain pipe sizes shall not be reduced in the direction of flow.

EXCEPTION: Drain pipe sizes for individual fixtures shall be not less than the minimum trap size required in Section 5.2.

Table 11.5.1A
BUILDING DRAINS AND SEWERS[1]

Maximum Number of Drainage Fixture Units (DFU) That May Be Connected to Any Portion of the Building Drain or the Building Sewer.

Pipe Size- Inches	Slope Per Foot 1/16-Inch	Slope Per Foot 1/8-Inch	Slope Per Foot 1/4-Inch	Slope Per Foot 1/2-Inch
2			21	26
3		36[2]	42[2]	50[2]
4		180	216	250
5		390	480	575
6		700	840	1,000
8	1,400	1,600	1,920	2,300
10	2,500	2,900	3,500	4,200
12	3,900	4,600	5,600	6,700
15	7,000	8,300	10,000	12,000

NOTES FOR TABLE 11.5.1A:
1. On-site sewers that serve more than one building may be sized according to the current standards and specifications of the Authority Having Jurisdiction for public sewers.
2. See Section 11.5.6.c & d for the restrictions on the number of water closets installed on 3" building drains and sewers.

Table 11.5.1B
HORIZONTAL FIXTURE BRANCHES AND STACKS

Maximum Number of Drainage Fixture Units (DFU) That May Be Connected to Any Horizontal Fixture Branch, a Stack of Three Branch Intervals or Less, or Stacks of more than Three Branch Intervals

Pipe Size- Inches	Any Horizontal Fixture Branch[1]	Stacks with Three Branch Intervals or Less	Stacks with more than Three Branch Intervals	
			Total for Stack	Discharged Into One Branch Interval
1-1/4	1	1	1	1
1-1/2	3	4	8	2
2	6	10	24	6
3	20[2]	48[3]	72[3]	20[3]
4	160	240	500	90
5	360	540	1,100	200
6	620	960	1,900	350
8	1,400	2,200	3,600	600
10	2,500	3,800	5,600	1,000
12	3,900	6,000	8,400	1,500
15	7,000	10,500	13,000	2,300

NOTES FOR TABLE 11.5.1B:
1. Does not include branches of the building drain.
2. See Section 11.5.6.a & d for the restrictions on the number of water closets installed on 3" horizontal fixture branches.
3. See Section 11.5.6.b & d for the restrictions on the number of water closets on 3" stacks and discharged into any branch interval of a 3" stack.

11.5.2 Size of Soil and Waste Stacks

The sizing of sanitary drain stacks shall be based on the maximum vertical height from their base to their highest branch drain connection and the total number of their branch intervals in accordance with either Section 11.5.2.1 or Section 11.5.2.2.

11.5.2.1 Less than Five Branch Intervals and Forty (40) Feet Maximum Total Stack Height

a. Sanitary drain stacks with 40 feet or less height shall be sized according to Table 11.5.1B based on less than five branch intervals and the number of connected drainage fixture units (DFU).

b. Sections of drain stacks shall be sized according to the total accumulated drainage fixture unit load (DFU) within each branch interval.

EXCEPTION: Sections of drain stacks shall not be smaller than their largest branch connection. Lower sections of the drain stack shall be increased to that size if necessary.

c. No portion of the drain stack shall be less than one-half of the pipe size of the drain stack at its base.

d. The top of the drain stack shall be connected to a stack vent that is not less than the size of the drain stack at that point.

11.5.2.2 Five or More Branch Intervals with No Limit on Total Stack Height

a. The entire drain stack shall be sized according to Table 11.5.1B based on the total drainage fixture unit (DFU) load at the base of the drain stack.

EXCEPTION: The drain stack size shall be not less than its largest branch connection.

b. The top of the drain stack shall be connected to a stack vent that is not less than the size of the drain stack.

c. A vent stack shall be provided in accordance with Section 12.3.1.

11.5.3 Size of Branch Drains and the Building Drain

a. Horizontal fixture branches, which connect to drain stacks, shall be sized in accordance with Table 11.5.1B.

b. Horizontal branch drains, which do not connect to drain stacks, shall be sized as branches of the building drain in accordance with Table 11.5.1A.

c. The building drain, branches of the building drain, and branch connections to the building drain shall be sized in accordance with Table 11.5.1A.

11.5.4 Provision for Future Fixtures

When provision is made for the future installation of fixtures, those provided for shall be considered in determining the required sizes of drain and vent pipes. Construction to provide for such future installation shall be terminated with a plugged fitting or fittings.

11.5.5 Minimum Size of Underground Drain Piping

No portion of the drainage system installed underground shall be less than two inch pipe size.
EXCEPTION: Condensate waste, tub and shower traps and trap arms, and piping that receives the discharge from relief valves after an air gap.

11.5.6 Restrictions on the Number of Water Closets on 3" Drains

a. 3" Horizontal Fixture Branches

No more than four water closets or bathroom groups shall be installed on a 3" horizontal fixture branch. EXCEPTION: Where the water closets are rated 3.5 gallons or more per flush, no more than two water closets or bathroom groups shall be permitted.

b. 3" Stacks

No more than four water closets or bathroom groups shall be installed within any branch interval of a 3" stack, and no more than a total of twelve on the stack.

EXCEPTION: Where the water closets are rated 3.5 gallons or more per flush, no more than two water closets or bathroom groups shall be permitted in any branch interval, and no more than a total of six on the stack.

 c. 3" Building Drains and Sewers

  1. In single dwelling units, no more than six water closets or bathroom groups shall be installed on a 3" building drain or building sewer, or branches thereof.

 EXCEPTION: Where the water closets are rated 3.5 gallons or more per flush, no more than three water closets or bathroom groups shall be permitted.

  2. In other than single dwelling units, no more than four water closets or bathroom groups shall be installed on a 3"building drain or building sewer, or branches thereof.

 EXCEPTION: Where the water closets are rated 3.5 gallons or more per flush, no more than two water closets or bathroom groups shall be permitted.

 d. Mixed Water Closets on 3" Drains

Where 3" drain piping serves a mixture of 1.6 GPF water closets and 3.5 (or higher) GPF water closets, the 3.5 (or higher) GPF water closets shall be counted as two water closets for the purpose of determining the total number of water closets on the 3" drain piping. The drainage fixture unit (DFU) load for each 3.5 (or higher) GPF water closet shall be as indicated in Table 11.4.1.

11.6 EFFECT OF OFFSETS IN SANITARY DRAIN STACKS

11.6.1 Vertical Offsets

An offset in a drain stack that is 45 degrees or more from horizontal shall be sized as a straight vertical stack in accordance with Table 11.5.1B.

11.6.2 Horizontal Offsets

Horizontal offsets in sanitary drain stacks shall be sized according to Table 11.5.1A for building drains and sewers. Drain piping downstream from a horizontal offset shall be not less than the pipe size of the offset.

11.6.3 Sanitary Drain Stacks with Horizontal Offsets

 a. Where a sanitary drain stack includes a horizontal offset, the upper and lower portions of the drain stack shall be sized and vented based on their individual number of branch intervals.

 b. Where a vertical portion of a drain stack has less than five branch intervals, it shall be sized according to Section 11.5.2.1 and vented by a stack vent in accordance with Section 12.3.1.

 c. Where a vertical portion of a drain stack has five or more branch intervals, it shall be sized according to Section 11.5.2.2 and vented by a stack vent and vent stack in accordance with Section 12.3.1.

 d. Where a drain stack with horizontal offsets is designed as a single drain stack and any vertical portion requires a vent stack, the size of all drain stack piping, including horizontal offsets, shall be not less than the pipe size at the base of the drain stack. The drain stack shall be vented according to Section 12.3.1.

See Figure 11.6.3

11.6.4 Offsets Above the Highest Branch

An offset in a drain stack above the highest horizontal branch drain connection shall not affect the size of the drain stack or stack vent.

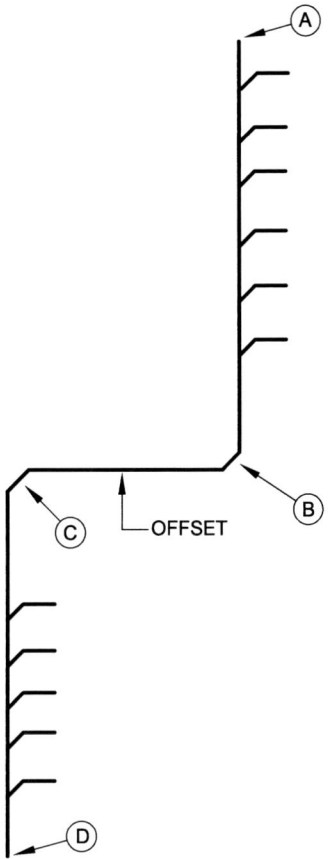

NOTES:
1. The drain stack contains five or more branch intervals in its upper portion and five or more branch intervals in its lower portion. It must be sized according to Section 11.5.2.
2. The upper portion of the stack (A-B) must be sized based on its total DFU load at its base, using Table 11.5.1B.
3. The offset (B-C) is horizontal and must be sized as a building drain, using Table 11.5.1A.
4. The lower portion of the stack (C-D) must be sized based on the total DFU at its base, using Table 11.5.1B, or the size of the offset, whichever is larger.
5. The size of the entire drain stack must not be smaller than (1) the base of the upper portion (A-B), (2) the hoizontal offset (B-C), (3) the base of the lower portion (C-D), or (4) its largest branch connection.
6. The upper and lower portions of the drain stack must have vent stacks and stack vents in accordance with Section 12.3.1.
7. If either the upper portion or the lower portion of the drain stack has ten or more branch intervals, relief vents must be provided for that portion in accordance with Section 12.3.2 and Figure 12.3.2.
8. If one stack is provided for both the upper and lower portions of the drain stack, it should be arranged as indicated in Figure12.3.3A.
9. If separated stacks are provided for the upper and lower portions of the drain stack, they should arranged as indicated in Figure 12.3.3B.

Figure 11.6.3
HORIZONTAL OFFSETS IN DRAIN STACKS

11.7 Sewage Pumping

11.7.1 General

   a. Equipment for pumping or lifting sewage shall be in accordance with the following:

    11.7.2 Building subdrains.

    11.7.3 Sewage pumps and ejectors.

    11.7.4 Pneumatic sewage ejectors.

    11.7.5 Grinder pumps.

    11.7.6 Macerating toilet systems.

    11.7.7 Sump pits, basins, and receptors.

    11.7.8 Discharge piping.

    11.7.9 Controls.

    11.7.10 High level alarm.

    11.7.11 Individual fixture pumps and ejectors.

   b. Pumping equipment shall be installed in accordance with the manufacturer's instructions and this Code.

11.7.2 Building Subdrains

   a. Sanitary drainage below the elevation of the building drain that cannot flow to the building drain by gravity shall be connected to a subdrain from which the contents shall be pumped up to a point in the building drainage system that will flow to the building drain by gravity.

   b. Only drainage that must be lifted for gravity flow to the building drain shall be connected to building subdrains. All other drainage shall flow to the building drain by gravity.

   EXCEPTION: Renovations and additions to an existing system.

   c. Building subdrain piping shall comply with the applicable requirements of Chapter 11.

   d. Building subdrain piping shall be vented in accordance with Chapter 12. Vents from subdrain piping may be combined with vents from building gravity drain piping or may be vented separately to outdoors.

See Figure 11.7.2

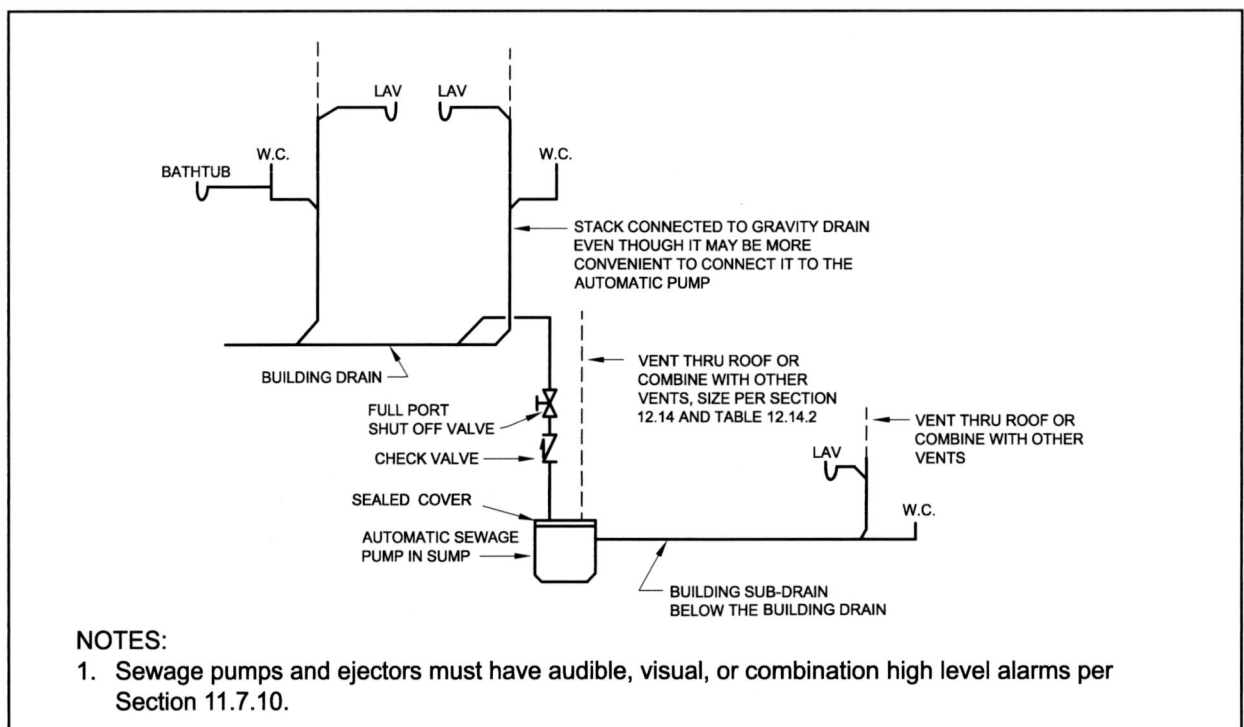

Figure 11.7.2
BUILDING SUBDRAIN AND SEWAGE PUMP

11.7.3 Sewage Pumps and Ejectors

a. Sewage pumps and ejectors shall:

1. be installed in a sump pit, basin, or receptor in accordance with Section 11.7.7.

2. in single dwelling units, be capable of passing a solid at least 1-1/2 inches in diameter and have a 2" minimum discharge pipe size.

3. in other than single dwelling units, be capable of passing a solid at least 2 inches in diameter and have a 3" minimum discharge pipe size.

4. have a minimum capacity of 20 gallons per minute if they receive drainage from a water closet or urinal.

b. Discharge piping shall comply with Section 11.7.8.

EXCEPTION: Individual fixture pumps and ejectors shall comply with Section 11.7.11.

11.7.4 Pneumatic Sewage Ejectors

a. Discharge piping shall comply with Section 11.7.8.

b. If water closets, urinals, or other fixtures are close enough to a pneumatic sewage ejector that they will overflow if flushed or discharged while the ejector is discharging under pressure, a surge tank shall be provided to provide temporary holding capacity and gravity flow to the ejector. Surge tanks shall be vented in accordance with Section 12.14.3.a.

c. Pressure release vents for pneumatic sewage ejectors shall not be connected to vent piping for drain piping. Pneumatic ejectors shall be vented separately to the outdoors in accordance with Section 12.14.3.b.

11.7.5 Grinder Pumps

a. Grinder pumps that discharge to a low pressure sewer system shall:

1. be approved by the authority for the sewer system.

2. be installed outdoors underground or indoors in a vented sump pit, basin, or receptor.

3. have discharge pressure that is coordinated with the pressure in the sewer system.

4. have discharge piping that complies with Section 11.7.8.

5. include materials for the underground discharge piping that are approved by the sewage authority.

b. Discharge pipe sizes shall be as follows:

1-1/4" size for up to 25 gallons per minute

1-1/2" size for 15 to 35 gallons per minute

2" size for 25 to 65 gallons per minute

11.7.6 Macerating Toilet Systems

a. Macerating toilet systems shall comply with ASME A112.3.4/CSA B45.9.

b. Units shall:

1. be either self-contained within a water closet or be an external unit.

2. be permitted to drain bathroom groups, kitchen groups, or individual fixtures.

3. have clothes washers drained indirectly through a laundry sink if necessary.

4. be complete including receptor, macerating pump, and automatic controls.

5. be suitable for the application and be installed in accordance with the manufacturer's instructions and this Code.

6. include connections for drain inlets, pumped discharge and vent.

7. be installed within the manufacturer's distance limits for vertical lift and horizontal discharge.

c. Discharge piping shall comply with the applicable requirements of Section 11.7.8 with 3/4" minimum pipe size. Horizontal piping beyond the vertical rise from the pump shall be sloped in accordance with the unit manufacturer's installation instructions.

d. Except for water closets, fixtures connected to macerating toilet units shall be trapped and vented before connection to the pumping unit.

11.7.7 Sump Pits, Basins, and Receptors

a. Sump pits, basins, and receptors shall comply with the following:
  1. be concrete, steel, fiberglass, or other suitable material.
  2. be suitable for indoor or outdoor installation, as required.
  3. be liquid-tight and gas-tight with a sealed cover.
  4. be not less than the minimum size recommended by the pump manufacturer.
  5. include necessary pipe connections for drains, discharge, and vent.
  6. provide access to the pump, piping, and controls.
  7. have provisions for electrical power and control wiring.
  8. shall have a solid bottom adequate to support the pump.

11.7.8 Discharge Piping

a. Discharge piping from pumps and ejectors shall comply with the following unless specified otherwise for specific applications.

b. The drainage load that sewage pumping units place on building gravity drain piping shall be based on two (2) drainage fixtures units (DFU) for each gallon per minute of pump discharge.

c. The discharge piping from sewage pumping units shall consist of rigid pipe and fittings that are suitable for conveying sewage and pressure rated for the maximum discharge pressure of the pumping unit.

d. Discharge pipe shall be:
  1. ASTM A53 galvanized steel
  2. ASTM B88 copper tube, Type L or K
  3. ASTM A377 ductile iron
  4. ASTM D1785 PVC, schedule 40 or 80
  5. ASTM F441 CPVC, schedule 40 or 80
  6. ASTM F442 CPVC, SDR
  7. ASTM D2846 CPVC, SDR 11, CTS
  8. ASTM F2389 PP, IPS schedule 80

e. Piping shall be sized for a flow velocity of not less than 2 feet per second.

f. The inside diameter of piping shall be not less than the solid passing capacity of the pump that it serves.

g. Fittings shall not reduce the inside diameter of the pipe.

h. Discharge piping from pumps shall include a check valve and full-way shutoff valve that are accessible.

11.7.9 Controls

Pump units shall have automatic liquid level controls. Units with multiple pumps shall include multi-stage lead-lag control.

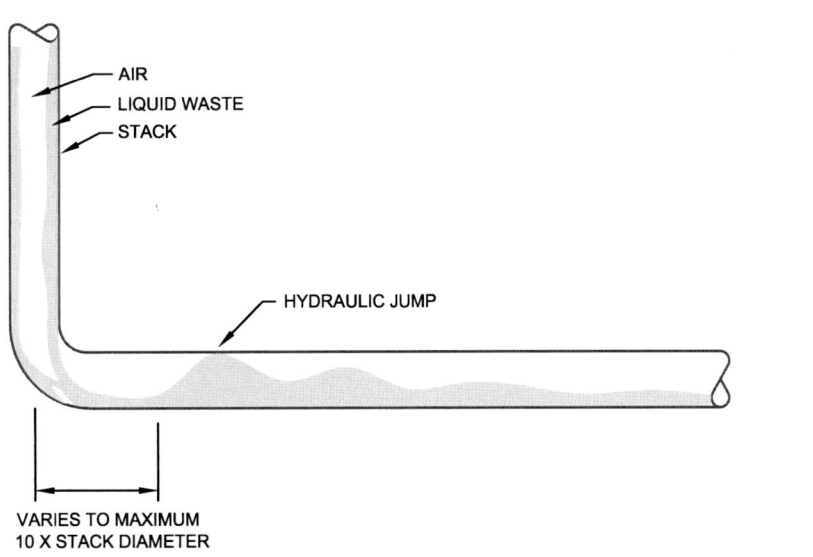

NOTES:
1. Because of "hydraulic jump", stable flow does not occur in the horizontal drain until a distance of up to 10 times the pipe size "D" from the base of the stack.
2. "Hydraulic jump" can completely close the horizontal drain.
3. "Hydraulic jump" can occur in stacks sized for as few as 4 drainage fixture units (DFU).
4. Branch drain connections are prohibited within 10 x "D" of the base of the stack where hydraulic jump can occur.

Figure 11.9
HYDRAULIC JUMP AT THE BASE OF DRAINAGE STACKS

11.7.10 High Level Alarm

Sewage pumping units shall include an audible, visual, or combination high liquid level alarm.
EXCEPTIONS:
(1) Existing sewage pumping units.
(2) Sewage ejectors and sewage pumps serving individual fixtures.
(3) Macerating toilet systems.

11.7.11 Individual Fixture Pumps and Ejectors

a. Individual fixtures other than water closets, urinals, and similar fixtures may discharge directly into an approved fixture-mounted pump or ejector, or into a receptor having a pump or ejector.

b. The discharge piping from a sewage pump or ejector for an individual fixture shall be sized to suit the discharge rate of the fixture and include a backwater valve and full-way shutoff valve.

c. Direct-mounted equipment may be manually or automatically controlled.

d. The installation of manually or automatically operated equipment shall not be subject to the venting requirements of this Code, but shall be vented only as required for proper operation of the equipment in accordance with the manufacturer's instructions.

e. If the equipment provides a proper liquid trap for the fixture, additional traps are not required.

f. A vent on the fixture side of a trap may terminate locally in the area served.

11.8 RESERVED

11.9 BRANCH CONNECTIONS NEAR THE BASE OF DRAIN STACKS

a. Branch drain connections near the base of drain stacks that do not have vent stacks shall be not less than 2 feet above the base of the stack and not less than 10 pipe diameters downstream from the base of the stack.

b. Where drain stacks have vent stacks in accordance with Section 12.3.1, branch drain connections near the base of the drain stack shall be coordinated with the location of its pressure relief connection by the vent stack.

1. If the pressure relief connection by the vent stack is above the base of the drain stack, there shall be no branch drain connections to the drain stack between its pressure relief connection and its base.

2. If the pressure relief connection by the vent stack is to the building drain beyond the base of the drain stack, there shall be no branch drain connections to the building drain between its pressure relief connection and the base of the drain stack.

See Figures 11.9 and 12.3.1

11.10 BRANCH CONNECTIONS TO OFFSETS IN DRAIN STACKS

a. Branch drains shall be permitted to connect to a horizontal stack offset, provided that the connection is not within 10 pipe diameters downstream from the upper portion of the stack.

b. Where stacks have five or more branch intervals above a horizontal offset, there shall be no branch connections to the stack within 2 feet above or below the offset.

c. Where stacks having five or more branch intervals above a vertical offset have branch connections to the stack within 2 feet above or below the offset, the offset shall be vented as required for a horizontal offset.

11.11 SUDS PRESSURE ZONES

11.11.1 General

Where suds-producing fixtures on upper floors discharge into a sanitary drain stack, suds pressure zones shall exist as described in Section 11.11.2. Fixture or branch drain connections shall not be made to such stacks in the suds pressure zones except where relief vents complying with Section 12.15 are provided. Suds-producing fixtures include kitchen sinks, laundry sinks, automatic clothes washers, dishwashers, and other fixtures that could discharge sudsy detergents.

> *Comment #1: The most likely fixtures to create suds pressure problems in drain stacks are clothes washers, dishwashers, kitchen sinks, and laundry sinks. Where liquid wastes in tall buildings include high-sudsing detergents, the detergent is vigorously mixed with the liquid waste and air in the stack. The liquid waste is heavier than the suds and does not carry them along with the flow. The suds will settle in the lower portions of the drainage system, including any horizontal offsets. The air that is flowing with the liquid waste compresses the suds and builds up pressure in the stack that can blow trap seals if not relieved. Zones of the drainage and vent piping where suds pressure can exist are described in Section 11.11.2 and illustrated in Figure 11.11.2.*
>
> *Comment #2: Suds pressure relief vents in Section 12.15 are larger than ordinary vents for drainage systems because the suds are heavier than air. Suds can weight from 2 to as much as 19 pounds per cubic foot.*

11.11.2 Locations in Stacks Serving Suds-Producing Fixtures

a. Zone 1 - at offsets greater than 45 degrees from vertical. A suds pressure zone shall extend 40 pipe diameters up the stack above the offset, 10 pipe diameters downstream from the base of the upper portion of the stack, and in the horizontal offset, 40 pipe diameters upstream from the top of the lower portion of the stack.

b. Zone 2 - at the base of a sanitary drain stack. A suds pressure zone shall extend 40 pipe diameters up the stack above its base.

c. Zone 3 - in the horizontal drain beyond the base of a sanitary drain stack. A suds pressure zone shall extend 10 pipe diameters from the base of the stack. Also, if a turn greater than 45 degrees occurs in the horizontal drain less than 50 feet from the base of the stack, suds pressure zones shall exist 40 pipe diameters upstream and 10 pipe diameters downstream from the horizontal turn.

d. Zone 4 - in a vent stack at the base of a sanitary drain stack. Where a vent stack connects above or beyond the base of a sanitary drain stack, a suds pressure zone shall extend up the vent stack to a level equal to the level of the suds pressure zone in the sanitary drain stack.

See Figure 11.11.2

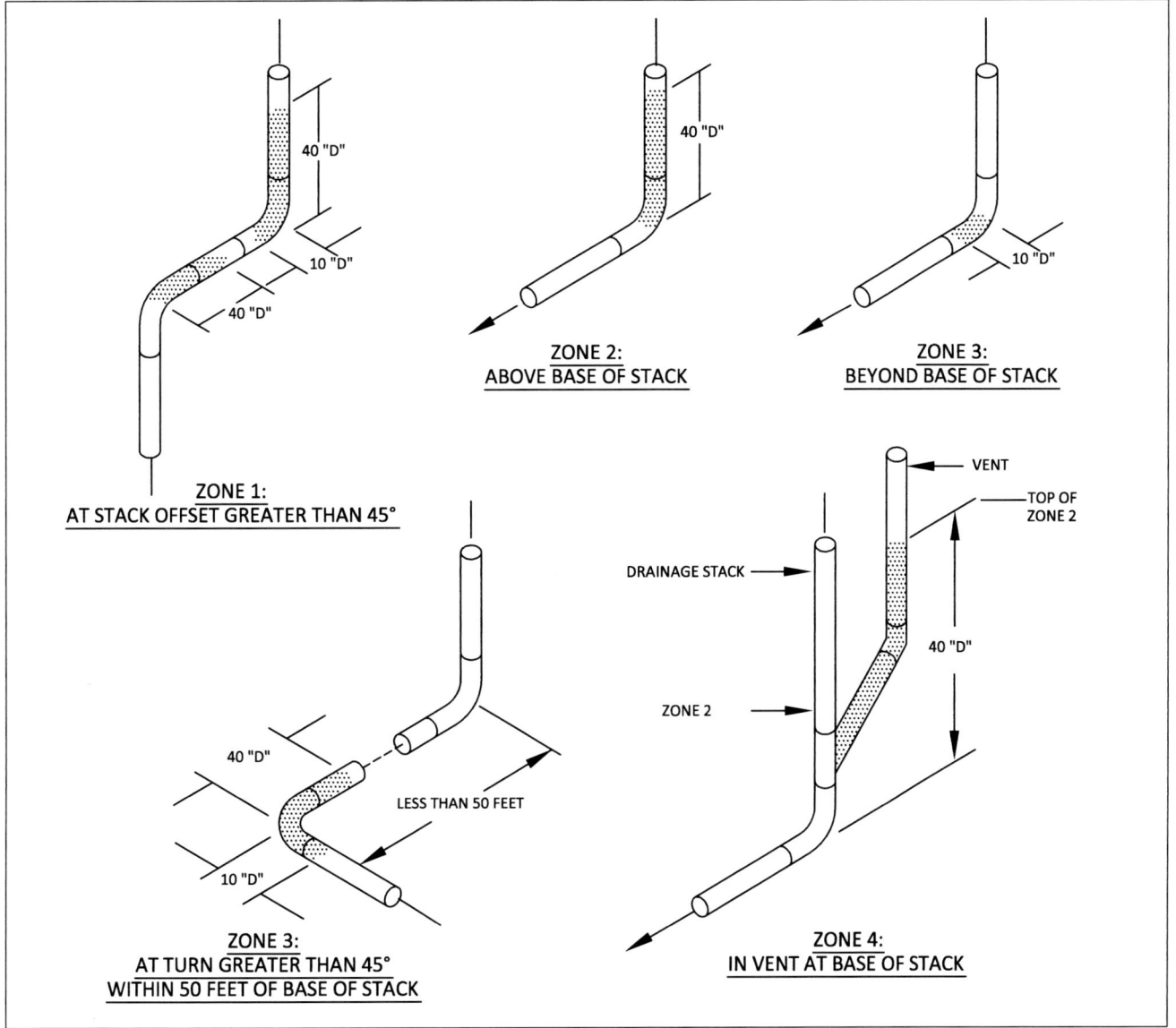

Figure 11.11.2
THE LOCATIONS OF SUDS PRESSURE ZONES IN DRAIN AND VENT PIPING

11.11.3 Separate Stacks

Where sanitary drain stacks serving suds-producing fixtures extend six or more floors above the base of the stack or above a horizontal offset in the stack, the lowest four floors above the base or horizontal offset shall be drained by a separate stack. In the case of a horizontal offset, the separate stack for the four floors above the offset may be reconnected to the main stack below the offset, provided that the point of connection is not a suds pressure zone in either stack.
See Figure 11.11.3

11.11.4 Exceptions

a. The requirements of Sections 11.11 and 12.15 shall not apply to the following:
   1. Stacks that are less than three stories in height.
   2. Stacks in individual dwellings having their own building sewer.

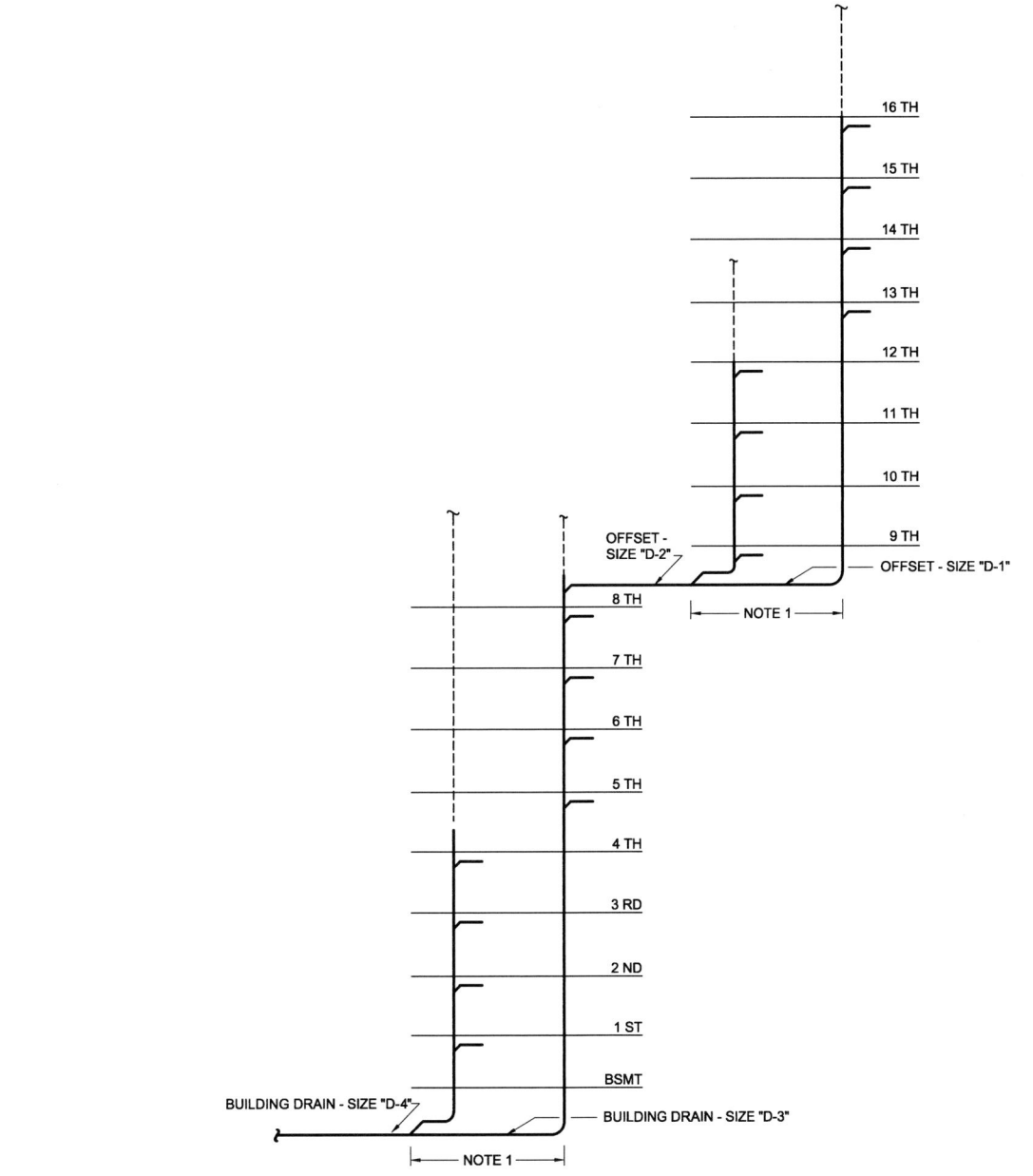

NOTES:
1. There should be no branch drain connections within 10 pipe diameters of the base of a drainage stack.
2. Portion D-1 of the upper floor offset between the 8th and 9th floors must be sized as a building drain, based on the DFU load for the upper floors above that portion of the offset.
3. Portion D-2 of the upper floor offset between the 8th and 9th floors must be sized as a building drain based on the DFU load for all floors above the offset.
4. Portion D-3 of the building drain must be sized based on the DFU load for the drainage stack above that portion of the building drain.
5. Portion D-4 of the building drain must be sized based on the DFU for the entire drainage stack.

Figure 11.11.3
SEPARATE DRAINAGE STACKS FOR LOWER FLOORS
WHERE THERE ARE SUDS PRODUCING FIXTURES ABOVE

Chapter 12

Vents and Venting

12.1 MATERIALS

See Section 3.6.

12.2 PROTECTION OF TRAP SEALS

12.2.1 Protection Required

a. The protection of trap seals from siphonage, aspiration, or back pressure shall be accomplished by the appropriate use of sanitary drain stacks with adequate venting in accordance with the requirements of this Code.

b. Venting systems shall be designed and installed so that at no time will trap seals be subjected to a pneumatic pressure differential of more than one inch of water column under design load conditions.

c. If a trap seal is subject to loss by evaporation, means shall be provided to prevent the escape of sewer gas. (See Section 5.3.6.)
See Figures 12.2.1-A and 12.2.1-B

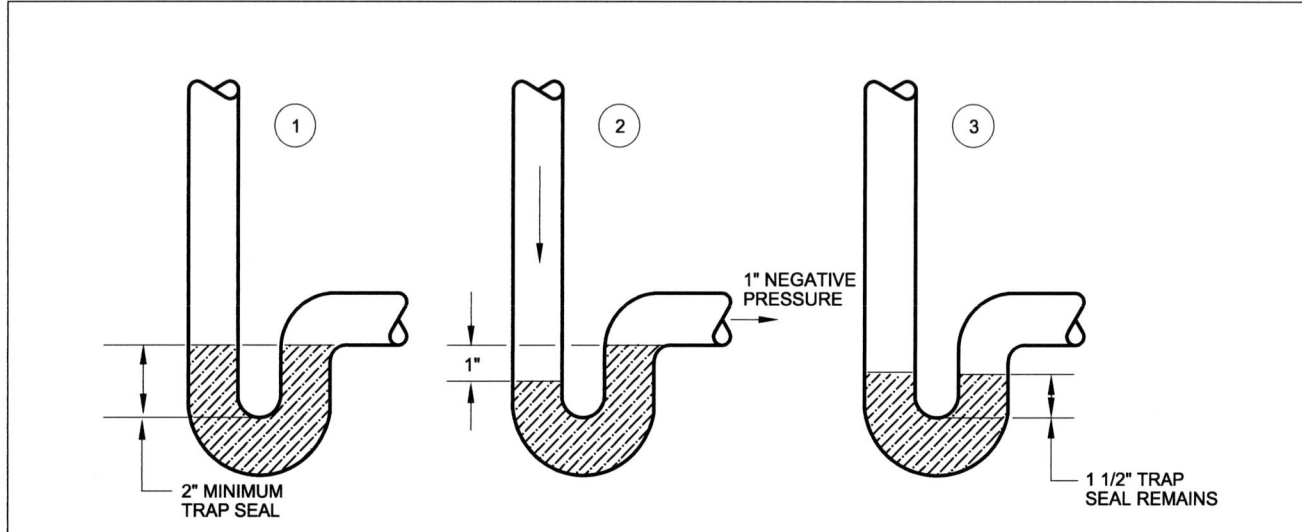

NOTES:
1. Trap at rest with 2" trap seal.
2. Trap subjected to 1" suction from building drain piping.
3. Trap at rest with 1/2" loss of trap seal. Trap will continue to spillover and lose trap seal when subjected to 1" suction until the trap seal is reduced to 1". The 2" initial trap seal permits the trap to withstand 1" suction and still maintain a trap seal of at least 1".

Figure 12.2.1 - A
TRAP SEAL REDUCTION FROM 1" NEGATIVE PRESSURE

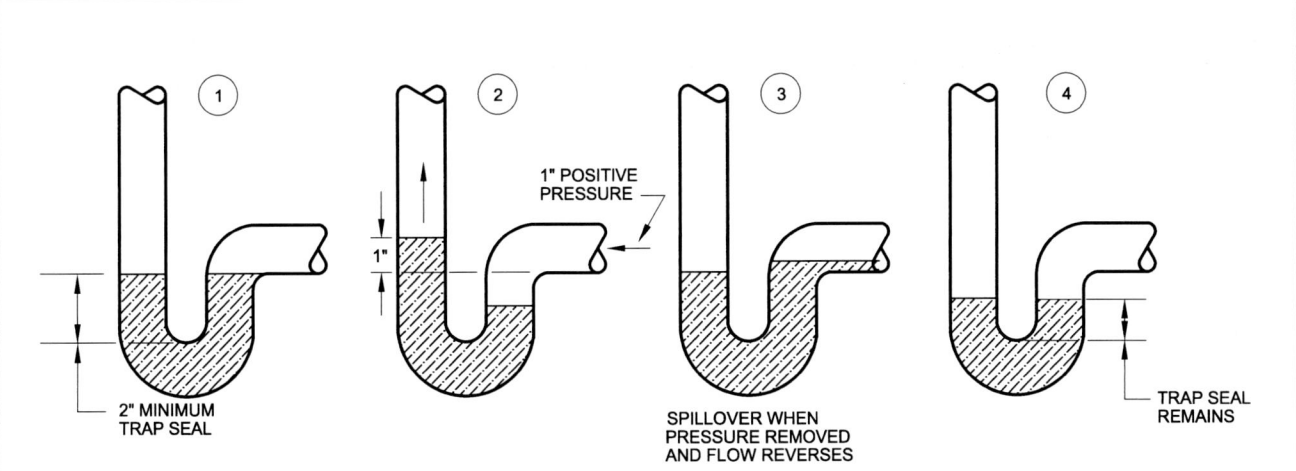

Figure 12.2.1 - B
TRAP SEAL REDUCTION FROM 1" POSITIVE PRESSURE

NOTES:
1. Trap at rest with 2" trap seal.
2. Trap subjected to 1" positive pressure from building drain piping.
3. When pressure is removed, some spillover occurs from the momentum of the trap legs equalizing.
4. Trap will continue to spillover and lose trap seal when subjected to 1" positive pressure until the trap seal is reduced to 1". The 2" initial trap seal permits the trap to withstand 1" positive pressure and still maintain a trap seal of at least 1".

12.3 VENTING OF DRAIN STACKS

12.3.1 Stack Vents and Vent Stacks

a. Stack vents shall extend from the top of a drain stack to an outdoor vent terminal.

b. A vent stack shall be provided for drain stacks having five or more branch intervals.

c. Where vent stacks are provided, their lower connection near the base of the drain stack for pressure relief, shall be either:

    1. to the drain stack at or below its lowest branch connection or,

    2. to the building drain within 10 pipe diameters downstream from the base of the drain stack.

d. The upper connection of vent stacks shall be to the stack vent for the drain stack or to an outdoor vent terminal.

e. Vent stacks shall be permitted to be provided for drain stacks having less than five branch intervals.

f. Stack vents and vent stacks shall be sized in accordance with Section 12.16.4.

See Figure 12.3.1

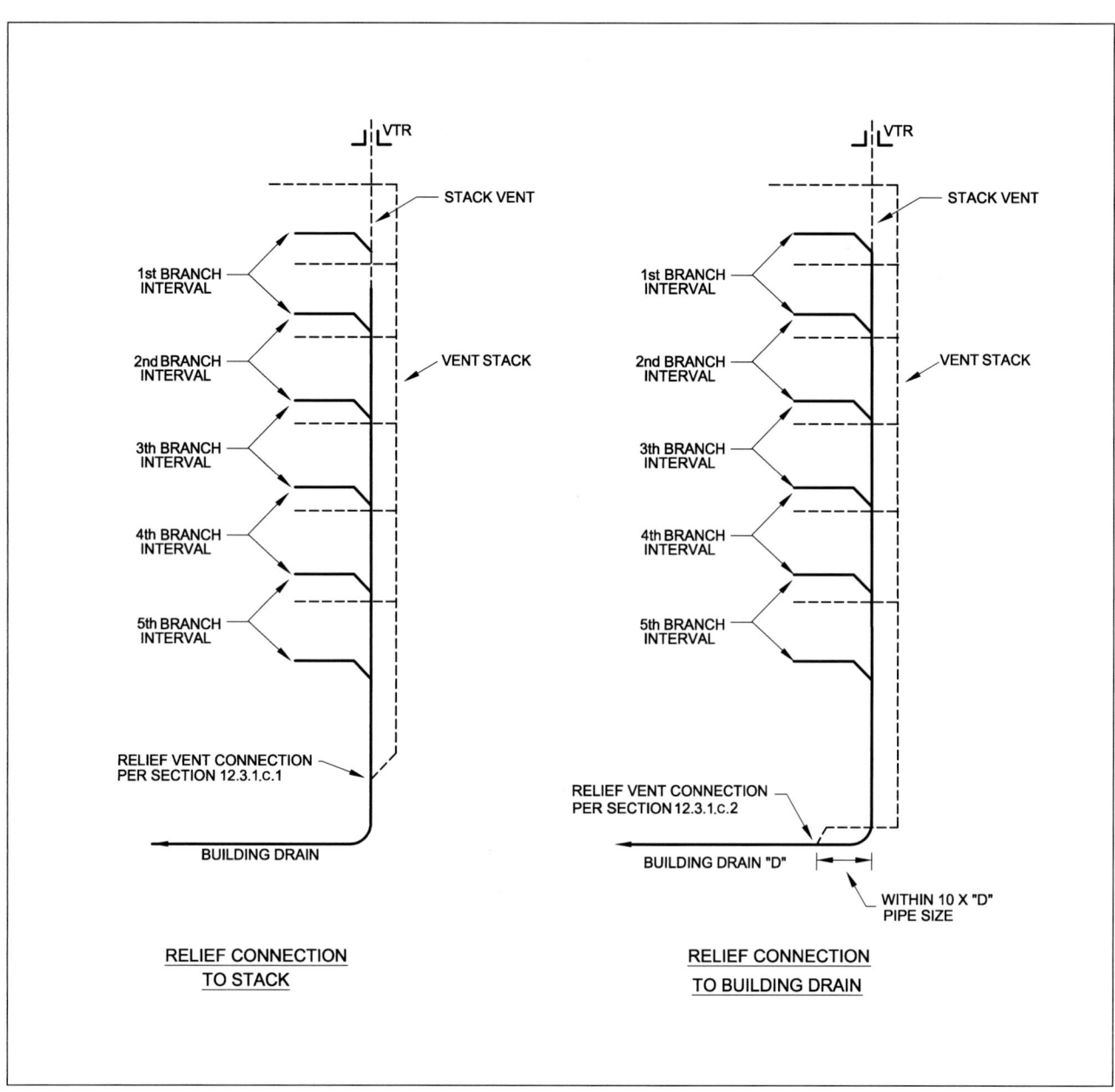

**Figure 12.3.1
VENT STACKS FOR DRAINAGE STACKS
HAVING 5 OR MORE BRANCH INTERVALS**

12.3.2 Relief Vents for Stacks Having Ten or More Branch Intervals

a. Where drain stacks have ten or more branch intervals, a relief vent shall be provided for each ten branch intervals, starting at the top of the stack.

b. The lower end of each relief vent shall connect to the drain stack as a yoke vent below its tenth branch interval.

c. The upper end of the relief vent shall connect to the vent stack at an elevation not less than 3 feet above the floor level served by the branch interval.
See Figure 12.3.2

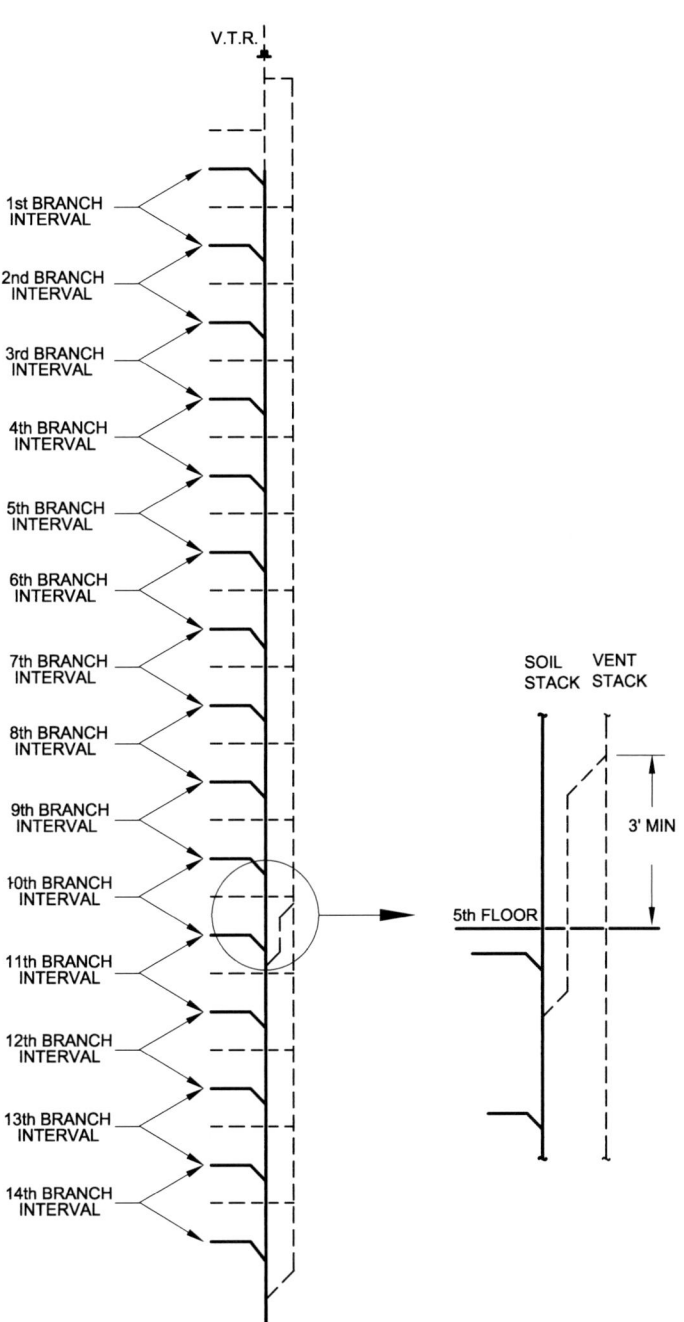

NOTES:
1. A branch interval is the vertical distance between branch connections to a drain stack, generally a story height, but never less than 8 feet. The drain stack has 14 branch intervals.
2. A relief vent is required for the upper ten branch intervals.
3. The relief vent connects as a yoke vent below the 5th Floor branch drain connection and connects to the vent stack at least 3 feet above the floor on the 5th Floor.
4. The 3-foot vertical rise in the yoke vent keeps pressure surges in the drain stack from causing spill-over of drainage into the vent stack.
5. Stack relief vents must be the same size as the vent stack. See Section 12.16.4 for sizing vent stacks.

**Figure 12.3.2
RELIEF FOR DRAIN STACKS HAVING
10 OR MORE BRANCH INTERVALS**

12.3.3 Horizontal Offsets

a. Horizontal offsets in stacks having five or more branch intervals discharging above the offset shall be vented either:

   1. by considering the stack as two separate stacks, one above and one below the offset, and venting each separately.

   2. by providing a yoke vent from the drain stack below the offset to the vent stack required by Section 12.3.1 not less than 3 feet above the offset. This relief vent may be a stack vent for the lower portion of the drain stack.

See Figures 12.3.3-A and 12.3.3-B

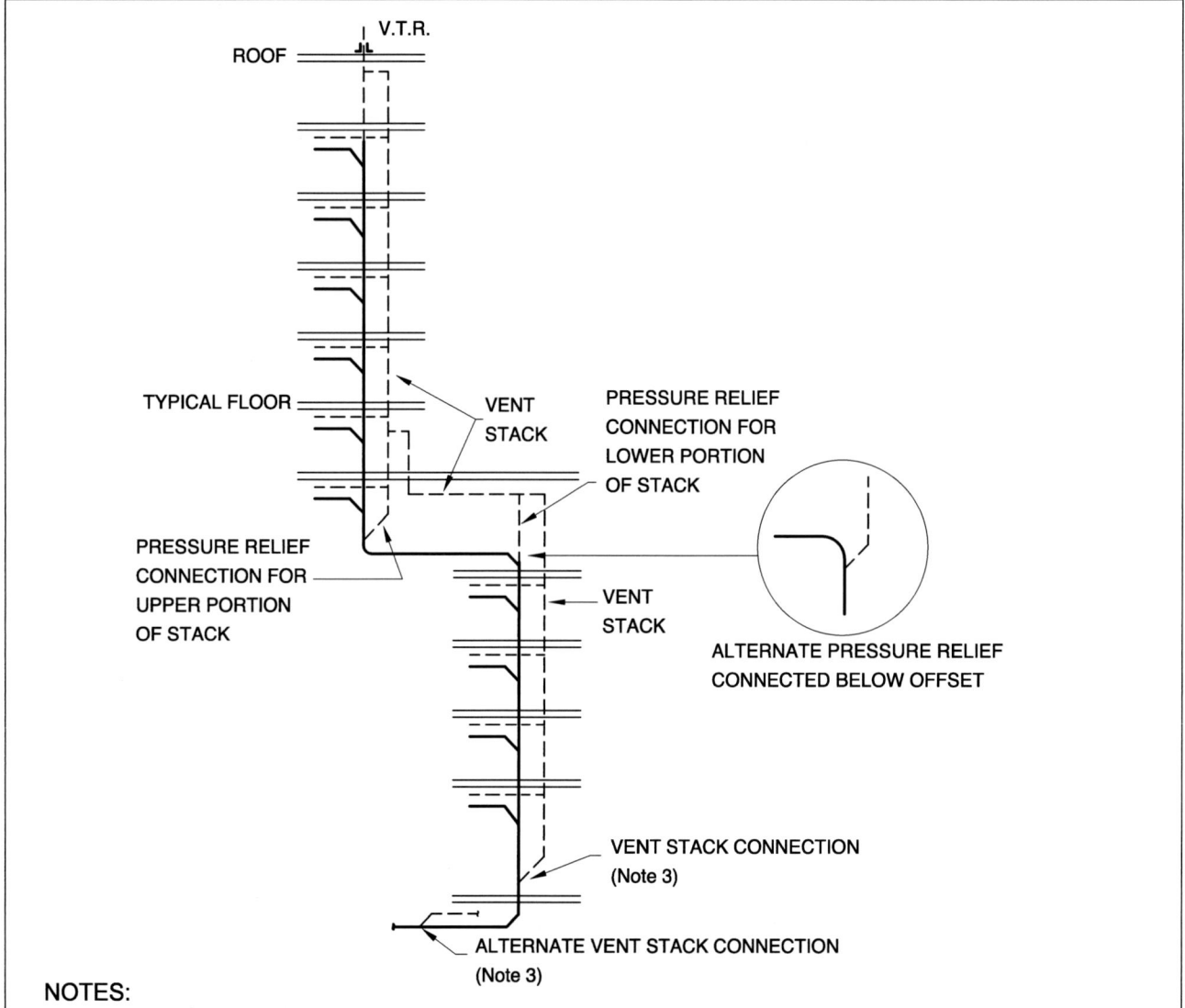

NOTES:
1. The vent stack must be sized for the entire DFU load on the drain stack per Table 12.16.4.
2. The pressure relief vents for the upper and lower portions of the drain stack are required to relieve pressure from the hydraulic effect of the offset. The relief vents must be the same size as the vent stack.
3. The vent stack must connect to the base of the drain stack, either below the lowest branch connection above the base of the stack or within 10 pipe diameters downstream from the base of the stack.

**Figure 12.3.3 - A
ONE STACK VENT FOR HORIZONTAL OFFSETS
IN DRAIN STACKS**

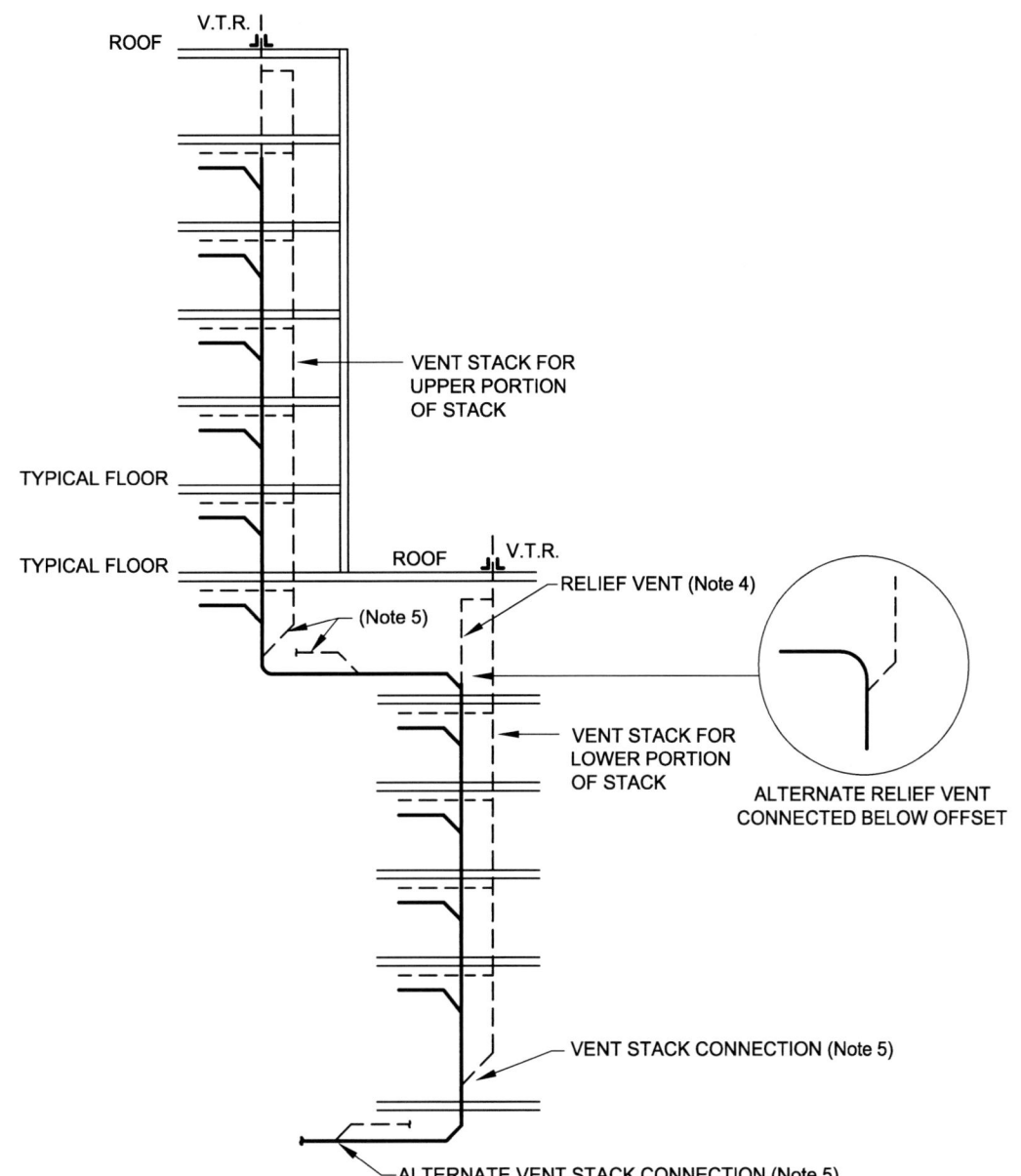

NOTES:
1. The drain stack has separate vent stacks for the upper and lower portions of the stack.
2. The vent stack for the upper portion of the drain stack can be sized for only the DFU load on the upper portion of the stack. Only vents for upper floor fixtures can be connected to the upper vent stack.
3. The lower portion of the drain stack carries the DFU load of the upper and lower portions of the stack. Its vent stack must be sized for the entire DFU load on the drain stack.
4. The relief vent for the lower portion of the drain stack must be the same size as the vent stack for the lower portion of the stack.
5. The vent stacks must connect to the base of each drainage stack, either below the lowest branch connection above the base of the stack or within 10 pipe diameters downstream from the base of the stack.

Figure 12.3.3 - B
TWO VENT STACKS FOR HORIZONTAL OFFSETS IN DRAIN STACKS

12.3.4 Vertical Offsets

Where vertical offsets in drain stacks having five or more branch intervals above the offset have branch connections within 2 feet above or below the offset, a relief vent shall be provided for the lower portion of the stack below the offset.

12.3.5 Vent Headers

Vents may be connected into a common header at the top of one or more stacks and then be extended to the open air at one point.

12.3.6 Other Use Prohibited

The plumbing vent system shall not be used for purposes other than venting of the plumbing system.

12.4 VENT TERMINALS

12.4.1 Extension Above Roofs

Vent pipes shall terminate not less than 6 inches above the roof, measured from the highest point where the vent intersects the roof.

EXCEPTION: Where a roof is used for any purpose other than weather protection and maintaining equipment, vents within ten (10) feet of the area being utilized shall extend at least seven (7) feet above the roof, shall be properly supported, and shall also comply with Section 12.4.4 for location.

See Figure 12.4.1

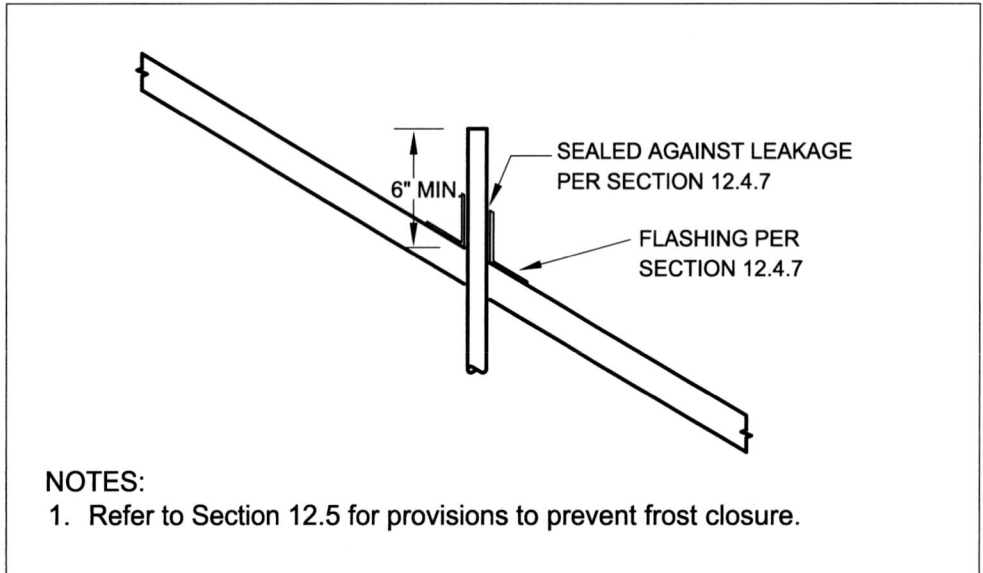

Figure 12.4.1
VENT EXTENSIONS THROUGH THE ROOF

12.4.2 Waterproof Flashings

Vent terminals shall be made watertight with the roof by proper flashing.

12.4.3 Flag Poling Prohibited

Vent terminals shall not be used for the purpose of flag poling, TV aerials, or similar purposes.

12.4.4 Location of Vent Terminal

a. Vent terminals shall not be located where vapors can enter the building.

b. No vent terminal shall be located directly beneath any door, window, or other ventilating opening of a building or of another building, nor shall any such vent terminal be within 10 feet horizontally of such opening unless it is at least 2 feet above the top of such opening.

c. Where a vent terminal is within 10 feet horizontally and less than 2 feet above a ventilation opening described in Section 12.4.4b and the line-of-sight from the vent terminal to the ventilation opening is interrupted by the continuous ridge of a roof, the ridge shall be at least 2 feet above the top of the opening. Otherwise, the vent terminal shall comply with Section 12.4.4b.

d. Where a vent terminal is within 10 feet horizontally and less than 2 feet above a ventilation opening described in Section 12.4.4b and the line of sight from the vent terminal to such ventilation opening is interrupted by a solid wall or solid barrier, the top of the wall or barrier shall be at least 2 feet above the top of the ventilation opening and the shortest travel distance around the wall or barrier from the vent terminal to the nearest edge of the ventilation opening shall be at least 10 feet. Otherwise, the vent terminal shall comply with Section 12.4.4b.

See Figure 12.4.4

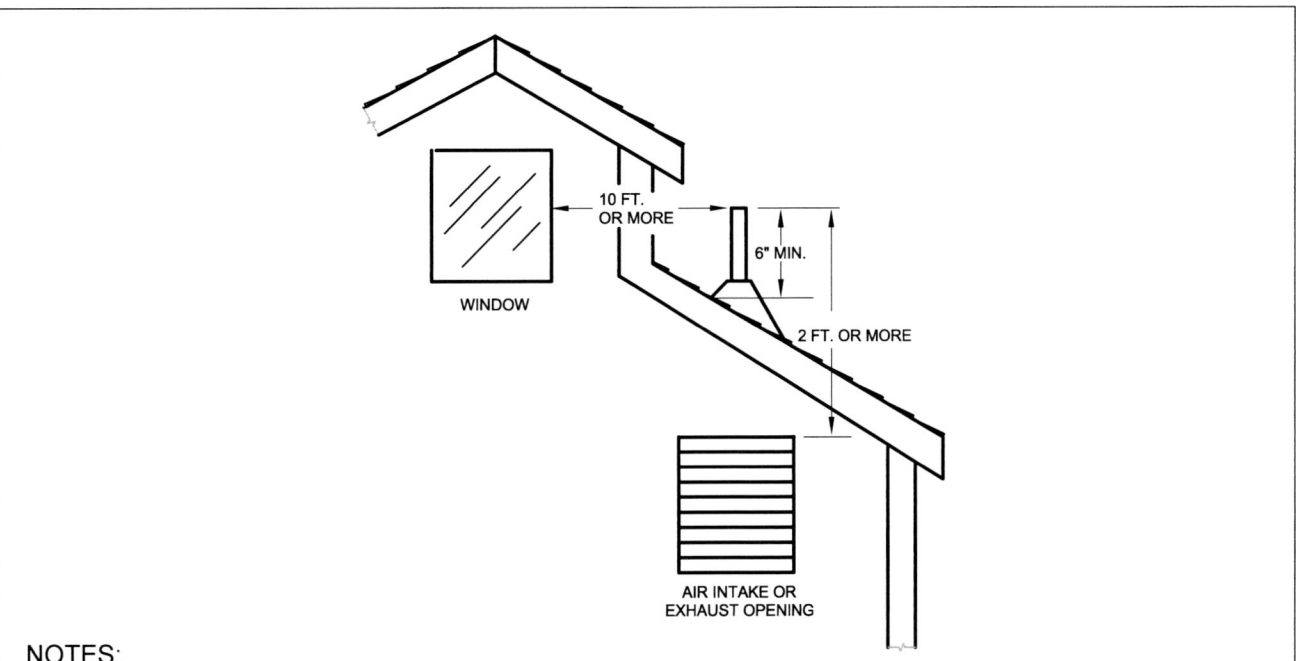

NOTES:
1. The separation of vent terminals from doors, windows, and air intake and exhaust openings keeps foul odors from entering the building.
2. The vent terminal is less than 2 feet above the window, but the window is 10 feet or more from the vent terminal horizontally.
3. The air intake or exhaust opening can be within 10 feet horizontally from the vent terminal if the vent terminal is 2 feet or more above the top of the opening.

Figure 12.4.4
ALLOWABLE LOCATIONS FOR VENT TERMINALS

12.4.5 Sidewall Venting

Vent terminals shall be permitted to extend through a wall on an existing building. They shall be at least 10 feet horizontally from any lot line, 10 feet above existing grade, and terminate with a corrosion-resistant bird screen. Vent terminals shall not terminate under an overhang of a building. They shall be located in accordance with Section 12.4.4.

12.4.6 Extensions Outside Building

No sanitary drain or vent pipe extension shall be installed on the outside of a wall of any new building, but shall be carried up inside the building
EXCEPTION: In those localities where the outdoor temperature does not drop below 32°F, the Authority Having Jurisdiction may approve the installation outside the building.

12.4.7 Flashing Roof Vent Terminals

a. Vent terminals through the roof shall be made watertight to the roof by sealing the flashing to either the exterior or interior of the vent terminal.

b. Vent terminals that are externally sealed shall employ manufactured vent stack flashing sleeves, roof couplings, or no-caulk roof vent flashings.

c. Where vent terminals are sealed by counter-flashing over the top of the vent terminal, the counter flashing shall not decrease the interior free area of the minimum required vent terminal size. Vent terminals shall be increased at least one pipe size when counter-flashed. Interior counter flashing shall be sealed gas-tight to prevent the entrance of sewer gas into the building through the flashing.

12.5 FROST CLOSURE

Where the Authority Having Jurisdiction requires protection against frost closure, vent terminals less than 3" pipe size shall be increased at least one pipe size to not less than 3" size. Where an increase is necessary, the increase in size shall be made inside the building at least one foot below a roof or ceiling that is thermally insulated and in an area not subject to freezing temperatures.

12.6 VENT SLOPES AND CONNECTIONS

12.6.1 Vent Slope

Vent and branch vent pipes shall be free from drops and sags and be sloped and connected as to drain by gravity to the drainage system.
See Figure 12.6.1

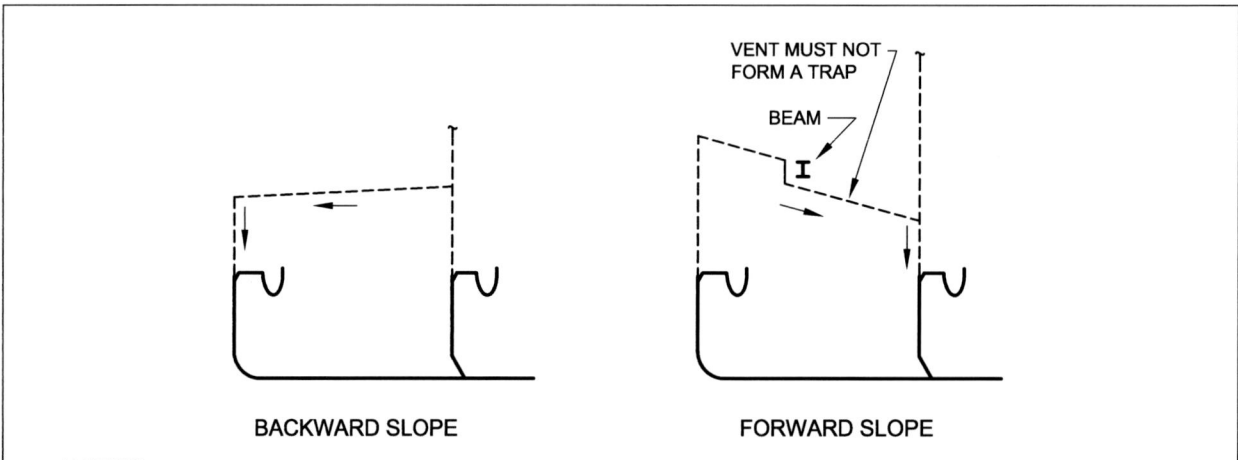

**Figure 12.6.1
VENT PIPING SLOPE**

12.6.2 Vertical Rise

Every vent shall rise vertically to a minimum of 6 inches above the flood level of the rim of the fixture being served before connecting to another vent.

EXCEPTIONS:

(1) Horizontal portions of a vent below the flood level rim of the fixture served that are installed in accordance with Sections 12.6.2.1, 12.6.2.2, and 12.6.2.3.

(2) Island sink vents in accordance with Section 12.18.

See Figures 12.6.2-A and 12.6.2-B

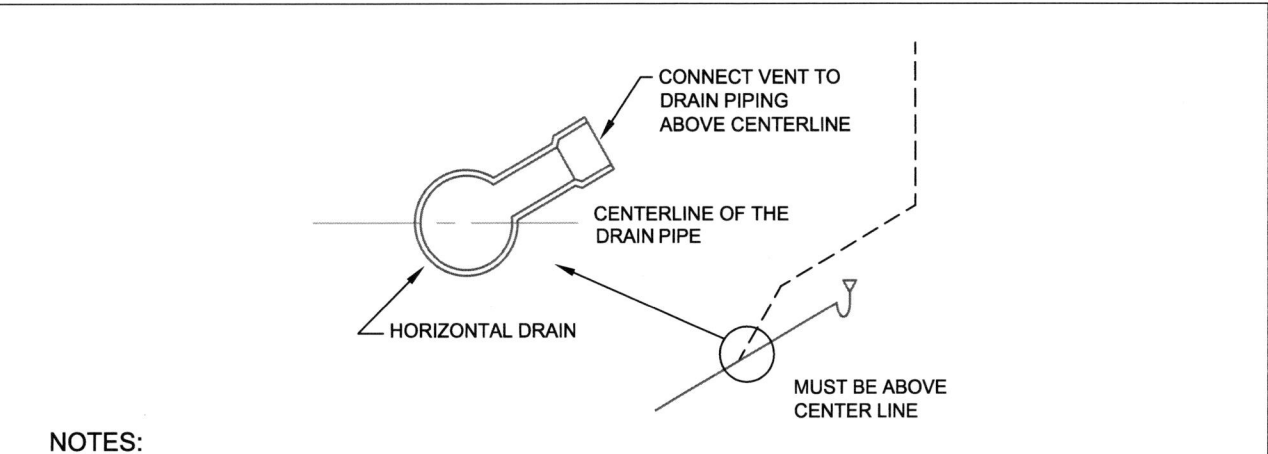

NOTES:
1. Connecting the vent above the centerline of a horizontal drain reduces the possibility of the vent connection being fouled by the flow in the drain.

Figure 12.6.2 - A
VENT PIPE CONNECTIONS TO HORIZONTAL DRAINS

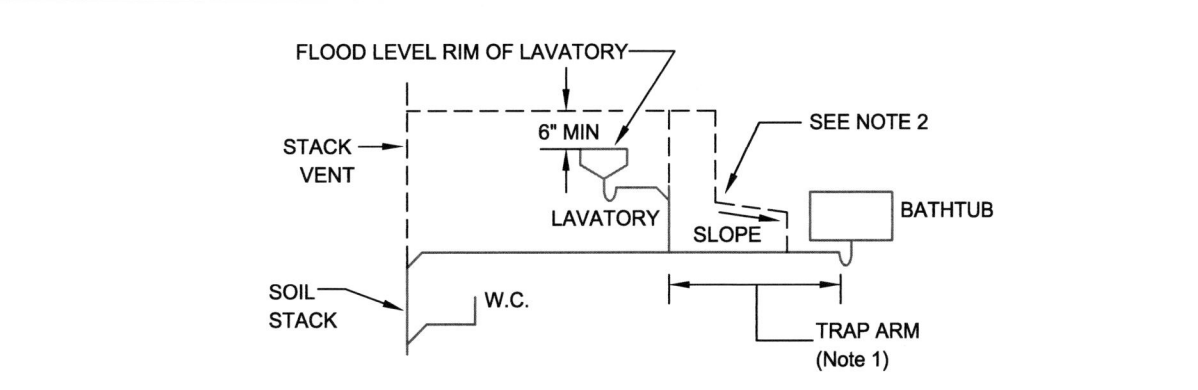

NOTES:
1. A separate vent will be required for the bathtub if the developed length of its trap arm exceeds the maximum allowable length in Table 12.8.1.
2. If a vent is required for the bathtub, it can have a horizontal offset below the flood level rim of the bathtub, but it must be 6 inches above the flood level rim of the lavatory before it connects to another vent. The horizontal offset below the flood level rim must be sloped so that it drains to the drainage system.

Figure 12.6.2 - B
HORIZONTAL VENT PIPING BELOW THE FLOOD LEVEL RIM OF FIXTURES

12.6.2.1 Horizontal Vent Below Fixture Flood Level Rim

Where a vent pipe connects to a horizontal fixture drain branch, and conditions require a horizontal offset in the vent below the flood level rim of the fixture served, the vent shall be taken off so that the invert of the horizontal portion of the vent pipe is at or above the centerline of the horizontal sanitary drain pipe.

12.6.2.2 Slope of Horizontal Vent

The portion of the horizontal vent installed below the flood level rim as permitted in Section 12.6.2.1 shall be installed with the required slope to drain by gravity to the drainage system. *See Figure 12.6.2-B*

12.6.2.3 Cleanouts

Cleanouts shall be provided in the vent piping so that any blockages in the vent piping below the flood level rim of the fixture served can be cleared into the drainage system.

12.6.3 Vent Connection Height Above Fixtures

Connections between any horizontal vent pipe, including individual vents, branch vents, relief vents, circuit vents or loop vents, and a vent stack or stack vent shall be made at least 6 inches above the flood level rim of the highest fixture on the floor level.

12.6.4 Side-Inlet Closet Bends

    a. Side-inlet closet bends shall be permitted only in cases where the fixture connection thereto is vented.
    b. In no case shall the side-inlet be used to vent a bathroom group without being washed by a fixture.
EXCEPTION: As allowed in Sections 12.10 and 12.11.

12.7 ADJACENT FIXTURES

Two fixtures set adjacent within the distance allowed between a trap and its vent, may be served with one common vent, provided that each fixture connects separately into an approved double fitting having inlet openings at the same level. (See Section 12.9.2 for inlet openings at different levels.)
See Figure 12.7

> *Comment: Suitable double fittings for connecting adjacent fixtures such as lavatories, sinks, tubs, showers, urinals, or floor-outlet water closets to a stack are double sanitary tees, double fixture fittings (plastic), and sanitary crosses. These fittings have high branch openings into the stack that keep the top of the vent opening above the weir of the traps to prevent self-siphonage.*

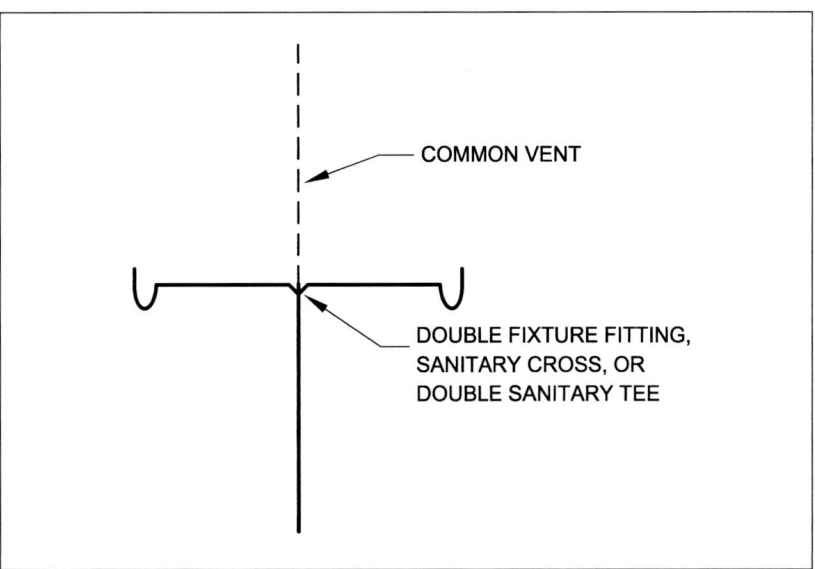

Figure 12.7
VENTING ADJACENT FIXTURES
CONNECTED TO A VERTICAL DRAIN AT THE SAME LEVEL

12.8 FIXTURE VENTS

12.8.1 Venting of Fixture Drains

Fixture drains shall have a vent so located that the vent connects above the top weir of the trap and the developed length of the trap arm is within the limits set forth in Table 12.8.1.

EXCEPTIONS:

(1) Water closets and similar siphonic fixtures.

(2) Combination waste and vent systems. (see Section 12.17)

(3) Vents may be connected below the top weir of the fixture trap if the following conditions are met:

   a) The vertical section of the drain pipe shall be at least one pipe size larger than the trap inlet size.

   b) The horizontal pipe connected to the trap outlet shall be at least two pipe diameters long.

   c) The developed length of the trap arm shall not exceed the values in Table 12.8.1.

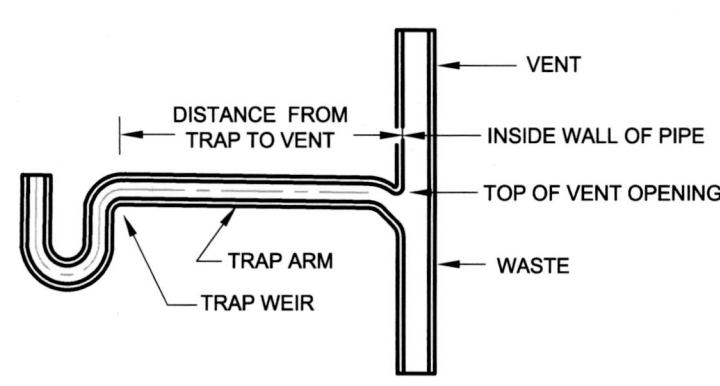

NOTES:
1. Trap arm lengths are limited by Table 12.8.1 to prevent self-siphonage of the fixture trap.
2. The top of the opening at the vent pipe must not be below the elevation of the trap weir. The maximum trap arm lengths in Table 12.8.1 are based on using sanitary tees or other short turn fittings to connect trap arms to the drainage system. Long turn fittings should not be used with Table 12.8.1.
3. The trap arm length is the developed length along its centerline, including any changes in direction.

Figure 12.8.1 - A
THE MAXIMUM LENGTH OF TRAP ARMS

Table 12.8.1 MAXIMUM LENGTH OF TRAP ARM		
Size of Trap Arm (Inches)	Length – Trap Arm to Vent	Slope – Inches per Foot
1-1/4	3' 6"	1/4
1-1/2	5'	1/4
2	8'	1/4
3	10'	1/8
4	12'	1/8

NOTES FOR TABLE 12.8.1:

This table has been expanded in the "length" requirements to reflect expanded application of the wet venting principles. Slope shall not exceed 1/4" per foot.

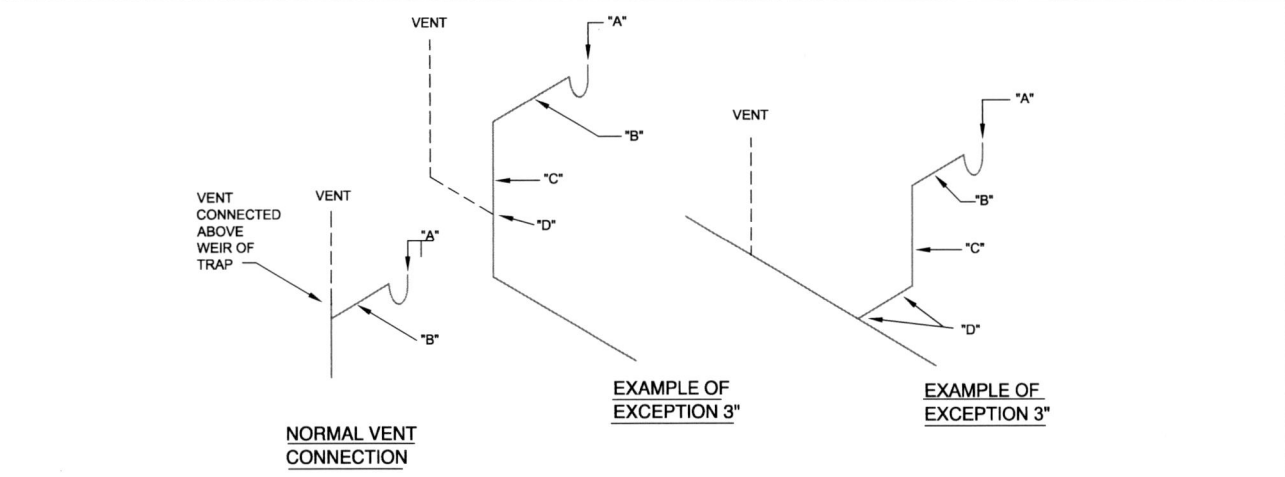

NOTES:
1. "A" is the trap pipe size.
2. "B" is the horizontal pipe between the trap outlet and the vertical leg. Its length must be at least two times the trap pipe size to avoid creating an "S" trap.
3. "C" is the vertical drop. It must be one size larger than the trap pipe size.
4. "D" is the connection to a vented drain line. The horizontal portion of "D" must be the same size as "C". The distance from the weir of the trap to the vent connection at "D" must be within the limits of Table 12.8.1, based on the size of the trap.

Figure 12.8.1 - B
EXAMPLES OF EXCEPTION #3 TO SECTION 12.8.1
(VENT CONNECTIONS BELOW THE WEIR OF A TRAP)

12.8.2 Provision for Venting Future Fixtures

On new construction of residential dwelling units with basements, a 2" minimum size vent shall be installed between the basement and attic or tied into an existing, properly sized vent and capped for future use.

12.8.3 Crown Venting Limitation

A vent shall not be installed within two pipe diameters of the trap weir. *See Figure 12.8.3*

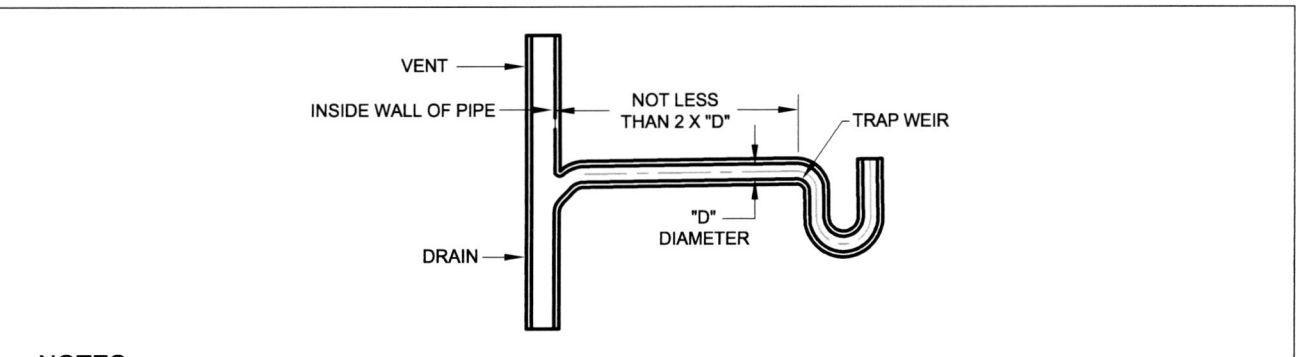

NOTES:
1. Vents connected too close to the crown or weir of a trap are subject to clogging, thereby rendering the vent ineffective. Vent connections must be at least two pipe diameters from the weir of the trap.

Figure 12.8.3
CROWN VENTS ARE PROHIBITED

12.8.4 Water Closets and Other Siphonic Fixtures

For water closets and other fixtures that operate by siphonic action, the distance between the outlet of the fixture and its vent connection shall not exceed 3 feet vertically and 9 feet horizontally.

12.9 COMMON VENTS

12.9.1 Individual Vent as Common Vent

An individual vent, installed vertically, may be used as a common vent for two fixture traps when both fixture drains connect with a vertical drain at the same level.

12.9.2 Fixtures Drains Connected at Different Levels

A common vent may be used for two fixtures installed on the same floor but connecting to a vertical drain at different levels, provided that the vertical drain is one pipe size larger than the upper fixture drain but in no case smaller than the lower fixture drain. *See Figure 12.9.2*

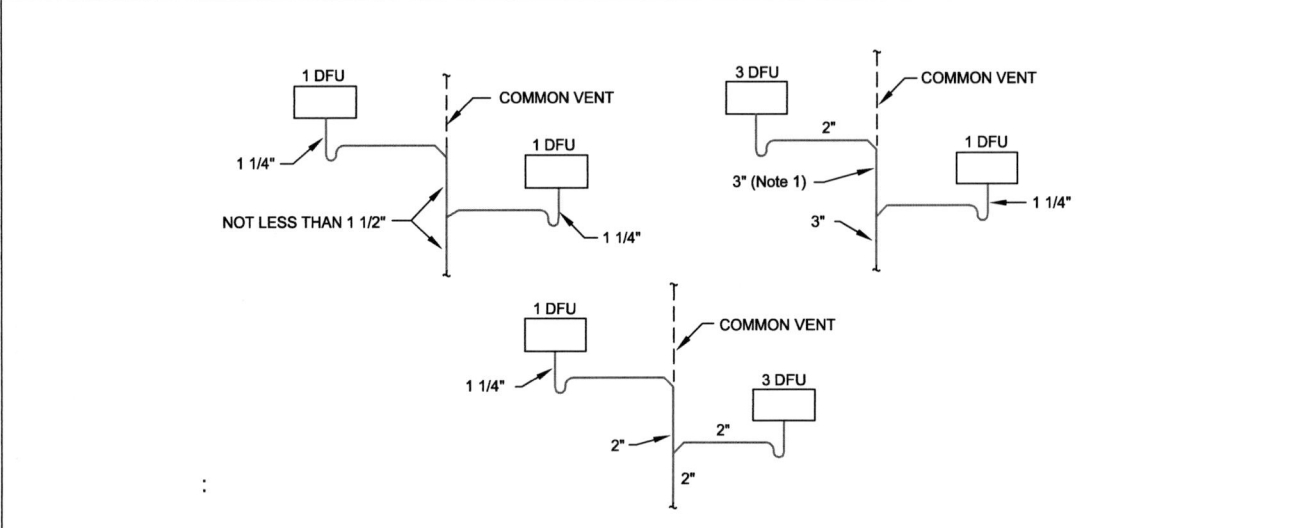

NOTES:
1. When the upper fixture is 3 DFU and the lower is 1 DFU, the vertical drain must be 3" instead of 2" so that the 1 DFU fixture is wet vented during the heavier discharge from the 3 DFU fixture.

Figure 12.9.2
VENTING FIXTURES CONNECTED TO A
VERTICAL DRAIN AT DIFFERENT LEVELS

12.10 WET VENTING

12.10.1 Single Bathroom Groups

   a. An individually vented lavatory in a single bathroom group shall be permitted to serve as a wet vent for the water closet, the bathtub or shower stall, or the water closet and bathtub/shower if all of the following conditions are met.

     1. The wet vent is 1-1/2" minimum pipe size if the water closet bend is 3" size or it shall be 2" minimum pipe size if the water closet bend is 4" pipe size.

     2. A horizontal branch drain serving both the lavatory and the bathtub or shower stall is 2" minimum pipe size.

3. The length of the trap arm for the bathtub or shower stall is within the limits of Table 12.8.1. If not, the bathtub or shower stall shall be individually vented.

4. The distance from the outlet of the water closet to the connection of the wet vent is within the limits established by Section 12.8.4. Otherwise, the water closet shall be individually vented.

5. A horizontal branch serving the lavatory and the bathtub or shower stall shall connect to the stack at the same level as the water closet, or it may connect to the water closet bend, or the lavatory and bathtub or shower stall may individually connect to the water closet bend.

6. When the bathroom group is the topmost load on a stack, a horizontal branch serving the lavatory and the bathtub or shower stall may connect to the stack below the water closet bend, or the lavatory and the bathtub or shower stall may individually connect to the stack below the water closet bend.
See Figures 12.10.1-A through -C

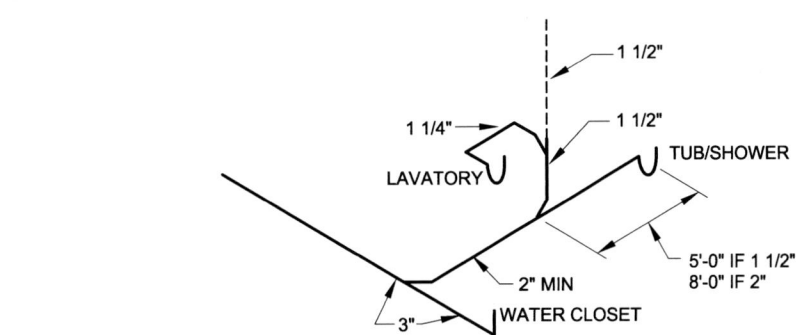

NOTES:
1. The vent for the bathroom group is 1-1/2", which is 1/2 the size of the 3" drain for the bathroom group.

Figure 12.10.1 - A
A WET VENTED BATHROOM GROUP WITH A 3" CLOSET BEND

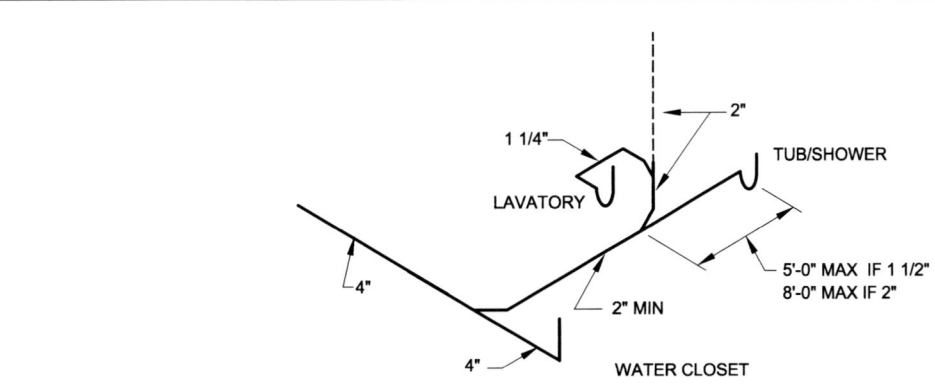

NOTES:
1. The vent for the bathroom group is 2", which is 1/2 the size of the 4" drain for the bathroom group. Even though the bathroom group could have had a 3" drain with a 1-1/2" vent, installing a 1-1/2" vent on the 4" drain would cause unnecessary confusion for the installer and inspector.

Figure 12.10.1 - B
A WET VENTED BATHROOM GROUP WITH A 4" CLOSET BEND

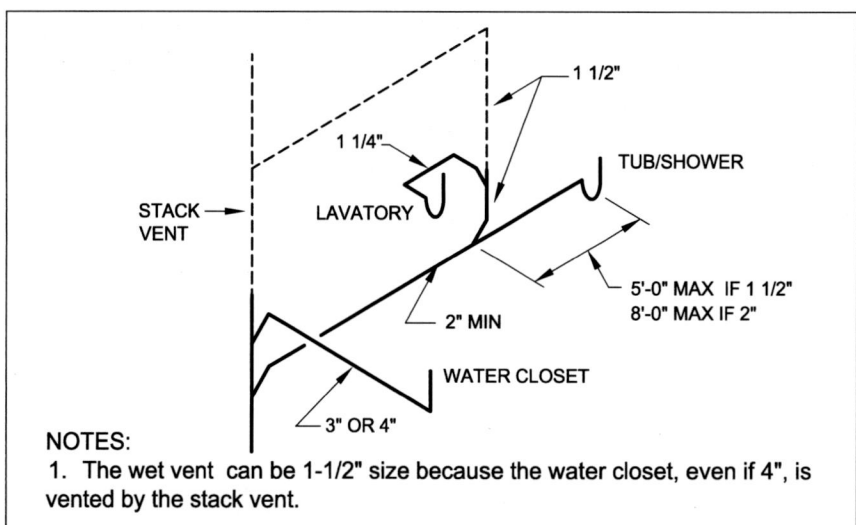

Figure 12.10.1 - C
A WET VENTED BATHROOM GROUP WITH 3" OR 4" CLOSET BEND
(SINGLE STORY OR TOP FLOOR)

12.10.2 Double Bathtubs and Lavatories

Two lavatories and two bathtubs or showers back-to-back may be installed on the same horizontal branch with a common vent for the lavatories and with no back vent for the bathtubs or shower stalls provided the wet vent is 2" in size and the lengths of the tub/shower drains conform to Table 12.8.1. *See Figure 12.10.2*

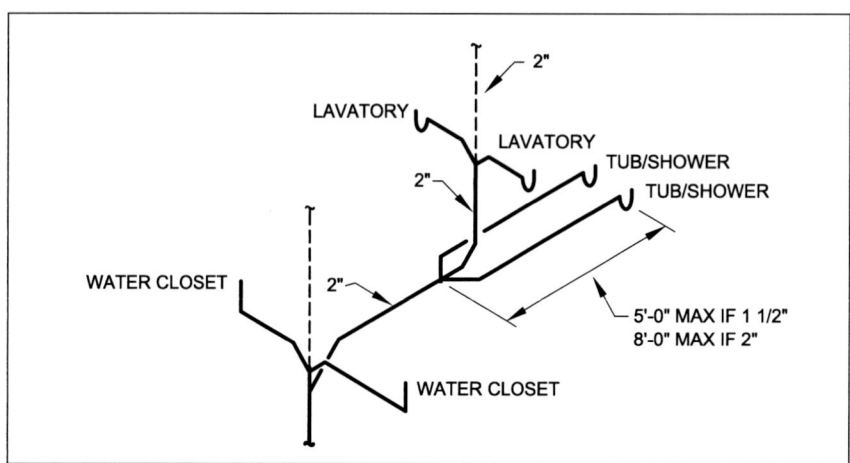

Figure 12.10.2
WET VENTED BACK-TO-BACK BATHROOM GROUPS
(SINGLE STORY OR TOP FLOOR)

12.10.3 Multi-Story Bathroom Groups

   a. On the lower floors of a stack, the waste pipe from one or two lavatories may be used as a wet vent for one or two bathtubs or showers as provided in Section 12.10.2.
   b. Each water closet below the top floor shall be individually back vented.
   EXCEPTION: The water closets in bathroom groups shall not be required to be back vented if the following conditions are met:
     (1) The 2" waste serving the tubs/showers and lavatories connect directly into the water closet bend with a 45° wye tap in the direction of flow or,

(2) A special stack fitting is used that consists of a 3" or 4" closet opening and two side inlets each 2" in size and the inverts of which are above the center, and below the top of the water closet opening; and one of the 2" inlets is connected to the tub/shower drains, and the other is connected to the waste pipe from a maximum of two lavatories that are vented to a vent stack or stack vent; or,

(3) In lieu of the special stack fitting of Section 12.10.3b(2) above, 4" closet bends with two 2" wye taps may be used.

See Figures 12.10.3-A through -E

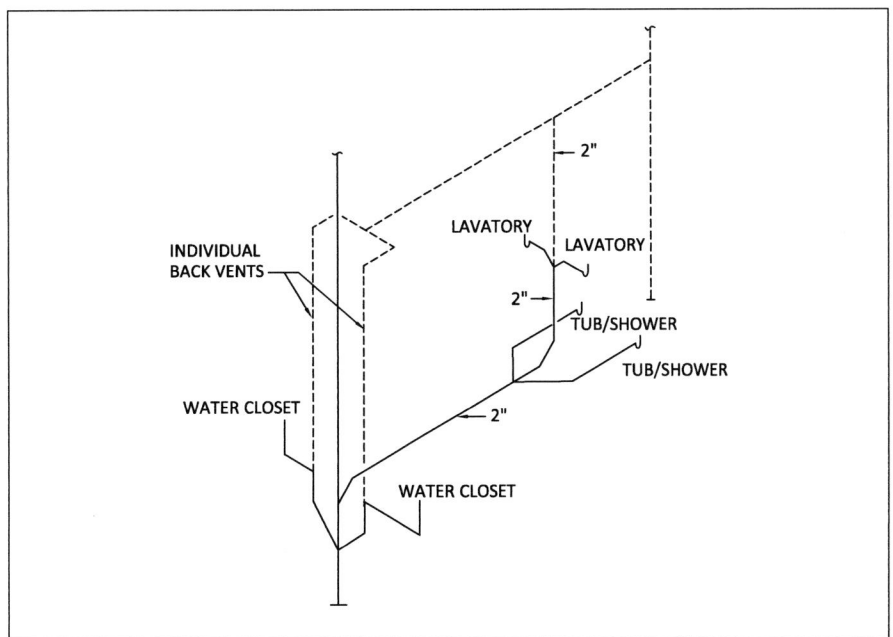

Figure 12.10.3 - A
WET VENTED BACK-TO-BACK BATHROOM GROUPS ON A LOWER FLOOR WITH BACK VENTED WATER CLOSETS (ONE ARRANGEMENT)

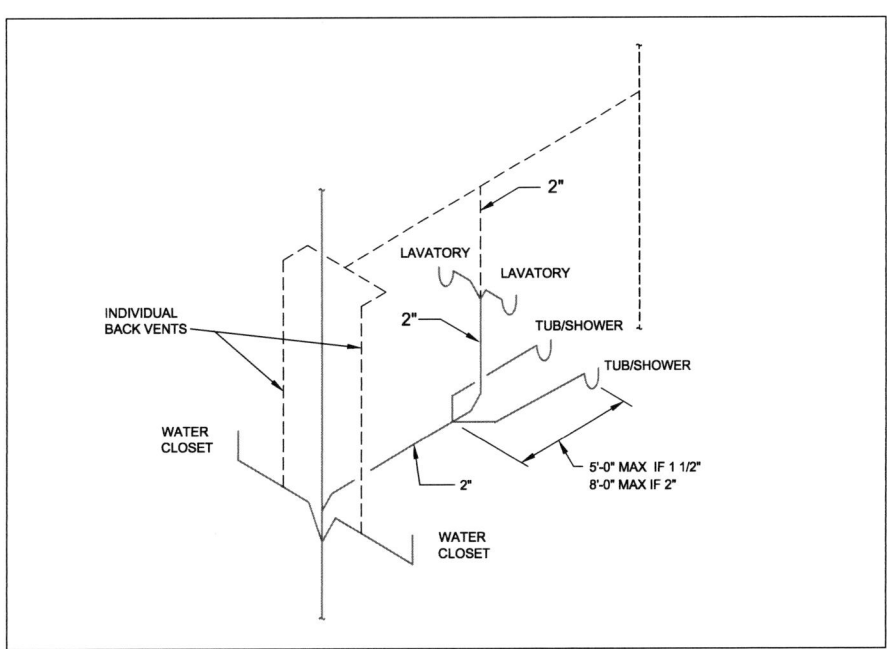

Figure 12.10.3 - B
WET VENTED BACK-TO-BACK BATHROOM GROUPS ON A LOWER FLOOR WITH BACK VENTED WATER CLOSETS (ANOTHER ARRANGEMENT)

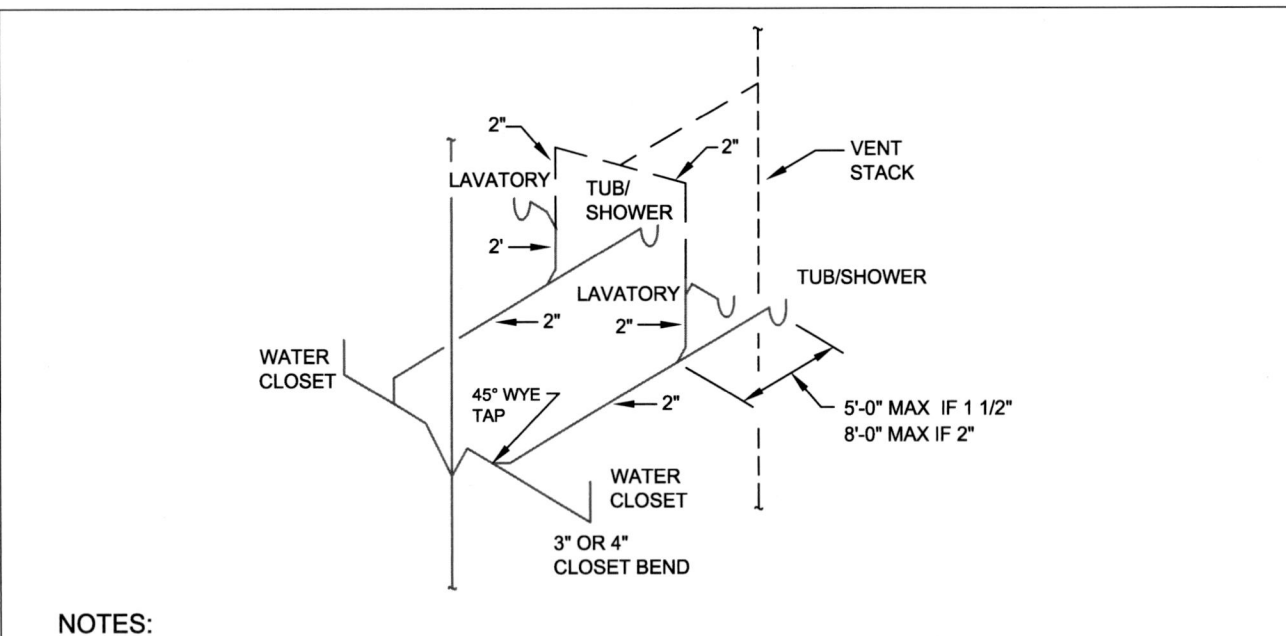

NOTES:
1. This is an example of Exception #1 to Section 12.10.3.
2. Two adjacent bathroom groups are shown but individual bathroom groups can also be wet vented in this manner.
3. A lower floor is shown, but the arrangement can be used at the top of a drain stack.

Figure 12.10.3 - C
ADJACENT BATHROOM GROUPS WET VENTED WITH 45 DEGREE WYE TAPS

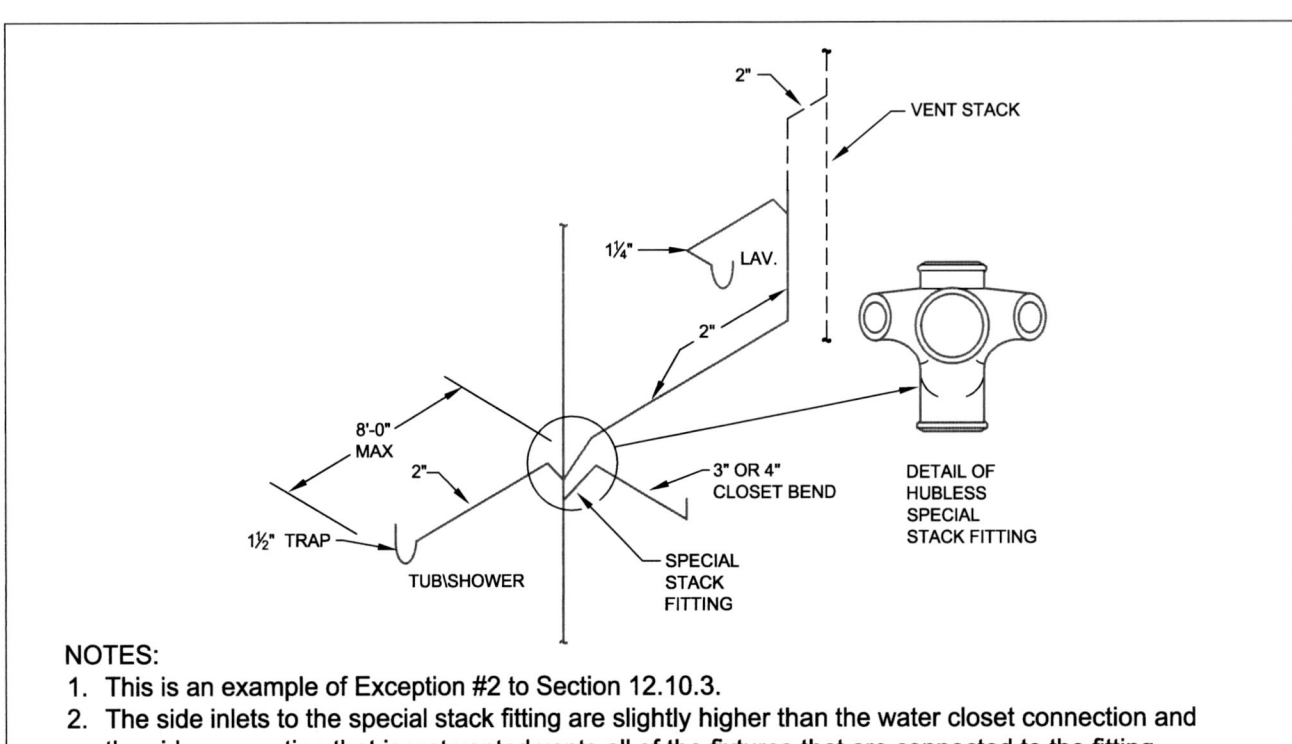

NOTES:
1. This is an example of Exception #2 to Section 12.10.3.
2. The side inlets to the special stack fitting are slightly higher than the water closet connection and the side connection that is wet vented vents all of the fixtures that are connected to the fitting.
3. A lower floor is shown, but the special stack fitting can be used at the top of a stack.

Figure 12.10.3 - D
A BATHROOM GROUP WET VENTED WITH A SPECIAL STACK FITTING

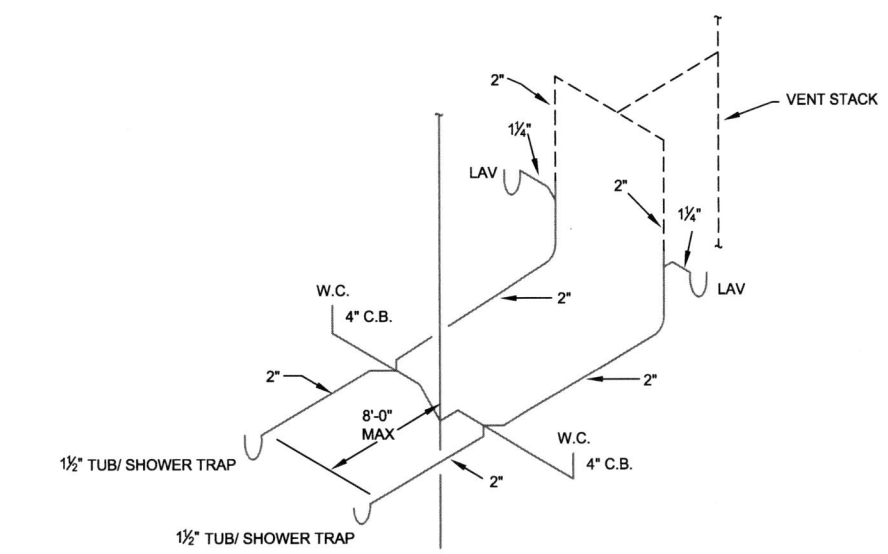

NOTES:
1. This is an example of Exception #3 to Section 12.10.3.
2. The oversized 4" water closet fixture drain permits the tub/shower on one side of the drain to be wet vented by the lavatory on the other side of the drain.
3. Two adjacent bathroom groups are shown, but individual bathroom groups can be wet vented in this manner.
4. A lower floor is shown, but the arrangement can be used at the top of a drain stack.

Figure 12.10.3 - E
ADJACENT BATHROOM GROUPS WET VENTED WITH 4" CLOSET BENDS HAVING TWO 2" WYE TAPS

12.10.4 Bathtubs and Water Closets

a. An individually-vented bathtub in a single bathroom group shall be permitted to serve as a wet vent for the water closet if all of the following conditions are met:

1. The wet vent is 2" minimum size.

2. The distance from the outlet of the water closet to the connection of the wet vent is within the limits established by Section 12.8.4. Otherwise, the water closet shall be individually vented.

See Figure 12.10.4

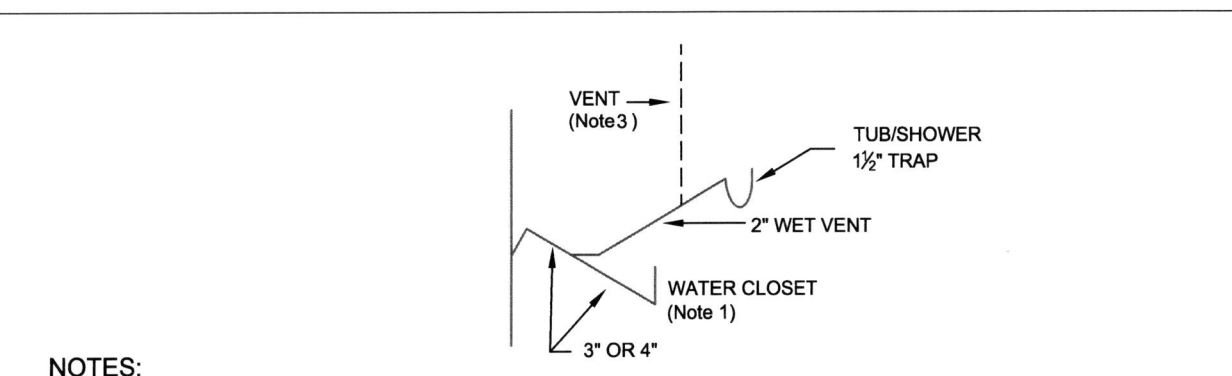

NOTES:
1. The maximum distance from the water closet outlet to its vent connection is 9 feet horizontal and 3 feet vertical.
2. A lower floor is shown, but the arrangement can be used at top of stack.
3. The vent must be 1 1/2" for a 3" closet bend or 2" for a 4" closet bend.

Figure 12.10.4
A WATER CLOSET WET VENTED BY A BATHTUB

12.10.5 Reserved

12.10.6 Floor Drains and Floor Sinks

a. A lavatory or sink shall be permitted to serve as a wet vent for a floor drain or floor sink if all of the following conditions are met:

1. The wet vent shall be not less than 1-1/2" size for a 1 DFU lavatory or 2" for a 2 DFU sink.
2. The wet vent shall be larger than 1/2 the size of the drain for the floor drain or floor sink.
3. The distance from the outlet of the floor drain or floor sink to the connection of the wet vent shall be within the limits established by Table 12.8.1.

12.11 STACK VENTING

12.11.1 Fixture Groups

a. A single bathroom group and a kitchen sink (with or without a disposer and/or dishwasher) located back-to-back, or two bathroom groups back-to-back may be installed without individual fixture vents in a one-story building or on the highest branch of a stack in a multi-story building provided that the following conditions are met:

1. Each fixture drain connects independently to the stack.
2. The tub and/or shower and water closet enter the stack at the same level.
3. The requirements of Table 12.8.1 are met.
4. A side inlet connection into a 4" closet bend shall be considered to be an independent connection to the stack.

See Figures 12.11.1-A and 12.11.1-B

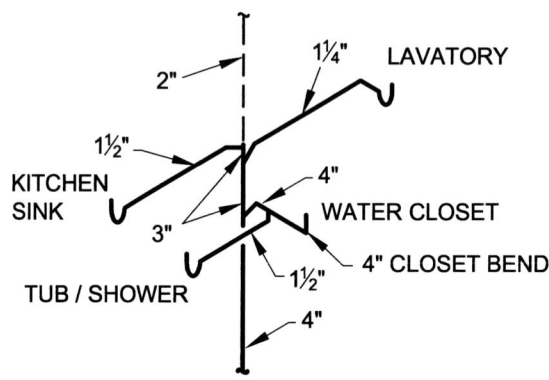

NOTES:
1. The arrangement shown is permitted in a one-story building or at the top of a stack.
2. Each fixture must connect independently to the stack except that the tub/shower connection to the 4" water closet fixture drain is considered to be a connection to the stack because the 4" drain is oversized for the water closet.
3. The length of all trap arms from their trap weir to their vent opening at the stack must not exceed the limits in Table 12.8.1. The vent distance for the water closet must be in accordance with Section 12.8.4.
4. Bathroom groups without kitchen sinks can also be stack vented as shown.
5. If the drain stack has 5 or more branch intervals, the drain stack and stack vent must be full size of the base of the drain stack or 4" minimum for the branch connection from the water closet and tub/shower.

Figure 12.11.1 - A
STACK VENTING A BATHROOM GROUP AND AN ADJACENT KITCHEN SINK

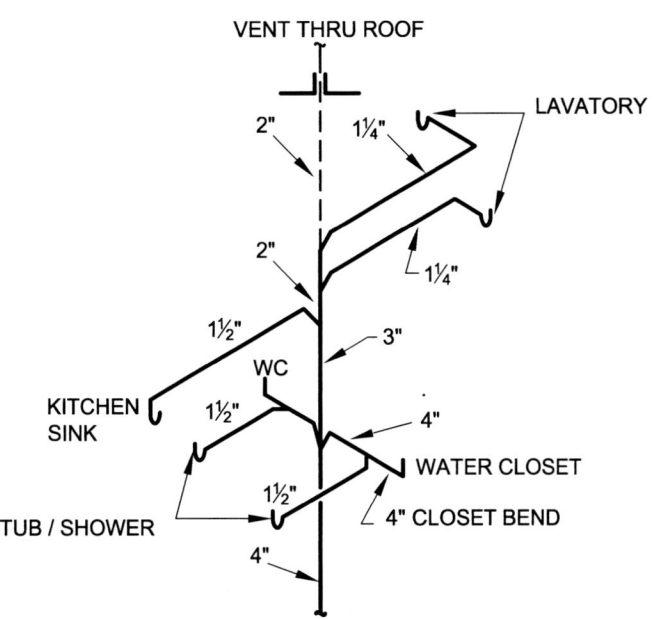

NOTES:
1. The arrangement shown is permitted in a one-story building or at the top of a stack.
2. Each fixture must connect independently to the stack except that the tub/shower connections to the 4" water closet fixture drains are considered to be connections to the stack because the 4" drains are over-sized for the water closets.
3. The length of all trap arms from their trap weir to their vent opening at the stack must not exceed the limits in Table 12.8.1. The vent distance for the water closets must be in accordance with Section 12.8.4.
4. Bathroom groups without a kitchen sink can also be stack vented as shown.
5. If the drain stack has 5 or more branch intervals, the drain stack and stack vent must be full size of the base of the base of the drain stack or 4" minimum for the branch connection from the water closet and tub/shower.

Figure 12.11.1 - B
STACK VENTING ADJACENT BATHROOM GROUPS
AND AN ADJACENT KITCHEN SINK

12.11.2 Lower Floors

a. Lower floor bathroom groups may be vented as provided in Section 12.11.1, provided the following conditions are met:

1. A wye is installed in the stack with an upright one-eighth bend continuing from the wye branch to serve the stack group.
2. A 2" relief vent is connected to the wye branch at least 6 inches above the flood level rim of the highest fixture on the wye branch.

See Figures 12.11.2-A and 12.11.2-B

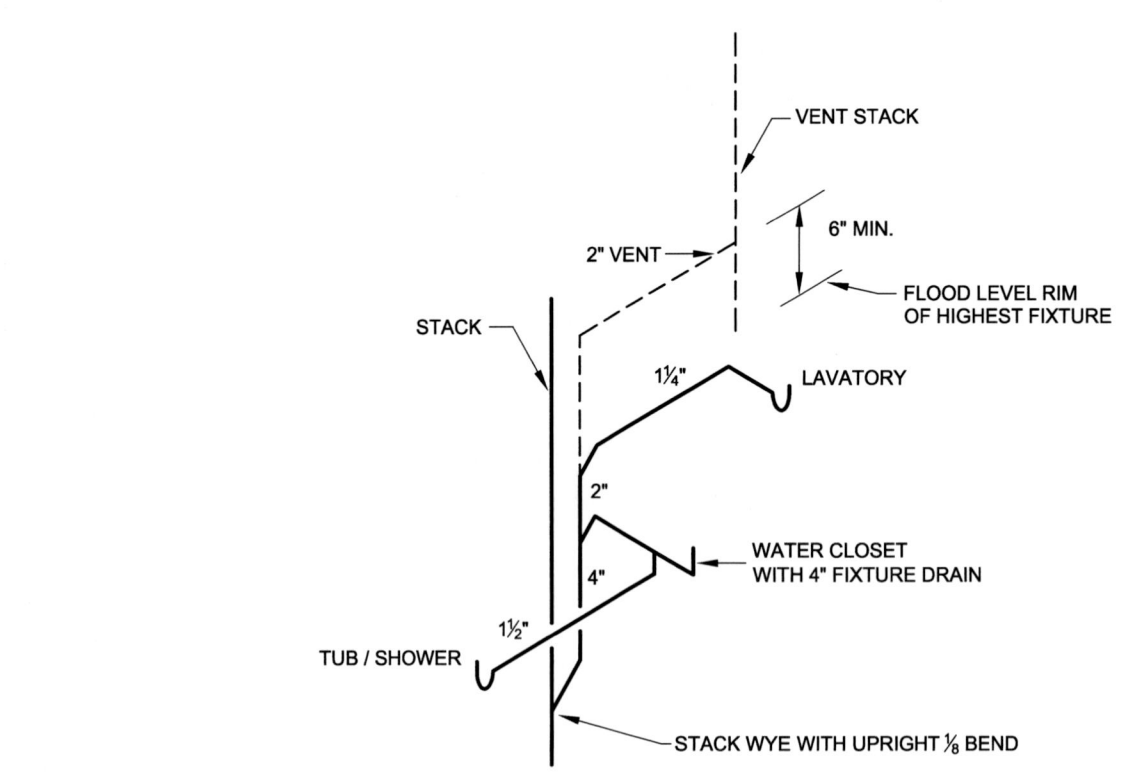

NOTES:
1. Each fixture in the stack group must connect independently to the sub-stack except that the connection to the 4" water closet fixture drain is considered to be a connection to the sub-stack because it is over-sized for the water closet.
2. The length of all trap arms from their trap weir to their stack must not exceed the limits in Table 12.8.1. The vent distance for the water closet must be in accordance with Section 12.8.4.

Figure 12.11.2 - A
STACK VENTING A BATHROOM GROUP ON A LOWER FLOOR OF A DRAIN STACK

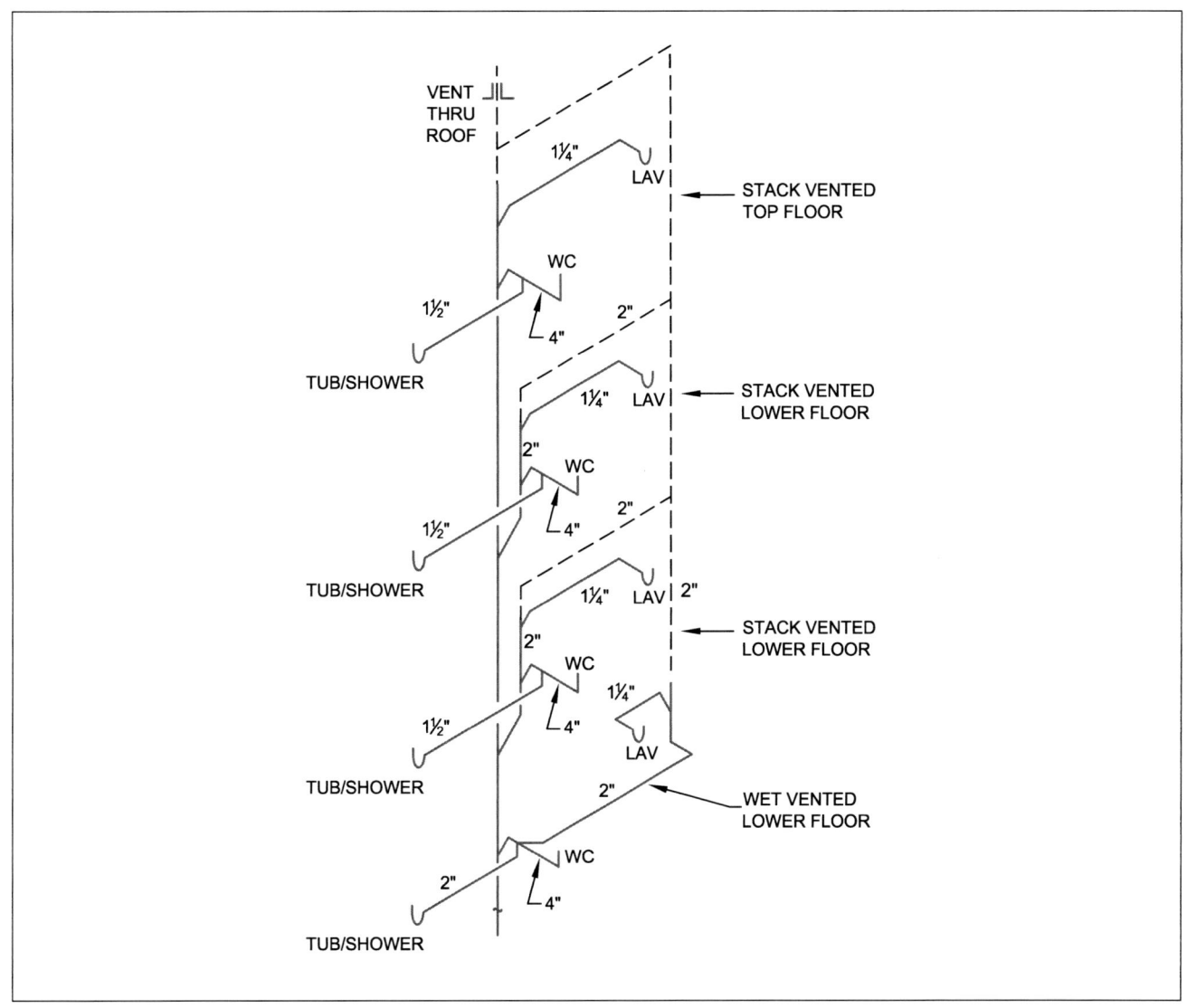

**Figure 12.11.2 - B
STACK VENTED AND WET VENTED BATHROOM GROUPS ON A DRAIN STACK**

12.12 FIXTURE REVENTING

12.12.1 Reserved

12.12.2 Horizontal Branches

Three lavatories or one sink within 8 feet developed length of a main-vented line may be installed on a 2" horizontal waste branch without reventing, provided the branch is not less than 2 inches in diameter throughout its length, and provided the wastes are connected into the side of the branch and the branch leads to its stack connection with a grade of not more than 1/4 inch per foot.

See Figure 12.12.2

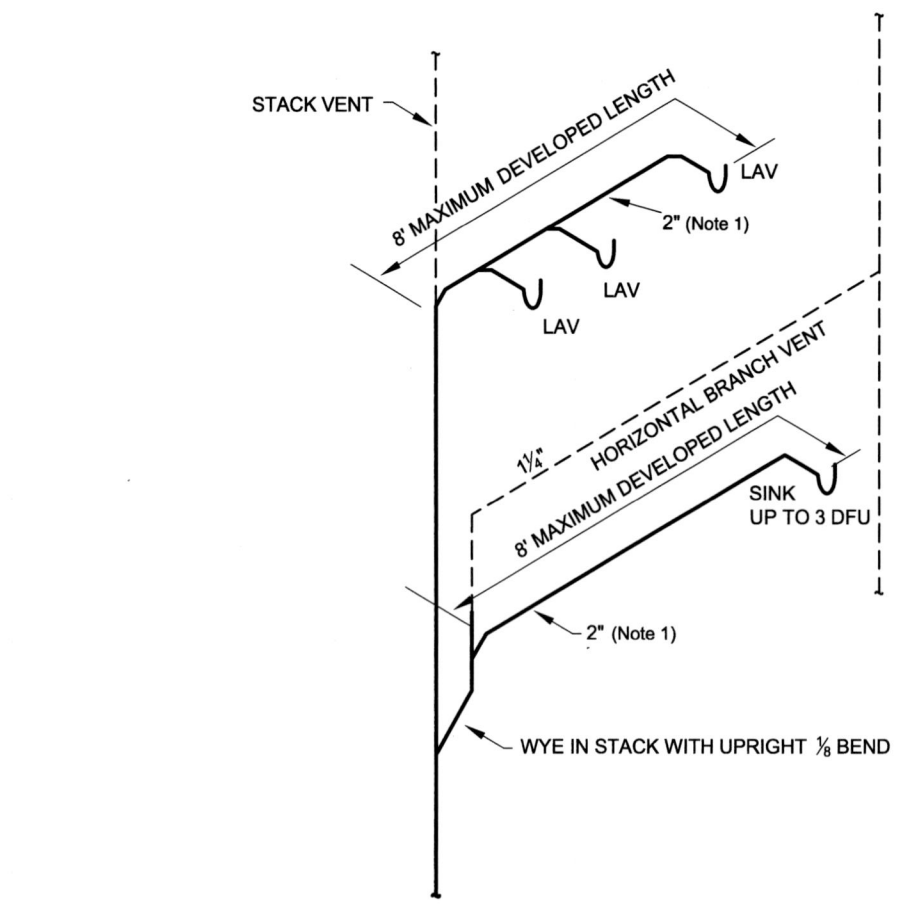

NOTES:
1. The horizontal branch must have not more than 8 feet developed length, be 2" minimum size, and be sloped not more than 1/4" per foot to avoid self-siphonage of the fixture trap.

Figure 12.12.2
FIXTURES ON HORIZONTAL BRANCHES OF A STACK WITHOUT REVENTS

12.12.3 Fixtures without Revents Above Highest Bathtubs and Water Closets

a. Fixtures without revents may be connected to a sanitary drain stack above the highest water closet or bathtub connection if all the following conditions are met:
   1. The total load does not exceed 3 dfu's.
   2. The soil or waste stack is 3" or larger.
   3. The total load on the stack is in accordance with Table 11.5.1B.
   4. The waste piping of the fixture above the water closet or bathtub connection is in accordance with Sections 12.8.1 and 12.12.2.

See Figure 12.12.3

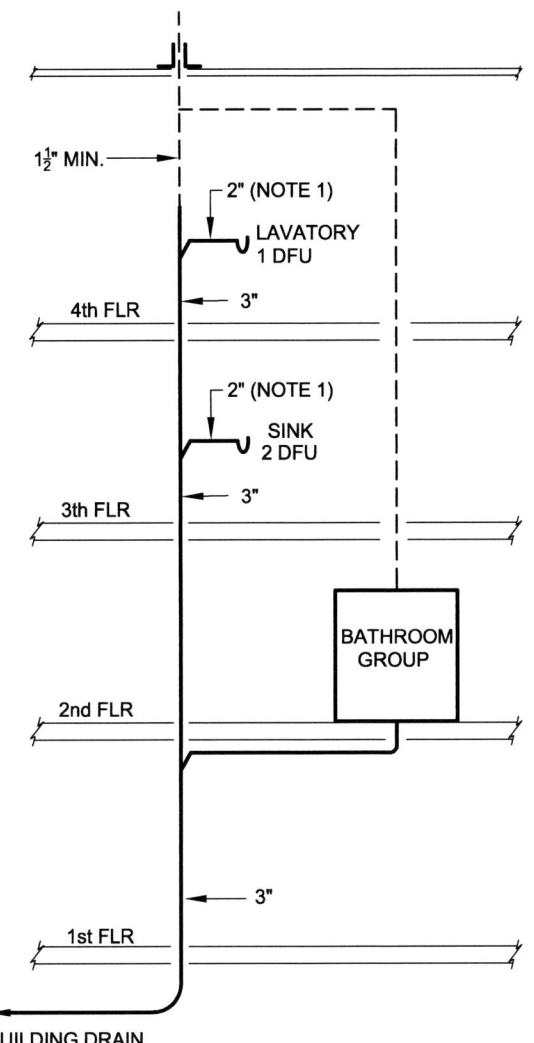

NOTES:
1. One or more fixtures totaling up to 3 DFU can be connected to a 3" minimum stack above the highest bathtub or water closet without individual vents (revents) if the horizontal waste branch(s) from the fixture(s) is 2" size with 8 feet maximum developed length and sloped no more than 1/4" per foot.
2. The stack must be larger than 3" size if required by Table 11.5.1.B.

Figure 12.12.3
UNVENTED FIXTURES ABOVE THE HIGHEST BATHTUBS AND WATER CLOSETS

12.12.4 Vent Washdown

a. Fixtures other than kitchen sinks or food waste disposers shall be permitted to wash down a vertical loop vent, circuit vent or relief vent associated with a battery-vented drain branch without reventing, provided that:

    1. Not more than 2 drainage fixture units (DFU) are drained to a 2" vent, nor more than 4 drainage fixture units (DFU) are drained to a 3" vent;

    2. The fixture trap arm lengths comply with Section 12.8.1;

    3. The fixtures drained to the vent are within the same branch as the other fixtures served by the vent; and

    4. No other fixtures are drained to the vent.

12.13 CIRCUIT AND LOOP VENTING

12.13.1 Battery Venting

a. A maximum of eight floor-outlet water closets, showers, bathtubs, or floor drains connected in battery on a horizontal drain shall be permitted to be battery vented.
EXCEPTION: Blowout type water closets.
b. Each fixture drain shall connect horizontally to the horizontal drain being so vented.
c. The horizontal battery-vented drain shall be considered as a vent extending from the most downstream fixture drain connection to the most upstream fixture drain connection.
d. Back-outlet water closets and wall-mounted urinals shall be permitted to be battery vented provided that no floor-outlet fixtures are connected to the same horizontal battery-vented drain.
EXCEPTION: Back-outlet blowout type water closets.
e. The battery vent shall be a circuit or loop vent connected to the horizontal drain between the two most upstream fixture drains and shall be installed in accordance with Section 12.6.
f. The entire length of the vented section of the horizontal battery-vented drain shall be uniformly sized for the total drainage fixture units (DFU) connected thereto.
g. The maximum slope of the horizontal battery-vented drain shall be 1 inch per foot.
h. A relief vent shall be provided on battery-vented horizontal drains that have four or more water closets connected on the lower floors of a drain stack or connect to the building drain or a branch of the building drain.
i. The relief vent shall connect to the horizontal battery-vented drain between its most downstream fixture drain connection and its connection to a drain stack, the building drain, or branch of the building drain.
j. Relief vents shall be installed in accordance with Section 12.6.
k. Circuit, loop, and relief vents shall be permitted to connect to a fixture drain, common vent, or continuous vent for fixtures located within the same branch interval as the battery-vented horizontal drain.
EXCEPTION: Fixture vents for more than four drainage fixture units (DFU) shall not connect to a battery vent.
l. Lavatories and similar fixtures shall be permitted to connect to the horizontal battery-vented drain, either horizontally or vertically, provided that,
    (1) the fixtures are on the same floor as the battery-vented drain, and
    (2) the fixtures have individual, common, or continuous vents.
m. Batteries of more than eight battery-vented fixtures shall have a circuit or loop vent for each group of eight or less fixtures.
n. Where there are two or more groups of battery-vented fixtures, the horizontal drain for each downstream group shall be sized for the total drainage fixture units (DFU) into that group, including any upstream groups and the fixtures within the group being sized.
o. Horizontal battery-vented drains that connect to drain stacks shall be sized as horizontal fixture branches in accordance with Table 11.5.1B.
p. Horizontal battery-vented drains that connect to the building drain or a branch of the building drain shall be sized as a branch of the building drain in accordance with Table 11.5.1A.

See Figure 12.13.1 and the definitions of "Battery of Fixtures", "Vent, Circuit", and "Vent, Loop".

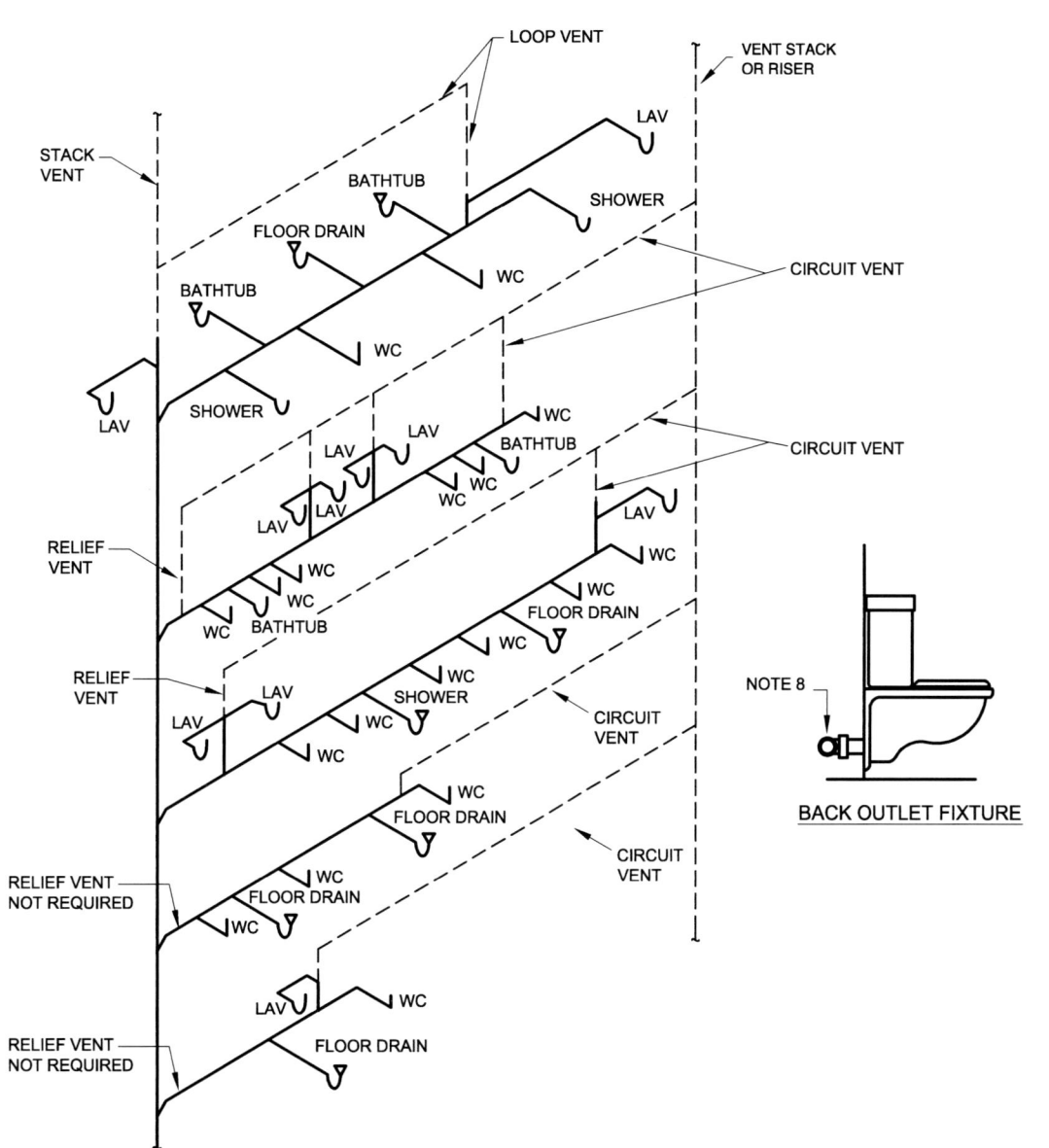

NOTES:
1. Battery vented horizontal drains can serve a mixture of two to eight water closets, floor drains, bathtubs, shower stalls, and other floor-outlet fixtures.
2. The fixtures must connect horizontally to the horizontal battery-vented drain.
3. The entire length of the horizontal battery-vented drain must be sized for the total connected DFU load. The slope of the drain must not exceed 1 inch per foot.
4. The developed length of fixture trap arms must be within the limits of Table 12.8.1 for the size of the trap arm.
5. Lavatories and sinks can connect horizontally or vertically to the horizontal battery vented drain, but they must be independently vented.
6. Relief vents are required where horizontal battery-vented drains serve more than four water closets on the lower floors of a drain stack or connect to the building drain or a branch of the building drain.
7. Fixtures are permitted to drain into circuit, loop and relief vents. A 2" vent can drain 2 DFU. A 3" vent can drain 4 DFU.
8. Back-outlet water closets and wall-mounted urinals must connect horizontally to the horizontal battery-vented drain with no vertical piping.

Figure 12.13.1
CIRCUIT AND LOOP VENTING BATTERIES OF FIXTURES

12.13.2 Joining Parallel Branches

Where parallel branches of up to eight battery-vented fixtures each are joined prior to connecting to a stack or building drain, the common downstream piping shall be sized for the combined total fixture unit (DFU) load of both branches. A relief vent shall be provided on the common downstream piping when the parallel branches serve a combined total of four or more water closets and connect to a stack receiving drainage from an upper floor.

12.13.3 Vent Connections

Circuit, loop, and relief vent connections to horizontal battery-vented drains shall be taken off at a vertical angle or from the top of the horizontal drain. *See Figure 12.16.2*

12.13.4 Fixtures Back-to-Back in Battery

When fixtures are connected to one horizontal battery-vented drain through a double wye or a sanitary tee in a vertical position, a common vent for each two fixtures back-to-back or double connected shall be provided. The common vent shall be installed in a vertical position as a continuation of the double drain connection.

12.14 VENTING OF BUILDING SUBDRAIN SYSTEMS

12.14.1 Fixture Venting

Fixtures and gravity drain piping in a building subdrain system shall be vented in the same manner as a conventional gravity drainage system and shall be permitted to connect to vent piping for fixtures and gravity drain piping that are not part of the subdrain system.

12.14.2 Subdrain Sump Pits for Atmospheric Pumps

a. The vent pipe for an atmospheric subdrain sump pit shall be sized based on the discharge capacity of its sewage pump or ejector and the drainage load (DFU) connected to the pit, whichever is larger.

1. The required vent pipe size and length based on the sewage pump or ejector capacity (GPM) shall be determined from Table 12.14.2.

2. The required vent pipe size and length based on the connected drainage load (DFU) shall be determined from Table 12.16.4.

b. The atmospheric vent pipe from a sump pit shall be permitted to be connected to gravity vent piping for fixtures other than those served by the sump pit. The sump pit vent shall carry the DFU load on the pit if it is connected to other vent piping and combined with their DFU loads.

Table 12.14.2
SIZE AND LENGTH[1] OF SUMP VENTS

Discharge Capacity of Sump Pump (gpm)	Diameter of Vent (inches)					
	1-1/4	1-1/2	2	2-1/2	3	4
10	NL[2]	NL[2]	NL[2]	NL[2]	NL[2]	NL[2]
20	270	NL[2]	NL[2]	NL[2]	NL[2]	NL[2]
40	72	160	NL[2]	NL[2]	NL[2]	NL[2]
60	31	75	270	NL[2]	NL[2]	NL[2]
80	16	41	150	380	NL[2]	NL[2]
100	<10[3]	25	97	250	NL[2]	NL[2]
150	NP[4]	<10[3]	44	110	370	NL[2]
200	NP[4]	NP[4]	20	60	210	NL[2]
250	NP[4]	NP[4]	10	36	132	NL[2]
300	NP[4]	NP[4]	<10[3]	22	88	380
400	NP[4]	NP[4]	NP[4]	<10[3]	44	210
500	NP[4]	NP[4]	NP[4]	NP[4]	24	130

NOTES FOR TABLE 12.14.2:
1. The lengths in the table are the developed lengths of the vent pipes in feet. An allowance has been made for entrance losses and friction for the pipe and fittings.
2. No Limit; actual values greater than 500 feet.
3. Less than 10 feet.
4. Not Permitted.

12.14.3 Pneumatic Sewage Ejectors

a. When a pneumatic sewage ejector requires a surge tank in Section 11.7.4, the tank shall be vented the same as an atmospheric sump pit. The vent shall be sized based on its connected DFU load.

b. Pressure release vents for pneumatic sewage ejectors shall extend to an outdoor vent terminal that is separate from any gravity system vents. Such pressure release vents shall be of sufficient size to reduce the ejector tank to atmospheric pressure within 10 seconds after discharge, but not less than 1-1/4" pipe size.

12.15 SUDS PRESSURE VENTING

12.15.1 Relief Venting

Where fixture or branch drains connect to a sanitary drain stack within a suds pressure zone as described in Section 11.11.2, a suds relief vent shall be provided for the fixture or branch drain. Suds relief vents shall be 2" minimum size but not less than one pipe size smaller than the drain branch that they serve. Such relief vents shall connect to the drain branch between the suds pressure zone and the first fixture trap on the branch.

Table 12.15.1 SUDS PRESSURE RELIEF VENTS	
Drain Size (inches)	Relief Vent Size (inches)
1-1/2	2
2	2
3	2
4	3
5	4
6	5
8	6

Comment: Suds pressure relief vents in Section 12.15 are larger than ordinary vents for drainage systems because the suds are heavier than air. Suds can weight from 2 to as much as 19 pounds per cubic foot.

12.15.2 Prohibited Vent Connections

Connections shall not be made within the suds pressure zone of a vent stack that connects at or downstream from the base of a sanitary drain stack, as described in Section 11.11.2.d.

12.16 SIZE AND LENGTH OF VENTS

12.16.1 Size of Fixture Vents

Vents for individual fixtures shall be not less than 1-1/4" size nor less than one-half the size of the fixture drain that they serve.
See Figure 12.16.1

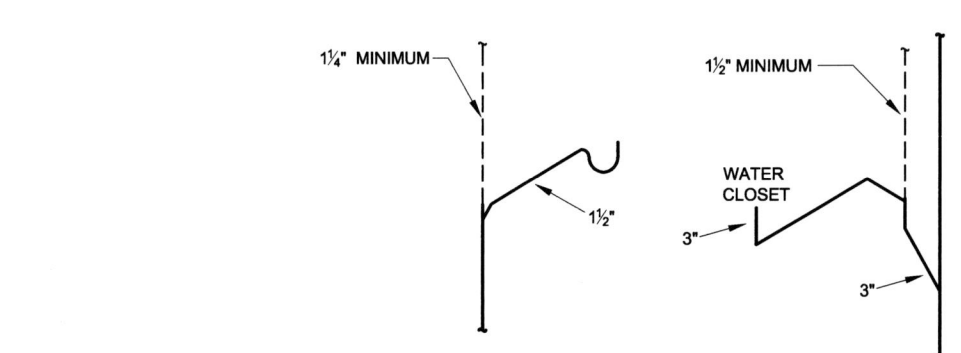

NOTES:
1. Vents for individual fixtures must be at least 1/2 of the size of the drain that they serve, but not less that 1-1/4"

**Figure 12.16.1
THE SIZE OF INDIVIDUAL FIXTURE VENTS**

12.16.2 Size of Circuit or Loop Vents

Circuit or loop vents shall be not less than one-half the size of the horizontal drain that they serve. *See Figure 12.16.2*

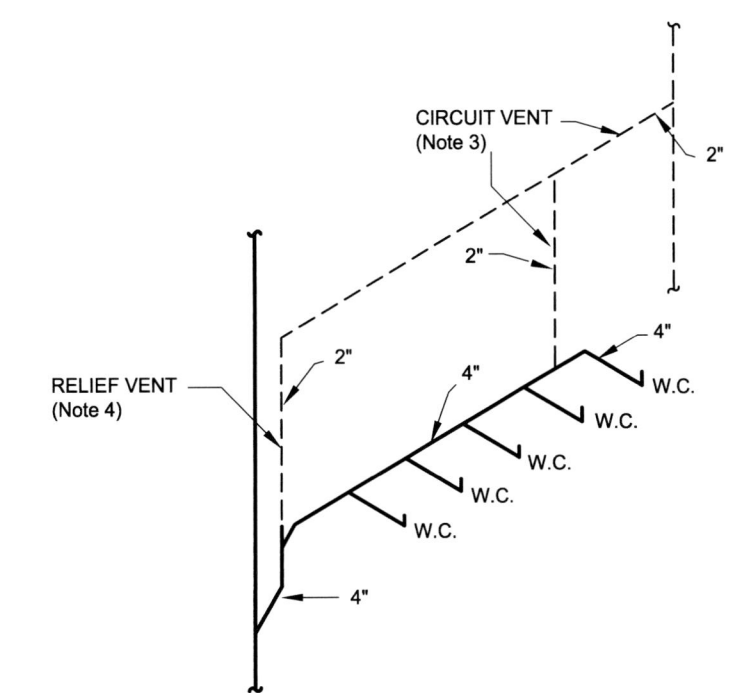

NOTES:
1. A lower floor is shown. It requires a relief vent because there are more than 4 water closets on the branch.
2. The horizontal battery-vented drain is 4" size.
3. The circuit vent must be 2" minimum size per Section 12.16.2.
4. The relief vent must be 2" minimum size per Section 12.16.3.

**Figure 12.16.2
THE SIZE OF CIRCUIT AND LOOP VENTS**

12.16.3 Size of Relief Vents

a. The size of relief vents for circuit or loop vented branches of the drainage system shall be not less than one-half the size of the horizontal battery-vented drain being served.

b. Relief vents for drain stacks having ten or more branch intervals and relief vents for horizontal offsets in such drain stacks shall be the same size as the vent stack to which they connect.

12.16.4 Size of Stack Vents and Vent Stacks

a. Stack vents shall be not less than the pipe size of the top of the drain stack that they are connected to.

b. Vent stacks, where required, shall be sized in accordance with Table 12.16.4, based on the total drainage fixture unit load (DFU) at the base of the drain stack that they serve. Connections of branch vents to the vent stack from fixtures on floor levels above the base of the drain stack shall not affect the size of the vent stack.

> *COMMENT: Where a vent stack is required for a drain stack, the drain stack and its stack vent must be the full size of the drain stack at its base.*

Table 12.16.4
SIZE AND LENGTH OF VENT STACKS

Size of Drain Stack (inches)	Drainage Fixture Units (DFU) Connected	Maximum Developed Length of Pipe (feet) for Vent Stack Size									
		1-1/4"	1-1/2"	2"	3"	4"	5"	6"	8"	10"	12"	15"
2"	6	129	314	1324								
2"	12	90	217	890								
2"	18	75	188	739								
2"	24	68	168	667								
3"	10		61	252	1830							
3"	25		35	141	1021							
3"	50		25	99	716							
3"	72		22	86	625							
4"	50			71	533	2101						
4"	200			30	220	931						
4"	350			23	169	704						
4"	500			20	149	616						
5"	350				81	330	993					
5"	600				61	249	747					
5"	850				52	215	642					
5"	1,100				48	197	595					
6"	700				29	119	361	881				
6"	1,100				23	95	286	701				
6"	1,500				20	83	250	618				
6"	1,900				19	77	231	572				
8"	1,500					21	76	187	775			
8"	2,200					18	63	155	641			
8"	2,900					16	56	138	571			
8"	3,600					15	52	129	534			
10"	2,600						23	56	232	701		
10"	3,600						20	48	199	600		
10"	4,600						18	43	180	543		
10"	5,600						17	41	169	509		
12"	5,400							19	77	235	571	
12"	6,400							17	72	219	535	
12"	7,400							16	68	208	507	
12"	8,400							16	66	200	487	
15"	7,000								26	79	193	583
15"	9,000								23	71	173	528
15"	11,000								22	65	160	494
15"	13,000								20	62	152	455

NOTES FOR TABLE 12.16.4:
1. The maximum developed lengths of pipe in the table are based on a 1" wg pressure drop for a single vent stack from its connection at the base of its drain stack to its connection to its stack vent or to its outdoor vent terminal.
2. The maximum developed length of pipe is the total length of the pipe alone. The pipe lengths in the table include an allowance for pipe fittings. The total equivalent length of pipe used in the calculations for 1" wg pressure drop is 2/3 pipe and 1/3 fittings.

c. For the purpose of sizing vent stacks, the length of a vent stack in Table 12.16.4 shall be its developed length from its pressure relief connection near the base of the drain stack being vented to its connection to the stack vent at the top of the drain stack. If not connected to the stack vent, the vent stack shall extend to an outdoor vent terminal.

12.16.5 Size of Vent Headers

a. Vent headers and portions thereof shall be sized according to Table 12.16.6.

b. The number of drainage fixture units (DFU) used to size sections of vent headers shall be the sum of all fixture units (DFU) for all drain stacks served by that section of the header.

c. The total developed length of all sections of a vent header shall be the longest vent length from the vent stack connection at the base of the most distant drain stack to the outdoor vent terminal.

d. Where two or more vent headers are connected before reaching an outdoor vent terminal, each vent header shall have its own total developed length to the outdoor vent terminal with regard to sizing per Table 12.16.6.

Table 12.16.6
SIZE OF BRANCH VENTS (1) AND VENT HEADERS (2)

DFU Being Vented	Vent Pipe Size (minimum) For 40 Feet Maximum Total Developed Length	Vent Pipe Size (minimum) For Greater than 40 Feet Total Developed Length
1	1-1/4"	1-1/2"
3	1-1/4"	1-1/2"
6	1-1/4"	1-1/2"
20	1-1/2"	2"
160	2"	3"
360	3"	4"
620	3"	4"
1400	4"	5"
2500	5"	6"
3900	6"	8"
7000	8"	10"

NOTES FOR TABLE 12.16.6:
1. The total developed length of branch vents shall comply with Section 12.16.6.c.
2. The total developed length of vent headers shall comply with Section 12.16.5.c or d.

12.16.6 Size of Branch Vents

a. Branch vents and portions thereof shall be sized according to Table 12.16.6.

b. The number of drainage fixture units (DFU) used to size sections of branch vents shall be the sum of all drainage fixture units (DFU) for the individual fixture vents or branch vents served by that section of the branch vent.

c. The total developed length of all sections of a branch vent shall be the longest length from its connection to the most distant individual fixture vent being vented to its connection to an outdoor vent terminal.

12.16.7 Underground Vent Piping

The minimum size of vent piping installed underground shall be 1-1/2".

12.16.8 Aggregate Size of Vent Terminals

a. Each building sewer shall be vented by one or more vents extending from the drainage system, or branches thereof, to the outside air.

b. The aggregate cross-sectional area of all vent terminals serving a sewer shall be not less than the cross-sectional area of the minimum required size of the building drain that they serve, at the point where it connects to the building sewer. (See Table 12.16.8 for the cross-sectional areas of pipes).

EXCEPTION: The aggregate cross-sectional area requirement shall be exclusive of any requirements to increase the size of a vent terminal to prevent frost closure under Section 12.5.

Table 12.16.8
NOMINAL PIPE CROSS SECTIONAL AREA (Sq. Inches)

Nominal Pipe Size (ID)	Cross Sectional Area (sq in.)
1-1/4"	1.2
1-1/2"	1.8
2"	3.1
2-1/2"	4.9
3"	7.1
4"	12.6
5"	19.6
6"	28.3
8"	50.3
10"	78.5
12"	113.1
15"	176.7

c. One or more vent terminals having the aggregate cross-sectional area of a 3" vent terminal shall be permitted to vent a 4" building drain if the drainage fixture unit load (DFU) and number of bathroom groups served by the building drain does not exceed the maximum number allowed on a 3" building drain, as permitted by Section 11.5.6.c.

12.17 COMBINATION WASTE AND VENT SYSTEM

12.17.1 Where Permitted

a. A combination waste and vent system shall be permitted only where conditions preclude the installation of a conventionally vented drainage system as otherwise required by this Code.

b. Combination waste and vent systems shall be limited to floor drains and other floor receptors, sinks, lavatories, and standpipes.

12.17.2 Trap Size

Traps in a combination waste and vent system shall be the normal size for the particular fixture. See Table 5.2.

12.17.3 Trap Arms

a. Fixtures shall be considered to be vented at the point that they connect to a combination waste and vent system.

b. Where fixtures have conventionally sized trap arms, the maximum length of the trap arm from the weir of its trap to the point of connection to the combination waste and vent systems shall be as limited in Table 12.8.1.

c. In the case of fixtures with above-the-floor outlets, the vertical drop at the end of the trap arm shall be one size larger than the trap arm and be considered as the beginning of the combination waste and vent system.

d. Floor-outlet fixtures shall also be permitted to drop into a vertical combination waste and vent that is at least one size larger than the trap arm.

e. Where a fixture trap arm is sized as a combination waste and vent, its length shall not be limited and it shall be considered as a branch of the combination waste and vent system.

f. The maximum vertical drop from a fixture trap arm to a horizontal drain below shall be 6 feet.

12.17.4 Pipe Sizing

The piping in a combination waste and vent system shall be sized according to Table 12.17.4, based on the number of drainage fixture units (DFU) served and the slope of the piping.
EXCEPTION: No pipe shall be smaller than any section of piping upstream, including vertical drops from trap arms.

Table 12.17.4
PIPE SIZING FOR COMBINATION WASTE AND VENT SYSTEMS

Load	slope 1/8" per ft	slope 1/4" per ft	slope 3/8" per ft	slope 1/2" per ft
3 dfu	4"	2"	2"	2"
12 dfu	4"	4"	3"	3"
20 dfu	5"	4"	4"	4"
180 dfu	5"	5"	4"	4"
218 dfu	6"	5"	5"	5"
390 dfu	8"	8"	5"	5"
480 dfu	8"	8"	6"	6"
700 dfu	8"	8"	6"	6"
840 dfu	10"	8"	8"	8"
1600 dfu	10"	10"	8"	8"
1920 dfu	12"	10"	10"	10"

12.17.5 Maximum Slope

All piping in a combination waste and vent system shall be horizontal and sloped at not greater than 1/2 inch per foot.

EXCEPTIONS:

(1) Vertical drops at the end of trap arms.

(2) Vertical drops of not greater than 45 degrees from horizontal where the vertical drop is not greater than 6 feet and the offset is at least 10 pipe diameters from any turn or branch connection.

12.17.6 Branch Connections

a. Connections to mains and branches within combination waste and vent systems shall be made horizontally at a slope not greater than 1/2 inch per foot.

b. Branch connections shall not be made in vertical drops or offsets.

12.17.7 Minimum Distances

The distance between turns, offsets, and branch connections in combination waste and vent piping shall be not less than 10 pipe diameters.

12.17.8 Connections to Conventional Drainage Systems

a. Combination waste and vent systems shall extend to the point of connection to a conventionally sized and conventionally vented drainage system.

b. At the point of connection, the pipe size of the conventional system shall be at least as large as the combination system, and sized to accept the added drainage load from the combination system.

c. Such connection from the combination system to the conventional system shall be made at an angle above horizontal of not less than 22-1/2 degrees nor more than 45 degrees.

12.17.9 Connection of Individual Fixtures

Where drains from individual fixtures are designed as a combination waste and vent and are connected to a conventional drainage system, the connection from the fixture to the conventional system shall be made according to Section 12.17.8.

12.18 ISLAND SINK VENTING

12.18.1 Where Permitted

Island sink venting shall be permitted for sinks and lavatories where the vent pipe cannot rise 6 inches above the flood level rim of the fixture before turning horizontal. Kitchen sinks in dwelling units with dishwasher connections, food waste disposer connections, or both, shall be permitted to be island vented.
See Figure 12.18

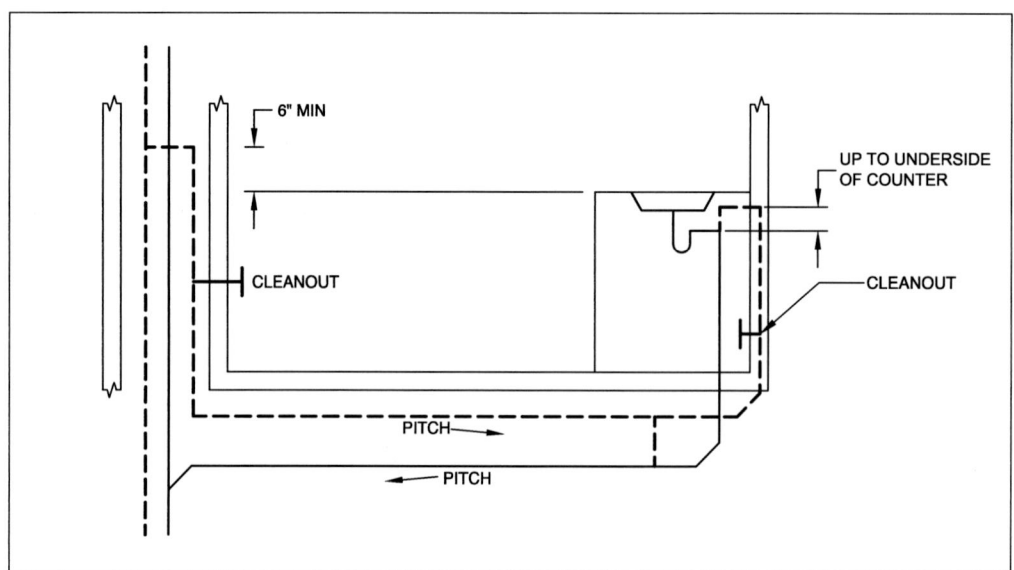

**Figure 12.18
ISLAND SINK VENTING**

12.18.2 Arrangement of Vent Piping

The island vent pipe shall rise vertically under the sink at the end of the fixture trap arm up to the underside of the counter top for the sink. The vent shall then turn downward and connect to the horizontal drain line below the floor downstream from the fixture drain connection so that the vertical vent drop will drain by gravity to the drainage system. A horizontal vent pipe shall be extended under the floor from the vertical vent drop to a point where it can rise vertically. The vertical rise at the end of the horizontal vent portion shall extend upward to at least 6 inches above the flood level rim of the fixture being vented before turning horizontal and connecting to a vent to the outdoors. The horizontal portion of the vent under the floor shall pitch back toward the sink so that it will drain by gravity through the vertical vent drop connection to the drainage system.

12.18.3 Size of Island Vent Pipes

Island vent pipes shall be sized as individual or common vents in accordance with Section 12.16.1.

12.18.4 Cleanouts Required

Cleanouts shall be provided in the vertical vent drop under the sink and in the vertical rise beyond the horizontal portion of the vent so that any blockages in the vent piping can be rodded into the drainage system.

12.19 WASTE STACK VENTING

12.19.1 Permitted Fixtures

Lavatories, bathtubs, showers, kitchen sinks with and without food waste disposers and dishwashers, laundry sinks, clothes washer standpipes, drinking fountains, floor drains, and similar fixtures shall be permitted to be vented by a waste stack that is sized and installed in accordance with the requirements of this Section.

12.19.2 Prohibited Fixtures

Water closets and urinals shall not be vented by waste stacks.

12.19.3 Waste Stacks

a. Waste stacks shall be uniformly sized from top to bottom according to the total connected fixture drainage load in accordance with Table 12.9.5.

b. Waste stacks shall be vertical for their entire height without offsets of any degree, except for the base of the stack below the lowest fixture connection.

12.19.4 Connections to the Stack

a. Each fixture shall individually connect to the stack through a single or double sanitary tee.

b. The maximum length of the trap arms from the individual fixtures shall be in accordance with Table 12.8.1.

12.19.5 Waste Stack Sizes

Waste stack sizes shall be in accordance with Table 12.19.5.

Table 12.19.5 WASTE STACK SIZES FOR WASTE STACK VENTING		
Stack Size (inches)	Total DFU into one Branch Interval	Total DFU into the Stack
1-1/2	1	2
2	2	4
3	6	24
4	6	50
5	6	75
6	6	100

12.19.6 Waste Stack Vents

a. Each waste stack shall be vented by a stack vent that is the same size as the waste stack.

b. Offsets shall be permitted in stack vents for waste stacks above the connection of the top-most fixture on the stack.

See Figure 12.19.6

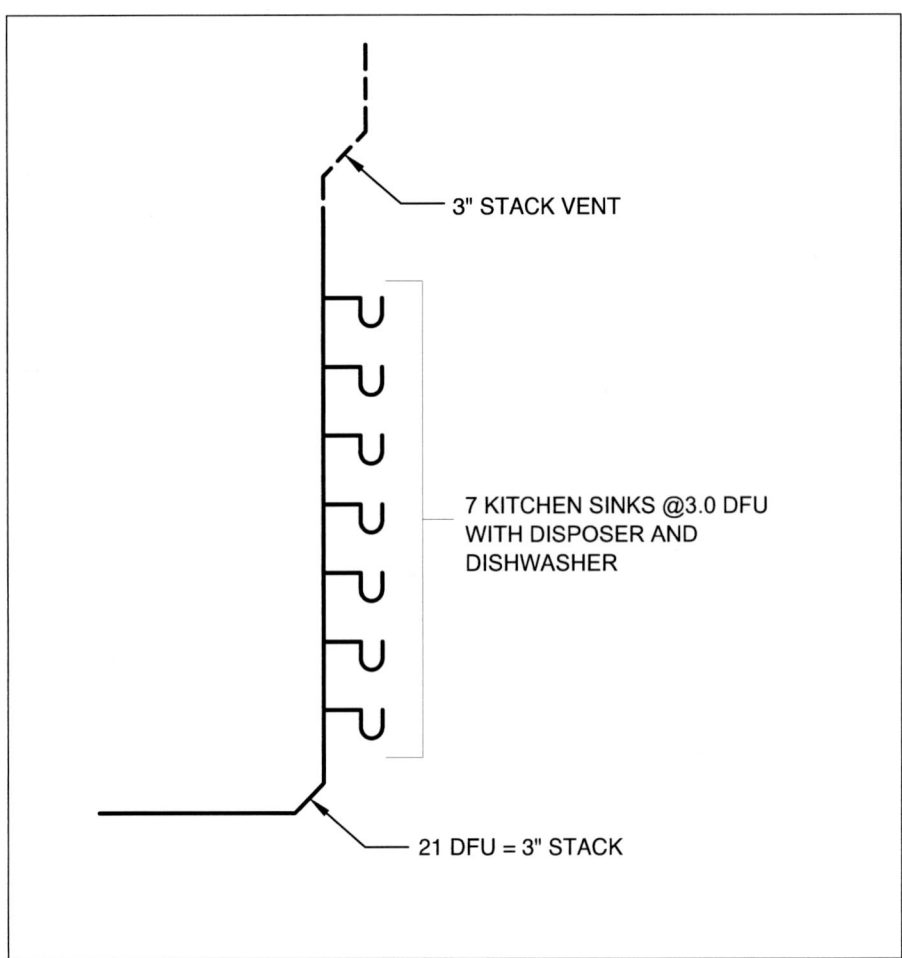

**Figure 12.19.6
WASTE STACK VENTING**

12.20 OTHER DESIGNS

Venting systems not described in this Code may be permitted by the Authority Having Jurisdiction if they provide the protection required by Section 12.2.1 and are individually designed by a licensed professional engineer. (See Appendix E - SPECIAL DESIGN PLUMBING SYSTEMS.)

Chapter 13

Storm Water Drainage

13.1 GENERAL

13.1.1 Where Required

Roofs, paved areas, yards, courts, and courtyards shall be drained to either a storm sewer where available, a combined sewer where necessary, or to a place of disposal satisfactory to the Authority Having Jurisdiction. EXCEPTION: Storm water from one- and two-family dwellings may be discharged on lawns or streets provided that the storm water flows away from the dwelling and does not otherwise create a nuisance.

13.1.2 Storm Water Drainage to Sanitary Sewer Prohibited

Storm water shall not be drained into sewers intended for sewage only, except as approved by the Authority Having Jurisdiction.

> *Comment #1: The peak storm water flows in combined sanitary and storm sewer systems can overload the sewage treatment facility, causing it to bypass untreated sewage into its point of discharge.*
>
> *Comment #2: In many cities with combined sewers, it is impractical to install separate sewers in the downtown areas because of the number of existing utilities under the streets.*
>
> *Comment #3: Some jurisdictions with combined sewers require that new or renovated buildings have separate sanitary and storm building drains so that they could be connected to separate sewers in the future.*

13.1.3 Sanitary and Storm Sewers

Where separate systems of sanitary drainage and storm water are installed in the same property, the storm and sanitary building sewers and drains may be laid side by side in the same trench.

13.1.4 Reserved

13.1.5 Foundation Drains

a. Foundation drains shall be provided around the perimeter of basements, cellars, crawl spaces, or any building space below grade. The drains shall be positioned either inside or outside of the footings, and shall be of perforated or open-joint approved drain tile or pipe not less than 3" pipe size. The top of foundation drains shall be not less than 2 inches below the underside of the floor slab being protected.

  b. Weep holes

    1. Where foundation drains are located on the interior side of hollow-core concrete masonry units, 1/2"- 3/4" diameter weep holes shall be located through the inside face of the foundation wall at the footing on 16 inch centers.

    2. Where foundation drains are located on the interior side of a poured concrete foundation wall, 1-1/2" pipes shall be installed through the footing on six foot centers.

c. Foundation drains shall be laid in a filter bed of gravel, crushed stone, slag, approved 3/4" crushed recycled glass aggregate, or other approved porous materials. The bottom of the filter bed shall be no higher than the bottom of the base course beneath the floor slab. There shall be not less than 2 inches of filter bed beneath the foundation drain. Where foundation drains are located outside of the footings, there shall be at least 6 inches of filter bed above the top of the pipe.

d. Drainage from foundations shall be discharged to a storm drain, street, alley, approved water course, or at grade. When discharged at grade, the point of discharge shall be at least 10 feet from any property line, where possible, and shall not create a nuisance.

e. Where foundation drains are below the required point of discharge, one or more automatic sump pumps shall be provided. The pump or pumps shall have adequate capacity to convey all drainage to its point of discharge. The minimum pump capacity shall be 15 gallons per minute at the required discharge head. Sump pits shall be sized to accommodate the pump(s), as recommended by the pump manufacturer, but shall be not less than 15 inches in diameter nor less than 18 inches deep. Sump pits shall be provided with fitted covers. Pits shall be located to avoid foot traffic where their covers do not have sufficient strength to carry such weight. Discharge lines from sump pumps shall be sized according to the design pump capacity and shall be not less than 1-1/4" pipe size. A check valve shall either be incorporated into each sump pump or be installed in the discharge line from each sump pump, except that check valves may be eliminated where the discharge pipe would be subject to freezing. Under such conditions, the sump pit shall be adequately sized to prevent short cycling of the pump.

f. Where sump pumps discharge at grade on unpaved surfaces, the discharge pipe shall extend to a splash block or equivalent, which shall be designed to contain the discharge, reduce its velocity, and avoid disturbing adjacent areas. Where necessary, the discharge pipe shall terminate with an elbow to direct the flow along the splash block. Splash blocks shall be at least 24 inches long.

g. Water-operated sump pumps shall comply with Section 13.1.13.

See Figure 13.1.5-A and 13.1.5-B

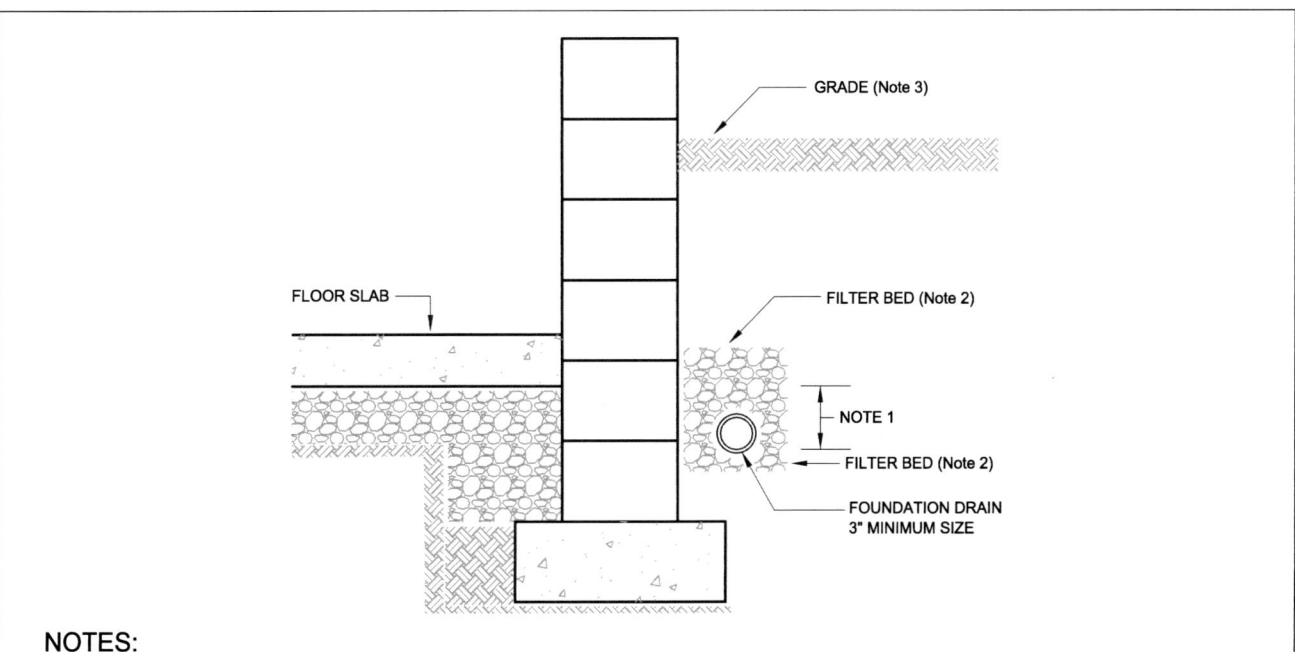

NOTES:
1. The invert of the foundation drain must be at least 2 inches below the underside of the floor slab.
2. The filter bed must be at least 2" beneath the bottom of the drain pipe and at least 6" above the top of the pipe. Installing a filter fabric over the filter bed will keep fine soil particles from clogging the porous openings in the bed.
3. Foundation drains require the same earth cover as drainage piping.

Figure 13.1.5 - A
A FOUNDATION DRAIN OUTSIDE OF A FOOTING

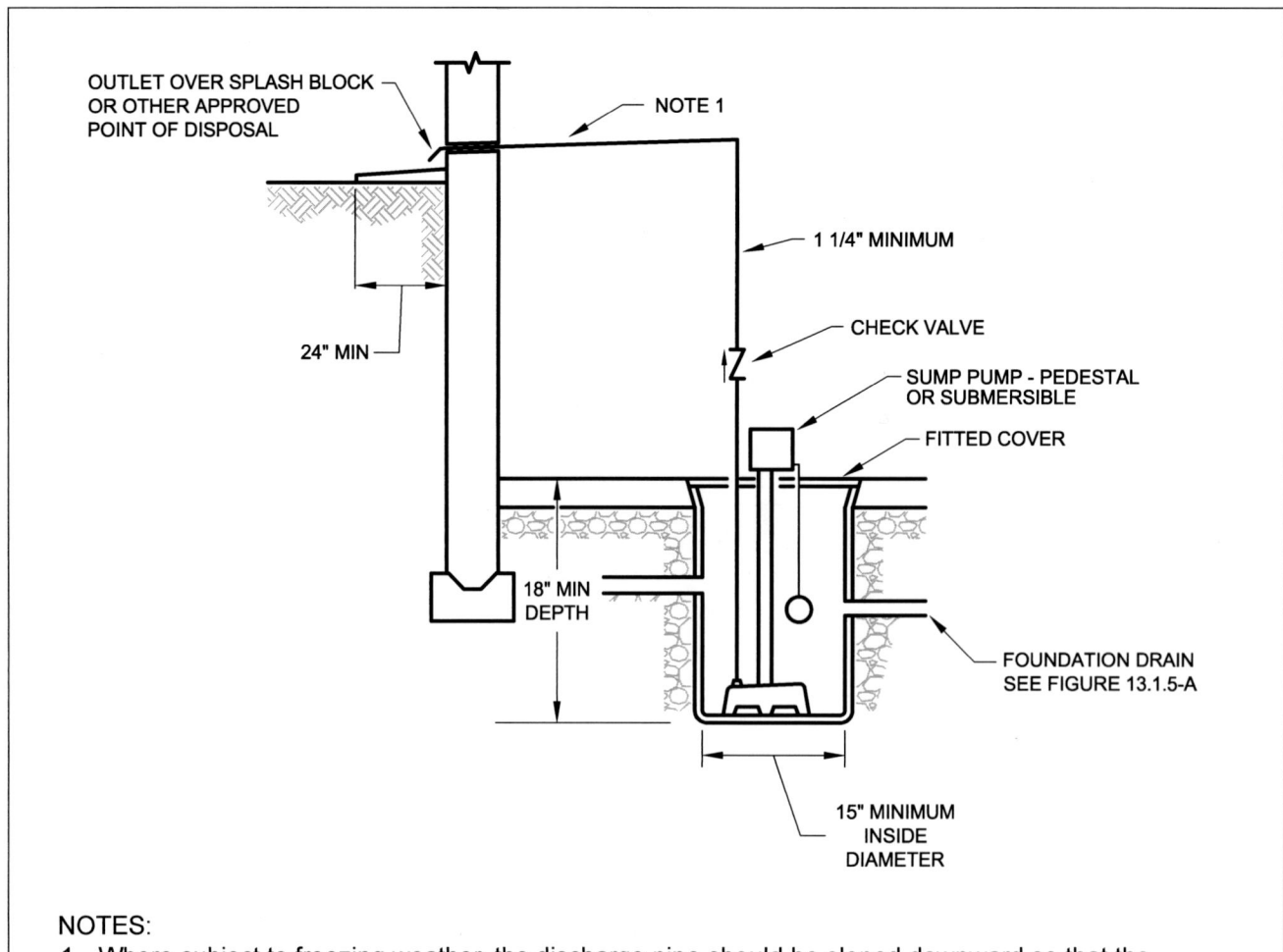

Figure 13.1.5 - B
A SUMP PUMP FOR FOUNDATION DRAINS

13.1.6 Areaway Drains

a. Drainage shall be provided for open areaways below grade where storm water may accumulate. Areaways include outdoor spaces that provide access to basements or floor levels of a building that are below grade. Drains in such areas shall be sized according to Table 13.6.2 and shall include strainers as required for roof drains or floor drains.

b. Drains for individual areaways that collect rainwater from more than 100 square feet of space shall not be connected to a foundation drain.

See Figures 13.1.6 and 1.2.6

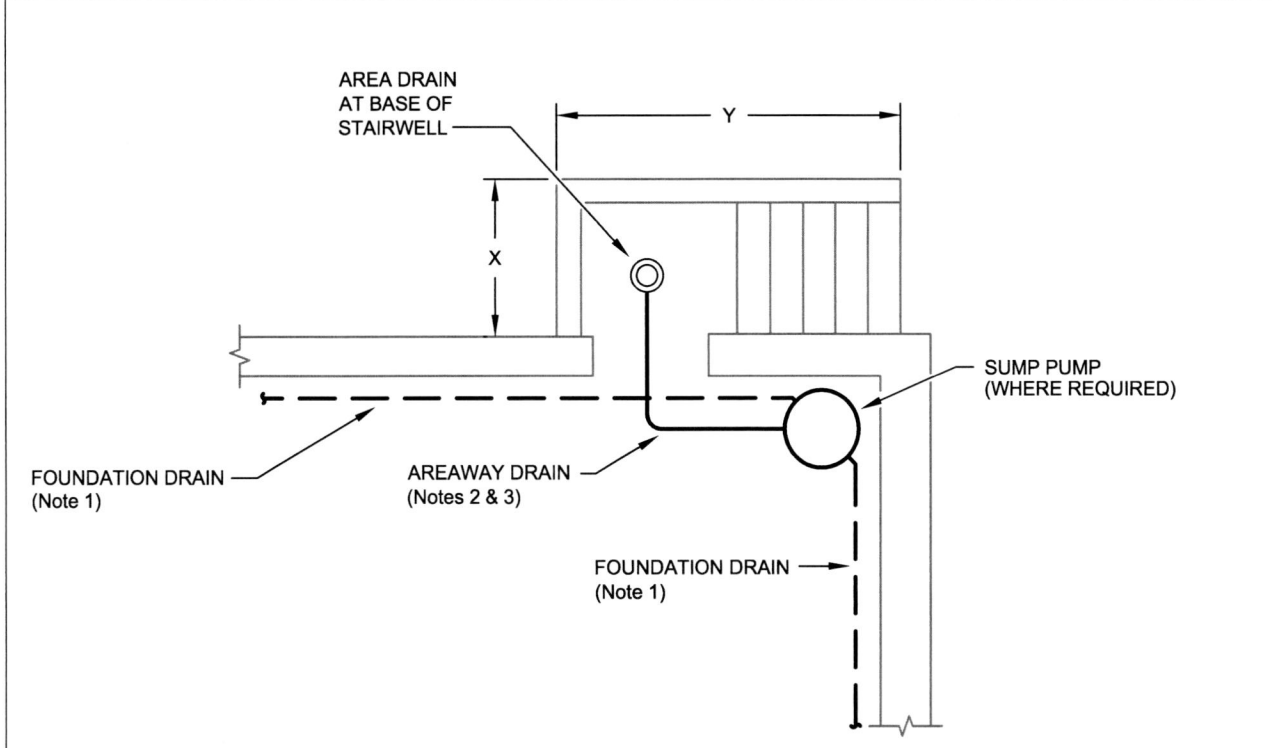

Figure 13.1.6
THE ARRANGEMENT OF AN AREAWAY DRAIN

13.1.7 Window Well Drains

Window wells shall be drained as required for areaways, except that window wells not greater than 10 square feet in area shall be permitted to drain into a foundation drain, either directly by means of a 2" minimum size drain, or indirectly through a porous filter bed.
See Figure 13.1.7-A and 13.1.7-B

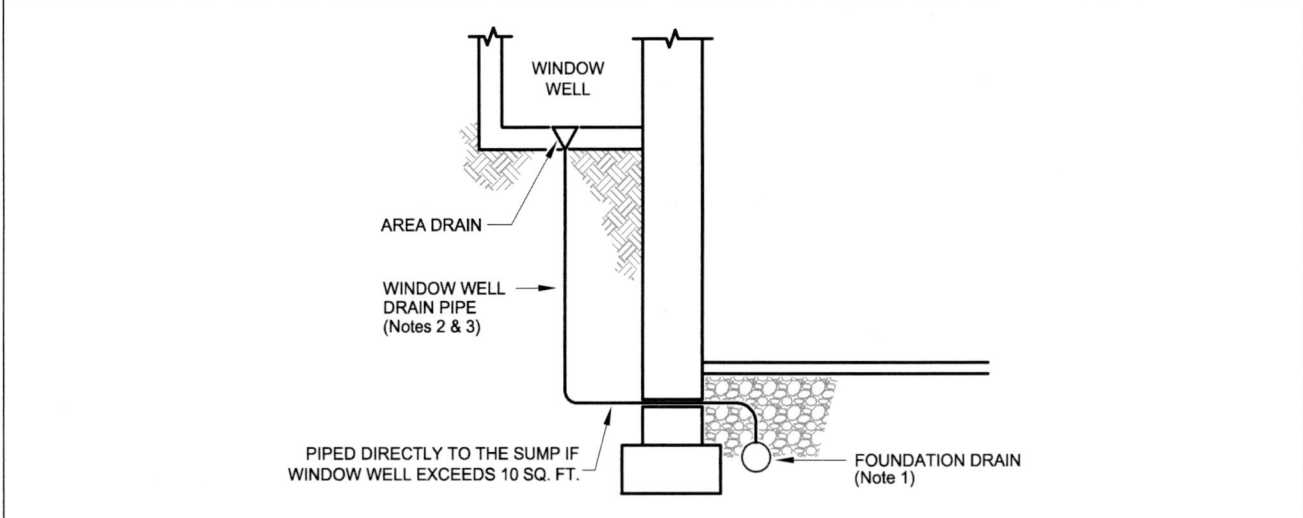

NOTES:
1. The foundation drains can be inside or outside of the footings.
2. The window well drain can connect to a foundation drain if the size of the window well (and any adjacent areas that drain into the window well) is 10 square feet or less. It must discharge separately if the area being drained is larger than 10 square feet.
3. Drain pipes for window wells must be sized according to Table 13.6.2, accounting for the rainfall rate and the size of the window well and any adjacent areas that may drain into the window well. The minimum drain pipe sizes are 2" if sloped 1/4 in./ft or more and 3" if sloped 1/8 in./ft.

Figure 13.1.7 - A
A WINDOW WELL DIRECTLY DRAINED

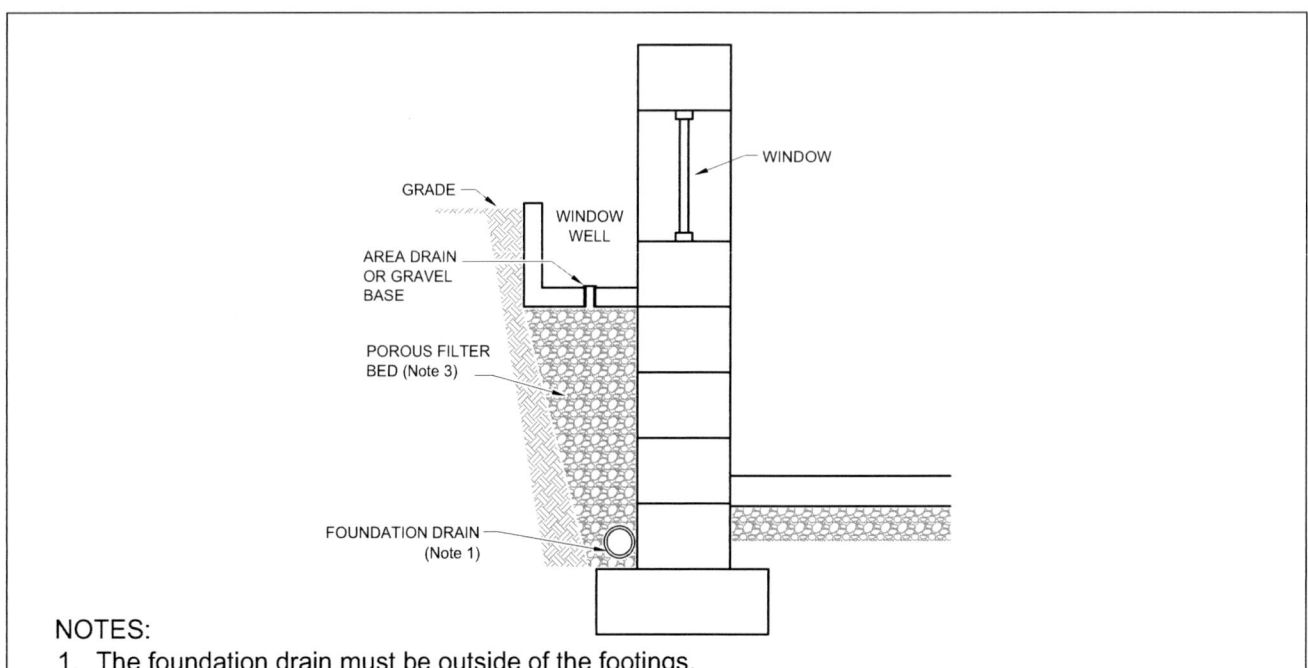

NOTES:
1. The foundation drain must be outside of the footings.
2. The window well and any surrounding drained area must not be greater than 10 square feet.
3. The porous filter bed must extend from the area drain in the window well to the foundation drain. Installing a filter fabric between the earth and the filter bed will keep the porous bed from becoming clogged with fine soil particles.

Figure 13.1.7 - B
A WINDOW WELL INDIRECTLY DRAINED

13.1.8 Parking and Service Garages

Storm water drainage from parking and service garages shall be in accordance with Sections 6.3.1.d and 6.3.1.e.

13.1.9 Trench Drains

Trench drains shall be installed in accordance with the manufacturer's instructions for excavation, anchoring, assembly, and the connection of piping.

13.1.10 Roof Drainage

13.1.10.1 Primary Roof Drainage

Roof areas of buildings shall be drained by roof drains or scuppers unless gutters and downspouts or other non-plumbing drainage is provided. The location and sizing of roof drains and scuppers shall be coordinated with the structural design and slope of the roof. Roof drains, scuppers, vertical conductors or leaders, and horizontal storm drainage piping for primary drainage shall be sized based on a storm of 60 minutes duration and a 100-year return period. (See Appendix A).

13.1.10.2 Secondary Roof Drainage

   a. Where parapet walls or other construction extend above the roof and create areas where storm water would become trapped if the primary roof drainage system failed to provide sufficient drainage, an independent secondary roof drainage system consisting of scuppers, standpipes, or roof drains shall be provided. Secondary roof drainage shall be sized for a 100-year, 15-minute storm (see Appendix A). The capacity of the primary system shall not be considered in the sizing of the secondary system.

   b. Where secondary drainage is provided by means of roof drains or standpipes, the secondary system shall be separate from the primary system and shall discharge independently at grade or other approved point of discharge.

   c. Where secondary roof drainage is provided, the overflow level(s) into the secondary system shall be established by the amount of ponding that is allowed in the structural design of the roof, including roof deflection. An allowance shall be made to account for the required overflow head of water above the secondary inlets. The elevation of the secondary inlet plus the required overflow head shall not exceed the maximum allowable water level on the roof.

   d. Scuppers shall be sized as rectangular weirs, using hydraulic principles to determine the required length and resulting overflow head (see Appendix A). Secondary roof drains and standpipes shall be sized according to Table 13.6.1 Where standpipes are used, the head allowance required under Section 13.1.10.2.c shall be not less than 1-1/2 inches.

   e. Strainers shall not be required on open standpipes when used for secondary inlets.

   f. Where secondary roof drainage is provided by roof drains or standpipes, they shall be permitted to discharge horizontally, similar to scuppers, but below the roof level.

13.1.10.3 Vertical Walls

Where vertical walls drain onto roofs, an allowance based on 50% of the maximum projected wall area shall be added to the roof area onto which each wall drains.

13.1.10.4 Equivalent Systems

When approved by the Authority Having Jurisdiction, the requirements of Sections 13.1.10.1 and 13.1.10.2 shall not preclude the installation of an engineered roof drainage system that has sufficient capacity to prevent water from ponding on the roof in excess of that allowed in the roof structural design during a 100-year, 15-minute storm.

13.1.11 Continuous Flow

Where continuous flow from a spring or ground water is encountered in a foundation drainage system or other subsoil drain, the discharge shall be piped to a storm sewer or approved water course.

13.1.12 Backwater Valves

Where foundation drains, areaway drains, window well drains, or other storm water drains discharge by gravity and are subject to backflow from their point of discharge, a backwater valve shall be provided in the discharge line. Backwater valves shall comply with the requirements of Section 5.5.

13.1.13 Water-Operated Sump Pumps

a. Water-operated sump pumps shall not be used as a primary sump pump. They shall be secondary to an electric-powered sump pump.

b. Backflow protection for the water supply to a water-operated sump pump shall comply with Section 10.5.13.d.

13.2 MATERIALS

See Section 3.7. See Table 3.7 for approved materials for storm water drainage.

13.3 TRAPS IN STORM DRAINAGE SYSTEMS

13.3.1 General

a. Traps shall be installed in a storm drainage system if it connects to a combined sewer conveying both sewage and storm water.

EXCEPTION: Traps shall not be required where roof drains, rain leaders, and other storm water inlets are at locations allowed under Section 12.4.4 for outdoor vent terminals.

b. Floor drains or other receptors within a building shall be individually trapped and vented if they are connected to a storm drainage system, regardless of whether or not the sewer is combined.

c. Traps required under this section shall comply with the requirements of Sections 5.3.1, 5.3.2, 5.3.3, 5.3.5 and 13.3.2. Traps shall have an accessible cleanout for their inlet, or other means of clearing the trap.

> *Comment: Floor drains or other receptors that are connected to a storm drainage system or combined sewer and are subject to evaporation of the trap seal must be provided with a deep seal trap or other means of maintaining the trap seal. See Section 5.3.6.*

13.3.2 Location of Traps

Where traps are required under Section 13.3.1.a, they shall be installed either on individual branches of the storm drainage system or in the building storm drain or building storm sewer before it connects to the combined sewer. Traps shall not be installed in locations where they will be subject to freezing. Where traps are required for rain leaders, the minimum earth cover shall be as required in Section 2.16.a.
See Figure 13.3.2

> *Comment #1: The "10 feet downstream" requirement allows the sanitary discharge to stabilize before mixing with the storm water flow. It also reduces the back pressure on the sanitary drainage system that may be caused by heavy storm water discharges.*
>
> *Comment #2: The horizontal wye connection between the two drainage systems permits the flows to mix more uniformly without blocking the drain.*

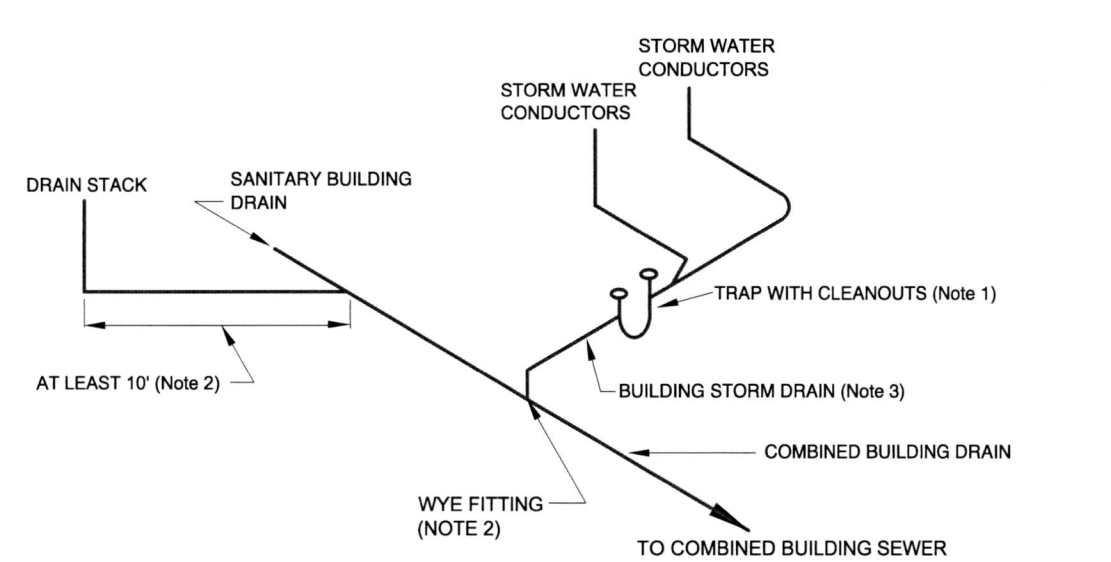

NOTES:
1. The trap is required when any of the storm water inlets are at locations where sanitary vent terminals would not be permitted in Section 12.4.4. If all storm water inlets are at locations where sanitary vent terminals would be permitted, the trap is not required. Only those inlets in prohibited locations for sanitary vent terminals need to be trapped.
2. The connection between the building storm drain and the building sanitary drain needs to be a horizontal wye fitting that is at least 10 feet downstream from any sanitary drain connection per Section 13.4.3.
3. The building storm drain piping must be increased one pipe size at least 10 feet from its connection to combined building drain piping.

Figure 13.3.2
A TRAP IN A STORM DRAINAGE SYSTEM CONNECTED TO A COMBINED SEWER

13.4 CONDUCTORS OR LEADERS AND CONNECTIONS

13.4.1 Not to be Improperly Connected

Except for combined building drains, stormwater conductors shall not be connected to sanitary drain or vent pipes nor shall sanitary drain or vent pipes be connected to stormwater conductors.

13.4.2 Protection of Rain Water Leaders

Rain water leaders installed along alleyways, driveways, or other locations where they may be exposed to damage shall be protected by metal guards, shall be recessed into the wall, or shall be constructed from ferrous alloy pipe to a point 5 foot above grade.
See Figure 13.4.2

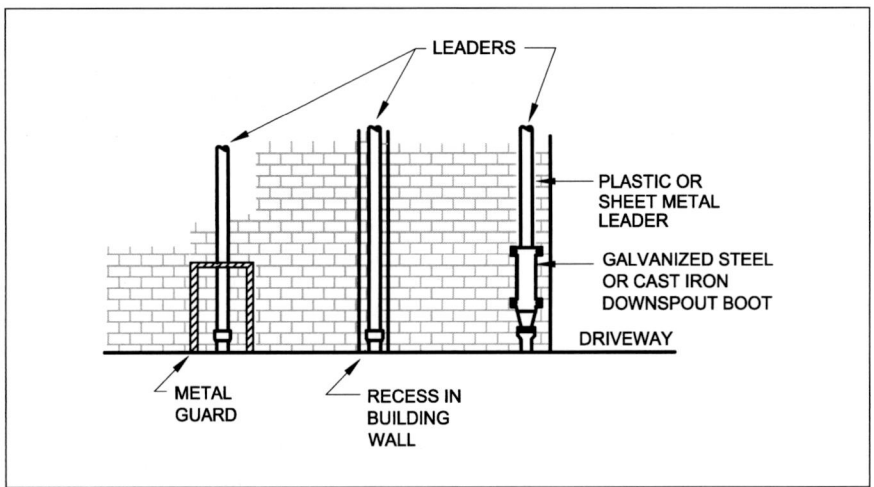

Figure 13.4.2
PROTECTION OF RAIN WATER LEADERS

13.4.3 Combining Storm with Sanitary Drainage

a. Storm and sanitary drainage shall not be combined unless connection to a combined sewer is required by the Authority Having Jurisdiction. The sanitary and storm building drains shall be entirely separate, except that where a combined building drain is required for connection to a combined building sewer, the storm drains shall connect horizontally to the combined building drain through single wye fittings that are at least 10 feet downstream from the base of any drain stack.

b. Horizontal storm drain piping from vertical storm water conductors shall be not less than 10 feet in length and shall be increased one pipe size larger than required for horizontal storm drains in Table 13.6.2 when it is 10 feet from its connection to a combined building drain.
See Figure 1.2.15

13.4.4 Double Connections of Storm Drains to Combined Building Drains

Double pattern findings shall not be used to connect storm drains to combined building drains.

> *Comment: Double pattern fittings are not permitted to connect storm drains to a combined drain. This is to prevent a heavy flow in one drain branch from causing a back pressure in the other drain branch.*

13.5 ROOF DRAINS

13.5.1 Compliance and Materials

Roof drains shall comply with ASME A112.6.4 and be constructed of coated or galvanized cast iron, bronze, stainless steel, plastic, or other corrosion-resisting materials. Drains shall include any deck clamps or other appurtenances necessary for installation and coordination with the roofing system.

13.5.2 Dome Strainers

Roof areas shall be drained to roof drains having raised dome strainers with dome free areas complying with ASME A112.6.4 for general purpose roof drains and for gutter or cornice roof drains.
EXCEPTIONS:
(1) Pitched roofs draining to hanging gutters.
(2) Roof areas subject to pedestrian and/or vehicular traffic.

13.5.3 Flat Grates

Roof drains on patios, sun decks, parking decks, and other areas subject to pedestrian and/or vehicular traffic shall have flat grates with a free inlet area complying with ASME A112.6.4 for parapet roof drains and for promenade or deck roof drains. Such drains shall not be located where they cannot be readily inspected and maintained on a regular basis.

13.5.4 Roof Drain Flashings

The connection between roofs and roof drains that pass through the roof and into the interior of the building shall be made watertight by the use of proper flashing material.

Table 13.6.1
SIZE OF VERTICAL STORM DRAINS

Nominal Diameter (inches)	Flow Capacity (GPM)	Allowable Projected Roof Areas (sq ft) at Various Rates of Rainfall per Hour (inches)					
		1"	2"	3"	4"	5"	6"
2"	23	2,180	1,090	727	545	436	363
3"	67	6,426	3,213	2,142	1,607	1,285	1,071
4"	144	13,840	6,920	4,613	3,460	2,768	2,307
5"	261	25,094	12,547	8,365	6,273	5,019	4,182
6"	424	40,805	20,402	13,602	10,201	8,161	6,801
8"	913	87,878	43,939	29,293	21,970	17,576	14,646
10"	1655	159,334	79,667	53,111	39,834	31,867	26,556
12"	2692	259,095	129,548	86,365	64,774	51,819	43,183
15"	4880	469,771	234,886	156,590	117,443	93,954	78,295
		7"	8"	9"	10"	11"	12"
2"	23	311	272	242	218	198	182
3"	67	918	803	714	643	584	536
4"	144	1,977	1,730	1,538	1,384	1,258	1,153
5"	261	3,585	3,137	2,788	2,509	2,281	2,091
6"	424	5,829	5,101	4,534	4,080	3,710	3,400
8"	913	12,554	10,985	9,764	8,788	7,989	7,323
10"	1655	22,762	19,917	17,704	15,933	14,485	13,277
12"	2692	37,014	32,387	28,788	25,910	23,554	21,591
15"	4880	67,110	58,721	52,197	46,977	42,706	39,146

NOTES FOR TABLE 13.6.1
1. Flow capacities are based on stacks flowing 7/24 full.
2. Interpolation between rainfall rates is permitted.

Table 13.6.2 Part 1
SIZE OF HORIZONTAL STORM DRAINS (for 1"/hr to 6"/hr rainfall rates)

Size of Drain (inches)	Design Flow of Drain	Allowable Projected Roof Area at Various Rates of Rainfall per Hour (Square Feet of Roof)					
		\multicolumn{6}{c}{Slope 1/16 inch/foot}					
	GPM	1"/hr	2"/hr	3"/hr	4"/hr	5"/hr	6"/hr
2							
3							
4	53	5,101	2,551	1,700	1,275	1,020	850
5	97	9,336	4,668	3,112	2,334	1,867	1,556
6	157	15,111	7,556	5,037	3,778	3,022	2,519
8	339	32,629	16,314	10,876	8,157	6,526	5,438
10	615	59,194	29,597	19,731	14,798	11,839	9,866
12	999	96,154	48,077	32,051	24,039	19,231	16,026
15	1812	174,405	87,203	58,135	43,601	34,881	29,068
\multicolumn{8}{c}{Slope 1/8 inch/foot}							
Size	GPM	1"/hr	2"/hr	3"/hr	4"/hr	5"/hr	6"/hr
2							
3	35	3,369	1,684	1,123	842	674	561
4	75	7,219	3,609	2,406	1,805	1,444	1,203
5	137	13,186	6,593	4,395	3,297	2,637	2,198
6	223	21,464	10,732	7,155	5,366	4,293	3,577
8	479	46,104	23,052	15,368	11,526	9,221	7,684
10	869	83,641	41,821	27,880	20,910	16,728	13,940
12	1413	136,002	68,001	45,334	34,000	27,200	22,667
15	2563	246,689	123,345	82,230	61,672	49,338	41,115
\multicolumn{8}{c}{Slope 1/4 inch/foot}							
Size	GPM	1"/hr	2"/hr	3"/hr	4"/hr	5"/hr	6"/hr
2	17	1,636	818	545	409	327	273
3	50	4,813	2,406	1,604	1,203	963	802
4	107	10,299	5,149	3,433	2,575	2,060	1,716
5	194	18,673	9,336	6,224	4,668	3,735	3,112
6	315	30,319	15,159	10,106	7,580	6,064	5,053
8	678	65,258	32,629	21,753	16,314	13,052	10,876
10	1229	118,292	59,146	39,431	29,573	23,658	19,715
12	1999	192,404	96,202	64,135	48,101	38,481	32,067
15	3625	348,907	174,454	116,302	87,227	69,781	58,151
\multicolumn{8}{c}{Slope 1/2 inch/foot}							
Size	GPM	1"/hr	2"/hr	3"/hr	4"/hr	5"/hr	6"/hr
2	24	2,310	1,155	770	578	462	385
3	70	6,738	3,369	2,246	1,684	1,348	1,123
4	151	14,534	7,267	4,845	3,633	2,907	2,422
5	274	26,373	13,186	8,791	6,593	5,275	4,395
6	445	42,831	21,416	14,277	10,708	8,566	7,139
8	959	92,304	46,152	30,768	23,076	18,461	15,384
10	1738	167,283	83,641	55,761	41,821	33,457	27,880
12	2827	272,099	136,050	90,700	68,025	54,420	45,350
15	5126	493,379	246,689	164,460	123,345	98,676	82,230

NOTES FOR TABLE 13.6.2 – Part 1:
1. Design flows are based on fairly rough pipe with a Manning friction coefficient of n = 0.015

Table 13.6.2 Part 2
SIZE OF HORIZONTAL STORM DRAINS (for 7"/hr to 12"/hr rainfall rates)

Size of Drain (inches)	Design Flow of Drain	Allowable Projected Roof Area at Various Rates of Rainfall per Hour (Square Feet of Roof)					
		Slope 1/16 inch/foot					
Size	GPM	7"/hr	8"/hr	9"/hr	10"/hr	11"/hr	12"/hr
2							
3							
4	53	729	638	567	510	464	425
5	97	1,334	1,167	1,037	934	849	778
6	157	2,159	1,889	1,679	1,511	1,374	1,259
8	339	4,661	4,079	3,625	3,263	2,966	2,719
10	615	8,456	7,399	6,577	5,919	5,381	4,933
12	999	13,736	12,019	10,684	9,615	8,741	8,013
15	1812	24,915	21,801	19,378	17,441	15,855	14,534
		Slope 1/8 inch/foot					
Size	GPM	7"/hr	8"/hr	9"/hr	10"/hr	11"/hr	12"/hr
2							
3	35	481	421	374	337	306	281
4	75	1,031	902	802	722	656	602
6	223	3,066	2,683	2,385	2,146	1,951	1,789
8	479	6,586	5,763	5,123	4,610	4,191	3,842
10	869	11,494	10,455	9,293	8,364	7,604	6,970
12	1413	12,429	17,000	15,111	13,600	12,364	11,334
15	2563	35,241	30,836	27,410	24,669	22,426	20,557
		Slope 1/4 inch/foot					
Size	GPM	7"/hr	8"/hr	9"/hr	10"/hr	11"/hr	12"/hr
2	17	234	205	182	164	149	136
3	50	688	602	535	481	438	401
4	107	1,471	1,287	1,144	1,030	936	858
5	194	2,668	2,334	2,075	1,867	1,698	1,556
6	315	4,331	3,790	3,369	3,032	2,756	2,527
8	678	9,323	8,157	7,251	6,526	5,933	5,438
10	1229	16,899	14,787	13,144	11,829	10,754	9,858
12	1999	27,486	24,051	21,378	19,240	17,491	16,034
15	3625	49,844	43,613	38,767	34,891	31,719	29,076
		Slope 1/2 inch/foot					
Size	GPM	7"/hr	8"/hr	9"/hr	10"/hr	11"/hr	12"/hr
2	24	330	289	257	231	210	193
3	70	963	842	749	674	613	562
4	151	2,076	1,817	1,615	1,453	1,321	1,211
5	274	3,768	3,297	2,930	2,637	2,398	2,198
6	445	6,119	5,354	4,759	4,283	3,894	3,569
8	959	13,186	11,538	10,256	9,230	8,391	7,692
10	1738	23,898	20,910	18,587	16,728	15,208	13,940
12	2827	38,871	34,012	30,233	27,210	24,736	22,675
15	5126	70,483	61,672	54,820	49,338	44,835	41,115

NOTES FOR TABLE 13.6.2 – Part 2:
1. Design flows are based on fairly rough pipe with a Manning friction coefficient of = 0.015.

13.5.5 Roof Drain Restrictions

The roof drain size shall not be restricted by insertion of any roofing material or other objects to insure water flow into the drain.

13.5.6 Roof Drain Outlet Pipe Size

a. The outlet pipe size of roof drains having vertical outlets shall be not less than the size required for vertical storm drains in Table 13.6.1.

b. The outlet pipe size for roof drains having horizontal outlets shall be not less than the pipe size required for horizontal storm drains in Table 13.6.2.

13.6 SIZE OF VERTICAL AND HORIZONTAL STORM DRAIN PIPING

13.6.1 Size of Vertical Storm Drain Piping

a. Vertical storm drain piping shall be sized according to Table 13.6.1.

b. Where a vertical pipe section is smaller than the following horizontal pipe section, the increase in size shall be made in the vertical pipe section.

c. Where a vertical pipe section is smaller than the preceding horizontal pipe section, the reduction in size shall be made in the vertical pipe section.

13.6.2 Size of Horizontal Storm Drain Piping

a. Horizontal storm drain piping shall be sized according to Table 13.6.2 – Part 1 or Part 2. Such piping includes horizontal offsets in vertical storm drains, building storm drains, building storm sewers, and branches thereof.

b. The design flows in Table 13.6.2 – Parts 1 and 2 are based on fairly rough pipe with a Manning friction coefficient "n" = 0.015.

13.7 AREA DRAINS AND TRENCH DRAINS

Area drains and trench drains shall comply with Section 7.16, including the required free area of the grates.

13.8 SIZING FOR CONTINUOUS OR INTERMITTENT FLOWS

Continuous or intermittent flows from a sump pump, air conditioning condensate drain, or other approved discharge into a storm drainage system shall be determined in gallons per minute flow. Air conditioning condensate drainage shall be based on not less than 0.006 gpm/ton of cooling capacity. Such flows shall be added to the stormwater load on the storm drainage system, which shall also be determined on the basis of gallons per minute according to Section A-3 and Table A-1 in Appendix A.

13.9 CONTROLLED FLOW STORM WATER SYSTEM

13.9.1 Application

In lieu of sizing the storm drainage system on the basis of actual maximum projected roof areas required in this Chapter, the roof drainage system, or parts thereof may be sized on equivalent or adjusted maximum projected roof areas that result from controlled flow and storage of storm water on the roof where flow control devices are used.

13.9.2 Design

A controlled flow storm water system shall be designed, installed, inspected and certified as an engineered special design plumbing system as outlined in Appendix E of this Code.

Blank Page

Chapter 14

Special Requirements For Health Care Facilities

14.1 GENERAL

This Chapter applies to special fixtures and systems that occur in health care facilities and to the special plumbing requirements in such facilities. Ordinary plumbing in such facilities shall comply with the other applicable Chapters of this Code.

14.2 WATER SERVICE

Where required by the Authority Having Jurisdiction, hospitals and similar health care facilities shall have dual water service lines to maintain a water supply in the event of a water main failure. Where possible, the service pipelines shall be connected to different water mains so that a single water main break can be isolated and repaired without shutting off all water service to the facility.

14.3 MEDICAL GAS AND VACUUM PIPING SYSTEMS

14.3.1 General

The installation of medical gas and vacuum piping systems shall be in accordance with the gas and vacuum system requirements of NFPA 99 - Health Care Facilities Code.

14.3.2 Professional Qualifications of Installers

Installers, including brazers, of medical gas and vacuum piping systems shall meet the qualification requirements of ASSE 6000 or the equivalent.

14.4 PROTRUSIONS FROM WALLS

Drinking fountains, control valves, medical gas station outlets, vacuum inlet stations, risers, cleanout covers, and other devices shall be fully-recessed in corridors and other areas where patients may be transported on a gurney, hospital bed, or wheelchair. Protective guards shall be provided where necessary.

14.5 MENTAL PATIENT ROOMS

Piping and drain traps in mental patient rooms shall be concealed. Fixtures and fittings shall be vandal-proof.

> *Comment: Plumbing that is accessible to mental patients must be suicide-proof, vandal-proof, and sound-proof. Piping is installed in chases that are not accessible to the patients. Special fixtures, such as combination lavatory/toilets, are available for such applications. The fixtures are pushbutton operated to avoid faucet handles. Most fixtures are constructed of stainless steel. The plumbing fixtures used in mental facilities are similar to those used in correctional institutions.*

14.6 PROHIBITED LOCATIONS FOR ICE STORAGE

Ice makers or ice storage chests shall not be located in a Soiled Utility Room or similar areas where subject to possible contamination.

14.7 CROSS CONNECTION CONTROL AND BACKFLOW PREVENTION

a. Backflow prevention shall be in accordance with Section 10.5.

b. Vacuum breakers for bedpan washers shall be not less than 5 feet above the floor.

14.8 CLINICAL SINKS AND BEDPAN WASHERS

14.8.1 General

a. Clinical sinks and bedpan washers, and flushing-rim service sinks shall be installed in the same manner as water closets. Where such fixtures have a vent connection on the inlet side of their trap, a local vent shall be provided in accordance with Section 14.9.

b. Clinical sinks shall not be used as a substitute for non-flushing service sinks, nor shall a non-flushing service sink be utilized to clean bedpans.

c. Vacuum breakers for bedpan washers shall be installed in accordance with Section 14.7.b.

See Figure 14.8

14.9 LOCAL VENTS AND STACKS FOR CLINICAL SINKS OR BEDPAN WASHERS

14.9.1 General

Where clinical sinks or bedpan washers have provisions for a local vent, a local vent shall be extended to the outdoors above the roof. Local vents shall terminate in accordance with Section 12.4. Local vents from clinical sinks or bedpan washers shall not be connected to vapor vents for sterilizers or to any drainage system vent.

14.9.2 Material

Local vent piping shall be of a material acceptable for sanitary vents in accordance with Section 3.6.

14.9.3 Required Size and Arrangement

A local vent serving a single clinical sink or bedpan washer shall be not less than 2" pipe size. Where such fixtures are installed back-to-back or are located above each other on more than one floor, a local vent stack may be provided to serve multiple fixtures. A 2" local vent stack may serve up to three fixtures. A 3" local vent stack may serve up to six fixtures. A 4" local vent stack may serve up to twelve fixtures. In multiple installations, the connections to the local vent stack shall be made using sanitary tee or tee-wye fittings oriented for upward flow from the branch. A branch connection to a local vent stack shall extend not more than 5 feet horizontally and shall be sloped not less than 1/4 inch per foot back towards the fixture served.

14.9.4 Provisions for Drainage

Provisions shall be made for the drainage of vapor condensation within local vent piping. A local vent serving a single fixture may drain back to the fixture served. The base of a local vent stack serving one or more fixtures shall be directly connected to a trapped and vented waste branch of the sanitary drainage system. The trap and waste branch shall be the same size as the local vent stack. The trap seal depth shall be not less than 3 inches. The vent for the waste branch shall be 1-1/4" minimum size, but not less than one-half the size of the waste branch.

14.9.5 Trap Priming

The waste trap required under Section 14.9.4 shall be primed by at least one clinical sink or bedpan washer on each floor served by the local vent stack. A priming line not less than 1/4" OD size shall be extended from the discharge or fixture-side of the vacuum breaker protecting the fixture water supply to the local vent stack. A trap having not less than a 3-inch water seal shall be provided in the priming line. The line shall prime the trap at the base of the local vent stack each time that a fixture is flushed.

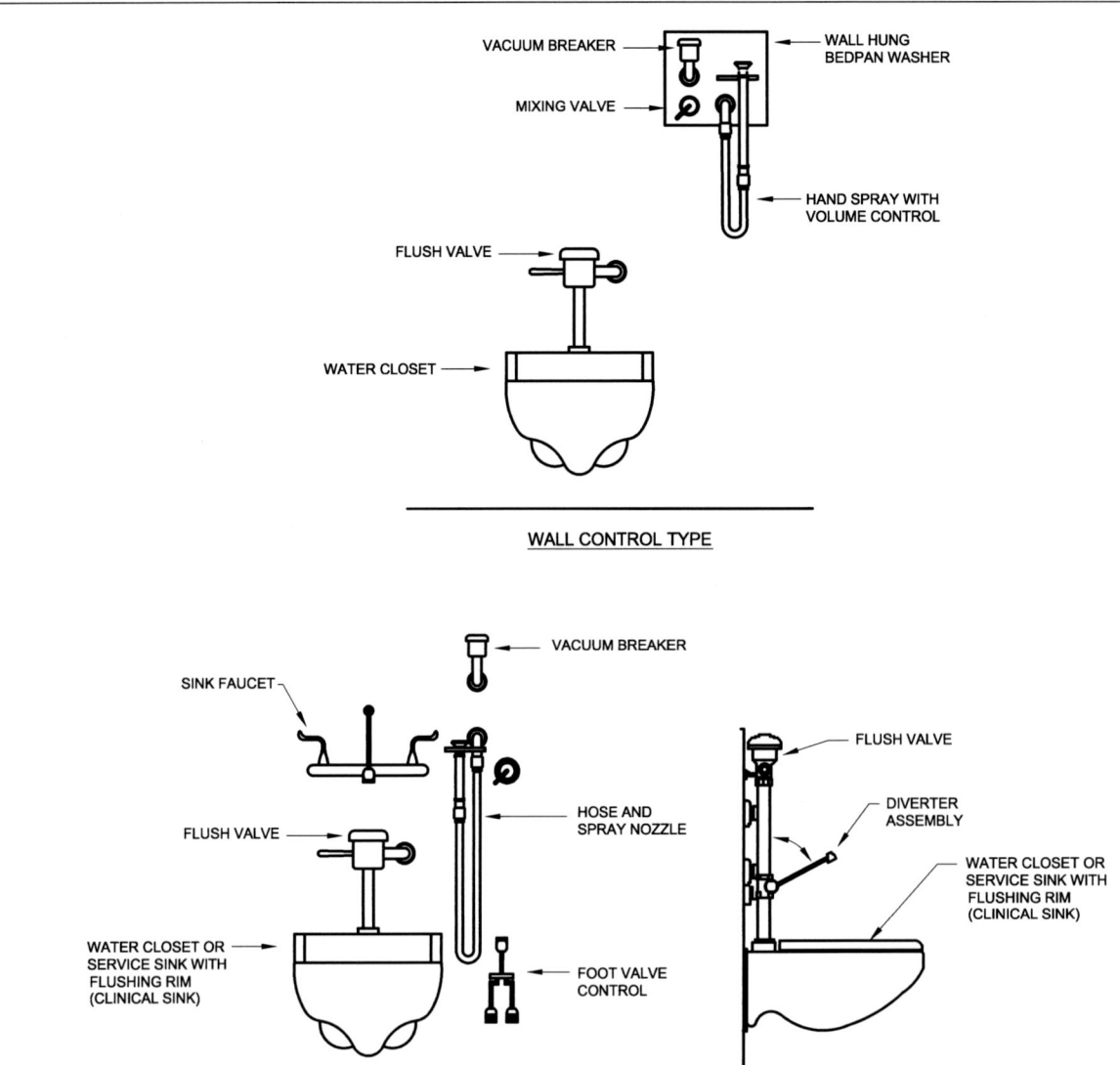

NOTES:
1. The divertor type of bedpan washer is typically installed in a patient toilet room. When the divertor is fully raised, the flush valve supplies the water closet bowl in the normal manner. When the divertor is lowered, the flush valve supplies the divertor to rinse the bedpan. After the bedpan has been rinsed into the water closet, the divertor can be raised and the water closet flushed. The elevation of the flush valve vacuum breaker above the overflow rim level of the water closet should be as recommended by the manufacturer.
2. The wall control type of bedpan washer is typically installed in a patient toilet room. The bedpan is rinsed into the water closet bowl using the hand spray from the wall control unit. The water closet is then flushed using the flush valve. Section 14.7.b requires that the vacuum breaker in the wall control unit be at least 5 feet above the floor to prevent a cross connection
3. The foot control type of bedpan washer is typically installed in a soiled utility room with a clinical sink. The bedpan is rinsed into the water closet bowl using the hand spray and foot control valve. The clinical sink is then flushed using the flush valve. Section 14.7.b requires that the vacuum breaker for the hose spray be at least 5 feet above the floor to prevent a cross connection. The sink faucet allows the clinical sink to be used as a service sink for housekeeping purposes.

Figure 14.8
TYPES OF BEDPAN WASHERS

14.10 STERILIZERS

14.10.1 General

The requirements of this Section apply to sterilizers and bedpan steamers. Such equipment shall be installed in accordance with this Code and the manufacturer's instructions.

14.10.2 Indirect Waste Connections

All waste drainage from sterilizers and bedpan steamers shall be indirectly connected to the sanitary drainage system through an air gap in accordance with Chapter 9. Indirect waste pipes shall be not less than the size of the drain connection on the fixture. Separate waste pipes shall be provided for each fixture, except that up to three sterilizers may have a common indirect waste pipe if its developed length does not exceed 8 feet. The size of such common indirect waste pipes shall be not less than the aggregate cross-sectional area of the individual sterilizer drain connections. Except for bedpan steamers, indirect waste pipes shall not require traps.

14.10.3 Floor Drains

a. A trapped and vented floor drain, not less than 3" pipe size, shall be provided in each recess room or space where recessed or concealed portions of sterilizers are located. The floor drain shall drain the entire floor area and shall receive the indirect waste from at least one sterilizer. Where an air gap fitting is provided, the waste pipe from the fitting may connect to the body of the floor drain above its trap seal.

b. Where required by the sterilizer manufacturer, a floor drain shall be located directly beneath the sterilizer within the area of its base.

14.10.4 Cooling Required

Waste drainage from condensers or steam traps shall be cooled below 140°F before being discharged indirectly to the sanitary drainage system.

14.10.5 Traps Required for Bedpan Steamers

A trap having a minimum seal of 3 inches shall be provided in the indirect waste pipe for a bedpan steamer, located between the fixture and the air gap at the indirect waste receptor.

14.11 VAPOR VENTS AND STACKS FOR STERILIZERS

14.11.1 General

Where sterilizers have provisions for a vapor vent and such a vent is required by their manufacturer, a vapor vent shall be extended to the outdoors above the roof. Sterilizer vapor vents shall terminate in accordance with Section 12.4 and shall not be connected to local vents for clinical sinks or bedpan washers or to any drainage system vent.

14.11.2 Material

Sterilizer vapor vent piping shall be of a material acceptable for sanitary vents in accordance with Section 3.6.

14.11.3 Required Size and Arrangement

a. Sterilizer vapor vents and stacks for individual sterilizers shall be not less than the size of the sterilizer vent connection, except that stacks shall be not less than 1-1/2" pipe size. Where vapor vent stacks serve more than one sterilizer, the cross-sectional area of the stack shall be not less than the aggregate cross-sectional areas of the vapor vents for all of the sterilizers served.

b. In single and multiple installations, the connections to the vapor vent stack shall be made using sanitary tee or tee-wye fittings oriented for upward flow from the branch. A branch connection to a sterilizer vapor vent stack shall extend not more than 5 feet horizontally and shall be sloped not less than 1/4 inch per foot away from the sterilizer and toward the vent stack.

14.11.4 Provisions for Drainage

Provisions shall be made for the drainage of vapor condensation within sterilizer vapor vent piping. The base of stacks shall drain indirectly through an air gap to a trapped and vented waste receptor connected to the sanitary drainage system.

14.12 DRAINAGE FROM CENTRAL VACUUM SYSTEMS

14.12.1 General

Provisions for drainage from medical, surgical, dental, and similar central vacuum systems shall be as required by NFPA 99 - Health Care Facilities Code. In addition, drainage from dental and other vacuum systems that collect fluid waste centrally shall comply with Sections 14.12.2 through 14.12.4.

14.12.2 Positive Pressure Drainage from Air/Waste Separators in Dental Vacuum Systems

a. The waste outlet from an air/waste separator on the discharge side of a vacuum pump or blower shall be direct-connected to the sanitary drainage system through a deep-seal trap that is conventionally vented within the plumbing system. The trap vent shall extend vertically to not less than 6 inches above the top of the separator before making any horizontal turns. The vacuum exhaust air flow from the separator shall be separately vented to outdoors as required under NFPA 99.

b. The trap and drain branch size shall be at least two pipe sizes larger than the waste pipe from the separator, but not less than 1-1/2" pipe size. The vent shall be the full size of the trap and drain. The trap seal shall be at least two times the exhaust back pressure in the separator, but not less than 4 inches deep.

14.12.3 Gravity Drainage from Waste Holding Tanks in Dental Vacuum Systems

a. The drainage from waste holding tanks shall extend from the vacuum check valve on the waste outlet of the tank and be direct-connected to the sanitary drainage system through a deep-seal trap that is conventionally vented within the plumbing system. In addition, a vent shall be installed between the vacuum check valve and the drain trap, on the inlet side of the trap, to seal the check valve when the holding tank is operating under vacuum and collecting waste. This vent shall be connected to the plumbing system vents. Both vents shall extend vertically to not less than 6 inches above the top of the holding tank before making any horizontal turns.

b. The trap and drain size shall be at least two pipe sizes larger than the waste outlet and vacuum check valve, but not less than 2" pipe size. The trap shall be not less than 4 inches deep. The vent for the vacuum check valve shall be not less than the size of the check valve. The trap vent shall be not less than one-half the size of the trap and drain branch.

14.12.4 Protection from Sewage Backup in Dental Vacuum Systems

A floor drain or other trapped and vented receptor shall be provided near the connection of the drain from a dental vacuum air/waste separator or waste holding tank to the sanitary drainage system that will overflow in the event of a backup in the sanitary drainage system and prevent the backup from reaching the level of the trap for the air/waste separator or the drain check valve for the waste holding tank. The trap of the floor drain or receptor shall be primed if it does not receive an indirect waste discharge

14.13 ASPIRATORS

Provisions for aspirators or other water-supplied suction devices shall be installed only with the specific approval of the Authority Having Jurisdiction. Where aspirators are used for removing body fluids, they shall include a collection bottle or similar fluid trap. Aspirators shall indirectly discharge to the sanitary drainage system through an air gap in accordance with Chapter 9. The potable water supply to an aspirator shall be protected by a vacuum breaker or equivalent, in accordance with Sections 14.7 and 10.5.3.

Blank Page

Chapter 15

Tests and Maintenance

15.1 EXPOSURE OF WORK

New, altered, extended or replaced plumbing shall be left uncovered and unconcealed until it has been inspected, tested and approved. Where such work has been covered or concealed before it is inspected, tested and approved, it shall be exposed as required.

15.2 EQUIPMENT, MATERIAL AND LABOR FOR TESTS

Equipment, material and labor required for testing a plumbing system or part thereof shall be provided by the installing contractor.

15.3 TESTING OF PLUMBING SYSTEMS

15.3.1 General

a. New plumbing systems shall be tested as prescribed hereinafter to disclose leaks and defects.

b. Where additions or changes are made to existing plumbing systems, the new work shall be tested or inspected to the degree necessary to disclose leaks and defects in the new work.

15.4 METHODS OF TESTING THE DRAINAGE AND VENT SYSTEMS

15.4.1 Rough Plumbing

a. Except for perforated or open jointed drain tile, the piping of plumbing drain and vent systems shall be tested upon completion of the rough piping installation by water or, for piping systems other than plastic, by air and proved watertight. The Authority Having Jurisdiction may require the removal of any cleanout plugs to ascertain if the pressure has reached all parts of the system. One of the following test methods shall be used:

1. The water test shall be applied to the drainage system either in its entirety or in sections after rough piping has been installed. If applied to the entire system, all openings in the piping shall be tightly closed, except the highest opening, and the system filled with water to point of overflow. If the system is tested in sections, each opening shall be tightly plugged except the highest opening of the section under test, and each section shall be filled with water, but no section shall be tested with less than a 10-foot head of water. In testing successive sections, at least the upper 10 feet of the next preceding section shall be tested, so that no joint or pipe in the building (except the uppermost 10 feet of the system) shall have been submitted to a test of less than 10-foot head of water. The water shall be kept in the system or in the portion under test for at least 15 minutes before inspection starts; the system shall then be tight at all points.

2. The air test shall be made by attaching an air compressor testing apparatus to any suitable opening and after closing all other inlets and outlets to the system, forcing air into the system until there is a uniform gauge pressure of 5 pounds per square inch or sufficient to balance a column of mercury 10 inches in height. This pressure shall be held without introduction of additional air for a period of at least 15 minutes.

15.4.2 Finished Plumbing

a. When the rough plumbing has been tested in accordance with Section 15.4.1, a final test of the finished plumbing system may be required to insure that the final fixture connections to the drainage system are gas-tight.

b. After the plumbing fixtures have been set and their traps filled with water, their connections shall be tested and proved gas and watertight. A final smoke or peppermint test shall be required, except in the case of a previous on site-inspected water or air tested system. If a smoke or peppermint test is required, the following test methods shall be employed:

    1. A smoke test shall be made by filling all traps with water and then reintroducing into the entire system a pungent, thick smoke produced by one or more smoke machines. When the smoke appears at stack openings on the roof, they shall be closed and a pressure equivalent to a one-inch water column shall be developed and maintained for the period of the inspection.

    2. Where the Authority Having Jurisdiction, due to practical difficulties or hardships, finds that a smoke test cannot be performed, a peppermint test shall be substituted in lieu thereof. Such peppermint test shall be conducted by the introduction of two ounces of oil of peppermint into the roof terminal of every line or stack to be tested. The oil of peppermint shall be followed at once by ten quarts of hot (140°F) water whereupon all roof vent terminals shall be sealed. A positive test, which reveals leakage, shall be the detection of the odor of peppermint at any trap or other point on the system. Oil of peppermint or persons whose person or clothes have come in contact with oil of peppermint shall be excluded from the test area.

> *Comment: The smoke or peppermint tests on the finished plumbing can be waived by the Authority Having Jurisdiction when they are considered to be unnecessary. However, the finished plumbing tests should be performed if the installation appears to be improper or there is the odor of sewer gas.*

15.5 METHOD OF TESTING BUILDING SEWERS

The building sewer shall be tested by insertion of a test plug at the point of connection with the public sewer, private sewer, individual sewage disposal system, or other point of disposal. It shall then be filled with water under a head of not less than 10 feet. The water level at the top of the test head of water shall not drop for at least 15 minutes. Where the final connection of the building sewer cannot reasonably be subjected to a hydrostatic test, it shall be visually inspected.

15.6 METHODS OF TESTING WATER SUPPLY SYSTEMS

    a. Upon completion of a section or the entire water supply system, it shall be tested and proved tight under a water pressure not less than the working pressure under which it is to be used or 80 pounds per square inch, whichever is greater.

    b. For metallic pipe and where the Authority Having Jurisdiction determines that providing potable water for the test represents a hardship or practical difficulty, the system may be tested with air to the pressures noted above, as allowed by the pipe manufacturer.

    c. For plastic pipe, testing by compressed gas or air pressure shall be prohibited.

    d. After testing, piping shall be flushed and disinfected as required by Section 10.9.

15.7 DEFECTIVE PLUMBING

Where there is reason to believe that the plumbing system of any building has become defective, it shall be subjected to test or inspection and any defects found shall be corrected.

15.8 MAINTENANCE

15.8.1 General

The plumbing systems shall be maintained at all times in compliance with the provisions of this Code.

15.8.2 Exception

Existing plumbing installed under prior regulations or lack thereof, may remain unchanged unless immediate hazards to health, life, or property are evident.

Chapter 16

Regulations Governing Individual Sewage Disposal Systems for Homes and Other Establishments Where Public Sewage Systems Are Not Available

16.1 GENERAL PROVISIONS

16.1.1 General

In the absence of State or other local laws governing the installation, use and maintenance of private sewage disposal systems, the provisions of this Chapter shall apply.

16.1.2 Sewage Disposal

"Sewage disposal" under this section shall mean all private methods of collecting and disposing of domestic sewage, including septic tanks.

16.1.3 Domestic Sewage

Domestic sewage shall be disposed of by an approved method of collection, treatment and effluent discharge. Domestic sewage or sewage effluent shall not be disposed of in any manner that will cause pollution of the ground surface, ground water, bathing areas, lakes, ponds, watercourses, tidewater, or create a nuisance. It shall not be discharged into any abandoned or unused well, or into any crevice, sink hole, or other opening either natural or artificial in a rock formation.

> *Comment: Federal and local environmental agencies have criteria for the allowable contaminant levels that can be discharged into a water course or other point of discharge. Improper disposal of sewage can result in; (1) contamination of public or private water supplies, (2) spread of disease by insects or vermin, (3) creation of objectionable odors, (4) pollution of public water resources, or (5) other conditions that are detrimental to public health and safety.*

16.1.4 Non-Water-Carried Sewage

When water under pressure is not available, all human body wastes shall be disposed of by depositing them in approved privies, chemical toilets, or such other installations acceptable to the Authority Having Jurisdiction.

16.1.5 Water-Carried Sewage

Water-carried sewage from bathrooms, kitchens, laundry fixtures and other household plumbing shall pass through a septic or other approved sedimentation tank prior to its discharge into the soil or into a sand filter. Where underground disposal or sand filtration is not feasible, consideration shall be given to special methods of collection and disposal.

16.1.6 Responsibility

The installing contractor is responsible for compliance with these regulations.

16.1.7 Abandoned Disposal Systems

Abandoned disposal systems shall be disconnected from the buildings, pumped out and filled with earth.

> *Comment: This requirement applies to all abandoned dry wells, septic tanks, cesspools, distribution boxes, seepage pits, and other structures that have handled sewage or sewage by-products.*

16.1.8 Absorption Capacity

No property shall be improved in excess of its capacity to properly absorb sewage effluent in the quantities and by means provided for in this Code. *See Sections 16.5 and 16.6*

> *Comment: The development of properties must be limited by the ability of the soil to absorb the effluent of the required sewage disposal system. Bacterial action within the soil provides secondary treatment of the effluent from septic tanks.*

16.2 RESERVED

16.3 DESIGN OF INDIVIDUAL SEWAGE DISPOSAL SYSTEMS

16.3.1 Design

The design of the individual sewage disposal system must take into consideration location with respect to wells or other sources of water supply, topography, water table, soil characteristics, area available, and maximum occupancy of the building.

> *Comment #1: The primary factors in the design of private sewage disposal systems are; (1) the location of the disposal system, (2) the absorption capacity of the soil, (3) the size of the drainage field, (4) the size of the septic tank, (5) the elevation of the ground water table, (6) the location of water supplies for the property and neighboring properties, (7) the use and population of the building(s) served, (8) the topography of the property.*
>
> *Comment #2: Problems with topography (elevation differences) can be overcome by sewage pumps, which can pump to elements of the disposal system that are at higher elevations than others.*

16.3.2 Type of System

The type of system to be installed shall be determined on the basis of location, soil permeability, and groundwater elevation. *See Sections 16.4 and 16.5*

16.3.3 Sanitary Sewage

The system shall be designed to receive all sanitary sewage, including laundry waste, from the building. Drainage from footings or roofs shall not enter the system.

16.3.4 Discharge

The system shall consist of a septic tank discharging into either a subsurface disposal field or one or more seepage pits or into a combination of both, if found adequate as such and approved by the Authority Having Jurisdiction. *Refer to the definition of "Septic Tank" and "Leaching Well or Pit".*

> *Comment: Where plumbing fixtures connected to a private sewage disposal system are subject to backflow from a blockage in the disposal system, they must be protected by a backwater valve. Sewage pumps and ejectors having check valves provide this protection for the fixtures that they serve.*

16.3.5 Backflow

Plumbing fixtures connected to a private sewage disposal system that are subject to backflow shall be protected by a backwater valve or a sewage ejector.

16.3.6 Reserved

16.3.7 Design Criteria

Design criteria for sewage flows shall be selected according to type of establishment. *See Table 16.3.7.*

16.4 LOCATION OF INDIVIDUAL SEWAGE DISPOSAL SYSTEMS

16.4.1 Reserved

16.4.2 Reserved

16.4.3 Minimum Distances

The minimum distances that shall be observed in locating the various components of the disposal system shall be as given in Table 16.4.3.

16.4.4 General

All sewage disposal systems shall conform with the following general principles regarding the site:

16.4.4.1 Location

Sewage disposal systems shall be located at the lowest point on the premises consistent with the general layout topography and surroundings, including abutting lots. Locations at a higher elevation through employment of a forced system may be used with the specific approval of the Authority Having Jurisdiction.

16.4.4.2 Watersheds

Sewage disposal facilities shall not be located on any watershed for a public water supply system.

16.4.4.3 Septic Tanks and Underground Disposal

Septic tanks and underground disposal means shall not be within 200 feet measured horizontally from the high water level in a reservoir or the banks of tributary streams when situated less than 3,000 feet upstream from an intake structure.

Table 16.3.7
SEWAGE FLOWS ACCORDING TO TYPE OF ESTABLISHMENT

Schools (toilets and lavatories only)	15 Gal. per day per person
Schools (with above plus cafeteria)	25 Gal. per day per person
Schools (with above plus cafeteria and showers)	35 Gal. per day per person
Day workers at schools and offices	15 Gal. per day per person
Day camps	25 Gal. per day per person
Trailer parks or tourist camps (with built-in bath)	50 Gal. per day per person
Trailer parks or tourist camps (with central bathhouse)	35 Gal. per day per person
Work or construction camps	50 Gal. per day per person
Public picnic parks (toilet wastes only)	5 Gal. per day per person
Public picnic parks (bathhouse, showers and flush toilets)	10 Gal. per day per person
Swimming pools and beaches	10 Gal. per day per person
Country clubs	25 Gal. per day per person
Luxury residences and estates	150 Gal. per day per person
Rooming houses	40 Gal. per day per person
Boarding schools	50 gal. per day per person
Hotels (with connecting baths)	50 Gal. per day per person
Hotels (with private baths–2 persons per room)	100 Gal. per day per person
Boarding schools	100 Gal. per day per person
Factories (gallons per person per shift–exclusive of industrial waste)	25 Gal. per day per person
Nursing homes	75 Gal. per day per person
General hospitals	150 Gal. per day per person
Public Institutions (other than hospitals)	100 Gal. per day per person
Restaurants (toilet and kitchen wastes per unit of serving capacity)	25 Gal. per day per person
Kitchen wastes from hotels, camps, boarding houses, etc. serving three meals per day	10 Gal. per day per person
Motels	50 Gal. per day per bed space
Motels with bath, toilet, and kitchen wastes	60 Gal. per day per bed space
Drive-in theaters	5 Gal. per day per car space
Stores	400 Gal. per day per toilet room
Service stations	10 Gal. per day per vehicle served
Airport terminals	3-5 Gal. per day per passenger
Assembly halls	2 Gal. per day per seat
Bowling alleys	75 Gal. per day per lane
Churches (small)	3-5 Gal. per day per sanctuary seat
Churches (large with kitchen)	5-7 Gal. per day per sanctuary seat
Dance halls	2 Gal. per day per person
Laundries (coin operated)	400 Gal. per day per machine
Service stations	1000 Gal. (first bay) per day 500 Gal. (each add'l bay)
Sub-divisions or individual homes	75 Gal. per day per person
Marinas–Flush toilets	36 Gal. per fixture per hr
Urinals	10 Gal. per fixture per hr
Wash basins	15 Gal. per fixture per hr
Showers	150 Gal. per fixture per hr

Table 16.4.3
MINIMUM DISTANCE BETWEEN COMPONENTS
OF AN INDIVIDUAL SEWAGE DISPOSAL SYSTEM (in feet)[1]

	Shallow Well	Deep Well	Single Suction Line	Septic Tank	Distribution Box	Disposal Field	Seepage Pit	Dry Well	Property Line	Building
Bldg. Sewer other than cast-iron	50	50	50	-	-	-	-	-	-	-
Bldg. Sewer cast-iron	10	10	10	-	-	-	-	-	-	-
Septic Tank	100	50	50	-	5	10	10	10	10	10
Distribution Box	100	50	50	5	-	-	5	5	10	20
Disposal Field	100	50	50	10	5	-	-	-	10	20
Seepage Pit	100	50	50	10	5	-	-	-	10	20
Dry Well	100	50	50	10	5	-	-	-	10	20
Shallow Well	-	-	-	100	100	100	100	100	-	-
Deep Well	-	-	-	50	50	50	50	50	-	-
Suction Line	-	-	-	50	50	50	50	50	-	-

[1] Minimum distances may be reduced if approved by the Authority Having Jurisdiction.

16.4.4.4 Beyond 3,000 feet

Sewage disposal facilities situated beyond 3,000 feet upstream from intake structures shall be located no less than 100 feet measured horizontally from the high water level in the reservoir or the banks of tributary streams.

16.4.4.5 Percolation Test

Prior to approval, the soil must prove satisfactory by the standard percolation test when underground disposal is used.

16.5 PERCOLATION TEST

Percolation tests to determine the absorption capacity of soil for septic tank effluent shall be conducted in the following manner:

16.5.1 Subsurface Irrigation

When subsurface irrigation is contemplated, a test pit shall be prepared 2 feet square and not less than 1 foot deep. At the time of conducting the percolation test, a hole 1 foot square and 1 foot deep shall be prepared in the test pit.

16.5.2 Water Depth

The hole shall be filled with water to a depth of 7 inches. For pre-wetting purposes, the water level shall be allowed to drop to 6 inches before time of recording is started.

16.5.3 Time Expired

The time required for the water level to drop 1 inch from 6 inches to 5 inches in depth shall be noted and the required length of tile in the subsurface irrigation system shall be obtained from Section 16.5.4. In no case, however, shall less than 100 feet of tile be installed when 1 foot trenches are used.

16.5.4 Trench Length in Disposal Fields

The trench length, in feet for each 100 gallons of sewage per day, shall comply with Table 16.5.4, based on the time expired during the percolation test in Section 16.5.3.

> *Comment: Determine the total daily sewage flow for the system from Table 16.3.7. Divide this number by 100 and determine the required length of trenches in the disposal field from Table 16.5.4. Tile lengths are limited to 100 feet by Section 16.9.4.*

Table 16.5.4
REQUIRED LENGTH OF TRENCHES IN SEWAGE DISPOSAL FIELDS IN FEET FOR EACH 100 GALLONS OF SEWAGE PER DAY

Time in Minutes for 1-inch Drop	For 1- foot Trench Width	For 2- foot Trench Width	For 3- foot Trench Width
1	25 feet/100 gpd	13 feet/100 gpd	9 feet/100 gpd
2	30 feet/100 gpd	15 feet/100 gpd	10 feet/100 gpd
3	35 feet/100 gpd	18 feet/100 gpd	12 feet/100 gpd
5	42 feet/100 gpd	21 feet/100 gpd	14 feet/100 gpd
10	59 feet/100 gpd	30 feet/100 gpd	20 feet/100 gpd
15	74 feet/100 gpd	37 feet/100 gpd	25 feet/100 gpd
20	91 feet/100 gpd	46 feet/100 gpd	31 feet/100 gpd
25	105 feet/100 gpd	53 feet/100 gpd	35 feet/100 gpd
30	125 feet/100 gpd	63 feet/100 gpd	42 feet/100 gpd

16.5.5 Seepage Pits

When seepage pits are contemplated, test pits approximately 5 feet in diameter to permit a man entering the pit by means of a ladder and to such depth as to reach a porous soil shall be prepared. In the bottom of this pit, a 1 foot square by 1 foot deep hole shall be made at the time of testing and the percolation test conducted as indicated under Sections 16.5.1, 16.5.2, and 16.5.3. **See the definition of "Leaching Well or Pit" and Figure 1.2.41.**

> *Comment: Observe safety requirements regarding sheeting, shoring, and bracing for deep excavations.*

16.5.6 Seepage Pit Absorption Area

The absorption area of a seepage pit required shall be obtained from Table 16.5.6. In no case, however, shall the absorption area in the porous soil be less than 125 square feet. The bottom of the pit shall not be considered part of the absorption area.

> *Comment: The absorption area of a seepage or leaching pit is the square foot area of the side walls. The sidewall area = 3.1416 x diameter x depth (both in feet).*

Table 16.5.6
REQUIRED ABSORPTION AREA IN SEEPAGE PITS FOR EACH 100 GALLONS OF SEWAGE PER DAY

Time in Minutes for 1-inch Drop	Effective Absorption Area in Square Feet
1	32 sq ft/100 gpd
2	40 sq ft/100 gpd
3	45 sq ft/100 gpd
5	56 sq ft/100 gpd
10	75 sq ft/100 gpd
15	96 sq ft/100 gpd
20	108 sq ft/100 gpd
25	139 sq ft/100 gpd
30	167 sq ft/100 gpd

16.5.7 Thickness of Porous Soil

The thickness of the porous soil below the point of percolation test must be determined by means of digging a pit or using a soil auger. The effective absorption area shall be calculated only within this porous soil.

16.6 CAPACITY OF SEPTIC TANKS

16.6.1 Liquid Capacity

The liquid capacity of all septic tanks shall conform to Table 16.3.7 and Table 16.6.1 as determined by the number of bedrooms or apartment units in dwelling occupancies and the occupant load or the number of plumbing fixture units as determined from Table 11.4.1, (whichever is greater) in other building occupancies.

> *Comment: The septic tank sizes in Table 16.6.1 have allowances for sludge storage space and the use of domestic food waste disposal units.*

Table 16.6.1
CAPACITY OF SEPTIC TANKS

Single family dwelling-number of bedrooms	Multiple dwellings units or apartments-one bedroom each	Other uses, maximum fixture units served	Minimum septic tank capacity in gallons
1-3		20	1000
4	2 units	25	1200
5 or 6	3	33	1500
7 or 8	4	45	2000
	5	55	2250
	6	60	2500
	7	70	2750
	8	80	3000
	9	90	3250
	10	100	3500

Extra bedroom: 150 gallons each.
Extra dwelling units over 10: 250 gallons each.
Extra fixture units over 100: 25 gallons per fixture unit.

16.6.2 Reserved

16.6.3 Multiple Compartment

In a tank of more than one compartment, the inlet compartment shall have a capacity of not less than two-thirds of the total tank capacity.

> *Comment: Septic tanks with two compartments operate more efficiently than single compartment tanks at removing solids in the effluent. The liquid that enters the second compartment is already substantially clarified. There is less turbulence in the second chamber which permits finer suspended solids to settle out.*

16.6.4 Septic Tank Materials

See Sections 3.3.11, 16.6.5 and 16.6.6.

> *Comment: Modern septic tanks are typically constructed of concrete (pre-cast), polyethylene, or fiberglass. Steel septic tanks are prohibited by many jurisdictions because of leakage due to corrosion.*

16.6.5 Steel Tanks

> *Comment: Steel septic tanks are no longer approved by most jurisdictions because of leakage due to corrosion. The average life expectancy of a steel septic tank is only 7 years.*

16.6.5.1 Welding

All steel tanks shall be continuous welded. (No spot welding is permitted.)

16.6.5.2 Wall Thickness

The minimum wall thickness of any steel septic tank shall be No. 12 U.S. gauge (0.109").

16.6.5.3 Coatings

Metal tanks shall be coated inside and out with an approved coating.

16.6.5.4 Baffles

The inlet and outlet baffles shall be at least 12 inches in diameter at the point opposite the opening in the tank.

16.6.5.5 Pumpout Opening

The pumpout opening in the top shall be large enough to permit a 6-inch cast-iron pumpout pipe to be inserted with a shoulder to support this pipe.

16.6.5.6 Tank Opening

The tank opening shall not be smaller than 6 inches with a 3-inch collar.

16.6.5.7 Outside Diameter of Collar

The outside diameter of this collar shall be 8 inches.

16.6.5.8 Pumpout Pipe

The pumpout pipe shall terminate at the surface and a 6-inch iron body brass cleanout shall be caulked into the hub of this pipe with oakum and molten lead; the cleanout nut shall be solid brass no smaller than one inch.

16.6.5.9 Manhole

There shall be a 24 x 24-inch manhole held in position by four 3/8" bolts securely welded in place.

16.6.5.10 Partition

There shall be a supporting partition welded in the center of these tanks as per drawings.

16.6.5.11 Partition Openings

This partition shall have 2-inch openings at intervals at the top for air circulation.

16.6.5.12 Capacity, Gauge Metal and Weight

The capacity, gauge metal, and weight must be stamped on a brass plate and welded to the top of metal septic tanks.

16.6.6 Concrete Tanks

16.6.6.1 Baffles

Concrete tanks shall have the same size baffles and pumpout openings as for steel tanks.

16.6.6.2 Tops

The tops shall have a 24-inch manhole with handle to remove same, or be cast in three or four sections cemented in place.

16.6.6.3 Wall Thickness

The minimum thickness of the walls shall be 2-3/4 inches.

16.6.6.4 Tops and Bottoms

The tops and bottoms shall be 4 inches thick unless placed under a driveway, then they shall be a minimum of 6 inches.

16.6.6.5 Walls and Bottoms

All tank walls and bottoms shall be reinforced with approved reinforcing.

16.6.6.6 Top Reinforcing

The tops shall have 3/8 inch steel reinforcing on 6-inch centers.

16.6.6.7 Watertight

The tank shall be watertight.

16.6.7 Depth of Septic Tank

The top of the septic tank shall be brought to within 36 inches of the finished grade. Where a greater depth is permitted by the Authority Having Jurisdiction, the access manhole must be extended to the finished grade and the manhole shall have a concrete marker at grade.

> *Comment: The typical depth of cover on the top of a septic tank is 12" - 18". Where the depth of cover exceeds 36", the structural design of the tank must be adequate for the imposed earth loads.*

16.6.8 Limitation

No septic tank shall serve more than one property unless authorized by the Authority Having Jurisdiction.

16.6.9 Effluent

The effluent from all septic tanks shall be disposed of underground by subsurface irrigation or seepage pits or both.

16.7 DISTRIBUTION BOX

16.7.1 When Required

A distribution box shall be required when more than one line of subsurface irrigation or more than one seepage pit is used.

> *Comment: The minimum distances from distribution boxes to the septic tank, disposal field, seepage pit, dry well, potable water wells, or suction lines from potable water wells must be in accordance with Table 16.4.3.*

16.7.2 Connection

Each lateral line shall be connected separately to the distribution box and shall not be subdivided.

16.7.3 Invert Level

The invert of all distribution box outlets shall be at the same level and approximately 2 inches above the bottom of the box. The inlet invert shall be at least 1 inch above the invert of the outlets. The size of the distribution box shall be sufficient to accommodate the number of lateral lines.

16.7.4 Watertight

The distribution box shall be of watertight construction arranged to receive the septic tank effluent and have an outlet or connecting line serving each trench or seepage pit.

16.7.5 Baffle

A baffle at least 6 inches high and 12 inches long shall rest on the bottom of the box and be placed at right angles to the direction of the incoming tank effluent and 12 inches in front of it.

16.7.6 Reserved

16.7.7 Inspection

The sides of the box shall extend to within a short distance of the ground surface to permit inspection, and shall have a concrete marker at grade.

16.8 SEEPAGE PITS

16.8.1 Use

Seepage pits may be used either to supplement the subsurface disposal field or in lieu of such field where conditions favor the operation of seepage pits, as may be found necessary and approved by the Authority Having Jurisdiction. *See Figure 1.2.41 and the definition of "Leaching Well or Pit"*

16.8.2 Water Table

Seepage pits shall not penetrate the water table.

16.8.3 Septic Tank Effluent Disposal

Where seepage pits are used for septic tank effluent disposal, the number, diameter and depth of the pits shall be determined after percolation tests have been made to ascertain the porosity of the soil.

16.8.4 Excavation

The excavation for a seepage pit shall be greater in diameter than the outside diameter of the vertical sidewalls to allow for the footing.

16.8.5 Annular Space

The annular space between the outside of the vertical walls and the excavation shall be backfilled with broken stone, coarse gravel, or other suitable material.

16.8.6 Construction

Seepage pits shall be constructed with the bottom being open with an outer ring, or footing, to support the sidewalls.

16.8.7 Sidewalls

The sidewalls shall be made of pre-cast concrete, stone, concrete or cinder blocks, or brick laid in cement mortar for strength, with openings at sufficient intervals to permit the septic tank effluent to pass out through the wall to the surrounding porous soil.

16.8.8 Cover Strength

All septic tank tops and seepage pit covers shall be of sufficient strength to carry the load imposed. Seepage pit covers shall be at least as required in Sections 16.8.9, 16.8.10, and 16.8.11.

16.8.9 Pre-Cast Top

Seepage pit tops shall be pre-cast, reinforced concrete (2,500 pounds per square inch minimum compressive strength) not less than 5 inches thick and designed to support an earth load of not less than 400 pounds per square foot. Each such cover shall extend not less than 3 inches beyond the sidewalls of the pit, shall be provided with a 6-inch minimum inspection hole with pipe extended to the surface, and a 6-inch cast-iron standpipe with cleanout at grade.

16.8.10 Depth Below Grade

The top shall be at least 36 inches below finished grade, except where less is permitted by the Authority Having Jurisdiction.

16.8.11 Field Fabricated Slabs

Where field fabricated slabs are used, Table 16.8.11 indicates the requirements.

Table 16.8.11
DESIGN OF SEEPAGE PIT COVERS

Pit Diameter	Pit Wall Thickness	Cover Thickness	Cover Weight	Reinforcing Steel Required in Two Perpendicular Directions
5ft.	4"	5"	1230 lb	#5 @ 10-1/2" c/c
6ft.	8"	5"	1770 lb	#5 @ 9" c/c
8ft.	8"	6"	3780 lb	#5 @ 7-1/2" c/c
10ft.	8"	8"	7850 lb	#5 @ 6-1/2" c/c

16.9 ABSORPTION TRENCHES

16.9.1 General

Absorption trenches shall be designed and constructed on the basis of the required effective percolation area.

16.9.2 Filter Material

The filter material shall cover the tile and extend the full width of the trench and shall be not less than 6 inches deep beneath the bottom of the drain tile, and 2 inches above the top of the tile. The filter material may be washed gravel, crushed stone, slag, or clean bank-run gravel ranging in size from 1/2 to 2-1/2 inches. The filter material shall be covered with burlap, filter cloth, 2 inches of straw, or equivalent permeable material prior to backfilling the excavation.

16.9.3 Absorption Field

The size and minimum spacing requirements for absorption fields shall conform to those given in Table 16.9.3.

Table 16.9.3
SIZE AND SPACING FOR DISPOSAL FIELDS

Width of trench at bottm (in.)	Recommended depth of trench (in.)	Spacing of trenches[1] (feet)	Effective absorption area per lineal ft. of trench (sq. ft.)
18	18 to 30	6.0	1.5
24	18 to 30	6.0	2.0
30	18 to 36	7.5	2.5
36	24 to 36	9.0	3.0

[1] A greater spacing is desirable where available area permits.

16.9.4 Lateral Length

Length of laterals shall not exceed 100 feet.

> *Comment; Where the required length of trenches from Section 16.5.4 would require lateral tiles longer than 100 feet, two or more laterals must be provided.*

16.9.5 Absorption Lines

Absorption lines shall be constructed of 4" pipe of open jointed or perforated vitrified clay pipe, open jointed or horizontally split or perforated clay tile, perforated plastic pipe or open jointed cast iron soil pipe, all conforming to approved standards. In the case of clay tile, open jointed clay pipe, or open jointed cast-iron soil pipe, the sections shall be spaced not more than 1/2 inch apart, and the upper half of the joint shall be protected by asphalt-treated paper while the piping is being covered.

16.9.6 Grade

The trench bottom shall be uniformly graded to slope from a minimum of 2 inches to a maximum of 4 inches per 100 feet.

16.10 RESERVED

16.11 PIPING MATERIAL

See Chapter 3.

> *Comment: Refer to Section 3.5 and Table 3.5 for approved materials for drainage piping. Refer to Section 16.9.5 and Table 3.8 for approved materials for sub-soil drainage and absorption lines.*

16.12 SAND FILTERS

16.12.1 General Specifications for Design and Construction of a Sand Filter with Chlorination

16.12.1.1 General

A sand filter shall consist of a bed of clean, graded sand on which septic tank effluent is distributed by means of a siphon and pipe, with the effluent percolating through the bed to a series of underdrains through which it passes to the point of disposal.

16.12.1.2 Filter Size

The filter size shall be determined on the basis of 1.15 gallons per square foot per day if covered, and 2.3 gallons per square foot per day if an open filter is to be used.

16.12.1.3 Dosing Tank Size

The septic tank effluent shall enter a dosing siphon tank of a size to provide a 2-inch coverage of the sand filter.

16.12.1.4 Siphon

The siphon shall be of a commercial type and shall discharge the effluent to the sand filter intermittently. A surge tank shall be used to receive the pump discharge prior to dosing on the sand filter.

16.12.1.5 Surge Tank

The siphon shall be omitted if a pump is used to lift the septic tank effluent to the sand filter.

16.12.1.6 Underdrains

Four-inch diameter vitrified clay pipe in 2-foot lengths laid with 1/2 inch open joints or unglazed farm tile in 1-foot lengths laid with open joints, with the top half of each joint covered with 4-inch wide strips of tar paper, burlap, or copper screen, or perforated bituminized-fiber pipe or other approved material shall be used for the underdrains.

16.12.1.7 Underdrain Bed

The underdrains shall be laid at the bottom of the sand filter, surrounded by washed gravel, crushed stone, slag, or clean bank-run gravel ranging in size from 1/2 inch to 2-1/2 inches and free of fines, dust, ashes or clay. The gravel shall extend from at least 2 inches below the bottom of the tile to a minimum of 2 inches above the top of the tile.

16.12.1.8 Underdrain Slope and Spacing

The underdrains shall have a slope from 2 inches to 4 inches per 100 feet and shall be placed at 6-foot to 8-foot intervals.

16.12.1.9 Underdrain Fill

Above the gravel or other material surrounding the underdrain shall be placed 2 feet of washed and graded sand having an effective size of from 0.35-0.5 mm and a uniformity coefficient of not over 3.5. (The effective size of a sand filter is that size of which 10% by weight is smaller and the uniformity coefficient is the ratio of that size of which 60% by weight is smaller to the effective size.)

16.12.1.10 Distribution Pipes

The distribution pipes shall be laid at the surface of the sand filter, surrounded by gravel as specified for the underdrains.

16.12.1.11 Gravel Cover

The gravel should be covered with untreated building paper and the entire area covered with a minimum of 12 inches of earth if the filter is to be covered.

16.12.1.12 Open Filter

If the filter is an open one, the four sides shall be constructed of wood or concrete to prevent earth erosion from entering the sand filter bed.

16.12.1.13 Chlorine Contact Tank

The chlorine contact tank for disinfection of sand filter effluent shall provide 20 minutes detention at average flow, but in no case shall it be smaller than 50 gallons capacity. Chlorine control should be provided by the use of hypochlorite or chlorine machines commercially available.

Chapter 17

Private Potable Water Supply Systems

17.1 GENERAL REGULATIONS

17.1.1 Applicability

The regulations in this chapter apply to any private potable water supply system where plumbing fixtures are installed for human occupancy.

> *Comment: This chapter applies to private water supply systems serving one or more buildings independent of any public water supply.*

17.1.2 Pumps

Pumps shall be installed only in wells, springs and cisterns that comply with the rules and regulations as determined by the Authority Having Jurisdiction.

17.2 QUANTITY OF WATER REQUIRED

17.2.1 Single Dwelling Units

The minimum capacity of the system in gallons per minute shall equal the number of fixtures installed.

> *Comment: A capacity of 1 GPM per fixture in a single dwelling unit is a rough rule of thumb. The peak demand should be verified using Tables 10.14.2A and 10.14.2B. See Section 17.2.4 where well yields do not satisfy the peak demand.*

17.2.2 Other Than a Single Dwelling Unit

In other than a single dwelling unit, the water system shall be designed in accordance with Tables 10.14.2A, and 10.14.2B and shall be capable of supplying the maximum demand to the system according to usage, but, in no case, less than for a minimum period of 30 minutes.

> *Comment: See Section 10.14.3 for determining the peak demand on the potable water supply system.*

17.2.3 Available Water

Total water available during any 24-hour period shall not be less than the requirements of Table 16.3.7.

> *Comment: Table 16.3.7 indicates the daily design sewage flows from various establishments. The available water supply must be able to satisfy this daily volume requirement, as well as the peak GPM demand.*

17.2.4 Secondary Sources of Water

a. When the available primary source of water does not meet the minimum requirement of Sections 17.2.1, 17.2.2, and 17.2.3, one of the following methods shall be used:

   1. Pressure tank of sufficient size.
   2. Gravity tank, see Section 10.8.
   3. Two pump system.

      (a) The capacity of the first pump shall not exceed the flow capacity of the well. It shall supply water to a tank that stores the water at atmospheric pressure and has a level control to start and stop the pump.

      (b) The second pump shall supply a hydro-pneumatic tank at the required pressure and volume to supply the water distribution system.

> *Comment: Where the primary potable water supply source cannot provide sufficient volume to satisfy the 24-hour demand of the water distribution system, water storage facilities must be provided.*

17.3 PRESSURE

Required water supply pressures are as indicated in Sections 10.14.2.a and 10.14.3.b.

> *Comment: Private potable water supply systems must provide adequate pressure for fixtures to deliver the flow rates for the various fixtures listed in Sections 10.14.2.a and 10.14.3.b.*

17.4 PIPING MATERIALS

Piping from a well or other water source to a building water distribution system shall be in accordance with Sections 3.4.1 and 3.4.2.

17.5 STORAGE TANKS

Storage equipment shall be as follows:

17.5.1 Certified Tanks

All tanks shall be certified for use with potable water.

17.5.2 Tank Material

All tanks shall be coated or made of material to resist corrosion.

17.5.3 Pressure Rating

Hydropneumatic tanks shall have a working pressure rating in excess of the maximum required system pressure.

17.5.4 Non-Toxic Materials

All tanks shall be constructed of materials and/or coatings that are non-toxic.

17.5.5 Drain Required

All tanks shall be provided with a means for draining.

17.5.6 Covers

Atmospheric storage tanks shall be provided with a cover as required in Section 10.8.4.

17.6 PUMPS

17.6.1 Pump Sizing

Pumps may be sized as recommended in the Water Systems Council's Water Systems Handbook.

> *Comment #1: Well pumps for private water supplies include shallow well jet pumps, deep well jet pumps, and submersible pumps.*
>
> *Comment #2: Jet pumps circulate water through an ejector fitting that develops a vacuum and draws water from the well. Shallow well jet pumps are used for wells less than 25 feet deep and have the ejector fitting located outside of the well. Deep well jet pumps can be used for wells up to 75 feet deep and have the ejector fitting located within the well casing. Some jet pumps are convertible and can be used for either one-pipe shallow well or two-pipe deep well installations, although their rated capacity is less for deep wells than for shallow wells.*
>
> *Comment #3: Submersible well pumps are located at the bottom of the drop pipe within the well casing. Submersible well pumps can lift water up to 500 feet or more.*

17.6.2 Pump Installation

Pumps shall be installed in accordance with the manufacturer's recommendations. and the recommendations in the Water Systems Council's Water Systems Handbook.

17.6.3 Equipment Installation

Pumping equipment shall be installed to prevent the entrance of contamination or objectionable material either into the well or into the water that is being pumped.

17.6.4 Pump Location

Pumps shall be located to facilitate necessary maintenance and repair. Vertical well pumps shall include overhead clearance for removal of drop pipes.

17.6.5 Pump Mounting

Pumps shall be suitably mounted to avoid objectionable vibration and noise, and to prevent damage to pumping equipment.

17.6.6 Pump Accessories

Pump controls and/or accessories shall be protected from weather.

17.7 PUMP DOWN CONTROL

17.7.1 Tailpipe

Thirty (30) feet of tailpipe shall be installed below the jet on deep well installation.

17.7.2 Switches

A low pressure cut-off switch and/or water level cut-off switch shall be installed.

17.7.3 Suction Pipe

Provide a vertical suction pipe of 30-foot length on shallow well jet installation.

17.8 CONTROLS

a. The following controls are required on all pump installations:
   1. Pressure switch
   2. Thermal overload switch
   3. Pressure relief valve on positive displacement pumps
   4. Low water cut-off switch where the pump capacity exceeds the source of water.

17.9 WELL TERMINAL

17.9.1 Upper Well Terminal

Well casings and curbs shall terminate not less than eight inches above the finished ground surface or pump house floor and at least 24 inches above the maximum high water level where flooding occurs. No casing shall be cut off or cut into below ground level except to install a pitless adapter.

17.9.2 Pitless Adapter

17.9.2.1 Design

Pitless adapters designed to replace a section of well casing or for attachment through the wall of the well casing shall be constructed of materials that provide strength and durability equal to the well casing.

17.9.2.2 Installation

Installation shall be by a threaded or welded adapter or a coupling connection to the casing or attachment through the wall of the casing and shall be watertight.

17.9.2.3 Adapter Units

Adapter units designed to replace a section of the well casing shall extend above the finished ground surface as provided in Section 17.9.1. The top of the adapter unit shall be capped with a cover having a downward flange that will overlap the edge of the unit. The cover shall be securely fastened to the unit and shall fit sufficiently snug to the unit to be vermin proof. The cover shall provide for watertight entrance of electrical cables and vent piping or air line if installed.

17.9.3 Hand Pumps

17.9.3.1 General

Hand pumps shall be of the force type equipped with a packing gland around the pump rod, a delivery spout that is closed and downward directed, and a one-piece bell type base that is part of the pump stand or is attached to the pump column in a watertight manner.

17.9.3.2 Installation

The bell base of the pump shall be bolted with a gasket to a flange that is securely attached to the casing or pipe sleeve.

17.9.4 Power Driven Pumps

17.9.4.1 General

The design and operating principles of each type of power driven pump determines where each may be located with respect to a well. The location selected for the pump determines what factors must be considered to make an acceptable installation.

17.9.4.2 Location Above Well

Any power driven pump located over a well shall be mounted on the well casing, pipe sleeve, pump foundation or pump stand such that a watertight closure is or can be made with the open top of the casing or sleeve.

17.9.4.3 Pump Base

The pump base bolted with a neoprene or rubber gasket or equivalent watertight seal to a foundation or plate provides an acceptable seal.

17.9.4.4 Large Pump Installation

On large pump installations, the bolting may be omitted when the weight of the pump and column is sufficient to make a watertight contact with the gasket.

17.9.4.5 Pump Location Other Than Over Casing

If the pump unit is not located over the casing or pipe sleeve, and the pump delivery or suction pipes emerge from the top of the well, a watertight expanding rubber seal or equivalent shall be installed between the well casing and piping to provide a watertight closure.

17.9.4.6 Seal Top

The top of the seal in Section 17.9.4.5 shall not extend below the uppermost edge of the casing or pipe sleeve.

17.9.5 Location in Well For Submersible Pumps

17.9.5.1 General

This type of location is permissible for submersible pumps only.

17.9.5.2 Top Discharge Line

When the discharge line leaves the well at the top of the casing, the opening between the discharge line and casing or pipe sleeve shall be sealed watertight with an expanding rubber seal or equivalent device.

17.9.5.3 Underground Discharge

When an underground discharge is desired, a properly installed pitless adapter shall be used. A check valve shall be installed in the discharge line above the pump in the well.

17.9.5.4 Top Discharge Line Sloped to Drain to Well

When the discharge pipe leaves the well at the top, remains above ground, and slopes to drain back to the well, the check valve can be located beyond the well.

17.9.6 Offset from Well

17.9.6.1 Location

Pumps offset from the well, if not located in an above-ground pump house or other building, may be located in an approved basement provided the pump and all suction pipes are elevated at least 12 inches above the floor.

17.9.6.2 Buried Lines

All portions of suction lines buried below the ground surface between the well and the pump and that are not enclosed in a protective pipe shall be located the same minimum distance from sources of contamination as are prescribed for the well in Section 16.4.3.

17.9.6.3 Protective Pipe

When the minimum distances in Section 16.4.3 cannot be obtained, the suction line shall be enclosed in a protective pipe of standard thickness from the well to the pump. The protective pipe shall be sealed watertight at both ends.

17.10 VENTS

17.10.1 Size

All vent piping shall be of adequate size to allow equalization of air pressure in the well and shall not be less than one-half inch in diameter.

17.10.2 Toxic or Flammable Gases

Particular attention shall be given to proper venting of wells and pressure tanks in areas where toxic or flammable gases are known to be a characteristic of the water. If determined that either of these types of gases are present, all vents when located in buildings shall be extended to discharge outside of the building at a height where they will not be a hazard.

17.10.3 Vent Extension

The vent shall extend above the upper terminal of the well with the end down turned and covered with not less than 16 mesh screen wire. The point of entry into the well shall be sealed watertight.

17.11 BEARING LUBRICATION

17.11.1 General

Lubrication of bearings of power driven pumps shall be with water or oil that will not adversely affect the quality of the water to be pumped.

17.11.2 Water Lubrication

If a storage tank is required for the lubrication water, it shall be designed to protect the water from contamination.

17.11.3 Oil Lubrication

The reservoir shall be designed to protect the oil from contamination. The oil shall not contain substances that will cause odor or taste to the water pumped.

17.12 WATER LEVEL MEASUREMENT

On wells of large capacity where access for measuring the water level in the well is provided, piping for this purpose shall terminate above the upper well terminal, be capped or otherwise closed, and all openings around the piping at the point of entry into the well sealed watertight.

17.13 PROHIBITED PUMPS

No pitcher or chain-bucket pump shall be installed on any water supply.

17.14 PUMP HOUSING

17.14.1 Watertight

A separate structure housing the water supply and pumping equipment shall have an impervious floor and rain-tight walls and roof.

17.14.2 Pump Pit

A pump pit shall be of watertight construction and provided with a positive drain or sump pump to keep the pit dry.

17.15 PRIVATE WATER SUPPLY SYSTEM RESTRICTIONS AND LIMITATIONS

17.15.1 Restriction

A private potable water supply system shall not be connected to a public water supply system unless approved in accordance with Section 10.4.4.

17.15.2 Limitations

No private potable water supply system shall serve more than one property unless approved by the Authority Having Jurisdiction.

Blank Page

Chapter 18

Mobile Home & Travel Trailer Park Plumbing Requirements

The primary objective of this Chapter is to assure sanitary plumbing installations in trailer parks. Reference should be made to the Authority Having Jurisdiction and the regulations promulgated by the Authority Having Jurisdiction governing the establishment and operation of trailer parks.

18.1 DEFINITIONS

> *Comment: The requirements of this chapter are intended as general requirements for parks and campgrounds that serve travel trailers, recreational vehicles, and mobile homes. Travel trailers and recreational vehicles are used as temporary dwellings for travel or recreational purposes. Mobile homes are movable structures or units that are designed as living quarters. The requirements of the Authority Having Jurisdiction may vary for the different types of parks and campgrounds.*

Service Buildings

A building housing toilet, laundry and any other such facilities as may be required.

Sewer Connection

Sewer connection is that portion of the drainage piping that extends as a single terminal under the trailer coach for connection with the trailer park drainage system.

Trailer Coach

Any camp-car, trailer, or other vehicle with or without motive power, designed and constructed to travel on the public thoroughfares in accordance with the provisions of the applicable State Vehicle Code and designed or used for human habitation.

Trailer Coach, Dependent

One which is not equipped with a water closet for sewage disposal.

Trailer Coach, Independent

One which is equipped with a water closet for sewage disposal.

Trailer Coach Drain Connection

The removable extension connecting the trailer coach drainage system to the trailer connection fixture.

Trailer Coach, Left Side

The side farthest from the curb when the trailer home is being towed or in transit.

Trailer Connection Fixture

A connection to a trap that is connected to the park drainage system, and receives the water, liquid or other waste discharge from a trailer coach.

Trailer Park Drainage System

The entire system of drainage piping used to convey sewage or other waste from a trailer connection fixture to the sewer.

Trailer Park Branch Line

That portion of drainage piping that receives the discharge from not more than two trailer connection fixtures.

Trailer Park

Any area or tract of land where space is rented or held-out for rent, or occupied by two or more trailer coaches.

Trailer Park Sewer System

That piping that extends from the public or private sewage disposal system to a point where the first trailer park drainage system branch fitting is installed.

Trailer Park Water Service Main

That portion of the water distribution system that extends from the street main, water meter, or other source of supply to the trailer site water service branch.

Trailer Site

That area set out by boundaries on which one trailer can be located.

Trailer Site Water Service Branch

That portion of the water distributing system extended from the park service main to a trailer site, and includes connections, devices, and appurtenances thereto.

Water Service Connection

That portion of the water supply piping that extends as a single terminal under the trailer coach for connection with the trailer coach park water supply system.

18.2 STANDARDS

18.2.1 General

Plumbing systems hereafter installed in trailer home parks shall conform to the provisions set forth in the preceding chapters of this Code, where applicable, and also to the provisions set forth in this Chapter. Trailer home park plumbing and drainage systems shall, in addition, conform to all other applicable Authority Having Jurisdiction regulations.

18.2.2 Plans and Specifications

Before any plumbing or sewage disposal facilities are installed or altered in any trailer park, plans and specifications shall be filed, and required permits obtained from the Authority Having Jurisdiction. Plans shall show the following in detail:

18.2.2.1 Plot Plan

Plot plan of the park, drawn to scale, indicating elevations, property lines, driveways, existing or proposed buildings, and sizes of trailer sites.

18.2.2.2 Plumbing Layout

Complete specifications and piping layout of proposed plumbing system or alteration.

18.2.2.3 Sewage Disposal Layout

Complete specifications and piping layout of proposed sewer system or alteration.

18.2.2.4 Conformance

Trailer park plumbing systems shall be designed and installed in accordance with the requirements of this Code and shall, in addition, conform to all other pertinent local ordinances and State regulations.

18.2.3 Materials

Materials shall conform to the approved standards set forth in other sections of this Code. *See Sections 3.4, 3.5, 3.6, and 3.7*

18.3 DRAINAGE SYSTEM

18.3.1 Design and Installation

The trailer park drainage system shall be designed and installed in accordance with the requirements of this Code.

18.3.2 Alternate

The trailer park drainage system may be installed by the use of a combination waste and vent drainage system (see Section 12.17), which shall consist of an installation of waste piping, as hereinafter provided in this Section, in which the traps for one or more trailer connection fixtures are not separately or independently vented, but which is vented through the waste piping of such size to provide free circulation of air therein.

18.3.3 Each Independent Trailer Site

Each independent trailer site shall be provided with a trapped trailer connection that shall consist of a three-inch horizontal iron pipe-size threaded connection, installed a minimum of three inches and a maximum of six inches (from the bottom of the connection), above the finished grade. The vertical connection to the trailer connection fixture shall be anchored in a concrete slab four inches thick, and 18" x 18" square.

18.3.4 Above Ground

Any part of the plumbing system extending above the ground shall be protected from damage when deemed necessary by the Authority Having Jurisdiction.

18.3.5 Trap Connections

Each trailer site shall be provided with a three-inch I.P.S. male or female threaded connection, extended above the surrounding grade, from a three- inch minimum size vented P-trap.

18.3.5.1 Location

Traps shall be located with reference to the immediate boundary lines of the designated space or area within each trailer site that will actually be occupied by the trailer. Each such trap shall be located in the rear third-quarter section along the left boundary line of the trailer parking area not less than one foot or more than three feet from the road side of the trailer and shall be a minimum of five feet from the rear boundary of the trailer site. This location may be varied by permission of the Authority Having Jurisdiction when unusual conditions are encountered.

18.3.5.2 Material

All traps, tail pipes, vertical vents, the upper five feet of any horizontal vent, and the first five feet of any trap branch shall be fabricated from materials approved for use within a building.

18.3.6 Reserved

18.3.7 Drain Connections

Mobile home and travel trailer drain connections shall be of approved semi-rigid or flexible reinforced hose having smooth interior surfaces and not be less than a 3-inch inside diameter. Main connections shall be equipped with a standard quick-disconnect screw or clamp type fitting, not smaller than the outlet. Main connections shall be gas-tight and no longer than necessary to make the connection between the trailer coach drain connection and the trailer connector fixture on the site.

18.3.8 Cleanouts

Cleanouts shall be provided as required by Chapter 5 of this Code, except cleanouts shall be provided in vent terminals one foot above grade.

18.3.9 Drainage Fixture Unit (DFU) Loading

For the purpose of determining pipe sizes, each trailer site connection shall be assigned a drainage loading value of six drainage fixture units (DFU) and each trailer park drainage system shall be sized in accordance with Table 18.3.12.

18.3.10 Slope of Sewers

The grade on sewers shall provide a minimum velocity of two feet per second when the pipe is flowing half full.

18.3.11 Drainage System Discharge

The discharge of the park drainage system shall be connected to a public sewer. Where a public sewer is not available within 300 feet for use, an individual sewage disposal system of a type that is acceptable and approved by the Authority Having Jurisdiction shall be installed.

> *Comment: It is preferred that trailer parks be connected to public sewers, even if more than 300 feet away. Trailer sites are typically densely spaced to maximize the use of the property, which leaves little space for an adequate on-site sewage disposal system.*

18.3.12 Minimum Drainage Pipe Size

Minimum pipe sizes in the drainage system shall be in accordance with Table 18.3.12.

Table 18.3.12
DRAIN PIPE SIZING

Maximum Number of Trailers, Individually Vented Systems	Maximum Number of Trailers, Wet-Vented Systems	Size of Drain
2	1	3"
30	10	4"
100	50	6"
400	-	8"
1000	-	10"

18.3.13 Trailer Connections

Each trailer connection fixture outlet shall be provided with a screw-type plug or cap, and be effectively capped when not in use.

18.4 VENTING

18.4.1 Location

Each wet-vented drainage system shall be provided with a vent not more than 15 feet downstream from its upper trap, and long mains shall be provided with additional relief vents at intervals of not more than 100 feet thereafter. The minimum size of each vent serving a wet-vented system shall be as set forth in Table 18.4.1.

Table 18.4.1 VENT SIZING	
Size of Wet-Vented Drain	Minimum Size of Vent
3"	2"
4"	3"
5"	4"
6"	5"

18.4.2 Reserved

18.4.3 Reserved

18.4.4 Reserved

18.4.5 Vent Connections

All vent intersections shall be taken off above the center line of the horizontal pipe. All vent stacks shall be supported by a four-inch by four-inch redwood post, set in at least two feet of concrete extending at least four inches above the ground, or supported by another approved method.

> *Comment: Treated lumber is considered to be the equivalent of redwood.*

18.4.6 Galvanized Steel Vent Pipe

Galvanized steel vent pipe may extend below the ground vertically, and may directly intersect a drainage line with an approved fitting, if the entire section around both the drain and the galvanized pipe is encased in concrete to prevent any movement. Galvanized steel pipe encased in concrete shall be first coated with bituminous paint, or equivalent protective material.

18.4.7 Location of Vent Pipes

Vent pipes shall terminate at least 10 feet above grade and be at least 10 feet from any property line. No vent shall terminate directly beneath any door, window, or ventilation opening of any building, nor shall any vent terminal be within 10 feet horizontally from such openings unless it is at least 2 feet above the top of such opening.

18.4.8 Reserved

18.4.9 Wet Vented Branch Drain Lines

No three-inch branch drain shall exceed six feet in length, and no four-inch branch drain shall exceed 15 feet in length, unless they are properly vented.

18.5 WATER DISTRIBUTION SYSTEM

18.5.1 Conformance

Each trailer park water distribution system shall conform to the requirements of Chapter 10 of this Code and shall be so designed and maintained as to provide a residual pressure of not less than 20 psi at each trailer site under normal operating conditions.

18.5.2 Individual Water Service Branch

Every trailer site shall be provided with an individual water service branch line that shall not be less than 3/4" size, delivering safe, potable water.

18.5.3 Water Service Branch Components

A manual shutoff valve shall be installed on the water service branch, followed by an ASSE 1024 or CSA B64.6 dual check backflow preventer on the discharge side of the shutoff valve, with a pressure relief valve located on the discharge side of the backflow preventer, and with a hose connection or other approved trailer attachment on the trailer side of the relief valve. The pressure relief valve shall be set for 125 psi and have a full-size discharge pipe with the end of the pipe not more than two feet or less than six inches above the ground and pointing downward. No part of the discharge pipe shall be trapped. No shut-off valve shall be installed between the pressure relief valve and the trailer that it serves. The backflow device and relief valve shall be located not less than 12 inches above the ground.

18.5.4 Connection Details

The service connection shall not be rigid. Flexible metal tubing is permitted. Fittings at either end shall be of a quick disconnect type not requiring any special tools or knowledge to install or remove.

18.5.5 Water Fixture Units

Each trailer outlet on the water distribution system shall be rated as six water supply fixture units (WSFU).

18.5.6 Location of Water Connection

The trailer park water outlet for each trailer coach space shall be located near the center of the left side of each trailer coach.

18.5.7 Fire Protection

In the design of the water distribution system in a trailer park, consideration for fire outlet stations throughout the park should be made relative to the location and quantity of water necessary during an emergency period.

18.5.8 Backflow Prevention

Backflow prevention shall be in accordance with Section 10.5.

18.6 RESERVED

18.7 TESTING

Installations shall be tested in accordance with Chapter 15.

18.8 SANITARY FACILITIES

18.8.1 Public Water Closets, Showers, and Lavatories

Separate public water closets, showers, and lavatories shall be installed and maintained for each sex in accordance with the following ratio of trailer sites:

18.8.1.1 Dependent Trailer

Trailer parks constructed and operated exclusively for dependent trailers shall have one water closet, one shower, and one lavatory for each 10 sites or fractional part thereof.

> *Comment: Dependent trailers are not equipped with a water closet and lavatory and are dependent on the trailer park for such facilities.*

18.8.1.2 Independent Trailer

Trailer parks constructed and operated exclusively for independent trailers shall have one water closet, one shower, and one lavatory for each 100 sites or fractional part thereof.

> *Comment: Independent trailers are equipped with a water closet and lavatory and are not dependent on the trailer park for such facilities.*

18.8.1.3 Combined Trailer Use

Trailer parks constructed and operated for the combined use of dependent and independent trailers shall have facilities as shown in Table 18.8.1.3.

Table 18.8.1.3 FACILITIES REQUIRED FOR COMBINED TRAILER USE			
Sites	Water Closets	Showers	Lavatories
2-25	1	1	1
26-70	2	2	2

18.8.1.4 Additional Water Closets

For combined trailer use, one additional water closet shall be provided for each 100 sites or fractional part thereof in excess of 70 sites.

> *Comment: Section 18.8.1.4 applies to trailer parks with combined trailer use, as in Section 18.8.1.3.*

18.8.2 Exclusivity

Each toilet facility shall be for the exclusive use of the occupants of the trailer sites in the trailer park.

18.8.3 Showers

In every trailer park, shower bathing facilities with hot and cold running water shall be installed in separate compartments. Every compartment shall be provided with a self-closing door or otherwise equipped with a waterproof draw curtain.

18.8.4 Laundry Facilities

Every trailer park shall be provided with an accessory utility building containing at least one clothes washer or laundry tray equipped with hot and cold running water for every 20 trailer sites or fractional part thereof, but in no case shall there be less than two laundry trays in any trailer park.

18.8.5 Shower Compartments

The inner face of walls of all shower compartments shall be finished with concrete, metal, tile or other approved waterproof materials extending to a height of not less than six feet above the floor. Floors or shower compartments shall be made of concrete or other similar impervious material. Floors shall be waterproof and slope 1/4 inch per foot to the drains.

18.9 MAINTENANCE

All required devices or safeguards shall be maintained in good working order. The owner, operator, or lessee of the trailer park, or his designated agent shall be responsible for the maintenance.

18.10 OPERATOR'S RESPONSIBILITY - VIOLATIONS

When it is evident that there exists, or may exist, a violation of any pertinent regulation, the owner, operator, lessee, person in charge of the park, or any other person causing a violation shall immediately disconnect the trailer water supply and sewer connections from the park systems and shall employ such other corrective measures as may be ordered by the Authority Having Jurisdiction.

Chapter 19

Referenced Standards

19.1 REFERENCED STANDARDS

Table 19.1 lists the standards that are referenced within the requirements of this Code. Table 19.1 shows the designation number of the standard, its edition date, a date if reaffirmed, any addendums or amendments, its title, and the Sections and/or Tables in the Code where the standard is referenced.

19.2 STANDARDS ORGANIZATIONS

AHAM — Association of Home Appliance Manufacturers
1111 19th Street, NW, Suite 402
Washington, DC 20036 USA
202-872-5955

ANSI — American National Standards Institute
1899 L Street, NW, 11th Floor
Washington, DC 20036 USA
202-293-8020   fax: 202-293-9287

ASHRAE — ASHRAE Headquarters
1791 Tullie Circle, NE
Atlanta, GA 30329 USA
404-636-8400   fax: 404-321-5478

ASME — ASME
Two Park Avenue
New York, NY 10016-5990 USA
800-843-2763

ASSE — ASSE International
18927 Hickory Creek Drive, Suite 220
Mokena, IL 60448 USA
708-995-3019   fax: 708-479-6139

ASTM — ASTM International
100 Barr Harbor Drive
P.O. Box C700
West Conshohocken, PA 19428-2959 USA
877-909-2786

AWWA — American Water Works Association
6666 W. Quincy Avenue
Denver, CO 80235 USA
800-926-7337   fax: 303-347-0804

CISPI — Cast Iron Soil Pipe Institute
3008 Preston Station Drive
Hixson, TN 37343 USA
423-842-2122

CSA — CSA Group
178 Rexdale Boulevard
Toronto, ON, Canada, M9W 1R3
800-463-6727

FM — FM Global
270 Central Avenue
Johnston, RI 02919-4949 USA
401-275-3000   fax: 401-275-3029

IAPMO — IAPMO
4755 E. Philadelphia Street
Ontario, CA 91761 USA
909-472-4100   fax: 909-472-4150

ISEA — International Safety Equipment Association
1901 North Moore Street
Arlington, VA 22209-1762 USA
703-525-1695   fax: 703-528-2148

MSS — Manufacturers Standardization Society
127 Park Street, NE
Vienna, VA 22180-4602 USA
703-281-6613

NFPA — NFPA
1 Batterymarch Park
Quincy, MA 02169-7471 USA
617-770-3000   fax: 617-770-0700

NSF — NSF International
789 N. Dixboro Road
P.O. 130140
Ann Arbor, MI 48105 USA
800-673-6275   fax: 734-769-0109

PDI — Plumbing and Drainage Institute
800 Turnpike Street - Suite 300
North Andover. MA 01845 USA
800-589-8956

UL — Underwriters Laboratories Inc.
333 Pfingsten Road
Northbrook, IL 60062-2096 USA
847-272-8800

Table 19.1 REFERENCED STANDARDS

Standard Number	Standard Title	2015 NSPC
AHAM DW-1 - 2010	Dishwashers	Table 3.1.3-VII, 7.15.1
AHAM FWD-1 - 2009	Food Waste Disposers	Table 3.1.3-VII, 7.14.1
AHAM HLW-1 - 2010	Performance Evaluation Procedures for Household Clothes Washers	Table 3.1.3-VII, 7.13.1
ANSI Z4.3 - 1995 (R2005)	Sanitation - Non-sewered Waste-Disposal Systems - Minimum Requirements	2.24, Table 3.1.3-X
ANSI Z21.10.1 - 2013	Gas Water Heaters - Volume I, Storage Water Heaters with Input Ratings of 75,000 Btu per Hour or Less	Table 3.1.3-VII, 10.15.11
ANSI Z21.10.1/CSA 4.1	Refer to ANSI Z21.10.1 or CSA 4.1	Table 3.1.3-VII, 10.15.11
ANSI Z21.10.3 - 2013	Gas Water Heaters - Volume III, Storage Water Heaters with Input Ratings above 75,000 Btu Per Hour, Circulating and Instantaneous	Table 3.1.3-VII, 10.15.11
ANSI Z21.10.3/CSA 4.3	Refer to ANSI Z21.10.3 or CSA 4.3	Table 3.1.3-VII, 10.15.11
ANSI Z21.22 - 1999 (R2008)	Relief Valves for Hot Water Supply Systems	Table 3.1.3-VIII, 10.16.7, Figure 10.16.7
ANSI Z21.22/CSA 4.4	Refer to ANSI Z21.22 or CSA 4.4	Table 3.1.3-VIII, 10.16.7, Figure 10.16.7
ANSI/ISEA Z358.1 - 2009	American National Standard for Emergency Eyewash and Shower Equipment	Table 3.1.3-V, Table 7.21.1 Note (17), 7.24
ASHRAE Standard 18 – 2008 (R2013)	Methods of Testing for Rating Drinking-Water Coolers with Self-Contained Mechanical Refrigeration	7.12.1
ASME A13.1 - 2007 (R2013)	Scheme for the Identification of Piping Systems	2.27, 10.21
ASME A112.1.2 - 2012	Air Gaps in Plumbing Systems (for Plumbing Fixtures and Water-Connected Receptors)	Table 3.1.3-IX, 10.5.1, 10.5.3
ASME A112.1.3 - 2000 (R2010)	Air Gap Fittings for Use with Plumbing Fixtures, Appliances, and Appurtenances	Table 3.1.3-IX, 10.5.1, 10.5.3
ASME A112.3.1 - 2007	Stainless Steel Drainage Systems for Sanitary DWV, Storm, and Vacuum Applications, Above- and Below-Ground	Table 3.1.3-I, Table 3.5, Table 3.6, Table 3.7, 7.16.1
ASME A112.3.4 - 2013	Plumbing Fixtures with Pumped Waste and Macerating Toilet Systems	Table 3.1.3-X, 7.4.8, 11.7.6
ASME A112.3.4/CSA B45.9	Refer to ASME A112.3.4 or CSA B45.9	Table 3.1.3-X, 7.4.8, 11.7.6
ASME A112.4.1 - 2009	Water Heater Relief Valve Drain Tubes	Table 3.1.3-VIII
ASME A112.4.2 - 2009	Water Closet Personal Hygiene Devices	Table 3.1.3-VI, 7.7.1
ASME A112.4.3 - 1999 (R2010)	Plastic Fittings for Connecting Water Closets to the Sanitary Drainage System	Table 3.1.3-IV
ASME A112.4.14 - 2004 (R2010)	Manually Operated, Quarter-Turn Shutoff Valves for Use in Plumbing Systems	Table 3.1.3-VIII
ASME A112.6.1M - 1997 (R2012)	Floor-Affixed Supports for Off-the-Floor Plumbing Fixtures for Public Use	Table 3.1.3-VI, 3.3.12, 7.3.5
ASME A112.6.2 - 2000 (R2010)	Framing-Affixed Supports for Off-the Floor Water Closets with Concealed Tanks	Table 3.1.3-VI, 3.3.12, 7.3.5

Table 19.1 REFERENCED STANDARDS (continued)

Standard Number	Standard Title	2015 NSPC
ASME A112.6.3 - 2001 (R2007)	Floor and Trench Drains	Table 3.1.3-V, 5.3.6, 7.16.1, Figure 7.16.2-A, 7.16.3
ASME A112.6.4 - 2003 (R2012)	Roof, Deck, and Balcony Drains	Table 3.1.3-V, 13.5.1, 13.5.2, 13.5.3
ASME A112.6.7 - 2010	Sanitary Floor Sinks	Table 3.1.3-V, 7.11.5
ASME A112.14.1 - 2003 (R2012)	Backwater Valves	Table 3.1.3-VIII, 5.5.2
ASME A112.14.3 - 2000 (R2004)	Grease Interceptors	Table 3.1.3-X, 6.1.4.2, 6.2.1.1, 6.2.2
ASME A112.14.4 - 2001 (R2012)	Grease Removal Devices	Table 3.1.3-X, 6.1.4.2, 6.2.1.2, 6.2.2
ASME A112.14.6 - 2010	FOG (Fats, Oils, and Greases) Disposal Systems	Table 3.1.3-X, 6.2.1.4
ASME A112.18.1 - 2012	Plumbing Supply Fittings	Table 3.1.3-VI, 7.6.2, 7.7.1, 7.8.3, 7.10.2, 7.11.1
ASME A112.18.1/CSA B125.1	Refer to ASME A112.18.1 or CSA B125.1	Table 3.1.3-VI, 7.6.2, 7.7.1, 7.8.3, 7.10.2, 7.11.1
ASME A112.18.2 - 2011	Plumbing Waste Fittings	Table 3.1.3-VI, 3.2.1, 7.11.1
ASME A112.18.2/CSA B125.2	Refer to ASME A112.18.2 or CSA B125.2	Table 3.1.3-VI, 3.2.1, 7.11.1
ASME A112.18.3 - 2002 (R2012)	Performance Requirements for Backflow Protection Devices and Systems in Plumbing Fixture Fittings	Table 3.1.3-IX
ASME A112.18.6 - 2009	Flexible Water Connectors	Table 3.1.3-VI, 10.13
ASME A112.18.6/CSA B125.6	Refer to ASME A112.18.6 or CSA B125.6	Table 3.1.3-VI, 10.13
ASME A112.18.9 - 2011	Protectors/Insulators for Exposed Wastes and Supplies on Accessible Fixtures	Table 3.1.3-VI, 7.2
ASME A112.19.1 - 2013	Enameled Cast Iron and Enameled Steel Plumbing Fixtures	Table 3.1.3-V, 7.6.1, 7.8.1, 7.11.1
ASME A112.19.1/CSA B45.2	Refer to ASME A112.19.1 or CSA B45.2	Table 3.1.3-V, 7.6.1, 7.8.1, 7.11.1
ASME A112.19.2 - 2013	Ceramic Plumbing Fixtures	Table 3.1.3-V, 7.4.1, 7.4.2, 7.5.1, 7.5.2, 7.6.1, 7.7.1, 7.11.1
ASME A112.19.2/CSA B45.1	Refer to ASME A112.19.2 or CSA B45.1	Table 3.1.3-V, 7.4.1, 7.4.2, 7.5.1, 7.5.2, 7.6.1, 7.7.1, 7.11.1
ASME A112.19.3 - 2008 (R2013)	Stainless Steel Plumbing Fixtures	Table 3.1.3-V, 7.6.1, 7.11.1
ASME A112.19.3/CSA B45.4	Refer to ASME A112.19.3 or CSA B45.4	Table 3.1.3-V, 7.6.1, 7.11.1
ASME A112.19.5 - 2011 w/update 1	Flush Valves and Spuds for Water Closets, Urinals, and Tanks	Table 3.1.3-VI
ASME A112.19.7 - 2012	Hydromassage Bathtub Systems	Table 3.1.3-V, 7.9.2

Table 19.1 REFERENCED STANDARDS (continued)		
Standard Number	**Standard Title**	**2015 NSPC**
ASME A112.19.7/CSA B45.10	Refer to ASME A112.19.7 or CSA B45.10	Table 3.1.3-V, 7.9.2
ASME A112.19.10 - 2003 (R2008)	Dual Flush Devices for Water Closets	Table 3.1.3-VI
ASME A112.19.12 - 2006 (R2011)	Wall Mounted, Pedestal Mounted, Adjustable, Elevating, Tilting, and Pivoting Lavatory, Sink, and Shampoo Bowl Carrier Systems and Drain Waste Systems	Table 3.1.3-VI, 3.3.12, 7.2
ASME A112.19.14 - 2013	Six Liter Water Closets Equipped with a Dual Flushing Device	Table 3.1.3-V
ASME A112.19.15 - 2012	Bathtubs/Whirlpool Bathtubs with Pressure Sealed Doors	Table 3.1.3-V, 7.8.1
ASME A112.19.17 - 2010	Manufactured Safety Vacuum Release Systems (SVRS) for Residential and Commercial Swimming Pool, Spa, Hot Tub, and Wading Pool Suction Systems	Table 3.1.3-VI
ASME A112.19.19 - 2006 (R2011)	Vitreous China Non-water Urinals	Table 3.1.3-V, 7.5.1
ASME A112.36.2M - 1991 (R2012)	Cleanouts	Table 3.1.3-VIII
ASME A112.1016 - 2011 w/updates 1 & 2	Performance Requirements for Automatic Compensating Valves for Individual Shower and Tub/Shower Combinations	Table 3.1.3-VI, Table 3.1.3-VIII, 10.15.6
ASME A112.1016/ASSE 1016/ CSA B125.16	Refer to ASME A112.1016, ASSE 1016, or CSA B125.16	Table 3.1.3-VI, Table 3.1.3-VIII, 10.15.6
ASME B1.20.1 - 2013	Pipe Threads, General Purpose (Inch)	Table 3.1.3-IV, 4.2.2
ASME B16.3 - 2011	Malleable Iron Threaded Fittings: Classes 150 and 300	Table 3.1.3-I, Table 3.4
ASME B16.4 - 2011	Gray Iron Threaded Fittings: Classes 125 and 250	Table 3.1.3-I, Table 3.4
ASME B16.5 - 2013	Pipe Flanges and Flanged Fittings: NPS 1/2 through NPS 24 Metric/Inch Standard	Table 3.1.3-I, Table 3.4
ASME B16.12 - 2009	Cast Iron Threaded Drainage Fittings	Table 3.1.3-I, Table 3.5, Table 3.6, Table 3.7
ASME B16.14 - 2013	Ferrous Pipe Plugs, Bushings, and Locknuts with Pipe Threads	Table 3.1.3-I
ASME B16.15 - 2013	Cast Copper Alloy Threaded Fittings: Classes 125 and 250	Table 3.1.3-II, Table 3.4
ASME B16.18 - 2012	Cast Copper Alloy Solder Joint Pressure Fittings	Table 3.1.3-II, Table 3.4, 4.2.4, 4.2.6, 10.20.4
ASME B16.22 - 2013	Wrought Copper and Copper Alloy Solder-Joint Pressure Fittings	Table 3.1.3-II, Table 3.4, 4.2.4, 4.2.6, 4.2.8.2, 10.20.4
ASME B16.23 - 2011	Cast Copper Alloy Solder Joint Drainage Fittings: DWV	Table 3.1.3-II, Table 3.5, Table 3.6, Table 3.7, 4.2.4
ASME B16.24 - 2011	Cast Copper Alloy Pipe Flanges and Flanged Fittings, Classes 150, 300, 400, 600, 800, 1500, and 2500	Table 3.1.3-II, Table 3.4, 4.2.4
ASME B16.26 - 2013	Cast Copper Alloy Fittings for Flared Copper Tubes	Table 3.1.3-II, Table 3.4, 4.2.5

Table 19.1 REFERENCED STANDARDS (continued)

Standard Number	Standard Title	2015 NSPC
ASME B16.29 - 2012	Wrought Copper and Wrought Copper Alloy Solder-Joint Drainage Fittings - DWV	Table 3.1.3-II, Table 3.5, Table 3.6, Table 3.7, 4.2.4
ASME B16.50 - 2013	Wrought Copper and Copper Alloy Braze-Joint Pressure Fittings	Table 3.1.3-II, Table 3.4, 4.2.4, 4.2.8.2
ASME B16.51 - 2013	Copper and Copper Alloy Press-Connect Pressure Fittings	Table 3.1.3-II, Table 3.4
ASSE 1001 - 2008	Performance Requirements for Atmospheric Type Vacuum Breakers	Table 3.1.3-IX, 7.8.4, 10.5.3, 10.5.10
ASSE 1002 - 2008	Performance Requirements for Anti-Siphon Fill Valves for Water Closet Tanks	Table 3.1.3-VI, 7.19.3
ASSE 1003 - 2009	Performance Requirements for Water Pressure Reducing Valves for Domestic Water Distribution Systems	Table 3.1.3-VIII, 10.14.6
ASSE 1004 - 2008	Backflow Prevention Requirements for Commercial Dishwashing Machines	Table 3.1.3-IX, 7.15.1
ASSE 1008 - 2006	Performance Requirements for Plumbing Aspects of Residential Food Waste Disposer Units	Table 3.1.3-VII, 7.14.1
ASSE 1010 - 2004	Performance Requirements for Water Hammer Arresters	Table 3.1.3-VIII, 10.14.7
ASSE 1011 - 2004	Performance Requirements for Hose Connection Vacuum Breakers	Table 3.1.3-IX, 10.5.3
ASSE 1012 - 2009	Performance Requirements for Backflow Preventer with Intermediate Atmospheric Vent	Table 3.1.3-IX, 10.5.3, 10.5.8
ASSE 1013 - 2011	Performance Requirements for Reduced Pressure Principle Backflow Preventers and Reduced Pressure Fire Protection Principle Backflow Preventers	Table 3.1.3-IX, 10.5.3, 10.5.6, 10.5.9, 10.5.10, 10.5.13
ASSE 1014 - 2005	Performance Requirements for Backflow Prevention Devices for Hand-Held Showers	Table 3.1.3-IX, 10.5.3
ASSE 1015 - 2011	Performance Requirements for Double Check Backflow Prevention Assemblies and Double Check Fire Protection Backflow Prevention Assemblies	Table 3.1.3-IX, 10.5.3, 10.5.6, 10.5.9
ASSE 1016 - 2011	Performance Requirements for Automatic Compensating Valves for Individual Shower and Tub/Shower Combinations	Table 3.1.3-VI, Table 3.1.3-VIII, 10.15.6
ASSE 1016/ASME A112.1016/ CSA B125.16	Refer to ASSE 1016, ASME A112.1016, or CSA B125.16	Table 3.1.3-VI, Table 3.1.3-VIII, 10.15.6
ASSE 1017 - 2009	Performance Requirements for Temperature Actuated Mixing Valves for Hot Water Distribution Systems	Table 3.1.3-VIII, 10.15.6, 10.15.10
ASSE 1018 - 2001	Performance Requirements for Trap Seal Primer Valves - Potable Water Supplied	Table 3.1.3-VIII, 5.3.6, 7.16.2
ASSE 1019 - 2011	Performance Requirements for Wall Hydrants with Backflow Protection and Freeze Resistance	Table 3.1.3-IX, 10.5.3
ASSE 1020 - 2004	Performance Requirements for Pressure Vacuum Breaker Assembly	Table 3.1.3-IX, 10.5.3, 10.5.5, Figure 10.5.5-G, 10.5.6, 10.5.10
ASSE 1021 - 2001	Performance Requirements for Drain Air Gaps for Domestic Dishwasher Applications	Table 3.1.3-IX

\multicolumn{3}{	c	}{**Table 19.1 REFERENCED STANDARDS** (continued)}
Standard Number	**Standard Title**	**2015 NSPC**
ASSE 1022 - 2003	Performance Requirements for Backflow Preventer for Beverage Dispensing Equipment	Table 3.1.3-IX, 10.5.3, 10.5.8
ASSE 1023 - 1979	Performance Requirements for Hot Water Dispensers Household Storage Type - Electrical	Table 3.1.3-VII
ASSE 1024 - 2004	Performance Requirements for Dual Check Backflow Preventers	Table 3.1.3-IX, 10.5.3, 10.5.9, 18.5.3
ASSE 1032 - 2004 (R2011)	Performance Requirements for Dual Check Valve Type Backflow Preventers for Carbonated Beverage Dispensers - Post Mix Type	Table 3.1.3-IX, 10.5.8
ASSE 1035 - 2008	Performance Requirements for Laboratory Faucet Backflow Preventers	Table 3.1.3-IX, 10.5.3, 10.5.13
ASSE 1037 - 1990	Performance Requirements for Pressurized Flushing Devices (Flushometers) for Plumbing Fixtures	Table 3.1.3-VI, 7.19.4, 7.19.5
ASSE 1044 - 2001	Performance Requirements for Trap Seal Primer Devices - Drainage Types and Electronic Design Types	Table 3.1.3-VIII, 5.3.6, 7.16.2
ASSE 1047 - 2011	Performance Requirements for Reduced Pressure Detector Fire Protection Backflow Prevention Assemblies	Table 3.1.3-IX, 10.5.3, 10.5.6, 10.5.9
ASSE 1048 - 2011	Performance Requirements for Double Check Detector Fire Protection Backflow Prevention Assemblies	Table 3.1.3-IX, 10.5.3, 10.5.6, 10.5.9
ASSE 1052 - 2004	Performance Requirements for Hose Connection Backflow Preventers	Table 3.1.3-IX, 10.5.3
ASSE 1053 - 2004	Performance Requirements for Dual Check Backflow Preventer Wall Hydrants - Freeze Resistant Type	Table 3.1.3-IX, 10.5.3
ASSE 1055 - 2009	Performance Requirements for Chemical Dispensing Systems	Table 3.1.3-IX, 10.5.13
ASSE 1056 - 2013	Performance Requirements for Spill Resistant Vacuum Breakers	Table 3.1.3-IX, 10.5.3, 10.5.5, Figure 10.5.5-G, 10.5.6, 10.5.10
ASSE 1057 - 2012	Performance Requirements for Freeze Resistant Sanitary Yard Hydrants with Backflow Protection	Table 3.1.3-IX, 10.5.3
ASSE 1060 - 2006	Performance Requirements for Outdoor Enclosures for Fluid Carrying Components	Table 3.1.3-IX, 10.5.5
ASSE 1061 - 2011	Performance Requirements for Push-Fit Fittings	Table 3.1.3-II, Table 3.1.3-III, Table 3.1.3-IV, 4.2.7
ASSE 1062 - 2006	Performance Requirements for Temperature Actuated Flow Reduction (TAFR) Valves for Individual Fixture Fittings	Table 3.1.3-VI, 10.15.6
ASSE 1069 - 2005	Performance Requirements for Automatic Temperature Control Mixing Valves	Table 3.1.3-VIII, 10.15.6
ASSE 1070 - 2004	Performance Requirements for Water Temperature Limiting Devices	Table 3.1.3-VIII, 10.15.6
ASSE 1071 - 2012	Performance Requirements for Temperature Actuated Mixing Valves for Plumbed Emergency Equipment	Table 3.1.3-VIII, 7.24
ASSE 1072 - 2007	Performance Requirements for Barrier Type Floor Drain Trap Seal Protection Devices	Table 3.1.3-VIII, 5.3.6, 7.16.2

| Table 19.1 REFERENCED STANDARDS (continued) |||
Standard Number	Standard Title	2015 NSPC
ASSE 1079 - 2012	Performance Requirements for Dielectric Pipe Unions	Table 3.1.3-IV, 4.3.7
ASSE 6000 - 2012	Professional Qualification Standard for Medical Gas Systems Personnel	14.3.2
ASTM A53 - 2012	Standard Specification for Pipe, Steel, Black and Hot-Dipped, Zinc Coated, Welded and Seamless	Table 3.1.3-I, Table 3.4, Table 3.5, Table 3.6, Table 3.7, 11.7.8
ASTM A74 - 2013a	Standard Specification for Cast Iron Soil Pipe and Fittings	Table 3.1.3-I, Table 3.5, Table 3.6, Table 3.7
ASTM A377 - 2003 (R2008)e1	Specifications for Ductile-Iron Pressure Pipe	Table 3.1.3-I, Table 3.4, 11.7.8
ASTM A716 - 2008	Standard Specification for Ductile Iron Culvert Pipe	Table 3.1.3-I, Table 3.5, Table 3.7
ASTM A746 - 2009	Standard Specification for Ductile Iron Gravity Sewer Pipe	Table 3.1.3-I, Table 3.5, Table 3.7
ASTM A888 - 2013a	Standard Specification for Hubless Cast Iron Soil Pipe and Fittings for Sanitary and Storm Drain, Waste, and Vent Piping Applications	Table 3.1.3-I, Table 3.5, Table 3.6, Table 3.7
ASTM A1056 - 2012	Standard Specification for Cast Iron Couplings for Joining Hubless Cast Iron Soil Pipe and Fittings	Table 3.1.3-IV, 4.3.8
ASTM B29 - 2003 (R2009)	Standard Specification for Refined Lead	Table 3.1.3-IV
ASTM B32 - 2008	Standard Specification for Solder Metal	Table 3.1.3-IV, 4.2.4, 10.20.4
ASTM B43 - 2009	Standard Specification for Red Brass Pipe, Standard Sizes	Table 3.1.3-II, Table 3.4
ASTM B88 - 2009	Standard Specification for Seamless Copper Water Tube	Table 3.1.3-II, Table 3.4, Table 3.5, Table 3.6, Table 3.7, 4.2.6, 10.20.4, 11.7.8
ASTM B302 - 2012	Standard Specification for Threadless Copper Pipe, Standard Sizes	Table 3.1.3-II
ASTM B306 - 2013	Standard Specification for Copper Drainage Tube (DWV)	Table 3.1.3-II, Table 3.5, Table 3.6, Table 3.7
ASTM B813 - 2010	Standard Specification for Liquid and Paste Fluxes for Copper and Copper Alloy Tube	Table 3.1.3-IV, 4.2.4, 10.20.4
ASTM B828 - 2002 (R2010)	Standard Practice for Making Capillary Joints by Soldering of Copper and Copper Alloy Tube and Fittings	Table 3.1.3-XI, 4.2.4, 10.20.4
ASTM C4 - 2004 (R2009)	Standard Specification for Clay Drain Tile and Perforated Clay Drain Tile	Table 3.1.3-III, Table 3.8
ASTM C12 – 2013a	Standard Practice for Installing Vitrified Clay Pipe Lines	Table 3.1.3-XI
ASTM C14 - 2014	Standard Specification for Non-reinforced Concrete Sewer, Storm Drain, and Culvert Pipe	Table 3.1.3-III, Table 3.5, Table 3.7
ASTM C76 - 2014	Standard Specification for Reinforced Concrete Culvert, Storm Drain, and Sewer Pipe	Table 3.1.3-III, Table 3.5, Table 3.7
ASTM C412 - 2011	Standard Specification for Concrete Drain Tile	Table 3.1.3-III, Table 3.8
ASTM C425 - 2004 (R2013)	Standard Specification for Compression Joints for Vitrified Clay Pipe and Fittings	Table 3.1.3-IV

Table 19.1 REFERENCED STANDARDS (continued)

Standard Number	Standard Title	2015 NSPC
ASTM C443 - 2012	Standard Specification for Joints for Concrete Pipe and Manholes, Using Rubber Gaskets	Table 3.1.3-IV
ASTM C444 - 2003 (R2009)	Standard Specification for Perforated Concrete Pipe	Table 3.1.3-III, Table 3.8
ASTM C564 - 2012	Standard Specification for Rubber Gaskets for Cast Iron Soil Pipe and Fittings	Table 3.1.3-IV
ASTM C700 - 2013	Standard Specification for Vitrified Clay Pipe, Extra Strength, Standard Strength, and Perforated	Table 3.1.3-III, Table 3.5, Table 3.6, Table 3.7, Table 3.8
ASTM C1173 - 2010e1	Standard Specification for Flexible Transition Couplings for Underground Piping Systems	Table 3.1.3-IV, 4.2.11.6, 4.3.8
ASTM C1277 - 2012	Standard Specification for Shielded Couplings Joining Hubless Cast Iron Soil Pipe and Fittings	Table 3.1.3-IV, 4.3.8
ASTM C1460 - 2012	Standard Specification for Shielded Transition Couplings for Use With Dissimilar DWV Pipe and Fittings Above Ground	Table 3.1.3-IV, 4.3.8
ASTM C1461 - 2008 (R2013)	Standard Specification for Mechanical Couplings Using Thermoplastic Elastomeric (TPE) Gaskets for Joining Drain, Waste, and Vent (DWV), Sewer, Sanitary, and Storm Plumbing Systems for Above and Below Ground Use	Table 3.1.3-IV, 4.3.8
ASTM C1540 - 2011	Standard Specification for Heavy Duty Shielded Couplings Joining Hubless Cast Iron Soil Pipe and Fittings	Table 3.1.3-IV, 4.3.8
ASTM D1330 - 2004 (R2010)	Standard Specification for Rubber Sheet Gaskets	Table 3.1.3-IV
ASTM D1785 - 2012	Standard Specification for Poly(Vinyl Chloride) (PVC) Plastic Pipe, Schedules 40, 80, and 120	Table 3.1.3-III, Table 3.4, Table 3.4.2, 11.7.8
ASTM D2235 - 2004 (R2011)	Standard Specification for Solvent Cement for Acrylonitrile-Butadiene-Styrene (ABS) Plastic Pipe and Fittings	Table 3.1.3-IV
ASTM D2239 - 2012a	Standard Specification for Polyethylene (PE) Plastic Pipe (SIDR-PR) Based on Controlled Inside Diameter	Table 3.1.3-III, Table 3.4, Table 3.4.2
ASTM D2241 - 2009	Standard Specification for Poly(Vinyl Chloride) (PVC) Pressure-Rated Pipe (SDR Series)	Table 3.1.3-III, Table 3.4, Table 3.4.2
ASTM D2321 - 2011	Standard Practice for Underground Installation of Thermoplastic Pipe for Sewers and Other Gravity-Flow Applications	2.6.6, Table 3.1.3-XI
ASTM D2464 - 2013	Standard Specification for Threaded Poly(Vinyl Chloride) (PVC) Plastic Pipe Fittings, Schedule 80	Table 3.1.3-III, Table 3.4
ASTM D2466 - 2013	Standard Specification for Poly(Vinyl Chloride) (PVC) Plastic Pipe Fittings, Schedule 40	Table 3.1.3-III, Table 3.4, Table 3.5, Table 3.6, Table 3.7
ASTM D2467 - 2013a	Standard Specification for Poly(Vinyl Chloride) (PVC) Plastic Pipe Fittings, Schedule 80	Table 3.1.3-III, Table 3.4
ASTM D2564 - 2012	Standard Specification for Solvent Cements for Poly(Vinyl Chloride) (PVC) Plastic Piping Systems	Table 3.1.3-IV
ASTM D2609 - 2002 (R2008)	Standard Specification for Plastic Insert Fittings for Polyethylene (PE) Plastic Pipe	Table 3.1.3-III, Table 3.4

Table 19.1 REFERENCED STANDARDS (continued)

Standard Number	Standard Title	2015 NSPC
ASTM D2661 - 2011	Standard Specification for Acrylonitrile-Butadiene-Styrene (ABS) Schedule 40 Plastic Drain, Waste, and Vent Pipe and Fittings	Table 3.1.3-III, Table 3.5, Table 3.6, Table 3.7
ASTM D2665 - 2012	Standard Specification for Poly(Vinyl Chloride) (PVC) Plastic Drain, Waste, and Vent Pipe and Fittings	Table 3.1.3-III, Table 3.5, Table 3.6, Table 3.7
ASTM D2672 - 1996a (R2009)	Standard Specification for Joints for IPS PVC Pipe Using Solvent Cement	Table 3.1.3-IV
ASTM D2680 - 2001 (R2009)	Standard Specification for Acrylonitrile-Butadiene-Styrene (ABS) and Poly(Vinyl Chloride) (PVC) Composite Sewer Piping	Table 3.1.3-III, Table 3.5, Table 3.7
ASTM D2683 - 2010e2	Standard Specification for Socket-Type Polyethylene Fittings for Outside Diameter Controlled Polyethylene Pipe and Tubing	Table 3.1.3-III, Table 3.4
ASTM D2729 - 2011	Standard Specification for Poly(Vinyl Chloride) (PVC) Sewer Pipe and Fittings	Table 3.1.3-III, Table 3.8
ASTM D2737 - 2012a	Standard Specification for Polyethylene (PE) Plastic Tubing	Table 3.1.3-III, Table 3.4, Table 3.4.2
ASTM D2774 - 2012	Standard Practice for Underground Installation of Thermoplastic Pressure Piping	2.6.6, Table 3.1.3-XI
ASTM D2846 - 2009be1	Standard Specification for Chlorinated Poly(Vinyl Chloride) (CPVC) Plastic Hot- and Cold-Water Distribution Systems	Table 3.1.3-III, Table 3.4, Table 3.4.2, Table 3.4.3, 11.7.8
ASTM D2852 - 1995 (R2008)	Standard Specification for Styrene-Rubber (SR) Plastic Drain Pipe and Fittings	Table 3.1.3-III, Table 3.8
ASTM D2855 - 1996 (R2010)	Standard Practice for Making Solvent-Cemented Joints with Poly(Vinyl Chloride) (PVC) Pipe and Fittings	Table 3.1.3-XI
ASTM D2949 - 2010	Standard Specification for 3.25-in. Outside Diameter Poly(Vinyl Chloride) (PVC) Plastic Drain, Waste, and Vent Pipe and Fittings	Table 3.1.3-III, Table 3.5, Table 3.6, Table 3.7
ASTM D3034 - 2014	Standard Specification for Type PSM Poly(Vinyl Chloride) (PVC) Sewer Pipe and Fittings	Table 3.1.3-III, Table 3.5, Table 3.7
ASTM D3035 - 2014	Standard Specification for Polyethylene (PE) Plastic Pipe (DR-PR) Based on Controlled Outside Diameter	Table 3.1.3-III, Table 3.4, Table 3.4.2, 4.2.14.5
ASTM D3122 - 1995 (R2009)	Standard Specification for Solvent Cements for Styrene-Rubber (SR) Plastic Pipe and Fittings	Table 3.1.3-IV
ASTM D3138 - 2004 (R2011)	Standard Specification for Solvent Cements for Transition Joints Between Acrylonitrile-Butadiene-Styrene (ABS) and Poly(Vinyl Chloride) (PVC) Non-Pressure Piping Components	Table 3.1.3-IV, 4.3.9
ASTM D3139 - 1998 (R2011)	Standard Specification for Joints for Plastic Pressure Pipes Using Flexible Elastomeric Seals	Table 3.1.3-IV, Table 3.4
ASTM D3212 - 2007 (R2013)	Standard Specification for Joints for Drain and Sewer Plastic Pipes Using Flexible Elastomeric Seals	Table 3.1.3-IV, 4.2.11.6
ASTM D3261 - 2012e1	Standard Specification for Butt Heat Fusion Polyethylene (PE) Plastic Fittings for Polyethylene (PE) Plastic Pipe and Tubing	Table 3.1.3-III, Table 3.4, 4.2.18

Table 19.1 REFERENCED STANDARDS (continued)		
Standard Number	**Standard Title**	**2015 NSPC**
ASTM D3262 - 2011	Standard Specification for "Fiberglass" (Glass-Fiber-Reinforced Thermosetting-Resin) Sewer Pipe	Table 3.1.3-III, Table 3.5, Table 3.7
ASTM D3311 - 2011	Standard Specification for Drain, Waste, and Vent (DWV) Plastic Fitting Patterns	Table 3.1.3-III
ASTM D3517 - 2014	Standard Specification for "Fiberglass" (Glass-Fiber-Reinforced Thermosetting-Resin) Pressure Pipe	Table 3.1.3-III, Table 3.4
ASTM D3840 - 2014	Standard Specification for "Fiberglass" (Glass-Fiber-Reinforced Thermosetting-Resin) Pipe Fittings for Non-Pressure Applications	Table 3.1.3-III, Table 3.5, Table 3.7
ASTM D4068 - 2009	Standard Specification for Chlorinated Polyethylene (CPE) Sheeting for Concealed Water-Containment Membrane	Table 3.1.3-X
ASTM D4551 - 2012	Standard Specification for Poly(Vinyl Chloride) (PVC) Plastic Flexible Concealed Water-Containment Membrane	Table 3.1.3-X
ASTM F402 - 2005 (R2012)	Standard Practice for Safe Handling of Solvent Cements, Primers, and Cleaners Used for Joining Thermoplastic Pipe and Fittings	Table 3.1.3-XI
ASTM F405 - 2013	Standard Specification for Corrugated Polyethylene (PE) Pipe and Fittings	Table 3.1.3-III, Table 3.8
ASTM F409 - 2012	Standard Specification for Thermoplastic Accessible and Replaceable Plastic Tube and Tubular Fittings	2.4.5, Table 3.1.3-VI
ASTM F437 - 2009	Standard Specification for Threaded Chlorinated Poly(Vinyl Chloride) (CPVC) Plastic Pipe Fittings, Schedule 80	Table 3.1.3-III, Table 3.4
ASTM F438 - 2009	Standard Specification for Socket-Type Chlorinated Poly(Vinyl Chloride) (CPVC) Plastic Pipe Fittings, Schedule 40	Table 3.1.3-III, Table 3.4
ASTM F439 - 2013	Standard Specification for Chlorinated Poly(Vinyl Chloride) (CPVC) Plastic Pipe Fittings, Schedule 80	Table 3.1.3-III, Table 3.4
ASTM F441 - 2013e1	Standard Specification for Chlorinated Poly(Vinyl Chloride) (CPVC) Plastic Pipe, Schedules 40 and 80	Table 3.1.3-III, Table 3.4, Table 3.4.2, Table 3.4.3, 11.7.8
ASTM F442 - 2013e1	Standard Specification for Chlorinated Poly(Vinyl Chloride) (CPVC) Plastic Pipe (SDR-PR)	Table 3.1.3-III, Table 3.4, Table 3.4.2, Table 3.4.3, 11.7.8
ASTM F477 - 2010	Standard Specification for Elastomeric Seals (Gaskets) for Joining Plastic Pipe	Table 3.1.3-IV
ASTM F481 - 1997 (R2008)	Standard Practice for Installation of Thermoplastic Pipe and Corrugated Pipe in Septic Tank Leach Fields	Table 3.1.3-XI
ASTM F493 - 2010	Standard Specification for Solvent Cements for Chlorinated Poly(Vinyl Chloride) (CPVC) Plastic Pipe and Fittings	Table 3.1.3-IV
ASTM F628 - 2012e1	Standard Specification for Acrylonitrile-Butadiene-Styrene (ABS) Schedule 40 Plastic Drain, Waste, and Vent Pipe with a Cellular Core	Table 3.1.3-III, Table 3.5, Table 3.6, Table 3.7

Table 19.1 REFERENCED STANDARDS (continued)

Standard Number	Standard Title	2015 NSPC
ASTM F656 - 2010	Standard Specification for Primers for Use in Solvent Cement Joints of Poly(Vinyl Chloride) (PVC) Plastic Pipe and Fittings	Table 3.1.3-IV
ASTM F714 - 2013	Standard Specification for Polyethylene (PE) Plastic Pipe (DR-PR) Based on Outside Diameter	Table 3.1.3-III, Table 3.4, Table 3.4.2, Table 3.5, Table 3.7, 4.2.14.5, 4.2.18
ASTM F810 - 2012	Standard Specification for Smoothwall Polyethylene (PE) Pipe for Use in Drainage and Waste Disposal Absorption Fields	Table 3.1.3-III
ASTM F876 - 2013a	Standard Specification for Crosslinked Polyethylene (PEX) Tubing	Table 3.1.3-III, Table 3.4, Table 3.4.2, Table 3.4.3
ASTM F877 - 2011a	Standard Specification for Crosslinked Polyethylene (PEX) Hot- and Cold-Water Distribution Systems	Table 3.1.3-III, Table 3.4, Table 3.4.2, Table 3.4.3
ASTM F891 - 2010	Standard Specification for Coextruded Poly(Vinyl Chloride) (PVC) Plastic Pipe with a Cellular Core	Table 3.1.3-III, Table 3.5, Table 3.6, Table 3.7
ASTM F1055 - 2013	Standard Specification for Electrofusion Type Polyethylene Fittings for Outside Diameter Controlled Polyethylene and Crosslinked Polyethylene (PEX) Pipe and Tubing	Table 3.1.3-III, Table 3.1.3-IV, Table 3.4
ASTM F1281 - 2011	Standard Specification for Crosslinked Polyethylene/Aluminum/Crosslinked Polyethylene (PEX-AL-PEX) Pressure Pipe	Table 3.1.3-III, Table 3.4, Table 3.4.2, Table 3.4.3
ASTM F1282 - 2010	Standard Specification for Polyethylene/Aluminum/Polyethylene (PE-AL-PE) Composite Pressure Pipe	Table 3.1.3-III, Table 3.4, Table 3.4.2, Table 3.4.3
ASTM F1290 - 1998a (R2011)	Standard Practice for Electrofusion Joining Polyolefin Pipe and Fittings	Table 3.1.3-XI
ASTM F1336 - 2007	Standard Specification for Poly(Vinyl Chloride) (PVC) Gasketed Sewer Fittings	Table 3.1.3-III, Table 3.5, Table 3.7
ASTM F1476 - 2007 (R2013)	Standard Specification for Performance of Gasketed Mechanical Couplings for Use in Piping Applications	Table 3.1.3-IV
ASTM F1498 - 2008 (2012)e1	Standard Specification for Taper Pipe Threads 60° for Thermoplastic Pipe and Fittings	4.2.14.6
ASTM F1760 - 2001 (R2011)	Standard Specification for Coextruded Poly(Vinyl Chloride) (PVC) Non-Pressure Plastic Pipe Having Reprocessed-Recycled Content	Table 3.1.3-III, Table 3.5, Table 3.6, Table 3.7
ASTM F1807 - 2013a	Standard Specification for Metal Insert Fittings Utilizing a Copper Crimp Ring for SDR9 Cross-linked Polyethylene (PEX) Tubing and SDR9 Polyethylene of Raised Temperature (PE-RT) Tubing	Table 3.1.3-III, Table 3.4
ASTM F1865 - 2009	Standard Specification for Mechanical Cold Expansion Insert Fitting With Compression Sleeve for Cross-linked Polyethylene (PEX) Tubing	Table 3.1.3-III, Table 3.4

\multicolumn{3}{c}{**Table 19.1 REFERENCED STANDARDS** (continued)}
Standard Number	**Standard Title**	**2015 NSPC**
ASTM F1866 - 2013	Standard Specification for Poly(Vinyl Chloride) (PVC) Plastic Schedule 40 Drainage and DWV Fabricated Fittings	Table 3.5, Table 3.6, Table 3.7
ASTM F1960 - 2012	Standard Specification for Cold Expansion Fittings with PEX Reinforcing Rings for Use with Cross-linked Polyethylene (PEX) Tubing	Table 3.1.3-III, Table 3.4
ASTM F1961 - 2009	Standard Specification for Metal Mechanical Cold Flare Compression Fittings with Disc Spring for Crosslinked Polyethylene (PEX) Tubing	Table 3.1.3-III, Table 3.4
ASTM F1970 - 2012e1	Standard Specification for Special Engineered Fittings, Appurtenances or Valves for use in Poly(Vinyl Chloride) (PVC) or Chlorinated Poly(Vinyl Chloride) (CPVC) Systems	Table 3.4, Table 3.4.2, Table 3.4.3
ASTM F1974 - 2009	Standard Specification for Metal Insert Fittings for Polyethylene/Aluminum/Polyethylene and Crosslinked Polyethylene/Aluminum/Crosslinked Polyethylene Composite Pressure Pipe	Table 3.1.3-III, Table 3.4
ASTM F2014 - 2000 (R2013)	Standard Specification for Non-Reinforced Extruded Tee Connections for Piping Applications	Table 3.1.3-IV, 4.2.8.3
ASTM F2080 - 2012	Standard Specification for Cold-Expansion Fittings with Metal Compression Sleeves for Crosslinked Polyethylene (PEX) Pipe	Table 3.1.3-III, Table 3.4
ASTM F2098 - 2008	Standard Specification for Stainless Steel Clamps for Securing SDR9 Cross-linked Polyethylene (PEX) Tubing to Metal Insert and Plastic Insert Fittings	Table 3.1.3-III, Table 3.4
ASTM F2159 - 2011	Standard Specification for Plastic Insert Fittings Utilizing a Copper Crimp Ring for SDR9 Cross-linked Polyethylene (PEX) Tubing and SDR9 Polyethylene of Raised Temperature (PE-RT) Tubing	Table 3.1.3-III, Table 3.4
ASTM F2165 - 2013	Standard Specification for Flexible Pre-Insulated Piping	Table 3.1.3-III, Table 3.4
ASTM F2262 - 2009	Standard Specification for Crosslinked Polyethylene/ Aluminum/Crosslinked Polyethylene Tubing OD Controlled SDR9	Table 3.1.3-III, Table 3.4, Table 3.4.2, Table 3.4.3
ASTM F2389 - 2010	Standard Specification for Pressure-rated Polypropylene (PP) Piping Systems	Table 3.1.3-III, Table 3.4, 11.7.8
ASTM F2434 - 2009	Standard Specification for Metal Insert Fittings Utilizing a Copper Crimp Ring for SDR9 Cross-linked Polyethylene (PEX) Tubing and SDR9 Cross-linked Polyethylene/Aluminum/Cross-linked Polyethylene (PEX-AL-PEX) Tubing	Table 3.1.3-III, Table 3.4
ASTM F2618 - 2009	Standard Specification for Chlorinated Poly(Vinyl Chloride) (CPVC) Pipe and Fittings for Chemical Waste Drainage Systems	Table 3.1.3-III, Table 3.5, Table 3.6
ASTM F2620 - 2013	Standard Practice for Heat Fusion Joining of Polyethylene Pipe and Fittings	Table 3.1.3-XI, 4.2.14.5, 4.2.18
ASTM F2735 - 2009	Standard Specification for Plastic Insert Fittings for SDR9 Cross-linked Polyethylene (PEX) and Polyethylene of Raised Temperature (PE-RT) Tubing	Table 3.1.3-III, Table 3.4

Table 19.1 REFERENCED STANDARDS (continued)

Standard Number	Standard Title	2015 NSPC
ASTM F2769 - 2010	Standard Specification for Polyethylene of Raised Temperature (PE-RT) Plastic Hot and Cold Water Tubing and Distribution Systems	Table 3.1.3-III, Table 3.4, Table 3.4.2, Table 3.4.3
ASTM F2788 - 2013	Standard Specification for Metric and Inch-Sized Crosslinked Polyethylene (PEX) Pipe	Table 3.1.3-III, Table 3.4
ASTM F2855 - 2012	Standard Specification for Chlorinated Poly(Vinyl Chloride)/Aluminum/Chlorinated Poly(Vinyl Chloride) (CPVC-AL-CPVC) Composite Pressure Tubing	Table 3.1.3-III, Table 3.4, Table 3.4.2, Table 3.4.3
AWS A5.8 - 2011, w/amendment 1	Specification for Filler Metals for Brazing and Braze Welding	Table 3.1.3-IV
AWWA C104/A21.4 - 2013	Cement Mortar Lining for Ductile Iron Pipe and Fittings for Water	Table 3.1.3-I
AWWA C110/A21.10 - 2012	Ductile-Iron and Gray-Iron Fittings	Table 3.1.3-I, Table 3.4, Table 3.5, Table 3.7
AWWA C111/A21.11 - 2012	Rubber-Gasketed Joints for Ductile-Iron Pressure Pipe and Fittings	Table 3.1.3-IV
AWWA C151/A21.51 - 2009	Ductile-Iron Pipe, Centrifugally Cast	Table 3.1.3-I, Table 3.4
AWWA C153/A21.53 - 2011	Ductile Iron Compact Fittings	Table 3.1.3-I, Table 3.4, Table 3.5, Table 3.7
AWWA C500 - 2009	Metal-Seated Gate Valves for Water Supply Service	Table 3.1.3-VIII
AWWA C600 - 2010	Installation of Ductile-Iron Water Mains and Their Appurtenances	Table 3.1.3-XI
AWWA C606 - 2011	Grooved and Shouldered Joints	Table 3.1.3-IV, Table 3.4
AWWA C800 - 2012	Underground Service Line Valves and Fittings	Table 3.4
AWWA C900 - 2007	Polyvinyl Chloride (PVC) Pressure Pipe and Fabricated Fittings, 4 In. Through 12 In. for Water Transmission and Distribution	Table 3.1.3-III, Table 3.4, Table 3.4.2
AWWA C901 - 2008	Polyethylene (PE) Pressure Pipe and Tubing, 1/2 In. Through 3 In. for Water Service	Table 3.1.3-III, Table 3.4, Table 3.4.2, 4.2.14.5
AWWA C903 - 2005	Polyethylene-Aluminum-Polyethylene Crosslinked Polyethylene Composite Pressure Pipes, 1/2 In. Through 2 In. for Water Service	Table 3.1.3-III, Table 3.4, Table 3.4.2
AWWA C904 - 2006	Cross-Linked Polyethylene (PEX) Pressure Pipe, 1/2 In. Through 3 In. for Water Service	Table 3.1.3-III, Table 3.4, Table 3.4.2
AWWA C950 - 2007	Fiberglass Pressure Pipe	Table 3.1.3-III, Table 3.4
CISPI 301 - 2012	Standard Specification for Hubless Cast Iron Soil Pipe and Fittings for Sanitary and Storm Drain, Waste, and Vent Piping Applications	Table 3.1.3-I, Table 3.5, Table 3.6, Table 3.7
CISPI 310 - 2012	Specification for Coupling for Use in Connection with Hubless Cast Iron Soil Pipe and Fittings for Sanitary and Storm Drain, Waste, and Vent Piping Applications	Table 3.1.3-IV, 4.3.8
CSA 4.1 - 2013	Gas Water Heaters - Volume I, Storage Water Heaters with Input Ratings of 75,000 Btu Per Hour or Less	Table 3.1.3-VII, 10.15.11

| \multicolumn{3}{c}{**Table 19.1 REFERENCED STANDARDS** (continued)} |

Standard Number	Standard Title	2015 NSPC
CSA 4.3 - 2013	Gas Water Heaters - Volume III, Storage Water Heaters, With Input Ratings Above 75,000 Btu Per Hour, Circulating and Instantaneous	Table 3.1.3-VII, 10.15.11
CSA 4.4 - M1999 (R2008)	Relief Valves for Hot Water Supply Systems	Table 3.1.3-VIII, 10.16.7, Figure 10.16.7
CSA B45.1 - 2013	Ceramic Plumbing Fixtures	Table 3.1.3-V, 7.4.1, 7.4.2, 7.5.1, 7.5.2, 7.6.1, 7.7.1, 7.11.1
CSA B45.2 - 2013	Enameled Cast Iron and Enameled Steel Plumbing Fixtures	Table 3.1.3-V, 7.6.1, 7.8.1, 7.11.1
CSA B45.4 - 2008 (R2013)	Stainless Steel Plumbing Fixtures	Table 3.1.3-V, 7.6.1, 7.11.1
CSA B45.5 - 2011 w/update 1	Plastic Plumbing Fixtures	Table 3.1.3-V, 7.4.1, 7.4.2, 7.5.1, 7.5.2, 7.6.1, 7.8.1, 7.10.1, 7.11.1
CSA B45.5/IAPMO Z124	Refer to CSA B45.5 or IAPMO Z124	Table 3.1.3-V, 7.4.1, 7.4.2, 7.5.1, 7.5.2, 7.6.1, 7.8.1, 7.10.1, 7.11.1
CSA B45.9 - 2013	Plumbing Fixtures with Pumped Waste and Macerating Plumbing Systems	Table 3.1.3-X, 7.4.8, 11.7.6
CSA B45.10 - 2012	Hydromassage Bathtub Systems	Table 3.1.3-V, 7.9.2
CSA B45.15 - 2011 w/update 1	Flush Valves and Spuds for Water Closets, Urinals, and Tanks	Table 3.1.3-VI
CSA B64.6 - 2011	Dual Check Valve (DuC) Backflow Preventers	18.5.3
CSA B125.1 - 2012	Plumbing Supply Fittings	Table 3.1.3-VI, 7.6.2, 7.7.1, 7.8.3, 7.10.2, 7.11.1
CSA B125.2 - 2011	Plumbing Waste Fittings	Table 3.1.3-VI, 3.2.1, 7.11.1
CSA B125.3 - 2012	Plumbing Fittings	Table 3.1.3-VI, Table 3.1.3-VIII, 7.19.4, 7.19.5, 10.15.6
CSA B125.6 - 2009	Flexible Water Connectors	Table 3.1.3-VI, 10.13
CSA B125.16 - 2011, w/update 1	Performance Requirements for Automatic Compensating Valves for Individual Shower and Tub/Shower Combinations	Table 3.1.3-VI, Table 3.1.3-VIII, 10.15.6
CSA B137.6 - 2013	Chlorinated Polyvinylchloride (CPVC) Pipe, Tubing, and Fittings for Hot- and Cold-Water Distribution Systems	Table 3.1.3-III, Table 3.4
CSA B137.9 - 2013	Polyethylene/Aluminum/Polyethylene (PE-AL-PE) Composite Pressure-Pipe Systems	Table 3.1.3-III
CSA B137.10 - 2013	Crosslinked Polyethylene/Aluminum/Crosslinked Polyethylene (PEX-AL-PEX) Composite Pressure-Pipe Systems	Table 3.1.3-III
CSA B181.1 - 2011 w/updates 1 & 2	Acrylonitrile-Butadiene-Styrene (ABS) Drain, Waste, and Vent Pipe and Fittings	Table 3.1.3-III, Table 3.1.3-VIII

Table 19.1 REFERENCED STANDARDS (continued)		
Standard Number	**Standard Title**	**2015 NSPC**
CSA B181.2 - 2011 w/updates 1 & 2	Polyvinylchloride (PVC) and Chlorinated Polyvinylchloride (CPVC) Drain, Waste, and Vent Pipe and Pipe Fittings	Table 3.1.3-III, Table 3.1.3-VIII
CSA B483.1 - 2007 (R2012)	Drinking Water Treatment Systems	Table 3.1.3-VII, 10.18.1
CSA B602 - 2010	Mechanical Couplings for Drain, Waste, and Vent Pipe and Sewer	Table 3.1.3-IV, 4.2.11.6
FM 1680 - 1989	Couplings Used in Hubless Cast Iron Systems for Drain, Waste or Vent, Sewer, Rainwater or Storm Drain Systems Above and Below Ground Industrial/Commercial and Residential	Table 3.1.3-IV, 4.3.8
IAPMO Z124 - 2011 w/update 1	Plastic Plumbing Fixtures	Table 3.1.3-V, 7.4.1, 7.4.2, 7.5.1, 7.5.2, 7.6.1, 7.8.1, 7.10.1, 7.11.1
IAPMO/ANSI Z124.5 - 2013e1	Plastic Toilet Seats	Table 3.1.3-VI, 7.4.5
IAPMO/ANSI Z124.7 - 2013	Prefabricated Plastic Spa Shells	Table 3.1.3-V
IAPMO/ANSI Z124.8 - 2013e1	Plastic Liners for Bathtubs and Shower Receptors	Table 3.1.3-VI
IAPMO/ANSI Z1001 - 2013	Prefabricated Gravity Grease Interceptors	Table 3.1.3-X, 6.2.1.3
IAPMO/ANSI Z1088 - 2013	Pre-Pressurized Water Expansion Tanks	Table 3.1.3-VIII
MSS SP-58 - 2009	Pipe Hangers and Supports - Materials, Design, Manufacture, Selection, Application, and Installation	Table 3.1.3-X, 8.10
MSS SP-70 - 2011	Gray Iron Gate Valves, Flanged and Threaded Ends	Table 3.1.3-VIII
MSS SP-71 - 2011	Gray Iron Swing Check Valves, Flanged and Threaded Ends	Table 3.1.3-VIII
MSS SP-80 - 2013	Bronze Gate, Globe, Angle and Check Valves	Table 3.1.3-VIII
MSS SP-110 - 2010	Ball Valves Threaded, Socket-Welding, Solder Joint, Grooved and Flared Ends	Table 3.1.3-VIII
NFPA 13 - 2013	Standard for the Installation of Sprinkler Systems	10.5.9
NFPA 13D - 2013	Standard for the Installation of Sprinkler Systems in One- and Two-Family Dwellings and Manufactured Homes	2.28, 10.5.9, 10.20.1, 10.20.2, 10.20.3, 10.20.4, 10.20.5, 10.20.7
NFPA 13R - 2013	Standard for the Installation of Sprinkler Systems in Low-Rise Residential Occupancies	10.5.9
NFPA 99 - 2015	Health Care Facilities Code	4.2.8.1, 4.2.8.2, 14.3.1, 14.12.1, 14.12.2
NSF 3 - 2012	Commercial Warewashing Equipment	Table 3.1.3-VII, 7.15.1
NSF 14 - 2013	Plastics Piping System Components and Related Materials	3.1.3, Table 3.1.3-III, 3.4.1
NSF 42 - 2013	Drinking Water Treatment Units - Aesthetic Effects	Table 3.1.3-VII, 10.18.1
NSF 44 - 2013	Residential Cation Exchange Water Softeners	Table 3.1.3-VII, 10.18.1
NSF 53 - 2013	Drinking Water Treatment Units - Health Effects	Table 3.1.3-VII, 10.18.1
NSF 55 - 2013	Ultraviolet Microbiological Water Treatment Systems	Table 3.1.3-VII, 10.18.1
NSF 58 - 2013	Reverse Osmosis Drinking Water Treatment Systems	Table 3.1.3-VII, 10.18.1
NSF 61 - 2013	Drinking Water System Components - Health Effects	1.2 Lead Content, 3.1.5, 3.4.6, 3.4.7, 4.2.6, 7.1.2
NSF 372 - 2011	Drinking Water System Components - Lead Content	3.4.7, 7.1.2

Table 19.1 REFERENCED STANDARDS (continued)		
Standard Number	**Standard Title**	**2015 NSPC**
PDI G 101 - 2012	Testing and Rating Procedure for Hydro Mechanical Grease Interceptors	Table 3.1.3-X, 6.1.4.2, 6.2.1.1, 6.2.2, 6.2.10
PDI WH 201 - 2010	Water Hammer Arrestors	Table 3.1.3-VIII, 10.14.7, Figure 10.14.7
UL 174 - 11th Edition, 2004	Standard for Household Electric Storage Tank Water Heaters	Table 3.1.3-VII, 10.15.11
UL 399 - 7th Edition, 2008	Standard for Drinking Water Coolers	Table 3.1.3-V, 7.12.1
UL 430 - 7th Edition, 2009	Standard for Waste Disposers	Table 3.1.3-VII, 7.14.1
UL 499 - 13th Edition, 2005	Standard for Electric Heating Appliances	Table 3.1.3-VII, 10.15.11
UL 732 - 5th edition, 1995	Standard for Oil-Fired Storage Tank Water Heaters	Table 3.1.3-VII, 10.15.11
UL 749 - 9th Edition, 2013	Household Dishwashers	Table 3.1.3-VII, 7.15.1
UL 921 - 6th Edition, 2006	Commercial Electric Dishwashers	Table 3.1.3-VII, 7.15.1
UL 1206 - 4th Edition, 2003	Standard for Electric Commercial Clothes-Washing Equipment	Table 3.1.3-VII, 7.13.1
UL 1453 - 5th Edition, 2004	Standard for Electric Booster and Commercial Storage Tank Water Heaters	Table 3.1.3-VII, 10.15.11
UL 2157 - 2nd Edition, 1997	Electrical Clothes Washing Machines and Extractors	Table 3.1.3-VII, 7.13.1

Appendix A

Sizing Storm Drainage Systems

A.1 Rainfall Rates for Cities 388

       Table A.1 Rainfall Rates for Cities 389

A.2 Roof Drainage 388

A.3 Sizing by Flow 388

A.4 Sizing by Roof Area 388

A.5 Capacity of Rectangular Scuppers 388

       Table A.5 Discharge from Rectangular Scuppers - GPM 393

A.1 Rainfall Rates for Cities

The rainfall rates in Table A.1, RAINFALL RATES FOR CITIES, are based on U.S. Weather Bureau Technical Paper No. 40, specifically Chart 14: 100-YEAR 1 HOUR RAINFALL (inches) and Chart 7: 100-YEAR 30-MINUTE RAINFALL (inches). The data in Chart 7 were multiplied by 0.72 to determine the rainfall for a 15-minute period, then multiplied by 4 to establish the corresponding rainfall rate in inches per hour. The flow rates in gallons per minute (gpm) were established by dividing the inches per hour by 12 to determine cubic feet per hour per square foot, then multiplying by 7.48 gallons per cubic foot to determine gallons per hour per square foot, then dividing by 60 minutes per hour to determine the equivalent gallons per minute.

A.2 Roof Drainage

Primary roof drainage systems are sized for a 100-year, 60-minute storm. Secondary roof drainage systems are sized for a more severe 100-year, 15-minute storm. The rainfall rates in Table A.1 should be used for design unless higher rates are established locally.

A.3 Sizing by Flow Rate

Storm drainage systems may be sized by stormwater flow rates, using the appropriate GPM/SF of rainfall listed in Table A.1 for the local area. Multiplying the listed GPM/SF by the roof area being drained (in square feet) produces the gallons per minute (gpm) of required flow for each drain inlet. The flow rates (gpm) can then be added to determine the flows in each section of the drainage system. Required pipe sizes can be determined from Table 13.6.1 and Table 13.6.2.

A.4 Sizing by Roof Area

Storm drainage systems may be sized using the roof area served by each section of the drainage system. Required pipe sizes can be determined from Table 13.6.1 and Table 13.6.2. Using this method, it may be necessary to interpolate between the various listed rainfall rates (inches per hour). To determine the allowable roof area for a listed size pipe at a listed slope, divide the allowable square feet of roof area for a 1" rainfall rate by the listed rainfall rate for the local area. For example, the allowable roof area for a 6" drain at 1/8" slope with a rainfall rate of 3.2 inches/hour is 21400/3.2 = 6688 square feet.

A.5 Capacity of Rectangular Scuppers

Table A.5 lists the discharge capacity of various width rectangular roof scuppers with various heads of water. The maximum allowable level of water on the roof should be obtained from the structural engineer, based on the design of the roof.

Table A.1 RAINFALL RATES FOR CITIES

STATES AND CITIES	PRIMARY STORM DRAINAGE 60-MIN. DURATION 100-YR RETURN		SECONDARY STORM DRAINAGE 15-MIN. DURATION 100-YR RETURN	
	IN/HR	GPM/SF	IN/HR	GPM/SF
ALABAMA				
Birmingham	3.7	0.038	7.8	0.081
Huntsville	3.3	0.034	7.5	0.078
Mobile	4.5	0.047	10.1	0.105
Montgomery	3.8	0.039	8.4	0.087
ALASKA				
Aleutian Islands	1.0	0.010	2.5	0.026
Anchorage	0.6	0.006	1.5	0.016
Bethel	0.8	0.008	2.0	0.021
Fairbanks	1.0	0.010	2.5	0.026
Juneau	0.6	0.006	1.5	0.016
ARIZONA				
Flagstaff	2.3	0.024	5.2	0.054
Phoenix	2.2	0.023	4.9	0.051
Tucson	3.0	0.031	5.8	0.060
ARKANSAS				
Eudora	3.8	0.039	8.6	0.089
Ft. Smith	3.9	0.041	8.9	0.092
Jonesboro	3.5	0.036	7.5	0.078
Little Rock	3.7	0.038	8.6	0.089
CALIFORNIA				
Eureka	1.5	0.016	3.7	0.038
Lake Tahoe	1.3	0.014	2.9	0.030
Los Angeles	2.0	0.021	4.3	0.045
Lucerne Valley	2.5	0.026	4.3	0.045
Needles	1.5	0.016	3.7	0.038
Palmdale	3.0	0.031	7.2	0.075
Redding	1.5	0.016	3.7	0.038
San Diego	1.5	0.016	3.7	0.038
San Francisco	1.5	0.016	3.5	0.036
San Luis Obispo	1.5	0.016	3.7	0.038
COLORADO				
Craig	1.5	0.016	3.5	0.036
Denver	2.2	0.023	4.6	0.048
Durango	1.8	0.019	4.3	0.045
Stratton	3.0	0.031	6.6	0.069
CONNECTICUT				
Hartford	2.8	0.029	6.6	0.069
New Haven	3.0	0.031	7.2	0.075
DELAWARE				
Dover	3.5	0.036	7.8	0.081
Rehobeth Beach	3.6	0.037	8.6	0.089
DISTRICT OF COLUMBIA				
Washington	4.0	0.042	8.6	0.089
FLORIDA				
Daytona Beach	4.0	0.042	8.6	0.089
Ft. Myers	4.0	0.042	10.1	0.105
Jacksonville	4.3	0.045	8.6	0.089
Melbourne	4.0	0.042	8.6	0.089
Miami	4.5	0.047	11.5	0.119
Palm Beach	5.0	0.052	11.5	0.119
Tampa	4.2	0.044	10.1	0.105
Tallahassee	4.1	0.043	9.2	0.096
GEORGIA				
Atlanta	3.5	0.036	7.8	0.081
Brunswick	4.0	0.042	8.6	0.089
Macon	3.7	0.038	8.1	0.084
Savannah	4.0	0.042	8.6	0.089
Thomasville	4.0	0.042	8.6	0.089
HAWAII				
Hawaiian Islands	(1)			

Table A.1 RAINFALL RATES FOR CITIES, continued

STATES AND CITIES	PRIMARY STORM DRAINAGE 60-MIN. DURATION 100-YR RETURN		SECONDARY STORM DRAINAGE 15-MIN. DURATION 100-YR RETURN	
	IN/HR	GPM/SF	IN/HR	GPM/SF
IDAHO				
Boise	1.0	0.010	2.3	0.024
Idaho Falls	1.2	0.012	3.2	0.033
Lewiston	1.0	0.010	2.9	0.030
Twin Falls	1.1	0.011	2.3	0.024
ILLINOIS				
Chicago	2.7	0.028	6.3	0.065
Harrisburg	3.1	0.032	6.9	0.072
Peoria	2.9	0.030	6.6	0.069
Springfield	3.0	0.031	6.6	0.069
INDIANA				
Evansville	3.0	0.031	6.9	0.072
Indianapolis	2.8	0.029	6.3	0.065
Richmond	2.7	0.028	6.3	0.065
South Bend	2.7	0.028	6.0	0.062
IOWA				
Council Bluffs	3.7	0.038	8.1	0.084
Davenport	3.0	0.031	7.2	0.075
Des Moines	3.4	0.035	7.8	0.081
Sioux City	3.6	0.037	7.8	0.081
KANSAS				
Goodland	3.5	0.036	7.5	0.078
Salina	3.8	0.039	8.6	0.089
Topeka	3.8	0.039	8.6	0.089
Wichita	3.9	0.041	8.9	0.092
KENTUCKY				
Bowling Green	2.9	0.030	6.9	0.072
Lexington	2.9	0.030	6.6	0.069
Louisville	2.8	0.029	6.3	0.065
Paducah	3.0	0.031	6.9	0.072
LOUISIANA				
Monroe	3.8	0.039	8.9	0.092
New Orleans	4.5	0.047	10.1	0.105
Shreveport	4.0	0.042	9.5	0.099
MAINE				
Bangor	2.2	0.023	4.9	0.051
Kittery	2.4	0.025	5.8	0.060
Millinocket	2.0	0.021	4.3	0.045
MARYLAND				
Baltimore	3.6	0.037	8.6	0.089
Frostburg	2.9	0.030	6.6	0.069
Ocean City	3.7	0.038	8.6	0.089
MASSACHUSETTS				
Adams	2.6	0.027	6.0	0.062
Boston	2.7	0.028	6.3	0.065
Springfield	2.7	0.028	6.3	0.065
MICHIGAN				
Cheboygan	2.1	0.022	4.6	0.048
Detroit	2.5	0.026	5.5	0.057
Grand Rapids	2.6	0.027	5.5	0.057
Kalamazoo	2.7	0.028	6.0	0.062
Traverse City	2.2	0.023	4.9	0.051
MINNESOTA				
Duluth	2.6	0.027	6.0	0.062
Grand Forks	2.5	0.026	6.0	0.062
Minneapolis	3.0	0.031	6.9	0.072
Worthington	3.4	0.035	7.5	0.078
MISSISSIPPI				
Biloxi	4.5	0.047	10.1	0.105
Columbus	3.5	0.036	7.8	0.081
Jackson	3.8	0.039	8.6	0.089

Table A.1 RAINFALL RATES FOR CITIES, continued

STATES AND CITIES	PRIMARY STORM DRAINAGE 60-MIN. DURATION 100-YR RETURN		SECONDARY STORM DRAINAGE 15-MIN. DURATION 100-YR RETURN	
	IN/HR	GPM/SF	IN/HR	GPM/SF
MISSOURI				
Independence	3.7	0.038	8.4	0.087
Jefferson City	3.4	0.035	7.8	0.081
St. Louis	3.2	0.033	7.2	0.075
Springfield	3.7	0.038	8.1	0.084
MONTANA				
Billings	1.8	0.019	3.7	0.038
Glendive	2.5	0.026	5.8	0.060
Great Falls	1.8	0.019	3.7	0.038
Missoula	1.3	0.014	2.9	0.030
NEBRASKA				
Omaha	3.6	0.037	8.1	0.084
North Platte	3.5	0.036	7.5	0.078
Scotts Bluff	2.8	0.029	6.0	0.062
NEVADA				
Las Vegas	1.5	0.016	3.5	0.036
Reno	1.2	0.012	2.9	0.030
Winnemucca	1.0	0.010	2.3	0.024
NEW HAMPSHIRE				
Berlin	2.2	0.023	5.2	0.054
Manchester	2.5	0.026	5.8	0.060
NEW JERSEY				
Atlantic City	3.4	0.035	8.1	0.084
Paterson	3.0	0.031	6.9	0.072
Trenton	3.2	0.033	7.2	0.075
NEW MEXICO				
Albuquerque	2.0	0.021	4.0	0.042
Carlsbad	2.6	0.027	6.0	0.062
Gallup	2.1	0.022	4.9	0.051
NEW YORK				
Binghamton	2.4	0.025	5.5	0.057
Buffalo	2.3	0.024	5.2	0.054
New York	3.1	0.032	6.9	0.072
Schenectady	2.5	0.026	5.8	0.060
Syracuse	2.4	0.025	5.2	0.054
NORTH CAROLINA				
Ashville	3.2	0.033	7.2	0.075
Charlotte	3.4	0.035	8.1	0.084
Raleigh	4.0	0.042	8.9	0.092
Wilmington	4.4	0.046	9.5	0.099
NORTH DAKOTA				
Bismarck	2.7	0.028	6.3	0.065
Fargo	2.9	0.030	6.6	0.069
Minot	2.6	0.027	5.8	0.060
OHIO				
Cincinnati	2.8	0.029	6.3	0.065
Cleveland	2.4	0.025	5.5	0.057
Columbus	2.7	0.028	6.3	0.065
Toledo	2.6	0.027	5.8	0.060
Youngstown	2.4	0.025	5.8	0.060
OKLAHOMA				
Boise City	3.4	0.035	7.8	0.081
Muskogee	4.0	0.042	9.2	0.096
Oklahoma City	4.1	0.043	9.2	0.096
OREGON				
Medford	1.3	0.014	3.2	0.033
Portland	1.3	0.014	3.2	0.033
Ontario	1.0	0.010	2.3	0.024

Table A.1 RAINFALL RATES FOR CITIES, *continued*

STATES AND CITIES	PRIMARY STORM DRAINAGE 60-MIN. DURATION 100-YR RETURN		SECONDARY STORM DRAINAGE 15-MIN. DURATION 100-YR RETURN	
	IN/HR	GPM/SF	IN/HR	GPM/SF
PENNSYLVANIA				
Erie	2.4	0.025	5.5	0.057
Harrisburg	2.9	0.030	6.6	0.069
Philadelphia	3.2	0.033	7.2	0.075
Pittsburgh	2.5	0.026	5.8	0.060
Scranton	2.8	0.029	6.0	0.062
RHODE ISLAND				
Newport	3.0	0.031	7.2	0.075
Providence	2.9	0.030	6.9	0.072
SOUTH CAROLINA				
Charleston	4.1	0.043	7.8	0.081
Columbia	3.5	0.036	8.4	0.087
Greenville	3.3	0.034	9.2	0.096
SOUTH DAKOTA				
Lemmon	2.7	0.028	6.3	0.065
Rapid City	2.7	0.028	6.3	0.065
Sioux Falls	3.4	0.035	7.5	0.078
TENNESSEE				
Knoxville	3.1	0.032	7.2	0.075
Memphis	3.5	0.036	7.5	0.078
Nashville	3.0	0.031	7.2	0.075
TEXAS				
Corpus Christi	4.6	0.048	10.7	0.111
Dallas	4.2	0.044	9.5	0.099
El Paso	2.0	0.021	4.9	0.051
Houston	4.6	0.048	10.7	0.111
Lubbock	3.3	0.034	7.5	0.078
San Antonio	4.4	0.046	9.8	0.102
UTAH				
Bluff	2.0	0.021	4.3	0.045
Cedar City	1.5	0.016	3.5	0.036
Salt Lake City	1.3	0.014	2.6	0.027
VERMONT				
Bennington	2.5	0.026	5.8	0.060
Burlington	2.3	0.024	5.2	0.054
Rutland	2.4	0.025	5.5	0.057
VIRGINIA				
Charlottesville	3.4	0.035	7.8	0.081
Richmond	4.0	0.042	8.9	0.092
Roanoke	3.3	0.034	7.8	0.081
Norfolk	4.0	0.042	9.5	0.099
WASHINGTON				
Seattle	1.0	0.010	2.3	0.024
Spokane	1.0	0.010	2.6	0.027
Walla Walla	1.0	0.010	2.9	0.030
WEST VIRGINIA				
Charleston	2.9	0.030	6.6	0.069
Martinsburg	3.0	0.031	7.2	0.075
Morgantown	2.7	0.028	6.3	0.065
WISCONSIN				
La Cross	2.9	0.030	6.9	0.072
Green Bay	2.5	0.026	5.8	0.060
Milwaukee	2.7	0.028	6.3	0.065
Wausau	2.5	0.026	5.8	0.060
WYOMING				
Casper	1.9	0.020	4.3	0.045
Cheyenne	2.5	0.026	5.5	0.057
Evaston	1.3	0.014	2.9	0.030
Rock Springs	1.4	0.015	3.5	0.036

1. Rainfall rates in Hawaiian Islands vary from 1.5 in/hr to 8.0 in/hr depending on location and elevation. Consult local data.

Table A.5
DISCHARGE FROM RECTANGULAR SCUPPERS - GALLONS PER MINUTE

WATER HEAD (Inches)	WIDTH OF SCUPPER - (Inches)					
	6	12	18	24	30	36
0.5	6	13	19	25	32	38
1	17	35	53	71	89	107
1.5	31	64	97	130	163	196
2		98	149	200	251	302
2.5		136	207	278	349	420
3		177	271	364	458	551
3.5			339	457	575	693
4			412	556	700	844

NOTES:
1. Table A.5 is based on discharge over a rectangular weir with end contractions.
2. Head is depth of water above bottom of scupper opening.
3. Height of scupper opening should be 2 times the design head.
4. Coordinate the allowable head of water with the structural design of the roof.

Blank Page

Appendix B

Sizing the Building Water Supply and Distribution Piping Systems

B.1 GENERAL 397

B.2 PRELIMINARY INFORMATION 397
    B.2.1 General 397
    B.2.2 Materials for System 397
    B.2.3 Characteristics of the Water Supply 397
    B.2.4 Location and Size of Water Supply Source 397
    B.2.5 Developed Length of System 398
    B.2.6 Pressure Data Relative to Source of Supply 398
    B.2.7 Elevations 398
    B.2.8 Minimum Pressure Required at Water Outlets 398
    B.2.9 Provision of Necessary Information on Plans 398

B.3 DEMAND AT INDIVIDUAL OUTLETS 398
          Table B.3 Maximum Demand at Individual Water Outlets 399

B.4 RESERVED 399

B.5 ESTIMATING DEMAND 399
    B.5.1 Standard Method 399
    B.5.2 Water Supply Fixture Units (WSFU) Assigned to Fixtures 400
          Table B.5.2 Water Supply Fixture Units (WSFU) and Minimum Fixture Branch Pipe
          Size for Individual Fixtures 401
    B.5.3 Water Supply Fixture Units for Groups of Fixtures 402
          Table B.5.3 Water Supply Fixture Units (WSFU) for Groups of Fixtures 403
    B.5.4 Demand (GPM) Corresponding to Fixture Load (WSFU) 402
          Table B.5.4 Table For Converting Demand in WSFU to GPM 404
    B.5.5 Total Demand Including Continuous Flow 402

B.6 LIMITATION OF VELOCITY 405
    B.6.1 Consideration of Velocity in Design 405
    B.6.2 Good Engineering Practice 405
    B.6.3 Manufacturer's Recommendations for Avoiding Erosion/Corrosion 405

B.7 SIMPLIFIED METHOD FOR SIZING SYSTEMS IN RELATIVELY LOW BUILDINGS 406
    B.7.1 Application 406
    B.7.2 Simplified Method Based on Velocity Limitations 406
    B.7.3 Sizing Tables Based on Velocity Limitations 406
          Table B.7.3.A - Galvanized Steel Pipe - Std Wt 407
          Table B.7.3.B - Type K Copper Tube 407
          Table B.7.3.C - Type L Copper Tube 407
          Table B.7.3.D - Type M Copper Tube 408

Table B.7.3.E - CPVC, PVC, ABS, PE Plastic Pipe - Schedule 40	408
Table B.7.3.F - CPVC, PVC, ABS, PE Plastic Pipe - Schedule 80	408
Table B.7.3.G - CPVC Plastic Tubing (Copper Tube Size) - SDR11	409
Table B.7.3.H - PEX Plastic Tubing (Copper Tube Size) SDR9	409
Table B.7.3.I - Composite Plastic Pipe (PE-AL-PE and PEX-AL-PEX)	409
B.7.4 Step-by-Step Procedure of Simplified Sizing Method	410
B.8 ILLUSTRATION OF SIMPLIFIED SIZING METHOD APPLICATION	410
B.8.1 Example	410
Figure B.8.1 Water Supply Fixture Units (WSFU)	411
Data for Figure B.8.1 – Water Supply Fixture Units (WSFU).	412
B.8.2 Solution	413
Figure B.8.2 Design Flow (GPM) and Pipe Sizes	414
Table B.8.2 Pressure Drops in the Basic Design Circuit in Figure B.8.2	415
B.8.3 Supplementary Check of Friction Loss in Main Lines and Risers	413
B.8.4 Application to Systems in High Buildings	413
B.9 LIMITATION OF FRICTION	413
B.9.1 Basic Criterion	413
B.9.2 Maximum Permissible Friction Loss	416
B.9.3 Basic Design Circuit	416
B.9.4 Friction Loss in Equipment	416
B.9.5 Estimating Pressure Loss in Displacement Type Cold-Water Meters	417
B.9.6 Uniform Pipe Friction Loss	417
B.9.7 Total Equivalent Length of Piping	417
Table B.9.7.A Equivalent Length of Pipe for Friction Loss in Threaded Fittings and Valves	418
Table B.9.7.B Equivalent Length of Pipe for Friction Loss in Copper Tube Fittings and Valves	418
Table B.9.7.C Equivalent Length of Pipe for Friction Loss in Schedule 40 CPVC Fittings	418
Table B.9.7.D Equivalent Length of Pipe for Friction Loss in Schedule 80 CPVC Fittings	419
Table B.9.7.E Equivalent Length of Pipe for Friction Loss in CPVC SDR11 (CTS) Tubing Fittings	419
B.9.8 Determination of Flow Rates Corresponding to Uniform Pipe Friction Loss	419
B.9.8.1 Chart B.9.8.1 - Flow Vs. Pressure Loss – Galvanized Steel ASTM A53	420
B.9.8.2 Chart B.9.8.2 - Flow Vs. Pressure Loss – Type K Copper Tube	421
B.9.8.3 Chart B.9.8.3 - Flow Vs. Pressure Loss – Type L Copper Tube	422
B.9.8.4 Chart B.9.8.4 - Flow Vs. Pressure Loss – Type M Copper Tube	423
B.9.8.5 Chart B.9.8.5 - Flow Vs. Pressure Loss – CPVC, PVC, ABS, PE Schedule 40 Pipe	424
B.9.8.6 Chart B.9.8.6 - Flow Vs. Pressure Loss – CPVC, PVC, ABS, PE Schedule 80 Pipe	425
B.9.8.7 Chart B.9.8.7 - Flow Vs. Pressure Loss – CPVC Tubing (Copper Tube Size) SDR11	426
B.10 DETAILED SIZING METHOD FOR SYSTEMS IN BUILDINGS OF ANY HEIGHT	427
B.11 ILLUSTRATION OF DETAILED SIZING METHOD APPLICATION	428
B.11.1 Example	428
B.11.2 Solution	428
Table B.11.2 Type L Copper Tubing for "Fairly Smooth" Condition	429
B.12 MANIFOLD TYPE PARALLEL WATER DISTRIBUTION SYSTEMS	429
B.12.1 Manifolds	429
Table B.12.1 Manifold Sizing	429
B.12.2 Distribution Lines	430

B.1 GENERAL

Note that there are two questions regarding water supply to a building: first, total consumption of water (hot or cold or both) over a period of time, or second, peak flow at any instant of time. This appendix considers only the second question.

Proper design of the water distribution system in a building is necessary to avoid excessive installed cost and in order that the various fixtures may function properly under normal conditions. The instantaneous flow of either hot or cold water in any building is variable, depending on the type of structure, usage, occupancy, and time of day. The correct design results in piping, water heating, and storage facilities of sufficient capacity to meet the probable peak demand without wasteful excess in either piping or maintenance cost.

For additional information on this subject, the reader is referred to:

National Bureau of Standards Building Materials and Structures Report BMS 65 (1940), Methods of Estimating Loads in Plumbing Systems, by R. B. Hunter

National Bureau of Standards Building Materials and Structures Report BMS 79 (1941), Water-Distributing Systems for Buildings, by R. B. Hunter

New York State Division of Housing and Community Renewal Building Codes Bureau Technical Report No. 1, (1964), A Simplified Method for Checking Sizes of Building Water Supply Systems, by Louis S. Nielsen

B.2 PRELIMINARY INFORMATION

B.2.1 General

The information necessary for sizing the building water supply and distribution systems are described in B.2.2 through B.2.9. Correct sizing is contingent upon accuracy and reliability of the information applied. Thus, such information should be obtained from responsible parties and appropriate local authorities recognized as sources of the necessary information.

B.2.2 Materials for System

Determine what kind or kinds of piping materials are to be installed in the system. This is a matter of selection by the owner of the building or his authorized representative, who may be the architect, engineer, or contractor, as the case may be.

B.2.3 Characteristics of the Water Supply

The corrosivity and the scale-forming tendency of a given water supply with respect to various kinds of piping materials is information that most officials, architects, engineers, and contractors in a water district normally have at their fingertips as a result of years of experience. For anyone without such experience and knowledge, significant characteristics of the water supply, such as its pH value, CO content, dissolved air content, carbonate hardness, Langelier Index, and Ryznar Index, may be applied to indicate its corrosivity and scale-forming tendency. The most appropriate source of such information is the local water authority having jurisdiction over the system supplying the water, or over the wells from which water is pumped from the underground water table.

B.2.4 Location and Size of Water Supply Source

Location and size of the public water main, where available, should be obtained from the local water authority. Where a private water supply source, such as a private well system, is to be used, the location and size as designed for the premises should be determined.

B.2.5 Developed Length of System

Information should be obtained regarding the developed length of the piping run from the source of water supply to the service shutoff valve of the building (i.e., the developed length of the water service pipe as shown on site plans). Also, determine the developed length of the piping run from the service shutoff valve to the highest and/or the most remote water outlet on the system. This may be established by measurement of the piping run on the plans of the system.

B.2.6 Pressure Data Relative to Source of Supply

Maximum and minimum pressures available in the public main at all times should be obtained from the water authority, as it is the best source of accurate and reliable information on this subject. Where a private well water supply system is to be used, the maximum and minimum pressures at which it will be adjusted to operate may be applied as appropriate in such cases.

B.2.7 Elevations

The relative elevations of the source of water supply and the highest water supply outlets to be supplied in the building must be determined. In the case of a public main, the elevation of the point where the water service connection is to be made to the public main should be obtained from the local water authority. It has the most authoritative record of elevations of the various parts of the public system, and such elevations are generally referred to a datum as the reference level, usually related to curb levels established for streets.

Elevation of the curb level directly in front of the building should be obtained from building plans, as such information is required to be shown on the building site plans. Elevations of each floor on which fixtures are to be supplied also may be determined from the building plans.

B.2.8 Minimum Pressure Required at Water Outlets

Information regarding the minimum flowing pressure required at water outlets for adequate, normal flow conditions consistent with satisfactory fixture usage and equipment function may be deemed to be as follows: 15 psig flowing for all water supply outlets at common plumbing fixtures, except 20 psig for flushometer valves on siphon jet water closets and 25 psig for flushometer valves for blowout water closets and blowout urinals. Flushometer tank (pressure assisted) water closets require a minimum of 25 psig static pressure. For other types of water supplied equipment, the minimum flow pressure required should be obtained from the manufacturer.

B.2.9 Provision of Necessary Information on Plans

The basis for designing sizes of water piping should be provided on plans of the water supply and distribution systems when submitted to plumbing plan examiners for proposed installations. Provision of such information permits the examiner to quickly and efficiently check the adequacy of sizes proposed for the various parts of the water supply and distribution systems.

B.3 DEMAND AT INDIVIDUAL OUTLETS

Maximum possible flow rates at individual fixtures and water outlets have become generally accepted as industry practice, which have since become maximum rates set by law. Recognized flow rates at individual water outlets for various types of typical plumbing fixtures and hose connections are given in Table B.3.

For older faucets, if the applied pressure is more than twice the minimum pressure required for satisfactory water supply conditions, an excessively high discharge rate may occur. Such rates may cause the actual flow in the piping to exceed greatly the estimated probable peak demand rate determined in accordance with the standard method discussed in Section B.5. Such excessive velocity of flow and friction loss in piping may adversely affect performance and durability of the system.

More recent faucets, however, are equipped with flow limiting devices that control the discharge rate at a nearly constant value over a large range of pressures.

Where necessary, it is recommended that means to control the rate of supply should be provided in the fixture supply pipe (or otherwise) wherever the available pressure at an outlet is more than twice the minimum pressure required for satisfactory supply. For this purpose, individual regulating valves, variable orifice flow control devices, or fixed orifices may be provided. They should be designed or adjusted to control the rate of supply to be equal to or less than the maximum rates set by law.

Table B.3
MAXIMUM DEMAND AT INDIVIDUAL WATER OUTLETS

Type of Outlet	Maximum Demand, (gpm)
Metering lavatory faucet	0.25 gal/cycle
Public lavatory faucet	0.5 @ 60 psi
Drinking fountain jet	0.75
Private lavatory faucet	2.2 @ 60 psi
Kitchen sink faucet	2.2 @ 60 psi
Shower head	2.5 @ 80 psi
Ballcock in water closet flush tank	3.0
Dishwashing machine (domestic)	4.0
Laundry machine (8 or 16 lbs.)	4.0
Laundry sink faucet	5.0
Service sink faucet	5.0
Bath faucet, 1/2"	5.0
Hose bibb or sillcock (1/2")	5.0
1/2" flush valve (15 psi flow pressure)	15.0
1" flush valve (15 psi flow pressure)	27.0
1" flush valve (25 psi flow pressure)	35.0

B.4 RESERVED

B.5 ESTIMATING DEMAND

B.5.1 Standard Method

A standard method for estimating the maximum probable demand in building water supply systems has evolved and become recognized as generally acceptable. In 1923, the fixture unit method of weighting fixtures in accordance with their load-producing effects was proposed by Roy B. Hunter, of the National Bureau of Standards. After studying application of the method in the design of federal buildings over a period of years, the method was revised by Hunter in 1940[1], and then recommended for general application. With appropriate modifications recently made for modern fixtures, the method fills the need for a reliable, rational way to estimate probable peak demand in water supply and distribution systems for all types of building occupancy.

Note that the concept of maximum probable demand is one of probability. We are saying, in effect, that the calculated flow rate at any point in a water piping system will not be exceeded more than, say, 0.1% of the time. For most systems designed by the method described herein, the design flow rates are never reached. Therefore, the method gives a conservative approach that still does not result in wasteful oversizing.

1. National Bureau of Standards Building Materials and Structures Report BMS 6, Methods of Estimating Loads in Plumbing Systems, by R. B. Hunter.

B.5.2 Water Supply Fixture Units (WSFU) Assigned to Fixtures

Individual fixture branch piping should be sized to provide the flow rates listed in Table B.3 for the particular fixture. Minimum fixture branch pipe sizes are listed in Table B.5.2.

Peak demand in building piping systems serving multiple fixtures cannot be determined exactly. The demand imposed on a system by intermittently used fixtures is related to the number, type, time between uses, and probable number of simultaneous uses of the fixtures installed in the building. In the standard method, fixtures using water intermittently under several conditions of service are assigned specific load values in terms of water supply fixture units. The water supply fixture unit (WSFU) is a factor so chosen that the load-producing effects of different kinds of fixtures under their conditions of service can be expressed approximately as multiples of that factor. WSFUs for two or more fixtures can then be added to determine their combined effect on the water piping system.

Values assigned to different kinds of fixtures and different types of occupancies are shown in Table B.5.2.

The total WSFUs represent the fixture's demand on the domestic water service to the building. For fixtures having both hot and cold water supplies, the values for separate hot and cold water demands are taken as being three-quarters (3/4) of the total value assigned to the fixture in each case, rounded to the nearest tenth of a WSFU. As an example, since the value assigned to a kitchen sink in an individual dwelling unit is 1.5 WSFU, the separate demands on the hot and cold water piping thereto are taken as being 1.1 WSFU.

Another consideration, added in 1994, is the nature of the application of the plumbing fixture. Table B.5.2 includes columns for Individual Dwelling Units, More Than 3 Dwelling Units, Other Than Dwelling Units, and Heavy-Use Assembly. The concept behind these added classifications is that the maximum probable demand created by plumbing fixtures varies depending on the type of occupancy in which they are installed.

TABLE B.5.2
WATER SUPPLY FIXTURE UNITS (WSFU) AND MINIMUM FIXTURE BRANCH PIPE SIZE FOR INDIVIDUAL FIXTURES

INDIVIDUAL FIXTURES	Minimum Branch Pipe Size		In Individual Dwelling Units			In 3 or More Dwelling Units			In Other Than Dwelling Units			In Heavy-Use Assembly		
	Cold	Hot	Total	Cold	Hot	Total	Cold	Hot	Total	Cold	Hot	Total	Cold	Hot
Bar Sink	3/8"	3/8"	1.0	0.8	0.8	0.5	0.4	0.4						
Bathtub or Combination Bath/Shower	1/2"	1/2"	4.0	3.0	3.0	3.5	2.6	2.6						
Bidet	1/2"	1/2"	1.0	0.8	0.8	0.5	0.4	0.4						
Clothes Washer, Domestic	1/2"	1/2"	4.0	3.0	3.0	2.5	1.9	1.9	4.0	3.0	3.0			
Dishwasher, Domestic		1/2"	1.5		1.5	1.0		1.0	1.5		1.5			
Drinking Fountain or Water Cooler	3/8"								0.5	0.5		0.8	0.8	
Hose Bibb (first)	1/2"		2.5	2.5		2.5	2.5		2.5	2.5				
Hose Bibb (each additional)	1/2"		1.0	1.0		1.0	1.0		1.0	1.0				
Kitchen Sink, Domestic	1/2"	1/2"	1.5	1.1	1.1	1.0	0.8	0.8	1.5	1.1	1.1			
Laundry Sink	1/2"	1/2"	2.0	1.5	1.5	1.0	0.8	0.8	2.0	1.5	1.5			
Lavatory	3/8"	3/8"	1.0	0.8	0.8	0.5	0.4	0.4	1.0	0.8	0.8	1.0	0.8	0.8
Service Sink Or Mop Sink	1/2"	1/2"							3.0	2.3	2.3			
Shower	1/2"	1/2"	2.0	1.5	1.5	2.0	1.5	1.5	2.0	1.5	1.5			
Shower, Continuous Use	1/2"	1/2"							5.0	3.8	3.8			
Urinal, 1.0 GPF	3/4"								4.0	4.0		5.0	5.0	
Urinal, Greater Than 1.0 GPF	3/4"								5.0	5.0		6.0	6.0	
Water Closet, 1.6 GPF Gravity Tank	1/2"		2.5	2.5		2.5	2.5		2.5	2.5		4.0	4.0	
Water Closet, 1.6 GPF Flushometer Tank	1/2"		2.5	2.5		2.5	2.5		2.5	2.5		3.5	3.5	
Water Closet, 1.6 GPF Flushometer Valve	1"		5.0	5.0		5.0	5.0		5.0	5.0		8.0	8.0	
Water Closet, 3.5 GPF (or higher) Gravity Tank	1/2"		3.0	3.0		3.0	3.0		5.5	5.5		7.0	7.0	
Water Closet, 3.5 GPF (or higher) Flushometer Valve	1"		7.0	7.0		7.0	7.0		8.0	8.0		10.0	10.0	
Whirlpool Bath or Combination Bath/Shower	1/2"	1/2"	4.0	3.0	3.0	4.0	3.0	3.0						

NOTES:
1. The fixture branch pipe sizes in Table B.5.2 are the minimum allowable. Larger sizes may be necessary if the water supply pressure at the fixture will be too low due to the available building supply pressure or the length of the fixture branch and other pressure losses in the distribution system.
2. Gravity tank water closets include the pump assisted and vacuum assisted types.

B.5.3 Water Supply Fixture Units for Groups of Fixtures

Table B.5.3 lists water supply fixture unit values for typical groups of fixtures in bathrooms, kitchens, and laundries in dwelling units. There is more diversity in the use of the fixtures in these groups than is reflected by WSFU values for the individual fixtures. The "Total WSFU" represents the demand that the group places on the domestic water service to the building. The separate cold and hot WSFUs for the group are each taken as 3/4 of the WSFU values for the individual fixtures in the group according to Table B.5.3, but not greater than the "Total WSFU" for the group. An exception is that the hot WSFU values for bathroom groups having 3.5 GPF (or greater) water closets are the same as those having 1.6 GPF water closets, since the hot WSFUs are not affected by the demand of the water closet.

B.5.4 Demand (GPM) Corresponding to Fixture Load (WSFU)

To determine the maximum probable demand in gallons per minute corresponding to any given load in water supply fixture units, reference should be made to Table B.5.4, in which the values have been arranged for convenient conversion of maximum probable demand from terms of water supply fixture units of load to gallons per minute of flow. Refer to the increased number of WSFU to GPM listings in Appendix M to avoid the need to interpolate between the values in Table B.5.4.

Note in the table that the maximum probable demand corresponding to a given number of water supply fixture units is generally much higher for a system in which water closets are flushed by means of direct-supply flushometer valves than for a system in which the water closets are flushed by other types of flushing devices.

The difference in maximum probable demand between the two systems diminishes as the total number of fixture units of load rises. At 1,000 water supply fixture units, the maximum probable demand in both types of systems is the same, 210 gpm.

Where a part of the system does not supply flushometer water closets, such as in the case with hot water supply piping and some cold water supply branches, the maximum probable demand corresponding to a given number of water supply fixture units may be determined from the values given for a system in which water closets are flushed by flush tanks.

B.5.5 Total Demand Including Continuous Flow

To estimate the maximum probable demand in gpm in any given water supply pipe that supplies outlets at which demand is intermittent and also outlets at which demand is continuous, the demand for outlets that pose continuous demand during peak periods should be calculated separately and added to the maximum probable demand for plumbing fixtures used intermittently. Examples of outlets that impose continuous demand are those for watering gardens, washing sidewalks, irrigating lawns, and for air conditioning or refrigeration apparatus.

Note that some continuous-flow outlets may be controlled to be used only during low-flow periods in the system. Such time-controlled loads should not be added to the maximum probable demand for intermittently used fixtures, since they will not occur at the same times. In such cases, it will be necessary to consider both situations and size the piping for the worse case.

Table B.5.3
WATER SUPPLY FIXTURE UNITS (WSFU) FOR GROUPS OF FIXTURES

	In Individual Dwelling Units			In 3 or More Dwelling Units		
	Total WSFU	Cold WSFU	Hot WSFU	Total WSFU	Cold WSFU	Hot WSFU
BATHROOM GROUPS HAVING 1.6 GPF GRAVITY-TANK WATER CLOSETS						
Half-Bath or Powder Room	3.5	3.3	0.8	2.5	2.5	0.4
1 Bathroom Group	5.0	5.0	3.8	3.5	3.5	3.0
1-1/2 Bathroom Groups	6.0	6.0	4.5	4.0	4.0	4.0
2 Bathroom Groups	7.0	7.0	7.0	4.5	4.5	4.5
2-1/2 Bathroom Groups	8.0	8.0	8.0	5.0	5.0	5.0
3 Bathroom Groups	9.0	9.0	9.0	5.5	5.5	5.5
Each Additional Half-Bath	0.5	0.5	0.5	0.5	0.5	0.5
Each Additional Bathroom Group	1.0	1.0	1.0	1.0	1.0	1.0
BATHROOM GROUPS HAVING 3.5 GPF (or higher) GRAVITY-TANK WATER CLOSETS						
Half-Bath or Powder Room	4.0	3.8	0.8	3.0	3.0	0.4
1 Bathroom Group	6.0	6.0	3.8	5.0	5.0	3.0
1-1/2 Bathroom Groups	8.0	8.0	4.5	5.5	5.5	4.0
2 Bathroom Groups	10.0	10.0	7.0	6.0	6.0	4.5
2-1/2 Bathroom Groups	11.0	11.0	8.0	6.5	6.5	5.0
3 Bathroom Groups	12.0	12.0	9.0	7.0	7.0	5.5
Each Additional Half-Bath	0.5	0.5	0.5	0.5	0.5	0.5
Each Additional Bathroom Group	1.0	1.0	1.0	1.0	1.0	1.0
OTHER GROUPS OF FIXTURES						
Bathroom Group with 1.6 GPF Flushometer Valve	6.0	6.0	3.8	4.0	4.0	3.0
Bathroom Group with 3.5 GPF (or higher) Flushometer Valve	8.0	8.0	3.8	6.0	6.0	3.0
Kitchen Group with Sink and Dishwasher	2.0	1.1	2.0	1.5	0.8	1.5
Laundry Group with Sink and Clothes Washer	5.0	4.5	4.5	3.0	2.6	2.6

NOTES:
1. The "Total WSFU" values for fixture groups represent their load on the water service. The separate cold and hot water supply fixture units for the group are each taken as 3/4 of the WSFU values for the individual fixtures in the group according to Table B.5.2, but not greater than the "Total WSFU" for the group in TableB.5.3, except that the hot WSFU for groups having 3.5 GPF water closets are the same as those having 1.6 GPF water closets.
2. The WSFU values for tank-type water closets apply to gravity tanks and pressurized tanks, flushometer tanks (pressure assisted), pump assisted tanks, and vacuum assisted tanks.

Table B.5.4
TABLE FOR CONVERTING DEMAND IN WSFU TO GPM[1,4]

WSFU	GPM Flush Tanks[2]	GPM Flush Valves[3]	WSFU	GPM Flush Tanks[2]	GPM Flush Valves[3]
3	3		120	49	74
4	4		140	53	78
5	4.5	22	160	57	83
6	5	23	180	61	87
7	6	24	200	65	91
8	7	25	225	70	95
9	7.5	26	250	75	100
10	8	27	300	85	110
11	8.5	28	400	105	125
12	9	29	500	125	140
13	10	29.5	750	170	175
14	10.5	30	1000	210	210
15	11	31	1250	240	240
16	12	32	1500	270	270
17	12.5	33	1750	300	300
18	13	33.5	2000	325	325
19	13.5	34	2500	380	380
20	14	35	3000	435	435
25	17	38	4000	525	525
30	20	41	5000	600	600
40	25	47	6000	650	650
50	29	51	7000	700	700
60	33	55	8000	730	730
80	39	62	9000	760	760
100	44	68	10,000	790	790

NOTES:
1. This table converts water supply demands in water supply fixture units (WSFU) to required water flow in gallons per minute (GPM) for the purpose of pipe sizing.
2. This column applies to the following portions of piping systems:
   (a). Hot water piping;
   (b). Cold water piping that serves no water closets; and
   (c). Cold water piping that serves water closets other than flush valve type.
3. This column applies to portions of piping systems where the water closets are the flush valve type.
4. Refer to Appendix M for WSFU to GPM listings between those in Table B.5.4 to avoid the need to interpolate between the values in Table B.5.4.

B.6 LIMITATION OF VELOCITY

B.6.1 Consideration of Velocity in Design

Velocity of flow through water supply piping during periods of peak demand is an important factor that must be considered in the building water supply and distribution systems. Limitation of water velocity should be observed in order to avoid objectionable noise effects in systems, shock damage to piping, equipment, tanks, coils, and joints, and accelerated deterioration and eventual failure of piping from corrosion. (Also see Section 10.14.1)

B.6.2 Good Engineering Practice

In accordance with good engineering practice, it is recommended generally that maximum velocity at maximum probable demand in water supply piping be limited to 8 fps. This is deemed essential in order to avoid such objectionable effects as the production of whistling line noise, the occurrence of cavitation, and associated excessive noise in fittings and valves.

Note that this velocity is too great for systems where the flow is continuous, as in the case of recirculated hot water piping. The continuous flow rate for hot water with modest chemical content should be limited to not more than 2 fps for such continuous systems. That is, verify that the flow rate in the system as a result of the circulation pump only does not exceed 2 fps at any point.

It is also recommended that maximum velocity be limited to 4 fps in water piping that supplies a quick-closing device, such as a solenoid valve, pneumatic valve, or a quick-closing valve or faucet of the self-closing, push-pull, push-button, or other similar type. This limitation is necessary in order to avoid excessive and damaging shock pressures in the piping and equipment when flow is suddenly shut off. Plumbing equipment and systems are not designed to withstand the very high shock pressures that may occur as the result of sudden cessation of high velocity flow in piping. (Also see Section 10.14.1)

B.6.3 Manufacturer's Recommendations for Avoiding Erosion/Corrosion

Velocity limits recommended by pipe manufacturers to avoid accelerated deterioration of their piping materials due to erosion/corrosion should be observed. Recent research studies have shown that turbulence accompanying even relatively low flow velocities is an important factor in causing erosion/corrosion, and that this is especially likely to occur where the water supply has a high carbon dioxide content (i.e., in excess of 10 ppm), and where it has been softened to zero hardness. Another important factor is elevated water temperature (i.e., in excess of 110°F).

To control erosion/corrosion effects in copper water tube and in copper and brass pipe, pipe manufacturers' recommendations are as follows:

(1) Where the water supply has a pH value higher than 6.9 and a positive scale-forming tendency, such as may be shown by a positive Langelier Index, peak velocity should be limited to 8 fps;

(2) Where the water supply has a pH value lower than 6.9 and may be classified as aggressively corrosive, or where the water supply has been softened to zero hardness by passage through a water softener, peak velocity should be limited to 4 fps; and

(3) The velocity in copper tube conveying hot water at up to 140°F should be limited to 5 fps because of the accelerated corrosion rate with hot water. Velocities should be limited to 2-3 fps for temperatures above 140°F.

Note that the above values apply to velocities at maximum probable demand. For continuous flow circulating systems, do not exceed 2 fps flow rate for the flow produced by the circulator.

B.7 SIMPLIFIED METHOD FOR SIZING SYSTEMS IN RELATIVELY LOW BUILDINGS

B.7.1 Application

A simplified method for sizing building water supply piping systems in accordance with the maximum probable demand load, in terms of water supply fixture units (WSFU), has been found to constitute a complete and proper method for adequately sizing the water piping systems for a specific category of buildings. In this category are all buildings supplied from a source at which the minimum available water pressure is adequate for supplying the highest and most remote fixtures satisfactorily during peak demand. Included are almost all one- and two-family dwellings, most multiple dwellings up to at least three stories in height, and a considerable portion of commercial and industrial buildings of limited height and area, when supplied from a source with a minimum available pressure of not less than 50 psi. Under such conditions, the available pressure generally is more than enough for overcoming static head and ordinary pipe friction losses, so that pipe friction is not an additional factor to consider in sizing.

B.7.2 Simplified Method Based on Velocity Limitations

This method is based solely on the application of velocity limitations that are:

(1) Recognized as good engineering practice; and

(2) Authoritative recommendations issued by manufacturers of piping materials regarding proper use of their products in order to achieve durable performance and avoid failure in service, especially in water areas where the supply is aggressively corrosive. These limitations have been detailed in Section B.6. (Also see Section 10.14.1.)

B.7.3 Sizing Tables Based on Velocity Limitations

Tables B.7.3.A through G provide a means of sizing water supply piping on the basis of flow velocities ranging from 4 fps to 8 fps. The velocity in copper water tube for hot water up to 140°F should not exceed 5 fps. The water flow rates, flow velocities, and pressure loss rates are based on Tables B.9.8.1 through B.9.8.7 for the various piping materials. The allowable water supply fixture unit (WSFU) fixture loadings are based on Table B.5.4.

The pressure loss data in the B.7.3 tables is based on friction for straight pipe and tube and does not include allowances for fittings, valves, and appurtenances. The equivalent length of the piping can be determined by adding the equivalent length of fittings and valves in Tables B.9.7.A, B, C, D, and E. If the exact layout of the piping systems cannot be determined, allowances for fittings and valves range up to 50% of the pipe length for smooth bore piping such as copper and solvent cement joint plastic piping and up to 75% of the pipe length for steel and plastic piping with threaded joints.

In Tables B.7.3.A through B.7.3.I, the columns headed "WSFU" (tanks) apply to piping that serves water closets having gravity or pressure-type flush tanks and no fixtures that are flushed by flushometer valves. The columns headed "WSFU (valves)" apply to piping that serves fixtures that are flushed by flushometer valves.

Table B.7.3.A - GALVANIZED STEEL PIPE - STD WT

PIPE SIZE	4 FPS VELOCITY				8 FPS VELOCITY				PIPE SIZE
	WSFU (tanks)	WSFU (valves)	FLOW (gpm)	PD psi/100ft	WSFU (tanks)	WSFU (valves)	FLOW (gpm)	PD psi/100ft	
1/2"	4		3.8	7.6	9		7.6	27.3	1/2"
3/4"	8		6.6	5.5	19		13.3	19.7	3/4"
1"	15		10.8	4.1	33	5	21.5	14.9	1"
1-1/4"	28		18.6	3.0	74	24	37.3	10.8	1-1/4"
1-1/2"	41	8	25.4	2.5	129	49	50.8	9.1	1-1/2"
2"	91	31	41.8	1.9	293	163	83.7	6.8	2"
2-1/2"	174	73	59.7	1.5	472	363	119.4	5.5	2-1/2"
3"	336	207	92.2	1.2	840	817	184.4	4.3	3"
4"	687	634	158.7	0.9	1925	1925	317.5	3.1	4"
5"	1329	1329	249.4	0.7	3710	3710	498.9	2.4	5"
6"	2320	2320	360.2	0.5	7681	7681	720.4	1.9	6"

Table B.7.3.B - TYPE K COPPER TUBE

TUBE SIZE	4 FPS VELOCITY				5 FPS VELOCITY				8 FPS VELOCITY				TUBE SIZE
	WSFU (tanks)	WSFU (valves)	FLOW (gpm)	PD psi/100ft	WSFU (tanks)	WSFU (valves)	FLOW (gpm)	PD psi/100ft	WSFU (tanks)	WSFU (valves)	FLOW (gpm)	PD psi/100ft	
3/8"			1.6	8.4			2.0	12.7	3		3.2	30.4	3/8"
1/2"			2.7	6.1	3		3.4	9.3	6		5.4	22.1	1/2"
3/4"	6		5.4	4.1	8		6.8	6.2	15		10.9	14.8	3/4"
1"	13		9.7	2.9	16		12.1	4.4	29		19.4	10.6	1"
1-1/4"	22		15.2	2.3	28		19.0	3.4	53	14	30.4	8.1	1-1/4"
1-1/2"	33	5	21.5	1.8	45	10	26.9	2.8	96	33	43.0	6.7	1-1/2"
2"	75	24	37.6	1.3	112	40	47.0	2.0	251	126	75.2	4.8	2"
2-1/2"	165	69	58.1	1.0	238	115	72.6	1.6	456	341	116.1	3.7	2-1/2"
3"	289	159	82.8	0.8	392	267	103.4	1.3	725	682	165.5	3.0	3"
4"	615	541	145.7	0.6	826	801	182.1	0.9	1678	1678	291.4	2.2	4"
5"	1134	1134	226.1	0.5	1605	1605	282.6	0.7	3191	3191	452.2	1.7	5"
6"	1978	1978	322.8	0.4	2713	2713	403.5	0.6	5910	5910	645.5	1.4	6"

Table B.7.3.C - TYPE L COPPER TUBE

TUBE SIZE	4 FPS VELOCITY				5 FPS VELOCITY				8 FPS VELOCITY				TUBE SIZE
	WSFU (tanks)	WSFU (valves)	FLOW (gpm)	PD psi/100ft	WSFU (tanks)	WSFU (valves)	FLOW (gpm)	PD psi/100ft	WSFU (tanks)	WSFU (valves)	FLOW (gpm)	PD psi/100ft	
3/8"			1.8	7.8	2		2.3	11.7	4		3.6	28.0	3/8"
1/2"	3		2.9	5.9	4		3.6	8.9	7		5.8	21.3	1/2"
3/4"	7		6.0	3.9	9		7.5	5.8	16		12.1	13.9	3/4"
1"	14		10.3	2.8	18		12.9	4.3	31		20.6	10.2	1"
1-1/4"	23		15.7	2.2	29		19.5	3.3	56	15	31.3	8.0	1-1/4"
1-1/2"	34	5	22.2	1.8	47	11	27.7	2.7	101	36	44.4	6.5	1-1/2"
2"	79	26	38.6	1.3	117	43	48.2	2.0	261	136	77.2	4.7	2"
2-1/2"	173	73	59.5	1.0	247	120	74.4	1.5	470	360	119.0	3.7	2-1/2"
3"	300	170	84.9	0.8	406	281	106.2	1.3	749	713	169.9	3.0	3"
4"	635	567	149.3	0.6	854	833	186.7	0.9	1739	1739	298.7	2.2	4"
5"	1189	1189	232.7	0.5	1674	1674	290.9	0.7	3338	3338	465.5	1.7	5"
6"	2087	2087	334.6	0.4	2847	2847	418.2	0.6	6382	6382	669.1	1.4	6"

Table B.7.3.D - TYPE M COPPER TUBE

TUBE SIZE	4 FPS VELOCITY				5 FPS VELOCITY				8 FPS VELOCITY				TUBE SIZE
	WSFU (tanks)	WSFU (valves)	FLOW (gpm)	PD psi/100ft	WSFU (tanks)	WSFU (valves)	FLOW (gpm)	PD psi/100ft	WSFU (tanks)	WSFU (valves)	FLOW (gpm)	PD psi/100ft	
3/8"			2.0	7.3			2.5	11.1	4		4.0	26.6	3/8"
1/2"	3		3.2	5.6	4		4.0	8.5	7		6.3	20.2	1/2"
3/4"	7		6.4	3.7	10		8.1	5.6	18		12.9	13.4	3/4"
1"	15		10.9	2.7	19		13.6	4.1	34	5	21.8	9.9	1"
1-1/4"	24		16.3	2.2	30		20.4	3.3	59	17	32.6	7.8	1-1/4"
1-1/2"	36	6	22.8	1.8	49	12	28.5	2.7	107	38	45.7	6.4	1-1/2"
2"	82	28	39.5	1.3	122	46	49.4	2.0	270	144	79.1	4.7	2"
2-1/2"	180	77	61.0	1.0	256	131	76.2	1.5	485	380	121.9	3.6	2-1/2"
3"	310	180	87.0	0.8	419	294	108.8	1.2	775	743	174.0	3.0	3"
4"	648	583	151.6	0.6	872	854	189.5	0.9	1783	1783	303.3	2.1	4"
5"	1215	1215	235.8	0.5	1706	1706	294.8	0.7	3407	3407	471.6	1.7	5"
6"	2125	2125	338.7	0.4	2894	2894	423.4	0.6	6548	6548	677.4	1.3	6"

Table B.7.3.E - CPVC, PVC, ABS, PE PLASTIC PIPE - SCHEDULE 40

PIPE SIZE	4 FPS VELOCITY				8 FPS VELOCITY				PIPE SIZE
	WSFU (tanks)	WSFU (valves)	FLOW (gpm)	PD psi/100 ft	WSFU (tanks)	WSFU (valves)	FLOW (gpm)	PD psi/100ft	
1/2"	4		3.6	5.2	9		7.3	18.7	1/2"
3/4"	7		6.4	3.7	18		12.9	13.4	3/4"
1"	14		10.5	2.8	32	4	20.9	10.1	1"
1-1/4"	27		18.2	2.0	71	22	36.4	7.3	1-1/4"
1-1/2"	40	8	24.9	1.7	124	47	49.7	6.1	1-1/2"
2"	89	30	41.1	1.3	286	157	82.2	4.6	2"
2-1/2"	168	70	58.5	1.0	460	347	117.1	3.7	2-1/2"
3"	328	198	90.6	0.8	820	795	181.2	2.9	3"
4"	675	618	156.5	0.6	1881	1881	313.1	2.1	4"
5"	1303	1303	246.4	0.5	3642	3642	492.8	1.6	5"
6"	2284	2284	356.2	0.4	7413	7413	712.4	1.3	6"

Table B.7.3.F - CPVC, PVC, ABS, PE PLASTIC PIPE - SCHEDULE 80

PIPE SIZE	4 FPS VELOCITY				8 FPS VELOCITY				PIPE SIZE
	WSFU (tanks)	WSFU (valves)	FLOW (gpm)	PD psi/100ft	WSFU (tanks)	WSFU (valves)	FLOW (gpm)	PD psi/100ft	
1/2"	3		2.7	6.1	7		5.5	22.1	1/2"
3/4"	6		5.1	4.2	14		10.3	15.3	3/4"
1"	11		8.6	3.1	25		17.2	11.3	1"
1-1/4"	22		15.4	2.2	55	15	30.8	8.1	1-1/4"
1-1/2"	33	4	21.3	1.9	95	33	42.7	6.7	1-1/2"
2"	70	21	35.8	1.4	233	112	71.7	4.9	2"
2-1/2"	132	51	51.4	1.1	389	264	102.7	4.0	2-1/2"
3"	277	149	80.3	0.9	698	648	160.7	3.1	3"
4"	585	503	140.4	0.6	1590	1590	280.7	2.2	4"
5"	1105	1105	222.6	0.5	3114	3114	445.3	1.7	5"
6"	1942	1942	319.2	0.4	5767	5767	638.3	1.4	6"

Table B.7.3.G - CPVC PLASTIC TUBING (Copper Tube Size) - SDR11

TUBE SIZE (CTS)	4 FPS VELOCITY				8 FPS VELOCITY				TUBE SIZE (CTS)
	WSFU (tanks)	WSFU (valves)	FLOW (gpm)	PD psi/100ft	WSFU (tanks)	WSFU (valves)	FLOW (gpm)	PD psi/100ft	
1/2"	2		2.4	6.6	6		4.8	23.9	1/2"
3/4"	6		4.9	4.4	13		9.8	15.8	3/4"
1"	10		8.1	3.3	24		16.2	11.8	1"
1-1/4"	16		12.1	2.6	38	7	24.2	9.3	1-1/4"
1-1/2"	25		16.9	2.1	62	18	33.7	7.7	1-1/2"
2"	50	12	28.9	1.6	164	68	57.8	5.6	2"

Table B.7.3.H - PEX PLASTIC TUBING (Copper Tube Size) - SDR9

TUBE SIZE (CTS)					8 FPS VELOCITY				TUBE SIZE (CTS)
					WSFU (tanks)	WSFU (valves)	FLOW (gpm)	PD psi/100 ft	
3/8"					1		2.4	35.3	3/8"
1/2"					5		4.5	24.8	1/2"
3/4"					11		8.8	16.8	3/4"
1"					21		14.5	12.5	1"
1-1/4"					33		21.6	9.9	1-1/4"
1-1/2"					53	14	30.2	8.2	1-1/2"
2"					134	52	51.7	6.0	2"

Table B.7.3.I - COMPOSITE PLASTIC PIPE (PE-AL-PE and PEX-AL-PEX)

PIPE SIZE	4 FPS VELOCITY				8 FPS VELOCITY				PIPE SIZE
	WSFU (tanks)	WSFU (valves)	FLOW (gpm)	PD psi/100 ft	WSFU (tanks)	WSFU (valves)	FLOW (gpm)	PD psi/100 ft	
3/8"	1		1.1	10.3	1		2.2	37.0	3/8"
1/2"	2		2.4	6.6	5		4.7	23.9	1/2"
3/4"	7		6.2	3.8	17		12.4	13.7	3/4"
1"	13		10.1	2.9	31		20.3	10.3	1"
1-1/4"	22		15.5	2.2	55	15	30.9	8.1	1-1/4"
1-1/2"	41	5	25.3	1.7	128	49	50.7	6.0	1-1/2"
2"	81	27	39.2	1.3	267	142	78.4	4.7	2"
2-1/2"	147	58	54.2	1.1	417	292	108.4	3.9	2-1/2"

B.7.4 Step-by-Step Procedure of Simplified Sizing Method

For sizing systems in relatively low buildings, the simplified sizing method consists of the following seven steps:

1. Obtain all information necessary for sizing the system. Such information should be obtained from responsible parties and appropriate local authorities recognized as sources of the necessary information. (See Section B.2.)

2. Provide a schematic elevation of the complete water piping system. Show all piping connections in proper sequence and all fixture supplies. Identify all fixtures and risers by means of appropriate letters, numbers, or combinations thereof. Identify all piping conveying water at a temperature above 150°F, and all branch piping to such water outlets as solenoid valves, pneumatic valves, or quick-closing valves or faucets. Provide on the schematic elevation all the necessary information obtained in Step 1. (See Section B.2.9.)

3. Mark on the schematic elevation, for each section of the complete system, the hot and cold water loads conveyed thereby in terms of water supply fixture units (WSFU) in accordance with Table B.5.2.

4. Mark on the schematic elevation, adjacent to all fixture unit notations, the demand in gallons per minute corresponding to the various fixture unit loads in accordance with Table B.5.3.

5. Mark on the schematic elevation, for appropriate sections of the system, the demand in gallons per minute for outlets at which demand is considered continuous, such as outlets for watering gardens, irrigating lawns, air conditioning apparatus, refrigeration machines, and similar equipment using water at a relatively continuous rate during peak demand periods. Add the continuous demand to the demand for intermittently used fixtures, and show the total demand at those sections where both types of demand occur. (See Section B.5.4.)

6. Size all individual fixture supply pipes to water outlets in accordance with the minimum sizes permitted by regulations. Minimum fixture supply pipe sizes for typical plumbing fixtures are given in Table B.5.2.

7. Size all other parts of the water piping system in accordance with velocity limitations recognized as good engineering practice, and with velocity limitations recommended by pipe manufacturers for avoiding accelerated deterioration and failure of their products under various conditions of service. (Sizing tables based on such velocity limitations and showing permissible loads terms of water supply fixture units (WSFU) for each size and kind of piping material have been provided and may be applied in this step.) (See Section B.6.)

B.8 ILLUSTRATION OF SIMPLIFIED SIZING METHOD APPLICATION

B.8.1 Example

A three-story, nine-family multiple dwelling fronts on a public street and is supplied by direct street pressure from a public main in which the certified minimum pressure is 50 psi. The building has a full basement and three above-grade stories, each of which is 10 feet in height from floor to floor. The first floor is 2 feet above the curb level in front of the building. The public water main is located under the street: 5 feet out from and 4 feet below the curb.

On each of the above-grade stories there are three dwelling units. Each dwelling unit has a sink and dishwasher, tank-type water closet, lavatory, and bathtub/shower combination.

The basement contains two automatic clothes washing machines, two service sinks, and a restroom with a flush-tank water closet and lavatory.

Two lawn faucets are installed, one on the front of the building and one in the rear. Hot water is to be supplied from a central storage-tank water heater. The water supply to the building will be metered at the water service entry point to the building. An isometric drawing of the water piping layout is shown in Figure B.8.1.

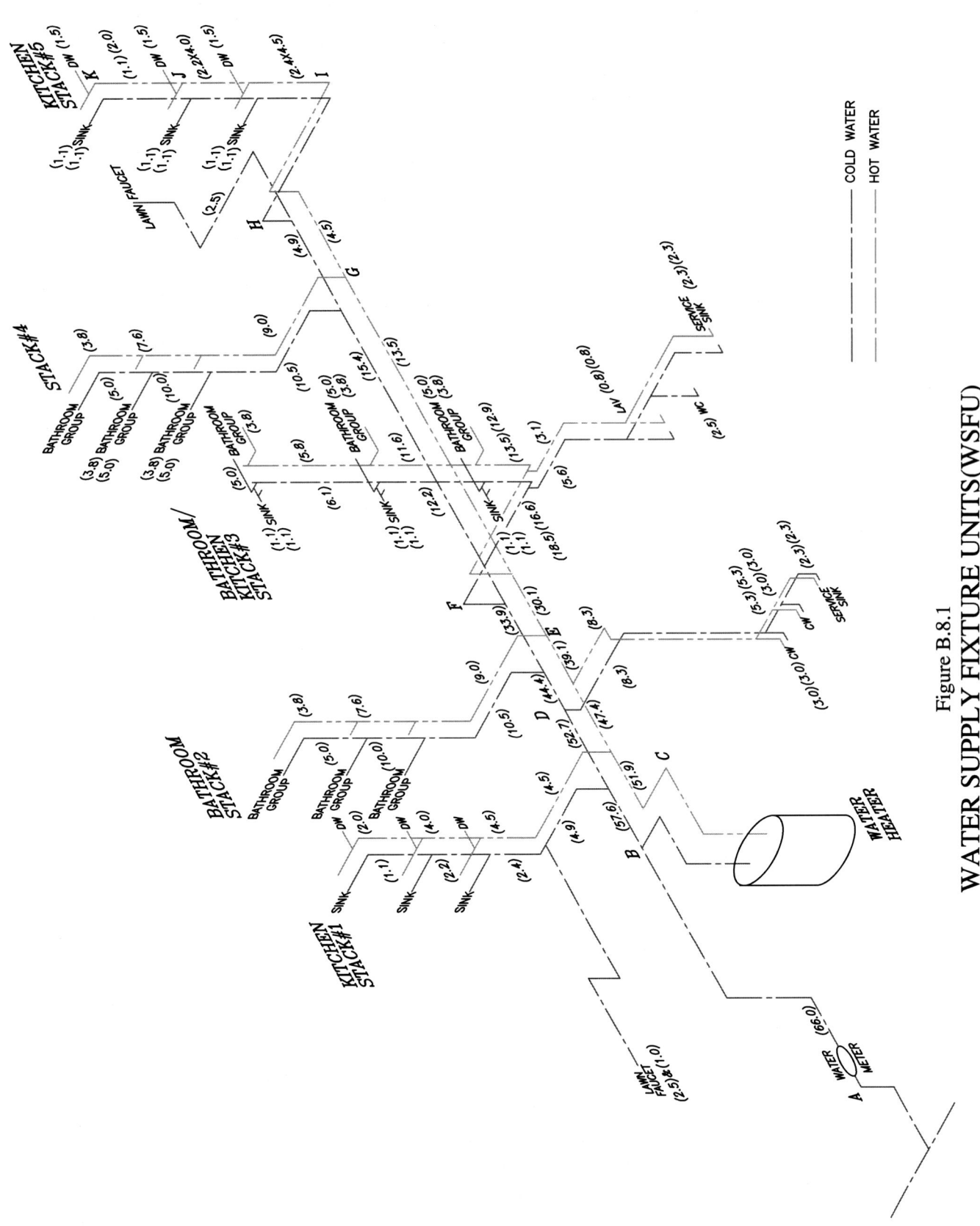

Figure B.8.1
WATER SUPPLY FIXTURE UNITS (WSFU)

DATA for Figure B.8.1 - WATER SUPPLY FIXTURE UNITS (WSFU)

WATER SUPPLY FIXTURE UNITS (WSFU)

Fixtures	Piping Serving Fixtures in 1 or 2 Dwellings			Piping Serving Fixtures in 3 or more Dwellings			Piping Serving Fixtures in Other than Dwelling Units		
	Total	Cold	Hot	Total	Cold	Hot	Total	Cold	Hot
Bathroom Group (1.6 GPF tank-type WC)	5.0	5.0	3.8	3.5	3.5	3.0			
Half Bath (1.6 GPF tank-type WC)	3.5	3.3	0.8	2.5	2.5	0.4			
Kitchen Group (sink & dishwasher)	2.0	1.1	2.0	1.5	0.8	1.5			
Clothes Washer							4.0	3.0	3.0
Service Sink							3.0	2.3	2.3
Hose Bibb							2.5	2.5	
Hose Bibb, each additional							1.0	1.0	

WATER SERVICE PIPE SIZING

Fixtures	Qty	WSFU	Total
Bathroom Groups (1.6 tank-type WC) (1)	9	3.5	31.5
Kitchen Groups (sink + dishwasher) (1)	9	1.5	13.5
Clothes Washers (2)	2	4.0	8.0
Service Sinks (2)	2	3.0	6.0
Half Bath (2)	1	3.5	3.5
Hose Bibb (2)	1	2.5	2.5
Hose Bibb (each additional) (2)	1	1.0	1.0
TOTAL WSFU			66.0
DEMAND (GPM)			34.8

(1) Fixtures in 3 or more dwellings
(2) Fixtures in other than dwelling units.

DESIGN BASIS

PIPING
Water service, Type K copper
Water distribution, Type L copper
wrought fittings and lead-free solder

PUBLIC WATER SUPPLY
10 inch main, 50 psig minimum pressure

WATER CHARACTERISTICS
No significant fouling or corrosive agents

ELEVATIONS
Curb as datum   10.0 ft
Water main   6.0 ft
Basement floor   2.0 ft
First floor   12.0 ft
Second floor   22.0 ft
Third floor   32.0 ft
Highest outlet @ "K"   35.0 ft

MINIMUM OUTLET PRESSURE
15 psig required

VELOCITY LIMITATIONS
8 fps
except 4 fps for branches with quick-closing valves
and 5 fps for hot water up to 140 deg F

LENGTH OF RUN TO FARTHEST OUTLET
Main - A   50 ft
A - B   12 ft
B - C   8 ft
C - D   10 ft
D - E   8 ft
E - F   8 ft
F - G   10 ft
G - H   4 ft
H - I   10 ft
I - J   10 ft
J - K   10 ft
Total = 140 ft plus fitting allowance

Hot water temperature is 140 deg F
controlled by water heater thermostat

B.8.2 Solution

1. All information necessary to develop the design must be obtained from appropriate sources.
2. After the information is known, the isometric drawing (Figure B.8.1) is marked up with general water supply information, and the mains, risers, and branches are suitably identified.
3. The water supply fixture unit (WSFU) loads are marked on the drawing next to each section of the system. These values are obtained from Tables B.5.2 and B.5.3. Many designers use parentheses marks for WSFU to distinguish them from gpm values.
4. The maximum probable demand in gpm is marked on the drawing for each section next to the WSFU values. These values are obtained from Table B.5.4, using the columns for flush-tank systems.
5. Where a section of piping serves a single hose bibb, it adds a demand of 2.5 WSFU. Where sections of the piping serve more than one hose bibb, each additional hose bibb adds a demand of 1.0 WSFU to those sections of the piping.
6. All individual fixture supply pipes to water outlets are sized in Figure B.8.1 in accordance with the minimum sizes shown in Table B.5.2.
7. All other parts of the system are sized in accordance with the velocity or pressure limitations established for this system as the basis of design. Piping is sized in accordance with the maximum probable demand for each section of the system. Sizing is done using Table B.7.3.A through Table B.7.3.I, and specifically the tables dealing with Copper Water Tube - Type K for sizing the water service pipe and with Copper Water Tube - Type L, for sizing piping inside the building since these are the materials of choice as given in the general information on the drawing.

B.8.3 Supplementary Check of Friction Loss in Main Lines and Risers

A supplementary check of the total friction loss in the main lines and risers is made for the longest run of piping from the public water outlet to be sure that the sizes determined were adequate. This run has been shown with letters noted at various points.

In Table B.8.2, the sum of all friction losses due to flow through pipe, valves, and fittings is found to be 7.9 psi, whereas the amount of excess pressure available for such friction loss is 19.4 psi. Thus, the sizes determined on the basis of velocity limitations exclusively are proven adequate. Checking of friction loss in this case is performed following steps 8 through 15 of the Detailed Sizing Method for Building of Any Height presented in Section B.10.

B.8.4 Application to Systems in High Buildings

This method of sizing, based upon the velocity limitations that should be observed in design of building water supply systems, has much broader application than just to systems in one-, two-, and three-story buildings where ample excess pressure is available at the source of supply. These velocity limitations should be observed in all building water supply systems. Thus, the sizes determined by this method are the minimum sizes recommended for use in any case. Where pipe friction is an additional factor to be considered in design, larger sizes may be required.

B.9 LIMITATION OF FRICTION

B.9.1 Basic Criterion

The design of a building water supply and distribution system must be such that the highest water outlets will have available, during periods of peak demand, at least the minimum pressure required at such outlets for satisfactory water supply conditions at the fixture or equipment.

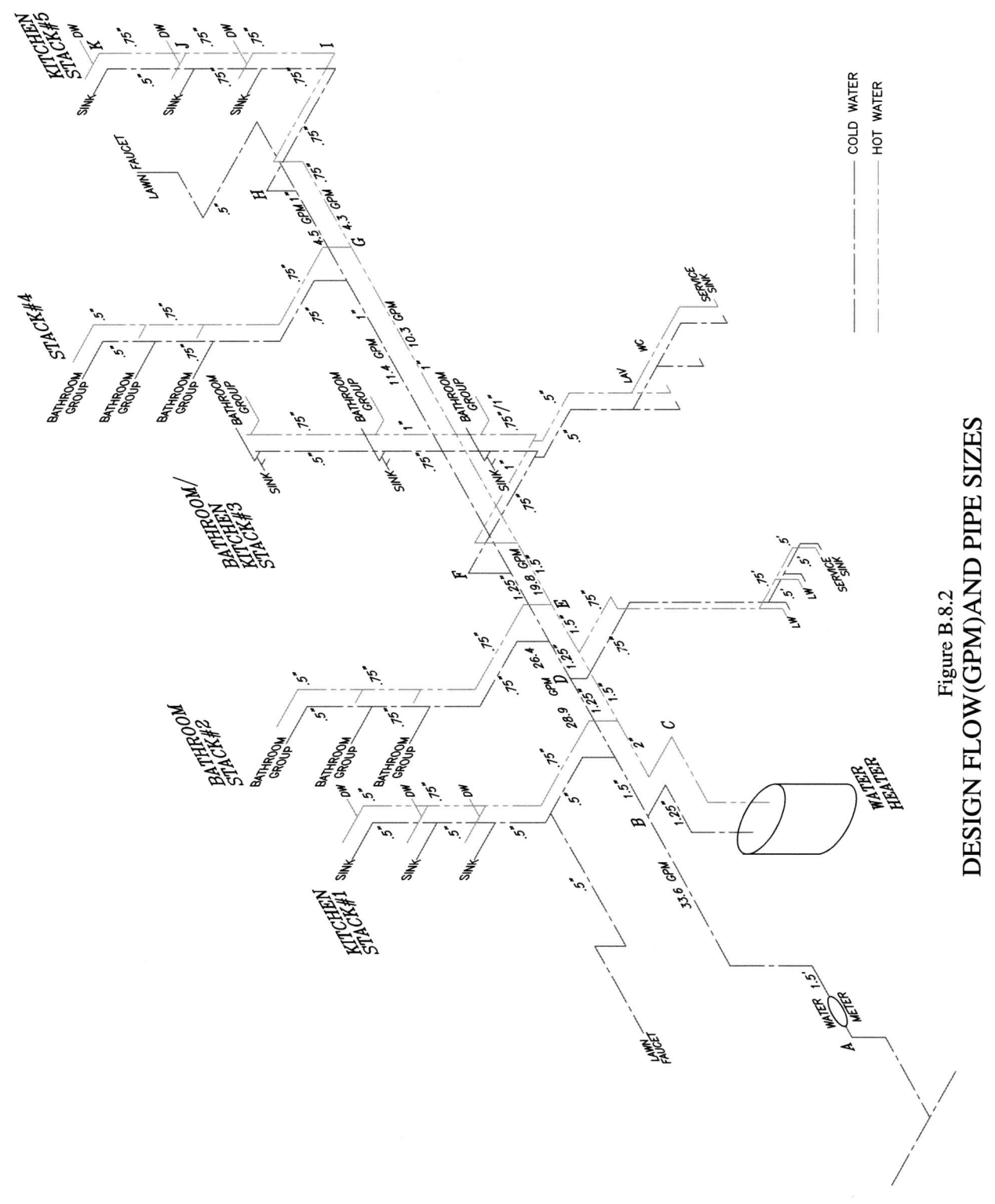

Figure B.8.2
DESIGN FLOW(GPM) AND PIPE SIZES

Table B.8.2
PRESSURE DROPS IN THE BASIC DESIGN CIRCUIT IN FIGURE B.8.2

COLD WATER FRICTION PRESSURE DROP FROM MAIN TO "K"						
SECTION	WSFU (1)	FLOW (gpm)	LENGTH (feet)	PIPE SIZE	VELOCITY (feet/second)	PD (psi/100 ft)	PRESSURE DROP (psi)
MAIN - A	66.0	34.8	50	1-1/2"	6.3	4.0	2.0
A - B	66.0	34.8	12	1-1/2"	6.3	4.0	0.5
B - C	57.6	32.0	8	1-1/2"	5.8	3.6	0.3
C - D	52.7	30.1	4	1-1/4"	7.7	8.0	0.3
D - E	44.4	26.8	8	1-1/4"	6.8	6.0	0.5
E - F	33.9	22.0	8	1-1/4"	5.6	4.0	0.3
F - G	15.4	11.4	10	1"	4.4	3.2	0.3
G - H	4.9	10.0 (6)	4	1" (4)	3.9	2.7	0.1
H - I	2.4	5.0 (7)	10	3/4" (4)	3.3	2.7	0.3
I - J	2.2 (2)	5.0 (7)	10	3/4" (4)	3.3	2.7	0.3
J - K	1.1 (2)	2.5 (8)	10	1/2" (4)	3.4	4.3	0.4
Total Pipe Pressure Drop (psig)							5.3
Fitting Allowance (50% of pipe loss, psig)							2.6
Total Pressure Drop Due to Pipe Friction (psig)							7.9

HOT WATER FRICTION PRESSURE DROP FROM MAIN TO "K"						
SECTION	WSFU (1)	FLOW (gpm)	LENGTH (feet)	PIPE SIZE	VELOCITY (feet/second)	PD (psi/100 ft)	PRESSURE DROP (psi)
MAIN - A	66.0	34.8	50	1-1/2"	6.3	4.0	2.0
A - B	66.0	34.8	12	1-1/2"	6.3	4.0	0.5
B - HWH	51.9	29.8	4	1-1/4"	7.6	7.5	0.3
HWH - C	51.9	29.8	4	2" (5)	3.1	0.8	0.0
C - D	47.4	28.0	10	1-1/2" (5)	5.0	2.7	0.3
D - E	39.1	24.6	8	1-1/2" (5)	4.4	2.1	0.2
E - F	30.1	20.1	8	1-1/2" (5)	3.6	1.6	0.1
F - G	13.5	10.3	10	1" (5)	4.0	3.0	0.3
G - H	4.5	4.3	4	3/4" (4)	2.9	2.1	0.1
H - I	4.5	4.3	10	3/4" (4)	2.9	2.1	0.2
I - J	4.0 (2)	4.0	10	3/4" (4)	2.7	1.8	0.2
J - K	2.0 (2)	4.0 (3)	10	3/4" (4)	2.7	1.8	0.2
Total Pipe Pressure Drop (psig)							4.3
Fitting Allowance (50% of pipe loss, psig)							2.2
Total Pressure Drop Due to Pipe Friction (psig)							6.5

NOTES FOR PRESSURE DROP CALCULATIONS
(1) Water supply fixture units (WSFU) are for sections of piping serving 3 or more dwelling units except as noted by (2).
(2) Water supply fixture units (WSFU) are for sections of piping serving fixtures in less than 3 dwelling units.
(3) Water flow (gpm) for the dishwasher.
(4) Velocity limited to 4 fps because of dishwashers and quick-closing sink faucets.
(5) Velocity in copper tube with 140 deg F domestic hot water is limited to 5 fps using Chart B.9.8.3.
(6) Allowance of 5 gpm for the hose bibb and two sinks at 2.5 gpm each.
(7) Allowance for two sinks at 2.5 gpm each.
(8) Allowance for one sink at 2.5 gpm.

Table B.8.2 (continued)
SUMMARY OF PRESSURE DROP CALCULATIONS

Minimum pressure in water main = 50.0 psig

Water meter pressure drop = 3.0 psig

Total cold water friction pressure drop from water main to "K" = 7.9 psig

Total hot water friction pressure drop from water main to "K" = 6.5 psig

Elevation pressure drop = (35 ft - 6 ft)(0.433) = 12.6 psig

Cold water pressure available at "K" = 50 - 3 - 7.9 -12.6 = 26.5 psig

Minimum required water pressure at "K" = 15 psig

Therefore, the pipe sizing is satisfactory.

If this calculation had shown that the pressure drop was excessive at "K", it would be necessary to examine the design for sections of the Basic Design Circuit that had the highest pressure drops and then increase those segment pipe sizes.

B.9.2 Maximum Permissible Friction Loss

The maximum allowable pressure loss due to friction in the water lines and risers to the highest water outlets is the amount of excess static pressure available above the minimum pressure required at such outlets when no-flow conditions exist. This may be calculated as the difference between the static pressure existing at the highest water outlets during no-flow conditions and the minimum pressure required at such outlets for satisfactory supply conditions.

Where water is supplied by direct pressure from a public main, to calculate the static pressure at the highest outlet, deduct from the certified minimum pressure available in the public main the amount of static pressure loss corresponding to the height at which the outlet is located above the public main (i.e., deduct 0.433 psi pressure for each foot of rise in elevation from the public main to the highest outlet).

Where supplied under pressure from a gravity water supply tank located at an elevation above the highest water outlet, the static pressure at that outlet is calculated as being equal to 0.433 psi pressure for each foot of difference in elevation between the outlet and the water level in the tank. In this case, the minimum static pressure at the outlet should be determined as that corresponding to the level of the lowest water level at which the tank is intended to operate.

B.9.3 Basic Design Circuit

Of all the water outlets on a system, the one at which the least available pressure will prevail during periods of peak demand is the critical outlet that controls the design. Normally, it is the highest outlet that is supplied through the longest run of piping extending from the source of supply.

This circuit is called the Basic Design Circuit (BDC) for sizing the main water lines and risers.

In most systems, the BDC will be found to be the run of cold water supply piping extending from the source of supply to the domestic hot water vessel plus the run of hot water supply piping extending to the highest and most remote hot water outlet on the system. However, in systems supplied directly from the public main and having flushometer-valve water closets at the topmost floor, the BDC may be found to be the run of cold water supply piping extending from the public main to the highest and most remote flushometer valve in the system.

B.9.4 Friction Loss in Equipment

Where a water meter, water filter, water softener, strainer, or instantaneous or tankless water heating coil is located in the BDC, the friction loss corresponding to the maximum probable demand through such equipment must be determined and included in pressure loss calculations. Manufacturers' charts and data sheets on their products provide such information generally, and should be used as a guide in selecting the best type and size of equipment to use with consideration for the limit to which pressure loss due to friction may be permitted to occur in the BDC. The rated pressure loss through such equipment should be deducted from the friction loss limit to establish the amount of pressure that is available to be dissipated by friction in pipe, valves, and fittings of the BDC.

B.9.5 Estimating Pressure Loss in Displacement Type Cold-Water Meters

The American Water Works Association standard for cold-water meters of the displacement type with bronze main cases is designated AWWA C700. It covers displacement meters known as nutating-disk or oscillating-piston or disc meters, which are practically positive in action. The standard establishes maximum capacity or delivery classification for each meter size as follows:

5/8"	20 gpm
3/4"	30 gpm
1"	50 gpm
1-1/2"	100 gpm
2"	160 gpm
3"	300 gpm

Also, the standard establishes the maximum pressure loss corresponding to these maximum capacities as follows:

15 psi for the 5/8", 3/4" and 1" meter sizes
20 psi for the 1-1/2", 2", 3", 4" and 6" sizes.

B.9.6 Uniform Pipe Friction Loss

To facilitate calculation of appropriate pipe sizes corresponding to the permissible friction loss in pipe, valves, and fittings, it is recommended that the BDC be designed in accordance with the principle of uniform pipe friction loss throughout its length. In this way, the friction limit for the piping run may be established in terms of pounds per square inch per 100 feet of piping length. The permissible uniform pipe friction loss in psi/100' is calculated by dividing the permissible friction loss in pipe, valves, and fittings by the total equivalent length of the basic design circuit, and multiplying by 100.

B.9.7 Total Equivalent Length of Piping

The total equivalent length of piping is its developed length plus the equivalent pipe length corresponding to the frictional resistance of all fittings and valves in the piping. When size of fittings are known, or has been established in accordance with sizes based upon appropriate limitation of velocity, corresponding equivalent lengths may be determined directly from available tables. Five such tables are included herein for various piping materials. See Tables B.9.7.A through B.9.7.E.

As a general finding, it has been shown by experience that the equivalent length to be added for pipe fittings and valves as a result of such calculations is approximately fifty percent of the developed length of the BDC in the case of copper water tube and plastic piping, and approximately seventy-five percent for standard threaded piping. The total equivalent length of copper and plastic piping is approximately 67% pipe and 33% pipe fittings and valves. The total equivalent length of standard threaded piping is approximately 57% pipe and 43% pipe fittings and valves.

Table B.9.7.A
EQUIVALENT LENGTH OF PIPE FOR FRICTION LOSS IN THREADED FITTINGS & VALVES

Fitting or Valve	Equivalent Feet of Pipe for Various Pipe Sizes										
	1/2"	3/4"	1"	1-1/4"	1-1/2"	2"	2-1/2"	3"	4"	5"	6"
45 deg Elbow	0.8	1.1	1.4	1.8	2.2	2.8	3.3	4.1	5.4	6.7	8.1
90 deg Elbow, std	1.6	2.1	2.6	3.5	4.0	5.2	6.2	7.7	10.1	12.6	15.2
Tee, Run	1.0	1.4	1.8	2.3	2.7	3.5	4.1	5.1	6.7	8.4	10.1
Tee, Branch	3.1	4.1	5.3	6.9	8.1	10.3	12.3	15.3	20.1	25.2	30.3
Gate Valve	0.4	0.6	0.7	0.9	1.1	1.4	1.7	2.0	2.7	3.4	4.0
Globe Valve	17.6	23.3	29.7	39.1	45.6	58.6	70.0	86.9	114	143	172
Angle Valve	7.8	10.3	13.1	17.3	20.1	25.8	30.9	38.4	50.3	63.1	75.8
Butterfly Valve						7.8	9.3	11.5	15.1	18.9	22.7
Swing Check Valve	5.2	6.9	8.7	11.5	13.4	17.2	20.6	25.5	33.6	42.1	50.5

NOTES FOR TABLE B.9.7.A
1) Equivalent lengths for valves are based on the valves being wide open.

Table B.9.7.B
EQUIVALENT LENGTH OF PIPE FOR FRICTION LOSS IN COPPER TUBE FITTINGS & VALVES

Fitting or Valve	Equivalent Feet of Pipe for Various Tube Sizes										
	1/2"	3/4"	1"	1-1/4"	1-1/2"	2"	2-1/2"	3"	4"	5"	6"
45 deg Elbow	0.5	0.5	1.0	1.0	1.5	2.0	2.5	3.5	5.0	6.0	7.0
90 deg Elbow, std	1.0	2.0	2.5	3.0	4.0	5.5	7.0	9.0	12.5	16.0	19.0
Tee, Run	0.0	0.0	0.0	0.5	0.5	0.5	0.5	1.0	1.0	1.5	2.0
Tee, Branch	2.0	3.0	4.5	5.5	7.0	9.0	12.0	15.0	21.0	27.0	34.0
Gate Valve	0.0	0.0	0.0	0.0	0.0	0.5	1.0	1.5	2.0	3.0	3.5
Globe Valve	17.6	23.3	29.7	39.1	45.6	58.6	70.0	86.9	114.0	143.0	172.0
Angle Valve	7.8	10.3	13.1	17.3	20.1	25.8	30.9	38.4	50.3	63.1	75.8
Butterfly Valve						7.5	10.0	15.5	16.0	11.5	13.5
Swing Check Valve	2.0	3.0	4.5	5.5	6.5	9.0	11.5	14.5	18.5	23.5	26.5

NOTES FOR TABLE B.9.7.B
1) Equivalent lengths for valves are based on the valves being wide open.
2) Data based in part on the 2004 Copper Tube Handbook by the Copper Development Association.

Table B.9.7.C
EQUIVALENT LENGTH OF PIPE FOR FRICTION LOSS IN SCHEDULE 40 CPVC FITTINGS

Fitting	Equivalent Feet of Pipe for Various Pipe Sizes										
	1/2"	3/4"	1"	1-1/4"	1-1/2"	2"	2-1/2"	3"	4"	5"	6"
45 deg Elbow	0.8	1.1	1.4	1.8	2.1	2.7	3.3	4.1	5.3	6.7	8.0
90 deg Elbow	1.5	2.0	2.6	3.4	4.0	5.1	6.1	7.6		12.5	15.1
Tee, Run	1.0	1.4	1.7	2.3	2.7	3.4	4.1	5.1	6.7	8.4	10.1
Tee, Branch	3.0	4.1	5.2	6.8	8.0	10.2	12.2	15.2		25.1	30.2

Table B.9.7.D
EQUIVALENT LENGTH OF PIPE FOR FRICTION LOSS IN SCHEDULE 80 CPVC FITTINGS

Fitting	Equivalent Feet of Pipe for Various Pipe Sizes										
	1/2"	3/4"	1"	1-1/4"	1-1/2"	2"	2-1/2"	3"	4"	5"	6"
45 deg Elbow	0.7	1.0	1.2	1.7	2.0	2.6	3.1	3.8	5.0	6.4	7.6
90 deg Elbow	1.3	1.8	2.3	3.1	3.7	4.8	5.7	7.2	9.5	11.9	14.3
Tee, Run	0.9	1.2	1.6	2.1	2.5	3.2	3.8	4.8	6.3	7.9	9.5
Tee, Branch	2.6	3.6	4.7	6.3	7.4	9.6	11.5	14.3	18.9	23.8	28.5

Table B.9.7.E
EQUIVALENT LENGTH OF PIPE FOR FRICTION LOSS CPVC SDR11 (CTS) TUBING FITTINGS

Fitting	Equivalent Feet of Pipe for Various Pipe Sizes					
	1/2" CTS	3/4" CTS	1" CTS	1-1/4" CTS	1-1/2" CTS	2" CTS
45 deg Elbow	0.8	1.1	1.4	1.8	2.2	2.8
90 deg Elbow	1.6	2.1	2.6	3.5	4.0	5.2
Tee, Run	1.0	1.4	1.8	2.3	2.7	3.5
Tee, Branch	3.1	4.1	5.3	6.9	8.1	10.3

B.9.8 Determination of Flow Rates Corresponding to Uniform Pipe Friction Loss

Flow rates corresponding to any given uniform pipe friction loss may be determined readily for each nominal size of the kind of pipe selected for the system. Pipe friction charts (B.9.8.1 through B.9.8.7) are presented herewith for each of the standard piping materials used for water supply systems in buildings. The appropriate chart to apply in any given case depends upon the kind of piping to be used and the effect the water to be conveyed will produce within the piping after extended service.

These charts are based on piping in average service. If piping is used in adverse service or in retrofit applications, conservative practice suggests selecting lower flow rates for a given pipe, or larger pipe for a given required flow rate.

For new work, with the range of materials now available, select a piping material that will not be affected by the water characteristics at the site.

CHART B.9.8.1
GALVANIZED STEEL - ASTM A53

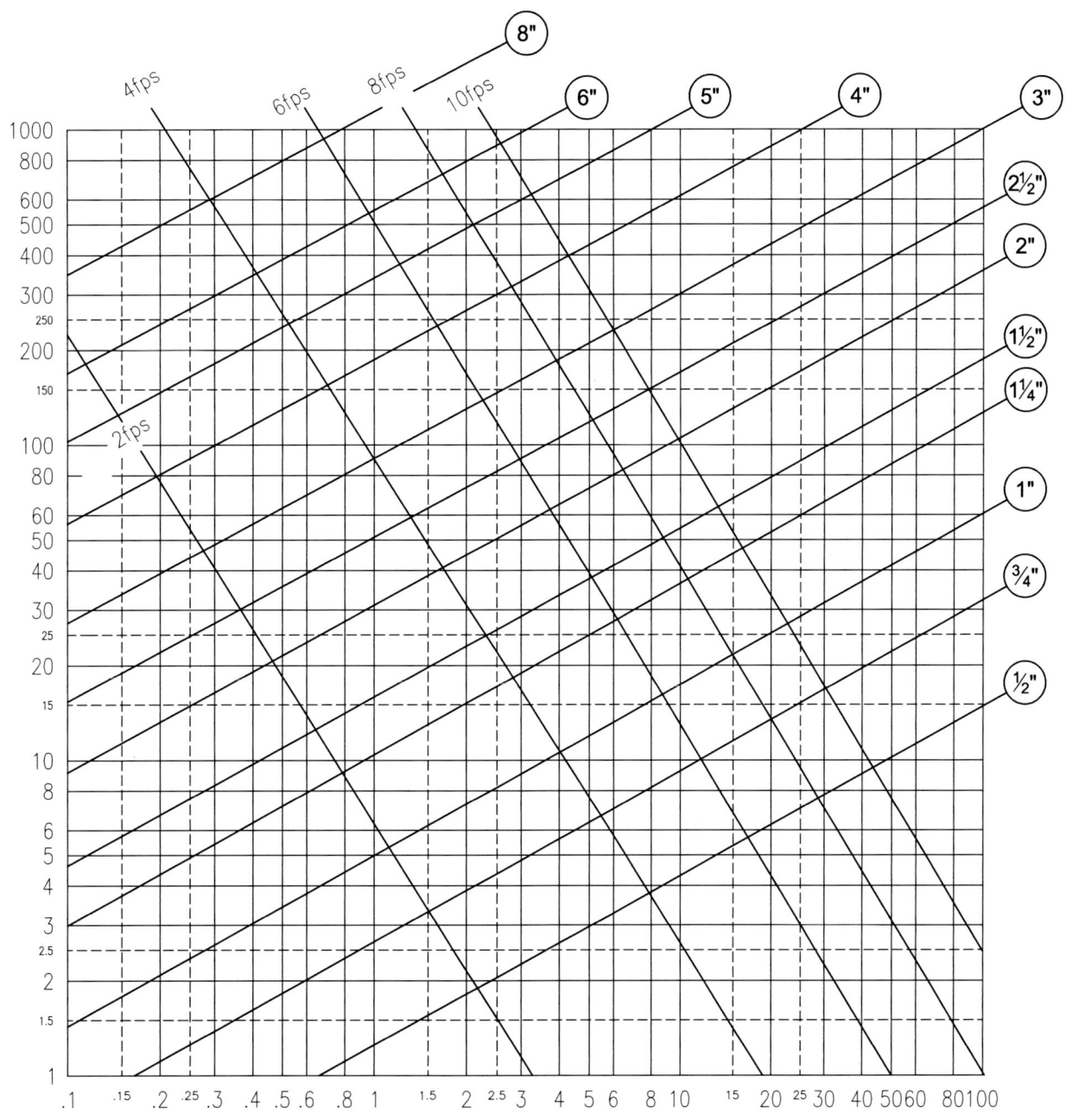

PRESSURE DROP (psi/100')

FLOW VS. PRESSURE LOSS

CHART B.9.8.2
TYPE K COPPER TUBE

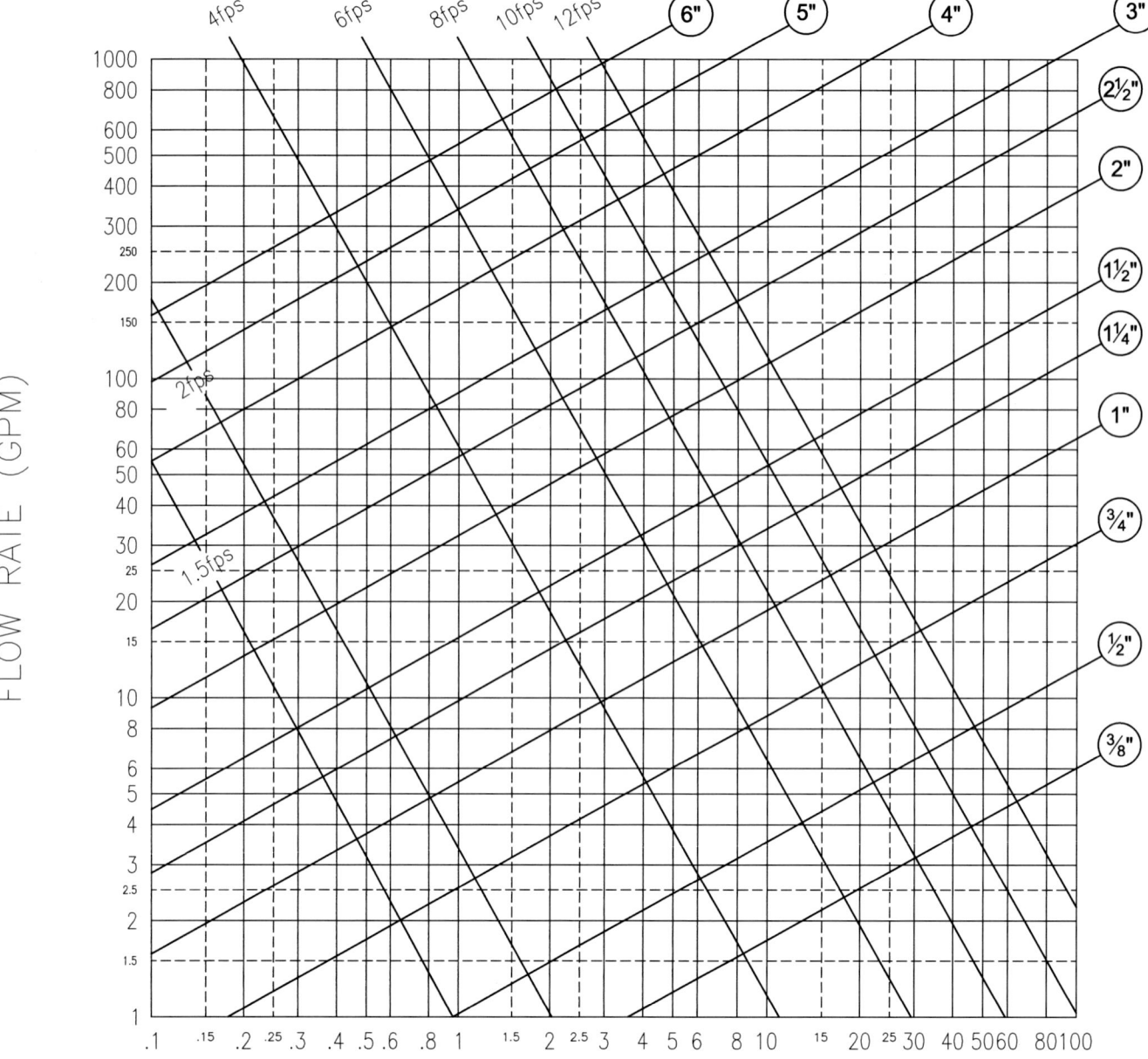

PRESSURE DROP (psi/100')

FLOW VS. PRESSURE LOSS

CHART B.9.8.3
TYPE L COPPER TUBE

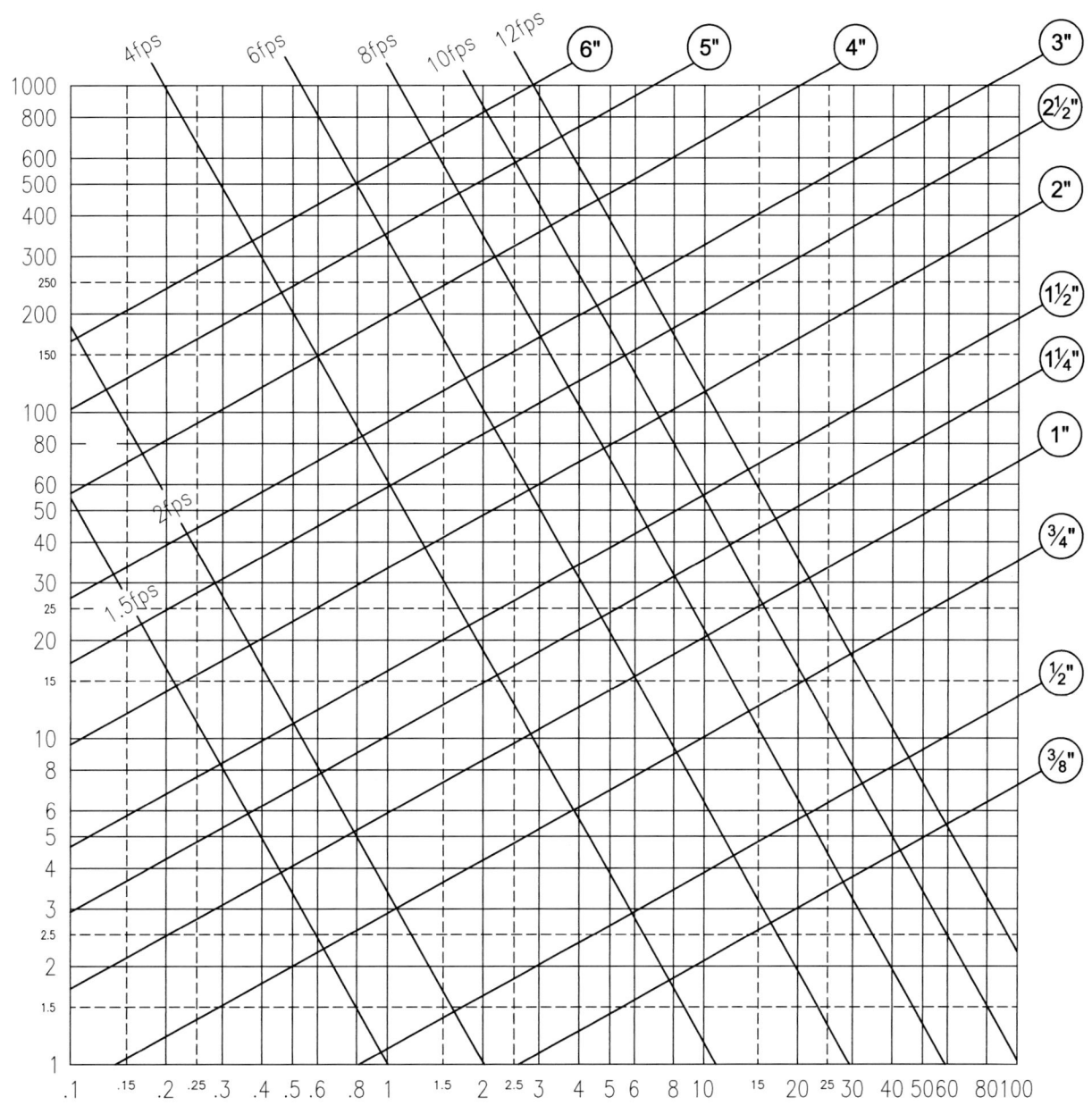

PRESSURE DROP (psi/100')

FLOW VS. PRESSURE LOSS

CHART B.9.8.4
TYPE M COPPER TUBE

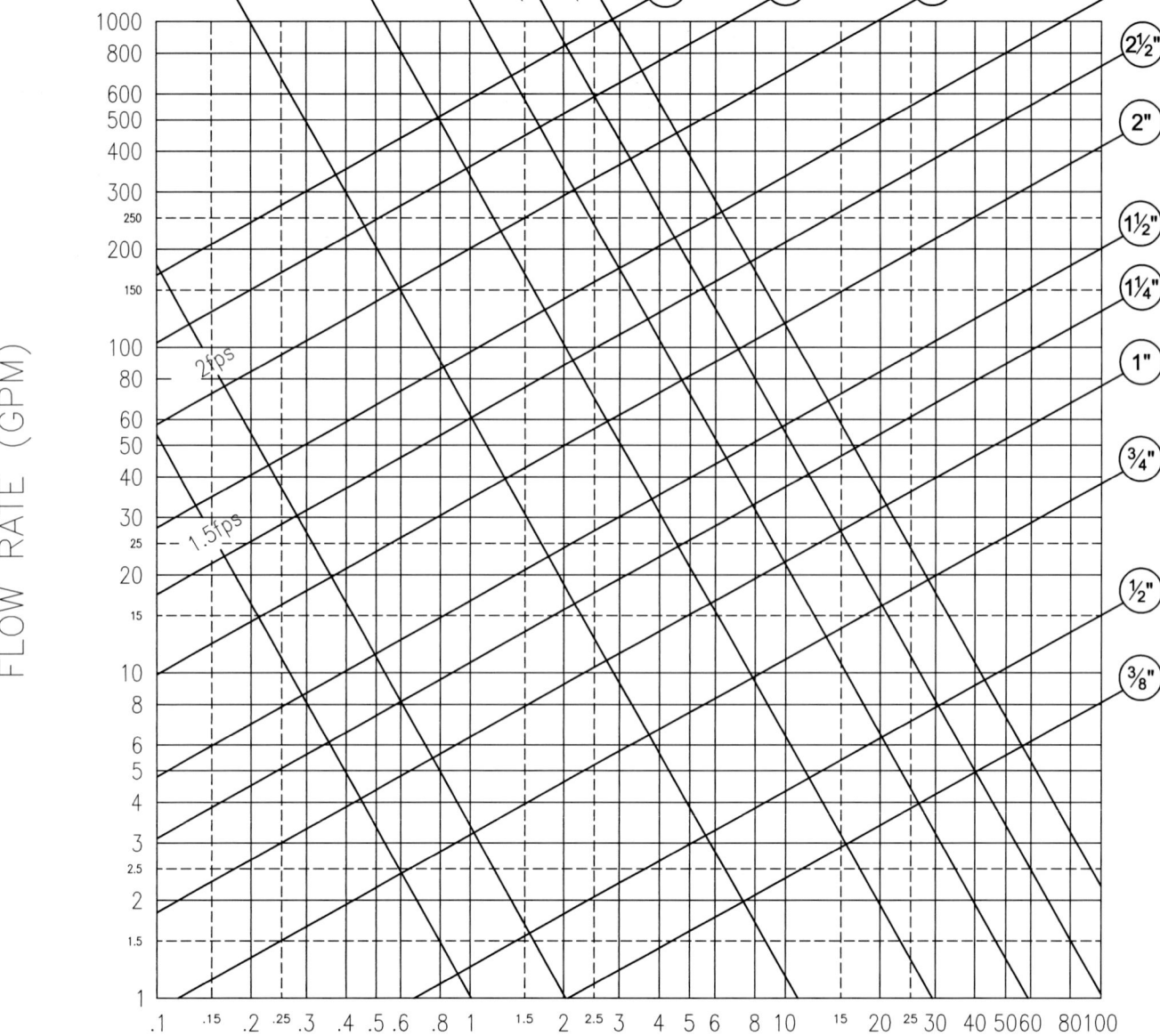

PRESSURE DROP (psi/100')

FLOW VS. PRESSURE LOSS

CHART B.9.8.5
CPVC, PVC, ABS, PE SCHEDULE 40 PIPE

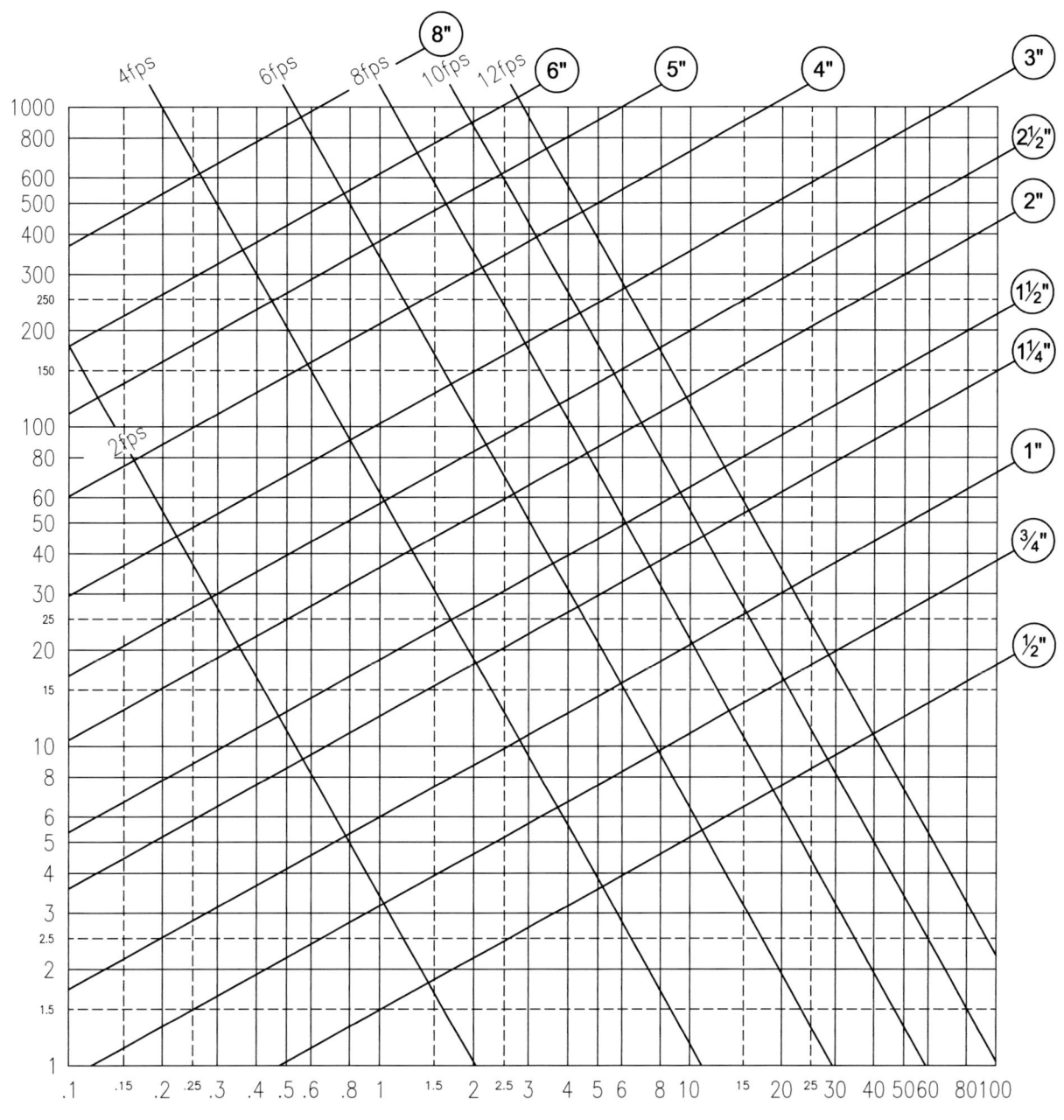

PRESSURE DROP (psi/100')

FLOW VS. PRESSURE LOSS

CHART B.9.8.6
CPVC, PVC, ABS, PE SCHEDULE 80 PIPE

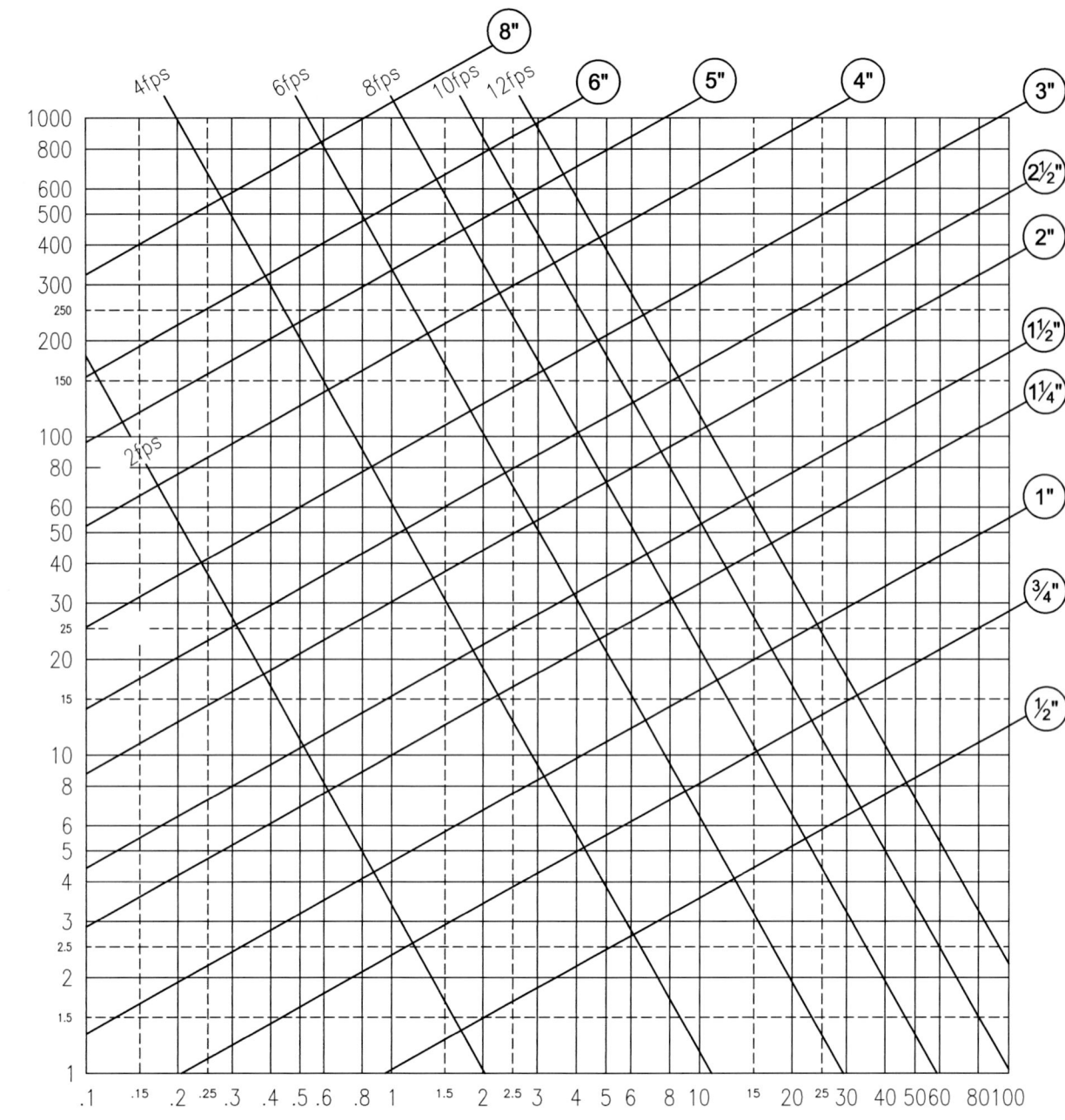

PRESSURE DROP (psi/100')

FLOW VS. PRESSURE LOSS

CHART B.9.8.7
CPVC TUBING (Copper Tube Size) SDR11

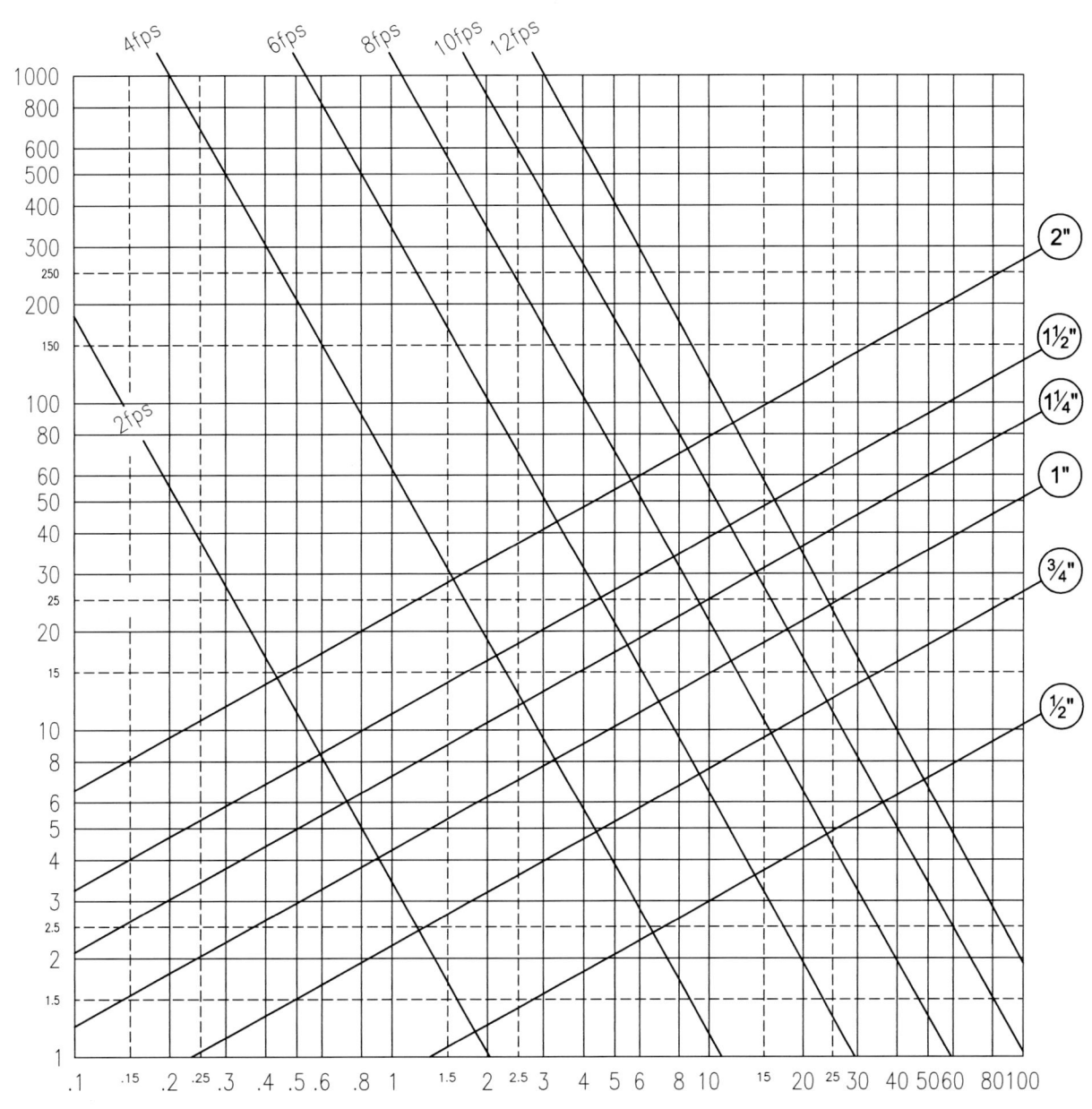

PRESSURE DROP (psi/100')

FLOW VS. PRESSURE LOSS

B.10 DETAILED SIZING METHOD FOR SYSTEMS IN BUILDINGS OF ANY HEIGHT

For sizing water supply systems in buildings of any height, a detailed method may be applied in the design of modern buildings. The procedure consists of sixteen (16) steps, as follows:

1. Obtain all information necessary for sizing the system. Such information shall be obtained from responsible parties and appropriate local authorities recognized as sources of the necessary information. (See Section B.2.)

2. Provide a schematic elevation of the complete water supply system. Show all piping connections in proper sequence and all fixture supplies. Identify all fixtures and risers by means of appropriate letters, numbers, or combinations thereof. Identify all piping conveying water at a temperature above 150°F, and all branch piping to such water outlets as solenoid valves, pneumatic valves, or quick-closing valves or faucets. Provide on the schematic elevation all the general information obtained per step 1. (See Section B.2.9.)

3. Mark on the schematic elevation, for each section of the complete system, the hot and cold water loads served in terms of water supply fixture units (WSFU) in accordance with Table B.5.2.

4. Mark on the schematic elevation, adjacent to all fixture unit notations, the probable maximum demand in gallons per minute corresponding to the various fixture unit loads in accordance with Table B.5.4.

5. Mark on the schematic elevation, for appropriate sections of the system, the demand in gallons per minute for outlets at which demand is considered continuous, such as outlets for watering gardens, irrigating lawns, air conditioning apparatus, refrigeration machines, and similar equipment. Add the continuous demand to the demand for intermittently used fixtures, and show the total demand at those sections where both types of demand occur. (See Section B.5.4.)

6. Size all individual fixture supply pipes to water outlets in accordance with the minimum sizes permitted by regulations. Minimum fixture supply pipe sizes for typical plumbing are given in Table B.5.2.

7. Size all other parts of the water supply system in accordance with velocity limitations recognized as good engineering practice, and with velocity limitations recommended by pipe manufacturers for avoiding accelerated deterioration and failure of their products under various conditions of service. Sizing tables based on such velocity limitations and showing permissible loads in terms of water supply fixture units (WSFU) for each size and kind of piping material have been provided and may be applied as a convenient and simplified method of sizing in this step. (See Section B.6)
Note: These sizes are tentative until verified in Steps 12, 13, 14, 15.

8. Assuming conditions of no-flow in the system, calculate the amount of pressure available at the topmost fixture in excess of the minimum pressure required at such fixtures for satisfactory supply conditions. This excess pressure is the limit for friction losses for peak demand in the system (1 foot of water column = 0.433 psi pressure). (See Section B.9.2.)

9. Determine which piping circuit of the system is the basic one for which pipe sizes in main lines and riser should be designed in accordance with friction loss limits. This circuit is the most extreme run of piping through which water flows from the public main, or other source of supply, to the highest and most distant water outlet. This basic design circuit (BDC) should be specifically identified on the schematic elevation of the system. (See Section B.9.3.)

10. Mark on the schematic elevation the pressure loss due to friction corresponding to the maximum probable demand through any water meter, water softener, or instantaneous or tankless water heating coil that may be provided in the BDC. (See Sections B.9.4 and B.9.5.)

11. Calculate the amount of pressure remaining and available for dissipation as friction loss during peak demand through pipe, valves, and fittings in the BDC. Deduct from the excess static pressure available at the topmost fixtures (determined in step 8), the friction losses for any water meters, softeners, and water heating coils provided in the BDC determined in step 10. (See Section B.9.4.)

12. Calculate the total equivalent length of the BDC. Pipe sizes established on the basis of velocity limitation in step 7 for main lines and risers must be considered just tentative at this stage, but may be deemed appropriate for determining corresponding equivalent lengths of fittings and valves in this step. (See Section B.9.7.)

13. Calculate the permissible uniform pressure loss for friction in piping of the BDC. The amount of pressure available in the circuit for dissipation as friction loss due to pipe, fittings, and valves (determined in step 11), is divided by the total equivalent length of the circuit (determined in step 12). This establishes the pipe friction limit for the circuit in terms of pressure loss in psi per foot of total equivalent pipe length. Multiply this value by 100 in order to express the pipe friction limit in terms of psi per 100 feet of length. (See Section B.9.6.)

14. Set up a sizing table showing the rates of flow for various sizes of the kind of piping to be used, corresponding to the permissible uniform pressure loss for pipe friction calculated for the BDC (determined in step 13). Such rates may be determined from a pipe friction chart appropriate for the piping to be used and for the effects upon the piping of the quality of the water to be conveyed thereby for extended service. (See Sections B.9.8 and B.2.3.)

15. Check the sizes of all parts of the BDC, and all other main lines and risers that supply water upward to the highest water outlets on the system, in accordance with the sizing table set up in step 14. Where sizes determined in this step are larger than those previously established in step 7 (based on velocity limitation), the increased sizes are applicable for limitation of friction.

16. Due consideration must be given to the action of the water on the interior of the piping, and proper allowance must be made where necessary as a design consideration, such as, where the kind of piping selected and the characteristics of the water conveyed are such that an appreciable buildup of corrosion products or hard-water scale may be anticipated to cause a significant reduction in bore of the piping system and inadequate capacity for satisfactory supply conditions during the normal service life of the system. A reasonable allowance in such cases is to select at least one standard pipe size larger than the sizes determined in the preceding steps. Where the water supply is treated in such manner as to avoid buildup of corrosion products or hard-water scale, no allowance need be made in sizing piping conveying such treated water. (See Sections B.2.3 and B.9.8.)

B.11 ILLUSTRATION OF DETAILED SIZING METHOD APPLICATION

B.11.1 Example

A seven-story building is supplied by direct street pressure from a public water main in which the minimum available pressure is 60 psi. The highest fixture supplied is 64'-8" above the public main, and requires 12 psi flow pressure at the fixture for satisfactory supply conditions.

The water supply is to be metered by a meter through which flow at the maximum probable demand rate will produce a pressure drop of 5.6 psi. Copper tubing, Type L, is to be used for the entire system. Quality of the water supply is known to be non-corrosive to copper tubing in the water district, and is recognized as being non-scaling in characteristic.

The entire system has been initially sized in accordance with the simplified method based solely on velocity limitations. Applying these sizes, the total equivalent length of piping from the public main to the highest and most remote fixture outlet has been calculated to be 600 feet.

B.11.2 Solution

Steps 1-7. The first seven steps of the detailed sizing method have already been performed. These steps constitute the simplified sizing method based solely on velocity limitations established as the design basis. All that remains is to perform steps 8 through 15 of the detailed sizing method which relate to sizing in accordance with the frictional limitation which must be observed for this particular system, and with allowances which may be necessary in view of the water characteristics.

Step 8. Assuming conditions of no-flow in the system, the amount of excess pressure available at the topmost fixture in excess of the minimum required at the fixture for satisfactory supply conditions is determined as follows: Excess pressure available = 60 psi - 12 psi - (64.67 x 0.4333 psi/ft) = 20 psi

Step 9. The BDC should be specifically identified on the schematic elevation provided as per step 2.

Step 10. The pressure loss through the water meter selected for this system for flow at maximum probable demand is given in the example as being 5.6 psi. No other items of equipment through which significant friction losses may occur have been noted in the example.

Step 11. The amount of pressure remaining for dissipation as friction loss during peak demand through pipes, valves, and fittings in the basic design circuit is determined as follows: Pressure available for friction in piping = 20 psi - 5.6 psi = 14.4 psi

Step 12. The total equivalent length of the basic design circuit has been given in the example as being 600 feet, based on the sizes determined in accordance with velocity limitations as per step 7.

Step 13. The permissible uniform pressure loss for friction in piping of the basic design circuit is determined as follows: Permissible uniform pipe friction loss = 14.4 psi x (100 ft/600 ft) = 2.4 psi per 100 ft pipe length.

Step 14. A sizing table showing the rates of flow through various sizes of copper tubing corresponding to a pipe friction loss rate of 2.4 psi per 100 feet of pipe length is given in Table B.11.2. These flow rates were determined from the chart applicable to such pipe with a "fairly smooth" surface condition after extended service conveying water having the effect stated in the example.

Step 15. All sections of the BDC should be selected and sized in accordance with the flow rates shown in the table established in step 14. Usually, all other parts of the system are sized using the same pressure drop limitation.

Table B.11.2
TYPE L COPPER TUBING FOR "FAIRLY SMOOTH" CONDITION

Nominal Pipe Size (in)	Flow Rate (gpm) Corresponding to Friction Loss of 2.4 psi per 100 feet
1/2	1.4
3/4	3.9
1	7.5
1-1/4	14.0
1-1/2	21.0
2	47.0
2-1/2	78.0
3	130.0
4	270.0

B.12 MANIFOLD TYPE PARALLEL WATER DISTRIBUTION SYSTEMS

B.12.1 Manifolds

The total water supply demand for the dwelling shall be determined in accordance with Appendix B.5. Manifolds shall be sized according to Table B.12.1 based on the total supply demand.

Table B.12.1
MANIFOLD SIZING[1]

Nominal Size Inches	Maximum GPM Available @ Velocity		
	@ 4 fps	@ 8 fps	@ 10 fps
1/2	2	5	6
3/4	6	11	14
1	10	20	25
1-1/4	15	31	38
1-1/2	22	44	55

1. Refer to Appendix B.6 for maximum velocity permitted.

B.12.2 Distribution Lines

a. The water pressure available for distribution pipe friction shall be determined from the minimum supply pressure available at the source, the developed length and size of the water service, the pressure drop through the water meter (if provided), the pressure drop through the manifold, the pressure drop through any other equipment or appurtenances in the system, the elevation of each distribution line, and the minimum pressure required at each fixture.

b. The water flow required at each fixture shall be in accordance with Table B.3. Where fixtures require both hot and cold water, the individual flow rates shall each be three-quarters (3/4) of the total flow rate in Table B.3. Distribution line sizes shall be in accordance with the system manufacturer's line sizing procedure.

c. The system manufacturer shall provide sizing data for the individual runs of tubing to each fixture based on the water pressure available for pipe friction and static elevation, the GPM required at each fixture, the tubing material, the tube size, and its maximum allowable length from the manifold to the fixture. Tube sizes for parallel water distribution systems include 3/8" nominal, 1/2" nominal, and 3/4" nominal.

Appendix C

Conversions: Metric to U.S. Customary Units

LENGTH

1 millimeter = 0.03937 inches
100 millimeters = 3.937 inches
1000 millimeters = 39.37 inches
1 meter = 3.2808 feet

AREA

1 square meter = 10.764 square feet

VOLUME

1 liter (L) = 0.2642 gallons (U.S.)
10 liters (L) = 2.642 gallons (U.S.)

VELOCITY

1 meter per second (m/s) = 3.2808 feet per second (ft/s)

FLOW

1 liter per minute (L/min) = 0.2642 gallons per minute (gpm)
10 liters per minute (L/min) = 2.642 gallons per minute (gpm)

PRESSURE

1 kilopascal (kPa) = 0.145 pounds per square inch (psi)
10 kilopascals (kPa) = 1.45 pounds per square inch (psi)

TEMPERATURE

deg F = 9/5 deg C +32
deg C = 5/9 (deg F -32)

TEMPERATURE	
Deg C	Deg F
0 C	32 F
5 C	41 F
10 C	50 F
15 C	59 F
20 C	68 F
25 C	77 F
30 C	86 F
35 C	95 F
40 C	104 F
45 C	113 F
50 C	122 F
55 C	131 F
60 C	140 F
65 C	149 F
70 C	158 F
75 C	167 F
80 C	176 F
85 C	185 F
90 C	194 F
95 C	203 F
100 C	212 F

CONVERTING WSFU DEMAND TO LITERS PER MINUTE (L/min) FLOW

WSFU	Flush Tanks L/min	Flush Valves L/min	WSFU	Flush Tanks L/min	Flush Valves L/min
3	11		120	186	280
4	15		140	201	295
5	17	83	160	216	314
6	19	87	180	231	329
7	23	91	200	246	345
8	27	95	225	265	360
9	28	98	250	284	379
10	30	102	300	322	416
11	32	106	400	398	473
12	34	110	500	473	530
13	38	112	750	644	662
14	40	114	1000	795	795
15	42	117	1250	909	909
16	45	121	1500	1022	1022
17	47	125	1750	1136	1136
18	49	127	2000	1230	1230
19	51	129	2500	1438	1438
20	53	133	3000	1647	1647
25	64	144	4000	1987	1987
30	76	155	5000	2271	2271
40	95	178	6000	2460	2460
50	110	193	7000	2650	2650
60	125	208	8000	2763	2763
80	148	235	9000	2877	2877
100	167	257	10,000	2990	2990

PIPE SIZES	
mm	NPS
15	1/2
20	3/4
25	1
32	1-1/4
40	1-1/2
50	2
65	2-1/2
80	3
100	4
125	5
150	6
200	8
250	10
300	12
380	15

Appendix D

Determining the Minimum Number of Required Plumbing Fixtures

The determination of the minimum number of required plumbing fixtures is a complex issue as many buildings are unique and the Authority Having Jurisdiction is called upon to use good judgment in applying this procedure.

1. DETERMINE THE SERVICE POPULATION

The population for the building or facility should be given on the plans. In the event the population of the building or service area is not given on the plans, three approaches may be utilized in determining the population to be served by the restroom facilities.

(a) Actual: In some instances the actual population, male and female, of the building service area may be known and this value may be used in the calculations.

(b) Engineering Estimate: The population of many buildings, especially those owner-occupied, may be determined on the basis of population densities. Typically, office building floor areas range from 200 to 400 square feet per person. In the absence of other data, the gender ratio should be 50/50.

(c) Legal Limit: Many building codes establish a legal occupancy limit based on the means of egress. Where the occupant load is based on the egress requirements of a building, the number of occupants for plumbing purposes shall be permitted to be reduced to two-thirds of that required for fire or life safety purposes.

2. MINIMUM NUMBER OF REQUIRED PLUMBING FIXTURES

The minimum number of required plumbing fixtures shall be determined based on building classification, user group and population as given in Table 7.21.1. The building classifications and user groups are consistent with nationally recognized building codes.

EXAMPLE A
The Building Classification, Occupancy Group, and the total population must be known or established.

Building Classification:	Assembly (A)
Occupancy Group:	A-3: Auditorium without permanent seating
Floor area:	20,000 net square feet, 1 story
Population:	2,857 based on egress per the Building Code at 7 sq ft per person

Determine the Plumbing Population
Since the population is based on egress requirements, the population for plumbing purposes can be 2/3 of that value per Section 7.21.2.b. The ratio of male and female occupants can be assumed to be 50% each per Section 7.21.2.c.

Population for plumbing purposes = 2857 x 2/3 = 1905 total
Male population = 1905 x 50% = 953
Female population = 1905 x 50% = 953

Determine the Minimum Required Number of Plumbing Fixtures
The minimum required number of plumbing fixtures is determined from Table 7.21.1 under Occupancy Group A-3: Assembly

The total numbers of males and females are greater than the three listed numerical groups listed in Table 7.21.1 for Group A-3 Assembly (with auditoriums). To determine the total number of the different fixtures required, calculate the number required by the three listed numerical groups and then determine how many additional groups of 300 there are over the first 300.

Male Water Closets

first 50 males (1-50)   = 1 water closet
next 100 males (51-150)   = 1 water closet
next 150 males (151-300)   = 1 water closet
add 1 water closet for each group of 300 over 300 (or parts thereof)

The three listed numerical groups account for the first 300 males.
Additional groups of 300 = (953-300) = 653 divided by 300 per group = 2.18 = 3 groups

The total number of water closets for males = 1 (first 50)
                                            = 1 (next 100)
                                            = 1 (next 150)
                                            = 3 (for three additional groups at 1 WC each)

The minimum number of water closets required for males = 6
Section 7.21.5a permits 50% of the required water closets for males to be urinals.
The male toilet room could contain 3 water closets and 3 urinals.

Female Water Closets

first 50 females (1-50)   = 1 water closet
next 100 females (51-150)   = 1 water closet
next 150 females (151-300)   = 1 water closet
add 1 water closet for each group of 300 over 300 (or parts thereof)

The three listed numerical groups account for the first 300 females.
Additional groups of 300 = (953-300) = 653 divided by 300 per group = 2.18 = 3 groups
The total number of water closets for females = 1 (first 50)
                                              = 1 (next 100)
                                              = 1 (next 150)
                                              = 3 (for three additional groups at 1 WC each)
The minimum number of water closets required for females = 6

Male Lavatories
first 50 males (1-50)   = 1 lavatory
next 100 males (51-150)   = 1 lavatory
next 150 males (151-300)   = 1 lavatory
add 1 lavatory for each group of 300 over 300 (or parts thereof)

The three listed numerical groups account for the first 300 males.
Additional groups of 300 = (953-300) = 653 divided by 300 per group = 2.18 = 3 groups

The total number of lavatories for males = 1 (first 50)
                                         = 1 (next 100)
                                         = 1 (next 150)
                                         = 3 (for three additional groups at 1 LAV each)
The minimum number of lavatories required for males = 6

Female Lavatories
first 50 females (1-50)   = 1 lavatory
next 100 females (51-150)   = 1 lavatory
next 150 females (151-300)   = 1 lavatory
add 1 lavatory for each group of 300 over 300 (or parts thereof)

The three listed numerical groups account for the first 300 females.
Additional groups of 300 = (953-300) = 653 divided by 300 per group = 2.18 = 3 groups

The total number of lavatories for females = 1 (first 50)
                                           = 1 (next 100)
                                           = 1 (next 150)
                                           = 3 (for three additional groups at 1 LAV each)
The minimum number of lavatories required for females = 6

Drinking Water Facilities
Table 7.21.1 requires 1 per 500 people (or parts thereof)
The minimum number of drinking water facilities = 1905 divided by 500 = 3.81 = 4 required

Service Sinks
Table 7.21.1 requires a minimum of 1 service sink per floor = 1 required

EXAMPLE B

The Building Classification, Occupancy Group, and the total population must be known or established.

Building Classification: Business (B)
Occupancy Group: B: Office building
Floor area: 48,300 net square feet, 1 story
Population: 386 based on expected occupancy

Determine the Plumbing Population

Population for plumbing purposes = 386
Male population = 386 x 50% = 193
Female population = 386 x 50% = 193

Determine the Minimum Required Number of Plumbing Fixtures

The minimum required number of plumbing fixtures is determined from Table 7.21.1 under Occupancy Group B Business.

The total numbers of males and females are greater than the total of the two listed numerical groups in Table 7.21.1 for Group B Business. To determine the total number of the different fixtures required, calculate the number required by the two listed numerical groups and then determine how many additional groups of 50 (or parts thereof) there are over the first 50.

Male Water Closets

first 15 males (1-15) = 1 water closet
next 35 males (16-50) = 1 water closet
add 1 water closet for each group of 50 over 50 (or parts thereof)

The two listed numerical groups account for the first 50 males.
Additional groups of 50 = (193-50) = 143 divided by 50 per group = 2.86 = 3 groups

The total number of water closets for males = 1 (first 15)
= 1 (next 35)
= 3 (for three additional groups of 50 at 1 WC each)

The minimum number of water closets required for males = 5
Section 7.21.5a permits 50% of the required water closets for males to be urinals.
The male toilet room could contain 3 water closets and 2 urinals.

Female Water Closets

first 15 females (1-15) = 1 water closet
next 35 females (16-50) = 1 water closet
add 1 water closet for each group of 50 over 50 (or parts thereof)

The two listed numerical groups account for the first 50 females.
Additional groups of 50 = (193-50) = 143 divided by 50 per group = 2.86 = 3 groups

The total number of water closets for females = 1 (first 15)
= 1 (next 35)
= 3 (for three additional groups at 1 WC each)

The minimum number of water closets required for females = 5.

Male Lavatories

first 15 males (1-15)   = 1 lavatory
next 35 males (16-50) = 0 lavatory
add 1 lavatory for each group of 50 over 50 (or parts thereof)

The two listed numerical groups account for the first 50 males.
Additional groups of 50 = (193-50) = 143 divided by 50 per group = 2.86 = 3 groups

The total number of lavatories for males = 1 (first 15)
                                         = 0 (next 35)
                                         = 3 (three additional groups of 50 at 1 LAV each)
The minimum number of lavatories required for males = 4.

Female Lavatories

first 15 females (1-15)   = 1 lavatory
next 35 females (16-50) = 0 lavatory
add 1 lavatory for each group of 50 over 50 (or parts thereof)

The two listed numerical groups account for the first 50 females.
Additional groups of 50 = (193-50) = 143 divided by 50 per group = 2.86 = 3 groups

The total number of lavatories for females = 1 (first 15)
                                           = 0 (next 35)
                                           = 3 (three additional groups of 50 at 1 LAV each)
The minimum number of lavatories required for females = 4.

Drinking Water Facilities

Table 7.21.1 requires 1 per 100 people (or parts thereof)
The minimum number of drinking water facilities = 386 divided by 100 = 3.86 = 4 required

Service Sinks

Table 7.21.1 requires a minimum of 1 service sink per floor = 1 required

Appendix E

Special Design Plumbing Systems

E.1 GENERAL REQUIREMENTS

E.1.1 Special design plumbing systems shall include all systems that vary in detail from the requirements of this Code.

E.1.2 The provisions of this Appendix shall control the design, installation, and inspection of special design plumbing systems.

E.1.3 Special design plumbing systems shall conform to the Basic Principles of this Code.

E.1.4 Special design plumbing systems shall be designed by a registered design professional who is licensed to practice in the particular jurisdiction.

E.1.5 The requirements of Section E.1, E.2, E.3, and E.4 shall apply to all special design plumbing systems in this Appendix E.

E.2 PLANS, SPECIFICATIONS, AND COMPUTATIONS

E.2.1 Plans, specifications, computations, and other related data for special design plumbing systems, prepared by the registered design professional, shall be submitted to the Authority Having Jurisdiction for review and approval prior to installation.

E.2.2 The design plans shall indicate that the plumbing system (or portions thereof) is a special design system.

E.3 INSTALLATION OF SPECIAL DESIGN PLUMBING SYSTEMS

E.3.1 Special design plumbing systems shall be installed according to established, tested and approved criteria, including manufacturer's instructions.

E.3.2 The installation shall comply with Chapter 2 - *General Regulations* and other applicable requirements of this Code.

E.4 CERTIFICATION OF COMPLIANCE

E.4.1 Inspections shall be made by the Authority Having Jurisdiction to ensure conformance with data submitted for approval and the applicable requirements of this Code.

E.4.2 The complete installation and performance of the special design plumbing system shall be certified by the registered design professional as complying with the requirements of the special design.

E.5 VACUUM DRAINAGE SYSTEMS

E.5.1 General Requirement

E.5.1.1 System Design

Vacuum drainage systems shall be designed in accordance with manufacturer's recommendations. The system layout, including piping layout, tank assemblies, vacuum pump assembly and other components/designs necessary for proper function of the system shall be per manufacturer's recommendations. Plans, specifications and other data for such systems shall be submitted to the Authority Having Jurisdiction for review and approval prior to installation.

E.5.1.2 Fixtures

Gravity type fixtures used in vacuum drainage systems shall comply with Chapter 7 of this Code.

E.5.1.3 Drainage Fixture Units

Fixture units for gravity drainage systems that discharge into or receive discharge from vacuum drainage systems shall be based upon values in Chapter 11 of this Code.

E.5.1.4 Water Supply Fixture Units

Water supply fixture units shall be based upon values in Chapter 10 of this Code with the addition that the fixture unit of a vacuum type water closet shall be 1.0 WSFU.

E.5.1.5 Traps and Cleanouts

Gravity type fixtures shall be provided with traps and cleanouts per Chapter 5 of this Code.

E.5.1.6 Materials

Vacuum drainage pipe, fittings and valve materials shall be as recommended by the vacuum drainage system manufacturer and as permitted by this Code.

E.5.2 Tests and Demonstrations

After completion of the entire system installation, the system shall be subjected to a vacuum test of 19 inches of mercury and shall be operated to function as required by the Authority Having Jurisdiction and the manufacturer. Recorded proof of all tests shall be submitted to the Authority Having Jurisdiction.

E.5.3 Written Instructions

Written instructions for the operation, maintenance, safety and emergency procedures shall be provided to the Building Owner as verified by the Authority Having Jurisdiction.

E.5.4 Requirements for Special Design Plumbing Systems

The requirements of Sections E.1, E.2, E.3, and E.4 apply to this special design plumbing system.

E.6 ONE-PIPE SANITARY DRAINAGE SYSTEMS EMPLOYING AERATOR AND DEAERATOR STACK FITTINGS

E.6.1 Compliance

a. One-pipe sanitary drainage systems employing aerator and deaerator stack fittings shall be permitted to be installed in accordance with (1) the fitting manufacturer's current piping design manual and technical bulletins, and (2) the applicable requirements of this Code.

b. The requirements of this Code shall supersede the fitting manufacturer's piping design manual with regard to acceptable piping materials, drainage fixture unit (DFU) values, and minimum drainage pipe sizes.

c. Complete detailed layout drawings shall be prepared prior to the installation of such a sanitary drainage system.

E.6.2 Piping Materials

Piping materials for drainage stacks and branches shall be in accordance with Section 3.5 of this Code. Piping materials for vents shall be in accordance with Section 3.6 of this Code.

E.6.3 Drainage Fixture Unit (DFU) Values

Drainage fixture unit values for bathroom groups and individual fixtures shall be in accordance with Table 11.4.1 of this Code.

E.6.4 Drainage Pipe Sizing

Piping in such a sanitary drainage system shall be sized according to the fitting manufacturer's current piping design manual.
EXCEPTION: Drainage pipe sizes shall not be less than those required by Section 11.5 of this Code.

E.6.5 Arrangement of Piping

Piping shall be arranged as illustrated in the fitting manufacturer's current design manual.

E.6.6 Requirements for Special Design Plumbing Systems

The requirements of Sections E.1, E.2, E.3, and E.4 apply to this special design plumbing system.

E.7 SINGLE STACK VENT SYSTEMS

E.7.1 Where permitted.

A drainage stack shall be permitted to serve as a single stack vent system when sized and installed in accordance with Sections E.7.2 through E.7.9. The drainage stack and branch piping in a single stack vent system shall provide for the flow of liquids, solids, and air without the loss of fixture trap seals.

E.7.2 Stack Size

Drainage stacks shall be sized according to Table E.7.2. A maximum of two water closets shall be permitted to discharge to a 3 inch stack. Stacks shall be uniformly sized based on the total connected drainage fixture unit load with no reductions in size.

Table E.7.2
SINGLE STACK SIZE

Stack Size (inches)	Maximum Connected Drainage Fixture Units (DFU)		
	Stacks Less than 75 Feet in Height	Stacks 75 Feet to Less than 160 Feet in Height	Stacks 160 Feet or Greater in Height
3	24 (2)	NP	NP
4	225	24	NP
5	480	225	24
6	1015	480	225
8	2320	1015	480
10	4500	2320	1015
12	8100	4500	2320
15	13,600	8100	4500

1. NP = not permitted
2. Not more than two (2) water closets are permitted on a 3" stack.

E.7.3 Branch Size

Horizontal branches connecting to a single stack vent system shall be sized according to Table 11.5.1 B.
EXCEPTIONS:
  (1) No more than one water closet within 18" of the stack horizontally shall be permitted on a 3" horizontal branch.
  (2) A water closet within 18" of a stack horizontally and one other fixture with up to 1-1/2 inch fixture drain size shall be permitted on a 3" horizontal branch when connected to the stack through a sanitary tee.

E.7.4 Length of Horizontal Branches

  a. Water closets shall be no more than four (4) feet horizontally from the stack.
EXCEPTION: Water closets shall be permitted to be up to eight (8) feet horizontally from the stack when connected to the stack through a sanitary tee.
  b. Fixtures other than water closets shall be no more than twelve (12) feet horizontally from the stack.
  c. The length of any vertical piping from a fixture trap to a horizontal branch shall not be considered it computing the fixture's horizontal distance from the stack.

E.7.5 Maximum Vertical Drops From Fixtures

Vertical drops from fixture traps to horizontal branch piping shall be one pipe size larger than the trap but not less than two (2) inch pipe size. Vertical drops shall be four (4) feet maximum length. Fixture drains that are not increased in size, or have a vertical drop exceeding 4 feet shall be individually vented.

E.7.6 Additional Venting Required

Additional venting shall be provided when more than one water closet is on a horizontal branch and where the distance from a fixture trap to the stack exceeds the limits in Section 12.18.1. Where additional venting is required, the fixture(s) shall be vented by an individual vent, common vent, wet vent, circuit vent, or a combination waste and vent pipe. The dry vent extensions for the additional venting shall connect to a branch vent, vent stack, stack vent, air admittance valve, or be extended outdoors and terminate to the open air.

E.7.7 Stack Offsets

Where there are no fixture drain connections below a horizontal offset in a stack, the offset does not need to be vented. When there are fixture drain connections below a horizontal offset in a stack, the offset shall be vented in accordance with Section 12.3.3. There shall be no fixture connections to a stack within 2 feet above and below a horizontal offset.

E.7.8 Separate Stacks Required

Where stacks are more than two stories high, a separate stack shall be provided for the fixtures on the lower two stories. The stack for the lower two stories may be connected to the branch of the building drain that serves the stack for the upper stories at a point that is at least 10 pipe diameters downstream from the base of the upper stack.

E.7.9 Sizing Building Drains and Sewers

The building drain and branches thereof, and the building sewer in a single stack vent system shall be sized in accordance with Table 11.5.1A.

E.7.10 Requirements for Special Design Plumbing Systems

The requirements of Sections E.1, E.2, E.3, and E.4 apply to this special design plumbing system.

E.8 AIR ADMITTANCE VALVES

E.8.1 Definition

Air admittance valve: A one-way valve designed to allow air to enter the plumbing drainage system when negative pressures develop in the system. The device closes by gravity, without springs or other mechanical means, and seals the vent terminal at zero differential pressure (no flow conditions) and also under positive internal pressure. The purposes of an air admittance valve are (1) to provide a method of allowing air to enter the plumbing drainage system without the need for a vent extended outdoors to open air, and (2) to prevent sewer gases from escaping into the building.

E.8.2 Where Permitted

E.8.2.1 Branch, circuit, common, continuous, and individual vents shall be permitted to terminate with a connection to an individual or branch type air admittance valve complying with ASSE 1051. Individual and branch type air admittance valves shall only vent fixtures that are on the same floor level and connect to a horizontal branch drain.

E.8.2.2 Vent stacks and stack vents shall be permitted to terminate at a stack type air admittance valve complying with ASSE 1050.
EXCEPTIONS
    (1) Vent stacks and stack vents serving drainage stacks that exceed six (6) branch intervals in height.
    (2) Vent stacks and stack vents that serve relief vents in Section E.8.3.

E.8.2.3 Air admittance valves shall not be permitted in the following applications:
    a. vents for special waste drainage systems (Sections 9.4.1 and 9.4.2).
    b. vents for sewage pump or ejector sump pits.
    c. vents for pneumatic sewage ejectors.
    d. suds pressure zone venting.
    e. relief vents required by Section E.8.3.1.
    f. in attics and other inaccessible locations.

E.8.3 Relief Vents

E.8.3.1 A relief vent shall be provided where a horizontal branch drain that is vented by one or more air admittance valves connects to a drainage stack more than four (4) branch intervals from the top of the stack. The relief vent shall connect vertically to the horizontal branch drain between the drainage stack and the most downstream fixture drain connection on the horizontal branch drain. Relief vents shall extend from the horizontal branch drain to a vent stack, stack vent, or other vent that terminates outdoors in open air.

E.8.3.2 Relief vents shall be the full size of the horizontal branch drain that they serve, up to 3" maximum required size.

E.8.3.3 Relief vents shall be permitted to vent fixtures other than those on the horizontal branch drain being relieved.

E.8.4 Installation

E.8.4.1 Air admittance valves shall be installed in accordance with the manufacturer's instructions and Section E.8.

E.8.4.2 Air admittance valves shall connect to fixture trap arms within the maximum allowable trap arm lengths in Table 12.8.1.

E.8.4.3 Individual and branch type air admittance valves shall be installed at least 4 inches above the top of the trap arm or horizontal branch drain that they serve.

E.8.4.4 Stack type air admittance valves shall be installed at least 6 inches above the flood level rim of the highest fixture served by the valve.

E.8.4.5 Air admittance valves shall be installed in accessible locations having free movement of air to enter the valve.

E.8.4.6 Air admittance valves shall not be installed in HVAC supply or return air plenums or other areas subject to other than atmospheric pressure.

E.8.4.7 Air admittance valves shall be the same size as the vent pipe to which they are connected.

E.8.4.8 Air admittance valves shall not be installed until all required pressure tests of the drainage and vent piping are successfully completed.

E.8.5 Vent to Outdoors Required

E.8.5.1 Where a plumbing drainage system is vented by one or more air admittance valves, at least one vent pipe shall extend to an outdoor vent terminal complying with Section 12.4. Outdoor venting shall comply with either Section E.8.5.2 or E.8.5.3.

E.8.5.2 The aggregate size of the outdoor vent terminals shall comply with Section 12.16.8 unless the building drain is vented in accordance with Section E.8.5.3.

E.8.5.3 Where the aggregate size of outdoor vent terminals does not comply with Section E.8.5.2, a dry vent, sized according to Table 12.16.4, shall be provided downstream from the last fixture connection, branch connection, or stack connection to the building drain before the connection of the building drain to the building sewer.

E.8.6 Referenced Standards

a. ASSE 1050-2009 Stack Air Admittance Valves for Sanitary Drainage Systems.
b. ASSE 1051-2009 Individual and Branch Type Air Admittance Valves for Plumbing Drainage Systems.

E.8.7 Requirements for Special Design Plumbing Systems

The requirements of Sections E.1, E.2, E.3, and E.4 apply to this special design plumbing system.

E.9 SIPHONIC ROOF DRAINAGE

E.9.1 General Requirements

a. Siphonic roof drainage systems for primary roof drainage shall be designed, installed, inspected, and tested in accordance with the requirements of ASPE Standard 45 - Siphonic Roof Drainage, its Appendices, and the other requirements of Section E.9 in this Appendix.

b. Systems shall be permitted to be designed using computer programs that produce results equivalent to ASPE Standard 45.

c. Design plans shall show all details of the system installation, including pipe sizing, routing, elevations, fittings, and orientation. The system installation shall be coordinated with all other aspects of the facility construction.

d. The requirements of Sections E.1, E.2, E.3, and E.4 apply to this special design plumbing system.

E.9.2 Roof Drains

Siphonic roof drains shall comply with ASME A112.6.9.

E.9.3 Rainfall Rates

The rainfall rates used for sizing primary and secondary roof drainage systems shall be in accordance with Table A.1 in NSPC Appendix A.

E.9.4 Secondary Roof Drainage

Secondary roof drainage shall be provided in accordance with NSPC Section 13.1.10.2.

E.9.5 Referenced Standards

ASPE Standard 45 - Siphonic Roof Drainage (current edition)
ASME A112.6.9-2005 Siphonic Roof Drains (current edition)

Blank Page

Appendix F

Requirements of the Adopting Agency

F.1 GENERAL

This appendix lists those sections of this Code where the Adopting Agency must establish specific requirements or confirm whether or not they have certain requirements. It also includes listings where the Authority Having Jurisdiction may need to have requirements. Adopting agencies should review the sections listed below under Section F.2. REFERENCES.

F.2 REFERENCES

ADM 1.4.8:	Appendices
2.16.a.1:	Minimum earth cover for water service pipe
2.16.a.2:	Minimum earth cover for building sewers
2.19.1:	Distance for required connections to public water supplies and sewers
2.19.2:	Standards and requirements for private water and sewage disposal systems
2.26	Draining elevator sump pits
2.28	NFPA 13D multipurpose residential fire sprinkler systems
Table 3.5:	SDR and PS ratings of PVC building sewers
Table 3.7:	SDR and PS ratings of PVC building storm sewers
5.3.4:	Building traps
5.4.6.b:	Cleanouts in sanitary sewers and storm sewers.
6.1.4.1	Design and installation of liquid waste treatment equipment
6.1.7	Point of discharge of liquid waste treatment equipment
6.2.1.a:	Grease interceptors
6.2.1.3.b:	Gravity grease interceptors
6.2.7:	Indoor grease interceptors
6.2.10.b:	Interceptor sizing (commercial kitchens)
6.2.12:	Combination grease interceptor systems
6.3.2.e:	Waste oil tanks for oil separators
7.5.6	Non-water urinals
10.1:	Use of reclaimed water, graywater, and harvested rainwater
10.4.3:	Cross connection control
10.20.1	NFPA 13D multipurpose residential fire sprinkler systems
12.5:	Frost closure of vent terminals
Chapter 16:	Private sewage disposal systems
Chapter 17:	Private potable water supply systems
Chapter 18:	Mobile home and travel trailer park plumbing requirements
Appendix E.5	Vacuum drainage systems
Appendix E.6	One pipe drainage systems with aerators and deaerators
Appendix E.7	Single stack vent systems
Appendix E.8	Air admittance valves
Appendix E.9	Siphonic roof drainage

F.3 EXCEPTIONS, WAIVERS, APPROVALS

There are instances where this Code specifically permits exceptions or waivers of its requirements and the approval of alternative materials and methods by the Adopting Agency. These occur throughout this Code.

Appendix G

Green Plumbing and Mechanical Code Supplement

G.1 FOREWORD 454

G.2 DEFINITIONS 455

G.3 WATER EFFICIENCY AND CONSERVATION 458
    G.3.1 General 458
        G.3.1.1 Scope 458
    G.3.2 Water-Conserving Plumbing Fixtures and Fittings 458
        G.3.2.1 General 458
        G.3.2.2 Water Closets 458
        G.3.2.3 Urinals 459
        G.3.2.4 Residential Kitchen Faucets 459
        G.3.2.5 Lavatory Faucets 459
        G.3.2.6 Showers 460
        G.3.2.7 Commercial Pre-Rinse Spray Valves 460
    G.3.3 Appliances 460
        G.3.3.1 Dishwashers 460
        G.3.3.2 Clothes Washers 460
    G.3.4 Water Pressure 461
        G.3.4.1 Installation 461
    G.3.5 Water Softeners and Treatment Devices 461
        G.3.5.1 Water Softeners 461
        G.3.5.2 Water Softener Limitations 461
        G.3.5.3 Point-of-Use Reverse Osmosis Water Treatment Systems 461
    G.3.6 Occupancy Specific Water Efficiency Requirements 461
        G.3.6.1 Commercial Food Service 461
        G.3.6.2 Medical and Laboratory Facilities 462
    G.3.7 Leak Detection and Control 462
        G.3.7.1 General 462
    G.3.8 Fountains and Other Water Features 462
        G.3.8.1 Use of Alternate Water Source for Special Water Features 462
    G.3.9 Meters 462
        G.3.9.1 Required 462
        G.3.9.2 Consumption Data 463
        G.3.9.3 Access 463
    G.3.10 HVAC Water Efficiency 463
        G.3.10.1 Once-Through Cooling 463
        G.3.10.2 Cooling Towers and Evaporative Coolers 463
        G.3.10.3 Cooling Tower Makeup Water 463
        G.3.10.4 Evaporative Cooler Water Use 463
        G.3.10.5 Use of Reclaimed (Recycled) and On-Site Treated Non-Potable Water for Cooling 464

G.3.11 Condensate Recovery	464
G.3.12 Water-Powered Sump Pumps	465
G.3.13 Landscape Irrigation Systems	465
G.3.13.1 General	465
G.3.13.2 Backflow Protection	465
G.3.13.3 Use of Alternate Water Sources for Landscape Irrigation	465
G.3.13.4 Irrigation Control Systems	465
G.3.13.5 Low Flow Irrigation	466
G.3.13.6 Mulched Planting Areas	466
G.3.13.7 System Performance Requirements	466
G.3.13.8 Narrow or Irregularly Shaped Landscape Areas	466
G.3.13.9 Sloped Areas	466
G.3.13.10 Sprinkler Head Installations	466
G.3.13.11 Irrigation Zone Performance Criteria	467
G.3.13.12 Qualifications	467
G.3.14 Trap Seal Protection	467
G.3.14.1 Water Supplied Trap Primers	467
G.3.14.2 Drainage Type Trap Seal Primer Devices	467
G.3.15 Vehicle Wash Facilities	467
G.4 ALTERNATE WATER SOURCES FOR NON-POTABLE APPLICATIONS	**468**
G.4.1 General	468
G.4.1.1 Scope	468
G.4.1.2 System Design	468
G.4.1.3 Permit	468
G.4.1.4 Component Identification	468
G.4.1.5 Maintenance and Inspection	468
G.4.1.6 Operation and Maintenance Manual	469
G.4.1.7 Minimum Water Quality Requirements	469
G.4.1.8 Material Compatibility	469
G.4.1.9 System Controls	470
G.4.1.10 Commercial, Industrial, and Institutional Restroom Signs	470
G.4.1.11 Inspection and Testing	471
G.4.1.12 Separation Requirements	472
G.4.1.13 Abandonment	473
G.4.1.14 Sizing	473
G.4.2 Gray Water Systems	473
G.4.2.1 General	473
G.4.2.2 Gray Water System	473
G.4.2.3 Connections to Potable and Reclaimed (Recycled) Water Systems	473
G.4.2.4 Location	474
G.4.2.5 Plot Plan Submission	474
G.4.2.6 Prohibited Location	474
G.4.2.7 Drawings and Specifications	474
G.4.2.8 Procedure for Estimating Gray Water Discharge	475
G.4.2.9 Gray Water System Components	475
G.4.2.10 Subsurface Irrigation System Zones	477
G.4.2.11 Subsurface and Subsoil Irrigation Field, and Mulch Basin Design and Construction	478
G.4.2.12 Gray Water System Color and Marking Information	480
G.4.2.13 Special Provisions	480

G.4.2.14 Testing	481
G.4.2.15 Maintenance	481
G.4.3 Reclaimed (Recycled) Water Systems	481
G.4.3.1 General	481
G.4.3.2 Permit	481
G.4.3.3 System Changes	481
G.4.3.4 Connections to Potable or Reclaimed (Recycled) Water Systems	481
G.4.3.5 Initial Cross-Connection Test	481
G.4.3.6 Reclaimed (Recycled) Water System Materials	482
G.4.3.7 Reclaimed (Recycled) Water System Color and Marking Information	482
G.4.3.8 Valves	482
G.4.3.9 Installation	482
G.4.3.10 Signs	483
G.4.3.11 Inspection and Testing	483
G.4.4 On-Site Treated Non-Potable Water Systems	483
G.4.4.1 General	483
G.4.4.2 Plumbing Plan Submission	483
G.4.4.3 System Changes	483
G.4.4.4 Connections to Potable or Reclaimed (Recycled) Water Systems	483
G.4.4.5 Initial Cross-Connection Test	483
G.4.4.6 On-Site Treated Non-Potable Water System Materials	483
G.4.4.7 On-Site Treated Non-Potable Water Devices and Systems	483
G.4.4.8 On-Site Treated Non-Potable Water System Color and Marking Information	484
G.4.4.9 Valves	484
G.4.4.10 Design and Installation	484
G.4.4.11 Signs	484
G.4.4.12 Inspection and Testing	484
G.4.5 Non-Potable Rainwater Catchment Systems	485
G.4.5.1 General	485
G.4.5.2 Plumbing Plan Submission	485
G.4.5.3 System Changes	485
G.4.5.4 Connections to Potable or Reclaimed (Recycled) Water Systems	485
G.4.5.5 Initial Cross-Connection Test	485
G.4.5.6 Sizing	485
G.4.5.7 Rainwater Catchment System Materials	485
G.4.5.8 Rainwater Catchment Water System Color and Marking Information	486
G.4.5.9 Design and Installation	486
G.4.5.10 Signs	489
G.4.5.11 Inspection and Testing	489
G.5 WATER HEATING DESIGN, EQUIPMENT AND INSTALLATION	**490**
G.5.1 General	490
G.5.1.1 Scope	490
G.5.1.2 Insulation	490
G.5.1.3 Recirculation Systems	490
G.5.2 Service Hot Water – Low-Rise Residential Buildings	491
G.5.2.1 General	491
G.5.2.2 Water Heaters and Storage Tanks	491
G.5.2.3 Recirculation Systems	491
G.5.2.4 Central Water Heating Equipment	492
G.5.2.5 Insulation	492

- G.5.2.6 Hard Water 492
- G.5.2.7 Maximum Volume of Hot Water 492
- G.5.3 Service Hot Water – Other Than Low-Rise Residential Buildings 493
  - G.5.3.1 General 493
  - G.5.3.2 Service Water Heating 493
  - G.5.3.3 Compliance Path(s) 493
  - G.5.3.4 Mandatory Provisions 494
  - G.5.3.5 Prescriptive Path 497
  - G.5.3.6 Submittals 498
- G.5.4 Solar Water Heating Systems 498
  - G.5.4.1 General 498
  - G.5.4.2 Annual Inspection and Maintenance 498
- G.5.5 Hard Water 498
  - G.5.5.1 Softening and Treatment 498
- G.5.6 Drain Water Heat Exchangers 498

G.6 METHOD OF CALCULATING WATER SAVINGS 499
- G.6.1 Water Savings Calculation 499
  - G.6.1.1 Purpose 499
  - G.6.1.2 Calculation of Water Savings 499

G.7 POTABLE RAIN WATER CATCHMENT SYSTEMS 503
- G.7.1 General 503
  - G.7.1.1 Scope 503
  - G.7.1.2 System Design 503
  - G.7.1.3 Permit 503
  - G.7.1.4 Product and Material Approval 503
  - G.7.1.5 Maintenance and Inspection 503
  - G.7.1.6 Operation and Maintenance Manual 504
  - G.7.1.7 Minimum Water Quality Requirements 505
  - G.7.1.8 Material Compatibility 505
  - G.7.1.9 System Controls 505
- G.7.2 Connection 505
  - G.7.2.1 General 505
  - G.7.2.2 Connections to Public or Private Potable Water Systems 506
  - G.7.2.3 Backflow Prevention 506
- G.7.3 Potable Rainfall Catchment System Materials 506
  - G.7.3.1 Collections Surfaces 506
  - G.7.3.2 Rainwater Catchment System Drainage Materials 506
  - G.7.3.3 Storage Tanks 506
  - G.7.3.4 Water Supply and Distribution Materials 506
- G.7.4 Design and Installation 506
  - G.7.4.1 Collection Surfaces 506
  - G.7.4.2 Minimum Water Quality 507
  - G.7.4.3 Overhanging Tree Branches and Vegetation 507
  - G.7.4.4 Rainwater Storage Tanks 507
  - G.7.4.5 Pumps 509
  - G.7.4.6 Roof Drains 509
  - G.7.4.7 Water Quality Devices and Equipment 509
  - G.7.4.8 Freeze Protection 509

  G.7.4.9 Roof Washer or Pre-Filtration System 509
  G.7.4.10 Filtration and Disinfection Systems 510
  G.7.4.11 Roof Gutters 510
  G.7.4.12 Drains, Conductors, and Leaders 510
  G.7.4.13 Size of Potable Water Piping 510
 G.7.5 Cleaning 510
  G.7.5.1 General 510
G.7.6 Supply System Inspection and Test 510

G.1 FOREWORD

Appendix G serves as a resource for code officials, plumbers, contractors, engineers, and manufacturers in designing, installing, inspecting and maintaining sustainable plumbing systems. Appendix G is intended to provide a comprehensive set of technically sound provisions that encourage sustainable practices and works towards enhancing the design of plumbing systems that result in a positive long-term environmental impact. Appendix G consists of excerpts from the 2012 Green Plumbing and Mechanical Code Supplement (GPMCS) copyrighted by the International Association of Plumbing and Mechanical Officials (IAPMO). No part of this Appendix may be reproduced without the permission of IAPMO.

G.2 DEFINITIONS

Alternate Water Source: Non-potable source of water that includes but not limited to gray water, on-site treated non-potable water, rainwater, and reclaimed (recycled) water.

Debris Excluder: A device installed on the rainwater catchment conveyance system to prevent the accumulation of leaves, needles, or other debris in the system.

Dry Weather Runoff: Water that flows along a surface, in a channel or sub-surface including groundwater seepage, and is not associated with a rainwater catchment system or stormwater catchment system.

Energy Star: A joint program of the U.S. Environmental Protection Agency and the U.S. Department of Energy. Energy Star is a voluntary program designed to identify and promote energy-efficient products and practices.

Evapotranspiration (ET): The combination of water transpired from vegetation and evaporated from the soil, water, and plant surfaces. Evapotranspiration rates are expressed in inches per day, week, month, or year. Evapotranspiration varies by climate and time of year. Common usage includes Evapotranspiration as the base rate (water demand of 4-6 inch tall cool season grass), with coefficients for specific plant types. Evapotranspiration rates are used as a factor in estimating the irrigation water needs of landscapes. Local agriculture extension, state departments of agriculture, water agencies, irrigation professionals, and internet websites are common sources for obtaining local Evapotranspiration rates.

Gang Showers (non-residential): Shower compartments designed and intended for use by multiple persons simultaneously in non-residential occupancies.

Geothermal: Renewable energy generated by deep-earth.

Gray Water: Untreated waste water that has not come into contact with toilet waste, kitchen sink waste, dishwasher waste or similarly contaminated sources. Gray water includes waste water from bathtubs, showers, lavatories, clothes washers and laundry tubs. Also known as grey water, graywater, and greywater.

Gray Water Diverter Valve: A valve that directs gray water to the sanitary drainage system or to a subsurface irrigation system.

Hydrozone: A grouping of plants with similar water requirements that are irrigated by the same irrigation zone.

Irrigation Demand: The amount of irrigation water not supplied by natural precipitation that is needed to maintain landscape plant life in good condition. Irrigation demand is calculated by subtracting natural effective precipitation from the ET rate adjusted by the crop coefficient of the plant being irrigated.

Irrigation Emission Device: The various landscape irrigation equipment terminal fittings or outlets that emit water for irrigating vegetation in a landscape.

Irrigation Zone: The landscape area that irrigated by a set of landscape irrigation emission devised installed on the same water supply line downstream of a single valve.

Kitchen and Bar Sink Faucets: A faucet that discharges into a kitchen or bar sink in domestic or commercial installations. Supply fittings that discharge into other type sinks, including clinic sinks, floor sinks, service sinks and laundry trays are not included.

Lavatory: 1) A basin or vessel, for washing. 2) A plumbing fixture, as above, especially placed for use in personal hygiene. Principally not used for laundry purposes and never used for food preparation or utensils, in food services. 3) A fixture designed for the washing of the hands and face. Sometimes called a wash basin.

Lavatory Faucet: A faucet that discharges into a lavatory basin in a domestic or commercial installation.

Low Application Rate Irrigation: A means of irrigation using Low Precipitation Rate Sprinkler Heads or Low Flow Emitters in conjunction with cycling irrigation schedules to apply water at a rate less than the soil absorption rate.

Low Flow Emitter: Low flow irrigation emission device designed to dissipate water pressure and discharge a small uniform flow or trickle of water at a constant flow rate. To be classified as a Low Flow Emitter: drip emitters shall discharge water at less than 4 gallons (15 L) per hour per emitter; micro-spray, micro-jet, and misters shall discharge water at a maximum of 30 gallons (113 L) per hour per nozzle.

Low Precipitation Rate Sprinkler Heads: Landscape irrigation emission devices or sprinkler heads with maximum precipitation rate of 1 inch per hour over the applied irrigation area.

Metering Faucet: A self-closing faucet that dispenses a specific volume of water for each actuation cycle. The volume or cycle duration can be fixed or adjustable.

Mulch: Organic materials, such as wood chips and fines, tree bark chips, and pine needles that are used in a mulch basin to conceal gray water outlets and permit the infiltration of gray water.

Mulch Basin: A subsurface catchment area for gray water that is filled with mulch and of sufficient depth and volume to prevent ponding, surfacing or runoff.

On-Site Treated Non-Potable Water: Non-potable water, including gray water that has been collected, treated, and intended to be used on-site and is suitable for direct beneficial use. The level of treatment and quality of the on-site treated non-potable water shall be approved by the public health Authority Having Jurisdiction.

Precipitation Rate: The sprinkler head application rate of water applied to landscape irrigation zone, measured as inches (millimeters) per hour. Precipitation rates of sprinkler heads are calculated according to the flow rate, pattern and spacing of the sprinkler heads.

Pre-Rinse Spray Valve: A handheld device for use with commercial dishwashing and ware washing equipment that sprays water on dishes, flatware, and other food service items for the purpose of removing food residue before cleaning and sanitizing the items.

Quick-Disconnect Device: A hand-operated device that provides a means for connecting and disconnecting a hose to a water supply and that is equipped with a means to shut off the water supply when the device is disconnected.

Rainwater: Natural precipitation that has contacted a rooftop or other man-made above ground surface and has not been put to beneficial use.

Rainwater Catchment System: A system that collects and stores rainwater for the intended purpose of beneficial use. Also known as Rain Water Harvesting System.

Recirculation System: A system of hot water supply and return piping with shutoff valves, balancing valves, circulating pumps, and a method of controlling the circulating system.

Reclaimed (Recycled) Water: Non-potable water provided by a water/wastewater utility that, as a result of treatment of domestic wastewater, meets requirements of the Authority Having Jurisdiction for its intended uses.

Roof Washer: A device or method for removal of sediment and debris from a collection surface by diverting initial rainfall from entry into the cistern(s). Also known as a first flush device.

Run Out: The developed length of pipe that extends away from the circulating loop system to a fixture(s).

Self Closing Faucet: A faucet that closes itself after the actuation or control mechanism is deactivated. The actuation or control mechanism can be mechanical or electronic.

Soil Absorption Rate: The rate of the soil's ability to allow water to percolate or infiltrate the soil and be retained in the root zone of the soil, expressed as inches (millimeters) per hour.

Sprinkler Head: Landscape irrigation emission device discharging water in the form of sprays or rotating streams, not including Low Flow Emitters.

Storage Tank: The central component of the rainwater, stormwater, or dry weather runoff catchment system. Also known as a cistern or rain barrel.

Stormwater: Natural precipitation that has contacted a surface at grade or below grade and has not been put to beneficial use.

Stormwater Catchment System: A system that collects and stores stormwater for a beneficial use.

Submeter: A meter installed subordinate to a site meter. Also known as a Dedicated Meter.

Subsoil Irrigation Field: Gray water irrigation field installed in a trench within the layer of soil below the topsoil. This system is typically used for irrigation of deep rooted plants.

Subsurface Irrigation Field: Gray water irrigation field installed below finished grade within the topsoil.

Surge Tank: A reservoir to modify the fluctuation in flow rates to allow for uniform distribution of gray water to the points of irrigation.

WaterSense: A voluntary program of the U.S. Environmental Protection Agency designed to identify and promote water-efficient products and practices.

Water Closet: A fixture with a water-containing receptor that receives liquid and solid body waste and on actuation conveys the waste through an exposed integral trap into a drainage system. Also referred to as a toilet.

Water Factor (WF): A measurement and rating of appliance water efficiency, most often used for residential and light commercial clothes washers, as follows:

> **Clothes Washer (residential and commercial):** The quantity of water in gallons used to complete a full wash and rinse cycle per measured cubic foot capacity of the clothes container.

Water/Wastewater Utility: A public or private entity which may treat, deliver or do both functions to reclaimed (recycled) water, potable water, or both to wholesale or retail customers.

G.3 WATER EFFICIENCY AND CONSERVATION

G.3.1 General

G.3.1.1 Scope

The provisions of this chapter establish the means of conserving potable and non-potable water used in and around a building.

G.3.2 Water-Conserving Plumbing Fixtures and Fittings

G.3.2.1 General

The maximum water consumption of fixtures and fixture fittings shall comply with the flow rates specified in Table G.3.2.1 and Section G.3.2.2 through Section G.3.2.7.

TABLE G.3.2.1 MAXIMUM FIXTURE AND FIXTURE FITTINGS FLOW RATES	
FIXTURE TYPE	**FLOW RATE**
Showerheads	2.0 gpm @ 80 psi[1]
Kitchen faucets residential[5]	1.8 gpm @ 60 psi
Lavatory faucets residential	1.5 gpm @ 60 psi
Lavatory faucets other than residential	0.5 gpm @ 60 psi
Metering faucets	0.25 gallons/cycle
Metering faucets for wash fountains	0.25 [rim space (in.)/20 gpm @ 60 psi]
Wash fountains	2.2 [rim space (in.)/20 gpm @ 60 psi]
Water Closets - other than remote locations[4]	1.28 gallons/flush[2]
Water Closets - remote locations[4]	1.6 gallons/flush
Urinals	0.5 gallons/flush[3]
Commercial Pre-Rinse Spray Valves	1.3 gpm @ 60 psi

1. For multiple showerheads serving one shower compartment see Section G.3.2.6.2
2. Shall also be listed to EPA WaterSense Tank-Type High Efficiency Toilet Specification.
3. Shall also be listed to EPA WaterSense Flushing Urinal Specification. Nonwater urinals shall meet the specifications listed in Section G.3.2.3.1.
4. Remote location is where a water closet is located at least 30 feet upstream of the nearest drain line connections or fixtures, and is located where less than 1.5 drainage fixture units are upstream of the water closet's drain line connection.
5. See Section G.3.2.4.

G.3.2.2 Water Closets

No water closet shall have a flush volume exceeding 1.6 gallons per flush (gpf) (6.1 Lpf).

G.3.2.2.1 Gravity, Pressure Assisted and Electro-Hydraulic Tank Type Water Closets

Gravity, pressure assisted, and electro-hydraulic tank type water closets shall have a maximum effective flush volume of not more than 1.28 gallons (4.84 L) of water per flush in accordance with ASME A112.19.2/CSA B45.1 or ASME A112.19.14 and shall also be listed to the EPA WaterSense Tank-Type High Efficiency Toilet Specification. The effective flush volume for dual flush toilets is defined as the composite, average flush volume of two reduced flushes and one full flush.

G.3.2.2.2 Flushometer-Valve Activated Water Closets

Flushometer-valve activated water closets shall have a maximum flush volume of not more than 1.6 gallons (6.1 L) of water per flush in accordance with ASME A112.19.2/CSA B45.1.

G.3.2.2.3 Composting Toilets. Reserved

G.3.2.3 Urinals

Urinals shall have a maximum flush volume of not more than 0.5 gallon (1.9 L) of water per flush in accordance with ASME A112.19.2/CSA B45.1 or IAPMO Z124.9. Flushing urinals shall be listed to the EPA WaterSense Flushing Urinal Specification.

G.3.2.3.1 Nonwater Urinals

Nonwater urinals shall comply with ASME A112.19.3/CSA B45.4, ASME A112.19.19/CSA B45.4 or IAPMO Z124.9. Nonwater urinals shall be cleaned and maintained in accordance with the manufacturer's instructions after installation. Where nonwater urinals are installed they shall have a water distribution line roughed-in to the urinal location at a height not less than 56 inches (1422 mm) above finished floor to allow for the installation of an approved backflow prevention device in the event of a retrofit. Such water distribution lines shall be installed with shutoff valves located as close as possible to the distributing main to prevent the creation of dead ends. Where nonwater urinals are installed, not less than one water supplied fixture rated at not less than 1 drainage fixture unit (DFU) shall be installed upstream on the same drain line to facilitate drain line flow and rinsing.

G.3.2.4 Residential Kitchen Faucets

The maximum flow rate of residential kitchen faucets shall not exceed 1.8 gallons per minute (gpm) (0.11 L/s) at 60 pounds-force per square inch (psi) (414 kPa). Kitchen faucets are permitted to temporarily increase the flow above the maximum rate, but not to exceed 2.2 gpm (0.77 L/s) at 60 psi (414 kPa), and must revert to a maximum flow rate of 1.8 gpm (0.11 L/s) at 60 psi (414 kPa) upon valve closure.

G.3.2.5 Lavatory Faucets

The maximum water flow rate of faucets shall be in accordance with Section G.3.2.5.1 and Section G.3.2.5.2.

G.3.2.5.1 Lavatory Faucets in Residences, Apartments, and Private Bathrooms in Lodging Facilities, Hospitals, and Patient Care Facilities

The flow rate for lavatory faucets installed in residences, apartments, and private bathrooms in lodging, hospitals, and patient care facilities (including skilled nursing and long-term care facilities) shall not exceed 1.5 gpm (0.09 L/s) at 60 psi (414 kPa) in accordance with ASME A112.18.1/CSA B125.1 and shall be listed to the U.S. EPA WaterSense High-Efficiency Lavatory Faucet Specification.

G.3.2.5.2.1 Maximum Flow Rate

The flow rate shall not exceed 0.5 gpm (0.03 L/s) at 60 psi (414 kPa) in accordance with ASME A112.18.1/CSA B125.1.

G.3.2.5.2.2 Metering Faucets

Metering faucets shall deliver not more than 0.25 gallons (0.95 L) of water per cycle.

G.3.2.6 Showers

G.3.2.6.1 Showerheads

Showerheads shall comply with the requirements of the Energy Policy Act of 1992, except that the flow rate shall not exceed 2.0 gpm (0.13 L/s) at 80 psi (552 kPa), when listed to ASME A112.18.1/CSA B125.1.

G.3.2.6.2 Multiple Showerheads Serving One Shower Compartment

The total allowable flow rate of water from multiple showerheads flowing at any given time, with or without a diverter, including rain systems, waterfalls, bodysprays, and jets, shall not exceed 2.0 gpm (0.13 L/s) per shower compartment, where the floor area of the shower compartment is less than 1800 square inches (1.161 m^2). For each increment of 1800 square inches (1.161 m^2) of floor area thereafter or part thereof, additional showerheads are allowed, provided the total flow rate of water from all flowing devices shall not exceed 2.0 gpm (0.13 L/s) for each such increment.

EXCEPTIONS:

(1) Gang showers in non-residential occupancies. Singular showerheads or multiple shower outlets serving one showering position in gang showers shall not have more than 2.0 gpm (0.13 L/s) total flow.

(2) Where provided, accessible shower compartments shall not be permitted to have more than 4.0 gpm (0.25 L/s) total flow, where one outlet is the hand shower. The hand shower shall have a control with a nonpositive shutoff feature.

G.3.2.6.3 Bath and Shower Diverters

The rate of leakage out of the tub spout of bath and shower diverters while operating in the shower mode shall not exceed 0.1 gpm (0.006 L/s) in accordance with ASME A112.18.1/CSA B125.1.

G.3.2.6.4 Shower Valves

Shower valves shall meet the temperature control performance requirements of ASSE 1016 or ASME A112.18.1/CSA B125.1 when tested at 2.0 gpm (0.13 L/s).

G.3.2.7 Commercial Pre-Rinse Spray Valves

The flow rate for a pre-rinse spray valve installed in a commercial kitchen to remove food waste from cookware and dishes prior to cleaning shall not be more than 1.3 gpm (0.08 L/s) at 60 psi (414 kPa). Where pre-rinse spray valves with maximum flow rates of 1.0 gpm (0.06 L/s) or less are installed, the static pressure shall be not less than 30 psi (207 kPa). Commercial kitchen pre-rinse spray valves shall be equipped with an integral automatic shutoff.

G.3.3 Appliances

G.3.3.1 Dishwashers

Residential and commercial dishwashers shall be in accordance with the Energy Star program requirements.

G.3.3.2 Clothes Washers

Residential clothes washers shall be in accordance with the Energy Star program requirements. Commercial clothes washers shall be in accordance with Energy Star program requirements, where such requirements exist.

G.3.4 Water Pressure

G.3.4.1 Installation

Pressure regulators shall be installed in accordance with the plumbing code.

G.3.5 Water Softeners and Treatment Devices

G.3.5.1 Water Softeners

Actuation of regeneration of water softeners shall be by demand initiation. Water softeners shall be listed to NSF/ANSI Standard 44. Water softeners shall have a rated salt efficiency exceeding 3400 grains (gr) (0.2200 kg) of total hardness exchange per pound (lb) (0.5 kg) of salt, based on sodium chloride (NaCl) equivalency, and shall not generate more than 5 gallons (19 L) of water per 1000 grains (0.0647 kg) of hardness removed during the service cycle.

G.3.5.2 Water Softener Limitations

In residential buildings, where the supplied potable water hardness is equal to or less than 8 grains per gallon (gr/gal) (137 mg/L) measured as total calcium carbonate equivalents, water softening equipment that discharges water into the wastewater system during the service cycle shall not be allowed, except as required for medical purposes.

G.3.5.3 Point-of-Use Reverse Osmosis Water Treatment Systems

Reverse osmosis water treatment systems installed in residential occupancies shall be equipped with automatic shutoff valves to prevent discharge when there is no call for producing treated water. Reverse osmosis water treatment systems shall be listed to meet NSF/ANSI Standard 58.

G.3.6 Occupancy Specific Water Efficiency Requirements

G.3.6.1 Commercial Food Service

G.3.6.1.1 Ice Makers

Ice makers shall be air cooled and shall be in accordance with Energy Star for commercial ice machines.

G.3.6.1.2 Food Steamers

All steamers shall consume not more than 5.0 gallons (19 L) per hour per steamer pan in the full operational mode.

G.3.6.1.3 Combination Ovens

Combination ovens shall not consume more than 3.5 gph (13 L/h) per pan in the full operational mode.

G.3.6.1.4 Grease Interceptors

Grease interceptor maintenance procedures shall not include post-pumping/cleaning refill using potable water. Refill shall be by connected appliance accumulated discharge only.

G.3.6.1.5 Dipper Well Faucets

Where dipper wells are installed, the water supply to a dipper well shall have a shutoff valve and flow control. The flow of water into a dipper well shall be limited by at least one of the following methods:

(1) Maximum Continuous Flow: Water flow shall not exceed the water capacity of the dipper well in one minute at supply pressure of 60 psi (414 kPa), and the maximum flow shall not exceed 2.2 gpm (0.14 L/s) at a supply pressure of 60 psi (414 kPa). The water capacity of a dipper well shall be the maximum amount of water that the fixture can hold before water flows into the drain.

(2) Metered Flow: The volume of water dispensed into a dipper well in each activation cycle of a self closing fixture fitting shall not exceed the water capacity of the dipper well, and the maximum flow shall not exceed 2.2 gpm (0.14 L/s) at a supply pressure of 60 psi (414 kPa).

G.3.6.2 Medical and Laboratory Facilities

G.3.6.2.1 Steam Sterilizers

Controls shall be installed to limit the discharge temperature of condensate or water from steam sterilizers to 140°F (60°C) or less. Venturi-type vacuum system shall not be utilized with vacuum sterilizers.

G.3.6.2.2 X-Ray Film Processing Units

Processors for X-ray film exceeding 6 inches (152 mm) in any dimension shall be equipped with water recycling units.

G.3.6.2.3 Exhaust Hood Liquid Scrubber Systems

Liquid scrubber systems for exhaust hoods and ducts shall be of the recirculation type. Liquid scrubber systems for perchloric acid exhaust hoods and ducts shall be equipped with a timer-controlled water recirculation system. The collection sump for perchloric acid exhaust systems shall be designed to automatically drain after the wash down process has completed.

G.3.7 Leak Detection and Control

G.3.7.1 General

Where installed, leak detection and control devices shall comply with IAPMO IGC115. Note: Leak detection and control devices help protect property from water damage and also conserve water by shutting off the flow when leaks are detected.

G.3.8 Fountains and Other Water Features

G.3.8.1 Use of Alternate Water Source for Special Water Features

Special water features such as ponds and water fountains shall be provided with reclaimed (recycled) water, rainwater, or on-site treated non-potable water where the source and capacity is available on the premises and approved by the Authority Having Jurisdiction.

G.3.9 Meters

G.3.9.1 Required

A water meter shall be required for buildings connected to a public water system, including municipally supplied reclaimed (recycled) water. In other than single-family houses, multi-family structures of three stories or fewer above grade, and modular houses, a separate meter or submeter shall be installed in the following locations:

(1) The water supply for irrigated landscape with an accumulative area exceeding 2500 square feet (232 m^2).

(2) The makeup water supply to cooling towers, evaporative condensers, and fluid coolers.

(3) The makeup water supply to one or more boilers collectively exceeding 1,000,000 British thermal units per hour (Btu/h) (293 kW).

(4) The water supply to a water-using process where the consumption exceeds 1,000 gallons per day (gal/d) (0.0438 L/s), except for manufacturing processes.

(5) The water supply to each building on a property with multiple buildings where the water consumption exceeds 500 gal/d (0.021 L/s).

(6) The water supply to an individual tenant space on a property where any of the following applies:

(a) Water consumption exceeds 500 gal/d (0.021 L/s) for that tenant.

(b) Tenant space is occupied by a commercial laundry, cleaning operation, restaurant, food service, medical office, dental office, laboratory, beauty salon, or barbershop.

(c) Total building area exceeds 50,000 square feet (4645 m^2).

(7) A makeup water supply to a swimming pool.

(8) The makeup water supply to an evaporative cooler having an air flow exceeding 30,000 cubic feet per minute (ft3/min) (14 158.2 L/s).

G.3.9.2 Consumption Data

A means of communicating water consumption data from submeters to the water consumer shall be provided.

G.3.9.3 Access

Meters and submeters shall be accessible.

G.3.10 HVAC Water Efficiency

G.3.10.1 Once-Through Cooling

Once-through cooling using potable water is prohibited.

G.3.10.2 Cooling Towers and Evaporative Coolers

Cooling towers and evaporative coolers shall be equipped with makeup water and blow down meters, conductivity controllers and overflow alarms. Cooling towers shall be equipped with efficiency drift eliminators that achieve drift reduction to 0.002 percent of the circulated water volume for counterflow towers and 0.005 percent for cross-flow towers.

G.3.10.3 Cooling Tower Makeup Water

Not less than five cycles of concentration is required for air-conditioning cooling tower makeup water having a total hardness of less than 11 gr/gal (188 mg/L) expressed as calcium carbonate. Not less than 3.5 cycles of concentration is required for air-conditioning cooling tower makeup water having a total hardness equal to or exceeding 11 gr/gal (188 mg/L) expressed as calcium carbonate. EXCEPTION: Air-conditioning cooling tower makeup water having discharge conductivity range not less than 7 gr/gal (120 mg/L) to 9 gr/gal (154 mg/L) of silica measured as silicon dioxide.

G.3.10.4 Evaporative Cooler Water Use

Evaporative cooling systems (also known as swamp coolers) shall use less than 3.5 gallons (13.2 L) of water per ton-hour of cooling when system controls are set to maximum water use. Water use, expressed in maximum water use per ton-hour of cooling, shall be marked on the device and included in product user manuals, product information literature, and installation instructions. Water use information shall be readily available at the time of code compliance inspection.

G.3.10.4.1 Overflow Alarm

Cooling systems shall be equipped with an overflow alarm to alert building owners, tenants, or maintenance personnel when the water refill valve continues to allow water to flow into the reservoir when the reservoir is already full. The alarm shall have a minimum sound pressure level rating of 85 dBa measured at a distance of 10 feet.

G.3.10.4.2 Automatic Pump Shut-Off

Cooling systems shall automatically cease pumping water to the evaporation pads when airflow across evaporation pads ceases.

G.3.10.4.3 Cooler Reservoir Discharge

A water quality management system (either timer or water quality sensor) is required. Where timers are used, the time interval between discharge of reservoir water shall be set to 6 hours or greater of cooler operation. Where water quality sensors are used, the discharge of reservoir water shall be set for greater 800 ppm or greater of TDS. Continuous discharge or continuous bleed systems are prohibited.

G.3.10.4.4 Discharge Water Reuse

Discharge water shall be reused where appropriate applications exist on site. Where a nonpotable water source system exists on site, evaporative cooler discharge water shall be collected and discharged to such collection system. EXCEPTION: Where the reservoir water adversely affects the quality of the nonpotable water supply making the nonpotable water unusable for its intended purposes.

G.3.10.4.5 Discharge Water to Drain

Where discharge water is not recovered for reuse, the sump overflow line shall not be directly connected to a drain. Where the discharge water is put into a sanitary drain, a minimum 6 inch (152 mm) air gap is required between the termination of the discharge line and the drain opening. The discharge line shall terminate in a location that is readily visible to the building owner, tenants, or maintenance personnel.

G.3.10.5 Use of Reclaimed (Recycled) and On-Site Treated Non-Potable Water for Cooling

Where approved for use by the water/wastewater utility and the Authority Having Jurisdiction, reclaimed (recycled) or on-site treated non-potable water shall be permitted to be used for industrial and commercial cooling or air-conditioning.

G.3.10.5.1 Drift Eliminator

A drift eliminator shall be utilized in a cooling system, utilizing alternate sources of water, where the aerosolized water may come in contact with employees or members of the public.

G.3.10.5.2 Disinfection

A biocide shall be used to treat the cooling system recirculation water where the recycled water may come in contact with employees or members of the public.

G.3.11 Condensate Recovery

Condensate is permitted to be used as on-site treated non-potable water where collected, stored and treated in accordance with Section G.4.4.

G.3.12 Water-Powered Sump Pumps

Sump pumps powered by potable or reclaimed (recycled) water pressure shall only be used as an emergency backup pump. The water-powered pump shall be equipped with a battery powered alarm having a minimum rating of 85 dBa at 10 feet (3048 mm). Water-powered pumps shall have a water efficiency factor of pumping at least 1.4 gallons (5.3 L) of water to a height of 10 feet (3048 mm) for every gallon of water used to operate the pump, measured at a water pressure of 60 psi (414 kPa). Pumps shall be clearly labeled as to the gallons of water pumped per gallon of potable water consumed.

Water-powered stormwater sump pumps shall be equipped with a reduced pressure principle backflow prevention assembly.

G.3.13 Landscape Irrigation Systems

G.3.13.1 General

Where landscape irrigation systems are installed, they shall use low application irrigation methods and comply with Sections G.3.13.2 through G.3.13.13. Requirements limiting the amount or type of plant material used in landscapes shall be established by the Authority Having Jurisdiction. EXCEPTION: Plants grown for food production.

G.3.13.2 Backflow Protection

Potable water and reclaimed water supplies to landscape irrigation systems shall be protected from backflow in accordance with the plumbing code and Authority Having Jurisdiction.

G.3.13.3 Use of Alternate Water Sources for Landscape Irrigation

Where available by pre-existing treatment, storage or distribution network, and where approved by the Authority Having Jurisdiction, alternative water source(s) complying with Chapter 5 shall be utilized for landscape irrigation. Where adequate capacity and volumes of pre-existing alternative water sources are available, the irrigation system shall be designed to use minimum of 75 percent of alternative water for the annual irrigation demand before supplemental potable water is used.

G.3.13.4 Irrigation Control Systems

Where installed as part of a landscape irrigation system, irrigation control systems shall:

G.3.13.4.1 Automatically adjust the irrigation schedule to respond to plant water needs determined by weather or soil moisture conditions.

G.3.13.4.2 Utilize sensors to suspend irrigation during a rainfall.

G.3.13.4.3 Utilize sensors to suspend irrigation when adequate soil moisture is present for plant growth.

G.3.13.4.4 Have the capability to program multiple and different run times for each irrigation zone to enable cycling of water applications and durations to mitigate water flowing off of the intended irrigation zone.

G.3.13.4.5 The site specific settings of the irrigation control system affecting the irrigation and shall be posted at the control system location. The posted data, where applicable to the settings of the controller, shall include:

(1) Precipitation rate for each zone.

(2) Plant evapotranspiration coefficients for each zone.

(3) Soil absorption rate for each zone.

(4) Rain sensor settings.

(5) Soil moisture setting.

(6) Peak demand schedule including run times for each zone and the number of cycles to mitigate runoff and monthly adjustments or percentage.

G.3.13.5 Low Flow Irrigation

Irrigation zones using low flow irrigation shall be equipped with filters sized for the irrigation emission devices, and with a pressure regulator installed upstream of the irrigation emission devices as necessary to reduce the operating water pressure meeting manufacturers' equipment requirements.

G.3.13.6 Mulched Planting Areas

Only low volume emitters are allowed to be installed in mulched planting areas with vegetation taller than 12 inches (305 mm).

G.3.13.7 System Performance Requirements

The landscape irrigation system shall be designed and installed to:

(1) Prevent irrigation water from runoff out of the irrigation zone.

(2) Prevent water in the supply-line drainage from draining out between irrigation events.

(3) Not allow irrigation water to be applied onto or enter non-targeted areas including: adjacent property and vegetation areas, adjacent hydrozones not requiring the irrigation water to meet its irrigation demand, non-vegetative areas, impermeable surfaces, roadways, and structures.

G.3.13.8 Narrow or Irregularly Shaped Landscape Areas

Narrow or irregularly shaped landscape areas, less than 4 feet (1219 mm) in any direction across any opposing boundaries shall not be irrigated by any irrigation emission device except low flow emitters.

G.3.13.9 Sloped Areas

Where soil surface rises more than 1 foot (305 mm) per 4 feet (1219 mm) of length, the irrigation zone system average precipitation rate shall not exceed 0.75 inches (19 mm) per hour as verified through either of the following methods:

(a) manufacturer documentation that the precipitation rate for the installed sprinkler head does not exceed 0.75 inches (19 mm) per hour where the sprinkler heads are installed no closer than the specified radius and where the water pressure of the irrigation system is no greater than the manufacturer's recommendations.

(b) catch can testing in accordance with the requirements of the Authority Having Jurisdiction and where emitted water volume is measured with a minimum of 6 catchment containers at random places within the irrigation zone for a minimum of 15 minutes to determine the average precipitation rate, expressed as inches per hour.

G.3.13.10 Sprinkler Head Installations

All installed sprinkler heads shall be low precipitation rate sprinkler heads.

G.3.13.10.1 Sprinkler Heads in Common Irrigation Zones

Sprinkler heads installed in irrigation zones served by a common valve shall be limited to applying water to plants with similar irrigation needs, and shall have matched precipitation rates (identical inches of water application per hour as rated or tested, plus or minus 5 percent).

G.3.13.10.2 Sprinkler Head Pressure Regulation

Sprinkler heads shall utilize pressure regulating devices (as part of irrigation system or integral to the sprinkler head) to maintain manufacturer's recommended operating pressure for each sprinkler and nozzle type.

G.3.13.10.3 Pop-up Type Sprinkler Heads

Where pop-up type sprinkler heads are installed, the sprinkler heads shall rise to a height of not less than 4 inches (102 mm) above the soil level when emitting water.

G.3.13.11 Irrigation Zone Performance Criteria

Irrigation zones shall be designed and installed to ensure the average precipitation rate of the sprinkler heads over the irrigated area does not exceed 1.0 inch per hour as verified through either of the following methods:

(a) manufacturer's documentation that the precipitation rate for the installed sprinkler head does not exceed 1.0 inches per hour where the sprinkler heads are installed no closer that the specified radius and where the water pressure of the irrigation system is no greater than the manufacturer's recommendations.

(b) catch can testing in accordance with the requirements of the Authority Having Jurisdiction and where emitted water volume is measured with a minimum of 6 catchment containers at random places within the irrigation zone for a minimum of 15 minutes to determine the average precipitation rate, expressed as inches per hour.

G.3.13.12 Qualifications

The Authority Having Jurisdiction shall have the authority to require landscape irrigation contractors, installers, or designers to demonstrate competency. Where required by the Authority Having Jurisdiction, the contractor, installer, or designer shall be certified to perform such work.

G.3.14 Trap Seal Protection

G.3.14.1 Water Supplied Trap Primers

Water supplied trap primers shall be electronic or pressure activated and shall use no more than 30 gallons (114 L) per year per drain. Where an alternate water source, as defined by this code, is used for fixture flushing or other uses in the same room, the alternate water source shall be used for the trap primer water supply. EXCEPTION: Flushometer tailpiece trap primers complying with IAPMO PS 76 are exempted from the provisions of this section.

G.3.14.2 Drainage Type Trap Seal Primer Devices

Drainage type trap seal primer devices shall not be limited in the amount of water they discharge.

G.3.15 Vehicle Wash Facilities

The maximum make-up water use for automobile washing shall not exceed 40 gallons (151 L) per vehicle for in-bay automatic car washes and 35 gallons (132 L) for conveyor and express type car washes. Spray wands and foamy brushes shall use no more than 3.0 gpm (0.06 L/s). Spot-free reverse osmosis discharge (reject) water shall be recycled. Towel ringers shall have a positive shut-off valve. Spray nozzles shall be replaced annually. EXEMPTION: Bus and large commercial vehicles washes are exempt from the requirements in this section.

G.4 ALTERNATE WATER SOURCES FOR NON-POTABLE APPLICATIONS

G.4.1 General

G.4.1.1 Scope

The provisions of this chapter shall apply to the construction, alteration, and repair of alternate water source systems for non-potable applications.

G.4.1.1.1 Allowable Use of Alternate Water

Where approved or required by the Authority Having Jurisdiction, alternate water sources (reclaimed (recycled) water, rainwater, gray water and onsite treated non-potable water) shall be permitted to be used in lieu of potable water for the applications identified in this chapter.

G.4.1.2 System Design

Alternate water source systems complying with this chapter shall be designed by a person registered or licensed to perform plumbing design work or who demonstrates competency to design the alternate water source system as required by the Authority Having Jurisdiction. Components, piping, and fittings used in any alternate water source system shall be listed.

EXCEPTIONS:

(1) A person registered or licensed to perform plumbing design work is not required to design rainwater catchment systems used for irrigation with a maximum storage capacity of 360 gallons (1363 L).

(2) A person registered or licensed to perform plumbing design work is not required to design rainwater catchment systems for single family dwellings where all outlets, piping, and system components are located on the exterior of the building.

(3) A person registered or licensed to perform plumbing design work is not required to design gray water systems having a maximum discharge capacity of 250 gallons per day (gal/d) (15.77 L/s) for single family and multi-family dwellings.

(4) A person registered or licensed to perform plumbing design work is not required to design an on-site treated non-potable water system for single family dwellings having a maximum discharge capacity of 250 gal/d (15.77 L/s).

G.4.1.3 Permit

It shall be unlawful for any person to construct, install, alter, or cause to be constructed, installed, or altered any alternate water source system in a building or on a premise without first obtaining a permit to do such work from the Authority Having Jurisdiction.

EXCEPTIONS:

(1) A permit is not required for exterior rainwater catchment systems used for outdoor drip and subsurface irrigation with a maximum storage capacity of 360 gallons (1363 L).

(2) A plumbing permit is not required for rainwater catchment systems for single family dwellings where all outlets, piping, and system components are located on the exterior of the building. This does not exempt the need for permits if required for electrical connections, tank supports, or enclosures.

G.4.1.4 Component Identification

System components shall be properly identified as to the manufacturer.

G.4.1.5 Maintenance and Inspection

Alternate water source systems and components shall be inspected and maintained in accordance with Section G.4.1.5.1 through Section G.4.1.5.3.

G.4.1.5.1 Frequency

Alternate water source systems and components shall be inspected and maintained in accordance with Table G.4.1.5 unless more frequent inspection and maintenance is required by the manufacturer.

G.4.1.5.2 Maintenance Log

A maintenance log for gray water, rainwater, and on-site treated non-potable water systems is required to have a permit in accordance with Section G.4.1.3 and shall be maintained by the property owner and be available for inspection. The property owner or designated appointee shall ensure that a record of testing, inspection and maintenance as required by Table G.4.1.5 is maintained in the log. The log will indicate the frequency of inspection and maintenance for each system.

G.4.1.5.3 Maintenance Responsibility

The required maintenance and inspection of alternate water source systems shall be the responsibility of the property owner, unless otherwise required by the Authority Having Jurisdiction.

G.4.1.6 Operation and Maintenance Manual

An operation and maintenance manual for gray water, rainwater, and on-site treated water systems required to have a permit in accordance with Section G.4.1.3 shall be supplied to the building owner by the system designer. The operating and maintenance manual shall include the following:

(1) Detailed diagram of the entire system and the location of system components.

(2) Instructions on operating and maintaining the system.

(3) Details on maintaining the required water quality as determined by the Authority Having Jurisdiction.

(4) Details on deactivating the system for maintenance, repair, or other purposes.

(5) Applicable testing, inspection, and maintenance frequencies as required by Table G.4.1.5.

(6) A method of contacting the manufacturer(s).

G.4.1.7 Minimum Water Quality Requirements

The minimum water quality for alternate water source systems shall meet the applicable water quality requirements for the intended application as determined by the Authority Having Jurisdiction. Water quality for non-potable rainwater catchment systems shall comply with Section G.4.5.9.4. In the absence of water quality requirements for on-site treated non-potable water and reclaimed (recycled) water systems, the EPA/625/R-04/108 contains recommended water reuse guidelines to assist regulatory agencies develop, revise, or expand alternate water source water quality standards.
EXCEPTIONS:

(1) Water treatment is not required for rainwater catchment systems used for aboveground irrigation with a maximum storage capacity of 360 gallons (1363 L).

(2) Water treatment is not required for gray water used for subsurface irrigation.

(3) Water treatment is not required for rainwater catchment systems used for subsurface or drip irrigation.

G.4.1.8 Material Compatibility

Alternate water source systems shall be constructed of materials that are compatible with the type of pipe and fitting materials, water treatment, and water conditions in the system.

TABLE G.4.1.5
MINIMUM ALTERNATE WATER SOURCE TESTING, INSPECTION, AND MAINTENANCE FREQUENCY

DESCRIPTION	MINIMUM FREQUENCY
Inspect and clean filters and screens, and replace (if necessary)	Every 3 months
Inspect and verify that disinfection, filters and water quality treatment devices and systems are operational and maintaining minimum water quality requirements as determined by the Authority Having Jurisdiction	In accordance with manufacturer's instructions, and the Authority Having Jurisdiction
Inspect and clear debris from rainwater gutters, downspouts, and roof washers	Every 6 months
Inspect and clear debris from roof or other aboveground rainwater collection surfaces	Every 6 months
Remove tree branches and vegetation overhanging roof or other aboveground rainwater collection surfaces	As needed
Inspect pumps and verify operation	After initial installation and every 12 months thereafter
Inspect valves and verify operation	After initial installation and every 12 months thereafter
Inspect pressure tanks and verify operation	After initial installation and every 12 months thereafter
Clear debris from and inspect storage tanks, locking devices, and verify operation	After initial installation and every 12 months thereafter
Inspect caution labels and marking	After initial installation and every 12 months thereafter
Inspect and maintain mulch basins for gray water irrigation systems	As needed to maintain mulch depth and prevent ponding and runoff
Cross-connection inspection and test*	After initial installation and every 12 months thereafter
Test water quality of rainwater catchment systems required by Section G.4.5.9.4 to maintain a minimum water quality	Every 12 months. After system renovation or repair.
*The cross-connection test shall be performed in the presence of the Authority Having Jurisdiction in accordance with the requirements of this Chapter.	

G.4.1.9 System Controls

Controls for pumps, valves, and other devices that contain mercury that come in contact with alternate water source water supply shall not be permitted.

G.4.1.10 Commercial, Industrial, and Institutional Restroom Signs

A sign shall be installed in all restrooms in commercial, industrial, and institutional occupancies using reclaimed (recycled) water, on-site treated water, and non-potable rainwater for water closets, urinals, or both. Each sign shall contain 1/2 inch (12.7 mm) letters of a highly visible color on a contrasting background. The location of the sign(s) shall be such that the sign(s) shall be visible to all users. The location of the sign(s) shall be approved by the Authority Having Jurisdiction and shall contain the following text:

TO CONSERVE WATER, THIS BUILDING USES *_____* TO FLUSH TOILETS AND URINALS.

G.4.1.10.1 Equipment Room Signs

Each room containing reclaimed (recycled) water, on-site treated water, and non-potable rainwater equipment shall have a sign posted in a location that is visible to anyone working on or near non-potable water equipment with the following wording in 1 inch (25.4 mm) letters:
CAUTION: NON-POTABLE *_____*, DO NOT DRINK. DO NOT CONNECT TO DRINKING WATER SYSTEM. NOTICE: CONTACT BUILDING MANAGEMENT BEFORE PERFORMING ANY WORK ON THIS WATER SYSTEM.
_____ Shall indicate RECLAIMED (RECYCLED) WATER, ON-SITE TREATED WATER, or RAINWATER accordingly.

G.4.1.11 Inspection and Testing

Alternate water source systems shall be inspected and tested in accordance with Section G.4.1.11.1 and Section G.4.1.11.2.

G.4.1.11.1 Supply System Inspection and Test

Alternate water source systems shall be inspected and tested in accordance with the plumbing code for testing of potable water piping.

G.4.1.11.2 Annual Cross-Connection Inspection and Testing

An initial and subsequent annual inspection and test shall be performed on both the potable and alternate water source systems. The potable and alternate water source system shall be isolated from each other and independently inspected and tested to ensure there is no cross-connection in accordance with Section G.4.1.11.2.1 through Section G.4.1.11.2.4.

G.4.1.11.2.1 Visual System Inspection

Prior to commencing the cross-connection testing, a dual system inspection shall be conducted by the Authority Having Jurisdiction and other authorities having jurisdiction as follows:

(1) Meter locations of the alternate water source and potable water lines shall be checked to verify that no modifications were made, and that no cross-connections are visible.

(2) Pumps and equipment, equipment room signs, and exposed piping in equipment room shall be checked.

(3) Valves shall be checked to ensure that valve lock seals are still in place and intact. Valve control door signs shall be checked to verify that no signs have been removed.

G.4.1.11.2.2 Cross-Connection Test

The procedure for determining cross-connection shall be followed by the applicant in the presence of the Authority Having Jurisdiction and other authorities having jurisdiction to determine whether a cross-connection has occurred as follows:

(1) The potable water system shall be activated and pressurized. The alternate water source system shall be shut down, depressurized, and drained.

(2) The potable water system shall remain pressurized for a minimum period of time specified by the Authority Having Jurisdiction while the alternate water source system is empty. The minimum period the alternate water source system is to remain depressurized shall be determined on a case-by-case basis, taking into account the size and complexity of the potable and the alternate water source distribution systems, but in no case shall that period be less than 1 hour.

(3) The drain on the alternate water source system shall be checked for flow during the test and all fixtures, potable and alternate water source, shall be tested and inspected for flow. Flow from any alternate water source system outlet indicates a cross-connection. No flow from a potable water outlet shall indicate that it is connected to the alternate water source system.

(4) The potable water system shall then be depressurized and drained.

(5) The alternate water source system shall then be activated and pressurized.

(6) The alternate water source system shall remain pressurized for a minimum period of time specified by the Authority Having Jurisdiction while the potable water system is empty. The minimum period the potable water system is to remain depressurized shall be determined on a case-by-case basis, but in no case shall that period be less than 1 hour.

(7) All fixtures, potable and alternate water source, shall be tested and inspected for flow. Flow from any potable water system outlet indicates a cross-connection. No flow from an alternate water source outlet will indicate that it is connected to the potable water system.

(8) The drain on the potable water system shall be checked for flow during the test and at the end of the test.

(9) If there is no flow detected in any of the fixtures which would indicate a cross-connection, the potable water system shall be repressurized.

G.4.1.11.2.3 Discovery of Cross-Connection

In the event that a cross-connection is discovered, the following procedure, in the presence of the Authority Having Jurisdiction, shall be activated immediately:

(1) The alternate water source piping to the building shall be shut down at the meter, and the alternate water source riser shall be drained.

(2) Potable water piping to the building shall be shut down at the meter.

(3) The cross-connection shall be uncovered and disconnected.

(4) The building shall be retested following procedures listed in Section G.4.1.11.2.1 and Section G.4.1.11.2.2.

(5) The potable water system shall be chlorinated with 50 parts-per-million (ppm) chlorine for 24 hours.

(6) The potable water system shall be flushed after 24 hours, and a standard bacteriological test shall be performed. If test results are acceptable, the potable water system shall be permitted to be recharged.

G.4.1.11.2.4 Annual Inspection

An annual inspection of the alternate water source system, following the procedures listed in Section G.4.1.11.2.1 shall be required. Annual cross-connection testing, following the procedures listed in Section G.4.1.11.2.2 shall be required by the Authority Having Jurisdiction, unless site conditions do not require it. In no event shall the test occur less than once in 4 years. Alternate testing requirements shall be permitted by the Authority Having Jurisdiction.

G.4.1.12 Separation Requirements

All underground alternate water source service piping other than gray water shall be separated from the building sewer in accordance with the plumbing code. Treated non-potable water pipes shall be permitted to be run or laid in the same trench as potable water pipes with a 12 inch (305 mm) minimum vertical and horizontal separation when both pipe materials are approved for use within a building. Where horizontal piping materials do not meet this requirement the minimum separation shall be increased to 60 inches (1524 mm). The potable water piping shall be installed at an elevation above the treated non-potable water piping.

G.4.1.13 Abandonment

All alternate water source systems that are no longer in use or fails to be maintained in accordance with Section G.4.1.5 shall be abandoned.

G.4.1.14 Sizing

Unless otherwise provided for in this supplement, alternate water source piping shall be sized in accordance with the plumbing code for sizing potable water piping.

G.4.2 Gray Water Systems

G.4.2.1 General

The provisions of this section shall apply to the construction, alteration, and repair of gray water systems.

G.4.2.2 Gray Water System

G.4.2.2.1 Discharge

Gray water shall be permitted to be diverted away from a sewer or private sewage disposal system, and discharge to a subsurface irrigation or subsoil irrigation system. The gray water shall be permitted to discharge to a mulch basin for single family and multi-family dwellings. Gray water shall not be used to irrigate root crops or food crops intended for human consumption that come in contact with soil.

G.4.2.2.2 Surge Capacity

Gray water systems shall be designed to have the capacity to accommodate peak flow rates and distribute the total amount of estimated gray water on a daily basis to a subsurface irrigation field, subsoil irrigation field, or mulch basin without surfacing, ponding, or runoff. A surge tank is required for all systems that are unable to accommodate peak flow rates and distribute the total amount of gray water by gravity drainage. The water discharge for gray water systems shall be determined in accordance with Section G.4.2.8.1 or Section G.4.2.8.2.

G.4.2.2.3 Diversion

The gray water system shall connect to the sanitary drainage system downstream of fixture traps and vent connections through an approved gray water diverter valve. The gray water diverter shall be installed in an accessible location and clearly indicate the direction of flow.

G.4.2.2.4 Backwater Valves

Gray water drains subject to backflow shall be provided with a backwater valve so located as to be accessible for inspection and maintenance.

G.4.2.3 Connections to Potable and Reclaimed (Recycled) Water Systems

Gray water systems shall have no direct connection to any potable water supply, on-site treated non-potable water supply, or reclaimed (recycled) water systems. Potable, on-site treated non-potable, or reclaimed (recycled) water is permitted to be used as makeup water for a non-pressurized storage tank provided the connection is protected by an airgap in accordance with the plumbing code.

G.4.2.4 Location

No gray water system or part thereof shall be located on any lot other than the lot that is the site of the building or structure that discharges the gray water, nor shall any gray water system or part thereof be located at any point having less than the minimum distances indicated in Table G.4.2.4.

G.4.2.5 Plot Plan Submission

No permit for any gray water system shall be issued until a plot plan with appropriate data satisfactory to the Authority Having Jurisdiction has been submitted and approved.

G.4.2.6 Prohibited Location

Where there is insufficient lot area or inappropriate soil conditions for adequate absorption to prevent the ponding, surfacing or runoff of the gray water, as determined by the Authority Having Jurisdiction, no gray water system shall be permitted. A gray water system is not permitted on any property in a geologically sensitive area as determined by the Authority Having Jurisdiction.

TABLE G.4.2.4
LOCATION OF GRAY WATER SYSTEM

MINIMUM HORIZONTAL DISTANCE IN CLEAR REQUIRED FROM:	SURGE TANK (feet)	SUBSURFACE AND SUBSOIL IRRIGATION FIELD AND MULCH BED (feet)
Building structures[1]	5[2,9]	2[3,8]
Property line adjoining private property	5	5[8]
Water supply wells[4]	50	100
Streams and lakes[4]	50	50[5]
Sewage pits or cesspools	5	5
Sewage disposal field	5	4[6]
Septic tank	0	5
On-site domestic water service line	5	5
Pressurized public water main	10	10[7]

For SI units: 1 foot = 304.8 mm
Note: Where irrigation or disposal fields are installed in sloping ground, the minimum horizontal distance between any part of the distribution system and the ground surface shall be 15 feet (4572 mm).
1. Including porches and steps, whether covered or uncovered, breezeways, roofed carports, roofed patios, carports, covered walks, covered driveways, and similar structures or appurtenances.
2. The distance shall be permitted to be reduced to 0 feet for aboveground tanks when first approved by the Authority Having Jurisdiction.
3. Reference to a 45 degree (0.79 rad) angle from foundation.
4. Where special hazards are involved, the distance required shall be increased as directed by the Authority Having Jurisdiction.
5. These minimum clear horizontal distances shall also apply between the irrigation or disposal field and the ocean mean higher high tide line.
6. Add 2 feet (610 mm) for each additional foot of depth in excess of 1 foot (305 mm) below the bottom of the drain line.
7. For parallel construction or for crossings, approval by the Authority Having Jurisdiction shall be required.
8. The distance shall be permitted to be reduced to 1 1/2 feet (457 mm) for drip and mulch basin irrigation systems.
9. The distance shall be permitted to be reduced to 0 feet for surge tanks of 75 gallons (284 L) or less.

G.4.2.7 Drawings and Specifications

The Authority Having Jurisdiction shall require any or all of the following information to be included with or in the plot plan before a permit is issued for a gray water system, or at any time during the construction thereof:

(1) Plot plan drawn to scale and completely dimensioned, showing lot lines and structures, direction and approximate slope of surface, location of all present or proposed retaining walls, drainage channels, water supply lines, wells, paved areas and structures on the plot, number of bedrooms and plumbing fixtures in each structure, location of private sewage disposal system and expansion area or building sewer connecting to the public sewer, and location of the proposed gray water system.

(2) Details of construction necessary to ensure compliance with the requirements of this chapter, together with a full description of the complete installation, including installation methods, construction, and materials as required by the Authority Having Jurisdiction.

(3) Details for all holding tanks shall include all dimensions, structural calculations, bracings, and such other pertinent data as required.

(4) A log of soil formations and groundwater level as determined by test holes dug in proximity to any proposed irrigation area, together with a statement of water absorption characteristics of the soil at the proposed site as determined by approved percolation tests. Exception: The Authority Having Jurisdiction shall permit the use of Table G.4.2.10 in lieu of percolation tests.

(5) Distance between the plot and any surface waters such as lakes, ponds, rivers or streams, and the slope between the plot and the surface water, if in close proximity.

G.4.2.8 Procedure for Estimating Gray Water Discharge

Gray water systems shall be designed to distribute the total amount of estimated gray water on a daily basis. The water discharge for gray water systems shall be determined in accordance with Section G.4.2.8.1 or Section G.4.2.8.2.

G.4.2.8.1 Single Family Dwellings and Multi-Family Dwellings

The gray water discharge for single family and multi-family dwellings shall be calculated by water use records, calculations of local daily per person interior water use, or the following procedure:

(1) The number of occupants of each dwelling unit shall be calculated as follows:

First Bedroom: 2 occupants
Each additional bedroom: 1 occupant

(2) The estimated gray water flows of each occupant shall be calculated as follows:

Showers, bathtubs   25 gallons (95 L) per day/occupant and lavatories
Laundry   15 gallons (57 L) per day/occupant

(3) The total number of occupants shall be multiplied by the applicable estimated gray water discharge as provided above and the type of fixtures connected to the gray water system.

G.4.2.8.2 Commercial, Industrial, and Institutional Occupancies

The gray water discharge for commercial, industrial, and institutional occupancies shall be calculated by utilizing the procedure in Section G.4.2.8.1, water use records, or other documentation to estimate gray water discharge.

G.4.2.9 Gray Water System Components

Gray water system components shall be in accordance with Section G.4.2.9.1 through Section G.4.2.9.7.

G.4.2.9.1 Surge Tanks

Where installed, surge tanks shall comply with the following:

(1) Surge tanks shall be constructed of solid, durable materials not subject to excessive corrosion or decay and shall be watertight. Surge tanks constructed of steel shall be approved by the Authority Having Jurisdiction, provided such tanks comply with approved applicable standards.

(2) Each surge tank shall be vented as required by the plumbing code. The vent size shall be determined based on the total gray water fixture units as outlined in the plumbing code.

(3) Each surge tank shall have an access opening with lockable gasketed covers or approved equivalent to allow for inspection and cleaning.

(4) Each surge tank shall have its rated capacity permanently marked on the unit. In addition, a sign stating GRAY WATER, DANGER — UNSAFE WATER shall be permanently marked on the holding tank.

(5) Each surge tank shall have an overflow drain. The overflow drains shall have permanent connections to the building drain or building sewer, upstream of septic tanks, if any. The overflow drain shall not be equipped with a shutoff valve.

(6) The overflow drainpipes shall not be less in size than the inlet pipe. Unions or equally effective fittings shall be provided for all piping connected to the surge tank.

(7) Surge tank shall be structurally designed to withstand anticipated earth or other loads. Surge tank covers shall be capable of supporting an earth load of not less than 300 pounds per square foot (lb/ft^2) (1465 kg/m^2) when the tank is designed for underground installation.

(8) If a surge tank is installed underground, the system shall be designed so that the tank overflow will gravity drain to the existing sewer line or septic tank. The tank shall be protected against sewer line backflow by a backwater valve installed in accordance with the plumbing code.

(9) Surge tanks shall be installed on dry, level, well-compacted soil if underground or on a level 3 inch (76 mm) thick concrete slab if aboveground.

(10) Surge tanks shall be anchored to prevent against overturning when installed aboveground. Underground tanks shall be ballasted, anchored, or otherwise secured, to prevent the tank from floating out of the ground when empty. The combined weight of the tank and hold down system shall meet or exceed the buoyancy forces of the tank.

G.4.2.9.2 Gray Water Pipe and Fitting Materials

Aboveground and underground building drainage and vent pipe and fittings for gray water systems shall comply with the requirements for aboveground and underground sanitary building drainage and vent pipe and fittings in the plumbing code. These materials shall extend not less than 2 feet (610 mm) outside the building.

G.4.2.9.3 Subsoil Irrigation Field Materials

Subsoil irrigation field piping shall be constructed of perforated high-density polyethylene pipe, perforated ABS pipe, perforated PVC pipe, or other approved materials, provided that sufficient openings are available for distribution of the gray water into the trench area. Material, construction, and perforation of the pipe shall be in compliance with the appropriate absorption field drainage piping standards and shall be approved by the Authority Having Jurisdiction.

G.4.2.9.4 Subsurface Irrigation Field and Mulch Basin Supply Line Materials

Materials for gray water piping outside the building shall be polyethylene or PVC. Drip feeder lines shall be PVC or polyethylene tubing.

G.4.2.9.5 Valves

Valves shall be accessible.

G.4.2.9.6 Trap

Gray water piping discharging into the surge tank or having a direct connection to the sanitary drain or sewer piping shall be downstream of an approved water seal type trap(s). If no such trap(s) exists, an approved vented running trap shall be installed upstream of the connection to protect the building from any possible waste or sewer gases.

G.4.2.9.7 Backwater Valve

A backwater valve shall be installed on all gray water drain connections to the sanitary drain or sewer.

G.4.2.10 Subsurface Irrigation System Zones

Irrigation or disposal fields shall be permitted to have one or more valved zones. Each zone must be of adequate size to receive the gray water anticipated in that zone.

G.4.2.10.1 Required Area of Subsurface Irrigation Fields, Subsoil Irrigation Fields and Mulch Basins

The minimum effective irrigation area of subsurface irrigation fields, subsoil irrigation fields, and mulch basins shall be determined by Table G.4.2.10 for the type of soil found in the excavation, based upon a calculation of estimated gray water discharge pursuant to Section G.4.2.8. For a subsoil irrigation field, the area shall be equal to the aggregate length of the perforated pipe sections within the valved zone multiplied by the width of the proposed subsoil irrigation field.

G.4.2.10.2 Determination of Maximum Absorption Capacity

The irrigation field and mulch basin size shall be based on the maximum absorption capacity of the soil and determined using Table G.4.2.10. For soils not listed in Table G.4.2.10, the maximum absorption capacity for the proposed site shall be determined by percolation tests or other method acceptable to the Authority Having Jurisdiction. A gray water system shall not be permitted, where the percolation test shows the absorption capacity of the soil is unable to accommodate the maximum discharge of the proposed gray water irrigation system.

G.4.2.10.3 Groundwater Level

No excavation for an irrigation field, disposal field, or mulch basin shall extend within 3 feet (914 mm) vertical of the highest known seasonal groundwater level, nor to a depth where gray water contaminates the groundwater or surface water. The applicant shall supply evidence of groundwater depth to the satisfaction of the Authority Having Jurisdiction.

TABLE G.4.2.10
DESIGN OF SIX TYPICAL SOILS

TYPE OF SOIL	MINIMUM SQUARE FEET OF IRRIGATION AREA PER 100 GALLONS OF ESTIMATED GRAY WATER DISCHARGE PER DAY	MAXIMUM ABSORPTION CAPACITY IN GALLONS PER SQUARE FOOT OF IRRIGATION/ LEACHING AREA FOR A 24-HOUR PERIOD
Coarse sand or gravel	20	5.0
Fine sand	25	4.0
Sandy loam	40	2.5
Sandy clay	60	1.7
Clay with considerable sand or gravel	90	1.1
Clay with small amounts of sand or gravel	120	0.8

For SI units: 1 square foot = 0.0929 m^2, 1 gallon per day = 0.000043 L/s

G.4.2.11 Subsurface and Subsoil Irrigation Field, and Mulch Basin Design and Construction

Subsurface and subsoil irrigation field, and mulch basin design and construction shall be in accordance with Section G.4.2.11.1 through Section G.4.2.11.3. Where a gray water irrigation system design is predicated on soil tests, the subsurface or subsoil irrigation field or mulch basin shall be installed at the same location and depth as the tested area.

G.4.2.11.1 Subsurface Irrigation Field

A subsurface irrigation field shall be in accordance with Section G.4.2.11.1.1 through Section G.4.2.11.1.6.

G.4.2.11.1.1 Minimum Depth

Supply piping, including drip feeders, shall be not less than 2 inches (51 mm) below finished grade and covered with mulch or soil.

G.4.2.11.1.2 Filter

Not less than 140 mesh (115 micron) filter with a capacity of 25 gallons per minute (gpm) (1.58 L/s), or equivalent shall be installed. Where a filter backwash is installed, the backwash and flush discharge shall discharge into the building sewer or private sewage disposal system. Filter backwash and flush water shall not be used for any purpose.

G.4.2.11.1.3 Emitter Size

Emitters shall be installed in accordance with the manufacturer's installation instructions. Emitters shall have a flow path of not less than 1200 microns (µ) (1200 µm) and shall not have a coefficient of manufacturing variation (Cv) exceeding 7 percent. Irrigation system design shall be such that emitter flow variation shall not exceed 10 percent.

G.4.2.11.1.4 Number of Emitters

The minimum number of emitters and the maximum discharge of each emitter in an irrigation field shall be in accordance with Table G.4.2.11.1.

G.4.2.11.1.5 Controls

The system design shall provide user controls, such as valves, switches, timers, and other controllers, to rotate the distribution of gray water between irrigation zones.

G.4.2.11.1.6 Maximum Pressure

Where pressure at the discharge side of the pump exceeds 20 pounds-force per square inch (psi) (138 kPa), a pressure-reducing valve able to maintain downstream pressure not exceeding 20 psi (138 kPa) shall be installed downstream from the pump and before any emission device.

G.4.2.11.2 Mulch Basin

A mulch basin shall be in accordance with Section G.4.2.11.2.1 through Section G.4.2.11.2.4.

G.4.2.11.2.1 Single Family and Multi-Family Dwellings

The gray water discharge to a mulch basin is limited to single family and multi-family dwellings.

G.4.2.11.2.2 Size

Mulch basins shall be of sufficient size to accommodate peak flow rates and distribute the total amount of estimated gray water on a daily basis without surfacing, ponding or runoff. Mulch basins shall have a depth of not less than 10 inches (254 mm) below finished grade. The mulch basin size shall be based on the maximum absorption capacity of the soil and determined using Table G.4.2.10.

G.4.2.11.2.3 Minimum Depth

Gray water supply piping, including drip feeders, shall be a minimum 2 inches (51 mm) below finished grade and covered with mulch.

G.4.2.11.2.4 Maintenance

The mulch basin shall be maintained periodically to retain the required depth and area, and to replenish the required mulch cover.

G.4.2.11.3 Subsoil Irrigation Field

Subsoil irrigation fields shall be in accordance with Section G.4.2.11.3.1 through Section G.4.2.11.3.3.

G.4.2.11.3.1 Minimum Pipe Size

Subsoil irrigation field distribution piping shall be not less than 3 inches (80 mm) diameter.

G.4.2.11.3.2 Filter Material and Backfill

Filter material, clean stone, gravel, slag, or similar material acceptable to the Authority Having Jurisdiction, varying in size from 3/4 of an inch (19.1 mm) to 2 1/2 inches (64 mm) shall be placed in the trench to the depth and grade in accordance with Table G.4.2.11.3. The perforated section of subsoil irrigation field distribution piping shall be laid on the filter material in an approved manner. The perforated section shall then be covered with filter material to the minimum depth in accordance with Table G.4.2.11.3. The filter material shall then be covered with porous material to prevent closure of voids with earth backfill. No earth backfill shall be placed over the filter material cover until after inspection and acceptance.

G.4.2.11.3.3 Subsoil Irrigation Field Construction

Subsoil irrigation fields shall be constructed in accordance with Table G.4.2.11.3. Where necessary on sloping ground to prevent excessive line slopes, irrigation lines shall be stepped. The lines between each horizontal leaching section shall be made with approved watertight joints and installed on natural or unfilled ground.

TABLE G.4.2.11.1
SUBSURFACE IRRIGATION DESIGN CRITERIA FOR SIX TYPICAL SOILS

TYPE OF SOIL	MAXIMUM EMITTER DISCHARGE	MINIMUM NUMBER OF EMITTERS PER GALLON OF ESTIMATED GRAY WATER DISCHARGE PER DAY*
	gallon/day	gallon/day
Sand	1.8	0.6
Sandy loam	1.4	0.7
Loam	1.2	0.9
Clay loam	0.9	1.1
Silty clay	0.6	1.6
Clay	0.5	2.0

For SI units: 1 gallon per day = 0.000043 L/s
*The estimated gray water discharge per day shall be determined in accordance with Section G.4.2.8.

TABLE G.4.2.11.3
SUBSOIL IRRIGATION FIELD CONSTRUCTION

DESCRIPTION	MINIMUM	MAXIMUM
Number of drain lines per valved zone	1	-
Length of each perforated line	-	100 feet
Bottom width of trench	12 inches	18 inches
Spacing of lines, center to center	4 feet	-
Depth of earth cover of lines	10 inches	-
Depth of filter material cover of lines	2 inches	-
Depth of filter material beneath lines	3 inches	-
Grade of perforated lines level	level	3 inches per 100 feet

For SI units: 1 inch = 25.4 mm, 1 foot = 304.8 mm, 1 inch per foot = 83.3 mm/m

G.4.2.12 Gray Water System Color and Marking Information

Pressurized gray water distribution systems shall be identified as containing non-potable water in accordance with the plumbing code.

G.4.2.13 Special Provisions

G.4.2.13.1 Other Collection and Distribution Systems

Other collection and distribution systems shall be approved by the local Authority Having Jurisdiction.

G.4.2.13.2 Higher Requirements

Nothing contained in this chapter shall be construed to prevent the Authority Having Jurisdiction from requiring compliance with higher requirements than those contained herein, where such higher requirements are essential to maintain a safe and sanitary condition.

G.4.2.14 Testing

Building drains and vents for gray water systems shall be tested in accordance with the plumbing code. Surge tanks shall be filled with water to the overflow line prior to and during inspection. Seams and joints shall be left exposed, and the tank shall remain watertight. A flow test shall be performed through the system to the point of gray water discharge. Lines and components shall be watertight up to the point of the irrigation perforated and drip lines.

G.4.2.15 Maintenance

Gray water systems and components shall be maintained in accordance with Table G.4.1.5.

G.4.3 Reclaimed (Recycled) Water Systems

G.4.3.1 General

The provisions of this section shall apply to the installation, construction, alteration, and repair of reclaimed (recycled) water systems intended to supply uses such as water closets, urinals, trap primers for floor drains and floor sinks, aboveground and subsurface irrigation, industrial or commercial cooling or air conditioning and other uses approved by the Authority Having Jurisdiction.

G.4.3.2 Permit

It shall be unlawful for any person to construct, install, alter, or cause to be constructed, installed, or altered any reclaimed (recycled) water system within a building or on a premises without first obtaining a permit to do such work from the Authority Having Jurisdiction.

G.4.3.2.1 Plumbing Plan Submission

No permit for any reclaimed (recycled) water system shall be issued until complete plumbing plans, with appropriate data satisfactory to the Authority Having Jurisdiction, have been submitted and approved.

G.4.3.3 System Changes

No changes or connections shall be made to either the reclaimed (recycled) water system or the potable water system within any site containing a reclaimed (recycled) water system without approval by the Authority Having Jurisdiction.

G.4.3.4 Connections to Potable or Reclaimed (Recycled) Water Systems

Reclaimed (recycled) water systems shall have no connection to any potable water supply or alternate water source system. Potable water is permitted to be used as makeup water for a reclaimed (recycled) water storage tank provided the water supply inlet is protected by an airgap or reduced-pressure principle backflow preventer complying with the plumbing code.

G.4.3.5 Initial Cross-Connection Test

A cross-connection test is required in accordance with Section G.4.1.11.2. Before the building is occupied or the system is activated, the installer shall perform the initial cross-connection test in the presence of the Authority Having Jurisdiction and other authorities having jurisdiction. The test shall be ruled successful by the Authority Having Jurisdiction before final approval is granted.

G.4.3.6 Reclaimed (Recycled) Water System Materials

Reclaimed (recycled) water supply and distribution system materials shall comply with the requirements of the plumbing code for potable water supply and distribution systems, unless otherwise provided for in this section.

G.4.3.7 Reclaimed (Recycled) Water System Color and Marking Information

Reclaimed (recycled) water systems shall have a colored background in accordance with the plumbing code. Reclaimed (recycled) water systems shall be marked, in lettering in accordance with the plumbing code, with the words: "CAUTION: NON-POTABLE RECLAIMED (RECYCLED) WATER, DO NOT DRINK." Field marking of pipe meeting these requirements shall be permitted.

G.4.3.8 Valves

Valves, except fixture supply control valves, shall be equipped with a locking feature.

G.4.3.9 Installation

G.4.3.9.1 Hose Bibbs

Hose bibbs shall not be allowed on reclaimed (recycled) water piping systems located in areas accessible to the public. Access to reclaimed (recycled) water at points in the system accessible to the public shall be through a quick-disconnect device that differs from those installed on the potable water system. Hose bibbs supplying reclaimed (recycled) water shall be marked with the words: "CAUTION: NON-POTABLE RECLAIMED WATER, DO NOT DRINK," and the symbol in Figure G.4.3.9.

G.4.3.9.2 Required Appurtenances

The reclaimed (recycled) water system and the potable water system within the building shall be provided with the required appurtenances (valves, air/vacuum relief valves, etc.) to allow for deactivation or drainage as required for cross-connection test in Section G.4.1.11.2.

G.4.3.9.3 Same Trench as Potable Water Pipes

Reclaimed (recycled) water pipes shall be permitted to be run or laid in the same trench as potable water pipes with a 12 inches (305 mm) minimum vertical and horizontal separation when both pipe materials are approved for use within a building. When piping materials do not meet this requirement, the minimum horizontal separation shall be increased to 60 inches (1524 mm). The potable water piping shall be installed at an elevation above the reclaimed (recycled) water piping. Reclaimed (recycled) water pipes laid in the same trench or crossing building sewer or drainage piping shall be installed in accordance with the plumbing code for potable water piping.

Figure G.4.3.9

G.4.3.10 Signs

Rooms and water closet tanks in buildings using reclaimed (recycled) water shall be in accordance with Section G.4.1.10.

G.4.3.11 Inspection and Testing

Reclaimed (recycled) water systems shall be inspected and tested in accordance with Section G.4.1.11.

G.4.4 On-Site Treated Non-Potable Water Systems

G.4.4.1 General

The provisions of this section shall apply to the installation, construction, alteration, and repair of on-site treated non-potable water systems intended to supply uses such as water closets, urinals, trap primers for floor drains and floor sinks, above and below ground irrigation, and other uses approved by the Authority Having Jurisdiction.

G.4.4.2 Plumbing Plan Submission

No permit for any on-site treated non-potable water system shall be issued until complete plumbing plans, with appropriate data satisfactory to the Authority Having Jurisdiction, have been submitted and approved.

G.4.4.3 System Changes

No changes or connections shall be made to either the on-site treated non-potable water system or the potable water system within any site containing an on-site treated non-potable water system without approval by the Authority Having Jurisdiction.

G.4.4.4 Connections to Potable or Reclaimed (Recycled) Water Systems

On-site treated non-potable water systems shall have no connection to any potable water supply or reclaimed (recycled) water source system. Potable or reclaimed (recycled) water is permitted to be used as makeup water for a non-pressurized storage tank provided the makeup water supply is protected by an airgap in accordance with the plumbing code.

G.4.4.5 Initial Cross-Connection Test

A cross-connection test is required in accordance with Section G.4.1.11.2. Before the building is occupied or the system is activated, the installer shall perform the initial cross-connection test in the presence of the Authority Having Jurisdiction and other authorities having jurisdiction. The test shall be ruled successful by the Authority Having Jurisdiction before final approval is granted.

G.4.4.6 On-Site Treated Non-Potable Water System Materials

On-site treated non-potable water supply and distribution system materials shall comply with the requirements of the plumbing code for potable water supply and distribution systems, unless otherwise provided for in this section.

G.4.4.7 On-Site Treated Non-Potable Water Devices and Systems

Devices or equipment used to treat on-site treated non-potable water in order to maintain the minimum water quality requirements determined by the Authority Having Jurisdiction shall be listed or labeled (third-party certified) by a listing agency (accredited conformity assessment body) or approved by the Authority Having Jurisdiction and approved for the intended application. Devices or equipment used to treat on-site treated non-potable water for use in water closet and urinal flushing, surface irrigation and similar applications shall be listed and labeled to IAPMO IGC207-2009a, NSF 350-2011 or approved by the Authority Having Jurisdiction.

G.4.4.8 On-Site Treated Non-Potable Water System Color and Marking Information

On-site treated water systems shall have a colored background in accordance with the plumbing code. On-site treated water systems shall be marked, in lettering in accordance with the plumbing code, with the words: "CAUTION: ON-SITE TREATED NON-POTABLE WATER, DO NOT DRINK." Field marking of pipe meeting these requirements shall be acceptable.

G.4.4.9 Valves

Valves, except fixture supply control valves, shall be equipped with a locking feature.

G.4.4.10 Design and Installation

The design and installation of on-site treated non-potable systems shall be in accordance with Section G.4.4.10.1 through Section G.4.4.10.5.

G.4.4.10.1 Listing Terms and Installation Instructions

On-site treated non-potable water systems shall be installed in accordance with the terms of its listing and the manufacturer's installation instructions.

G.4.4.10.2 Minimum Water Quality

On-site treated non-potable water supplied to toilets or urinals or for other uses in which it is sprayed or exposed shall be disinfected. Acceptable disinfection methods shall include chlorination, ultraviolet sterilization, ozone, or other methods as approved by the Authority Having Jurisdiction. The minimum water quality for on-site treated non-potable water systems shall meet the applicable water quality requirements for the intended applications as determined by the Authority Having Jurisdiction.

G.4.4.10.3 Deactivation and Drainage

The on-site treated non-potable water system and the potable water system within the building shall be provided with the required appurtenances (valves, air/vacuum relief valves, etc.) to allow for deactivation or drainage as required for cross-connection test in accordance with Section G.4.1.11.2.

G.4.4.10.4 Near Underground Potable Water Pipe

On-site treated non-potable water pipes shall be permitted to be run or laid in the same trench as potable water pipes with a 12 inch (305 mm) minimum vertical and horizontal separation when both pipe materials are approved for use within a building. Where piping materials do not meet this requirement the minimum separation shall be increased to 60 inches (1524 mm). The potable water piping shall be installed at an elevation above the on-site treated non-potable water piping.

G.4.4.10.5 Required Filters

A filter permitting the passage of particulates no larger than 100 microns (100 μm) shall be provided for on-site treated non-potable water supplied to water closets, urinals, trap primers, and drip irrigation system.

G.4.4.11 Signs

Signs in buildings using on-site treated non-potable water shall be in accordance with Section G.4.1.10.

G.4.4.12 Inspection and Testing

On-site treated non-potable water systems shall be inspected and tested in accordance with Section G.4.1.11.

G.4.5 Non-Potable Rainwater Catchment Systems

G.4.5.1 General

The provisions of this section shall apply to the installation, construction, alteration, and repair of rainwater catchments systems intended to supply uses such as water closets, urinals, trap primers for floor drains and floor sinks, irrigation, industrial processes, water features, cooling tower makeup and other uses approved by the Authority Having Jurisdiction. Additional design criteria can be found in the ARCSA/ASPE Rainwater Catchment Design and Installation Standard.

G.4.5.2 Plumbing Plan Submission

No permit for any rainwater catchment system requiring a permit shall be issued until complete plumbing plans, with appropriate data satisfactory to the Authority Having Jurisdiction, have been submitted and approved. No changes or connections shall be made to either the rainwater catchment or the potable water system within any site containing a rainwater catchment water system without approval by the Authority Having Jurisdiction.

G.4.5.3 System Changes

No changes or connections shall be made to either the rainwater catchment system or the potable water system within any site containing a rainwater catchment system requiring a permit without approval by the Authority Having Jurisdiction.

G.4.5.4 Connections to Potable or Reclaimed (Recycled) Water Systems

Rainwater catchment systems shall have no direct connection to any potable water supply or alternate water source system. Potable or reclaimed (recycled) water is permitted to be used as makeup water for a rainwater catchment system provided the potable or reclaimed (recycled) water supply connection is protected by an airgap or reduced-pressure principle backflow preventer in accordance with the plumbing code.

G.4.5.5 Initial Cross-Connection Test

Where any portion of a rainwater catchment system is installed within a building, a cross-connection test is required in accordance with G.4.1.11.2. Before the building is occupied or the system is activated, the installer shall perform the initial cross-connection test in the presence of the Authority Having Jurisdiction and other authorities having jurisdiction. The test shall be ruled successful by the Authority Having Jurisdiction before final approval is granted.

G.4.5.6 Sizing

The design and size of rainwater drains, gutters, conductors, and leaders shall be in accordance with the plumbing code.

G.4.5.7 Rainwater Catchment System Materials

Rainwater catchment system materials shall be in accordance with Section G.4.5.7.1 through Section G.4.5.7.4.

G.4.5.7.1 Water Supply and Distribution Materials

Rainwater catchment water supply and distribution materials shall comply with the requirements of the plumbing code for potable water supply and distribution systems, unless otherwise provided for in this section.

G.4.5.7.2 Rainwater Catchment System Drainage Materials

Materials used in rainwater catchment drainage systems, including gutters, downspouts, conductors, and leaders shall comply with the requirements of the plumbing code for storm drainage.

G.4.5.7.3 Storage Tanks

Rainwater storage tanks shall be in accordance with Section G.4.5.9.5.

G.4.5.7.4 Collections Surfaces

The collection surface shall be constructed of a hard, impervious material.

G.4.5.8 Rainwater Catchment Water System Color and Marking Information

Rainwater catchment systems shall have a colored background in accordance with the plumbing code. Rainwater catchment systems shall be marked, in lettering in accordance with the plumbing code, with the words: "CAUTION: NON-POTABLE RAINWATER WATER, DO NOT DRINK."

G.4.5.9 Design and Installation

G.4.5.9.1 Outside Hose Bibbs

Outside hose bibbs shall be allowed on rainwater piping systems. Hose bibbs supplying rainwater shall be marked with the words: "CAUTION: NON-POTABLE WATER, DO NOT DRINK" and the symbol in Figure G.4.3.9.

G.4.5.9.2 Deactivation and Drainage for Cross-connection Test

The rainwater catchment system and the potable water system within the building shall be provided with the required appurtenances (e.g., valves, air or vacuum relief valves, etc.) to allow for deactivation or drainage as required for cross-connection test in Section G.4.1.11.2.

G.4.5.9.3 Collection Surfaces

G.4.5.9.3.1 Rainwater Catchment System Surfaces

Rainwater shall be collected from roof surfaces or other manmade, aboveground collection surfaces.

G.4.5.9.3.2 Other Surfaces

Natural precipitation collected from surface water runoff, vehicular parking surfaces or manmade surfaces at or below grade shall comply with the stormwater requirements for on-site treated non-potable water systems in Section G.4.4.

G.4.5.9.3.3 Prohibited Discharges

Overflows and bleed-off pipes from roof-mounted equipment and appliances shall not discharge onto roof surfaces that are intended to collect rainwater

G.4.5.9.4 Minimum Water Quality

The minimum water quality for harvested rainwater shall meet the applicable water quality requirements for the intended applications as determined by the Authority Having Jurisdiction. In the absence of water quality requirements determined by the Authority Having Jurisdiction, the minimum treatment and water quality shall also comply with Table G.4.5.9.4.

TABLE G.4.5.9.4
MINIMUM WATER QUALITY

APPLICATION	MINIMUM TREATMENT	MINIMUM WATER QUALITY
Car washing	Debris excluder or other approved means in compliance with Section G.4.5.9.10, and 100 Micron (100 μm) in compliance with Section G.4.5.9.11 for drip irrigation.	N/A
Subsurface and drip irrigation	Debris excluder or other approved means in compliance with Section G.4.5.9.10, and 100 Micron (100 μm) in compliance with Section G.4.5.9.11 for drip irrigation.	N/A
Spray irrigation where the maximum storage volume is less than 360 gallons (1363 L)	Debris excluder or other approved means in compliance with Section G.4.5.9.10, and Disinfection in accordance with Section G.4.5.9.8.	N/A
Spray irrigation where the maximum storage volume is equal to or greater than 360 gallons (1363 L)	Debris excluder or other approved means in compliance with Section G.4.5.9.10.	Escherichia coli: < 100 CFU/100 mL, and Turbidity: < 10 NTU
Urinal and water closet flushing, clothes washing, and trap priming	Debris excluder or other approved means in compliance with Section G.4.5.9.10, and 100 Micron (100 μm) in compliance with Section G.4.5.9.11.	Escherichia coli: < 100 CFU/100 mL, and Turbidity: < 10 NTU
Ornamental fountains and other water features	Debris excluder or other approved means in compliance with Section G.4.5.9.10.	Escherichia coli: < 100 CFU/100 mL, and Turbidity: < 10 NTU
Cooling tower make up water	Debris excluder or other approved means in compliance with Section G.4.5.9.10, and 100 Micron (100 μm) in compliance with Section G.4.5.9.11.	Escherichia coli: < 100 CFU/100 mL, and Turbidity: < 10 NTU

G.4.5.9.5 Rainwater Storage Tanks

Rainwater storage tanks shall be constructed and installed in accordance with Section G.4.5.9.5.1 through Section G.4.5.9.5.8.

G.4.5.9.5.1 Construction

Rainwater storage shall be constructed of solid, durable materials not subject to excessive corrosion or decay and shall be watertight. Storage tanks shall be approved by the Authority Having Jurisdiction, provided such tanks comply with approved applicable standards.

G.4.5.9.5.2 Location

Rainwater storage tanks shall be permitted to be installed above or below grade.

G.4.5.9.5.3 Above Grade

Above grade storage tanks shall be of an opaque material, approved for aboveground use in direct sunlight or shall be shielded from direct sunlight. Tanks shall be installed in an accessible location to allow for inspection and cleaning. The tank shall be installed on a foundation or platform that is constructed to accommodate all loads in accordance with the building code.

G.4.5.9.5.4 Below Grade

Rainwater storage tanks installed below grade shall be structurally designed to withstand all anticipated earth or other loads. Holding tank covers shall be capable of supporting an earth load of not less than 300 pounds per square foot (lb/ft^2) (1465 kg/m^2) when the tank is designed for underground installation. Below grade rainwater tanks installed underground shall be provided with manholes. The manhole opening shall be a minimum diameter of 20 inches (508 mm) and located a minimum of 4 inches (102 mm) above the surrounding grade. The surrounding grade shall be sloped away from the manhole. Underground tanks shall be ballasted, anchored, or otherwise secured, to prevent the tank from floating out of the ground when empty. The combined weight of the tank and hold down system should meet or exceed the buoyancy force of the tank.

G.4.5.9.5.5 Drainage and Overflow

Rainwater storage tanks shall be provided with a means of draining and cleaning. The overflow drain shall not be equipped with a shutoff valve. The overflow outlet shall discharge as required by the plumbing code for storm drainage systems. Where discharging to the storm drainage system, the overflow drain shall be protected from backflow of the storm drainage system by a backwater valve or other approved method.

G.4.5.9.5.5.1 Overflow Outlet Size

The overflow outlet shall be sized to accommodate the flow of the rainwater entering the tank and not less than the aggregate cross-sectional area of all inflow pipes.

G.4.5.9.5.6 Opening and Access Protection

G.4.5.9.5.6.1 Animals and Insects

Rainwater tank openings shall be protected to prevent the entrance of insects, birds, or rodents into the tank.

G.4.5.9.5.6.2 Human Access

Rainwater tank access openings exceeding 12 inches (305 mm) in diameter shall be secured to prevent tampering and unintended entry by either a lockable device or other approved method.

G.4.5.9.5.7 Marking

Rainwater tanks shall be permanently marked with the capacity and the language: "NON-POTABLE RAINWATER." Where openings are provided to allow a person to enter the tank, the opening shall be marked with the following language: "DANGER-CONFINED SPACE."

G.4.5.9.5.8 Storage Tank Venting

Where venting by means of drainage or overflow piping is not provided or is considered insufficient, a vent shall be installed on each tank. The vent shall extend from the top of the tank and terminate a minimum of 6 inches (152 mm) above grade and shall be a minimum of 1 1/2 inches (38 mm) in diameter. The vent terminal shall be directed downward and covered with a 3/32 inch (2.4 mm) mesh screen to prevent the entry of vermin and insects.

G.4.5.9.6 Pumps

Pumps serving rainwater catchment systems shall be listed. Pumps supplying water to water closets, urinals, and trap primers shall be capable of delivering not less than 15 psi (103 kPa) residual pressure at the highest and most remote outlet served. Where the water pressure in the rainwater supply system within the building exceeds 80 psi (552 kPa), a pressure reducing valve reducing the pressure to 80 psi (552 kPa) or less to all water outlets in the building shall be installed in accordance with the plumbing code.

G.4.5.9.7 Roof Drains

Primary and secondary roof drains, conductors, leaders, and gutters shall be designed and installed in accordance with the plumbing code.

G.4.5.9.8 Water Quality Devices and Equipment

Devices and equipment used to treat rainwater to maintain the minimum water quality requirements determined by the Authority Having Jurisdiction shall be listed or labeled (third-party certified) by a listing agency (accredited conformity assessment body) and approved for the intended application.

G.4.5.9.9 Freeze Protection

Tanks and piping installed in locations subject to freezing shall be provided with an adequate means of freeze protection.

G.4.5.9.10 Debris Removal

The rainwater catchment conveyance system shall be equipped with a debris excluder or other approved means to prevent the accumulation of leaves, needles, other debris and sediment from entering the storage tank. Devices or methods used to remove debris or sediment shall be accessible and sized and installed in accordance with manufacturer's installation instructions.

G.4.5.9.11 Required Filters

A filter permitting the passage of particulates no larger than 100 microns (100 μm) shall be provided for rainwater supplied to water closets, urinals, trap primers, and drip irrigation system.

G.4.5.9.12 Roof Gutters

Gutters shall maintain a minimum slope and be sized in accordance with the plumbing code.

G.4.5.10 Signs

Signs in buildings using rainwater water shall be in accordance with Section G.4.1.10.

G.4.5.11 Inspection and Testing

Rainwater catchment systems shall be inspected and tested in accordance with Section G.4.1.11.

G.4.5.11.1 Supply System Inspection and Test

Rainwater catchment systems shall be inspected and tested in accordance with Section G.4.1.11 and the applicable provisions of the plumbing code for testing of potable water and storm drainage systems. Storage tanks shall be filled with water to the overflow opening for a period of 24 hours and during inspection or by other means as approved by the Authority Having Jurisdiction. All seams and joints shall be exposed during inspection and checked for water tightness.

G.5 WATER HEATING DESIGN, EQUIPMENT AND INSTALLATION

G.5.1 General

G.5.1.1 Scope

The provisions of this chapter shall establish the means of conserving potable and non-potable water and energy associated with the generation and use of hot water in a building. This includes provisions for the hot water distribution system, which is the portion of the potable water distribution system between a water heating device and the plumbing fixtures, including all dedicated return piping and appurtenances to the water heating device in a recirculation system.

G.5.1.2 Insulation

Hot water supply and return piping shall be thermally insulated. The wall thickness of the insulation shall be equal to the nominal diameter of the pipe up to 2 inches (50 mm). The wall thickness shall be not less than 2 inches (50 mm) for nominal pipe diameters exceeding 2 inches (50 mm). The conductivity of the insulation [k-factor (Btu•in/(h•ft^2•°F))], measured radially, shall be less than or equal to 0.28 [Btu•in/(h•ft^2•°F)] [0.04 W/(m•k)]. Hot water piping to be insulated shall be installed such that insulation is continuous. Pipe insulation shall be installed to within 1/4 inch (6.4 mm) of all appliances, appurtenances, fixtures, structural members, or a wall where the pipe passes through to connect to a fixture within 24 inches (610 mm). Building cavities shall be large enough to accommodate the combined diameter of the pipe plus the insulation, plus any other objects in the cavity that the piping must cross. Pipe supports shall be installed on the outside of the pipe insulation.
EXCEPTIONS:

(1) Where the hot water pipe is installed in a wall that is not of sufficient width to accommodate the pipe and insulation, the insulation thickness shall be permitted to have the maximum thickness that the wall can accommodate and not less than 1/2 inch (12.7 mm) thick.

(2) Hot water supply piping exposed under sinks, lavatories, and similar fixtures.

(3) Where hot water distribution piping is installed within attic, crawlspace, or wall insulation.

(a) In attics and crawlspaces the insulation shall cover the pipe not less than 5 inches (140 mm) further away from the conditioned space.

(b) In walls, the insulation must completely surround the pipe with not less than 1 inch (25.4 mm) of insulation.

(c) If burial within the insulation will not completely or continuously surround the pipe, then these exceptions do not apply.

G.5.1.3 Recirculation Systems

G.5.1.3.1 Pump Operation

G.5.1.3.1.1 For Low-Rise Residential Buildings

Circulating hot water systems shall be arranged so that the circulating pump(s) can be turned off (automatically or manually) when the hot water system is not in operation. [ASHRAE 90.2:7.2]

G.5.1.3.1.2 For Pumps Between Boilers and Storage Tanks

When used to maintain storage tank water temperature, recirculating pumps shall be equipped with controls limiting operation to a period from the start of the heating cycle to a maximum of 5 minutes after the end of the heating cycle. [ASHRAE 90.1:7.4.4.4]

G.5.1.3.2 Recirculation Pump Controls

Pump controls shall include on-demand activation or time clocks combined with temperature sensing. Time clock controls for pumps shall not let the pump operate more than 15 minutes every hour. Temperature sensors shall stop circulation when the temperature set point is reached and shall be located on the circulation loop at or near the last fixture. The pump, pump controls and temperature sensors shall be accessible. Pump operation shall be limited to the building's hours of operation.

G.5.1.3.3 Temperature Maintenance Controls

For other than low-rise residential buildings, systems designed to maintain usage temperatures in hot-water pipes, such as recirculating hot-water systems or heat trace, shall be equipped with automatic time switches or other controls that can be set to switch off the usage temperature maintenance system during extended periods when hot water is not required. [ASHRAE 90.1:7.4.4.2]

G.5.1.3.4 System Balancing

Systems with multiple recirculation zones shall be balanced to uniformly distribute hot water, or they shall be operated with a pump for each zone. The circulation pump controls shall comply with the provisions of Section G.5.1.3.2.

G.5.1.3.5 Flow Balancing Valves

Flow balancing valves shall be a factory preset automatic flow control valve, a flow regulating valve, or a balancing valve with memory stop.

G.5.1.3.6 Air Elimination

Provision shall be made for the elimination of air from the return system.

G.5.1.3.7 Gravity or Thermosyphon Systems

Gravity or thermosyphon systems are prohibited.

G.5.2 Service Hot Water – Low-Rise Residential Buildings

G.5.2.1 General

The service water heating system for single-family houses, multi-family structures of three stories or fewer above grade, and modular houses shall be in accordance with Section G.5.2.2 through Section G.5.2.7. The service water heating system of all other buildings shall be in accordance with Section G.5.3.

G.5.2.2 Water Heaters and Storage Tanks

Residential-type water heaters, pool heaters, and unfired water heater storage tanks shall meet the minimum performance requirements specified by federal law. Unfired storage water heating equipment shall have a heat loss through the tank surface area of less than 6.5 British thermal units per hour per square foot (Btu/h•ft^2) (20.5 W/m^2). [ASHRAE 90.2:7.1]

G.5.2.3 Recirculation Systems

Recirculation systems shall meet the provisions in Section G.5.1.3.

G.5.2.4 Central Water Heating Equipment

Service water heating equipment (central systems) that does not fall under the requirements for residential-type service water heating equipment addressed in Section G.5.2 shall meet the applicable requirements for service water-heating equipment found in Section G.5.3. [ASHRAE 90.2:7.3]

G.5.2.5 Insulation

Insulation of hot water and return piping shall meet the provisions of Section G.5.1.2.

G.5.2.6 Hard Water

Where water has hardness equal to or exceeding 9 grains per gallon (gr/gal) (154 mg/L) measured as total calcium carbonate equivalents, the water supply line to water heating equipment in new one- and two-family dwellings shall be roughed-in to allow for the installation of water treatment equipment.

G.5.2.7 Maximum Volume of Hot Water

The maximum volume of water contained in the hot water distribution shall comply with Sections G.5.2.7.1 or G.5.2.7.2. The water volume shall be calculated using Table G.5.2.7.

G.5.2.7.1 Maximum Volume of Hot Water Without Recirculation or Heat Trace

The maximum volume of water contained in the hot water distribution pipe between the water heater and any fixture fitting shall not exceed 32 ounces (oz) (946 mL). Where a fixture fitting shut off valve (supply stop) is installed ahead of the fixture fitting, the maximum volume of water is permitted to be calculated between the water heater and the fixture fitting shut off valve (supply stop).

G.5.2.7.2 Maximum Volume of Hot Water With Recirculation or Heat Trace

The maximum volume of water contained in the branches between the recirculation loop or electrically heat traced pipe and the fixture fitting shall not exceed a 16 oz (473 mL). Where a fixture fitting shut off valve (supply stop) is installed ahead of the fixture fitting, the maximum volume of water is permitted to be calculated between the recirculation loop or electrically heat traced pipe and the fixture fitting shut off valve (supply stop).

EXCEPTION: Whirlpool bathtubs or bathtubs that are not equipped with a shower are exempted from the requirements of Section G.5.2.7.

G.5.2.7.3 Hot Water System Submeters

Where a hot water pipe from a circulation loop or electric heat trace line is equipped with a submeter, the hot water distribution system downstream of the submeter shall have either an end-of-line hot water circulation pump or shall be electrically heat traced. The maximum volume of water in any branch from the circulation loop or electric heat trace line downstream of the submeter shall not exceed 16 oz (473 mL).

If there is no circulation loop or electric heat traced line downstream of the submeter, the submeter shall be located within 2 feet (610 mm) of the central hot water system; or the branch line to the submeter shall be circulated or heat traced to within 2 feet of the submeter. The maximum volume from the submeter to each fixture shall not exceed 32 oz (946 mL).

The circulation pump controls shall comply with the provisions of Section G.5.1.3.2.

TABLE G.5.2.7
WATER VOLUME FOR DISTRIBUTION PIPING MATERIALS
OUNCES OF WATER PER FOOT LENGTH OF PIPING

NOMINAL SIZE (inch)	COPPER M	COPPER L	COPPER K	CPVC CTS SDR 11	CPVC SCH 40	PEX-AL-PEX	PE-AL-PE	CPVC SCH 80	PEX CTS SDR 9	PE-RT SDR 9	PP SDR 6	PP SDR 7.3	PP SDR 11
3/8	1.06	0.97	0.84	NA	1.17	0.63	0.63	NA	0.64	0.64	0.91	1.09	1.24
1/2	1.69	1.55	1.45	1.25	1.89	1.31	1.31	1.46	1.18	1.18	1.41	1.68	2.12
3/4	3.43	3.22	2.90	2.67	3.38	3.39	3.39	2.74	2.35	2.35	2.23	2.62	3.37
1	5.81	5.49	5.17	4.43	5.53	5.56	5.56	4.57	3.91	3.91	3.64	4.36	5.56
1 1/4	8.70	8.36	8.09	6.61	9.66	8.49	8.49	8.24	5.81	5.81	5.73	6.81	8.60
1 1/2	12.18	11.83	11.45	9.22	13.20	13.88	13.88	11.38	8.09	8.09	9.03	10.61	13.47
2	21.08	20.58	20.04	15.79	21.88	21.48	21.48	19.11	13.86	13.86	14.28	16.98	21.39

G.5.3 Service Hot Water – Other Than Low-Rise Residential Buildings

G.5.3.1 General

The service hot water, other than single-family houses, multi-family structures of three stories or fewer above grade, and modular houses, shall comply with this section.

G.5.3.2 Service Water Heating

G.5.3.2.1 New Buildings

Service water heating systems and equipment shall comply with the requirements of this section as described in Section G.5.3.3. [ASHRAE 90.1:7.1.1.1]

G.5.3.2.2 Additions to Existing Buildings

Service water heating systems and equipment shall comply with the requirements of this section.

EXCEPTION: When the service water heating to an addition is provided by existing service water heating systems and equipment, such systems and equipment shall not be required to comply with this supplement. However, any new systems or equipment installed must comply with specific requirements applicable to those systems and equipment. [ASHRAE 90.1:7.1.1.2]

G.5.3.2.3 Alterations to Existing Buildings

Building service water heating equipment installed as a direct replacement for existing building service water heating equipment shall comply with the requirements of Section G.5.3 applicable to the equipment being replaced. New and replacement piping shall comply with Section G.5.3.4.3.

EXCEPTION: Compliance shall not be required where there is insufficient space or access to meet these requirements. [ASHRAE 90.1:7.1.1.3]

G.5.3.3 Compliance Path(s)

G.5.3.3.1 General

Compliance shall be achieved by meeting the requirements of Section G.5.3.1, General; Section G.5.3.4, Mandatory Provisions; Section G.5.3.5, Prescriptive Path; and Section G.5.3.6, Submittals. [ASHRAE 90.1:7.2.1]

G.5.3.3.2 Energy Cost Budget Method

Projects using the Energy Cost Budget Method (Section 11 of ASHRAE 90.1) for demonstrating compliance with the standard shall meet the requirements of Section G.5.3.4, Mandatory Provisions, in conjunction with Section 11 of ASHRAE 90.1, Energy Cost Budget Method. [ASHRAE 90.1:7.2.2]

G.5.3.4 Mandatory Provisions

G.5.3.4.1 Load Calculations

Service water heating system design loads for the purpose of sizing systems and equipment shall be determined in accordance with manufacturers' published sizing guidelines or generally accepted engineering standards and handbooks acceptable to the adopting authority (e.g., ASHRAE Handbook – HVAC Applications). [ASHRAE 90.1:7.4.1]

G.5.3.4.2 Equipment Efficiency

Water heating equipment, hot-water supply boilers used solely for heating potable water, pool heaters, and hot-water storage tanks shall meet the criteria listed in Table G.5.3.4.2. Where multiple criteria are listed, all criteria shall be met. Omission of minimum performance requirements for certain classes of equipment does not preclude use of such equipment where appropriate. Equipment not listed in Table G.5.3.4.2 has no minimum performance requirements.

EXCEPTIONS: Water heaters and hot-water supply boilers having more than 140 gallons (530 L) of storage capacity are not required to meet the standby loss (SL) requirements of Table G.5.3.4.2 when:

(1) The tank surface is thermally insulated to R-12.5.

(2) A standing pilot light is not installed.

(3) Gas- or oil-fired storage water heaters have a flue damper or fan-assisted combustion. [ASHRAE 90.1:7.4.2]

G.5.3.4.3 Insulation

Insulation of hot water and return piping shall meet the provisions in Section G.5.1.2.

G.5.3.4.4 Hot Water System Design

G.5.3.4.4.1 Recirculation System

Recirculation systems shall meet the provisions in Section G.5.1.3.

G.5.3.4.4.4 Maximum Volume of Hot Water

The maximum volume of water contained in hot water distribution lines between the water heater and the fixture stop or connection to showers, kitchen faucets, and lavatories shall be determined in accordance with Section G.5.2.7.

G.5.3.4.5 Service Water Heating System Controls

G.5.3.4.5.1 Temperature Controls

Temperature controls shall be provided that allow for storage temperature adjustment from 120°F (49°C) or lower to a maximum temperature compatible with the intended use.

EXCEPTION: When the manufacturers' installation instructions specify a higher minimum thermostat setting to minimize condensation and resulting corrosion. [ASHRAE 90.1:7.4.4.1]

G.5.3.4.5.2 Outlet Temperature Controls

Temperature controlling means shall be provided to limit the maximum temperature of water delivered from lavatory faucets in public facility restrooms to 110°F (43°C). [ASHRAE 90.1:7.4.4.3]

G.5.3.4.6 Pools

G.5.3.4.6.1 Pool Heaters

Pool heaters shall be equipped with a readily accessible ON/OFF switch to allow shutting off the heater without adjusting the thermostat setting. Pool heaters fired by natural gas shall not have continuously burning pilot lights. [ASHRAE 90.1:7.4.5.1]

G.5.3.4.6.2 Pool Covers

Heated pools shall be equipped with a vapor retardant pool cover on or at the water surface. Pools heated to more than 90°F (32°C) shall have a pool cover with a minimum insulation value of R-12.

EXCEPTION: Pools deriving over 60 percent of the energy for heating from site-recovered energy or solar energy source. [ASHRAE 90.1:7.4.5.2]

G.5.3.4.6.3 Time Switches

Time switches shall be installed on swimming pool heaters and pumps.
EXCEPTIONS:

(1) Where public health standards require 24-hour pump operation.

(2) Where pumps are required to operate solar and waste heat recovery pool heating systems. [ASHRAE 90.1:7.4.5.3]

G.5.3.4.7 Heat Traps

Vertical pipe risers serving storage water heaters and storage tanks not having integral heat traps and serving a nonrecirculating system shall have heat traps on both the inlet and outlet piping as close as practical to the storage tank. A heat trap is a means to counteract the natural convection of heated water in a vertical pipe run. The means is either a device specifically designed for the purpose or an arrangement of tubing that forms a loop of 360 degrees (6.28 rad) or piping that from the point of connection to the water heater (inlet or outlet) includes a length of piping directed downward before connection to the vertical piping of the supply water or hot-water distribution system, as applicable. [ASHRAE 90.1:7.4.6]

Table G.5.3.4.2
PERFORMANCE REQUIREMENTS FOR WATER HEATING EQUIPMENT
(ASHRAE 90.1: TABLE 7.8)

EQUIPMENT TYPE	SIZE CATEGORY (INPUT)	SUBCATEGORY OR RATING CONDITION	PERFORMANCE REQUIRED[1]	TEST PROCEDURE[2,3]
Electric table top water heaters	≤12 kW	Resistance ≥20 gal	0.93–0.00132V EF	DOE 10 CFR Part 430
Electric water heaters	≤12 kW	Resistance ≥20 gal	0.97–0.00132V EF	DOE 10 CFR Part 430
	>12 kW	Resistance ≥20 gal	20 + 35√V SL, Btu/h	Section G.2 of ANSI Z21.10.3
	≤24 Amps and ≤250 Volts	Heat Pump	0.93–0.00132V EF	DOE 10 CFR Part 430
Gas storage water heaters	≤75 000 Btu/h	≥20 gal	0.62–0.0019V EF	DOE 10 CFR Part 430
	>75 000 Btu/h	<4000 (Btu/h)/gal	80% Et (Q/800 + 110√V) SL, Btu/h	Sections G.1 and G.2 of ANSI Z21.10.3
Gas instantaneous water heaters	>50 000 Btu/h and <200 000 Btu/h	≥4000 (Btu/h)/gal and <2 gal	0.62–0.0019V EF	DOE 10 CFR Part 430
	≥200 000 Btu/h[4]	≥4000 (Btu/h)/gal and <10 gal	80% Et	Sections G.1 and G.2 of ANSI Z21.10.3
	≥200 000 Btu/h	≥4000 (Btu/h)/gal and ≥10 gal	80% Et (Q/800 + 110√V) SL, Btu/h	
Electronic instantaneous water heaters[5]	≤ 12 kW	≥ 4000 (Btu/h)/gal and <2 gal	0.93 – (0.00132·V) EF	DOE 10 CFR Part 430
	> 12 kW	≥ 4000 (Btu/h)/gal and <2 gal	95% Et	Section G.2 of ANSI Z21.10.3
Oil storage water heaters	≤105 000 Btu/h	≥20 gal	0.59-0.0019V EF	DOE 10 CFR Part 430
	>105 000 Btu/h	<4000 (Btu/h)/gal	78% Et (Q/800 + 110√V) SL, Btu/h	Sections G.1 and G.2 of ANSI Z21.10.3
Oil instantaneous water heaters	≤210 000 Btu/h	≥4000 (Btu/h)/gal and <2 gal	0.59–0.0019V EF	DOE 10 CFR Part 430
	>210 000 Btu/h	≥4000 (Btu/h)/gal and <10 gal	80% Et	Sections G.1 and G.2 of ANSI Z21.10.3
	>210 000 Btu/h	≥4000 (Btu/h)/gal and ≥10 gal	78% Et (Q/800 + 110√V) SL, Btu/h	
Hot-water supply boilers, gas and oil	≥300 000 Btu/h and <12 500 000 Btu/h	≥4000 (Btu/h)/gal and <2 gal	80% Et	Sections G.1 and G.2 of ANSI Z21.10.3
Hot-water supply boilers, gas	———	≥4000 (Btu/h)/gal and <10 gal	80% Et (Q/800 + 110√V) SL, Btu/h	
Hot-water supply boilers, oil	———	≥4000 (Btu/h)/gal and ≥10 gal	78% Et (Q/800 + 110√V) SL, Btu/h	
Pool heaters, oil and gas	All	———	78% Et	ASHRAE 146
Heat pump pool heaters	All	50.0°F db 44.2°F wb outdoor air 80.0°F Entering Water	4.0 COP	AHRI 1160
Unfired storage tanks	All		R-12.5	(none)

G.5.3.5 Prescriptive Path

G.5.3.5.1 Space Heating and Water Heating

The use of a gas-fired or oil-fired space-heating boiler system otherwise complying with Section G.5.3 to provide the total space heating and water heating for a building is allowed when one of the following conditions is met:

(1) The single space-heating boiler, or the component of a modular or multiple boiler system that is heating the service water, has a standby loss in Btu/h (kW) not exceeding $(13.3 \times pmd + 400)/n$, where (pmd) is the probable maximum demand in gallons per hour, determined in accordance with the procedures described in generally accepted engineering standards and handbooks, and (n) is the fraction of the year when the outdoor daily mean temperature is greater than 64.9°F (18.28°C).

The standby loss is to be determined for a test period of 24 hours duration while maintaining a boiler water temperature of at least 90°F (32°C) above ambient, with an ambient temperature between 60°F (16°C) and 90°F (32°C). For a boiler with a modulating burner, this test shall be conducted at the lowest input.

(2) It is demonstrated to the satisfaction of the Authority Having Jurisdiction that the use of a single heat source will consume less energy than separate units.

(3) The energy input of the combined boiler and water heater system is less than 150 000 Btu/h (44 kW). [ASHRAE 90.1:7.5.1]

G.5.3.5.2 Service Water Heating Equipment

Service water heating equipment used to provide the additional function of space heating as part of a combination (integrated) system shall satisfy all stated requirements for the service water heating equipment. [ASHRAE 90.1:7.5.2]

G.5.3.5.3 Heat Recovery for Service Water Heating

G.5.3.5.3.1

Condenser heat recovery systems shall be installed for heating or preheating of service hot water provided all of the following are true:

(1) The facility operates 24 hours a day.

(2) The total installed heat rejection capacity of the water-cooled systems exceeds 6,000,000 Btu/h (1758 kW) of heat rejection.

(3) The design service water heating load exceeds 1,000,000 Btu/h (293 kW). [ASHRAE 90.1:6.5.6.2.1]

G.5.3.5.3.2

The required heat recovery system shall have the capacity to provide the smaller of:

(1) Sixty percent of the peak heat rejection load at design conditions.

(2) Preheat of the peak service hot water draw to 85°F (29°C).
[ASHRAE 90.1:6.5.6.2.2]
EXCEPTIONS:

(a) Facilities that employ condenser heat recovery for space heating with a heat recovery design exceeding 30 percent of the peak water-cooled condenser load at design conditions.

(b) Facilities that provide 60 percent of their service water heating from site-solar or site-recovered energy or from other sources.

G.5.3.6 Submittals

G.5.3.6.1 General

The Authority Having Jurisdiction shall require submittal of compliance documentation and supplemental information, in accordance with this supplement and the applicable mechanical and building codes.

G.5.4 Solar Water Heating Systems

G.5.4.1 General

The erection, installation, alteration, addition to, use or maintenance of solar water heating systems shall be in accordance with this section and the Uniform Solar Energy Code.

G.5.4.2 Annual Inspection and Maintenance

Solar energy systems that utilize a heat transfer fluid shall be inspected annually, unless inspections are required on a more frequent basis by the solar energy system manufacturer.

G.5.5 Hard Water

G.5.5.1 Softening and Treatment

Where water has hardness equal to or exceeding 10 gr/gal (171 mg/L) measured as total calcium carbonate equivalents, the water supply line to water heating equipment and the circuit of boilers shall be softened or treated to prevent accumulation of lime scale and consequent reduction in energy efficiency.

G.5.6 Drain Water Heat Exchangers

Drain water heat exchangers shall comply with IAPMO PS-92. The heat exchanger shall be accessible.

G.6 METHOD OF CALCULATING WATER SAVINGS

G.6.1 Water Savings Calculation

G.6.1.1 Purpose

The purpose of this appendix is to provide a means of estimating the water savings when installing plumbing and fixture fittings that use less water than the maximum required by Energy Policy Act of 1992 and 2005 and the plumbing code.

G.6.1.2 Calculation of Water Savings

Table G.6.1.2(1) and Table G.6.1.2(2) can be used to establish a water use baseline in calculating the amount of water saved as a result of using plumbing fixtures and fixture fittings that use less water than the required maximum. Water use is determined by the following equation:

Water use = (Flow rate or Consumption) x (Duration) x (Occupants) x (Daily uses)

TABLE G.6.1.2(1)
WATER USE BASELINE[5]

FIXTURE TYPE	MAXIMUM FLOW-RATE CONSUMPTION[2]	DURATION	ESTIMATED DAILY USES PER PERSON	OCCUPANTS[3,4]
Showerheads	2.5 gpm @ 80 psi	8 minutes	1	–
Private or Private Use Lavatory Faucets	2.2 gpm @ 60 psi	0.25 minutes	4	–
Residential Kitchen Faucets	2.2 gpm @ 60 psi	4 minutes	1	–
Wash Fountains	2.2 gpm / 20 [rim space (inches) @ 60 psi]	–	–	–
Lavatory Faucets in other than Residences, Apartments, and Private Bathrooms in Lodging Facilities (See Section 402.4.2)	0.5 gpm	0.25 minutes	46	–
Metering Faucets	0.25 gallons /cycle	0.25 minutes	3	–
Metering Faucets for Wash Fountains	0.25 gpm / 20 [rim space (inches) @ 60 psi]	0.25 minutes	–	–
Water Closets	1.6 gallons per flush	1 flush	1 male[1]	–
			3 female	
Urinals	1.0 gallons per flush	1 flush	2 male	–
Commercial Pre-Rinse Spray Valves	1.6 gpm @ 60 psi	-	-	-

For SI units: 1 gallon per minute = 0.06 L/s, 1 pound-force per square inch = 6.89 kPa, 1 gallon = 3.785 L
1. The daily use number shall be increased to three if urinals are not installed in the room.
2. The maximum flow rate or consumption is from the Energy Policy Act.
3. For residential occupancies, the number of occupants shall be based on two persons for the first bedroom, and one additional person for each additional bedroom.
4. For non-residential occupancies, refer to the plumbing code, for occupant load factors.
5. When determining calculations, assume one use per person for metering or self closing faucets.
Notes continued on next page.

Notes and instructions for Table G.6.1.2(2):

Table G.6.1.2(1) is an example of a calculator that can help estimate water savings in residential and nonresidential structures. The "Duration" of use and "Daily Uses" values that appear in the table are estimates only and based on previous studies. The first example shown below is for a commercial office building with 300 occupants, 150 females, and 150 males. The second example is for a 3 bedroom residential building. To obtain and use a working copy of this calculator, follow the download and use instructions below.

Instructions for download:

1. Go to the IAPMO web site at www.iapmogreen.org in order to download the water-savings calculator. The calculator is a Microsoft Office Excel file (1997 or later), your computer must be capable of running MS Excel.
2. Follow the instructions for downloading and running the file.

Instructions for use:

1. In the Baseline Case section, insert the number of total occupants, male occupants and female occupants that apply for the building in the "Occupants" column. Unless specific gender ratio values are provided, assume a 50/50 gender ratio.
2. Copy and paste these same values in the "Occupants" column of the Calculator section.
3. In the Calculator section only, insert the consumption values (flow rates in gpm or gallons per flush or per cycle) in the "Consumption" column.
4. Estimated water savings in terms of percent savings versus baseline values, gallons per day and gallons per year will be automatically calculated.

TABLE G.6.1.2(2)
WATER SAVINGS CALCULATOR
NON-RESIDENTIAL BUILDINGS
BASELINE CASE: CHANGE OCCUPANT VALUES TO REFLECT ANTICIPATED OCCUPANCY

FIXTURE TYPE	CONSUMPTION (gallons per minute)	DAILY USES	DURATION (minutes)	OCCUPANTS	DAILY WATER USES (gallons)
1.6 gpf toilet - male (gallons per flush)	1.6	1	1	150	240
1.6 gpf toilet - female (gallons per flush)	1.6	3	1	150	720
1.0 gpf urinal - male (gallons per flush)	1	2	1	150	300
Commercial Lavatory Faucet - 0.5 gpm (gallons per minute)	0.5	3	0.25	300	113
Kitchen sink - 2.2 gpm (gallons per minute)	2.2	1	0.25	300	165
Showerhead - 2.5 gpm (gallons per minute)	2.5	0.1	8	300	600
				Total Daily Volume	2,138
				Annual Work Days	260
				Total Annual Usage	555,750

SI units: 1 gallon per minute = 0.06 L/s, 1 gallon = 3.785 L
Calculator: To determine estimated savings, insert occupant values (same as Baseline) and consumption values based on fixtures and fixture fittings installed.

FIXTURE TYPE	CONSUMPTION (gallons per minute)	DAILY USES	DURATION (minutes)	OCCUPANTS	DAILY WATER USES (gallons)
1.6 gpf toilet - male (gallons per flush)	1.28	1	1	150	192
1.6 gpf toilet - female (gallons per flush)	1.28	3	1	150	576
1.0 gpf urinal - male (gallons per flush)	0.5	2	1	150	150
Commercial Lavatory Faucet - 0.5 gpm (gallons per minute)	0.5	3	0.25	300	113
Kitchen sink - 2.2 gpm (gallons per minute)	2.2	1	0.25	300	165
Showerhead - 2.5 gpm (gallons per minute)	2.5	0.1	8	300	600
			Total Daily Volume		1,796
			Annual Work Days		260
			Total Annual Usage		466,830
			Annual Savings		88,920
			% Reduction		16.0 percent

For SI units: 1 gallon per minute = 0.06 L/s, 1 gallon = 3.785 L

Notes:
(1) Consumption values shown as underlined reflect the maximum consumption values associated with the provisions called out in the IAPMO Green Plumbing and Mechanical Code Supplement.
(2) If metering faucets are used, insert the flow rate of the faucet in the "Consumption" column and insert the cycle time in the "Duration" column (assume 1 cycle per use).

RESIDENTIAL 3 BEDROOM STRUCTURE

BASELINE CASE: CHANGE OCCUPANT VALUES BASED ON NUMBER OF BEDROOMS (EXAMPLE SHOWN IS FOR 3 BEDROOMS)

FIXTURE TYPE	CONSUMPTION (gallons per minute)	DAILY USES	DURATION	OCCUPANTS	DAILY WATER USES (gallons)
1.6 gpf toilets	1.6	5	1	4	32
Lavatory Faucet - 2.2 gpm	2.2	8	0.25	4	18
Kitchen sink - 2.2 gpm	2.2	6	0.25	4	13
Showerhead - 2.5 gpm	2.5	0.75	8	4	60
			Total Daily Volume		123
			Annual Usage		44,822

For SI units: 1 gallon per minute = 0.06 L/s, 1 gallon = 3.785 L
Calculator: To determine estimated savings, insert occupant values (same as Baseline) and consumption, consumption values based on fixtures and fixture fittings installed.

FIXTURE TYPE	CONSUMPTION (gallons per minute)	DAILY USES	DURATION	OCCUPANTS	DAILY WATER USES (gallons)
1.6 gpf toilet - male	1.28	5	1	4	26
Lavatory Faucet - 1.5 gpm	1.5	8	0.25	4	12
Kitchen sink - 2.2 gpm	2.2	6	0.25	4	13
Showerhead - 2.5 gpm	2.5	0.75	8	4	60
			Total Daily Volume		111
			Annual Usage		40,442
			Annual Savings		4,380
			% Reduction		-9.8 percent

For SI units: 1 gallon per minute = 0.06 L/s, 1 gallon = 3.785 L

Calculator: To determine estimated savings, insert occupant values (same as Baseline) and consumption, consumption values based on fixtures and fixture fittings installed.

G.7 POTABLE RAIN WATER CATCHMENT SYSTEMS

G.7.1 General

G.7.1.1 Scope

The provisions of this appendix shall apply to the installation, construction, alteration, and repair of potable rainwater catchment systems.

G.7.1.2 System Design

Potable rainwater catchment systems complying with this appendix shall be designed by a person registered, licensed, or deemed competent by the Authority Having Jurisdiction to perform potable rainwater catchment system design work.

G.7.1.3 Permit

It shall be unlawful for any person to construct, install, or alter, or cause to be constructed, installed, or altered any potable rainwater catchment systems in a building or on a premise without first obtaining a permit to do such work from the Authority Having Jurisdiction.

G.7.1.3.1 Plumbing Plan Submission

No permit for any rainwater catchment system requiring a permit shall be issued until complete plumbing plans, with appropriate data satisfactory to the Authority Having Jurisdiction, have been submitted and approved. No changes or connections shall be made to either the rainfall catchment or the potable water system within any site containing a rainwater catchment water system without approval by the Authority Having Jurisdiction.

G.7.1.3.2 System Changes

No changes or connections shall be made to either the rainwater catchment system or the potable water system within any site containing a rainwater catchment system requiring a permit without approval by the Authority Having Jurisdiction.

G.7.1.4 Product and Material Approval

G.7.1.4.1 Component Identification

System components shall be properly identified as to the manufacturer.

G.7.1.4.2 Plumbing Materials and Systems

Pipe, pipe fittings, traps, fixtures, material, and devices used in a potable rainwater system shall be listed or labeled (third-party certified) by a listing agency (accredited conformity assessment body) and shall conform to approved applicable recognized standards referenced in this supplement and the plumbing code, and shall be free from defects. Unless otherwise provided for in this supplement, all materials, fixtures, or devices used or entering into the construction of plumbing systems, or parts thereof, shall be submitted to the Authority Having Jurisdiction for approval.

G.7.1.5 Maintenance and Inspection

Potable rainwater catchment systems and components shall be inspected and maintained in accordance with Section G.7.1.5.1 through Section G.7.1.5.3.

G.7.1.5.1 Frequency

Potable rainwater catchment systems and components shall be inspected and maintained in accordance with Table G.7.1.5.1 unless more frequent inspection and maintenance is required by the manufacturer.

G.7.1.5.2 Maintenance Log

A maintenance log for potable rainwater catchment systems shall be maintained by the property owner and be available for inspection. The property owner or designated appointee shall ensure that a record of testing, inspection and maintenance as required by Table G.7.1.5.1 is maintained in the log. The log will indicate the frequency of inspection, and maintenance for each system. A record of the required water quality tests shall be retained for not less than 2 years.

G.7.1.5.3 Maintenance Responsibility

The required maintenance and inspection of potable rainwater catchment systems shall be the responsibility of the property owner, unless otherwise required by the Authority Having Jurisdiction.

G.7.1.6 Operation and Maintenance Manual

An operation and maintenance manual for potable rainwater catchment systems shall be supplied to the building owner by the system designer. The operating and maintenance manual shall include the following:

(1) Detailed diagram of the entire system and the location of all system components.

(2) Instructions on operating and maintaining the system.

(3) Details on maintaining the required water quality as determined by the Authority Having Jurisdiction.

(4) Details on deactivating the system for maintenance, repair, or other purposes.

(5) Applicable testing, inspection and maintenance frequencies as required by Table G.7.1.5.1.

(6) A method of contacting the manufacturer(s).

TABLE G.7.1.5.1
MINIMUM POTABLE RAINWATER CATCHMENT SYSTEM
TESTING, INSPECTION, AND MAINTENANCE FREQUENCY

Description	Minimum Frequency
Inspect and clean filters and screens, and replace (if necessary)	Every 3 months
Inspect and verify that disinfection, filters and water quality treatment devices and systems are operational. Perform any water quality tests as required by the Authority Having Jurisdiction.	In accordance with the manufacturer's instructions, and the Authority Having Jurisdiction
Perform applicable water quality tests to verify compliance with Section G.7.4.2	Every 3 months
Perform a water quality test for E. Coli, Total Coliform, and Heterotrophic bacteria. For a system where 25 different people consume water from the system over a 60 day period, a water quality test for cryptosporidium shall also be performed.	After initial installation and every 12 months thereafter, or as directed by the Authority Having Jurisdiction
Inspect and clear debris from rainwater gutters, downspouts, and roof washers	Every 6 months
Inspect and clear debris from roof or other aboveground rainwater collection surface	Every 6 months
Remove tree branches and vegetation overhanging roof or other aboveground rainwater collection surface	As needed
Inspect pumps and verify operation	After initial installation and every 12 months thereafter
Inspect valves and verify operation	After initial installation and every 12 months thereafter
Inspect pressure tanks and verify operation	After initial installation and every 12 months thereafter
Clear debris and inspect storage tanks, locking devices, and verify operation	After initial installation and every 12 months thereafter
Inspect caution labels and marking	After initial installation and every 12 months thereafter

G.7.1.7 Minimum Water Quality Requirements

The minimum water quality for all potable rainwater catchment systems shall meet the applicable water quality requirements as determined by the Authority Having Jurisdiction. In the absence of water quality requirements, the guidelines EPA/625/R-04/108 contains recommended water reuse guidelines to assist regulatory agencies develop, revise, or expand alternate water source water quality standards.

G.7.1.8 Material Compatibility

In addition to the requirements of this appendix, potable rainwater catchment systems shall be constructed of materials that are compatible with the type of pipe and fitting materials and water conditions in the system.

G.7.1.9 System Controls

Controls for pumps, valves, and other devices that contain mercury that come in contact with the water supply shall not be permitted.

G.7.2 Connection

G.7.2.1 General

No water piping supplied by a potable rainwater catchment system shall be connected to any other source of supply without the approval of the Authority Having Jurisdiction, Health Department or other department having jurisdiction.

G.7.2.2 Connections to Public or Private Potable Water Systems

Potable rainwater catchment systems shall have no direct connection to any public or private potable water supply or alternate water source system. Potable water from a public or private potable water system is permitted to be used as makeup water to the rainwater storage tank provided the public or private potable water supply connection is protected by an airgap or reduced-pressure principle backflow preventer in accordance with the plumbing code.

G.7.2.3 Backflow Prevention

The potable rainwater catchment system shall be protected against backflow in accordance with the plumbing code.

G.7.3 Potable Rainfall Catchment System Materials

G.7.3.1 Collections Surfaces

The collection surface for potable applications shall be constructed of a hard, impervious material and shall be approved for potable water use. Roof coatings, paints, and liners shall comply with NSF Protocol P151.

G.7.3.1.1 Prohibited

Roof paints and coatings with lead, chromium, or zinc shall not be permitted. Wood roofing material and lead flashing shall not be permitted.

G.7.3.2 Rainwater Catchment System Drainage Materials

Materials used in rainwater catchment drainage systems, including gutters, downspouts, conductors, and leaders shall be in accordance with the requirements of the plumbing code for storm drainage.

G.7.3.3 Storage Tanks

Rainwater storage shall be in accordance with Section G.7.4.4.

G.7.3.4 Water Supply and Distribution Materials

Potable rainwater supply and distribution materials shall be in accordance with the requirements of the plumbing code for potable water supply and distribution systems.

G.7.4 Design and Installation

G.7.4.1 Collection Surfaces

Rainwater shall be collected from roof or other cleanable aboveground surfaces specifically designed for rainwater catchment. Rainwater catchment system shall not collect rainwater from:

(1) Vehicular parking surfaces.
(2) Surface water runoff.
(3) Bodies of standing water.

G.7.4.1.1 Prohibited Discharges

Overflows, condensate, and bleed-off pipes from roof-mounted equipment and appliances shall not discharge onto roof surfaces that are intended to collect rainwater.

G.7.4.2 Minimum Water Quality

Upon initial system startup, the quality of the water for the intended applications shall be verified at the point(s) of use, as determined by the Authority Having Jurisdiction. In the absence of water quality requirements determined by the Authority Having Jurisdiction, the minimum water quality shall comply with the following limits:

Escherichia coli (fecal coliform):	99.9% reduction
Protozoan Cysts:	99.99% reduction
Viruses:	99.99% reduction
Turbidity:	<0.3 NTU

Normal system maintenance will require system testing every 3 months. System shall comply with the following standards:

| Escherichia coli (fecal coliform): | 99.9% reduction |
| Turbidity: | <0.3 NTU |

a. Upon failure of the fecal coliform test, system shall be re-commissioned involving cleaning, and retesting in accordance with section G.7.4.2.

b. One sample shall be analyzed for applications serving up to 1,000 persons. When the treated water shall serve 1,000-2,500 persons two (2) samples shall be analyzed and for 2,501-3,300 persons three (3) samples shall be analyzed.

G.7.4.2.1 Filtration Devices

Potable water filters shall comply with NSF 53 and shall be installed in accordance with manufacturer's instructions.

G.7.4.2.2 Disinfection Devices

Chlorination, ozone, and ultraviolet or other disinfection methods approved by an Authority Having Jurisdiction, or the product is listed and certified according to a microbiological reduction performance standard for drinking water shall be used to treat harvested rainwater to meet the required water quality permitted. The disinfection devices and systems shall be installed in accordance with the manufacturer's installation instructions and the conditions of listing. Disinfection devices and systems shall be located downstream of the water storage tank.

G.7.4.3 Overhanging Tree Branches and Vegetation

Tree branches and vegetation shall not be located over the roof or other aboveground rainwater collection surface. Where existing tree branch and vegetation growth extends over the rainwater collection surface, it shall be removed as required in Section G.7.1.5.

G.7.4.4 Rainwater Storage Tanks

Rainwater storage tanks shall be installed in accordance with Section G.7.4.4.1 through Section G.7.4.4.7.

G.7.4.4.1 Location

Rainwater storage tanks shall be constructed of solid, durable materials not subject to excessive corrosion or decay and shall be watertight. Storage tanks shall be approved by the Authority Having Jurisdiction for potable water applications, provided such tanks comply with approved applicable standards.

G.7.4.4.2 Drainage and Overflow

Rainwater storage tanks shall be permitted to be installed above or below grade.

G.7.4.4.2.1 Overflow Outlet Size

Above grade storage tanks shall be of an opaque material, approved for aboveground use in direct sunlight, or shall be shielded from direct sunlight. Tanks shall be installed in an accessible location to allow for inspection and cleaning. The tank shall be installed on a foundation or platform that is constructed to accommodate all loads in accordance with the building code.

G.7.4.4.2.2 Below Grade

Rainwater storage tanks installed below grade shall be structurally designed to withstand all anticipated earth or other loads. Holding tank covers shall be capable of supporting an earth load of not less than 300 pounds per square foot (lb/ft2) (1465 kg/m2) when the tank is designed for underground installation. Below grade rainwater tanks installed underground shall be provided with manholes. The manhole opening shall be a minimum diameter of 20 inches (508 mm) and located not less than 4 inches (102 mm) above the surrounding grade. The surrounding grade shall be sloped away from the manhole. Underground tanks shall be ballasted, anchored, or otherwise secured, to prevent the tank from floating out of the ground when empty. The combined weight of the tank and hold down system should meet or exceed the buoyancy force of the tank.

G.7.4.4.3 Drainage and Overflow

Rainwater storage tanks shall be provided with a means of draining and cleaning. The overflow drain shall not be equipped with a shutoff valve. The overflow outlet shall discharge as required by the plumbing code for storm drainage systems. Where discharging to the storm drainage system, the overflow drain shall be protected from backflow of the storm drainage system by a backwater valve or other approved method.

G.7.4.4.3.1 Overflow Outlet Size

The overflow outlet shall be sized to accommodate the flow of the rainwater entering the tank and not less than the aggregate cross-sectional area of the inflow pipes.

G.7.4.4.4 Opening and Access Protection

G.7.4.4.4.1 Animals and Insects

Rainwater tank openings to the atmosphere shall be protected to prevent the entrance of insects, birds, or rodents into the tank.

G.7.4.4.4.2 Human Access

Rainwater tank access openings exceeding 12 inches (305 mm) in diameter shall be secured to prevent tampering and unintended entry by either a lockable device or other approved method.

G.7.4.4.4.3 Exposure to Sunlight

Rainwater tank openings shall not be exposed to direct sunlight.

G.7.4.4.5 Inlets

A device or arrangement of fittings shall be installed at the inlet of the tank to prevent rainwater from disturbing sediment as it enters the tank.

G.7.4.4.6 Primary Tank Outlets

The primary tank outlet shall be located not less than 4 inches (102 mm) above the bottom of the tank, or shall be provided with floating inlet to draw water from the cistern just below the water surface.

G.7.4.4.7 Storage Tank Venting

Where venting by means of drainage or overflow piping is not provided or is considered insufficient, a vent shall be installed on each tank. The vent shall extend from the top of the tank and terminate a minimum of 6 inches (152 mm) above grade and shall be a minimum of 1 1/2" (38 mm) in diameter. The vent terminal shall be directed downward and covered with a 3/32 inch (2.4 mm) mesh screen to prevent the entry of vermin and insects.

G.7.4.5 Pumps

Pumps serving rainwater catchment systems shall be listed for potable water use. Pumps supplying water to water closets, urinals, and trap primers shall be capable of delivering not less than 15 pounds-force per square inch (psi) (103 kPa) residual pressure at the highest and most remote outlet served. Where the water pressure in the rainwater supply system within the building exceeds 80 psi (552 kPa), a pressure reducing valve reducing the pressure to 80 psi (552 kPa) or less to water outlets in the building shall be installed in accordance with the plumbing code.

G.7.4.6 Roof Drains

Primary and secondary roof drains, conductors, leaders, overflows, and gutters shall be designed and installed as required by the plumbing code.

G.7.4.7 Water Quality Devices and Equipment

Devices and equipment used to treat rainwater to maintain the minimum water quality requirements determined by the Authority Having Jurisdiction shall be listed or labeled (third-party certified) by a listing agency (accredited conformity assessment body) and approved for the intended application.

G.7.4.7.1 Filtration and Disinfection Systems

Filtration and disinfection systems shall be located after the water storage tank. Where a chlorination system is installed, it shall be installed upstream of filtration systems. Where ultraviolet disinfection system is installed, a filter not greater than 5 microns (5 μm) shall be installed upstream of the disinfection system.

G.7.4.8 Freeze Protection

Tanks and piping installed in locations subject to freezing shall be provided with an adequate means of freeze protection.

G.7.4.9 Roof Washer or Pre-Filtration System

Collected rainwater shall pass through a roof washer or pre-filtration system before the water enters the rainwater storage tank. Roof washer systems shall comply with Section G.7.4.9.1 through Section G.7.4.9.4.

G.7.4.9.1 Size

The roof washer shall be sized to direct a sufficient volume of rainwater containing debris that has accumulated on the collection surface away from the storage tank. The ARCSA/ASPE rainwater catchment design and installation standard contains additional guidance on acceptable methods of sizing roof washers.

G.7.4.9.2 Debris Screen

The inlet to the roof washer shall be provided with a debris screen or other approved means that protects the roof washer from the intrusion of debris and vermin. Where the debris screen is installed, the debris screen shall be corrosion resistant and shall have openings no larger than 1/2 of an inch (12.7 mm).

G.7.4.9.3 Drain Discharge

Water drained from the roof washer or pre-filter shall be diverted away from the storage tank and discharged to a disposal area that does not cause property damage or erosion. Roof washer drainage shall not drain over a public way.

G.7.4.9.4 Automatic Drain

Roof washing systems shall be provided with an automatic means of self draining between rain events.

G.7.4.10 Filtration and Disinfection Systems

Filtration and disinfection systems shall be located after the water storage tank. Where a chlorination system is installed, it shall be installed upstream of filtration systems. Where ultraviolet disinfection system is installed, a filter not greater than 5 microns (5 μm) shall be installed upstream of the disinfection system.

G.7.4.11 Roof Gutters

Gutters shall maintain a minimum slope and be sized in accordance with the plumbing code.

G.7.4.12 Drains, Conductors, and Leaders

The design and size of rainwater drains, conductors, and leaders shall be in accordance with the plumbing code.

G.7.4.13 Size of Potable Water Piping

Potable rainwater system distribution piping shall be sized in accordance with the plumbing code for sizing potable water piping.

G.7.5 Cleaning

G.7.5.1 General

The interior surfaces of tanks and equipment shall be clean before they are put into service.

G.7.6 Supply System Inspection and Test

Rainwater catchment systems shall be inspected and tested in accordance with the applicable provisions of the plumbing code for testing of potable water and storm drainage systems. Storage tanks shall be filled with water to the overflow opening for a period of 24 hours and during inspection or by other means as approved by the Authority Having Jurisdiction. All seams and joints shall be exposed during inspection and checked for water tightness.

Appendix H

Combined Building Drains and Building Sewers

H.1 General

a. Where combined building drains and building sewers are required by the Authority Having Jurisdiction, the connections between the drainage systems shall comply with Sections 13.4.3 and 13.4.4. There shall be no sanitary drain connections downstream from the connection of the storm water piping.

b. Combined building drains and building sewers shall be sized based on the combined flow of the storm drain piping and sanitary drain piping.

c. Combined building drains and building sewers shall be sized to flow 1/2 full.

d. The gallons per minute (GPM) of flow for roof drainage shall be the square feet of roof area times the inches per hour of rainfall times 0.01039.

e. The gallons per minute (GPM) of sanitary drainage shall be determined from Table H-1. The GPM shall be interpolated between the values in the Table. Table H-1 is based on the maximum DFU allowances in Table 11.5.1A with the drains flowing 1/2 full.

f. The size of the combined building drain and sewer shall be based on Table H-2 or H-3, depending on the roughness of the drain piping.

Table H-1
Drainage Fixture Units (DFU) to Gallons per Minute (GPM)

DFU	GPM
36	22
42	31
50	44
180	47
216	67
250	71
390	86
480	105
575	126
700	153
840	184
1000	219
1400	306
1600	350
1920	420
2300	503
2500	547
2900	634
3500	765
3900	853
4200	918
4600	1006
5600	1224
6700	1465
7000	1530
8300	1815
10,000	2186
12,000	2624

Table H-2
GPM for COMBINED BUILDING DRAINS & BUILDING SEWERS

Pipe Size	Slope for Fairly Smooth Pipe (n = 0.012)			
	1/16in/ft	1/8in/ft	1/4in/ft	1/2in/ft
3"		22	31	44
4"	33	47	67	94
5"	61	86	121	171
6"	98	140	197	279
8"	212	300	424	600
10"	384	544	769	1087
12"	625	884	1250	1768
15"	1133	1603	2267	3206

Table H-3
GPM for COMBINED BUILDING DRAINS & BUILDING SEWERS

Pipe Size	Slope for Fairly Rough Pipe (n = 0.015)			
	1/16in/ft	1/8in/ft	1/4in/ft	1/2in/ft
3"		18	25	35
4"	27	38	54	76
5"	49	69	97	137
6"	79	112	158	223
8"	170	240	340	480
10"	308	435	615	870
12"	500	707	1000	1415
15"	907	1282	1814	2565

Appendix I

Fixture Unit Value
Curves for Water Closets

I.1 DRAINAGE FIXTURE UNITS — WATER CLOSETS

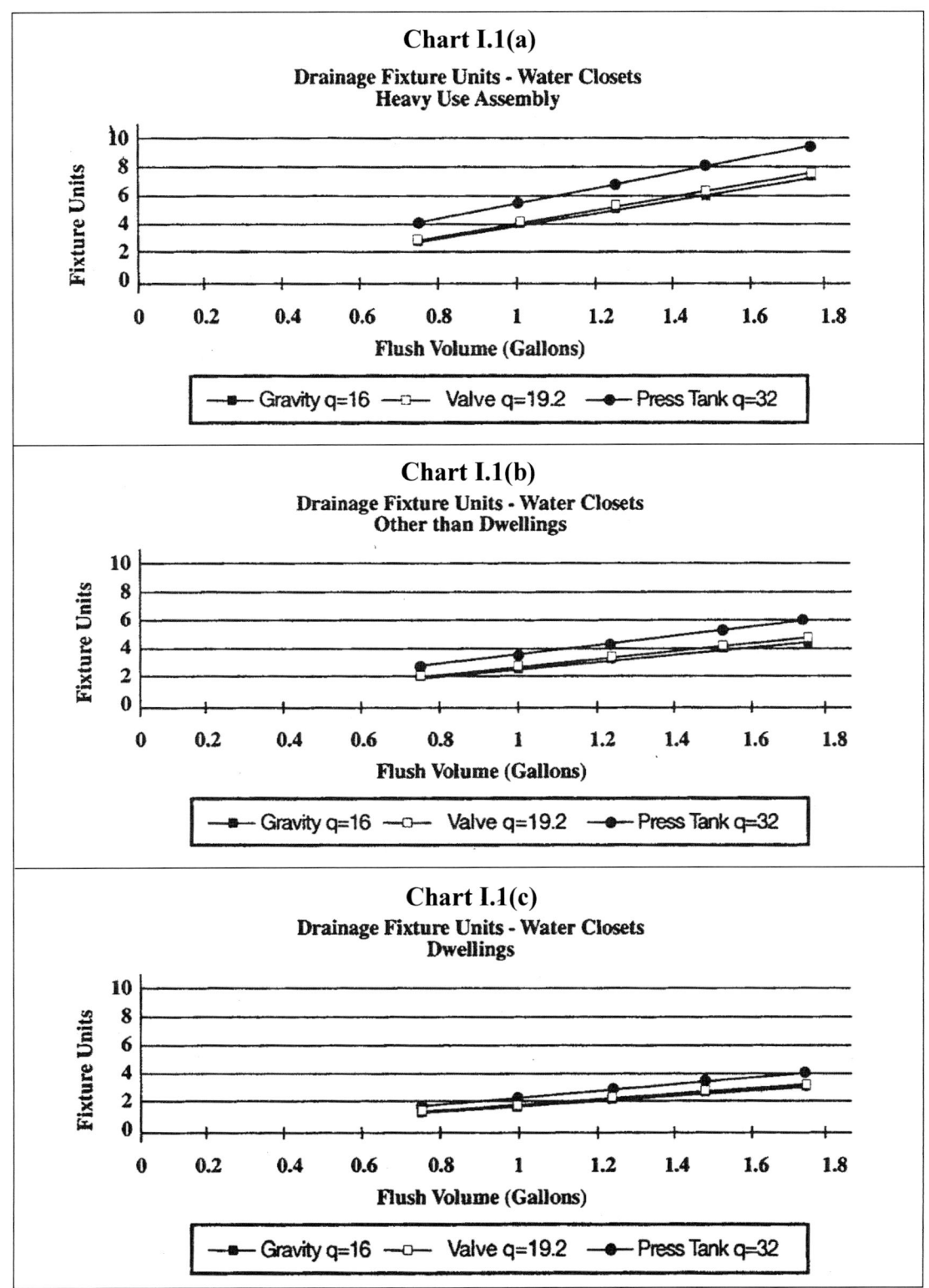

©1994 Stevens Institute of Technology

I.2 WATER SUPPLY FIXTURE UNITS — WATER CLOSETS

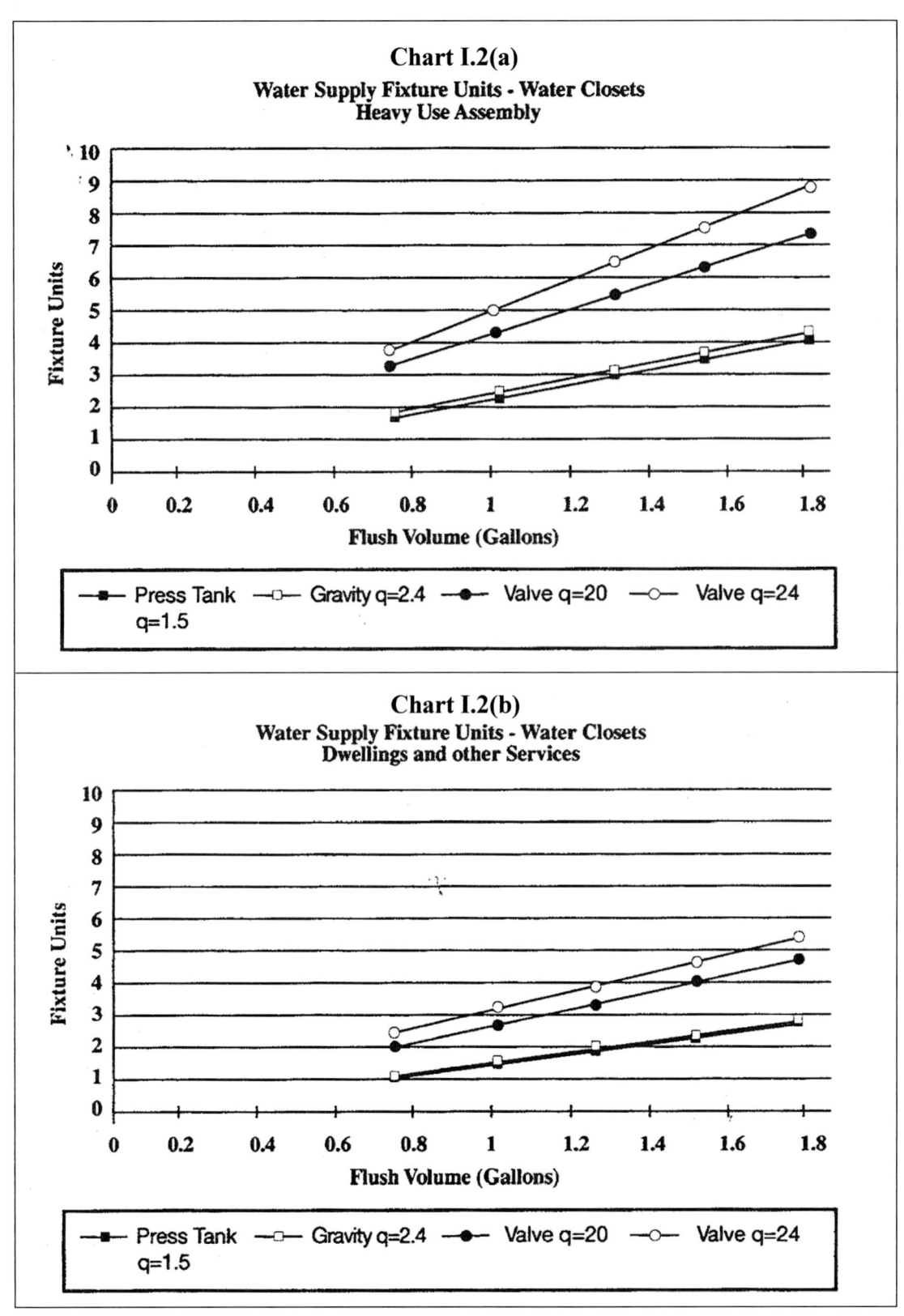

©1994 Stevens Institute of Technology

Blank Page

Appendix K

Flow In Sloping Drains

Flow in Sloping Drains

Tables K-1 and K-2 list flow rates in gallons per minute and velocities in feet per second for various size drains at various slopes. Table K-l is based on fairly rough pipe. Table K-2 is based on smooth pipe.

The minimum flow velocity to achieve scouring in horizontal sanitary drain lines is two (2) feet per second.

For this reason, based on Table K-2 for smooth pipe, drains that are 2 inches and smaller must be sloped at not less than 1/4 inch per foot. Drains that are 3" size and larger can be sloped at 1/8 inch per foot.

Even at 1/4" per foot slope, the uniform velocity in drains that are 1-1/4", 1-1/2", and 2" size is less than 2 feet per second. Either the slope should be increased or the length of such drains should be kept to a minimum so that the entrance velocity will provide scouring for the short distance involved.

Tables K-1 and K-2 are based on the Manning Formula for 1/2 full pipe. For full flow, multiply the flow by 2.00 and the velocity by 1.00. For 1/4 full flow, multiply the flow by 0.274 and the velocity by 0.701. For 3/4 full flow, multiply the flow by 1.82 and the velocity by 1.13.

In Table K-l, which is based on fairly rough pipe with "n" = 0.015, for smoother pipe, multiply the flow and velocity by 0.015 and divide by the "n" value of the smoother pipe.

Horizontal sanitary drain pipes are sized to be one-half (1/2) full under design loads. Horizontal storm drains are sized to run full under design loads.

Table 11.5.1A Building Drains and Sewers and the horizontal piping in Table 11.5.1B "Horizontal Fixture Branches and Stacks" is based on Table K-2 for smooth pipe.

Table 13.6.2 "Size of Horizontal Storm Drains" is based on Table K-1 for fairly rough pipe. The design flow of the drains is based on the pipes flowing full.

Table K-1
APPROXIMATE FLOW RATES AND VELOCITIES IN SLOPING DRAINS
(FOR FAIRLY ROUGH PIPE n = 0.015)
Flowing Half Full

Pipe Size (inches)	1/16 in./ft. slope		1/8 in./ft. slope		1/4 in./ft. slope		1/2 in./ft. slope	
	Flow gpm	Vel. fps	Flow gpm	Vel. fps	Flow gpm	Vel. fps	Flow gpm	Vel. fps
1.250					2.40	1.25	3.40	1.77
1.375					3.10	1.35	4.38	1.91
1.500					3.90	1.41	5.52	1.99
1.625					4.83	1.50	6.83	2.12
2					8.41	1.73	11.9	2.44
3			17.5	1.60	24.8	2.26	35.0	3.20
4	26.7	1.36	37.7	1.92	53.4	2.72	75.5	3.84
5	48.4	1.58	68.4	2.23	96.8	3.16	137	4.47
6	78.7	1.79	111	2.53	157	3.57	223	5.05
8	169	2.17	240	3.06	339	4.33	479	6.13
10	307	2.51	435	3.55	615	5.02	869	7.10
12	500	2.84	707	4.01	999	5.67	1413	8.02
15	906	3.29	1281	4.66	1812	6.59	2563	9.32

Table K-2
APPROXIMATE FLOW RATES AND VELOCITIES IN SLOPING DRAINS
(FOR SMOOTH PIPE n = 0.011)
Flowing Half Full

Pipe Size (inches)	1/16 in./ft. slope		1/8 in./ft. slope		1/4 in./ft. slope		1/2 in./ft. slope	
	Flow gpm	Vel. fps	Flow gpm	Vel. fps	Flow gpm	Vel. fps	Flow gpm	Vel. fps
1.250					3.27	1.71	4.63	2.42
1.375					4.22	1.84	5.97	2.60
1.500					5.32	1.92	7.53	2.72
1.625					6.86	2.04	9.32	2.89
2					11.5	2.35	16.2	3.33
3			23.9	2.18	33.8	3.08	47.8	4.36
4	36.4	1.85	51.5	2.62	72.8	3.71	103	5.24
5	66.0	2.15	93.3	3.05	132	4.31	187	6.09
6	107	2.44	152	3.44	215	4.87	303	6.89
8	231	2.95	327	4.18	462	5.91	654	8.36
10	419	3.42	593	4.84	838	6.84	1185	9.68
12	681	3.87	964	5.47	1363	7.73	1927	10.94
15	1236	4.49	1747	6.35	2471	8.99	3495	12.71

Appendix L

An Acceptable Brazing Procedure for General Plumbing

The following is extracted and edited with permission from Chapter VIII of The Copper Tube Handbook published by the Copper Development Association. This brazing procedure is acceptable for general plumbing work. Refer to NFPA 99 for brazing medical gas piping.

Introduction

Strong, leak-tight brazed connections for copper tube may be made by brazing with filler metals which melt at temperatures in the range between 1100 F and 1500 F, as listed in Table 12. Brazing filler metals are sometimes referred to as hard solders or silver solders. These confusing terms should be avoided.

The temperature at which a filler metal starts to melt on heating is the solidus temperature; the liquidus temperature is the higher temperature at which the filler metal is completely melted. The liquidus temperature is the minimum temperature at which brazing will take place.

The difference between solidus and liquidus is the melting range and may be of importance when selecting a filler metal. It indicates the width of the working range for the alloy and the speed with which the alloy will become fully solid after brazing. Filler metals with narrow ranges, with or without silver, solidify more quickly and, therefore, require more careful application of heat. The melting ranges of common brazing metals are shown in Figure 8a.

Brazing Filler Metals

Brazing filler metals suitable for joining copper tube are of two classes:
  (1) alloys that contain phosphorus (the BCuP series) and
  (2) alloys containing a high silver content (the BAg series)

The two classes differ in their melting, fluxing and flowing characteristics and this should be considered in selection of a filler metal. (See Table 12.) For joining copper tube, any of these filler metals will provide the necessary strength when used with standard solder-type fittings or commercially available short-cup brazing fittings.

Fluxes

The fluxes used for brazing copper joints are different in composition from soldering fluxes. The two types cannot be used interchangeably.

Brazing fluxes are water based, whereas most soldering fluxes are petroleum based. Similar to soldering fluxes, brazing fluxes dissolve and remove residual oxides from the metal surface, protect the metal from re-oxidation during heating and promote wetting of the surfaces to be joined by the brazing filler metal.

Fluxes also provide the craftsman with an indication of temperature. If the outside of the fitting and the heat-affected area of the tube are covered with flux (in addition to the end of the tube and the cup), oxidation will be prevented and the appearance of the joint will be greatly improved.

The fluxes best suited for brazing copper and copper alloy tube should meet AWS Classification FB3-A or FB3-C as listed in Table 4.1 of the AWS Brazing Handbook. Figure 9 illustrates the need for brazing flux with different types of copper and copper-alloy tube, fittings and filler metals when brazing.

Assembling

Assemble the joint by inserting the tube into the socket hard against the stop and turn if possible. The assembly should be firmly supported so that it will remain in alignment during the brazing operation.

Applying Heat and Brazing

Step one: Apply heat to the parts to be joined, preferably with an oxy-fuel flame. Air-fuel is sometimes used on smaller sizes. A neutral flame should be used. Heat the tube first, beginning about one inch from the edge of the fitting, sweeping the flame around the tube in short strokes at right angles to the axis of the tube.

It is very important that the flame be in motion continuously and not remain on any one point long enough to damage the tube. The flux may be used as a guide as to how long to heat the tube; continue heating the tube until the flux becomes quiet and transparent like clear water. The behavior of flux during the brazing cycle is described in Figure 8b.

Step two: Switch the flame to the fitting at the base of the cup. Heat uniformly, sweeping the flame from the fitting to the tube until the flux on the fitting becomes quiet. Avoid excessive heating of cast fittings.

Step three: When the flux appears liquid and transparent on both the tube and fitting, start sweeping the flame back and forth along the axis of the joint to maintain heat on the parts to be joined, especially toward the base of the cup of the fitting. The flame must be kept moving to avoid melting the tube or fitting.

Step four: Apply the brazing filler metal at a point where the tube enters the socket of the fitting. When the proper temperature is reached, the filler metal will flow readily into the space between the tube and fitting socket, drawn in by the natural force of capillary action.

Keep the flame away from the filler metal itself as it is fed into the joint. The temperature of the tube and fitting at the joint should be high enough to melt the filler metal.

Keep both the fitting and tube heated by moving the flame back and forth from one to the other as the filler metal is drawn into the joint.

When the joint is properly made, a continuous fillet of filler metal will be visible completely around the joint. Stop feeding as soon as you see that fillet. Table 11 is a guide to estimating how much filler metal will be consumed.

For 1-inch tube and larger it may be difficult to bring the whole joint up to heat at one time. It frequently will be found desirable to use a multiple-orifice torch tip to maintain a proper temperature over large areas. A mild preheating of the whole fitting is recommended for larger sizes. Heating then can proceed as outlined in the steps above.

Horizontal and Vertical Joints

When brazing horizontal joints, it is preferable to first apply the filler metal at the bottom, then the two sides, and finally the top, making sure the operations overlap. On vertical joints it is immaterial where the start is made. If the opening of the socket is pointing down, care should be taken to avoid overheating the tube, as this may cause the brazing filler metal to run down the outside of the tube. If this happens, take the heat away and allow the filler metal to set. Then reheat the cup of the fitting to draw up the filler metal.

Removing Residue

After the brazed joint has cooled, the flux residue should be removed with a clean cloth, brush, or swab using warm water. Remove all flux residue to avoid the risk of the hardened flux temporarily retaining pressure and masking an imperfectly brazed joint. Wrought fittings may be cooled more readily than cast fittings, but all fittings should be allowed to cool naturally before wetting.

General Hints and Suggestions

If the filler metal fails to flow or has a tendency to ball up, it indicates oxidation on the metal surfaces or insufficient heat on the parts to be joined. If the tube or fitting start to oxidize during heating there is too little flux. If the filler metal does not enter the joint and tends to flow over the outside of either member of the joint, it indicates that one member is overheated or the other is underheated.

Testing

Test all completed assemblies for joint integrity. Follow the testing procedure prescribed by applicable codes governing the intended service.

Table 12 Filler Metals for Brazing								
AWS Classification[1]	Principal Elements						Temperature F	
	Silver	Phosphorus	Zinc	Cadmium	Tin	Copper	Solidus	Liquidus
BCup-2	–	7.0-7.5	–	–	–	Remainder	1310	1460
BCup-3	4.8-5.2	5.8-6.2	–	–	–	Remainder	1190	1495
BCup-4	5.8-6.2	7.0-7.5	–	–	–	Remainder	1190	1325
BCup-5	14.5-15.5	4.8-5.2	–	–	–	Remainder	1190	1475
BAg-1[2]	44-46	–	14-18	23-25[2]	–	14-16	1125	1145
BAg-2[2]	34-36	–	19-23	17-19[2]	–	25-27	1125	1295
BAg-5	44-46	–	23-27	–	–	29-31	1225	1370
BAg-7	55-57	–	15-19	–	4.5-5.5	21-23	1145	1205

1- ANSI/AWS A5.8 Specification for Filler Metals for Brazing
2- WARNING: BAg1 and BAg2 contain cadmium. Heating when brazing can produce highly toxic fumes.
CAUTION: Avoid breathing fumes - use adequate ventilation. Refer to ANSI/ASC Z49.1 *Safety in Welding and Cutting*.

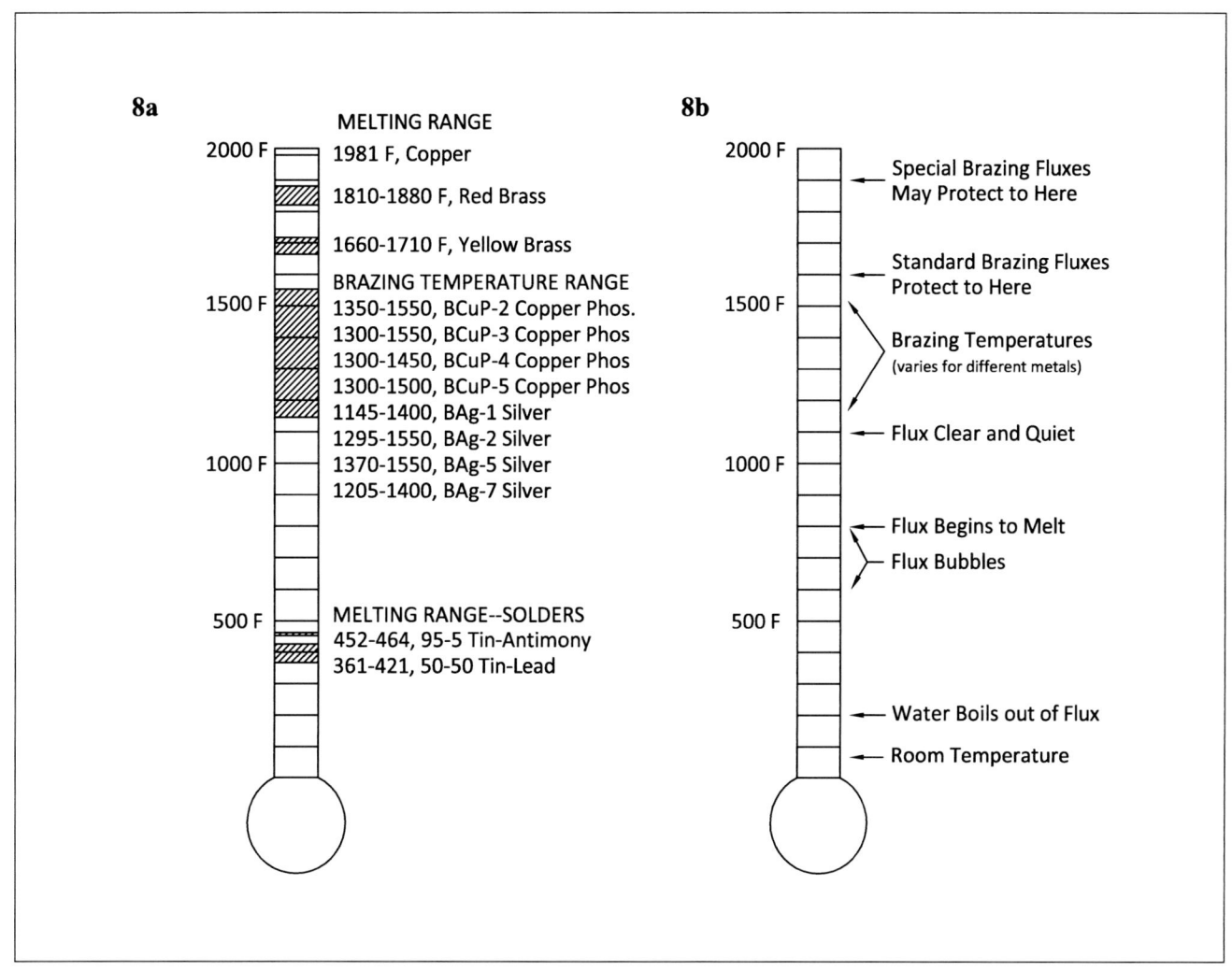

Figure 8
MELTING TEMPERATURE RANGES FOR COPPER AND COPPER ALLOYS, BRAZING FILLER METALS, FLUX AND SOLDERS

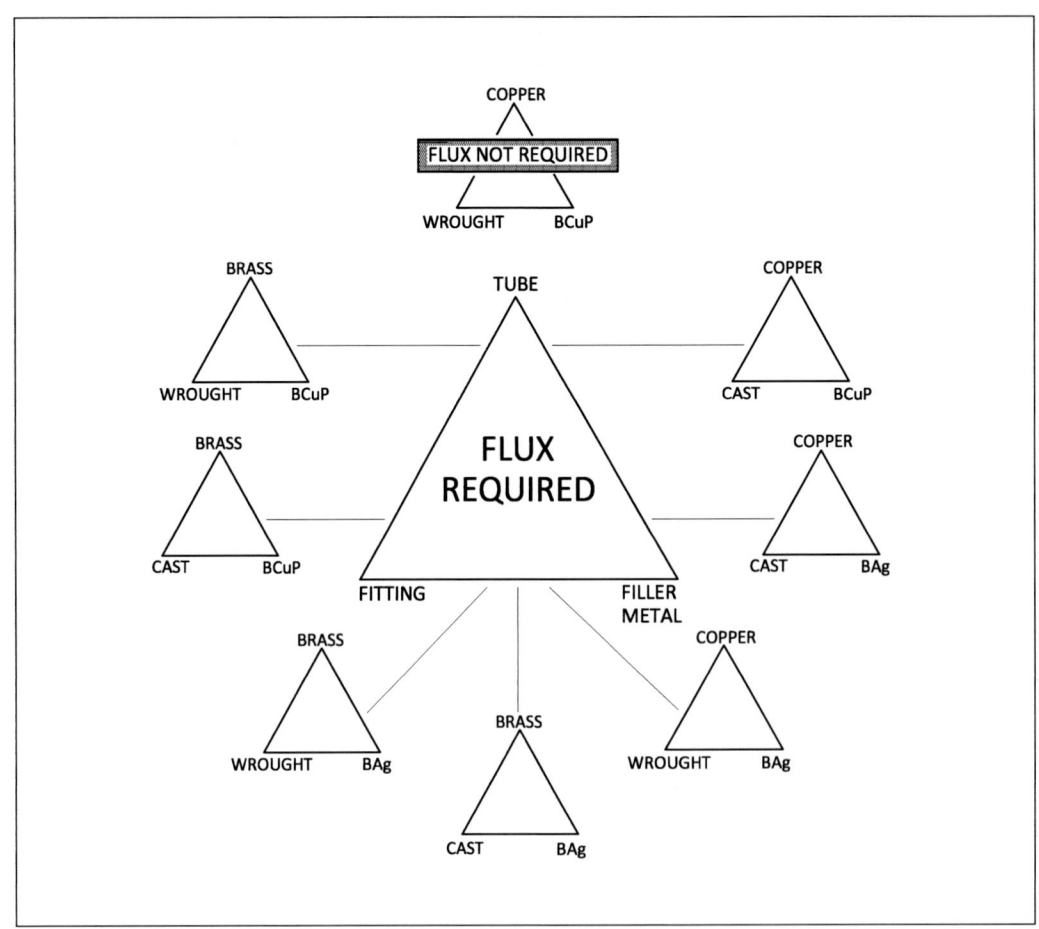

Figure 9
BRAZING FLUX RECOMMENDATIONS

Table 11
Typical Brazing Filler Metal Consumption

Tube, nominal or standard size, inches	Filler Metal Length, inches				Average weight per 100 joints, pounds[1]
	1/16 inch wire	1/8 inch x 0.050 inch rod	3/32 inch wire	1/8 inch wire	
1/4	1/2	1/4	1/4	1/8	.04
3/8	5/8	3/8	3/8	1/4	.06
1/2	1-1/8	5/8	1/2	3/8	.10
5/8	1-5/8	7/8	5/8	1/2	.15
3/4	2-1/4	1-1/8	1	5/8	.21
1	3-1/2	1-3/4	1-5/8	7/8	.32
1-1/4	4-1/2	2-1/4	2	1-1/4	.42
1-1/2	–	3	2-5/8	1-1/2	.56
2	–	4-3/4	4-3/8	2-1/2	.90
2-1/2	–	6-1/2	5-7/8	3-3/8	1.22
3	–	8-5/8	7-7/8	4-1/2	1.64
3-1/2	–	11-1/2	10-1/2	5-7/8	2.18
4	–	14-7/8	13-1/2	7-5/8	2.81
5	–	22-5/8	20-1/2	11-5/8	4.30
6	–	31-1/2	28-1/2	16	5.97
8	–	53-1/2	48-1/2	27-3/8	10.20
10	–	67-1/4	61	34-1/4	12.77
12	–	90-1/2	82	46-1/8	17.20

Footnote 1

The amount of filler material indicated is based on an average two-thirds penetration of the cup and with no provision for a fillet. For estimating purposes, actual consumption may be two to three times the amounts indicated in this table, depending on the size of the joints, method of application and level of workmanship.

NOTE:
1090 inches of 1/16 inch wire = 1 pound
534 inches of 1/8 inch x .050-inch wire = 1 pound
484 inches of 3/32 inch wire = 1 pound
268 inches of 1/8 inch wire = 1 pound

Appendix M

Converting Water Supply Fixture Units (WSFU) to Gallons Per Minute Flow (GPM)

INTRODUCTION

The water supply fixture unit (WSFU) values in this Appendix are interpolated between the values listed in Table 10.14.2.B to the degree that the increments between the WSFU listings are close enough that they do not produce GPM differences that will be significant in sizing the water supply distribution piping.

WSFU	GPM FLUSH TANKS	GPM FLUSH VALVES	WSFU	GPM FLUSH TANKS	GPM FLUSH VALVES	WSFU	GPM FLUSH TANKS	GPM FLUSH VALVES
			50	29.0	51.0	100	44.0	68.0
1	1.0		51	29.4	51.4	102	44.5	68.6
2	2.0		52	29.8	51.8	104	45.0	69.2
3	3.0		53	30.2	52.2	106	45.5	69.8
4	4.0		54	30.6	52.6	108	46.0	70.4
5	4.5	22.0	55	31.0	53.0	110	46.5	71.0
6	5.0	23.0	56	31.4	53.4	112	47.0	71.6
7	6.0	24.0	57	31.8	53.8	114	47.5	72.2
8	7.0	25.0	58	32.2	54.2	116	48.0	72.8
9	7.5	26.0	59	32.6	54.6	118	48.5	73.4
10	8.0	27.0	60	33.0	55.0	120	49.0	74.0
11	8.5	28.0	61	33.3	55.4	122	49.4	74.4
12	9.0	29.0	62	33.6	55.7	124	49.8	74.8
13	10.0	29.5	63	33.9	56.1	126	50.2	75.2
14	10.5	30.0	64	34.2	56.4	128	50.6	75.6
15	11.0	31.0	65	34.5	56.8	130	51.0	76.0
16	12.0	32.0	66	34.8	57.1	132	51.4	76.4
17	12.5	33.0	67	35.1	57.5	134	51.8	76.8
18	13.0	33.5	68	35.4	57.8	136	52.2	77.2
19	13.5	34.0	69	35.7	58.2	138	52.6	77.6
20	14.0	35.0	70	36.0	58.5	140	53.0	78.0
21	14.6	35.6	71	36.3	58.9	142	53.4	78.5
22	15.2	36.2	72	36.6	59.2	144	53.8	79.0
23	15.8	36.8	73	36.9	59.6	146	53.8	79.5
24	16.4	37.4	74	37.2	59.9	148	54.6	80.0
25	17.0	38.0	75	37.5	60.3	150	55.0	80.5
26	17.6	38.6	76	37.8	60.6	152	55.4	81.0
27	18.2	39.2	77	38.1	61.0	154	55.8	81.5
28	18.8	39.8	78	38.4	61.3	156	56.2	82.0
29	19.4	40.4	79	38.7	61.7	158	56.6	82.5
30	20.0	41.0	80	39.0	62.0	160	57.0	83.0
31	20.5	41.6	81	39.3	62.3	162	57.4	83.4
32	21.0	42.2	82	39.5	62.6	164	57.8	83.8
33	21.5	42.8	83	39.8	62.9	166	58.2	84.2
34	22.0	43.4	84	40.0	63.2	168	58.6	84.6
35	22.5	44.0	85	40.3	63.5	170	59.0	85.0
36	23.0	44.6	86	40.5	63.8	172	59.4	85.4
37	23.5	45.2	87	40.8	64.1	174	59.8	85.8
38	24.0	45.8	88	41.0	64.4	176	60.2	86.2
39	24.5	46.4	89	41.3	64.7	178	60.6	86.6
40	25.0	47.0	90	41.5	65.0	180	61.0	87.0
41	25.4	47.4	91	41.8	65.3	182	61.4	87.4
42	25.8	47.8	92	42.0	65.6	184	61.8	87.8
43	26.2	48.2	93	42.3	65.9	186	62.2	88.2
44	26.6	48.6	94	42.5	66.2	188	62.6	88.6
45	27.0	49.0	95	42.8	66.5	190	63.0	89.0
46	27.4	49.4	96	43.0	66.8	192	63.4	89.4
47	27.8	49.8	97	43.3	67.1	194	63.8	89.8
48	28.2	50.2	98	43.5	67.4	196	64.2	90.2
49	28.6	50.6	99	43.8	67.7	198	64.6	90.6

WSFU	GPM FLUSH TANKS	GPM FLUSH VALVES	WSFU	GPM FLUSH TANKS	GPM FLUSH VALVES	WSFU	GPM FLUSH TANKS	GPM FLUSH VALVES
200	65.0	91.0	450	115.0	132.5	900	194.0	196.0
205	66.0	91.8	455	116.0	133.3	910	195.6	197.4
210	67.0	92.6	460	117.0	134.0	920	197.2	198.8
215	68.0	93.4	465	118.0	134.8	930	198.8	200.2
220	69.0	94.2	470	119.0	135.5	940	200.4	201.6
225	70.0	95.0	475	120.0	136.3	950	202.0	203.0
230	71.0	95.8	480	121.0	137.0	960	203.6	204.4
235	72.0	96.6	485	122.0	137.8	970	205.2	205.8
240	73.0	97.4	490	123.0	138.5	980	206.8	207.2
245	74.0	98.2	495	124.0	139.3	990	208.4	208.6
250	75.0	100.0	500	125.0	140.0	1000	210.0	210.0
255	76.0	101.0	510	126.8	141.4	1020	212.4	212.4
260	77.0	102.0	520	128.6	142.8	1040	214.8	214.8
265	78.0	103.0	530	130.4	144.2	1060	217.2	217.2
270	79.0	104.0	540	132.2	145.6	1080	219.6	219.6
275	80.0	105.0	550	134.0	147.0	1100	222.0	222.0
280	81.0	106.0	560	135.8	148.4	1120	224.4	224.4
285	82.0	107.0	570	137.6	149.8	1140	226.8	226.8
290	83.0	108.0	580	139.4	151.2	1160	229.2	229.2
295	84.0	109.0	590	141.2	152.6	1180	231.6	231.6
300	85.0	110.0	600	143.0	152.6	1200	234.0	234.0
305	86.0	110.8	610	144.8	155.4	1220	236.4	236.4
310	87.0	111.5	620	146.6	156.8	1240	238.8	238.8
315	88.0	112.3	630	148.4	158.2	1260	241.2	241.2
320	89.0	113.0	640	150.2	159.6	1280	243.6	243.6
325	90.0	113.8	650	152.0	161.0	1300	246.0	246.0
330	91.0	114.5	660	153.8	162.4	1320	248.4	248.4
335	92.0	115.3	670	155.6	163.8	1340	250.8	250.8
340	93.0	116.0	680	157.4	165.2	1360	253.2	253.2
345	94.0	116.8	690	159.2	166.6	1380	255.6	255.6
350	95.0	117.5	700	161.0	168.0	1400	258.0	258.0
355	96.0	118.3	710	162.8	169.4	1420	260.4	260.4
360	97.0	119.0	720	164.6	170.8	1440	262.8	262.8
365	98.0	119.8	730	166.4	172.2	1460	265.2	265.2
370	99.0	120.5	740	168.2	173.6	1480	267.6	267.6
375	100.0	121.3	750	170.0	175.0	1500	270.0	270.0
380	101.0	122.0	760	171.6	176.4	1520	272.4	272.4
385	102.0	122.8	770	173.2	177.8	1540	274.8	274.8
390	103.0	123.5	780	174.8	179.2	1560	277.2	277.2
395	104.0	124.3	790	176.4	180.6	1580	279.6	279.6
400	105.0	125.0	800	178.0	182.0	1600	282.0	282.0
405	106.0	125.8	810	179.6	183.4	1620	284.4	284.4
410	107.0	126.5	820	181.2	184.8	1640	286.8	286.8
415	108.0	127.3	830	182.8	186.2	1660	289.2	289.2
420	109.0	128.0	840	184.4	187.6	1680	291.6	291.6
425	110.0	128.8	850	186.0	189.0	1700	294.0	294.0
430	111.0	129.5	860	187.6	190.4	1720	296.4	296.4
435	112.0	130.3	870	189.2	191.8	1740	298.8	298.8
440	113.0	131.0	880	190.8	193.2	1760	301.0	301.0
445	114.0	131.8	890	192.4	194.6	1780	303.0	303.0

WSFU	GPM FLUSH TANKS	GPM FLUSH VALVES	WSFU	GPM FLUSH TANKS	GPM FLUSH VALVES	WSFU	GPM FLUSH TANKS	GPM FLUSH VALVES
1800	305.0	305.0	2800	413.0	413.0	3800	507.0	507.0
1820	307.0	307.0	2820	415.2	415.2	3820	508.8	508.8
1840	309.0	309.0	2840	417.4	417.4	3840	510.6	510.6
1860	311.0	311.0	2860	419.6	419.6	3860	512.4	512.4
1880	313.0	313.0	2880	421.8	421.8	3880	514.2	514.2
1900	315.0	315.0	2900	424.0	424.0	3900	516.0	516.0
1920	317.0	317.0	2920	426.2	426.2	3920	517.8	517.8
1940	319.0	319.0	2940	428.4	428.4	3940	519.6	519.6
1960	321.0	321.0	2960	430.6	430.6	3960	521.4	521.4
1980	323.0	323.0	2980	432.8	432.8	3980	523.2	523.2
2000	325.0	325.0	3000	435.0	435.0	4000	525.0	525.0
2020	327.2	327.2	3020	436.8	436.8	4020	526.5	526.5
2040	329.4	329.4	3040	438.6	438.6	4040	528.0	528.0
2060	331.6	331.6	3060	440.4	440.4	4060	529.5	529.5
2080	333.8	333.8	3080	442.2	442.2	4080	531.0	531.0
2100	336.0	336.0	3100	444.0	444.0	4100	532.5	532.5
2120	338.2	338.2	3120	445.8	445.8	4120	534.0	534.0
2140	340.4	340.4	3140	447.6	447.6	4140	535.5	535.5
2160	342.6	342.6	3160	449.4	449.4	4160	537.0	537.0
2180	344.8	344.8	3180	451.2	451.2	4180	538.5	538.5
2200	347.0	347.0	3200	453.0	453.0	4200	540.0	540.0
2220	349.2	349.2	3220	454.8	454.8	4220	541.5	541.5
2240	351.4	351.4	3240	456.6	456.6	4240	543.0	543.0
2260	353.6	353.6	3260	458.4	458.4	4260	544.5	544.5
2280	355.8	355.8	3280	460.2	460.2	4280	546.0	546.0
2300	358.0	358.0	3300	462.0	462.0	4300	547.5	547.5
2320	360.2	360.2	3320	463.8	463.8	4320	549.0	549.0
2340	262.4	262.4	3340	465.6	465.6	4340	550.5	550.5
2360	364.6	364.6	3360	467.4	467.4	4360	552.0	552.0
2380	366.8	366.8	3380	469.2	469.2	4380	553.5	553.5
2400	369.0	369.0	3400	471.0	471.0	4400	555.0	555.0
2420	371.2	371.2	3420	472.8	472.8	4420	556.5	556.5
2440	373.4	373.4	3440	474.6	474.6	4440	558.0	558.0
2460	375.6	375.6	3460	476.4	476.4	4460	559.5	559.5
2480	377.8	377.8	3480	478.2	478.2	4480	561.0	561.0
2500	380.0	380.0	3500	480.0	480.0	4500	562.5	562.5
2520	382.2	382.2	3520	481.8	481.8	4520	564.0	564.0
2540	384.4	384.4	3540	483.6	483.6	4540	565.5	565.5
2560	386.6	386.6	3560	485.4	485.4	4560	567.0	567.0
2580	388.8	388.8	3580	487.2	487.2	4580	568.5	568.5
2600	391.0	391.0	3600	489.0	489.0	4600	570.0	570.0
2620	393.2	393.2	3620	490.8	490.8	4620	571.5	571.5
2640	395.4	395.4	3640	492.6	492.6	4640	573.0	573.0
2660	397.6	397.6	3660	494.4	494.4	4660	574.5	574.5
2680	399.8	399.8	3680	496.2	496.2	4680	576.0	576.0
2700	402.0	402.0	3700	498.0	498.0	4700	577.5	577.5
2720	404.2	404.2	3720	499.8	499.8	4720	579.0	579.0
2740	406.4	406.4	3740	501.6	501.6	4740	580.5	580.5
2760	408.6	408.6	3760	503.4	503.4	4760	582.0	582.0
2780	410.8	410.8	3780	505.2	505.2	4780	583.5	583.5

WSFU	GPM FLUSH TANKS	GPM FLUSH VALVES	WSFU	GPM FLUSH TANKS	GPM FLUSH VALVES	WSFU	GPM FLUSH TANKS	GPM FLUSH VALVES
4800	585.0	585.0	5800	640.0	640.0	6800	690.0	690.0
4820	586.5	586.5	5820	641.0	641.0	6820	691.0	691.0
4840	588.0	588.0	5840	642.0	642.0	6840	692.0	692.0
4860	589.5	589.5	5860	643.0	643.0	6860	693.0	693.0
4880	591.0	591.0	5880	644.0	644.0	6880	694.0	694.0
4900	592.5	592.5	5900	645.0	645.0	6900	695.0	695.0
4920	594.0	594.0	5920	646.0	646.0	6920	696.0	696.0
4940	595.5	595.5	5940	647.0	647.0	6940	697.0	697.0
4960	597.0	597.0	5960	648.0	648.0	6960	698.0	698.0
4980	598.5	598.5	5980	649.0	649.0	6980	699.0	699.0
5000	600.0	600.0	6000	650.0	650.0	7000	700.0	700.0
5020	601.0	601.0	6020	651.0	651.0	7020	700.6	700.6
5040	602.0	602.0	6040	652.0	652.0	7040	701.2	701.2
5060	603.0	603.0	6060	653.0	653.0	7060	701.8	701.8
5080	604.0	604.0	6080	654.0	654.0	7080	702.4	702.4
5100	605.0	605.0	6100	655.0	655.0	7100	703.0	703.0
5120	606.0	606.0	6120	656.0	656.0	7120	703.6	703.6
5140	607.0	607.0	6140	657.0	657.0	7140	704.2	704.2
5160	608.0	608.0	6160	658.0	658.0	7160	704.8	704.8
5180	609.0	609.0	6180	659.0	659.0	7180	705.4	705.4
5200	610.0	610.0	6200	660.0	660.0	7200	706.0	706.0
5220	611.0	611.0	6220	661.0	661.0	7220	706.6	706.6
5240	612.0	612.0	6240	662.0	662.0	7240	707.2	707.2
5260	613.0	613.0	6260	663.0	663.0	7260	707.8	707.8
5280	614.0	614.0	6280	664.0	664.0	7280	708.4	708.4
5300	615.0	615.0	6300	665.0	665.0	7300	709.0	709.0
5320	616.0	616.0	6320	666.0	666.0	7320	709.6	709.6
5340	617.0	617.0	6340	667.0	667.0	7340	710.2	710.2
5360	618.0	618.0	6360	668.0	668.0	7360	710.8	710.8
5380	619.0	619.0	6380	669.0	669.0	7380	711.4	711.4
5400	620.0	620.0	6400	670.0	670.0	7400	712.0	712.0
5420	621.0	621.0	6420	671.0	671.0	7420	712.6	712.6
5440	622.0	622.0	6440	672.0	672.0	7440	713.2	713.2
5460	623.0	623.0	6460	673.0	673.0	7460	713.8	713.8
5480	624.0	624.0	6480	674.0	674.0	7480	714.4	714.4
5500	625.0	625.0	6500	675.0	675.0	7500	715.0	715.0
5520	626.0	626.0	6520	676.0	676.0	7520	715.6	715.6
5540	627.0	627.0	6540	677.0	677.0	7540	716.2	716.2
5560	628.0	628.0	6560	678.0	678.0	7560	716.8	716.8
5580	629.0	629.0	6580	679.0	679.0	7580	717.4	717.4
5600	630.0	630.0	6600	680.0	680.0	7600	718.0	718.0
5620	631.0	631.0	6620	681.0	681.0	7620	718.6	718.6
5640	632.0	632.0	6640	682.0	682.0	7640	719.2	719.2
5660	633.0	633.0	6660	683.0	683.0	7660	719.8	719.8
5680	634.0	634.0	6680	684.0	684.0	7680	720.4	720.4
5700	635.0	635.0	6700	685.0	685.0	7700	721.0	721.0
5720	636.0	636.0	6720	686.0	686.0	7720	721.6	721.6
5740	637.0	637.0	6740	687.0	687.0	7740	722.2	722.2
5760	638.0	638.0	6760	688.0	688.0	7760	722.8	722.8
5780	639.0	639.0	6780	689.0	689.0	7780	723.4	723.4

WSFU	GPM FLUSH TANKS	GPM FLUSH VALVES	WSFU	GPM FLUSH TANKS	GPM FLUSH VALVES	WSFU	GPM FLUSH TANKS	GPM FLUSH VALVES
7800	724.0	724.0	8720	751.6	751.6	9640	779.2	779.2
7820	724.6	724.6	8740	752.2	752.2	9660	779.8	779.8
7840	725.2	725.2	8760	752.8	752.8	9680	780.4	780.4
7860	725.8	725.8	8780	753.4	753.4	9700	781.0	781.0
7880	726.4	726.4	8800	754.0	754.0	9720	781.6	781.6
7900	727.0	727.0	8820	754.6	754.6	9740	782.2	782.2
7920	727.6	727.6	8840	755.2	755.2	9760	782.8	782.8
7940	728.2	728.2	8860	755.8	755.8	9780	783.4	783.4
7960	728.8	728.8	8880	756.4	756.4	9800	784.0	784.0
7980	729.4	729.4	8900	757.0	757.0	9820	784.6	784.6
8000	730.0	730.0	8920	757.6	757.6	9840	785.2	785.2
8020	730.6	730.6	8940	758.2	758.2	9860	785.8	785.8
8040	731.2	731.2	8960	758.8	758.8	9880	786.4	786.4
8060	731.8	731.8	8980	759.4	759.4	9900	787.0	787.0
8080	732.4	732.4	9000	760.0	760.0	9920	787.6	787.6
8100	733.0	733.0	9020	760.6	760.6	9940	788.2	788.2
8120	733.6	733.6	9040	761.2	761.2	9960	788.8	788.8
8140	734.2	734.2	9060	761.8	761.8	9980	789.4	789.4
8160	734.8	734.8	9080	762.4	762.4	10,000	790.0	790.0
8180	735.4	735.4	9100	763.0	763.0			
8200	736.0	736.0	9120	763.6	763.6			
8220	736.6	736.6	9140	764.2	764.2			
8240	737.2	737.2	9160	764.8	764.8			
8260	737.8	737.8	9180	765.4	765.4			
8280	738.4	738.4	9200	766.0	766.0			
8300	739.0	739.0	9220	766.6	766.6			
8320	739.6	739.6	9240	767.2	767.2			
8340	740.2	740.2	9260	767.8	767.8			
8360	740.8	740.8	9280	768.4	768.4			
8380	741.4	741.4	9300	769.0	769.0			
8400	742.0	742.0	9320	769.6	769.6			
8420	742.6	742.6	9340	770.2	770.2			
8440	743.2	743.2	9360	770.8	770.8			
8460	743.8	743.8	9380	771.4	771.4			
8480	744.4	744.4	9400	772.0	772.0			
8500	745.0	745.0	9420	772.6	772.6			
8520	745.6	745.6	9440	773.2	773.2			
8540	746.2	746.2	9460	773.8	773.8			
8560	746.8	746.8	9480	774.4	774.4			
8580	747.4	747.4	9500	775.0	775.0			
8600	748.0	748.0	9520	775.6	775.6			
8620	748.6	748.6	9540	776.2	776.2			
8640	749.2	749.2	9560	776.8	776.8			
8660	749.8	749.8	9580	777.4	777.4			
8680	750.4	750.4	9600	778.0	778.0			
8700	751.0	751.0	9620	778.6	778.6			

Alphabetical Index

Accessible and Readily Accessible:
    Backflow preventers, 10.5.5
    Backwater valves, 5.5.2
    Building valves, 10.12.2
    Definition of, 1.2
    Expansion joints, 4.1.3
    Flexible water connectors, 10.13
    Flushometer valves, 7.19.5
    Individual fixture valves, 10.12.6
    In-line pressure balancing valves, 10.15.6
    Liquid waste treatment equipment, 6.1.6
    Manifold shutoff valves, 10.17.3
    Waste oil pumps, 6.3.2.e
    Water hammer arrestors, 10.14.7
    Water supply shutoff valves, 10.12.9
Adopting Agency:
    Definition of, 1.2
    Requirements of, Appendix F
Air Break:
    Definition of, 1.2
    Indirect waste pipes and receptors, 9.1.3
    Where permitted, 9.1.3
Air Break or Air Gap (drainage):
    Air conditioning equipment, 9.1.10
    Commercial dishwashing machines, 7.15.4
    Drinking fountains, 9.1.9
    Fixtures and appliances, 9.1.5
    Food display equipment, 2.25.4
    Food service equipment, 2.25.2
    Potable clear water wastes, 9.1.8
    Residential dishwashers, 7.15.2, 7.15.3
    Walk-in coolers and freezers, 9.1.6
    Water coolers, 9.1.9
Air Conditioning Condensate:
    Piping for, 9.4.3
Air Gap (drainage):
    Aspirators, 14.13
    Bedpan steamers, 14.10.2
    Definition of, 1.2
    Dishwasher air gap fitting, 7.15.2, 7.15.3
    Drinking water treatment units, 10.18.2
    Food prep and service equipment, 10.5.8.e
    Indirect waste pipes and receptors, 9.1.2
    Medical and sterile equipment, 9.1.7
    Sterilizers, 9.1.7, 14.10.2
    Swimming pools, 9.1.11
    Water heater relief valve discharge piping, 10.16.6
    Water treatment systems, 7.22
Air Gap (water supply):
    Automatic clothes washers, 7.13.1
    Bathtub filler spout, 7.8.4
    Bidets without flushing rims, 7.7.2
    Definition of, 1.2
    Dishwashers, 7.15.1
    Food prep and service equipment, 10.5.8
    Inlets to gravity water tanks, 10.8.5
    Minimum sizes, 10.5.2, Table 10.5.2
Alternate Materials and Methods:
    Requirements for, 3.12
Area Drain:
    Definition of, 1.2
Aspirator:
    Definition of, 1.2
Authority Having Jurisdiction:
    Definition of, 1.2
Automatic Compensating Valves:
    Definition of, 1.2
Backfilling:
    Bedding, 2.6.1
    Final backfill, 2.6.4
    For copper piping, 2.6.7
    For plastic piping, 2.6.6
    Initial backfill, 2.6.3
    Side-fill, 2.6.2, 2.6.3
    Supervision of, 2.6.9
    Trenching, 2.6.1
Backflow Preventers:
    Approval of, 10.5.4
    Definition of, 1.2
    Installation of, 10.5.5
    Required types, 10.5.3
    Standards for, Table 3.1.3-IX
    Testing and repair of, 10.5.6
Backflow Protection from:
    Baptisteries, 7.18.1
    Chemical dispensers, 10.5.13
    Cleaning units, portable, 10.5.13
    Decorative fountains, 7.18.1
    Dental pumping units, 10.5.13
    Fire protection systems, 10.5.9
    Food prep and service equipment, 10.5.8

Hose connections, 10.5.12
Irrigation systems, 10.5.10
Laboratory sink faucets, 10.5.13
Lawn sprinklers, 10.5.10
Ornamental pools, 7.18.1
Ornamental waterfalls, 7.18.1
Plumbing fixtures, 10.5.1
Special installations, 7.18.1
Sump pumps (water powered), 10.5.13
Swimming and wading pools, 7.18.1
Tanks and vats, 10.5.7
Water heat exchangers, 10.5.11
Water supply outlets, 10.5.1

Back Pressure Backflow:
Definition of, 1.2

Back Siphonage Backflow:
Definition of, 1.2

Backwater Valve:
Definition of, 1.2
Exception for installation, 5.5.1
Notice of installation, 5.5.3
Where required, 5.5.1

Bar Sinks:
For drinking water, 7.21.5.b
Requirements for, 7.11.2

Basic Design Circuit – BDC (for water):
Explanation of, B.9.3

Bathroom Groups:
Definition of, 1.2
DFUs for, Table 11.4.1
Stack venting of, 12.11
Wet venting of, 12.10
WSFUs for, Table 10.14.2A

Bathtubs:
Above food handling areas, 2.25.1
Requirements for, 7.8
Temperature control for, 10.15.6

Battery of Fixtures:
Definition of, 1.2

Bedpan Washers:
Local vents and stacks for, 14.9
Requirements for, 14.8
Vacuum breakers for, 14.7.b

Bidets:
Requirements for, 7.7

Branch Connections:
In drain stack offsets, 11.10

Branch Drain:
Definition of, 1.2

Branch Interval:
Definition of, 1.2

Building Classification:
Definition of, 1.2

Building Drain, Combined:
Definition of, 1.2

Building Drain, Sanitary:
Definition of, 1.2
Materials for, Table 3.5
Required size, Table 11.5.1A
Separation from water service, 10.6.1

Building Drain, Storm:
Definition of, 1.2
Materials for, Table 3.7
Required size, Table 13.6.2

Building Sewer, Combined:
Definition of, 1.2

Building Sewer, Sanitary:
Definition of, 1.2
Materials for, Table 3.5
Required size, Table 11.5.1A
Separation from water service, 10.6.1

Building Sewer, Storm:
Definition of, 1.2
Materials for, Table 3.7
Required size, Table 13.6.2

Building Subdrains:
Definition of, 1.2
Requirements for, 11.7.2
Venting, 11.7.2.d

Building Trap:
Definition of, 1.2
Where required, 5.3.4

Cesspool:
Definition of, 1.2

Chemical Waste:
Definition of, 1.2
Piping, 3.11
Treatment of, 9.4.1

Cleanouts:
At base of stacks, 5.4.5
At building traps, 5.3.4
At changes of direction, 5.4.3
Cleanout equivalent, 5.4.13
Clearances, 5.4.11
Direction of opening, 5.4.7
For floor drains, 5.4.14
For indirect waste piping, 9.2.4
For island sink vents, 12.18.4
For oil separators, 6.3.2.d
For sand interceptor outlets, 6.4.3
For storm drain traps, 13.3.1.c
For vent piping, 12.6.2.3
In building sewer piping, 5.4.2
In concealed piping, 5.4.4
Manholes, 5.4.10
Plugs and caps, 3.3.2

Prohibited connections to, 5.4.8
Required size, 5.4.9
Spacing, 5.4.1
Clear Water Waste:
Definition of, 1.2
Clinical Sinks:
Definition of, 1.2
Flushing devices for, 7.19.1
Local vents and stacks for, 14.9
Requirements for, 14.8.1
Clothes Washers:
Laundry sink receptor for, 9.3.1.c.2
Required compliance, 7.13.1
Standpipe receptor for, 9.3.4
Combination Fixture:
Definition of, 1.2
Combination Waste and Vent System:
Definition of, 1.2
Requirements for, 12.17
Combined Building Drains and Sewers:
Definitions of, 1.2
Requirements for, 13.4.3, 13.4.4
Common Laundry Room:
Definition of, 1.2
Floor drains required, 7.16.4
Condemned Equipment:
Not to be used, 2.15
Conductors (storm water):
Definition of, 1.2
Improper connections, 13.4.1
Continuous Waste Piping:
Definition of, 1.2
Conversions:
Metric to U.S. Units, Appendix C
Corrosive Fill:
Protection of pipes, 2.9.2
Corrosive Wastes:
Treatment of, 9.4.1
Critical Level:
Definition of, 1.2
Of backflow preventers, 10.5.5
Cross Connection:
Between water supplies, 10.4.4.b
Control and protection, 10.4.3
Definition of, 1.2
Cross Connection Control:
Program for, 10.4.3
Cutting and Notching:
Structural members, 2.9.3
Day Care Center:
Definition of, 1.2
Day Nursery:
Definition of, 1.2

Dead End:
Definition of, 1.2
Defective Plumbing:
Testing or inspection, 15.7
Definitions:
Code terms, Chapter 1
Demand (water):
Estimating procedure, B.5
Sizing distribution piping, 10.14.3
Dental Vacuum System Drainage:
Requirements for, 14.12
Developed Length (of piping):
Definition of, 1.2
DFU:
Definition of, 1.2
Dilution Tanks:
Requirements for, 6.6
Where required, 6.1.1
Directional Changes:
Of drainage piping, 2.3
Dishwashing Machines:
Requirements for, 7.15
Disinfection:
Of potable water piping, 10.9.2
Domestic Sewage:
Definition of, 1.2
Double Check Valve Assembly:
Application of, 10.5.3
Definition of, 1.2
Installation of, 10.5.5.d
Testing and maintenance of, 10.5.6
Drain Stack:
Definition of, 1.2
Sizing of, Table 11.5.1B
Sizing procedures, 11.5.2
Drainage System, Sanitary:
Branch drain piping, 11.5.3
Changes in direction, 2.3
Connections near stack bases, 11.9
Connections to stack offsets, 11.10
Diversity factors, 11.4.3
Drain stack sizing, 11.5.2
Effect of offsets, 11.6
Fittings and connections, 2.4
Fixture units (DFU), 11.4, Table 11.4.1
Pipe sizing, 11.5
Sewage pumping, 11.7
Slope of piping, 11.3.1
Suds pressure zones, 11.11
Drainage System, Storm Water:
Air conditioning condensate, 13.8
Areaway drains, 13.1.6
Backwater valves, 13.1.12

Combined bldg drain and sewer, 13.4.3
Controlled flow system, 13.9
Equivalent systems, 13.1.10.4
Foundation drains, 13.1.5
Parking and service garages, 13.1.8
Pipe sizing, 13.6
Piping materials, Table 3.7
Protection of leaders, 13.4.2
Roof drainage, primary, 13.1.10.1
Roof drainage, secondary, 13.1.10.2
Roof drains, 13.5
System sizing, Appendix A
Traps, 13.3
Vertical walls, 3.1.10.3
Where required, 13.1.1
Window well drains, 13.1.7

Drinking Water Facilities:
Compliance, 7.12.1
Indirect waste discharge, 9.1.9
Minimum number required, 7.12.2
Outdoor type, 7.12.4
Prohibited locations, 7.12.3

Drip Pans:
Construction, 10.15.9.2
Drainage from, 10.15.9.3
Where required, 10.15.9.1

DWV:
Definition of, 1.2

Effective Opening:
Definition of, 1.2

Ejector, Pump Type:
Definition of, 1.2

Elevator Pits:
Drainage of, 2.26.2
Requirements for, 2.26

Emergency Eyewash, Eye/Face Wash and Shower:
Requirements for, 7.24

Equivalent Length of Fittings and Valves:
Definition of, 1.2

Equivalent Length, Total:
Definition of, 1.2
With fittings and valves, B.9.7

Fixture Branch Drain:
Definition of, 1.2

Fixture Drain:
Definition of, 1.2

Fixture Traps:
General requirements, 5.3
Liquid seal, 5.3.2
Materials for, 3.2
Prohibited types, 5.3.5
Required size of, 5.2, Table 5.2

Seal maintenance, 5.3.6
Separate for each fixture, 5.1
With a grease interceptor, 6.2.3
With indirect waste piping, 9.2.3

Fixture Unit (Drainage-DFU):
Definition of, 1.2
Diversity factors, 11.4.3
For bathroom groups, Table 11.4.1
For individual fixtures, Table 11.4.1

Fixture Unit (Water Supply-WSFU):
Conversion to GPM, Appendix M
Conversion to GPM, Table 10.14.2B
Definition of, 1.2
For estimating GPM demand, B.5
For groups and fixtures, Table 10.14.2A

Flashing:
Materials for, 3.2
Roof drains, 13.5.4
Vent terminals, 12.4.2, 12.4.7
Vent through roof, 4.5

Flood Level Rim:
Definition of, 1.2

Floor Drains:
Above food handling areas, 2.25.1.g
Floor slope, 7.16.6
In areas with sterilizers, 14.10.3
Required locations, 7.16.4
Requirements for, 7.16
Size of, 7.16.3

Floor Flanges:
For water closets, 2.22, 3.3.4

Flow Pressure (water):
Definition of, 1.2

Flow Rates (drainage):
In sloping drains, Appendix K

Flow Rates (water):
Based on flow velocity, B.7.3
Based on friction pressure loss, B.9.8
For individual fixtures, 10.14.2

Flushing Devices:
Flush tanks, 7.19.3
Flushometer tanks, 7.19.4
Flushometer valves, 7.19.5

Food Handling Areas:
Area protection, 2.25.1
Exposed food products, 2.25.3
Food display equipment, 2.25.4
Sanitary floor sinks, 2.25.5
Service equipment and fixtures, 2.25.2
Vacuum condensate drainage, 2.25.6

Food Waste Disposers:
As indirect waste receptor, 9.3.1.c.1
Discharge to grease interceptors, 6.2.5

Discharge with sink and dishwasher, 7.15.3
If solids separator required, 6.2.5
Requirements for, 7.14
Vent washdown prohibited, 12.12.4
Footings:
Protection of, 2.17
Foundation Drains:
Definition of, 1.2
Installation of, 13.1.5
Pump discharge piping, 13.1.5.f
Pumps for, 13.1.5.e
Water-operated pumps for, 13.1.5.g
Freezing:
Foundation drain discharge pipe, 13.1.5.e
Protection of fixture traps, 5.3.3
Protection of piping, 2.16
Storm drain traps, 13.3.2
Frost Closure:
Vent terminals, 12.5
Garbage Can Washers:
Requirements for, 7.17
Grade:
Definition of, 1.2
Graywater:
Code coverage, 10.1.f
Definition of, 1.2
Permitted uses: 10.1.c, 10.1.d
Grease Interceptors:
Approval of, 6.1.4.1
Certification of, 6.1.4.2
Definition of, 1.2
Fats, oils, greases (FOG), 6.2.1.4
Gravity type, 6.2.1.3
Grease Removal Device (GRD), 6.2.1.2
Hydro-mechanical, 6.2.1.1, 6.2.2
Location of, 6.2.7
Prohibitions, 6.2.8, 6.2.9
Sizing, 6.2.10
Green Code Supplement:
Appendix G
Grinder Pumps:
Definition of, 1.2
Requirements for, 11.7.5
Ground Water:
Definition of, 1.2
Discharge of continuous flow, 13.1.11
Outdoor drinking fountains, 7.12.3
Protection of, Basic Principle No.22
Hangers and Supports:
Requirements for, Chapter 8

Harvested Rainwater:
Code coverage, 10.1.f
Definition of, 1.2
Permitted uses: 10.1.e
Health Care Facilities:
Aspirators, 14.13
Backflow prevention, 14.7
Bedpan washers, 14.8, 14.9
Central vacuum systems, 14.12
Clinical sinks, 14.8, 14.9
Ice storage locations, 14.6
Medical gas and vacuum piping, 14.3
Mental patient rooms, 14.5
Protrusions from walls, 14.4
Sterilizers, 14.10, 14.11
Water service, 14.2
Health Hazards:
Backflow preventers for, 10.5.3
Definition of, 1.2
Health and Safety Hazards:
Abatement of, 2.5
High Temperature Wastes:
Discharge of, 9.4.2
Horizontal Battery-Vented Drain:
Definition of, 1.2
Horizontal Branch Drain:
Definition of, 1.2
Sizing of, 11.5.3.b
Horizontal Fixture Branch:
Definition of, 1.2
Sizing of, 11.5.3.a
Horizontal Pipe:
Definition of, 1.2
Hot Water:
Definition of, 1.2
Plastic piping, 10.15.8
Temperature control, 10.15.6
Temperature maintenance, 10.15.2
Thermal expansion control, 10.15.7
Where required, 10.15.1
Hot Water Tanks:
Drainage from, 10.15.4
Minimum requirements for, 10.15.3
Pressure marking of, 10.15.5
Safety devices for, 3.3.10, 10.16
Storage tanks, 3.3.8
Hydraulic Jump:
At the base of drain stacks, Figure 11.9
Indirect Waste Connections:
Air breaks, 9.1.3
Air conditioning equipment, 9.1.10
Air gaps, 9.1.2
Bedpan steamers, 14.10.2

Definition of, 1.2
Drinking fountains, 9.1.9
Exposed food products, 2.25.3
Food handling equipment, 9.1.5
High temperature wastes, 9.4.2
Medical equipment, 9.1.7
Piping, 9.2
Potable clear-water waste, 9.1.8
Prevent contamination, Basic Principle 15
Requirements for, 9.1
Sterile equipment, 9.1.7
Sterilizers, 14.10.2
Swimming pools, 9.1.11
Walk-in coolers, freezers, 9.1.6
Waste receptors, 9.3
Water coolers, 9.1.9
Water heater relief valve
 discharge piping, 10.16.6
Where required, 9.1.4

Indirect Waste Pipe:
Definition of, 1.2
Requirements for, 9.2

Indirect Waste Receptors:
Requirements for, 9.3
Strainers or baskets for, 9.3.2

Individual Sewage Disposal Systems:
Absorption trenches, 16.9
Design of, 16.3
Distribution box, 16.7
General provisions, 16.1
Location of, 16.4
Percolation test, 16.5
Piping materials, 16.11
Sand filters, 16.12
Seepage pits, 16.8
Septic tanks, 16.6
Sewage flows, Table 16.3.7
Spacing of components, Table 16.4.3

Industrial Wastes:
Definition of, 1.2
Exclusion of, 2.10.2

Interceptor:
Definition of, 1.2

Invert:
Definition of, 1.2

Laundry Sinks:
Requirements for, 7.11.3

Lavatories:
Minimum number required, Table 7.21.1
Requirements for, 7.6
With integral overflow, 7.6.4

Laundries, Commercial:
Lint interceptors, 6.7.1

Lead-Free Plumbing:
Limit on lead content, 3.4.6

Leaders (rain water):
Definition of, 1.2
Protection of, 13.4.2

Liquid Waste Treatment Tanks:
Neutralizing and dilution, 6.6

Local Vents and Stacks:
For bedpan washers, 14.9
For clinical sinks, 14.9

Macerating Toilet Systems:
Definition of, 1.2
Requirements for, 11.7.6

May:
Definition of, 1.2

Medical Gas and Vacuum Systems:
Definitions of, 1.2
Installer qualifications, 14.3.2
NFPA 99 compliance, 14.3.1

Metric Units:
Metric to U.S. conventional, Appendix C

Mixed Water Temperature Control:
Animal washing, 10.15.6.m
Application of devices, 10.15.6.c
Bath/shower combinations, 10.15.6.e
Bathtubs, 10.15.6.h
Bidets, 10.15.6.i
Check valves, 10.15.6.d
Field adjustment of, 10.15.6.o
Hair/shampoo sprays, 10.15.6.k
Hand washing, 10.15.6.j
Hot water distribution, 10.15.6.b
In-line pressure balancing, 10.15.6.n
One-pipe tempered for lavatories, 10.15.6.g
One-pipe tempered for showers, 10.15.6.f
Showers, 10.15.6.e
TAFR flow reduction devices, 10.15.6.m
Whirlpool baths, 10.15.6.h

Mobile Home Parks:
Requirements for, Chapter 18

Multiple Dwelling:
Definition of, 1.2

NFPA 13D Fire Sprinkler Systems:
Multipurpose residential, 2.28, 10.20

Non-Health Hazards:
Backflow preventers for, 10.5.3
Definition of, 1.2

Non-Potable Water:
Definition of, 1.2
Identification of, 2.27, 10.21
Piping for, 10.21

Obstruction to Flow:
By fittings, devices, installation, 2.4.3

Offsets in Piping:
    Definition of, 1.2
    Drain stacks with offsets, 11.6.3
    Horizontal in drain stacks, 11.6.2
    Vertical in drain stacks, 11.6.1
Oil/Water Separators:
    Requirements for, 6.3
Pipe Joints:
    Bending copper, 4.2.19
    Between different materials, 4.3
    Brazed, copper, 4.2.8
    Burned (welded), lead, 4.2.10
    Caulked, cast iron soil, 4.2.1.1
    Caulked, cast iron water, 4.2.1.2
    Cement mortar, cast iron, 4.2.9
    Drain piping to certain fixtures, 4.4
    Ductile iron, 4.2.12
    Expansion, 4.2.16
    Flared, copper, 4.2.5
    Heat fusion, PE, 4.2.18
    Mechanical (flexible or slip), 4.2.11
    Mechanical, cast iron soil, 4.2.11.2
    Mechanical, cast iron water, 4.2.11.3
    Mechanical, clay, 4.2.11.4
    Mechanical, concrete, 4.2.11.5
    Mechanical, elastomeric sleeves, 4.2.11.6
    Mechanical, formed tee branch, 4.2.8.3
    Mechanical, pressed (crimped), 4.2.6
    Mechanical, stainless steel DWV, 4.2.11.1
    Plastic, 4.2.14
    Plastic DWV to others, 4.3.9
    Push-fit, 4.2.7
    Slip joints, 4.2.15
    Soldered, copper, 4.2.4
    Special for drain piping, 4.3.8
    Split couplings, 4.2.17
    Threaded, metal, 4.2.2
    Wiped, lead, 4.2.3
Pipe Sleeves:
    For fixtures above food areas, 2.25.1.e
    For water service pipes, 10.6.4
    Requirements for, 2.12
Piping:
    AC condensate, 3.9, 9.4.3
    Chemical waste, 3.11, 9.4.1
    Flexible water connectors, 10.13
    Foundation drainage, 3.8, Table 3.8
    Joints, 4.2
    Joints with different materials, 4.3
    Plastic water distribution, Table 3.4.3
    Plastic water service, Table 3.4.2
    Potable water, 3.4, Table 3.4
    Protection of, 2.9
    Sanitary waste & drain, 3.5, Table 3.5
    Standards for, Table 3.1.3
    Storm drainage, 3.7, Table 3.7
    Support spacing, 8.2, 8.3
    Support spacing, alternate, 8.10
    Valves, Table 3.1.3
    Vent, 3.6, Table 3.6
    Water hammer arresters, 10.14.7
Piping Materials:
    Air conditioning condensate drains, 3.9
    Approved, Table 3.1.3
    Chemical waste, 3.11
    Combustion condensate drains, 3.10
    Foundation drains, 3.8
    General requirements, 3.1.2
    Potable water, 3.4
    Sanitary drainage, 3.5
    Special materials, 3.2
    Storm drainage, 3.7
    Subsoil drains, 3.8
    Venting, 3.6
Plumbing:
    Definition of, 1.2
Plumbing Appliance:
    Definition of, 1.2
Plumbing Appurtenance:
    Definition of, 1.2
Plumbing Code:
    Definition of, 1.2
Plumbing Fixtures:
    Access for cleaning, 7.3.3
    Access for use, 7.21.3
    Air gaps for, 10.5.1, 10.5.2
    Approved, Table 3.1.3, Part V
    Area drains, 7.16
    Automatic clothes washers, 7.13
    Back-to-back, 2.3.3
    Bathtubs, 7.8
    Bidets, 7.7
    Clearances required, 7.3.2
    Combination bath/showers, 7.8.3
    Definition of, 1.2
    Dishwashing machines, 7.15
    Drainage fixture units (DFU), Table 11.4.1
    Drinking water, 7.12
    Emergency eye/face wash, 7.24
    Emergency eyewash, 7.24
    Emergency shower, 7.24
    Family and assisted-use, 7.21.9
    Floor drains, 7.16
    Floor sinks, 2.25.5
    Flushing devices, 7.19
    Food waste grinders, 7.14

For accessible use, 7.2
  For correctional institutions, 7.20
  For customers and employees, 7.21.7
  For detention centers, 7.20
  For special occupancies, 7.21.6
  Fractions of minimum required, 7.21.10
  Garbage can washers, 7.17
  In food service establishments, 7.21.8
  Individual pumps and ejectors, 11.7.11
  Installation of, 7.3
  Lavatories, 7.6
  Lavatory equivalents, 7.6.7
  Minimum number required, 7.21
  Non-water urinals, 7.5.6
  Occupant load for, 7.21.2
  Omission of, 7.21.5
  Prohibited urinals, 7.5.4
  Separate toilet facilities, 7.21.4
  Showers, 7.10
  Sinks, 7.11
  Spa and hot tub safety features, 7.23
  Special installations, 7.18
  Standards for, Table 3.1.3 – Part V
  Substitution of urinals, 7.21.5
  Trench drains, 7.16
  Urinals, 7.5
  Wash fountains, 7.6.5
  Wash sinks, 7.6.5
  Water closets, 7.4
  Water coolers, 7.12
  Water supply (WSFU) for, Table 10.14.2A
  Whirlpool baths, 7.9
Plumbing System:
  Definition of, 1.2
  Maintenance of, 15.8
  Required testing, 15.3
  Testing building sewer, 15.5
  Testing drain & vent piping, 15.4
  Testing water supply piping, 15.6
Pneumatic Sewage Ejectors:
  Discharge piping, 11.7.4
  Pressure release vent, 11.7.4, 12.14.3
  Surge tank, 11.7.4, 12.14.3
Pollution (of potable water):
  Definition of, 1.2
Potable Water:
  Definition of, 1.2
Pressure Balancing Compensating Valve:
  Definition of, 1.2
Pressure Relief Connection:
  Definition of, 1.2

Private Sewage Disposal Systems:
  Definition of, 1.2
  Requirements for, Chapter 16
Private Use:
  Definition of, 1.2
Private Water Supply Systems:
  Definition of, 1.2
  Requirements for, Chapter 17
Public Toilet Room:
  Definition of, 1.2
Public Use:
  Definition of, 1.2
Public Water Main:
  Availability, 2.19.1
  Interconnections, 10.4.2
Radon Gas:
  Definition of, 1.2
  Mitigation systems, 2.29
Rainfall Rates:
  For U.S. cities, Appendix A.1
Readily Accessible:
  Definition of, 1.2
Reclaimed Non-Potable Water:
  Code coverage, 10.1.f
  Definition of, 1.2
  Permitted uses, 10.1.b
Reduced Pressure Principle Backflow Preventer:
  Application of, 10.5.3
  Definition of, 1.2
  Installation of, 10.5.5
  Testing and maintenance, 10.5.6
Revent, Reventing:
  Definitions of, 1.2
Riser, Water Supply:
  Definition of, 1.2
Roof Drain:
  Definition of, 1.2
  Requirements for, 13.5
Rough Plumbing:
  Definition of, 1.2
Sand Interceptors:
  Requirements for, 6.4
Sanitary Sewer:
  Definition of, 1.2
SDR (standard dimension ratio):
  Definition of, 1.2
  For plastic sanitary drain piping, 3.5.4
  For plastic storm drain piping, 3.7.5
  For plastic vent piping, 3.6.3
Self-Draining:
  Definition of, 1.2

Septic Tanks:
  Definition of, 1.2
  Requirements for, 3.3.11, 16.6
Sewage:
  Definition of, 1.2:
Sewage Disposal Systems:
  Individual, Chapter 16
  Private, 2.19.2
Sewage Pump:
  Definition of, 1.2
Sewage Pumping:
  Building subdrains, 11.7.2
  Controls, 11.7.9
  Discharge piping, 11.7.8
  Grinder pumps, 11.7.5
  High level alarm, 11.7.10
  Individual fixtures, 11.7.11
  Macerating toilet systems, 11.7.6
  Pneumatic sewage ejectors, 11.7.4
  Pumps and ejectors, 11.7.3
  Sump pits, basins, receptors, 11.7.7
Shall:
  Definition of, 1.2
Short Term:
  Definition of, 1.2
Showers:
  Combination bath/shower, 7.8.3
  Compliance, 7.10.1
  Drainage fixture units (DFU), Table 11.4.1
  Emergency type, 7.24
  Floors and pan liners, 7.10.6
  Mixed water temperature control, 7.10.3
  Shower compartments, 7.10.5
  Shower head riser support, 7.10.7
  Trap sizes, 5.2, Table 5.2
  Waste outlet, 7.10.4
  Water conservation, 7.10.2
Sidewall Venting:
  Where permitted, 12.4.5
Sinks:
  Backflow prevention, 7.11.6
  Bar sinks, 7.11.2
  Clinical sinks, 14.8
  Compliance, 7.11.1
  Domestic kitchen sinks, 7.11.2
  Floor sinks, 7.11.5
  Laundry sinks, 7.11.3
  Minimum trap size, 5.2, Table 5.2
  Mop receptors, 7.11.4
  Service sinks, 7.11.4
  Shampoo sink traps, 6.7.4
Slaughterhouses:
  Drainage in, 6.7.3

Soil Pipe:
  Definition of, 1.2
Solids Interceptors:
  Requirements for, 6.5
Special Design Plumbing Systems:
  Aerators and deaerators, E.6
  Air admittance valves, E.8
  One-pipe sanitary drainage, E.6
  Requirements for, E.1, E.2, E.3, E.4
  Single stack vent systems, E.7
  Siphonic roof drainage, E.9
  Vacuum drainage, E.5
Stack:
  Definition of, 1.2
Stack Vent:
  Definition of, 1.2
Stack Venting:
  Definition of, 1.2
  Fixture groups, 12.11.1
  On lower floors, 12.11.2
Standards:
  List of referenced, Table 19.1
  Organizations, 19.2
Sterilizers:
  Drainage from, 14.10
  Vapor vents and stacks for, 14.11
Storm Drain:
  Definition of, 1.2
Storm Sewer:
  Definition of, 1.2
Storm Water:
  Definition of, 1.2
Storm Water Drainage:
  Areaway drains, 13.1.6
  Backwater valves, 13.1.12
  Combined with sanitary, 13.4.3
  Continuous flow, 13.1.11
  Controlled flow, 13.9
  Foundation drains, 13.1.5
  Primary roof drainage, 13.1.10.1
  Roof drainage, 13.1.10
  Roof drains, 13.5
  Secondary roof drainage, 13.1.10.2
  Size of horizontal piping, 13.6.2
  Size of vertical piping, 13.6.1
  Traps, 13.3
  Trench drains, 13.1.9
  Vertical walls, 13.1.10.3
  Water-operated pumps, 13.1.13
  Where required, 13.1.1
  Window well drains, 13.1.7
Sub-Stack:
  Definition of, 1.2

Suds Pressure Zones:
    Drain connections within, 11.11.1
    Exceptions to, 11.11.4
    Locations of, 11.11.2
    Separate stacks, 11.11.3
    Suds producing fixtures, 11.11.1
    Venting of, 12.15
Tempered Water:
    Definition of, 1.2
    For one-pipe lavatories, 10.15.6.g
    For one-pipe showers, 10.15.6.f
Thermostatic Compensating Valve:
    Definition of, 1.2
Thermostatic/Pressure Balancing Combination Compensating Valve:
    Definition of, 1.2
Toilet Facilities:
    Access to fixtures, 7.21.3
    Definition of, 1.2
    For business occupancies, 7.21.7
    For construction workers, 2.24
    For family and assisted-use, 7.21.9
    For food service establishments, 7.21.8
    For mercantile occupancies, 7.21.7
    For special occupancies, 7.21.6
    Minimum fixtures required, 7.21.1
    Separate for each sex, 7.21.4
    Toilet room requirements, 2.20
Traps:
    Building trap, 5.3.4
    Definition of, 1.2
    Design of, 5.3.1
    Level setting, 5.3.3
    Liquid seal required, 5.3.2
    Prohibited types, 5.3.5
    Protected from freezing, 5.3.3
    Required size, 5.2, Table 5.2
Trap Arm:
    Definition of, 1.2
    Maximum length of, Table 12.8.1
Trap Seal:
    Definition of, 1.2
    Maintenance of, 5.3.6
    Liquid seal required, 5.3.2
    Required depth of, 5.3.2
Travel Trailer Parks:
    Requirements for, Chapter 18
Trench Drains:
    For sanitary drainage, 7.16
    For storm water, 13.1.9, 13.7
Trenchless Pipe Replacement:
    Requirements for, 2.6.10

Tub/Shower:
    Definition of, 1.2
Tunneling:
    Pipe installation, 2.6.5
Urinals:
    Compliance, 7.5.1
    Non-water type, 7.5.6
    Prohibited types, 7.5.4
    Surrounding surfaces, 7.5.3
    Water conservation, 7.5.2
Used Material and Equipment:
    Where prohibited, 2.14
Vacuum Breakers:
    For backflow prevention, 10.5.3
    For chemical dispensing units, 10.5.13.a
    For cleaning equipment, 10.5.13.b
    For dental pump equipment, 10.5.13.c
    For food prep and service, 10.5.8
    For hose connections, 10.5.12
    For irrigation systems, 10.5.10
    For lawn sprinkler systems, 10.5.10
    Installation of, 10.5.5
    Testing and maintenance of, 10.5.6
Vacuum Condensate Drainage Systems:
    Requirements for, 2.25.6
Valves:
    Access to, 10.12.9
    Automatic compensating valves, 10.15.6.e
    Backflow prevention, 10.5
    Backwater valve, 5.5
    Building valve, 10.12.2
    Curb valve, 10.12.1
    Flush tank, 7.19.3
    Flushometer valve, 7.19.5
    For individual fixtures, 10.12.6
    In dwelling units, 10.12.4
    In other than dwelling units, 10.12.6
    In-line pressure balancing, 10.15.6
    Mixed water temperature control, 10.15.6
    Pressure reducing, 10.14.6
    Riser shutoff, 10.12.5
    TAFR flow reduction device, 10.15.6
    Temperature-actuated mixing, 10.15.6
    Temperature limiting device, 10.15.6
    Water heater shutoff, 10.12.7
    Water meter discharge shutoff, 10.12.8
    Water supply tank shutoff, 10.12.3
Valves for Pressure Vessels:
    Pressure relief, 10.16.2
    Pressure/temperature relief, 10.16.4
    Relief valve discharge piping, 10.16.6
    Temperature relief, 10.16.3
    Water heater shutoff, 10.12.7

Vent, Back:
 Definition of, 1.2
Vent, Branch:
 Definition of, 1.2
Vent, Circuit:
 Definition of, 1.2
Vent, Loop:
 Definition of, 1.2
Vent, Relief:
 Definition of, 1.2
Vent, Side:
 Definition of, 1.2
Vent, Wet:
 Definition of, 1.2
Vent, Yoke:
 Definition of, 1.2
Vent Piping:
 Aggregate size of terminals, 12.16.8
 Battery venting, 12.13.1, 12.13.2, 12.13.3
 Battery-vented back-to-back, 12.13.4
 Branch vent size, 12.16.6
 Building subdrain venting, 12.14
 Circuit and loop vent size, 12.16.2
 Circuit and loop venting, 12.13
 Combination waste and vent, 12.17
 Common vents, 12.9
 Connection height above fixtures, 12.6.3
 Extension of terminals above roof, 12.4.1
 Fixture reventing, 12.12
 Fixture vent, 12.8
 Fixture vent size, 12.16.1
 For horizontal offsets in stacks, 12.3.3
 For pneumatic sewage ejectors, 12.14.3
 For sump pits, 12.14.2
 For vertical offsets in stacks, 12.3.4
 Frost closure, 12.5
 Island sink venting, 12.18
 Location of vent terminals, 12.4.4
 Loop vent size, 12.16.2
 Loop venting, 12.13, 12.16.2
 Materials for, 3.6, Table 3.6
 On outside walls, 12.4.6
 Other designs for, 12.20
 Pipe slope, 12.6.1
 Prohibited use, 12.3.6, 12.4.3
 Protection of trap seals, 12.2.1
 Relief vent size, 12.16.3
 Relief vents for stacks, 12.3.2
 Revent, reventing, 1.2, 12.12
 Sidewall venting, 12.4.5
 Stack vent size, 12.16.4
 Stack venting, 12.11
 Stack vents, 12.3.1.a

 Suds pressure venting, 12.15
 Test methods, 15.4.1, 15.4.2
 Underground piping, 12.16.7
 Vent headers, 12.3.5, 12.16.5
 Vent stack size, 12.16.4, Table 12.16.4
 Vent stacks, 12.3.1
 Vent terminals, 12.4
 Vent washdown, 12.12.4
 Vertical rise above fixtures, 12.6.2
 Waste stack venting, 12.19
 Waterproof roof flashings, 12.4.2
 Wet venting, 12.10
Vent Riser:
 Definition of, 1.2
Vent Stack:
 Definition of, 1.2
Vertical Pipe:
 Definition of, 1.2
Washroom:
 Requirements for, 2.20
Waste:
 Definition of, 1.2
 Discharge of clear water, 9.1.8
 Industrial, 2.10.2
 Special wastes, 9.4
Water Bottle Filling Station:
 Compliance, 7.12.1
 Definition of, 1.2
Water Distribution Piping:
 Backflow prevention, 10.5
 Basic Design Circuit (BDC), B.9.3
 Cross connection control, 10.4.3
 Definition of, 1.2
 Disinfecting, 10.9.2
 Excessive pressures, 10.14.6
 Flushing, 10.9.1
 Hot water, 10.15
 Inadequate pressure, 10.14.4
 Limit on lead content, 3.4.6
 Materials for, 3.4, Table 3.4
 Minimum requirements for, 10.14
 Mixed water temperature control, 10.15.6
 Pipe sizing, 10.14.3
 Plastic distribution piping, Table 3.4.3
 Plastic hot water piping, 10.15.8
 Pressure booster systems, 10.8
 Pressure reducing valves, 10.14.6
 Protection of, 10.4
 Requirements for, 3.4.3, 3.4.4, 3.4.5
 Shutoff valves for, 10.12
 Temperature maintenance, 10.15.2
 Thermal expansion control, 10.15.7
 Variable street pressures, 10.14.5

Water hammer, 10.14.7
Water quality, 10.1
Water Hammer Arrestor:
    Definition of, 1.2
    Requirements for, 10.14.7
Water Heaters, Domestic:
    Combination relief valves for, 10.16.4
    Drain valves for, 10.15.4
    Drip pans for, 10.15.9
    Electric, commercial, 10.15.11
    Electric, household, 10.15.11
    Gas-fired storage, 10.15.11
    Oil-fired storage, 10.15.11
    Pressure relief valves for, 10.16.2
    Replacement of relief valves, 10.16.8
    Standards for, Table 3.1.3 - VII
    Storage tanks, 10.15.3
    Tankless, electric, 10.15.11
    Tankless, gas-fired, 10.15.11
    Temperature relief valves for, 10.16.3
    Thermal expansion control, 10.15.7
    Used for space heating, 10.15.10
    Vacuum relief valves for, 10.16.7
Water Piping Velocity Limitations:
    Good engineering practice, B.6.2
    In copper tubing, B.6.3
    Manufacturer's recommendations, B.6.3
    Maximum velocities, 10.14.1
Water Pressure:
    Excessive, 10.14.6
    Inadequate, 10.14.4
    Minimum required, 10.14.2
    Use of minimum street pressure, 10.14.5
Water Pressure Booster Systems:
    Requirements for, 10.8
    Where required, 10.8.1
Water Pressure Tanks:
    Construction of, 3.3.8
    Drains for, 10.8.6
    Pressure relief for, 10.8.9
    Safety devices for, 3.3.10
    Vacuum relief for, 10.8.8
Water Service Piping:
    Clearance from pollution, 10.6.2
    Definition of, 1.2
    Materials for, 3.4, Table 3.4
    Plastic piping for, Table 3.4.2
    Requirements for, 3.4.2, 10.6
Water Treatment Systems:
    Requirements for, 7.22
    Standards for, 10.18.1

Whirlpool Baths:
    Above food handling areas, 2.25.1
    Requirements for, 7.9
    Temperature control for, 10.15.6.h
WSFU:
    Definition of, 1.2

NOTES

NOTES